Anti-aging Drugs

From Basic Research to Clinical Practice

RSC Drug Discovery Series

Titles in the Series:

1: Metabolism, Pharmacokinetics and Toxicity of Functional Groups
2: Emerging Drugs and Targets for Alzheimer's Disease; Volume 1
3: Emerging Drugs and Targets for Alzheimer's Disease; Volume 2
4: Accounts in Drug Discovery
5: New Frontiers in Chemical Biology
6: Animal Models for Neurodegenerative Disease
7: Neurodegeneration
8: G Protein-Coupled Receptors
9: Pharmaceutical Process Development
10: Extracellular and Intracellular Signaling
11: New Synthetic Technologies in Medicinal Chemistry
12: New Horizons in Predictive Toxicology
13: Drug Design Strategies: Quantitative Approaches
14: Neglected Diseases and Drug Discovery
15: Biomedical Imaging
16: Pharmaceutical Salts and Cocrystals
17: Polyamine Drug Discovery
18: Proteinases as Drug Targets
19: Kinase Drug Discovery
20: Drug Design Strategies: Computational Techniques and Applications
21: Designing Multi-Target Drugs
22: Nanostructured Biomaterials for Overcoming Biological Barriers
23: Physico-Chemical and Computational Approaches to Drug Discovery
24: Biomarkers for Traumatic Brain Injury
25: Drug Discovery from Natural Products
26: Anti-Inflammatory Drug Discovery
27: New Therapeutic Strategies for Type 2 Diabetes: Small Molecules
28: Drug Discovery for Psychiatric Disorders
29: Organic Chemistry of Drug Degradation
30: Computational Approaches to Nuclear Receptors

31: Traditional Chinese Medicine
32: Successful Strategies for the Discovery of Antiviral Drugs
33: Comprehensive Biomarker Discovery and Validation for Clinical Application
34: Emerging Drugs and Targets for Parkinson's Disease
35: Pain Therapeutics; Current and Future Treatment Paradigms
36: Biotherapeutics: Recent Developments using Chemical and Molecular Biology
37: Inhibitors of Molecular Chaperones as Therapeutic Agents
38: Orphan Drugs and Rare Diseases
39: Ion Channel Drug Discovery
40: Macrocycles in Drug Discovery
41: Human-based Systems for Translational Research
42: Venoms to Drugs: Venom as a Source for the Development of Human Therapeutics
43: Carbohydrates in Drug Design and Discovery
44: Drug Discovery for Schizophrenia
45: Cardiovascular and Metabolic Disease: Scientific Discoveries and New Therapies
46: Green Chemistry Strategies for Drug Discovery
47: Fragment-Based Drug Discovery
48: Epigenetics for Drug Discovery
49: New Horizons in Predictive Drug Metabolism and Pharmacokinetics
50: Privileged Scaffolds in Medicinal Chemistry: Design, Synthesis, Evaluation
51: Nanomedicines: Design, Delivery and Detection
52: Synthetic Methods in Drug Discovery: Volume 1
53: Synthetic Methods in Drug Discovery: Volume 2
54: Drug Transporters: Role and Importance in ADME and Drug Development
55: Drug Transporters: Recent Advances and Emerging Technologies
56: Allosterism in Drug Discovery
57: Anti-aging Drugs: From Basic Research to Clinical Practice

How to obtain future titles on publication:
A standing order plan is available for this series. A standing order will bring delivery of each new volume immediately on publication.

For further information please contact:
Book Sales Department, Royal Society of Chemistry, Thomas Graham House, Science Park, Milton Road, Cambridge, CB4 0WF, UK
Telephone: +44 (0)1223 420066, Fax: +44 (0)1223 420247,
Email: booksales@rsc.org
Visit our website at www.rsc.org/books

Anti-aging Drugs
From Basic Research to Clinical Practice

Edited by

Alexander M. Vaiserman
Institute of Gerontology, Kiev, Ukraine
Email: vaiserman@geront.kiev.ua

ROYAL SOCIETY OF CHEMISTRY

THE QUEEN'S AWARDS FOR ENTERPRISE: INTERNATIONAL TRADE 2013

RSC Drug Discovery Series No. 57

Print ISBN: 978-1-78262-435-6
PDF eISBN: 978-1-78262-660-2
EPUB eISBN: 978-1-78262-985-6
ISSN: 2041-3203

A catalogue record for this book is available from the British Library

The Royal Society of Chemistry is a charity, registered in England and Wales, Number 207890, and a company incorporated in England by Royal Charter (Registered No. RC000524), registered office: Burlington House, Piccadilly, London W1J 0BA, UK, Telephone: +44 (0) 207 4378 6556.

Visit our website at www.rsc.org/books

Printed in the United Kingdom by CPI Group (UK) Ltd, Croydon, CR0 4YY, UK

Foreword

In 1970, the world's largest learned society focused on aging underwent a schism that persists to this day. Denham Harman, one of the foremost American gerontologists of that era, had become so incensed at the flight from translational work—or even, to judge from public pronouncements, translational aspirations—of nearly all his colleagues that he felt it necessary to found the American Aging Association in direct competition with the Gerontological Society of America, which had overseen the field for the previous quarter-century.

Was that a good move? This excellent volume provides a fitting affirmative answer. The American Aging Association languished in genuine obscurity and neglect for over 20 years, but by 2000 it had risen to a much greater degree of respect, and it has since become arguably the most prestigious society in the field worldwide, without ever losing sight of its intervention-focused roots. It has done so because of real progress in the laboratory: progress that has shifted other communities to a more translation-friendly stance rather than the other way around.

The pharmacological approach that dominates the following chapters is by no means the only option available to the biomedical gerontologist; in particular, my own work and that of SENS Research Foundation is focused mainly on stem cell and gene therapies. But it remains apparent that pharmacological interventions, simply by virtue of being so much easier to administer, are of immense value even if they only provide much lesser benefit to the average older person than more exotic alternatives, not only because even modest benefit is better than nothing, but also because the latter will not be available for a while and the former can act as a bridge to them.

The first and last sections of this book are no less important. Biogerontology runs the same risk as any science, of becoming an echo-chamber immune to the need for interaction with wider society. Biologists of aging

RSC Drug Discovery Series No. 57
Anti-aging Drugs: From Basic Research to Clinical Practice
Edited by Alexander M. Vaiserman

Published by the Royal Society of Chemistry, www.rsc.org

are perhaps even more duty-bound than any scientists, in consequence of the humanitarian importance of their field, to avoid falling into such a trap. It is therefore laudable that Vaiserman has chosen to invite chapters covering the pros and cons of both the feasibility and the desirability of significant, near-term success in the age-old quest to extend our youth. As one who has dedicated his life to that mission, I can attest that the best way to further it is to discuss it.

Enjoy these chapters as much as I have. They jointly constitute a comprehensive and invaluable primer in the current state of pharmacological anti-aging medicine.

Aubrey de Grey

Preface

Over the last few years, anti-aging medicine has received increasing attention in both public and scientific communities. Public interest in this area of research is largely driven by media attention related to recent developments in regenerative medicine and genome modification technologies. Probably the most famous example of that is the case of Elizabeth Parrish, the CEO of Seattle-based biotech firm BioViva, who claims that she had managed to reverse her own aging process with CRISPR gene editing technology by receiving a treatment targeting two gene loci, one a gene controlling telomere length and the other to protect against loss of muscle mass with age. Even though no confirmation has been received so far on whether or not this technology successfully changed her genome, many safety, ethical and regulatory issues are raised from this case. First of all, this concern is related to possible side effects associated with the use of this technology, primarily cancer. In this respect, using the more conventional pharmacologically based approach seems a reasonable alternative, particularly since many natural and synthetic agents have shown great potential for promoting health and longevity in numerous animal models. Among them, the most attention is currently drawn to rapamycin, resveratrol and the antidiabetic drug metformin. The last one was recently approved by the FDA to be examined in the Targeting Aging with Metformin (TAME) clinical trial to establish whether it may reduce the risk for aging-associated pathologies, such as cognitive impairments, cardiovascular disease and cancer, in non-diabetic persons. If successful, the TAME study would be the first demonstration that a particular drug can prevent or delay the onset of aging-associated chronic human disorders. It might provide a novel regulatory pathway for further clinical trials of pharmaceuticals specifically designed to slow the aging process.

The present volume is the first one devoted entirely to the pharmacological aspects of anti-aging medicine. It provides a comprehensive overview

RSC Drug Discovery Series No. 57
Anti-aging Drugs: From Basic Research to Clinical Practice
Edited by Alexander M. Vaiserman

Published by the Royal Society of Chemistry, www.rsc.org

of current research aimed to search for natural and synthetic compounds that can potentially be developed as drugs for treating aging-related chronic pathologies and, ultimately, for healthy life extension. In the first section of the book, the basic conceptual and methodological aspects of modern anti-aging medicine are described. The next sections are concerned with the main classes of lifespan-promoting agents, such as antioxidants, calorie restriction mimetics, epigenome-targeted drugs and phytochemicals with health-promoting properties. In the subsequent sections, the strategies for translation of research findings in the field of anti-aging medicine into clinical and healthcare practice as well as opportunities and challenges related to the implementation of such approaches are discussed. This volume constitutes a comprehensive collection of chapters written by leading experts in the field. It will be a relevant and useful resource not only for professional scientists and clinicians, but also for scientifically interested amateurs wishing to know more about the current research in anti-aging pharmacology.

Finally, I would like to acknowledge Dr Oksana Zabuga for the helpful assistance in preparing the manuscript of this volume, as well as the editorial staff at the Royal Society of Chemistry, especially Harriet Manning and Rowan Frame, with whom I had the good fortune to work on this project, for their patience and encouragement.

Alexander M. Vaiserman

Contents

Section I: Overview

RSC Drug Discovery Series No. 57
Anti-aging Drugs: From Basic Research to Clinical Practice
Edited by Alexander M. Vaiserman

Published by the Royal Society of Chemistry, www.rsc.org

Section III: Antioxidants

Section IV: Mimicking Caloric Restriction

Section V: Other Pharmacological Approaches

Section VI: Social Context

Section I

Overview

CHAPTER 1

Anti-Aging Drugs: Where are We and Where are We Going?

ALEXANDER M. VAISERMAN*[a] AND OLEH V. LUSHCHAK[b]

[a]Laboratory of Epigenetics, Institute of Gerontology, Kiev, Ukraine;
[b]Department of Biochemistry and Biotechnology, Vasyl Stefanyk Precarpathian National University, Ivano-Frankivsk, Ukraine
*E-mail: vaiserman@geront.kiev.ua

1.1 Introduction

Human longevity dramatically increased during the last century when implementation of vaccinations, disinfectants and antibiotics led to a substantial reduction of infectious diseases as a leading cause of death.[1] The decline in mortality among the elderly has continued over the past few decades. It is most probably owing to preventative factors, such as improved diets, as well as exercise and reduction in smoking.[2] If current demographic trends continue then 20% percent of the global population of 9 billion will be over the age of 60 by 2050.[3] As a consequence, most modern nations are undergoing rapid population aging. Although the life expectancy has enhanced dramatically in modern generations, this process has, nevertheless, not been accompanied by an equivalent increase in healthy life expectancy.[4] Since aging is a primary risk factor in most chronic disorders, the prevalence of age-associated disorders, such as type 2 diabetes, neurodegenerative disease, cardiovascular disease, osteoporosis and cancer, rises considerably with the increasing average age in populations of developed countries, representing a

RSC Drug Discovery Series No. 57
Anti-aging Drugs: From Basic Research to Clinical Practice
Edited by Alexander M. Vaiserman

Published by the Royal Society of Chemistry, www.rsc.org

great socio-economic challenge. It is estimated that there will be more than 30 million people over the age of 80 will be in the U.S. by 2050; about half of them will suffer from different forms of dementia, and at least 3 million of all adults will be diagnosed with Parkinson's disease.[5] The expected prevalence of age-associated conditions will have substantial consequences for future society, including increased financial and psychological burdens for families and greater pressure on government health care programs and entitlement budgets.[6,7] The demographic trend consisting of an increasing proportion of aged people in the populations of developed countries likely explains the dramatic increase in the interest of the lay public and country leaders in research in the field of biogerontology.[8]

1.2 Human Life Extension: Concerns and Considerations

Investigations aimed at human life extension have traditionally raised concerns that it can lead to the growth of the older population segment and, consequently, to the high prevalence of ageing-associated chronic pathologies. Numerous experimental studies have, however, demonstrated that life extension is usually accompanied by delayed or reduced morbidity, including cardiovascular disease, neurodegeneration, and tumors.[9] There is also increasing evidence from epidemiological studies, which is consistent with the findings from animal models. For example, centenarians, in particular those who live in so-called 'Blue Zones' (five regions in Europe, Latin America, Asia and the US with unusually high concentrations of centenarians), have been not only shown to exhibit exceptional longevity but also often remain free from disability and chronic diseases until very advanced age.[10]

The compression of morbidity has been the primary strategy in gerontology and geriatric research during the last few decades. This strategy claims that we may limit morbidity to a shorter period closer to the natural ending of life, thus reducing the burden of illness and disability by delaying the age at onset of major age-associated chronic disorders.[11] Geroscience, a novel branch of geriatric medicine, is centered on healthspan extension.[12] Extension of healthspan is a crucial component of achieving 'optimal longevity', defined as living long, but with good health and quality of life, including improved functioning, productivity and independence.[11] Attempts to increase healthspan are currently focused on slowing the basic biological processes accompanying aging, such as mitochondrial dysfunction, cellular senescence, age-related decline of stress resistance, dysregulated cellular energy sensing and growth pathways, impaired proteostasis, deteriorated stem cell function/bioavailability, as well as oxidative and inflammation stress.[13,14] All these processes interfere with the normal physiological cellular signaling pathways, demanding compensatory adjustments with aging to maintain homeostasis. At a certain age, however, these compensatory mechanisms become exhausted and different aspects of aging are manifested, thereby increasing the risk for functional decline and the onset and progression of chronic diseases.[15] Therapeutic strategies to combat aging and age-related diseases are a part of an investigation field commonly referred to as 'anti-aging medicine'. Anti-aging medicine has emerged as a new specialization in medical practice at the beginning of the 1990s. Over the past few years, it has become an increasingly discussed and debated topic.[16] Its main purpose is to prolong both healthspan and lifespan by specific regimes of exercise and dieting, as well as by advanced biomedical interventions aimed at slowing, stopping or reversing the aging process.[17,18]

Traditionally, the process of aging is believed to be 'natural' and therefore inevitable. However, in the view of many authors, the idea that aging is an indefeasible part of human nature is quite questionable.[19] In accordance with many modern evolutionary theories, aging has emerged as a by-product of evolutionary processes and does not have a specific function.[20] If aging is really not an intrinsic, irrevocable component of life, then it could be manipulated similarly to other processes that are generally deemed to be unnatural or pathological. The major assumption underlying anti-aging research is that age-associated senescence may be regarded as a pathophysiological phenomenon that might be prevented or even reversed.[21] Modern anti-aging medicine promotes biomedical technologies and approaches that have the potential to delay or postpone aging processes.[2] The success obtained in this research field is greatly attributed to the increasingly broad application of omics-based approaches, such as genomics, transcriptomics, proteomics and metabolomics.[22] Through the implementation of these technologies, a better understanding has been achieved regarding the key molecular and cellular pathways involved in the aging process, including inflammation, proteostasis, autophagy, mitochondrial efficiency and nutrient signaling, and regarding the most effective interventions to counteract age-related senescence.[23,24] The impetuous progress in highlighting the mechanisms underlying aging and longevity and first successful pharmacological interventions to extend healthy lifespan in different model organisms indicate that the aging process is malleable.

1.3 Anti-Aging Pharmacology: Promises and Pitfalls

The development of pharmacological agents targeting aging-related functional declines and pathological manifestations ('anti-aging drugs') is now in the spotlight in geroscience. An exponential growth of research in the field of geriatric pharmacology, including the study of prospective anti-aging drugs, has been observed over the past 20 years.[25] The first step in the process of drug development is known to involve the selection of druggable targets.[26] The situation when gene targets are determined by the study of genetic variations linked to either gain-of-function or loss-of-function phenotypes is especially useful because these targets can be considered as those that have been reliably validated.[27] Over the last two decades, a number of genetic pathways have been identified that play an unequivocal role in control of the aging process and longevity;[28–30] all these genes represent attractive drug targets. Currently, many pharmacological agents targeting the putative mechanisms of aging are under development.

Taking into account the extraordinary complexity of the mechanistic pathways underlying the aging process, the recognition of these pathways and development of anti-aging interventions seems a challenging task. Significant progress has, however, been achieved in the last few years in this research field. A number of pharmacological agents with the potential to target particular aging-associated pathways and to produce protective responses against age-related pathologies are currently under investigation. In recent years, several classes of bioactive chemical agents and nutraceuticals have been shown to have potential therapeutic efficacy in anti-aging medicine.[3,31] In experimental studies, many substances have been identified as having life-extending properties. Among them are calorie restriction mimetics, such as resveratrol, rapamycin and metformin,[32,33] antioxidants (vitamins A, C and E, quercetin, melatonin, coenzyme Q10, *etc.*),[34] autophagy inductors, such as spermidine,[35,36] senolytics,[37] phytochemicals, *e.g.*, curcumin, genistein, catechins and epigallocatechin gallate (EGCG),[38] and several other natural and chemical compounds. In recent years, modern biotechnological approaches have been used for developing novel anti-aging pharmaceutical applications. For example, the coupling of curcumin-based nanoparticles with the Tet-1 peptide, which has affinity for neurons and possess retrograde transportation properties,[39] as well as mitochondria-targeted antioxidant SkQ1,[40] have been recently explored as promising therapeutic applications for the treatment of Alzheimer's disease. Over the last decade, consistent evidence has also been reported for the role of epigenetic factors, including DNA methylation, histone modifications and microRNA regulation, in the aging process as well as in the pathogenesis and progression of age-related diseases.[41,42] A lot of hope is being pinned, therefore, on pharmacological agents targeted to the epigenetic regulation of gene activity, such as inhibitors of DNA methyltransferases and histone deacetylases, including sodium butyrate, trichostatin A, sodium 4-phenylbutyrate and suberoylanilide hydroxamic acid.[43]

It should, however, be noted that all agents that can be classified as potent anti-aging therapeutic compounds are multi-functional and targeted at multiple signaling pathways mediating aging. Moreover, the evidence remains limited regarding the overall health benefits of these substances, including epidemiological studies exploring the consequences of their long-term intake for human health. Furthermore, there is evidence that uncontrolled intake of some anti-aging agents can be useless or even harmful. For example, the consumption of antioxidants is considered as quite reasonable by many researchers, especially in the cardiovascular research area.[44] The appropriateness of antioxidant intake, however, still remains a matter of debate. Meta-analysis of observational studies and randomized controlled trials conducted in well-nourished and healthy populations demonstrated that antioxidant supplementation may be associated with undesirable consequences for health and all-cause mortality.[45] Another example is the fact that supplementation with several promising pro-healthspan compounds can in some cases trigger insulin resistance. This applies to substances such as rapamycin[46] and statins.[47] Therefore, people should use them with caution and only with careful medical monitoring.

Another method of anti-aging drug discovery is evaluating the pharmacological agents already approved by the FDA and other regulatory agencies for treatment of particular conditions associated with aging, such as statins, metformin, beta-blockers, renin-angiotensin-aldosterone system inhibitors, thiazolidinediones, and anti-inflammatory medications.[48] These classes of drugs are commonly used in the treatment of patients with various chronic medical conditions and their efficacy and safety have been proven in many clinical trials. They have also been shown to improve health, physiological functioning and well-being in middle to old age patients with chronic disorders.[49] Such agents are presently not used in the treatment of age-associated physiological dysfunctions in the absence of clinical manifestation of disease. However, these medications might theoretically be redirected to treating or preventing conditions or syndromes typically associated with aging.

Le Couteur *et al.*[50] noted in their review that 'despite the potential profits and the extraordinary capacity of drug discovery technology, there is a paucity of new drugs in the development pipeline, particularly for those medications that are likely to be highly profitable because they are used long term and by a large proportion of the population.' The longevity dividend, *i.e.* an idea that extending healthy life by slowing aging is the most efficient way to combat the fatal and disabling pathologies that plague us today,[51] may provide an opportunity to revitalize the drug development pipeline. Indeed, by delaying the aging process *per se*, it likely would be possible to prevent or delay all age-associated pathologies rather than to overcome them one by one, which is the current approach of the disease-based paradigm in drug development. Furthermore, prevention of a particular age-related chronic disorder, *e.g.*, cardiovascular disease, will apparently have only a modest effect on the population life expectancy because comorbidity, *e.g.*, cancer, will to a great extent substitute the reduction in mortality risk caused by preventing the targeted pathology. The main idea of geroscience is that preventing the clinical manifestations of all age-related diseases as a group by inhibiting the basic aging mechanisms can be far more effective than preventing the individual chronic disorders.[11,49] A recent analysis conducted by Goldman *et al.*[52] demonstrated that substantial socio-economic benefits might be derived from this approach in comparison with current public health strategy targeted to prevention of particular disorders. According to this analysis, the economic impact of delaying aging and increasing healthspan in the US is estimated at ~7 trillion dollars over the next fifty years. Hence, it is obvious that discovery of new drug targets based on biogerontological research represents an incredible opportunity for the pharmaceutical and healthcare industries.[53] Currently, the consensus among physicians and health professionals that the optimization of physiological and mental functioning throughout the life course should be a major emphasis of any contemporaneous biomedical policy addressing global aging. A healthy lifestyle comprising proper nutrition and physical activity represents the first-line function-preserving strategy. Pharmacological compounds, both existing and potential, can serve as a prospective complementary approach.[48]

1.4 Concluding Remarks and Future Directions

To summarize, it can be assumed that targeting aging *per se* can be a more effective approach to postponing or preventing age-related disorders than treatments targeted to specific pathological conditions. Because of the aging population, such a therapeutic strategy is undoubtedly an area of increasing relevance for the pharmaceutical industry and public health organizations. As has been recently emphasized by Longo *et al.*,[54] 'the time has come not only to consider several therapeutic options for the treatment of age-related comorbidities, but to initiate clinical trials with the ultimate goal of increasing the healthspan (and perhaps longevity) of human populations, while respecting the guiding principle of physicians *primum non nocere*.' In modern pharmacy, anti-aging is likely one of the most prospective markets because the target group can potentially include each person. Several supplements, such as resveratrol, are already advertised in the pharmaceutical market as "anti-aging pills".[55] Very promising in this regard is rapamycin (also known as sirolimus), which is already approved by the FDA as an antibiotic and immunosuppressant drug. Current marketing research demonstrates that most people are willing to pay for long-term pharmacological therapy to prevent or delay the aging-related decline in physical and mental functions.[50] Recent sociological surveys show a great desire for extended life and health in the US and worldwide. In most of the surveys conducted until now, the cautious attitude to life extension was a consequence of an erroneous equation of extended life with a prolonged period of age-related functional decline and frailty. When continued health was stipulated in the questionnaire design, responses significantly favored longer life. In the survey by Donner *et al.*,[56] 20% of respondents wished to die at the age of 85, whereas 42% wanted to have an unlimited lifespan. Despite the widespread misconception that implementation of anti-aging medicine would increase the proportion of chronic patients in modern societies, it in fact would lead to reducing the ratio of unhealthy to healthy population since it would result in delaying the onset of age-related pathological conditions. In other words, it may lead to a decrease of biological age (*i.e.*, old individuals will become biologically younger) and to an increase of the age of disability, thereby increasing the retirement age and enhancing revenues without enhancing taxes.[57] Optimistic predictions of the feasibility of health- and life-extending interventions, however, should certainly be critically discussed in terms of their ethical, economic and social implications. Only after in-depth examination and following comprehensive debates will the implementation of such approaches in clinical practice be possible.

References

1. J. P. de Magalhães, *Rejuvenation Res.*, 2014, **17**, 458.
2. J. Vijg and A. D. de Grey, *Gerontology*, 2014, **60**, 373.
3. B. K. Kennedy and J. K. Pennypacker, *Transl. Res.*, 2014, **163**, 456.
4. W. W. Hung, J. S. Ross, K. S. Boockvar and A. L. Siu, *BMC Geriatr.*, 2011, **11**, 47.
5. G. A. Petsko, *Genome Biol.*, 2008, **9**, 113.
6. S. Harper, *Science*, 2014, **346**, 587.
7. J. R. Beard and D. E. Bloom, *Lancet*, 2015, **385**, 658.
8. E. Le Bourg, *Biogerontology*, 2013, **14**, 221.
9. L. Fontana, L. Partridge and V. D. Longo, *Science*, 2010, **328**, 321.
10. B. J. Willcox, D. C. Willcox and L. Ferrucci, *J. Gerontol., Ser. A*, 2008, **63**, 1181.
11. D. R. Seals, R. E. Kaplon, R. A. Gioscia-Ryan and T. J. LaRocca, *Physiology*, 2014, **29**, 250.
12. B. K. Kennedy, S. L. Berger, A. Brunet, J. Campisi, A. M. Cuervo, E. S. Epel, C. Franceschi, G. J. Lithgow, R. I. Morimoto, J. E. Pessin, T. A. Rando, A. Richardson and E. E. Schadt, *et al.*, *Cell*, 2014, **159**, 709.
13. J. L. Kirkland, *Exp. Gerontol.*, 2013, **48**, 1.
14. L. Fontana, B. K. Kennedy, V. D. Longo, D. Seals and S. Melov, *Nature*, 2014, **511**, 405.
15. E. S. Epel and G. J. Lithgow, *J. Gerontol., Ser. A*, 2014, **69**, S10.
16. G. Barazzetti and M. Reichlin, *Swiss Med. Wkly.*, 2011, **141**, w13181.
17. R. Klatz, *Ann. N. Y. Acad. Sci.*, 2005, **1057**, 536.
18. M. Tosato, V. Zamboni, A. Ferrini and M. Cesari, *Clin. Interventions Aging*, 2007, **2**, 401.
19. A. L. Caplan, in *The Fountain of Youth. Cultural, Scientific, and Ethical Perspectives on a Biomedical Goal*, ed. S. G. Post and R. H. Binstock, Oxford University Press, Oxford, 2004, pp. 271–285.
20. J. F. Lemaître, V. Berger, C. Bonenfant, M. Douhard, M. Gamelon, F. Plard and J. M. Gaillard, *Proc. Biol. Sci.*, 2015, **282**, 2015020.
21. B. Anton, L. Vitetta, F. Cortizo and A. Sali, *Ann. N. Y. Acad. Sci.*, 2005, **1057**, 525.
22. E. Cevenini, E. Bellavista, P. Tieri, G. Castellani, F. Lescai, M. Francesconi, M. Mishto, A. Santoro, S. Valensin, S. Salvioli, M. Capri, A. Zaikin and D. Monti, *et al.*, *Curr. Pharm. Des.*, 2010, **16**, 802.
23. R. de Cabo, D. Carmona-Gutierrez, M. Bernier, M. N. Hall and F. Madeo, *Cell*, 2014, **157**, 1515.
24. E. K. Quarles, D. F. Dai, A. Tocchi, N. Basisty, L. Gitari and P. S. Rabinovitch, *Ageing Res. Rev.*, 2015, **23**, 101.
25. E. Verdaguer, F. Junyent, J. Folch, C. Beas-Zarate, C. Auladell, M. Pallàs and A. Camins, *Expert Opin. Drug Discovery*, 2012, **7**, 217.
26. Y. Zhou and N. Huang, *Methods Mol. Biol.*, 2015, **1289**, 13.
27. S. M. Paul, D. S. Mytelka, C. T. Dunwiddie, C. C. Persinger, B. H. Munos, S. R. Lindborg and A. L. Schacht, *Nat. Rev. Drug Discovery*, 2010, **9**, 203.

28. A. A. Moskalev, A. M. Aliper, Z. Smit-McBride, A. Buzdin and A. Zhavoronkov, *Cell Cycle*, 2014, **13**, 1063.
29. C. M. Lindborg, K. J. Property and R. J. Pignolo, *Mech. Ageing Dev.*, 2015, **146–148**, 23.
30. A. H. Shadyab and A. Z. LaCroix, *Ageing Res. Rev.*, 2015, **19**, 1.
31. M. V. Blagosklonny, *Cell Death Discovery*, 2014, **5**, e1552.
32. G. Testa, F. Biasi, G. Poli and E. Chiarpotto, *Curr. Pharm. Des.*, 2014, **20**, 2950.
33. D. K. Ingram and G. S. Roth, *Ageing Res. Rev.*, 2015, **20**, 46.
34. M. Wojcik, I. Burzynska-Pedziwiatr and L. A. Wozniak, *Curr. Med. Chem.*, 2010, **17**, 3262.
35. N. Minois, *Gerontology*, 2014, **60**, 319.
36. F. Madeo, A. Zimmermann, M. C. Maiuri and G. Kroemer, *J. Clin. Invest.*, 2015, **125**, 85.
37. M. Malavolta, E. Pierpaoli, R. Giacconi, L. Costarelli, F. Piacenza, A. Basso, M. Cardelli and M. Provinciali, *Curr. Drug Targets*, 2016, **17**, 447.
38. H. Si and D. Liu, *J. Nutr. Biochem.*, 2014, **25**, 581.
39. A. Mathew, T. Fukuda, Y. Nagaoka, T. Hasumura, H. Morimoto, Y. Yoshida, T. Maekawa, K. Venugopal and D. S. Kumar, *PLoS One*, 2012, **7**, e32616.
40. N. A. Stefanova, N. A. Muraleva, V. P. Skulachev and N. G. Kolosova, *J. Alzheimers Dis.*, 2014, **38**, 681.
41. A. Brunet and S. L. Berger, *J. Gerontol., Ser. A*, 2014, **69**, S17.
42. D. Ben-Avraham, *Adv. Exp. Med. Biol.*, 2015, **847**, 179.
43. A. M. Vaiserman and E. G. Pasyukova, *Front. Genet.*, 2012, **3**, 224.
44. U. Alehagen, J. Aaseth and P. Johansson, *PLoS One*, 2015, **10**, e0141641.
45. G. Bjelakovic, D. Nikolova and C. Gluud, *Curr. Opin. Clin. Nutr. Metab. Care*, 2014, **17**, 40.
46. M. V. Blagosklonny, *Aging (Albany NY)*, 2012a, **4**, 350.
47. B. D. Henriksbo and J. D. Schertzer, *Adipocyte*, 2015, **4**, 232.
48. D. R. Seals, J. N. Justice and T. J. LaRocca, *J. Physiol.*, 2016, **594**, 2001.
49. D. R. Seals and S. Melov, *Aging (Albany NY)*, 2014, **6**, 718.
50. D. G. Le Couteur, A. J. McLachlan, R. J. Quinn, S. J. Simpson and R. de Cabo, *J. Gerontol., Ser. A*, 2012, **67A**, 168.
51. S. J. Olshansky, *Public Policy Aging Rep.*, 2013, **23**, 3.
52. D. P. Goldman, D. Cutler, J. W. Rowe, P. C. Michaud, J. Sullivan, D. Peneva and S. J. Olshansky, *Health Aff.*, 2013, **32**, 1698.
53. D. G. Le Couteur and D. A. Sinclair, *J. Gerontol., Ser. A*, 2010, **65**, 693.
54. V. D. Longo, A. Antebi, A. Bartke, N. Barzilai, H. M. Brown-Borg, C. Caruso, T. J. Curiel, R. de Cabo, C. Franceschi, D. Gems, D. K. Ingram, T. E. Johnson and B. K. Kennedy, *et al.*, *Aging Cell*, 2015, **14**, 497.
55. J. Aschemann-Witzel and K. G. Grunert, *Ann. N. Y. Acad. Sci.*, 2015, **1348**, 171.
56. Y. Donner, K. Fortney, S. R. Calimport, K. Pfleger, M. Shah and J. Betts-LaCroix, *Front. Genet.*, 2015, **6**, 353.
57. M. V. Blagosklonny, *Aging (Albany NY)*, 2012b, **4**, 547.

CHAPTER 2

Aging: Natural or Disease? A View from Medical Textbooks

S. JANAC[a], B. CLARKE[b] AND D. GEMS*[a]

[a]University College London, Institute of Healthy Ageing, Gower Street, London, WC1E 6BT, UK; [b]University College London, Department of Science and Technology Studies, Gower Street, London, WC1E 6BT, UK
*E-mail: david.gems@ucl.ac.uk

2.1 Introduction

Whether a given condition is labelled as a disease or not can depend on a number of factors—including linguistics. For example, in one survey people were asked of 60 different conditions whether they considered them to be a disease or not.[1] The study found that alcoholism was seen as a disease, but smoking not. In some ways this is an odd finding since both—broadly speaking—elicit dependence symptoms, involve substance abuse and are detrimental to health in the long-term. Plausibly, this quirk reflects the choice of words employed in the survey. Perhaps if the terms used had instead been *drinking* and *nicotine addiction*, the classification would have come out the other way around.

Difficulties of classification also affect *aging*. For example, if one went to the doctor and asked for a prescription for anti-aging drugs, their response would likely be surprise, amusement or perhaps mild irritation. This is because aging, in the medical field, is not regarded as a disease.

RSC Drug Discovery Series No. 57
Anti-aging Drugs: From Basic Research to Clinical Practice
Edited by Alexander M. Vaiserman

Published by the Royal Society of Chemistry, www.rsc.org

The question of what exactly is meant by "anti-aging drugs" is complicated by several factors. First, linguistics, and the problem that the word "aging" has more than one meaning. Second, the question of whether aging is a disease. Thirdly, problems relating to what counts as an anti-aging intervention. These issues will be reviewed here briefly and a serving definition of the meaning of *anti-aging* suggested. This builds on previous work that attempts to define anti-aging interventions.[2–10] We will then present an attempt at a broad and general description of the biological basis of aging, to offer the beginnings of an etiological basis for the understanding of senescence as a disease syndrome. Then, in the main part of this chapter, we examine how the aging *vs.* disease question is presented in general medical textbooks.

2.1.1 What Does "Aging" Mean?

The word *aging* acts as a stumbling block in discussion because it has multiple meanings that are sometimes conflated. The main, distinct meanings are:

- The passage of time (*calendar aging*).
- Time-dependent alterations, usually in adult living organisms, but also inert objects (*age changes*).
- Cumulative deteriorative changes in adult organisms leading to pathology and death (*senescence*).[11–13] Senescence is one type of age change.

An unfortunate additional source of confusion is that the word senescence also has a second meaning, as introduced by Leonard Hayflick, that of *cellular senescence*. This refers to a specific type of cellular change where the proliferative capacity of cells is lost and a pathogenic hypertrophic phenotype appears. Confusion between these two meanings can, in some contexts, be avoided by use of the term *organismal senescence* to contrast with cellular senescence. However, it seems likely that the two meanings of senescence will continue to generate confusion. Replacement of *cellular senescence* with another term would solve this problem.

Thus, the English language is a hindrance in that the multiple meanings of aging impede understanding. Not all languages have this problem; for example the Russian *stareniye* (старение) means, essentially, senescence. For people, age changes include maturational changes, such as the attainment of wisdom, and character development. In this sense, an anti-aging drug would be highly undesirable; clearly, the interest is in anti-senescence (or geroprotectant) drugs, where senescence is meant in its original sense, not the sense of cellular senescence.

2.1.2 Is Aging a Disease?

Human senescence manifests as a wide range of deteriorative changes, including some that are debilitating and sometimes fatal (*e.g.* cardiovascular disease, cancer and dementia) and some that are not (*e.g.* greying of hair and wrinkling of skin). In medicine, a conceptual division is made between the former, as diseases for which aging is a risk factor, and the latter, which are not pathological but rather manifestations of *normal* aging.[14–16] Here, aging itself is viewed as a natural and non-pathological process. However, this division and the notion of normal aging is problematic in a number of respects. For example, the designation of particular senescent changes as normal or pathological has been controversial, as illustrated by the transfer of late-onset Alzheimer's disease and osteoporosis from the former to the latter category.[5] Moreover, from a biological perspective, senescence, a biological process whose defining characteristic is deterioration, is a fundamentally pathological process, identifiable as damage accumulation, degeneration, loss of function, and emergence of numerous disease states that can cause suffering and death. At present there exists some division between perspectives on aging in the medical and scientific domain. In the former the concept of normal aging is more prevalent, whereas in the latter there are more doubts about the existence (or meaning) of "non-pathological senescence".

As a contribution to this debate, we present here an attempt at a disease definition of aging. Ideally, a disease definition will include a full description of the disease etiology. In the case of aging this is not possible since the biological mechanisms that cause senescence are only partly understood. This definition does not pretend to encompass the views of all biogerontologists, and it surely will not do so. We hope that its faults will incite others to develop better definitions.

2.1.2.1 An Attempt at a Broad Account of the Etiology of Senescence

Organismal senescence manifests as diverse pathologies, including neurodegenerative diseases, cardiovascular disease and cancer, as well as minor pathologies such as skin wrinkling, and encompasses the etiologies of these conditions. There is no single etiology of organismal senescence, but rather multiple causes that generate a number of syndromes and unitary diseases. Thus, aging is a disease super-syndrome. These etiologies are predominantly the result of inherited predisposition, but environmental factors that promote damage and injury also play an important role, often through effects on the expression of predispositions (*e.g.* mechanical injury to joints can contribute to osteoarthritis).

Insofar as it is genetically determined, organismal senescence is a form of genetic disease, but of a special kind, as follows. According to contemporary medical understanding, a genetic disease is the result of a mutation in a gene that disrupts its evolved function, changing the gene from wild type to mutant, thereby disrupting biological function and causing pathology. By contrast, the inherited predisposition to organismal senescence is largely specified by wild-type genes. This seemingly paradoxical claim makes sense in the light of the evolution of aging.

Until the middle of the last century, aging was viewed as an adaptation that benefited the species by removing worn out, old individuals. This view is still quite common among the general public, but it is incorrect. According to modern evolution theory, natural selection moulds gene action to optimize reproductive success, not longevity. Natural selection can favour genes that increase reproductive success in early adulthood even though they might promote pathology in later ages—so-called antagonistic pleiotropy (AP).[13] This means that although organismal senescence is not an adaptation, it is genetically programmed: late-life action of genes that bring us into existence eventually cause our death.[17]

Evolutionary theory predicts that senescent pathologies arise from late-life action of many genes. Animal model research has shown that some genes specifying central regulators of growth and development (*i.e.* common to most cell types) are AP determinants of multiple age-related pathologies, including some that contribute to late-life mortality. Inhibiting late-life effects of these genes can lead to amelioration of a wide spectrum of late-life pathologies—typically delaying their onset (decelerated aging). Thus, within the broader AP genetic predisposition one can define discrete genetic etiologies leading to syndromes of age-related pathology. For example, late-life accumulation of senescent cells (*sensu* Hayflick) and, particularly, mTor-activated gene expression in these cells appears to contribute to development of multiple age-related pathologies; research in mice suggests that these pathologies include atherosclerosis, the three major classes of cancer (carcinomas, sarcomas, lymphomas), glomerulosclerosis, cardiomyocyte hypertrophy, cardiac dysfunction, lipodystrophy and cataractogenesis.[18,19]

In conclusion, organismal senescence is a disease multi-syndrome, a set of syndromes and unitary diseases. The main underlying cause is wild-type genetic pre-disposition, in which respect it is etiologically distinct from most other diseases. However, standard etiologies (*e.g.* microbial pathogens, injury, gene mutations) do play a major role in organismal senescence, particularly when programmed aspects of aging increase predisposition to their pathogenic effects.

2.1.3 What is an Anti-Aging Intervention?

If one rejects the traditional dichotomy between normal aging and aging-associated diseases, then the meaning of *anti-aging* becomes problematic. If the aging disease super-syndrome is understood as the sum of senescent pathologies and their causes, this suggests that any treatment of any senescent pathology could be understood to be an anti-aging treatment. This is problematic because treatments for existing senescent pathologies (*e.g.* chemotherapy for late-life cancer or hip replacement operations) do not conform with the objective of improving late-life health by intervening in aging itself. This critical, central aim of biogerontology seems diluted or lost within such a redefinition of anti-aging.

In response to this, a new definition of anti-aging has been proposed[9] to retain the utility of the term. Here, anti-aging refers specifically to the preventative approach to improving late-life health. By this view, anti-aging treatments are interventions that counteract any etiology of organismal senescence. Based on the above account, two basic types of anti-aging treatment can be distinguished: those that act upon late-life AP etiologies (*e.g.* rapamycin); and those that prevent causes of pathology for which AP-generated senescence is a prerequisite (*e.g.* sunscreen, to prevent pre-cancerous lesion formation).

A further difference in this proposed new understanding of anti-aging is that it is based on the understanding that there is no one etiology of senescence. This means that no treatments inhibit the totality of aging, only parts of it. Consistent with this, anti-aging treatments with efficacy in animal models can improve late-life health and extend lifespan but not prevent aging altogether. This type of outcome is sometimes described as *decelerated aging*, but this is likely to be an imprecise description; more exactly, interventions of this sort impact the etiology of a cluster of senescent pathologies that limit lifespan—not the aging process overall.

According to the new definition of anti-aging, any preventative approach to senescent pathologies is an anti-aging treatment, whether the etiologies involved generate a broad or a narrow range of pathologies (defining broad *vs.* narrow geroprotectants). This means that not only are (potential) broad-sense interventions such as dietary restriction and mTor inhibition anti-aging interventions, but so also are narrow-sense interventions, such as the use of sunscreen to prevent sun damage to skin and the use of toothbrushes to prevent dental decay. By this view, dentists and, particularly, dental hygienists are narrow-sense anti-aging practitioners.

2.1.4 Aims of this Study: How is the Aging *vs.* Disease Division Represented in Medical Textbooks?

Do healthcare professionals regard aging as a disease, as a normal process, or as something entirely different? How much emphasis does medical education put on the process of aging? To explore these issues, we have taken two approaches. Firstly, we examine several previous studies that examine attitudes of health care professionals towards aging. Secondly, we explore what medical students are taught about aging. One may suppose that the rejection of the aging *vs.* disease dichotomy by many biogerontologists is informed by their study of the biology of aging, including reading the views of other biogerontologists. Similarly, the belief in the aging *vs.* disease dichotomy common among doctors is presumably attributable, at least in part, to what they learn in medical school. Important determinants of the frameworks of ideas within scientific and professional fields are the reference textbooks that are used for undergraduate teaching.[20] We have conducted a preliminary investigation of what medical students are taught about the relationship between aging and disease, analyzing 14 widely used textbooks of general medicine. We wished to discover to what extent textbooks argue that aging is distinct from disease and, if so, to examine the arguments and evidence presented for such a claim. For reference and comparison, Table 2.1 presents a selection of quotes arguing against the aging *vs.* disease dichotomy, many from biogerontologists.

Table 2.1 Selected quotations arguing against the aging *vs.* disease dichotomy.

Charcot, 1881, p. 20 [43]	"The textural changes which old age induce in the organism sometimes attain such a point that the physiological and pathological states seem to mingle by an imperceptible transition and to be no longer sharply distinguishable."
Kleemeier, 1965, p. 55 [44]	"Can the effects of aging *per se* be distinguished from those of pathology? (...) to attribute to aging all time associated changes to which no specific cause can be found is at best a temporary holding tactic which will suffice only as long as we are ignorant of the mechanism involved. Time alone causes nothing."
Hall, 1984, p. 78f [45]	"Attempts have been made by numerous workers to separate physiological from pathological aging. The two are, however, so interrelated as to make attempts relatively abortive. It would be far more relevant to accept the existence of a continuum of ageing phenomena."
Rattan, 1991, p. 526 [46]	"Although it is well known that most diseases show marked increases with age, the connection between the ageing process and the incidence of age-related diseases is highly underestimated. Recent developments in gerontology are unearthing the molecular link between ageing and disease."
Holliday, 1995, p. 138 [47]	"...The distinction between so-called natural ageing and the pathologies that are common in old people is artificial. What we see is an increasing likelihood of many diseases in individuals as they age, which does not, of course, mean that all individuals develop all the pathologies."
Callahan and Topinkova, 1998, p. 94 [2]	"In short, not only does aging lend itself to be characterised as a disease, but the advantage of doing so is that, by rejecting the seeming fatalism of the label 'natural', it better legitimises medical efforts to either eliminate it or to get rid of those undesirable conditions associated with it."
Guarente and Kenyon, 2000, p. 261 [3]	"When single genes are changed, animals that should be old stay young. In humans, these mutants would be analogous to a ninety-year-old who looks and feels forty-five. On this basis we begin to think of ageing as a disease that can be cured, or at least postponed."
Caplan, 2005, p. S75 [6]	"(...) the common belief that ageing is a natural process is also mistaken. And if that is true, and if it is actually the case that what occurs during the ageing process parallels the changes that occur during paradigmatic examples of disease (...), then it would be reasonable to consider ageing as a disease."
Gems, 2009, p. 3 [48]	"The evolutionary theory adds insult to injury by telling us that it is a process without any kind of benign function in the cycle of life; moreover, it is, essentially, a form of genetic disease, that everybody has and that is invariably fatal. We, all of us, have inherited a horrible and invariably fatal genetic disease."
Bulterijs *et al.*, 2015 p. 3 [49]	"As aging appropriately fits the definition of disease, there is a shifting consensus that aging should be seen as a disease process in itself, and not a benign progression of age that increases the risk of disease."

2.2 How is Aging Viewed in the Medical Field?

2.2.1 Two Surveys of the Medical Perception of Aging

How is the relationship between aging and disease perceived in the medical establishment? We were unable to identify any studies addressing this issue specifically. However, two studies analyse the concept of disease more broadly and include the question of the status of aging, and therefore give some indication of the medical perception of aging.[1,21] It is worth noting that neither study deals with the linguistics, *i.e.* no study distinguishes the different meanings of aging (*e.g.* maturation *vs.* senescence).

2.2.1.1 BMJ Vote on the Top 'Non-Diseases'

In 2002, the British Medical Journal (BMJ) ran a poll to identify the most widely recognized *non-diseases*. Non-disease was defined as "a human process or problem that some have defined as a medical condition but where people may have better outcomes if the problem or process was not defined in that way."[21]

The BMJ is ranked fifth amongst general medical journals. It targets doctors, researchers and other health professionals,[22] thereby addressing the core medical field. For the survey, the editorial board and journal readers brainstormed nearly 200 conditions potentially qualifying as non-diseases. Then, 570 people voted on whether a particular condition was a non-disease. Among these, aging ranked first, constituting the top non-disease (Table 2.2). This is striking considering the presence of other, clearly non-pathological conditions like work (2rd place) or boredom (3rd place).

One may argue that the survey format is likely to miss differing opinions on the classification of aging for several reasons. Firstly, of the 570 participants only 271 (44%) believed aging was a non-disease. Perhaps the remaining 56% disagree with aging as the top non-disease. However, whilst keeping this possibility in mind, the fact that aging is the most frequently identified non-disease is a strong indication of the prevailing notions in the medical field. Secondly, and most importantly, the BMJ definition of non-disease is not saying "this state is not a disease". Instead, the poll asked for conditions

Table 2.2 Top 20 non-diseases in descending order of *non-diseaseness*.[21]

1 Aging	11 Childbirth
2 Work	12 Allergy to the 21st century
3 Boredom	13 Jet lag
4 Bags under eyes	14 Unhappiness
5 Ignorance	15 Cellulite
6 Baldness	16 Hangover
7 Freckles	17 Anxiety about penis size/penis envy
8 Big ears	18 Pregnancy
9 Grey or white hair	19 Road rage
10 Ugliness	20 Loneliness

that are *best* not labelled as diseases. This definition of non-disease does not prohibit the opinion that aging is a disease. However, it appears irrational to think of something as a disease whilst also thinking that it was best not labelled as one. Thus, despite these two limitations, the BMJ study can be used to demonstrate that aging is not classified alongside other recognized diseases in the medical field.

2.2.1.2 Surveying the Public, Health Professionals and Legislators on Disease

A study from Finland by Tikkinen *et al.* provides a clearer picture of the medical perception of aging.[1] Again an opinion poll was taken on about 60 different states of being, with participants evaluating two claims: "(this state of being) is a disease" and "(this state of being) should be treated with public tax revenue." The study consulted four groups: 1517 members of the general public, 56 members of parliament, 741 doctors and 966 nurses. Given our interest in aging and disease, we have focused on the results of the first claim; however, it is notable that a correlation exists between responses to the two claims.

Tikkinen and colleagues show that of the 60 conditions, there is considerable variation in opinion as to whether 43 of them constitute diseases. The classification of the remaining 17 cases is clearer, as more than 80% of respondants agree with each other. Here, twelve states are clearly seen as diseases, and five states are clearly not. Interestingly, aging is among the conditions that are clearly not seen as diseases, along with grief, homosexuality, wrinkles and smoking (Figure 2.1).

A strength of this study is the large sample size. Its results suggest that laypeople are slightly more likely than health professionals to see aging as pathological. This is despite the fact that health professionals are, if anything, more inclined than laypeople to classify states as diseases.[1,23] But if clinicians do not view aging as a disease, what do they see it as instead? This question is particularly interesting as some states associated with aging (*e.g.* breast cancer, prostate cancer, deafness, adult onset diabetes) are viewed as diseases while others (*e.g.* insomnia, night-time urination, menopause, wrinkles) are not. Do healthcare professionals distinguish between pathological and non-pathological aging? To try to address this, we turned our attention to medical textbooks.

2.2.2 Medical Textbook Analysis

The BMJ survey and the Finnish study suggest that aging is best not labelled as a disease in medicine. However, as pointed out previously, the terminology is vague and there are conflations between chronological aging, age changes and senescence, as well as what constitutes normal and pathological in each of these areas. We suspect that the results of these surveys may partially reflect linguistic confusion. In particular, we argue that whilst chronological aging and many age changes are not pathological, senescence is a disease

Figure 2.1 Laypeople (L), doctors (D), nurses (N) and members of parliament (P) evaluate the claim "[This state of being] is a disease". Although there is much variation in the perception of disease, aging is one of the five states that is clearly not seen as a disease. This view is stronger amongst doctors and nurses than laypeople.[1] Reproduced from *BMJ Open*, Tikkinen *et al.*, 2, e001632 (© 2012), with permission from BMJ Publishing Group.[1]

(or a disease syndrome). By means of a textbook analysis, we aim to trace the roots of this linguistic confusion as they grow in the soil of undergraduate medical education.

We have examined how aging is described in medical textbooks. Textbooks accompany the medical curriculum, represent the roots of medical education and build a foundation for the values and attitudes in medicine. How frameworks of ideas are maintained within different fields can be discovered by textbook analysis.[20] How textbooks present aging is likely to be a major determinant of the medical view of the aging *vs.* disease dichotomy. The textbook analysis also served several additional purposes.

- To supplement the findings of the Tikkinen *et al.*[1] study; in particular to probe whether they are representative of attitudes beyond Finland.
- To add a qualitative dimension to the Finnish study; if aging is not viewed as a disease, then how is it viewed?
- To test the claim[6] that medical textbooks do not sufficiently deal with aging. As far as we can ascertain, a formal medical textbook analysis to this end has not been conducted before.
- To create a foundation for future, more detailed investigations of this issue.

Textbook analyses have been used in research before, for example to look at multiple editions of the same textbook to understand how the presentation of obesity has changed,[24] and how the idea of giving medical prognoses has faded over the years.[25] Other studies have looked at a range of textbooks to assess whether they provide adequate factual information on specific topics[26–28] or adequate patient-orientated communication skills.[29]

Our main aim here is to discover how medical textbooks present the relationship between aging and disease. In particular, do they specifically argue the existence of a separate, non-pathological process of aging? If so, what is the justification for this separation? And what are the criteria for deciding which deteriorative age-changes are part of normal aging, and which are pathological changes?

2.2.2.1 Methodology

2.2.2.1.1 Textbook Selection. The study was conducted in University College London (UCL) libraries. For the final analysis, 14 textbooks were selected. Due to the great number and variety of medical textbooks, the selection process was not straightforward. Medicine is divided into more than 40 disciplines with separate textbooks.[30] An interesting question is how aging is understood in different medical disciplines, but this lies beyond the scope of this study. Instead, we focus on textbooks of general clinical medicine, also known as reference books. These textbooks include factual knowledge to practise medicine, explain basic science, research evidence and the context of underlying principles. Moreover, they outline how to apply this knowledge to manage patients. However, even within this niche, a plethora of textbooks exists. For instance, some may constitute multi-volume reference guides for professionals, whilst others are intended as pocket books for junior doctors and others as revision aids for a specific student exam. Which ones are most widely consulted amongst university students? It appears that a universal list identifying key textbooks for each medical discipline, compiled by asking medical schools for their recommendations, was last created in 1971.[30] Other studies employ one or a few subjective methods to create a selection of textbooks for analysis.[28,29] Therefore, we used a combination of approaches to assess the popularity of medical textbooks. Overlap in the following subjective sources indicates the frequent use of particular textbooks:

- A review of articles and blog posts recommending a list of medical textbooks (*e.g.* The Student Room Community, 2015).[31]
- A review of the number of holdings in the library shelf WB100 *Practice of Medicine*.
- A review of the short loans collection, shelving the most frequently borrowed library books.
- A review of textbooks used in previous medical textbook analyses.
- A review of the medical core collection for libraries with the tag *general medicine*, as specified by the Chartered Institute of Library and Information Professionals (CILIP).
- A review of readings lists for the Bachelor of Medicine and Surgery (MBBS) programme.

Table 2.3 shows our final selection of medical textbooks. The table includes the number of worldwide library holdings as an indicator of the relative popularity of the selected textbooks. This information was extracted from WorldCat, a global platform assembling library holdings and thereby creating a *collective collection* of worldwide libraries.[32]

2.2.2.1.2 The Research Process. To established how the selected textbooks deal with aging, we first reviewed the index for the term *aging/ageing* (US/UK spelling). For books where the index did not contain the term, no further analysis was conducted. If the index did contain the term, we checked whether: (a) the textbook dedicates an entire chapter or more on aging; and (b) the textbook deals with the aging *vs.* disease dichotomy. For the latter, the book had to show some acknowledgement of the complexity of the aging process and to put it into a medical context. This might include addressing some of the following questions. What is the relationship between aging and age-related diseases? Is the first normal and the second pathological? What is the current state of research into the biology of aging? What are the mechanisms and evolutionary origins of aging? What is the future of geriatric medicine? Can one intervene in the aging process? Textbooks do not deal with the aging *vs.* disease dichotomy if they only offer descriptive accounts, such as outlining changes or diseases with the highest prevalence amongst elderly, or discussing

Table 2.3 Medical textbooks selected for analysis. Listed by title in alphabetical order.

Name of textbook	Mention 'aging' in index (Yes/No)	Chapter on 'aging' (Yes/No)	Library holdings worldwide[b]
Blueprints Medicine[50]	N	N	232
Color Atlas and Text of clinical Medicine[51]	N	N	409
Davidson's Principles and Practise of Medicine[37]	Y[a]	Y	848
Goldman's Cecil Medicine[52]	Y[a]	Y	572
Harrison's Principles of Internal Medicine[36]	Y[a]	Y	2089
Kumar and Clark's Clinical Medicine[34]	Y[a]	N	485
Lecture Notes: Clinical Medicine[53]	N[a]	N	293
Medical Sciences[54]	N[a]	N	268
Medicine[33]	Y[a]	N	185
Medicine and Surgery – an Integrated Textbook[55]	N	N	240
Medicine at a Glance[56]	Y[a]	N	347
Oxford Handbook of Clinical Medicine[35]	Y[a]	N	488
Oxford Textbook of Medicine[57]	Y[a]	N[a]	14
Textbook of Medicine[38]	Y[a]	Y	348

[a]These textbooks mention a related term such as "elderly", "geriatrics" or "older adult".
[b]Across all editions. Extracted from WorldCat on 7th December 2015.

how to manage and treat older adults. For the purpose of this study, we are interested in textbook passages that discuss the aging-disease dichotomy.

To check whether accounts of the aging *vs.* disease dichotomy are tied to other terms, we reviewed the textbooks for the terms *elderly*, *geriatrics* and *older adults*. Additionally, in the final qualitative analysis, we selected the textbooks that touch on the aging *vs.* disease dichotomy, and used quotations and illustrations to analyse specifically whether aging is seen as a disease.

2.2.2.1.3 Strengths and Limitations of the Textbook Analysis. In terms of the textbook selection, this study is limited to textbooks of general clinical medicine. Moreover, it does not consider the specific audience each textbook is aimed at. For example, some books are clearly geared towards undergraduate students, while others may be written as handbooks for junior doctors or reference guides for professionals. Thus, it is not clear whether similar results would be obtained by analysing textbooks across different medical fields or textbooks targeting specific audiences.

Additionally, it is unlikely that undergraduate medical students will limit their readings to student textbooks, but access professional sources for better and more detailed understanding of particular areas. However, the sample of textbooks analyzed here does represent popular general clinical textbooks used by undergraduate students; several methods were used to define the most widely consulted general clinical textbooks, to try to reduce selection bias. Nonetheless, the selection was drawn from a study of University College London (UCL) libraries, and other universities may hold different types of clinical textbooks.

This limitation was somewhat balanced by including WorldCat ratings in the analysis. These offer a global basis for comparison of the popularity of the final textbook selection. It should be noted, however, that the WorldCat numbers refer to all editions of a particular textbook. Thus, it is likely that older textbooks will have more holdings than more recent textbooks, regardless of the popularity. Moreover, WorldCat search results do not necessarily correspond to all available items because some libraries may not have subscribed to their service. Additionally, it appears that WorldCat is biased, for example by excluding non-academic libraries, such as hospital libraries, from their search results. Nevertheless, WorldCat numbers are valuable in offering an objective measure of textbook usage.

2.2.2.2 Results

Of the 14 books reviewed, five (35.7%) do not mention aging and ten (71.4%) do not dedicate an entire chapter to the topic (Table 2.3). Searches for additional terms elderly, geriatrics or older adults were also performed but did not lead to discussions of the nature of aging. For example, the *Oxford Textbook of Medicine*, a comprehensive three volume reference guide dedicates an entire chapter to Gerontology, but does not discuss the nature of aging itself. Instead, the chapter focuses on the concept of frailty and the major problems which bring older people into hospital other than specific diseases (*e.g.* falls, pressure sores, incontinence). The chapter also includes a comprehensive guide to geriatric assessment and care.

Similarly, in other less detailed textbooks index entries for elderly, geriatrics or older adults typically refer to changes occurring in particular age groups. A typical example is the following passage from *Medicine*: "Many of the patients now on renal replacement therapy are elderly. In the elderly, most renal diseases are seen with greater frequency because of increased incidence of hypertension, diabetes mellitus, vascular disease and prostatic disease. Tumours are also more common in the elderly. Of the glomerular diseases, membranous nephropathy is more common in the elderly" (p. 503).[33]

2.2.2.2.1 Books Mentioning Aging but Without Dedicated Chapters. Nine textbooks (64.3%) mention aging in the index, of which four dedicate at least one chapter on the topic and are discussed later. In the remaining five, there are few index entries on the topic, and what they refer to varies considerably. For example, they examine aging under the topic headings cancer, drug side effects, haematological changes, hypogonadism and skin changes. Thus, aging is presented as a modulator and risk factor for disease. The nature of aging itself is not discussed.

Among these five, two—*Kumar and Clark's* and the *Oxford Handbook of Medicine*—stand out in how they depict aging and portray the elderly, acknowledging the different ways that aging is conceptualised in medicine.[34,35] In *Kumar and Clark's* aging is described in the chapter *Nutrition*. Here, there is a synthesis between theories of aging and nutrition as a key moderator of the aging process. There is no discussion of aging in relation to disease, though the elderly are occasionally portrayed in a somewhat depressing way. For example, there is a discussion of malnutrition due to "lack of cooking skills (particularly in widowers), depression and lack of motivation" (p. 215). Moreover, it is noted that elderly people "in institutions" commonly have multiple nutritional deficiencies and vitamin D supplements may be required because "often elderly people do not go into the sunlight" (p. 215).

By contrast, the *Oxford Handbook of Clinical Medicine* appears more at pains to counter negative stereotyping of the elderly. Here aging is mentioned in the chapter 'Thinking about Medicine' and presented as a disease-like state: "Any deterioration in an elderly patient is from treatable disease until proven otherwise. Find the cause; don't think: this is simply aging. Old age is associated with disease but doesn't cause it *per se*. Do not restrict treatment because of age—age alone is a poor predicator of outcome." (p. 12). Interestingly, despite this clear statement about aging, there is no separate chapter to deal with this issue. In summary, the index entries and contrasting depictions demonstrate the varied representation of aging in medical textbooks.

2.2.2.2.2 Books with Dedicated Chapters on Aging. Of the 14 textbooks, four (28.6%) have a specific chapter on aging. Looking at the number of editions and WorldCat library holdings, these textbooks are among the most established and popular of those examined here. They make the most reference to questions of what aging is and its relationship to disease. *Harrison's Principles of Internal Medicine* includes two chapters on aging[36] and *Davidson's Principles and Practise of Medicine* has information boxes throughout the book relating each condition to old age.[37] So is aging depicted as distinct from, similar to or the same as disease?

The general trend across these four textbooks is similar: aging is neither regarded as a disease, nor as something entirely normal, but has components of both. A recurrent term is *geriatric condition*, referring to deteriorative changes with age that are not regarded as diseases. For example, "a sudden onset of headaches or a recent change in bowel habit is never normal in old age, whereas gradually failing hearing and vision may be" (*Textbook of Medicine*, p. 191).[38] Table 2.4 lists more quotations that touch upon the distinction between disease and aging. Additionally, Figure 2.2 and Tables 2.5 and 2.6 show how various senescent changes are categorized into pathologies and non-pathologies.

How is it decided whether a given senescent change is to be viewed as pathological or normal? Overall, there seems to be a consensus agreement

Table 2.4 Quotations from four medical textbooks highlighting the distinction between normal and pathological aging.

Davidson's Principles and Practise of Medicine[37]	"The physiological features of normal ageing have been identified by examining disease-free populations of older people, to separate the effects of pathology from those due to time alone. However, the fraction of older people who age without disease ultimately decline to very low levels so that use of the term 'normal' becomes debatable." p. 167
	"The effects of ageing are usually not enough to interfere with organ function under normal conditions, but reserve capacity is significantly reduced." p. 167
	Information box on atherosclerosis and ageing: "Prevalence: related almost exponentially to age in developed countries, although atherosclerosis is not considered part of the normal ageing process" p. 602b
Textbook of Medicine[38]	"The resulting disability is not fixed or inevitable. For example, high-tone deafness and high blood pressure are common in elderly Britons, but are absent among elderly persons in the Eastern Islands. Osteoporosis is common in western Europe and the USA but rare in China. Thus descriptions of physical decline are too variable to be useful for defining ageing." p. 172
	"Degenerative changes occur throughout the body with increasing age, but these may become sufficiently marked to constitute a pathological process. The distinction between this and normal physiological ageing is often difficult to make, and there is increasing recognition that so-called normal ageing is the result of occult pathology." p. 174
	"A sudden onset of headaches or a recent change in bowel habit is never normal in old age, whereas gradually failing hearing and vision may be." p. 191
	"Occasional ectopic beats occur in about one-sixth of elderly persons, but any other arrhythmia should be regarded as abnormal and investigated by electrocardiography" p. 192
Goldman's Cecil Medicine[52]	"Health status in aging is a result of many factors, including the chronic diseases of aging and many other prevalent "geriatric" conditions that cannot be defined as classic "diseases" because they do not result from a single pathological cause." (p. 98)
	"Physiologic aging modulates the way in which illnesses cause signs and symptoms" p. 105
	"Geriatric syndromes emerge from these age-related changes" p. 105
	"Some physiological changes imitate illness when they may be a normal part of aging. Diabetes mellitus may 'appear' and 'disappear' in the elderly." p. 105
	"The major clinical impact of normal physiologic ageing in the lungs in an earlier appearance of shortness of breath as warning signal of underlying disease." p. 106
	"Normal ageing produces an obvious decrease in the size of the thymus gland" p. 106
	"Severe neuropsychiatric conditions are due to diseases that increase with age but are not part of the normal aging process." p. 114
	"The process of aging produces important physiologic changes in the central nervous system (...). These processes result in age-related symptoms and manifestations for many older persons (...) the decline may be modified by factors such as diet, exercise, environment, lifestyle, genetic predisposition, disability, disease and side effects of drugs. These changes can result in the common age-related symptoms of benign senescence, slowed reaction time, postural hypotension, vertigo (...). In the absence of disease, these physiologic changes usually result in relatively modest symptoms and little restriction in activities of daily living. The changes decrease physiologic reserve, however, and increase the susceptibility to challenges posed by disease-related, pharmacologic and environmental stressors." p. 114

(continued)

Table 2.4 (*continued*)

Harrison's Principles of Internal Medicine[36]	"The phenotype that results from the aging process is characterized by increased susceptibility to diseases, high risk of multiple coexisting diseases, impaired response to stress (including limited ability to heal or recover after an acute disease), emergence of "geriatric syndromes" (characterized by stereotyped clinical manifestations but multifactorial causes), altered response to treatment, high risk of disability, and loss of personal autonomy with all its psychological and social consequences. In addition, these key aging processes may interfere with the typical pathophysiology of specific diseases, thereby altering expected clinical manifestations and confounding diagnosis." p. 76 "The term geriatric syndrome encompasses clinical conditions that are frequently encountered in older persons; have a deleterious effect on function and quality of life; have a multifactorial pathophysiology, often involving systems unrelated to the apparent chief symptom; and are manifested by stereotypical clinical presentations. The list of geriatric syndromes includes incontinence, delirium, falls, pressure ulcers, sleep disorders, problems with eating or feeding, pain, and depressed mood. In addition, dementia and physical disability are sometimes considered to be geriatric syndromes." p. 79 "Normal aging is associated with a decline in food intake that is more marked in men than in women." p. 81 "Modest changes in balance function have been described in fit older individuals as a result of normal aging." p. 164 "The aging process is the major risk factor underlying disease and disability in developed nations, and older people respond differently to therapies developed for younger adults (usually with less effectiveness and more adverse reactions)." p. 94e-1 "The phenotypic components of aging include structural and functional changes that are separated, somewhat artificially, into either primary aging changes (*e.g.* sarcopenia, grey hair, oxidative stress, increased peripheral vascular resistance) or age-related disease (*e.g.* dementia, osteoporosis, arthritis, insulin resistance, hypertension)." p. 94 e-1 "Clinicians need to understand aging biology in order to better manage people who are elderly now. Moreover there is an urgent need to develop strategies based on aging biology that delay aging, reduce or postpone the onset of age-related disorders, and increase functional life and healthspan for future generations. Interventions related to nutritional interventions and drugs that act on nutrient-sensing pathways are being developed and, in some cases, are already being studied in humans. Whether these interventions are universally effective or species/individual specific needs to be determined." p. 94 e-7

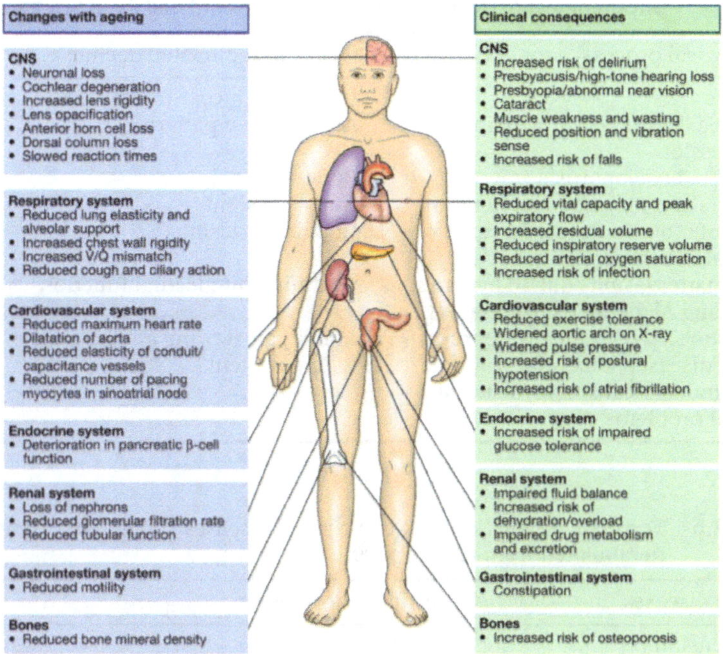

Features and consequences of normal ageing.

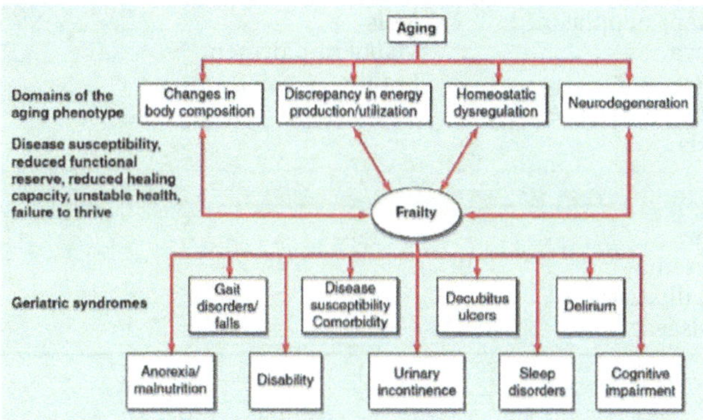

A unifying model of aging, frailty, and the geriatric syndromes.

Figure 2.2 Illustrations of the distinction between normal and pathological aging from two textbooks of clinical medicine. Top, Davidson's Principles and Practise of Medicine, ref. 37. Bottom, Kasper, *et al.*, Harrison's Principles of Internal Medicine 19/E, 2015, McGraw-Hill Education, ref. 36. Reproduced with permission of McGraw-Hill Education.

with respect to this division. For example, visual impairment, hearing impairment, falls and bone loss are attributed to normal aging. Diabetes and Alzheimer's disease, on the other hand, are referred to as diseases. However, for some conditions, such as sleep disturbances, classifications vary (Figure 2.2).

Table 2.5 Textbook distinction between normal *vs.* pathological decrements in physiological performance in old age. Retabulated from ref. 38, p. 177.

Pathological process	'Normal' age-related decline
Diabetes	Impaired glucose tolerance
Insomnia due to nocturia, pain and depression	Altered sleep pattern
Accelerated osteoporosis	Bone loss
Cataracts and macular degeneration	Impaired sight
Effects of ototoxic drugs, diuretics or vascular disorders	Impaired hearing
White matter lesions due to hypertension and Alzheimer's disease	Declining intellectual function
Parkinson's, Alzheimer's, cerebrovascular disease, dementia, vestibular lesions, cervical spondylosis and visual problems	Minor gait slowing and balance impairment

Table 2.6 Textbook distinction between clinical diseases and geriatric syndromes. Retabulated from ref. 52, p. 99.

Clinical diseases	Geriatric syndromes
Hypertension	Disability
Arthritis	Hearing impairment
Heart disease	Urinary incontinence
Malignant neoplasm	Falls
Influenza	Visual impairment
Diabetes	Frailty
Alzheimer's disease	
Sinusitis	
Ulcers	
Stroke	
Asthma	
Emphysema	
Kidney disease	
Liver disease	

While no clear criteria are presented to distinguish normal senescent changes from pathological ones, there is some recognition that this binary distinction can be problematic. For example, according to *Davidson's*, "the physiological features of normal aging have been identified by examining disease-free populations of older people, to separate the effects of pathology from those due to time alone. However, the fraction of older people who age without disease ultimately declines to very low levels so that use of the term 'normal' becomes debatable" (p. 167). *Harrison's* notes that the separation occurs "somewhat artificially" (p. 94 e-1). This problem is further outlined in the *Textbook of Medicine*: "There has been a long-running debate on

Figure 2.3 Representation of the conventional view that "aging is normal"; from Basic Pathology, Lakhani *et al.*, (© 2009), Taylor & Francis ref. 58. This excellent textbook on the biological basis of disease presents this nice cartoon without further explanation. Reproduced by permission of Taylor & Francis Books UK.

what constitutes normal gait in old age and the cause of senile gait disorder" (p. 180f.). Generally, the textbook recognises that "the distinction between this and normal physiological aging is often difficult to make, and there is increasing recognition that so-called normal aging is the result of occult pathology" (p. 174) (Figure 2.3).

2.2.2.2.3 Summary of Findings. There is considerable variability in the way that aging is represented in textbooks of general clinical medicine. Broadly, three categories exist:

(1) Textbooks that do not mention aging in the index. This is the case in a large proportion of textbooks. They do not address the question of what aging is, the division between aging and disease, *etc.* Related terms in the index, such as elderly, geriatrics, or older people, refer to descriptive passages, where changes in the elderly are outlined or examination procedures recommended.

(2) Textbooks that mention aging in the index, but do not have a specific chapter dedicated to the topic. Here, aging is referred to in multiple sections of the book. Aging is not the focus of these sections, but rather a risk factor for other diseases or a modifier for drug dosages. In most cases, the textbooks do not include discussions of aging itself.

(3) Textbooks that have at least one chapter on aging. This is the case in four clinical textbooks, which appear to be the more popular ones. They deal with aging in the most detail, but do not label it as a disease. Instead, all four textbooks present aging as partly normal and partly pathological. Problems arising from this distinction are recognized by all textbooks. However, none provide a rationale for viewing aging as a natural and non-pathological part of senescence, or define criteria for distinguishing pathological and non-pathological elements of senescence.

2.3 Discussion

The aim of this study is to examine the extent to which medicine still relies on the traditional distinction between normal aging and disease. Based on its etiology, we argue that senescence is a pathological process and phenomenon. By contrast, non-pathological senescence is a relic concept rooted in traditional ideas about aging, whose origins seem to lie as far back as the writings of the Roman physician Galen in the second century AD.[39] Interestingly, our analysis reveals considerable variety in the way that aging is presented in medical textbooks. It is at times presented as an underlying risk factor for disease, or a modulator for drug dosages, or looked at in biogerontological terms, or barely mentioned at all. Only four out of fourteen textbooks examined consider the nature of aging itself, and the relation between aging and disease. Here aging is seen not as a disease, but as something between a pathological and normal process. This diverse pattern of representation of aging across medical textbooks appears to reflect linguistic confusion caused by the multiple and easily conflated meanings of the English word aging. This confusion impacts on medical understanding and medical care.

What are the implications of this pattern of representation of aging? Given that some textbooks barely refer to aging, and those that do rarely discuss the nature of aging or its relationship to pathology, this suggests that many medical students are left in the dark about these critical issues. Moreover, regardless of which textbook medical students use, they will not be taught that aging is a disease. Not even the more popular textbooks that discuss aging support this notion. Instead, they explain that aging lies somewhere between normal and pathological processes. This distinction is artificial, confusing and problematic, especially when classifications vary, as seen with sleep disturbances. Therefore, one clinician may refrain from treatment, dismissing sleep disturbances as normal, whilst another clinician may seek treatment. More broadly, failing to understand senescence as pathology is not only inadequate in scientific terms, but also a barrier to delivering quality treatment to the elderly. Underlining this point, the surgeon Gawande (2014) acknowledges that "(...) scientific advances have turned the processes of aging and dying into medical experiences, matters to be managed by health care professionals. And we in the medical world have proved alarmingly unprepared for it" (p. 6).[40]

Our small scale textbook analysis raises several further questions and directions for future research. First, it leaves unanswered whether the content of the textbooks examined represent general medical views, which are also influenced by lectures, work placements, and personal experiences. Looking at the previous studies on medical conceptualisations of aging, there appears to be a correspondence between textbook content and the later attitude of medical professionals. For instance, like the textbooks, the study by Tikkinen *et al.* indicates that healthcare professionals do not see aging as a disease.[1] However, the study gives no indication of whether doctors see aging as part pathological and part normal—the view represented in more popular textbooks—or whether they have encountered arguments against the aging-disease false dichotomy at all. In-depth interviews or focus groups could yield more information about how medical professionals learn about concepts of aging, and how these influence their treatment of the elderly.

Second, the question arises as to why the view of aging not being a disease, but rather a normal occurrence, is so persistent in medicine. Gawande (2014) claims that "people naturally prefer to avoid the subject of their decrepitude" (p. 35) and doctors are turned off by geriatrics, because they do not have the faculties to cope with it.[40] This idea is discussed by Caplan (2005), who suggests that doctors employ the ideas of the *naturalness* of aging as a type of defense mechanism against despair when repeatedly dealing with chronically and incurably ill elderly patients.[6] There are other possible explanations for doctors' reluctance to view aging as a disease that could be investigated. For example, is it that they wish to avoid association with quack peddlers of anti-aging medication? Do they view interventions in the aging process as artificial enhancement technologies? Do they anticipate adverse economic consequences of an increasingly older population? Do they believe that the goals of treating aging are intangible? Or do they believe that experiencing the aging process has its benefits? These questions could be answered by structured interviews or focus groups with medical professionals.

It would also be interesting to explore whether the medical representation of aging has changed across the years, similar to that of obesity.[24] For example, osteoporosis was not regarded as a disease by the WHO until 1994.[41] Perhaps this is indicative of a broader reconceptualisation of aging (*i.e.* senescence) informed by biogerontological investigation. Considering the textbook analysis, the more popular and more established textbooks distinguish themselves by noting the artificial distinction between normal and pathological aging. Perhaps this acknowledgment is the first step into fully recognizing aging as a disease? A textbook analysis, looking at how aging is portrayed in previous editions of these volumes, could test for the existence of this transformation process.

Lastly, it must be noted that although moving towards a disease classification is reasonable and beneficial in several ways, it increases the risk of biomedicalizing aging. That is, there is a danger that a new medical model takes over and defines other non-biological processes of aging, including social and psychological ones.[42] Therefore, more research is needed to find out how to recognise aging as a disease without diverting resources away from understanding these other important phenomena of aging.

To conclude, our analysis of medical textbook content suggests a general neglect of the question of what aging is, unease about the somewhat arbitrary classification of different manifestations of senescence as normal or pathological, and the absence of any rationalization of the concept of normal aging. Some of these problems reflect linguistic confusion created by the word aging. These observations suggest that medicine remains in the dark about aging.

Acknowledgements

We thank E. Galimov for critical reading of the manuscript, A. Zhavoronkov for drawing our attention to the Tikkinen *et al.* (2012) study, and D. Khaltourina for pointing out the different meanings of *aging* and *stareniye*. This work was supported by a Wellcome Trust Strategic Award to D.G.

References

1. K. A. O. Tikkinen, J. S. Leinonen, G. H. Guyatt, S. Ebrahim and T. L. N. Jarvinen, *BMJ Open*, 2012, **2**, e001632.
2. D. Callahan and E. Topinkova, *Drugs Aging*, 1998, **13**, 93.
3. L. Guarente and C. Kenyon, *Nature*, 2000, **408**, 255.
4. N. Cutler, N. W. Whitelaw and B. L. Beattie, *American Perceptions of Aging in the 21st Century*, National Council on Aging, Washington, D.C., 2002.
5. G. J. Izaks and R. G. Westendorp, *BMC Geriatr.*, 2003, **3**, 7.
6. A. L. Caplan, *EMBO Rep.*, 2005, **6**, S72.
7. M. Ereshefsky, *Stud. Hist. Philos. Biol. Biomed. Sci.*, 2009, **40**, 221.
8. T. Schramme, *Theor. Med. Bioeth.*, 2013, **34**, 171.
9. D. Gems, *Exp. Gerontol.*, 2014, **58**, 14.
10. D. Gems, *Front. Genet.*, 2015, **6**, 212.
11. A. Comfort, *The Biology of Senescence*, Elsevier, New York, 1979.
12. C. E. Finch, *Longevity, Senescence and the Genome*, University of Chicago Press, Chicago, 1990.
13. G. C. Williams, *Evolution*, 1957, **11**, 398.
14. B. L. Strehler, *Time, Cells, and Aging*, Academic Press, New York, 1977.
15. L. Hayflick, *Ann. N. Y. Acad. Sci.*, 2007, **1100**, 1.
16. S. I. S. Rattan, *Aging Dis.*, 2014, **5**, 196.
17. M. V. Blagosklonny, *Cell Cycle*, 2006, **5**, 2087.
18. D. J. Baker, B. G. Childs, M. Durik, M. E. Wijers, C. J. Sieben, J. Zhong, R. A. Saltness, K. B. Jeganathan, G. C. Verzosa, A. Pezeshki, K. Khazaie, J. D. Miller and J. M. van Deursen, *Nature*, 2016, **530**, 184.
19. J. M. van Deursen, *Nature*, 2014, **509**, 439.
20. T. S. Kuhn, *The Structure of Scientific Revolutions*, University of Chicago Press, Chicago, 1962.
21. R. D. Smith, *BMJ*, 2002, **324**, 883.
22. BMJ, 2016, http://www.bmj.com/about-bmj.

23. E. J. M. Campbell, J. G. Scadding and R. S. Roberts, *BMJ*, 1979, **2**, 757.
24. V. W. Chang and N. A. Christakis, *Sociol. Health Illn.*, 2002, **24**, 151.
25. N. A. Christakis, *Soc. Sci. Med.*, 1997, **44**, 301.
26. S. A. Azer, *AJP: Adv. Physiol. Educ.*, 2004, **28**, 36.
27. M. Andrikopoulou, L. Michala, S. M. Creighton and L. M. Liao, *J. Obstet. Gynaecol.*, 2013, **33**, 648.
28. A. G. Powell, S. Paterson-Brown and G. B. Drummond, *BMC Med. Educ.*, 2014, **14**, 35.
29. A. M. Easson, J. A. Crosby and S. L. Librach, *Am. J. Surg.*, 2001, **182**, 34.
30. G. Inke, *Bull. Med. Libr. Assoc.*, 1971, **59**, 589.
31. The Student Room Community, 2015, http://thestudentroom.co.uk/content.php?r=16151-medicine-textbook-recommendations.
32. OCLC, 2016, https://www.oclc.org/worldcat.en.html.
33. J. S. Axford and C. A. O'Callaghan, *Medicine*, Wiley-Blackwell, Malden, MA, 2nd edn, 2004.
34. P. Kumar and M. L. Clark, *Kumar and Clark's Clinical Medicine*, Elsevier, Edinburgh, 8th edn, 2012.
35. J. M. Longmore, I. Wilkinson, A. Baldwin and E. Wallin, *Oxford Handbook of Clinical Medicine*, Oxford University Press, Oxford, 9th edn, 2010.
36. D. L. Kasper, A. S. Fauci, S. L. Hauser, D. L. Longo, J. L. Jameson and J. Loscalzo, *Harrison's Principles of Internal Medicine*, McGraw-Hill Education, New York, 19th edn, 2015.
37. B. R. Walker, N. R. Colledge, S. H. Ralston and I. D. Penman, *Davidson's Principles and Practice of Medicine*, Churchill Livingstone, Edinburgh, 22nd edn, 2014.
38. R. L. Souhami and J. Moxham, *Textbook of Medicine*, Churchill Livingstone, Edinburgh, 4th edn, 2002.
39. K. Cockayne, *Experiencing Old Age in Ancient Rome*, Taylor and Francis, 2003.
40. A. Gawande, *Being mortal: Illness, Medicine and What Matters in the End*, Profile Books, London, 2014.
41. WHO, *Assessment of Fracture Risk and its Application to Screening for Postmenopausal Osteoporosis*, WHO, Geneva, 1994.
42. C. L. Estes and E. A. Binney, *Gerontologist*, 1989, **29**, 587.
43. J. M. Charcot, *Clinical Lectures on the Diseases of Old Age*, Wood, 1881.
44. R. W. Kleemeier, *Contemp. Psychol.*, 1965, **10**, 53.
45. D. I. A. Hall, *The Biomedical Basis of Gerontology*, Wright–PSG, London, 1984.
46. S. I. S. Rattan, *Perspect. Biol. Med.*, 1991, **34**, 526.
47. R. Holliday, *Understanding Ageing*, Cambridge University Press, Cambridge, 1995.
48. D. Gems, in *Länger leben? Philosophische und biowissenschaftliche Perspektiven*, ed. M. W. S. Knell, Suhrkamp Taschenbuch Wissenschaft, 2009, pp. 25–45.
49. S. Bulterijs, R. S. Hull, V. C. E. Björk and A. G. Roy, *Front. Genet.*, 2015, **6**, 205.

50. V. B. Young, W. A. Kormos, D. A. Chick and A. H. Goroll, *Blueprints Medicine*, Lippincott Williams & Wilkins, Philadelphia, 5th edn, 2009.
51. C. D. Forbes and W. F. Jackson, *Color Atlas and Text of Clinical Medicine*, Mosby, Edinburgh, 3rd edn, 2003.
52. R. L. Cecil, L. Goldman and A. I. Schafer, *Goldman's Cecil Medicine*, Elsevier, Philadelphia, 24th edn, 2012.
53. J. Bradley, M. Gurnell, D. Wood and D. Rubenstein, *Lecture Notes. Clinical Medicine*, Wiley-Blackwell, Hoboken, NJ, 7th edn, 2012.
54. J. Naish, P. Revest and D. Syndercombe Court, *Medical Sciences*, Elsevier, Edinburgh, 1st edn, 2009.
55. E. K. S. Lim, Y. K. Loke and A. M. Thompson, *Medicine and Surgery: An Integrated Textbook*, Churchill Livingstone, Edinburgh, 1st edn, 2007.
56. P. Davey, *Medicine at a Glance*, Wiley-Blackwell, Hoboken, NJ, 3rd edn, 2010.
57. D. A. Warrell, T. M. Cox and J. D. Firth, *Oxford Textbook of Medicine*, Oxford University Press, Oxford, 5th edn, 2010.
58. S. Lakhani, S. A. Dilly and S. J. Finlayson, *Basic Pathology*, Hodder Arnold, London, 4th edn, 2009.

CHAPTER 3

The Search for the "Anti-Aging Pill": A Critical Viewpoint

ÉRIC LE BOURG[a]

[a]Centre de Recherches sur la Cognition Animale (CRCA), Centre de Biologie Intégrative, Université de Toulouse, CNRS, 31062 Toulouse cedex 9, France
*E-mail: eric.le-bourg@univ-tlse3.fr

3.1 Introduction

For centuries, not to say millennia, some people tried to discover (or claimed to have discovered) miraculous recipes providing a long and youthful life. For instance, a French book[1] reported in 1768 the recipe of an elixir of long life (Figure 3.1): this very *élixir de longue vie* is still sold today, under the names of Swedish bitters in the USA or UK and *Élixir du Suédois* in France (but mainly for digestive purposes, and no longer as an elixir of long life). These elixirs of long life had more to do with charlatanism than with basic science and were mocked as early as in 1749 by the famous French naturalist Georges Buffon, who wrote that "the universal panacea, the transfusion of blood, and other methods which have been proposed to render our bodies immortal, are as chimerical as the fountain of youth is fabulous".[2]

Up to recent times, infant mortality was so high (*e.g.* 33% in Ontario, Canada, in 1901[3]) and infectious diseases so widespread[4] that mean lifespan was rather low, *e.g. ca.* 50 years at the beginning of the twentieth century in Western Europe. Since that time, life expectancy has strongly increased in developed countries, but also in emergent countries, and the fate of an increasing

RSC Drug Discovery Series No. 57
Anti-aging Drugs: From Basic Research to Clinical Practice
Edited by Alexander M. Vaiserman

Published by the Royal Society of Chemistry, www.rsc.org

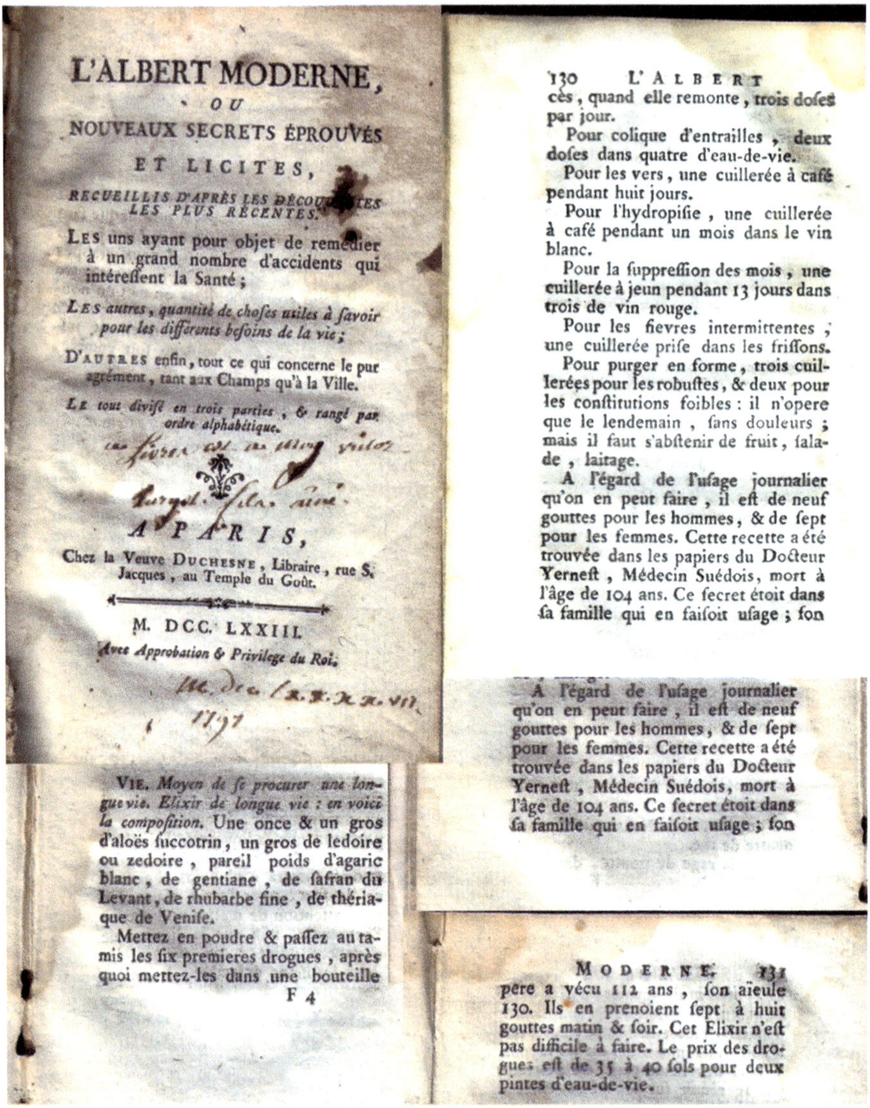

L'ALBERT MODERNE,
OU
NOUVEAUX SECRETS ÉPROUVÉS
ET LICITES,
RECUEILLIS D'APRÈS LES DÉCOUVERTES LES PLUS RÉCENTES.

LES uns ayant pour objet de remédier à un grand nombre d'accidents qui intéreſſent la Santé;

LES autres, quantité de choſes utiles à ſavoir pour les différents beſoins de la vie;

D'AUTRES enfin, tout ce qui concerne le pur agrément, tant aux Champs qu'à la Ville.

Le tout diviſé en trois parties, & rangé par ordre alphabétique.

A PARIS,
Chez la Veuve DUCHESNE, Libraire, rue S. Jacques, au Temple du Goût.

M. DCC. LXXIII.
Avec Approbation & Privilege du Roi.

130 L'ALBERT

cès, quand elle remonte, trois doſes par jour.

Pour colique d'entrailles, deux doſes dans quatre d'eau-de-vie.

Pour les vers, une cuillerée à café pendant huit jours.

Pour l'hydropiſie, une cuillerée à café pendant un mois dans le vin blanc.

Pour la ſuppreſſion des mois, une cuillerée à jeun pendant 13 jours dans trois de vin rouge.

Pour les fievres intermittentes, une cuillerée priſe dans les friſſons.

Pour purger en forme, trois cuillerées pour les robuſtes, & deux pour les conſtitutions foibles : il n'opere que le lendemain, ſans douleurs ; mais il faut s'abſtenir de fruit, ſalade, laitage.

A l'égard de l'uſage journalier qu'on en peut faire, il eſt de neuf gouttes pour les hommes, & de ſept pour les femmes. Cette recette a été trouvée dans les papiers du Docteur Yerneſt, Médecin Suédois, mort à l'âge de 104 ans. Ce ſecret étoit dans ſa famille qui en faiſoit uſage ; ſon

A l'égard de l'uſage journalier qu'on en peut faire, il eſt de neuf gouttes pour les hommes, & de ſept pour les femmes. Cette recette a été trouvée dans les papiers du Docteur Yerneſt, Médecin Suédois, mort à l'âge de 104 ans. Ce ſecret étoit dans ſa famille qui en faiſoit uſage ; ſon

VIE. Moyen de ſe procurer une longue vie. Elixir de longue vie : en voici la compoſition. Une once & un gros d'aloës ſuccotrin, un gros de ledoire ou zedoire, pareil poids d'agaric blanc, de gentiane, de ſafran du Levant, de rhubarbe fine, de thériaque de Veniſe.

Mettez en poudre & paſſez au tamis les ſix premieres drogues, après quoi mettez-les dans une bouteille
F 4

MODERNE. 131

pere a vécu 112 ans, ſon aïeule 130. Ils en prenoient ſept à huit gouttes matin & ſoir. Cet Elixir n'eſt pas difficile à faire. Le prix des drogues eſt de 35 à 40 ſols pour deux pintes d'eau-de-vie.

Figure 3.1 French book published in Paris in 1773 (first edition in 1768), *L'Albert moderne*,[1] containing a recipe for an elixir of long life (see the page below the front cover). Page 130 lists various indications, such as colics, intestinal worms or fever. Pages 130–131 assert that the recipe was found in the Dr Yernest's papers after his death at 104 years of age (a Swedish physician, said to have died by accident according to various unchecked sources), and that his father and grandmother, who drank the elixir twice daily, died respectively at 112 and 130 years of age. A part of the recipe has not been reproduced in the figure.

number of people is now to die at old age. In such conditions, it is not surprising that the search for remedies to improve healthspan has boomed in recent decades, particularly because there are now many scientists studying the aging process, beside modern charlatans still trying to fool the lay public with pseudo-anti-aging products.

However, some scientists relying on studies performed with animal models are of the opinion that the aging process could be delayed and longevity extended *via* a single genetic pathway or chemical product and that it is possible "to think of ageing as a disease that can be cured, or at least postponed".[5] One of the purposes of this chapter is to argue that it is not certain that results gathered on the classical animal models bear the promise that human aging and longevity can be modified as in these animal models, particularly because the life-history strategies of human beings and rodents, for instance, are very different.

Other problems described below prohibit expecting that many results reported in animal models can be observed in human beings, or concluding that a product affecting healthspan and lifespan truly affects the aging process. In addition, molecules improving health and lifespan in sick animals cannot be considered as real "anti-aging" drugs.

It is not to say that no product is (or will be) of therapeutic value, but simply that a very cautious attitude is required before making the hypothesis that what is efficient in an animal model could be too in humans, and that a product increasing lifespan or delaying some features of the aging process truly targets the aging process.

3.2 Diverse Life-History Strategies: Consequences for Lifespan Modulation

3.2.1 There Are Various Life-History Strategies in Mammals

Each species complies with a life-history strategy and these strategies differ among species.[6] In mammals,[7] there are on one side of a continuum short-lived species with a small body size, as for instance mice and rats, maturing quickly after a short gestation time and giving birth at short intervals to numerous offspring (Figure 3.2). However, they may have only one season of reproduction, if not a single reproduction episode, due to a high predatory load on these small-sized species. On the other side of this continuum are species with an opposite life-history strategy. They are thus longevous and of a large size, they need a long gestation time and an extended period with parental care to reach adulthood, and they give birth repeatedly to a few offspring during a long period, as is the case for instance in elephants or primates. These species, particularly because of their large size, do not suffer from a high predatory load, as small species do.

Therefore, some species need a long life to propagate and thrive while living long is not necessary for other species. As a consequence, mouse traps and poisoned baits are not a threat for the survival of mice as a species because

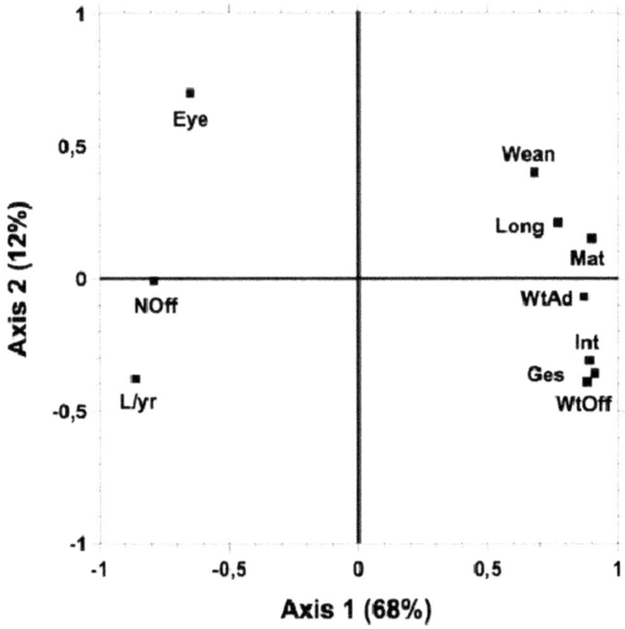

Figure 3.2 Principal component analysis of relationships between life-history variables in 65 mammalian species.[7] Signification of acronyms (as in the original article): Ges: gestation period, Eye: age at eye opening, WtOff: mass of offspring, Noff: number of offspring, Int: inter-litter interval, L/Yr: litters per year, Mat: age at maturity, Long: longevity, Wean: duration of lactation. The first and second axes explain, respectively, 68 and 12% of the variance and "the first principal component arranges the species along a dimension that varies from short gestation, small offspring, short inter-litter intervals, early maturity, small adults, short life, short lactation, late eye opening, many offspring, and many litters per year, to the opposite for each trait at the other extreme."[7]

mice reproduce heavily and quickly. By contrast, hunting adult elephants for their ivory is sufficient to make them an endangered species because hunters kill adults in charge of offspring.

One can thus understand why human beings *must* live for decades because of their low fecundity, long gestation time, inter-litter interval, and parental care: human beings living for *ca.* 25 years could simply not survive as a species. Saying that three centuries ago mean lifespan was 25 years can be misleading because this low life expectancy is explained by the very high infant mortality, *i.e.* the death before one year of age (at least 25%): life expectancy at 20 years of age was *ca.* 35 years, which means that people reaching their twenties could live up to *ca.* 55 years of age[8] or even longer, for instance *ca.* 65 years in the old Québec.[9] In contrast, mice and rats can live only a few weeks or months because they can reproduce at an early age, their offspring being able to live quickly on their own.

These various life-history strategies have very important consequences. Short-lived species can quickly exploit a new environment if food resources are plentiful and produce many offspring in a few weeks: farmers painfully know that innumerable mice can infest their silos very quickly. By contrast, long-lived species need a long time to colonise a land: despite their will, European pioneers needed decades, if not centuries, to populate America.

3.2.1.1 Only Some Species Increase Their Lifespan When Facing Food Shortage

A consequence of these diverging life-history strategies is that longevity cannot evolve independently from the other life-history traits because it "is the product of an evolutionary history that established the tempo of growth, development and maturation needed to survive and reproduce."[10] Mice live only for a very few years while humans can live for *ca.* 120 years at a maximum and these values have been selected during the course of evolution. This suffices to argue that the hypothesis that people born in 2000 could live for 5000 years[11] is mere fantasy, and that human lifespan cannot increase to a great extent because of these relationships between life-history traits.[12]

A second consequence, which is crucial if one wishes to extend to human beings the results showing a longevity increase in animal models, is that short- and long-lived mammals differ in their strategy to cope with food shortages. In short-lived mammals living in the wild, such as rats or mice, median lifespan can be less than half a year[13] and thus there is only one reproductive season. For these species plagued with a high predation, the best strategy to save reproduction in the event of famine is probably to stay in the same place and to live up to the end of the starvation period, or even to the next reproductive season, *i.e.* to the next year. Therefore, the mean lifespan of diet-restricted mice is expected to increase for a maximum of one year, *i.e.* the time until the next reproductive season.[14] In large and long-lived species facing a lower predatory load, another strategy can be emigration to discover new food sources and/or to delay reproduction. These species can afford this delay because they do not reproduce only once but repeatedly in successive years. For these species, there is thus no selective ground for an increased lifespan in the event of food scarcity because this increase is useless since other strategies are at hand.[15]

A way to test whether long-lived species live longer under diet restriction could be to observe the lifespan of long-lived primates subjected to diet restriction. Diet-restricted rhesus macaques have been reported not to live longer than control ones[16] or to live *ca.* 2 years longer.[17] It has been argued that the control group of the former study had a low weight when compared to usual results, and thus that this control group was maybe not *ad libitum*-fed but slightly diet-restricted, but it has also been stressed that the "question remains if *ad libitum* feeding is more an obesity model than a normal feeding state in primates."[18] Therefore, the evidence in favour of a lifespan increase in diet-restricted long-lived mammals is scarce.

3.2.1.2 *Can Modulating the Insulin–IGF1 Pathway Increase Lifespan in Human Beings?*

Another way to determine whether modulation of metabolism could increase lifespan would be to test whether the insulin–IGF1 pathway regulating metabolism and responses to a food shortage in very different species (*e.g.* nematodes, flies, rodents, humans) can modulate lifespan. Dwarf mice bearing mutations of this pathway live longer,[19] but no such increase has been observed in dwarf human mutants.[20] In addition, the effect on lifespan of genetic polymorphisms at loci governing metabolism has been tested in human beings and it has been reported that FOXO3A gene polymorphisms were linked to longevity. These studies compared very long-lived subjects (>90 years) to younger controls and a meta-analysis concluded that some variants were linked to very high longevity, at least in one sex.[21] However, these studies compare living persons, a first cohort being composed of very old subjects and the second control cohort comprising younger subjects, and there is no grounds to argue that persons of the second cohort will live less than those of the first one. In other words, the real longevity difference between the two cohorts, when they will be extinct, could be low or even absent. This conclusion is strengthened by the fact that a significant effect reported in a study comparing the oldest-old Danes to a 30–50 years younger cohort was not observed when linking the longevity of individuals of the oldest cohort (all persons were dead at this time but one) with the very same polymorphism (variant rs7762395).[22] In addition, a later study reported small differences in the prevalence of this rs7762395 variant between different birth cohorts observed at the same ages (>95 years, cohorts born from 1895 to 1915), which shows that factors other than longevity could explain polymorphism.[23] Could it be that differences between birth cohorts separated by several decades could give rise to differences not linked to age?

A direct test of a link between IGF-1 and remaining lifespan was tested in nonagenarians.[24] The plasmatic IGF-1 level was not linked to lifespan in men, but women with a level below the median survived longer. However, in the subgroup with a history of cancer (23% of the cohort: 34/151) subjects with a high IGF-1 level died before those with a low one, no such effect being observed in people without cancer (77% of the cohort). Because the whole cohort comprised *ca.* 75% women, it seems clear that the effect observed in people with cancer is mainly due to women. Thus, it can be said that in nonagenarian women with a history a cancer, those with a high IGF-1 level survived less than those with a low level, but such an effect is not observed in women without cancer and in all men. Because IGF-1 promotes metabolism, it is not surprising that people with a high IGF-1 level have a higher cancer risk (*i.e.* a higher risk for an anarchic cellular proliferation), but the main result of this study is the absence of a link between IGF-1 level and remaining lifespan in people without cancer. Obviously, it would be of interest to replicate this study with younger people.

Thus, these results do not clearly show that longevity is linked to the FOXO3A gene polymorphism and it is not certain that turning down the insulin–IGF1 genetic pathway can increase lifespan in long-lived species, such as human beings, as it does for instance in mice, because the lifespan of human beings is very less plastic than that of mice.

3.2.1.3 Conclusions

Because human beings are not giant mice and mice are not miniature humans, it is a flaw to expect that a treatment increasing lifespan in mice will have the same effect in human beings or other long-lived mammals. For some species, increasing lifespan when confronted with a food shortage has been selected because it is a valuable strategy for these species, but other species have not selected this response because they have other strategies at hand, such as fleeing or delaying reproduction. In such conditions, there is no reason to expect that, for instance, modulating the insulin–IGF1 pathway could have similar effects in, say, mice and human beings.

This conclusion goes beyond studies of aging because molecules used to treat amyotrophic lateral sclerosis increased survival of mice but failed in clinical trials.[25] A comment in *Nature* proposed recommendations for next translational research studies[25] but one could add that a treatment increasing the lifespan of mice will always fail to give a similar result in human beings because the lifespan of long-lived mammals is less plastic than that of mice. It has been argued that failing to reproduce in human beings the association observed in animal models between some genes and longevity could be explained by limitations of these animal models (*e.g.* limited genetic or environmental diversity) and that "pathways that extend lifespan in short-lived organisms may not work the same way in long-lived ones."[26] One may add that a sensible explanation of these discrepancies could lie with the different life-history strategies of short- and long-lived species: increasing lifespan does not appear to be a response to food shortage in long-lived species.

Therefore, showing that a treatment increases lifespan in mice definitely does not offer any clue for a positive result on human aging or longevity. This rationale could be extended to nematodes, flies, and other species with life-history strategies very different from that of humans. Let us consider the example of nematodes.

3.2.2 The Life-History Strategy of the Nematode Caenorhabditis Elegans Could Explain Why Its Longevity is Plastic

The nematode *C. elegans* is an animal model that has been widely used in research on aging for more than three decades.[27] This 1 mm worm lives in the soil where the main threats are drought, food scarcity, and temperature variation. Because this worm is unable to escape its environment, it can be understood why it can enter a very resistant Dauer larval stage to wait for 2 months for better times, before resuming its normal life cycle,[28] and why mutations of the insulin–IGF1-like signalling pathway regulating metabolism, responses to food shortage and Dauer formation strongly increase longevity in the laboratory.[29]

Another consequence of the features of this worm is that its only way to survive various threats could be to live longer, even if not entering the Dauer (duration in German) larval stage, because the lifespan in the soil is less than 2 days, *i.e.* 7-fold less than in the laboratory.[30] When subjected to a toxic chemical product in the soil, and because the worm cannot escape, an appropriate response could be to live longer, waiting for dilution (rain?) or destruction (bacteria?) of the product in the soil. It is thus not unexpected that many toxic molecules can increase lifespan in worms, even if some (too toxic or concentrated) also decrease lifespan. For instance, longevity is increased by the toxic products juglone (+6–29%),[31] hyperbaric oxygen (+22%),[32] hydrogen sulfide (+74%),[33] carbon dioxide (+26–44%),[34] plumbagin (+12%),[35] and dimethyl formamide (+30%).[36] These results are probably hormetic effects (beneficial effects of a low dose of a toxic product: see below), but other molecules not considered to be toxic also increase lifespan in worms. These increases could be explained by a real positive effect because worms are offered essential molecules (*e.g.* vitamin E: +22%[37]), but molecules *a priori* considered by the experimenter as beneficial (*e.g.* antioxidants, like trolox: +31%[38]) could indeed have a hormetic effect. Whatever the mechanism of the increased lifespan could be, this increase probably better reflects the life-history strategy of the worm, *i.e.* a very plastic lifespan when confronted with a threat rather than a real effect on the aging process.

Therefore, any experiment showing that a chemical product increases lifespan in *C. elegans* should be interpreted with caution because this beneficial effect is maybe only linked to the life-history strategy of this worm and thus could not have any beneficial effect on lifespan in other species with different life-history strategies. Thus, it is not certain that such studies set "the stage for future studies to investigate whether compounds that increase lifespan in the nematode may also have a beneficial effect on aging in mammals".[39] One could agree with the conclusion that "while *C. elegans* remains a valuable organism for the study of ageing, it is critical to consider its natural history when interpreting results from such studies".[30]

3.3 Toxic and Essential Molecules May Have the Same Effects at Low Doses

The effect on lifespan of many molecules has been tested in past decades, particularly in *Drosophila melanogaster*.[40] On the one hand, essential molecules such as vitamins may have positive effects at low doses, because deficiency is deleterious, but they can be toxic at a high dose (*e.g.* vitamin A on lifespan of *D. melanogaster* flies[41]). On the other hand, low doses of toxic molecules can have positive effects in organisms. This phenomenon is called hormesis:[42] a mild stress disturbs the homeostasis of the organism without inducing severe damage and provokes a general adaptive response of the organism enhancing the ability to resist other stresses. Mild stresses can also increase lifespan[43] (see above some examples of toxic products in *C. elegans*) and these chemical, physical (*e.g.* hypergravity[44]), or biological[45] mild stresses do not require the existence of specific cellular receptors to exert their effects. Therefore, toxic and essential molecules can have similar positive effects at low doses, the main difference between essential molecules and chemical stressors being that the former are necessary for the organism to thrive while the latter are not.

There is however a third category of molecules, those that are *a priori* considered by the experimenter as beneficial. For instance, the effect of antioxidants on lifespan has been investigated,[46] positive effects often being expected because the free radical theory of aging[47] has been accepted by many authors as a satisfactory explanation of the aging process. Thus, if the expected positive effect is observed one can be led to conclude that the results support the free radical theory. For instance, the antioxidant *N*-acetyl-cysteine has been added to the food of flies and their lifespan recorded: low doses (0.01–10 mg ml^{-1}) increased lifespan up to 25% while a higher dose (20 mg ml^{-1}) decreased it by 50%. The authors concluded that the "results give further support to the free radical theory of ageing".[48] Essential or toxic molecules and molecules that do not clearly fall in one of these two categories, but are postulated to be beneficial, can thus increase lifespan or, for instance, resistance to a severe stress,[32] but showing that this last kind of molecule has beneficial effects is not a proof for the postulated mechanism of action (*e.g.* "antioxidants increase lifespan and delay aging") because these products could also have hormetic effects.

Particularly, there is a debate on the mode of action of phytochemicals, which have been thought as beneficial for health because of their antioxidant capacity. However, they have also been considered either as toxins with hormetic effects[49] or as signalling molecules used by plants to resist various stresses: when the animal eats the plant, its organism implements a reaction similar to that of the plant and is better able to resist various stresses (xenohormesis hypothesis[50]). These phytochemicals can activate or inhibit transcription factors (*e.g.* respectively Nrf2 and NF-κB).[51]

In summary, while the mechanisms of action of the three kinds of molecules can be different, they can have similar effects, such as an increased lifespan. In the absence of a clear means to differentiate their mechanisms of action[52] it can be difficult to conclude that a tested molecule is essential for life and thus has a specific effect on the aging process, or rather has a hormetic effect, an antioxidant effect, and so on. However, in some cases, it could be possible to tell whether the tested product has a hormetic action or a specific effect on the aging process. Let us imagine that, if given to rodents, the hormone X (say for instance melatonin) increases lifespan and delays aging (for instance, a delay in cognitive and locomotor abilities): one could conclude that, because this hormone is normally synthesised by the animal and enters a well-known biochemical pathway, the extra dose probably enters the same biochemical pathway and modulates the aging process but, unfortunately, most tested molecules are not copies of those naturally produced by the organism. By contrast, let us imagine that an experimenter shows that a low dose of the highly toxic molecule Y has positive effects on lifespan and aging: one could conclude that its mechanism of action is probably hormetic but, unfortunately again, most tested molecules are not highly toxic.

Nevertheless, many drugs have hormetic effects even when they are not highly toxic. Some drugs are beneficial at low doses and toxic at high doses, like for instance paracetamol or aspirin. Other drugs have deleterious effects at low doses and beneficial ones at high doses, like antibiotics that stimulate growth of bacteria at low doses and kill them at high doses or anti-tumour drugs.[53] In both cases, these drugs display a hormetic dose–response curve, with opposite effects at low and high doses, and the physician relies on symmetrical strategies to treat the disease: low doses of paracetamol against pain and fever, and high doses of antibiotics to kill bacteria (Figure 3.3).

However, even if one is able to discover a molecule with positive effects on lifespan and health at old age in animal models, an issue is that this molecule could have an effect only in short-lived (see above) or compromised animals (see below) and thus would not be a real "anti-aging" drug, even if useful in therapy.

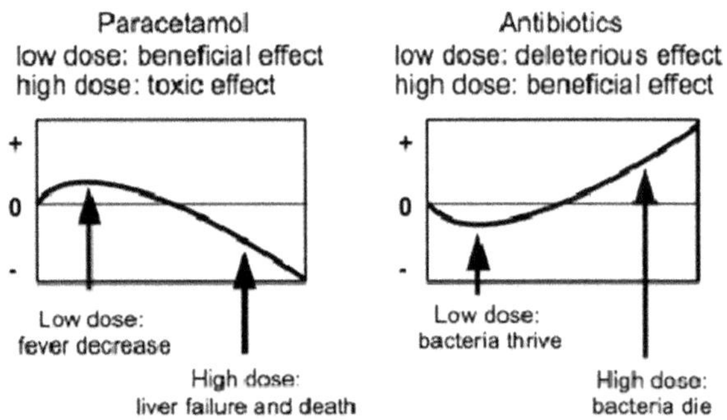

Figure 3.3 Hormetic dose–response curves showing opposite effects at low and high doses. The ordinates show beneficial (+), neutral (0), or deleterious (−) effects for the patient. Different drugs can have either beneficial or deleterious effects at a low dose and two examples are shown. Low doses of paracetamol decrease fever or pain, but very high doses can lead to death because of hepatotoxicity. Low doses of antibiotics can stimulate bacterial growth because they disturb the homeostasis of the bacteria without inducing severe damages and provoke a general adaptive response, enhancing the ability of bacteria to proliferate. By contrast, high doses are toxic to bacteria and kill them, which is the expected therapeutic effect of antibiotics.[53]

3.4 A Drug Treating an Age-Related Pathology is not an "Anti-Aging" Drug

All people age but they do not all suffer from the same age-related pathologies. For instance, obesity and cardiac diseases are often age-linked but many people are spared from these illnesses. Therefore, a treatment improving health and increasing lifespan in sick animals because it is a cure for this disease is not necessarily able to delay aging and increase longevity in healthy animals. This is what we can conclude from the story of resveratrol, a polyphenol found in grapes and red wine, for instance.

In 2004, an article reported that resveratrol increased lifespan in *C. elegans* and in *D. melanogaster* flies, provided these flies were diet-restricted, and the title of this article thus claimed that resveratrol delays aging.[54] The lifespan results on flies were confirmed[55] or not[56] and those on *C. elegans* were also poorly confirmed[56,57] or not at all.[58] In mice, resveratrol increased lifespan of animals living on a shortening lifespan high-calorie diet,[59] but not of those feeding on a normal diet.[60–62] Thus, resveratrol helped to recover normal longevity in animals living shorter because of a bad diet but had no effect in animals living in better conditions. A similar result was shown in *D. melanogaster* because resveratrol increased the lifespan of flies feeding on a shortening lifespan medium but had no effect if the medium provided a normal lifespan.[63] A possible consequence of these results is that resveratrol could eventually become a therapy for people living less because of obesity or cardiovascular diseases linked to an inappropriate diet, but would be of no help to other people with no metabolic diseases and a normal lifespan, as it seems to be the case.[64,65] This is exactly the definition of a drug: a molecule that fights a disease but should not be used by healthy people. Therefore, the biochemical pathways targeted by resveratrol probably have no role in the aging process and resveratrol is thus not the magic pill aggressively advertised on internet to delay aging in all people. In any case, even if resveratrol became a therapy against obesity, one might feel that eating a magic pill to circumvent the deleterious effects of a bad diet is not a good idea because the most efficient solution would obviously be modifying feeding habits.

To sum up, any product helping animals with short lives because of bad diet, disease, and so on to reach a normal lifespan should not be considered as an "anti-aging" drug able to delay aging and increase lifespan, for the same reason that the bacillus Calmette–Guérin vaccine that strongly increased lifespan during the last century, because people did not longer die at young or middle age from tuberculosis, was not an "anti-aging" vaccine.

3.5 Conclusions

Searching for new means to improve the lifespan of elderly people is a respectable endeavour. Some of these means, beside quitting smoking, taking exercise, avoiding junk food, and so on, could be new drugs helping either to protect from age-related ailments (*e.g.* hearing loss, cataracts) or diseases

(Alzheimer's). One cannot exclude that, one day, we will discover new means to delay the aging process more than we have been able to do up to now. Indeed, delaying aging has already been done during the last decades: thanks to sanitation, hygiene, medicine, social progress, 70 and 80 year-old people are much "younger" than they were, say, 50 years ago, and this point is very clear when looking at pictures of that time. For instance, one could imagine that a drug would be a mild stress with hormetic effects and could help to implement defences against severe stress, these defences not being mobilised if the mild stress were absent. In such conditions, maybe age-related diseases would be less severe or delayed because of a higher resistance to severe stress. Other paths will be pursued, with no doubt. In any case, in order not to be fooled by ourselves and our hopes, it is necessary to pay attention to traps on this road and this chapter has tried to delineate some of them.

Regarding drugs, it would obviously be the wrong attitude to conclude that a drug with positive effects on lifespan and the aging process because of hormetic effects should be disregarded because it does not truly target the aging process. This drug should be used because, as tells a Chinese aphorism, the colour of the cat does not matter as long as it catches mice, but it would be an error to conclude that one has been able to discover the secret of the aging process. This drug could be efficient not because of a specific action on the aging process, but rather because it stresses the organism and provokes a response to this stress.

Anyway, there are surely other traps on the road to the discovery of new means to improve the health of elderly people by relying on chemicals. For instance, some authors have attributed the aging process to a single cause and expect that a single molecule could modulate the whole process: this would be the definitive anti-aging drug. There are many examples of such theories of aging,[66] but the most famous one is probably the free radical theory of aging stating that "aging and the degenerative diseases associated with it are attributed basically to the deleterious side attacks of free radicals."[47] If this theory were valid, the obvious therapy would be to lower these attacks with antioxidants and many attempts were made over decades to discover efficient antioxidants, to no avail however,[67] and the free radical theory is now rejected by many authors.[68] One may think that reductionist views of the aging process have to be given up because a living being is a system with dialectical interactions among its components, and not the mere addition of molecules, cells and organs.[52,69]

Another trap lies with the old dream of humans to live longer, a trap that has caught many scientists during past decades and centuries. For instance, the double Nobel laureate Linus Pauling (chemistry and peace) promoted taking mega-doses of vitamin C and stressed "that vitamin C will have great value in controlling the problems associated with advancing age."[70] More recently, a biogerontologist claimed on an ABC broadcast that feeding mice with nicotinamide mono nucleotide "reversed aging completely within just a week of treatment in the muscle,"[71] but the published article stressed that "we did not observe an improvement in muscle strength…indicating that 1 week of treatment might not be sufficient to reverse whole-organism aging,"[72] which is not exactly the conclusion reported on the broadcast. Indeed, because there is often a strong desire of the mass-media to publish fabulous news, scientists should be very cautious before claiming to have discovered the miraculous recipe for living longer or delaying aging. While it can be fully understood that quacks are prone to make such claims for obvious reasons, colleagues should adopt less enthusiastic attitudes.[73,74] It has also been claimed in a book published by a prestigious academic publisher that people should consume starch and fat blockers in addition to a moderate caloric restriction:[75] starch and fat blockers are drugs with known side-effects that are prescribed to fight diabetes and obesity and should not be used in the absence of any disease.

The aim of this chapter was not to delineate the best methods to study the effects of drugs on the aging process but, obviously, the mandatory condition to claim that aging is delayed is improving the aging process, and not only increasing lifespan: various authors have warned against this flaw for decades.[76–78] However, some studies, particularly those using invertebrate models, can only observe longevity and conclude that the drug under study has an effect on aging.

References

1. Anonymous, *L'Albert moderne ou nouveaux secrets éprouvés et licites, recueillis d'après les découvertes les plus récentes (The Modern Albert or New Proven and Legitimate Secrets, Collected According to the Latest Findings)*, Chez la veuve Duchesne, Paris, 1st edn: 1768, 1773.
2. G. L. Buffon (G. L. Leclerc, earl of Buffon), *Histoire naturelle de l'Homme (Natural History of Man)*. Imprimerie royale, Paris, 1749. English translation in Natural History, General and Particular, W. Strahan and T. Cadell, London, 1785.
3. M. E. Mercier and C. G. Boone, *J. Hist. Geogr.*, 2002, **28**, 486.
4. C. E. Finch, *The Biology of Human Longevity. Inflammation, Nutrition, and Aging in the Evolution of Lifespans*, Academic Press, Burlington, 2007.
5. L. Guarente and C. Kenyon, *Nature*, 2000, **408**, 255.
6. E. R. Pianka, *Am. Nat.*, 1970, **102**, 592.
7. S. C. Stearns, *Oikos*, 1983, **41**, 173.
8. C. Masset, *Popul. Soc.*, 2002, (380).
9. E. Le Bourg, B. Thon, J. Légaré, B. Desjardins and H. Charbonneau, *Exp. Gerontol.*, 1993, **28**, 217.
10. B. A. Carnes, S. J. Olshansky and D. Grahn, *Biogerontology*, 2003, **4**, 31.
11. A. D. N. J. de Grey, in *Aging Interventions and Therapies*, ed. S. I. S. Rattan, World Scientific Publishers, Singapore, 2005, pp. 379–395.
12. E. Le Bourg, *Ageing Res. Rev.*, 2012, **11**, 325.
13. J. P. Phelan and S. N. Austad, *Growth, Dev. Aging*, 1989, **53**, 4.
14. A. D. N. J. de Grey, *Gerontology*, 2005, **51**, 73.
15. E. Le Bourg, *Ageing Res. Rev.*, 2010, **9**, 289.

16. J. A. Mattison, G. S. Roth, T. M. Beasley, E. M. Tilmont, A. M. Handy, R. L. Herbert, D. L. Longo, D. B. Allison, J. E. Young, M. Bryant, D. Barnard, W. F. Ward, W. Qi, D. K. Ingram and R. de Cabo, *Nature*, 2012, **489**, 318.
17. R. J. Colman, T. M. Beasley, J. W. Kemnitz, S. C. Johnson, R. Weindruch and R. M. Anderson, *Nat. Commun.*, 2014, **5**, 3557.
18. K. Verburgh, *Aging Cell*, 2015, **14**, 17.
19. A. Bartke, *Endocrinology*, 2005, **146**, 3718.
20. A. Bartke, *Gerontology*, 2012, **58**, 337.
21. J. M. Bao, X. L. Song, Y. Q. Hong, H. L. Zhu, C. Li, T. Zhang, W. Chen, S. C. Zhao and Q. Chen, *Asian J. Androl.*, 2014, **16**, 446.
22. M. Soerensen, S. Dato, K. Christensen, M. McGue, T. Stevnsner, V. A. Bohr and L. Christiansen, *Aging Cell*, 2010, **9**, 1010.
23. M. Nygaard, R. Lindahl-Jacobsen, M. Soerensen, J. Mengel-From, K. Andersen-Ranberg, B. Jeune, J. W. Vaupel, Q. Tan, L. Christiansen and K. Christensen, *Exp. Gerontol.*, 2014, **57**, 41.
24. S. Milman, G. Atzmon, D. M. Huffman, J. Wan, J. P. Crandall, P. Cohen and N. Barzilai, *Aging Cell*, 2014, **13**, 769.
25. S. Perrin, *Nature*, 2014, **507**, 423.
26. J. P. de Magalhães, *Cell Cycle*, 2014, **17**, 2671.
27. A. Z. Reznick and D. Gershon, in *Methods in Aging Research*, ed. B. P. Yu, Academic Press, Boca Raton, 1999, pp. 168–190.
28. M. R. Klass and D. Hirsh, *Nature*, 1976, **260**, 523.
29. H. A. Tissenbaum, *J. Gerontol., Ser. A*, 2012, **67**, 503.
30. W. A. Van Voorhies, J. Fuchs and S. Thomas, *Biol. Lett.*, 2005, **22**, 247.
31. T. Heidler, K. Hartwig, H. Daniel and U. Wenzel, *Biogerontology*, 2010, **11**, 183.
32. J. R. Cypser and T. E. Johnson, *J. Gerontol., Ser. A*, 2002, **57A**, B109.
33. D. L. Miller and M. B. Roth, *Proc. Natl. Acad. Sci. U. S. A.*, 2007, **104**, 20618.
34. K. Sharabi, A. Hurwitz, A. J. Simon, G. J. Beitel, R. I. Morimoto, G. Rechavi, J. I. Sznajder and Y. Gruenbaum, *Proc. Natl. Acad. Sci. U. S. A.*, 2009, **106**, 4024.
35. P. R. Hunt, T. G. Son, M. A. Wilson, Q. S. Yu, W. H. Wood, Y. Zhang, K. G. Becker, N. H. Greig, M. P. Mattson, S. Camandola and C. A. Wolkow, *PLoS One*, 2011, **6**, e21922.
36. H. Frankowski, S. Alavez, P. Spilman, K. A. Mark, J. D. Nelson, P. Mollahan, R. V. Rao, S. F. Chen, G. J. Lithgow and H. M. Ellerby, *Mech. Ageing Dev.*, 2013, **134**, 69.
37. L. A. Harrington and C. B. Harley, *Mech. Ageing Dev.*, 1988, **43**, 71.
38. M. G. Benedetti, A. L. Foster, M. C. Vantipalli, M. P. White, J. N. Sampayo, M. S. Gill, A. Olsen and G. J. Lithgow, *Exp. Gerontol.*, 2008, **43**, 882.
39. X. Ye, J. M. Linton, N. J. Schork, L. B. Buck and M. Petrascheck, *Aging Cell*, 2014, **13**, 206.
40. H. R. Massie, in *Drosophila as a Model Organism for Ageing Studies*, ed. F. A. Lints and M. H. Soliman, Blackie, Glasgow, 1988, pp. 59–70.
41. H. R. Massie, V. R. Aiello, T. R. Williams, M. B. Baird and J. L. Hough, *Exp. Gerontol.*, 1993, **28**, 601.

42. E. J. Calabrese, in *Hormesis in Health and Disease*, ed. S. I. S. Rattan and E. Le Bourg, CRC Press, Boca Raton, 2014, pp. 3–11.
43. E. Le Bourg, *Biochim. Biophys. Acta*, 2009, **1790**, 1030.
44. E. Le Bourg, in *Mild Stress and Healthy Aging. Applying Hormesis in Aging Research and Therapy*, ed. E. Le Bourg and S. I. S. Rattan, Springer, Dordrecht, 2008, pp. 43–63.
45. C. D. McClure, W. Zhong, V. L. Hunt, F. M. Chapman, F. V. Hill and N. K. Priest, *Evolution*, 2014, **68**, 2225.
46. E. Le Bourg, *FEBS Lett.*, 2001, **498**, 183.
47. D. Harman, *J. Gerontol.*, 1956, **11**, 298.
48. C. Brack, E. Bechter-Thüring and M. Labuhn, *Cell. Mol. Life Sci.*, 1997, **53**, 960.
49. M. P. Mattson, *Dose-Response*, 2014, **12**, 600.
50. B. J. Morris, in *Mild Stress and Healthy Aging. Applying Hormesis in Aging Research and Therapy*, ed. E. Le Bourg and S. I. S. Rattan, Springer, Dordrecht, 2008, pp. 115–138.
51. A. R. Weseler and A. Bast, in *Hormesis in Health and Disease*, ed. S. I. S. Rattan and E. Le Bourg, CRC Press, Boca Raton, 2014, pp. 79–92.
52. A. M. Vaiserman, *Front. Genet.*, 2014, **5**, 446.
53. E. J. Calabrese and M. P. Mattson, in *Hormesis. A Revolution in Biology, Toxicology and Medicine*, ed. M. P. Mattson and E. J. Calabrese, Springer, Dordrecht, 2010, pp. 177–198.
54. J. G. Wood, B. Rogina, S. Lavu, K. Howitz, S. L. Helfand, M. Tatar and D. Sinclair, *Nature*, 2004, **430**, 686.
55. J. H. Bauer, S. Goupil, G. G. Garber and S. L. Helfand, *Proc. Natl. Acad. Sci. U. S. A.*, 2004, **101**, 12980.
56. T. M. Bass, D. Weinkove, K. Houthoofd, D. Gems and L. Partridge, *Mech. Ageing Dev.*, 2007, **128**, 546.
57. K. Zarse, S. Schmeisser, M. Birringer, E. Falk, D. Schmoll and M. Ristow, *Horm. Metab. Res.*, 2010, **42**, 837.
58. W. Chen, L. Rezaizadehnajafi and M. Wink, *J. Pharm. Pharmacol.*, 2013, **65**, 682.
59. J. A. Baur, K. J. Pearson, N. L. Price, H. A. Jamieson, C. Lerin, A. Kalra, V. V. Prabhu, J. S. Allard, G. Lopez-Lluch, K. Lewis, P. J. Pistell, S. Poosala, K. G. Becker, O. Boss, D. Gwinn, M. Wang, S. Ramaswamy, K. W. Fishbein, R. G. Spencer, E. G. Lakatta, D. Le Couteur, R. J. Shaw, P. Navas, P. Puigserver, D. K. Ingram, R. de Cabo and D. A. Sinclair, *Nature*, 2006, **444**, 337.
60. K. J. Pearson, J. A. Baur, K. N. Lewis, L. Peshkin, N. L. Price, N. Labinskyy, W. R. Swindell, D. Kamara, R. K. Minor and E. Perez, *Cell Metab.*, 2008, **8**, 157.
61. R. A. Miller, D. E. Harrison, C. Astle, J. A. Baur, A. R. Boyd, R. de Cabo, E. Fernandez, K. Flurkey, M. A. Javors and J. F. Nelson, *J. Gerontol., Ser. A*, 2010, **66A**, 191.
62. R. Strong, R. A. Miller, C. M. Astle, J. A. Baur, R. de Cabo, E. Fernandez, W. Guo, M. Javors, J. L. Kirkland and J. F. Nelson, *J. Gerontol., Ser. A*, 2013, **68**, 6.

63. C. Wang, C. T. Wheeler, T. Alberico, X. Sun, J. Seeberger, M. Laslo, E. Spangler, B. Kern, R. de Cabo and S. Zou, *Age*, 2013, **35**, 69.
64. R. D. Semba, L. Ferrucci, B. Bartali, M. Urpí-Sarda, R. Zamora-Ros, K. Sun, A. Cherubini, S. Bandinelli and C. Andres-Lacueva, *JAMA Intern. Med.*, 2014, **174**, 1077.
65. M. G. Novelle, D. Wahl, C. Diéguez, M. Bernier and R. de Cabo, *Ageing Res. Rev.*, 2015, **21**, 1.
66. L. Hayflick, *Exp. Gerontol.*, 1985, **20**, 145.
67. J. R. Speakman and C. Selman, *BioEssays*, 2011, **33**, 255.
68. A. B. Salmon, A. Richardson and V. I. Pérez, *Free Radical Biol. Med.*, 2010, **48**, 642.
69. E. Le Bourg, *Biogerontology*, 2013, **14**, 221.
70. L. Pauling, *Agressologie*, 1983, **24**, 317.
71. S. Lannin, http://www.abc.net.au/news/2014-11-04/scientists-reverse-ageing-process-in-mice/5865714.
72. A. P. Gomes, N. L. Price, A. J. Y. Ling, J. J. Moslehi, M. K. Montgomery, L. Rajman, J. P. White, J. S. Teodoro, C. D. Wrann, B. P. Hubbard, E. M. Mercken, C. M. Palmeira, R. de Cabo, A. P. Rolo, N. Turner, E. L. Bell and D. A. Sinclair, *Cell*, 2013, **155**, 1624.
73. E. Le Bourg, *Biogerontology*, 2000, **1**, 89.
74. T. Kirkwood, *Time of Our Lives*, Weidenfeld and Nicolson, London, 1999.
75. R. Kurzweill and T. Grossman, in *The Future of Aging*, ed. G. M. Fahy, M. D. West, L. S. Coles and S. B. Harris, Springer, Dordrecht, 2010, pp. 3–22.
76. E. Le Bourg, F. A. Lints, N. Fresquet and P. Bullens, *Comp. Biochem. Physiol., Part A*, 1993, **105**, 389.
77. G. A. Rutledge and M. R. Rose, in *Life Extension: Lessons from Drosophila*, ed. A. M. Vaiserman, A. A. Moskalev and E. G. Pasuykova, Springer, Doredrecht, 2015, pp. 265–283.
78. E. Le Bourg and N. Minois, *Nature*, 1998, **395**, 538.

Section II

Basic Concepts, Models and Approaches

CHAPTER 4

Testing of Geroprotectors in Experiments on Cell Cultures: Pros and Cons

ALEXANDER N. KHOKHLOV*[a] AND GALINA V. MORGUNOVA[a]

[a]Evolutionary Cytogerontology Sector, School of Biology, Lomonosov Moscow State University, Leninskiye Gory 1-12, 119234 Moscow, Russia
*E-mail: khokhlov@mail.bio.msu.ru

4.1 Introduction

According to the classic definition of aging we share, it is a combination of changes in an organism leading to an increase in the probability of its death (rate of mortality).[1–4] It should be noted that the data on increasing or decreasing the life span affected by various factors are often interpreted in the studies as a modification of the aging process *per se*. However, aging and life span are not necessarily interrelated. If people did not age at all, they would not live eternally anyway. People would die because of random reasons, and the life expectancy would be increased "only" up to 700–800 years.[1,5]

It is known that there are both aging and non-aging organisms. The former can be distinguished from the latter only by the shape of the survival curves of respective cohorts.[5,6] The aging organisms die "according to Gompertz law", whereas the non-aging ones die "exponentially". In the

RSC Drug Discovery Series No. 57
Anti-aging Drugs: From Basic Research to Clinical Practice
Edited by Alexander M. Vaiserman

Published by the Royal Society of Chemistry, www.rsc.org

very rare cases of the complete absence of death, *e.g.*, in the case of freshwater hydra populations under certain conditions,[7–10] the survival curve is simply a horizontal line. The conclusion whether or not a given factor affects the aging process is made on the basis of the pattern of modification of such curves under the influence of this factor. It can be assumed that a "true" geroprotector (any agent that slows down the aging process) should cause a rightward shift of the survival curve without changing its shape (*i.e.*, it should increase both the average and maximum life span). And the survival curve must not be exponential! However, a hypothetical "immortalizer" that makes the survival curve horizontal (*i.e.*, virtually abolishes the death of the members of the cohort) could also be regarded as a geroprotector—in this case, an "ideal" one. It should also be emphasized that, in our opinion, it is not very important in this approach whether aging is a programmed process or whether it is only a "byproduct" of the program of development and is determined by the stochastic processes triggered after the completion of the program.[2,5,11–22]

The factors that increase the life span of the non-aging organisms, apparently, cannot be considered geroprotectors, because they do not affect the process of increasing the probability of death with age. Regarding the drugs that are used to combat the age-related diseases, formally, with their help we can slow down (or postpone) the age-associated increase in the probability of death but they may hardly affect the maximum (species-specific) life span. If such drugs are also regarded as geroprotectors, then this group should include almost everything that ensures the normal existence of an organism (water, food, vitamins, trace elements, *etc.*). We share the point of view that age-related diseases are *the result* of aging but not *vice versa*.

The growing interest in experimental gerontological research during recent years has, unfortunately, resulted in a paradoxical situation: although the number of publications in this field is increasing, only a minor part of them is actually devoted to the mechanisms of aging. In our opinion, this is due, among others, to the following methodological problems:

(1) As a rule, the authors ignore the above-mentioned classical definition of aging as a complex of age-related changes that increase the probability of death.
(2) The emphasis in such studies is on an increase or decrease in life span, although this often, as previously said, has no relation to modification of the aging process (in particular, it is possible to prolong the life span of non-aging organisms, while the fact of aging itself is not necessarily indicative of low longevity).
(3) The control group often consists of animals with certain abnormalities or genetic disorders, so that any favorable influence on the corresponding pathological processes results in life span prolongation.
(4) Too much significance is assigned to an increase or decrease in *the average* life span, which is largely determined by factors unrelated to aging.

(5) An increasing number of gerontological experiments are performed on model systems providing only indirect information on the mechanisms of aging, and its interpretation largely depends on the basic concept supported by a given research team. In particular, this concerns the usage of the term "cell/cellular senescence," which was originally introduced to designate a complex of various adverse changes occurring in normal cells due to the exhaustion of their proliferative potential.[23–27] Today, however, many authors apply it to the phenomenon of suppression of proliferative activity in cells (including transformed cells) under the effect of various DNA damaging factors, which is accompanied by a certain cascade of intracellular events.[28–31]

There are also some extra problems we will touch on in the next sections concerning various approaches to testing of geroprotectors (anti-aging compounds or physical factors) in experiments on cultured cells.

4.2 Cytogerontological Model Systems

Cytogerontology deals with analysis of aging mechanisms on cultured cells.[5,21,32–34] It is the cytogerontological approach that is increasingly being used to test potential geroprotectors (any physical or chemical factors retarding the increase in the probability of death with age). It should be emphasized that cytogerontology as a branch of gerontology cannot successfully develop in the absence of the correct general gerontological concepts and definitions described in Section 4.1. Because of this, we will review various approaches to the testing of geroprotectors in experiments on cultured cells keeping all our general considerations in mind.

There is the issue of what we call "the problem of reductionism." In the absolute majority of gerontological theories proposed in the past few decades, the mechanisms of both "normal" and accelerated or retarded aging of multicellular organisms are reduced to certain macromolecular changes (no matter stochastic or programmed) in their constituent cells. As a consequence, numerous model systems have been developed to study "age-related" changes in the cells relieved from "organismal noise" associated with the functioning of the neurohumoral system. Such reductionism in experimental gerontology ("it all depends on adverse changes in individual cells") has played its role, particularly in the development of the Hayflick model and also of some models used in our laboratory, such as the "stationary phase aging" model, the cell kinetic model for testing of geroprotectors and geropromoters (any factors that accelerate aging), and the model based on evaluation of cell colony-forming capacity.

What is not often remembered is that the foundations of this science were laid by August Weismann[35,36] as early as in the late 19th century. As for the term "cytogerontology," it was introduced by Leonard Hayflick[32,37] to describe research on aging *in vitro*, *i.e.*, "age-related" changes in cultures of normal cells that have exhausted their mitotic potential (in fact, it is this replicative senescence that was subsequently named the "Hayflick phenomenon"). The term "cytogerontology" has then been extrapolated to any studies on the mechanisms of aging in experiments on cell cultures.[34,38–41]

Weismann was the first to emphasize the essential distinction between germ line cells, whose population is basically immortal, and somatic cells, which age and die. Thus, the cornerstone of his concept is that there exist the mortal soma and the immortal "germ plasm" (*Keimplasma*). However, Weismann failed to give a clear definition of what cell aging/senescence is, and this probably accounted for the findings and conclusions made by Alexis Carrel,[42,43] who laid the experimental foundations of cytogerontology in the early 20th century.

Carrel was interested to test whether somatic cells isolated from higher animals would "senesce" and die instead of propagating indefinitely. To this end, he developed a procedure for culturing epithelial or fibroblast-like cells in special flasks, which is still used today with only minor modifications. However, the results of his experiments did not fit the "mortal soma" concept: some cell strains derived from chicken embryos could be maintained in culture almost indefinitely, without showing any signs of degradation. This is why gerontologists in the 20th century for almost 50 years considered somatic cells to be capable of unlimited replication, until the experiments performed in the 1950s and 1960s by Swim and Parker[23] and, subsequently, by Hayflick[24,25,27] showed that the results obtained by Carrel were apparently artifactual. In fact, almost all normal animal cells have proved to have a limited proliferation potential, being capable of no more than 100–120 divisions in culture (about 50 cell population doublings).

Unfortunately, the model based on the Hayflick limit concept (aging *in vitro*) is apparently not directly related to the mechanisms of aging, as has been repeatedly noted previously.[2–6,44–46] In other words, we cannot conclusively explain why we age by relying solely on the phenomenon of limited mitotic potential of normal cells, which is practically never fully utilized *in vivo*. However, owing to Olovnikov's theory of marginotomy,[47–49] we at least know today how this phenomenon is realized in the cells.

It is not excluded that, if the human life span were extended severalfold, some cell populations would eventually exhaust their mitotic potential (thereby reaching the Hayflick limit), which could have resulted in the "second wave" of aging, but this has not occurred so far. It should be noted, however, that some researchers still hold the opinion that the shortening of telomeres in the cells is the key mechanism of aging. In particular, according to the point of view described by Mikhelson,[50,51] a certain "mosaicism" in the proliferative parameters, observed in a highly organized multicellular organism, allows the shortening of telomeres to be considered as an important factor in aging and longevity.

The body of evidence for the gerontological value of the Hayflick phenomenon is based only on a series of *correlations*,[6,40] like reduced mitotic potential of fibroblasts from the patients with progeria, direct relationship of this parameter to the species lifespan, or its inverse relationship to the age of cell donor. When demonstrating that Hayflick's model is appropriate for studying aging mechanisms, it is usually emphasized that various changes, similar to those in the cells of an aging organism, take place in normal cultured cells as the number of cell population doublings increases.[24–26,33,52] In other words, cells either accumulate or lose something during aging *in vitro* in the same way as during aging *in vivo*. Therefore, it is again the case of *correlation*; this time it is correlation of the changes of certain biomarkers of aging (BA).

Despite its "correlativity," Hayflick's model has been widely used. Based on this model, a large body of data was obtained, which explained many problematic aspects in the life activity of organisms. In particular, it concerned the mechanisms of development and malignant transformation. However, the study of aging *in vitro*, in our opinion, practically did not help gerontologists to understand the fundamental mechanisms of aging and longevity.

Keeping in mind the main topic of the review, we should note that, when testing geroprotectors on the model system, researchers have followed either: (1) the proliferative potential of the cells studied, or (2) the process of the accumulation of various BA.

We developed another "correlative" model for testing of geroprotectors and geropromoters—the "cell kinetics model".[39,53,54] It is based on the well-known inverse correlation between the "age" of cultured cells (*i.e.*, age of their donor) and their saturation density.[55] This term is used for the maximum density (number of cells per square unit) of a cell culture in the stationary phase of growth when the cells stop propagating due to contact inhibition. It was assumed that the higher the saturation density is, the "younger" the cells studied are. The model allowed us to perform preliminary testing[33,39,56,57] of a lot of different compounds and factors (gamma-irradiation, DNA-alkylating agent thiophosphamide, low frequency electromagnetic field, antioxidants 2-ethyl-6-methyl-3-hydroxypyridine chlorohydrate and butylated hydroxytoluene, *etc.*) that are interesting from a gerontological point of view, but, unfortunately, it also revealed no information about the real mechanisms of aging or its modulation.

Unlike the Hayflick model and the cell kinetics model, which are based on a series of correlations, our model of "stationary phase aging" (accumulation of "age-dependent" injuries in cultured cells whose proliferation is restricted in a certain way, preferably by contact inhibition) is a "gist" model based on the assumption that processes taking place in this model system are essentially similar to those in an aging multicellular organism.[33,40,41,58–62] In fact, this assumption directly issues from our concept that the restriction of cell proliferation is the main mechanism providing for the accumulation of macromolecular defects in cells of aging multicellular organisms.[2,5,6,33]

Our numerous experiments provide evidence that changes in the cells occurring in our model system are indeed similar to those in the cells of aging multicellular organisms. They include accumulation of DNA single-strand breaks and DNA–protein crosslinks, DNA demethylation, changes in the level of spontaneous sister chromatid exchanges, structural defects in the cell nucleus, alterations in the plasma membrane, retardation of mitogen-stimulated proliferation, impairment of colony-forming capacity, changes in dealkylase activity of cytochrome P450, accumulation of 8-oxo-2′-deoxyguanosine (a known biomarker of aging) in the DNA, increase in the number of cells with senescence-associated beta-galactosidase activity (the most popular biomarker of cell senescence), and inhibition of poly(ADP-ribosyl)ation of chromatin proteins.[5,33,63–71]

It should be emphasized that such experiments can be performed with cells of different origin, including human and animal cells, bacteria,[72,73] yeasts (currently most widely used in experiments on "stationary phase aging"[74]), plant cells,[33] microalgae,[75] and mycoplasmas.[76,77] This provides a basis for the evolutionary approach to the analysis of experimental results.[78] Moreover, the "age-related" changes in cells of stationary cultures can be revealed within a relatively short time: as a rule in 2–3 weeks after the start of the experiment.

The stationary phase aging of yeasts is called "chronological aging" and is most frequently studied in *Saccharomyces cerevisiae*. It is observed in a population of yeast cells in the stationary phase of growth when their proliferation is stopped in one way or another.[79] In this case, the viability of cells is usually estimated by their ability to form colonies in a fresh growth medium.[80] The chronological aging of yeasts should be distinguished from their so-called "replicative aging". The latter is based on the phenomenon of a limited number of daughter cells that can be generated by one mother cell. This model is very similar to the Hayflick model. However, unlike the cultured human and animal cells, the daughter cell of the yeast *Saccharomyces cerevisiae*, which is typically much smaller than the mother cell, is formed as a result of asymmetric budding. In this case, the mother cell loses its ability for such budding after a certain number of divisions and then undergoes degradation and lysis, and the daughter cells "are born very young." This process is similar to the aging of the stem cell pool in higher organisms.[81] It should be noted that, for the yeast *Schizosaccharomyces pombe*, in which two identical daughter cells are formed as a result of symmetrical division (fission) of one mother cell, only the chronological aging model can be used.[82]

It is important that in studies on the Hayflick model it is fairly difficult to correctly perform repeated experiments with the same cell strain because the cells continuously change from passage to passage ("no man ever steps into the same river twice"), whereas the "stationary phase aging" model allows, as already mentioned above, experimentation with transformed (or normal but immortalized) animal and human cells with an unlimited mitotic potential so that multiple replication of an experiment is no longer a problem.[83]

4.3 Constructing of Survival Curves for Cultured Cells in Cytogerontological Experiments

During many years of research on the "stationary phase aging" model, our premise was that cultured cells whose proliferation is restricted in some way (preferably by contact inhibition) accumulate "age-related" defects similar to those in cells of aging multicellular organisms (and geroprotectors should postpone/retard the accumulation), with the kinetics of cell death in this model system remaining behind the scene. Our subsequent studies have shown that cells in this model die out in accordance with the Gompertz law; *i.e.*, they age in the true sense.[6,84] In other words, the probability of their death increases exponentially with age, as in aging animals and humans. Incidentally, similar results were obtained with the suspension cultures of *Acholeplasma laidlawii*,[77] and our previous experiments with this mycoplasma showed that its "stationary phase aging" could be successfully delayed by treatment with a geroprotective antioxidant 2-ethyl-6-methyl-3-hydroxypyridine chlorohydrate.[76]

It should be noted that most of the cell survival curves in our studies were obtained with transformed animal and human cells. Under appropriate conditions, most cancer cells are capable of proliferating indefinitely, with a given cell line (but not individual cells!) being "immortal." For example, the well-known HeLa cell line has been maintained in hundreds of laboratories over more than 60 years. However, when the growth of such a culture is restricted by certain physiological means (not causing cell death), various defects at different structural and functional levels begin to accumulate in the cells, and the probability of their death increases; *i.e.*, the cells, as already mentioned, age in the true sense.[83] At the same time, with regard to the reliability theory, it should be taken into account that an aging multicellular organism should not necessarily consist of senescing cells: the cells can simply die out "by exponent" (*i.e.*, without senescence), as in the case of radioactive decay.

Usually, no special methods of cell viability assessment were used by us in such experiments with human and animal cell cultures, and the proportion of cells survived by a given moment of time was determined visually, simply by counting the cells under a light microscope. Hence, the question has arisen as to how adequately the viability of an individual cell is evaluated using such an approach. This aspect is especially important for correctly constructing the survival curve's right tail, where the scattering of data points reaches a maximum because of significant reduction in the absolute number of the cell population.[62]

It should be noted that the correct assessment of cell viability is a problem for all specialists working with cell cultures, but it is especially acute in the case of cytogerontological experiments, where attention is focused on the temporal dynamics of the live/dead cell ratio in culture. It is such a parameter that should be determined in the first place in studies on cell aging both in the Hayflick model and in our model of stationary phase aging. However, this task is not as simple as it may seem at first glance. First, the cells may divide, thereby disrupting the integrity of the cell cohort; second, it is fairly difficult to correctly determine the time of death for a particular cell: the period of dying may be commensurate with cell life span, and it is tough to tell what stage in this long process is the point of no return,[85] after which the cell can be certainly considered dead.

Today, a variety of probes are available for assessing cell viability (*e.g.*, see Section 15.2 'Viability and Cytotoxicity Assay Reagents' in *Molecular Probes Handbook*[86]), but the results obtained with different probes unfortunately differ from each other. This is not surprising because the rationales for using certain probes are based on different concepts of what exactly is the main criterion of cell viability (the integrity of the plasmalemma, the ability to synthesize ATP, the level of dehydrogenase activity, cell respiration rate, *etc.*). In other words, this is a fairly common situation when a given cell is classified as live in one test and as dead in another.

In our experiments, we repeatedly determined the proportion of dead cells in the same "stationary aged" culture (not subcultured for 2–3 weeks) by directly examining the cells under a microscope and by taking digital images of the culture and taking cell counts on a computer display. In both cases, the cells were examined either "as is" (without any special treatment) or after adding dyes/probes commonly used for differential staining of live and dead cells (in particular, trypan blue, methylene blue, neutral red, and MTT). In many cases, the dead cell ratio detected by these methods proved to differ significantly, which casts doubt on the efficiency of such an approach to cell viability assessment in cytogerontological research. It should also be noted that some popular dyes have a number of side effects, which researchers often fail to mention. In particular, this concerns tetrazolium salts (MTT, XTT, *etc.*), which are inexpensive and can be used in experiments with cells of different origin, from bacteria to mammalian cells. However, some specialists consider that these probes are not optimal for assessing cell viability, even though they allow correct estimation of metabolic activity.[87] First, cell metabolic activity may change due to a variety of factors, even when the number of live cells in the population remains unchanged;[88] second, formazan crystals formed in the cells can damage the plasma membrane, thereby contributing to cell mortality.[87] The accuracy of analysis may be improved by using standard reagent kits containing several molecular probes each,[89] but this does not solve the problem in general.

Three groups of approaches to assessing the viability of cultured cells (Table 4.1) were diagrammatically represented in a paper of ours.[90] Table 4.1 does not cover all possible variants of live/dead cell tests but provide an idea of how broad the spectrum of such approaches can be. All methods have certain advantages and drawbacks. In particular, the occurrence of holes in the plasma membrane is not necessarily fatal for the cell, since sometimes the membrane can be repaired.[91]

Meanwhile, there is one method that usually gives a correct answer to the question about the proportion of dead cells in the test culture under study, in which the viability of cells is estimated from their colony-forming efficiency (CFE).[92–94] This method was widely introduced in cytogerontological experimentation in the 1970s, with the development of studies on the Hayflick phenomenon, *i.e.*, aging *in vitro*.[95,96] In particular, this was due to the fact that the proportion of colonies consisting of at least 64 cells (in some studies, at least 16 cells) proved to be a good indicator of the "biological age" of normal cell culture, well correlated

Table 4.1 A simplified classification of approaches to assessing the viability of cultured cells.[a]

Approach	Indices assessed	Comments/ specific parameters measured	Selected probes or dyes
Evaluation of cell proliferative activity	DNA synthesis	Label incorporation into nascent DNA	^{3}H-Thymidine; bromodeoxyuridine
	Colony-forming ability	AWNC; CFE	—
	Dilution of dye in daughter cells	Dye concentration decreases by half with each cell division	CFDA SE, CMFDA; SNARF-1 and its derivatives
Evaluation of plasma membrane integrity	Damaged plasmalemma	Dye penetration into the cell	Trypan blue; ethidium bromide
		Enzyme leakage from the cell	Bis-AAF-R110
	Intact plasmalemma	Dye retention in the cell	FDA, CFDA AM; SNARF-1
Evaluation of metabolic activity	Enzymatic activity	Redox reactions	Methylene blue; tetrazolium salts
		Esterase activity	FDA, SFDA; BCECF AM
	Transmembrane potential and other concentration gradients	Dye accumulation in mitochondria	Rhodamine 123; Di-4-ANEPPS; JC-1, JC-9
	ATP content in the cell	ATP-dependent transport	2-NBDG; $Na_2{}^{51}CrO_4$
		ATP-dependent enzyme activities	Luciferin/luciferase

[a]Abbreviations – 2-NBDG: 2-(*N*-(7-nitrobenz-2-*oxa*-1,3-diazol-4-yl)amino)-2-deoxyglucose; BCECF AM: 2′,7′-bis-(2-carboxyethyl)-5-(6)-carboxyfluorescein acetoxymethyl ester; bis-AAF-R110: bis-(alanyl-alanyl-phenylalanyl)-rhodamine 110; CFDA AM: 5-(6)-carboxyfluorescein diacetate acetoxymethyl ester; CFDA SE: 5-(6)-carboxyfluorescein diacetate *N*-succinimidyl ester; CMFDA: 5-chloromethylfluorescein diacetate; di-4-ANEPPS: 3-(4-(2-(6-(dibutylamino)naphthalene-2-yl) vinyl)pyridinium-1-yl)propane-1-sulfonate; FDA: fluorescein diacetate; JC-1: 5,5′,6,6′-tetrachloro-1,1′3,3′-tetraethylbenzimidazole carbocyanide iodide; JC-9: 3,3′-dimethyl-α-naphthoxazole iodide; SFDA: 5-sulfofluorescein diacetate; SNARF-1: seminaphtharhodafluor-1; AWNC: average weighted number of the class; CFE: colony-forming efficiency.

with the cell population doubling level. The CFE assay is also actively used to test the mitogenic or cytotoxic activity of various compounds.[92,97]

As a rule, this assay is performed by plating 100–200 cells from the test culture into Petri dishes and evaluating the number and size of colonies grown after several days. The same approach is suitable for determining the CFE of cells from donors of different ages in studies on their aging *in vivo*, but it is for obvious reasons inapplicable to postmitotic or very slowly dividing cells composing organs critically important for the aging process (neurons, cardiomyocytes, hepatocytes, *etc.*). Unfortunately, the viability of such cells can only be assessed using the aforementioned probes for measuring a certain functional parameter. However, the choice of such a parameter largely depends on what concept of aging is supported by the researcher, while the idea that if a cell divides, then this is certainly a live cell, is evident to all gerontologists. This is why, when possible, it is most expedient to rely on measurements of CFE as the best indicator of cell viability in the population studied. Unfortunately, this method in its routine variant often fails to reveal subtle modifications of CFE manifested as changes in the distribution of colonies by size rather than in their number. Hence, it is often difficult to compare histograms of size distribution for colonies formed in different Petri dishes, *e.g.*, in experiments on the effect on cell cultures of certain biologically active substances (in particular, potential geroprotectors or geropromoters). To facilitate this procedure, it would be desirable to have a certain numerical parameter providing an integrated characteristic of each histogram, which allows simple statistical analysis of the data.

To this end, we have modified the method for the CFE assay by distributing the colonies grown in a dish into size classes and calculating the average weighted number of the class (AWNC) for each distribution.[41,62] It has been assumed that an increase in AWNC (a shift of the distribution to larger colony size) is indicative of improvement in the functional status of the test culture (*i.e.*, a reduction of its "biological age"), while a decrease in this parameter is evidence that the culture is "getting older." This approach may be ineffective in some cases since it is theoretically possible that AWNC remains the same while the shape of the histogram changes, but we have never observed such a situation in our experiments. An additional advantage of this approach is that it provides for lower scattering of results obtained by different researchers for the same Petri dishes with cell colonies. Therefore, more researchers may be involved in the tedious process of cell counting in all the colonies in order to accelerate it, without any significant increase in the contribution of subjectivity to the dispersion of the results. To plot the distribution of colonies by size, we divide them into 17 classes with regard to the number of cells per colony (1–15, 16–31, 32–47···240–255, 256 and greater). Thus, all classes except the first and the last are of the same size (16 cells).

The above approach to the testing of biologically active compounds on cell cultures is well illustrated by the results of our study on the effect of hydrated C_{60}-fullerene on the CFE of transformed Chinese hamster cells.[62] These results confirm the geropromoter activity of the test agent, which has been revealed in previous experiments with the stationary phase aging model. As found in this study, the calculation of CFE and, especially, AWNC markedly simplifies the interpretation of experimental data and practically eliminates the problem of subjectivity in taking colony cell counts. To date, we have performed a number of cell culture experiments with various potential geroprotectors and geropromoters, and the results obtained provide conclusive evidence for the expediency of the proposed approach for rapid testing of such agents.[41,98]

To be objective, it should be noted that analysis of stationary phase cell aging by CFE assay may be complicated by the fact that cultured cells should be first removed from the growth substrate by treatment with special agents (usually a mixture of trypsin and Versene solutions) that disrupts the calcium–protein bridges attaching the cells to the surface and then plated at a very low density into Petri dishes or culture plates with fresh medium. This procedure is fairly traumatic and may even be fatal, especially for "elderly" cells, and the scattering of data on their CFE may sharply increase at the late stages of cell survival in this model system.[62]

4.4 Interpretation of Data About the Impact of Geroprotectors on Viability of Cultured Cells in Cytogerontological Studies

Based on the data reviewed in the former section it could be assumed that the solution of problems related to evaluating the viability of cultured cells in cytogerontological experiments, with special emphasis placed on the problems associated with constructing of the survival curves for cultured cells in the stationary phase aging model, should ensure successful testing of potential geroprotectors in experiments based on this model as well as on some other cytogerontological model systems. However, the following questions (in addition to the questions formulated in the Introduction) regarding the interpretation of data obtained in such studies in application to humans, whose aging is of primary interest to us, remain open:

(1) Whether the factors (chemical or physical) that improve the viability of cultured cells should always slow down the aging of a multicellular organism, and *vice versa?*
(2) How important is it which criteria of cell viability are used in testing geroprotectors in cytogerontological experiments?
(3) How can the interpretation of results obtained in a study depend on the origin of cells that were used in this study?

We will try to answer these questions below.

When studying potential geroprotectors in cytogerontological experiments (*i.e.*, in experiments on cell cultures), we usually evaluate their effect on cell viability. However, the criteria of this viability, as already mentioned, may be fundamentally different depending on the theory of aging to which a specific researcher adheres. In particular, the concept according to which the aging of a multicellular organism is caused by the limited mitotic potential of the normal cells constituting this organism has been very popular for many years. For this reason, the compounds that increase the proliferative potential (the "Hayflick limit") of such cells *in vitro* were automatically regarded as geroprotectors (it should be emphasized that we are talking about the proliferative *potential* of cells but not about their proliferative *activity*; unfortunately, these parameters are very often confused in the cytogerontological literature). At the same time, data according to which the aging of an organism is largely determined by its postmitotic or very slowly proliferating cells (neurons, cardiomyocytes, hepatocytes, egg cells, *etc.*), which never have enough time to realize even the "normal" proliferative potential during the lifetime of the "host", have been ignored.[6,99] The majority of human cells do not proliferate or proliferate very slowly because they *should not* do it rather than because they *cannot*. Therefore, the induction of telomerase activity in the normal cells, leading to a significant increase in their mitotic potential (possibly, even making it unlimited), cannot be realized in these cells. And for the cells of the organism that already have telomerase (stem and germ line cells), this induction is even more useless.

If a test compound has a positive effect on the proliferative *activity* of cells (which is manifested, for example, in increasing their CFE), the effects of this drug on the organism can be dual type. On the one hand, for some cells (for example, those involved in the regeneration processes), such stimulation can be useful. On the other hand, this effect can, firstly, stimulate the proliferation of those cells that, as mentioned above, should not divide, and secondly, increase the probability of a rapid propagation of the precancerous (or even cancerous) cells present in the organism. An increase in the incidence of benign tumors also cannot be ruled out. However, it should be emphasized that the evaluation of the CFE of cells is one of the few methods that provide data on the characteristics of *individual* cells rather than the cell population in general.[92] In the latter case, information about possible subpopulations of cells that may differently respond to the test compound is lost due to averaging. For example, under the influence of a test factor, the content of 8-oxo-2′-deoxyguanosine, a popular aging biomarker,[70] in DNA of different cells may increase, decrease, or remain unchanged. As a result, the estimation of the content of 8-oxo-2′-deoxyguanosine *on average* can lead to the conclusion about the absence of changes in this parameter.

Some parameters used to assess cell viability in cytogerontological experiments can be purely "correlative",[40] so that their interpretation becomes even more complicated. For example, this applies to the saturating density of a cell culture. It is known that, for normal diploid cells, this parameter is inversely well correlated with the age of the cell donor (in this case, the cause–effect relationships remain unclear). It was this parameter we used in our cell kinetics model (see Section 4.2) to assess potential geroprotectors. It was assumed that the factors that increase the saturating density of the culture and, thereby, reduce the "biological age" of cells should have a positive effect on the viability and aging of a multicellular organism. However, in this case, we may face the same problems as in interpreting the data of experiments on CFE. It is not obvious that an improved ability of cells to reach a high saturating density in culture will slow down the aging of a multicellular organism in all cases. It cannot be ruled out that it may have no effect on the aging process at all or may even accelerate it.

It is very important which cell types are used in cytogerontological experiments on testing potential geroprotectors—normal or transformed cells of multicellular organisms, unicellular eukaryotic or prokaryotic organisms, *etc.* As noted above, differences in interpreting the results of geroprotector testing that were obtained on normal and transformed human or animal cells can become quite apparent when these results are extrapolated to humans, many of whom die from cancer. In particular, the biologically active compounds that reduce the viability of cultured cancer cells can extend the life span of humans and experimental animals, similarly to the agents that increase the viability of normal cultured cells. The use of unicellular organisms, such as bacteria or yeast, makes it possible to estimate the effect of various agents on the cells that represent independent organisms. However, a bacterium, for example, is so dramatically different from a mammalian cell that the same compound can kill the former but havc hardly any effect on the viability of the latter (for example, this refers to antibiotics).

In our opinion, the use of the stationary phase aging model in many cases makes it possible to avoid many of these problems because the key factor that triggers the "aging" of all cells used in experiments is the restriction of cell proliferation with the help of various quite physiological impacts. A classic example is the chronological aging of yeast,[74,80] the results of studies of which are often pretty successfully used for studying the mechanisms of aging of humans and animals. In particular, experiments with the yeast *Saccharomyces cerevisiae* showed that rapamycin, a well-known mTOR inhibitor, in small doses that are sufficient for slowing down the proliferation of yeast cells but do not completely block this process, increases the culture life span in the chronological aging model.[100,101] Later this compound was shown to extend the life span of experimental animals—mice[102,103] and fruit flies.[104] It should be noted that, according to the ideas of some researchers,[101,105] the positive "gerontological" effect of rapamycin may be associated with the activation of autophagy. It also cannot be ruled out that the beneficial effect of rapamycin on the life span of animals may be due to its ability to suppress the emergence and development of malignant tumors.[17,106] As already mentioned above, in this case, it can hardly be considered a geroprotector. In addition, it is interesting to note that animals may develop tolerance to rapamycin over time. For this reason, some authors suggest that this drug should be used in combination with other active compounds, such as resveratrol.[107] Unfortunately, such problems are unlikely to be "caught" in cytogerontological studies.

4.5 Some Words About Biomarkers of Cell Aging/ Senescence

It appears that, today, the construction of the survival curves of the test animal/human cohorts is the most reliable way to estimate the efficiency of interventions in the aging process. Unfortunately, this method is inefficient in terms of labor, time, and finance expenditures. Because of this, overeager gerontologists currently rely mainly on so-called BA. Space limitations do not allow us to dwell on the essence of this term, but this is not necessary, since the relevant literature is available to any reader. It should only be noted that the researchers who use this term usually have in mind not so much the markers of aging itself as the markers of biological age. In other words, the markers (parameters) are well correlated with the chronological age of the test organisms but not with aging, *i.e.*, the time-dependent increase in the probability of death.

An illustrative example to this issue is the situation with human hair turning gray: the relative amount of gray hairs is well correlated with age but shows practically no correlation with mortality. Thus, relevant parameters in gerontology are those related to the basic mechanisms of aging, preferably in a cause-and-effect mode, and the majority of gerontologists consider that these are cellular or molecular mechanisms. The batteries of tests for determining the biological age (in other words, the degree of senescence) based on evaluation of various physiological parameters, which have been used on a wide scale, gradually recede in the past, giving way to studies with emphasis on "fundamental" BA—that is, on certain cellular or molecular characteristics. Moreover, these parameters are currently usually tied in with the phenomenon named cell/cellular senescence, which is central in cytogerontology.

It was initially considered that cell senescence takes place "by itself", *i.e.*, it is driven by an intrinsic mechanism, and all subsequent changes in the cells are mere *consequences* of this process. In fact, this fully applies to the mechanism of telomere shortening with every cell division, discovered by Alexey M. Olovnikov.[47] In the 1980s, one of us formulated the concept of aging,[33] according to which the restriction of cell proliferation imposed during development (due to the formation of populations of highly differentiated postmitotic or very slowly dividing cells) is the main cause of age-dependent accumulation of various macromolecular defects (mainly DNA damage) in the cells. This concept provides a simple explanation to "age-dependent" changes in senescent cell cultures: as cell proliferation at later passages is retarded, spontaneous DNA injuries are no longer "diluted" among newly emerging cells and their frequency in the population as a whole increases. The population aspect is very important since some cells fully retain the ability to divide, but their proportion decreases with passaging, so that cell senescence is manifested at the level of whole *cell population*. In essence, our "stationary phase aging" model[5,33,58–61] was based on 100% suppression of cell proliferation in culture by contact inhibition or some other physiological factor, with consequent accumulation of "age-dependent" defects in the cell population. In this case as well, we first made the cells "senesce" and only then analyzed them for certain biomarkers of *in vivo* aging (*e.g.*, DNA breaks). Thus, in the "classic" approach it was assumed that cell senescence is driven by a certain intrinsic mechanism, which leads to the emergence of various macromolecular defects (first of all, DNA damage) in the cells.

In recent years, however, cell senescence is understood primarily as the appearance or accumulation in the cells (most often, transformed cells not prone to replicative senescence) of certain "BA" (this time in quotation marks, because the situation is by no means related to real aging) under the impact of various external factors causing DNA damage (oxidative stress, H_2O_2, mitomycin C, doxorubicin, ethanol, ionizing radiation, *etc.*).[30,108–111] This phenomenon is referred to as DNA damage response (DDR). Within this definition, the "senescence" of cells takes place under the impact of DNA-damaging agents rather than on itself. It is also called "stress-induced premature senescence".[112] The aforementioned BA include senescence-associated beta-galactosidase (SA-β-Gal) activity, expression of p53 and p21 proteins as well as of regulators of inflammation such as IL-6 or IL-8, activation of oncogenes, *etc.* Therefore, cell "senescence" in the context of the above definition occurs not by itself but *because of* the impact of DNA-damaging agents. In our opinion, such an approach is very important for defining the strategy of cancer control but, yet again, leads away from the study of actual mechanisms of organismal aging.[31] A similar view was expressed by famous gerontologist Denham Harman in his brief comment published in the journal *Biogerontology*.[113] It should be emphasized that in our "stationary phase aging" model[5,33,59,60] we also observe certain BA in cell cultures, but in this case they appear *due to* restriction of proliferation by contact inhibition, *i.e.*, by a physiological factor that itself causes no damage to the cells. This situation is closely similar to what takes place in a multicellular organism.

The most popular biomarker of cellular senescence is SA-β-Gal (β-galactosidase pH 6.0). The enzyme β-galactosidase, a lysosomal hydrolase, cleaves off the terminal β-galactose from the compounds containing it (lactose, keratin sulfates, sphingolipids, *etc.*). It is involved in some "minor" metabolic reactions and is present in almost all tissues. This enzyme exhibits maximum activity at pH 4.0; however, the difference in this index between the "old" and "young" cells can be better detected by certain biochemical methods at pH 6.0. The feasibility of using SA-β-Gal activity as a BA was first postulated in 1995 by Dimri *et al.*,[114] who demonstrated that the expression of this enzyme increases with aging both *in vitro* and *in vivo*. In subsequent years, this BA was widely used in cytogerontological experiments to assess the "age" of cells and is currently the most common in the studies[29,115] based on the definition of cellular senescence that we do not accept. However, in parallel, several studies were published whose authors emphasized that SA-β-Gal activity in cells is not so good a BA, because, in many cases, it depends not so much on age (both *in vivo* and *in vitro*) as on the method of research and/or the presence of certain pathologies as well as, what is most important, on the proliferative status of the cells.[116–122] It seems that cell proliferation restriction, for whatever reason (differentiation, contact inhibition, DDR, some diseases, *etc.*), is the factor that causes stimulation of SA-β-Gal expression. In other words, SA-β-Gal appears even in the "young" cells if their proliferation is suppressed. Not long ago, we showed[71] that in the stationary phase culture of transformed Chinese hamster cells, the proportion of cells in which SA-β-Gal is detected by the method of Dimri *et al.* increases with time, and this is accompanied, on the one hand, by an increase in the level of poly(ADP-ribose) in the cells and, on the other hand, by a decline in their capacity to synthesize poly(ADP-ribose) in response to DNA damage induced by H_2O_2. Such data, in our opinion, provide further evidence of the viability of our concept of aging, which postulates the crucial role of cell proliferation restriction in the accumulation in cells of various macromolecular defects (the most important of which are DNA lesions), which, in turn, lead to deterioration in the functioning of organs and tissues and further increase in the probability of death of macroorganisms.[5,6,22]

It is also interesting to note that, in the experiments designed to compare the effects of "stationary-phase" or "stress-induced" (exposure to 4% ethanol for 2 h per day for 5 days) aging on the transformed Chinese hamster cells, we showed that the percentage of cells stained for SA-β-Gal by the method of Dimri *et al.* in a 14 day-old "stationary-phase-old" culture was much higher than in the "young" (7 day-old) control culture but comparable to that detected in 7 day-old cells incubated with ethanol.[123]

Finally, we would like to mention another study,[124] the authors of which showed that, both in the "stress-induced premature senescence" and in the replicative senescence "according to Hayflick," SA-β-Gal does not accumulate if the expression of the *GLB1* gene, which encodes the lysosomal β-galactosidase, is disrupted.

4.6 Conclusions

(1) We think that any "true" geroprotector should retard the age-related increase in the probability of death of aging organisms causing a rightward shift of the survival curve and increasing both the average and maximum life span.
(2) We do not think that the drugs that are used to combat age-related diseases could be considered geroprotectors, as well as the factors that increase the life span of the non-aging organisms.
(3) At present, there are several cytogerontological models that are used for testing of potential geroprotectors. The most popular among them are the Hayflick model, the stationary phase (chronological) aging model, and the cell kinetics model. In our opinion, the least number of problems associated with interpreting the results of testing potential geroprotectors in cytogerontological experiments arises when such studies are performed using the model of stationary phase aging (which is based on the concept of cell proliferation restriction as the main cause of accumulation of macromolecular lesions in cells of multicellular organisms with age, leading to the deterioration of the functioning of tissues and organs and, as a result, an increase in the probability of death) of normal cells. However, even this approach will not give the *final* answer to the question of whether or not the studied factor is a

"true" geroprotector. Answering this question will inevitably require both experiments in animals and clinical trials.

(4) The cytogerontological models mentioned can be effectively used to test various agents (drugs) or their combinations for their potential ability to accelerate or retard aging only if their effect is realized at the cell level. Unfortunately, we have recently got the impression that even the data obtained with "gist" cell culture models cannot be directly extrapolated to the organism as a whole. Our cytogerontological tests of various geroprotectors on the models of "stationary phase aging", cell kinetics, and cell CFE have shown that these factors fairly often have no favorable effect on the viability of cultured cells, even though they prolong the life span of experimental animals and improve the state of human health. This fact suggests that the effect of a geroprotector in many cases manifests itself only at the organismal level (probably due to activation/suppression of certain biochemical or neurophysiological processes) and is not limited to the improvement of viability of individual cells. Apparently, the same is also true of geropromoters. Thus, it was probably a serious mistake to perform experiments with cell cultures so as to exclude the influence of the endocrine and central nervous systems (which actually was the main purpose of gerontologists, beginning from studies by Alexis Carrel[42,43]). By all accounts, the results of cytogerontological experiments should be thoroughly verified in studies on laboratory animals and even in clinical trials (provided this complies with ethical principles of human subject research). Of course, this will lessen our chance for an early breakthrough in studies aimed at retarding the process of aging, but the reliability of the obtained data will be significantly higher.

(5) Regarding the approaches to cell viability testing in cytogerontological experiments, the choice of methods to this end depends mainly, apparently, on the researchers' ideas about molecular and cellular mechanisms of aging. The most appropriate method, the evaluation of CFE, though optimal for cell viability assessment, is not applicable to postmitotic or very slowly propagating cells. Unfortunately, many problems encountered when using popular molecular probes designed for live/dead cell viability assays remain open.

(6) When interpreting the results of geroprotector testing in experiments on cell cultures, the conclusions strongly depend on which cells types are used (see Section 4.4). In particular, they could vary greatly for normal and transformed animal cells.

(7) Instead of analyzing the effect of a potential anti-aging factor on the proliferative potential of cultured cells or their stationary phase life span we can follow some BA during cell aging/senescence *in vitro* or stationary phase (chronological) aging. If all that was said in Section 4.5 is true, then it may well be that canceling the aging process will not necessarily cause any significant changes in the age-dependent dynamics of those BA (regardless of whether they accumulate or disappear)

that are directly connected with the proliferative status of cells forming organs and tissues. This should apply at least to those BA that are not directly involved in the mechanisms responsible for the age-dependent increase in the probability of death. If certain BA are "gist" markers, *i.e.*, the aging process cannot be retarded without affecting these BA, then the postulated mechanism of aging canceling should provide an explanation as to how these BA will be continuously removed from postmitotic or very slowly proliferating cells.

Finally, here is the main conclusion. As already noted above, it appears that gerontologists analyzing the possibilities for retarding or even blocking the aging process currently have no fully adequate alternative to the construction of survival curves for the cohorts of animals or humans, even though this approach is highly expensive and requires great labor expenditures. Apparently, all the cytogerontological models reviewed provide only *preliminary* testing of potential anti-aging factors.

References

1. A. Comfort, *Ageing: The Biology of Senescence*, Routledge & Kegan Paul, London, 2nd edn, 1964.
2. A. N. Khokhlov, *Russ. J. Gen. Chem.*, 2010, **80**(7), 1507.
3. L. Wei, Y. Li, J. He and A. N. Khokhlov, *Moscow Univ. Biol. Sci. Bull.*, 2012, **67**(1), 13.
4. A. N. Khokhlov, L. Wei, Y. Li and J. He, *Adv. Gerontol.*, 2012, **25**(3), 513.
5. A. N. Khokhlov, *Curr. Aging Sci.*, 2013, **6**(1), 14.
6. A. N. Khokhlov, *Biophysics*, 2010, **55**(5), 859.
7. D. Martínez, *Exp. Gerontol.*, 1998, **33**(3), 217.
8. D. E. Martínez and D. Bridge, *Int. J. Dev. Biol.*, 2012, **56**(6–8), 479.
9. O. R. Jones, A. Scheuerlein, R. Salguero-Gómes, C. G. Camarda, R. Schaible, B. B. Casper, J. P. Dahlgren, J. Ehrlén, M. B. García, E. S. Menges, P. F. Quintana-Ascencio, H. Caswell, A. Baudisch and J. W. Vaupel, *Nature*, 2014, **505**(7482), 169.
10. A. N. Khokhlov, *Moscow Univ. Biol. Sci. Bull.*, 2014, **69**(4), 153.
11. C. S. Minot, *The Problem of Age, Growth, and Death; a Study of Cytomorphosis, Based on the Lectures at the Lowell Institute, March, 1907*, G. P. Patnam's Sons, New York–London, 1908.
12. I. I. Schmalhausen, *The Problem of Death and Immortality*, Gosizdat, Moscow–Leningrad, 1926 (in Russian).
13. G. P. Bidder, *Br. Med. J.*, 1932, **2**(3742), 583.
14. V. M. Dilman, *Lancet*, 1971, **297**(7711), 1211.
15. S. N. Austad, *Why We Age: What Science Is Discovering about The Body's Journey through Life*, John Wiley & Sons, Inc., Chichester, 1st edn, 1999.
16. S. N. Austad, *Aging Cell*, 2004, **3**(5), 253.
17. M. V. Blagosklonny, *Cell Cycle*, 2006, **5**(18), 2087.

18. L. Hayflick, *PLoS Genet.*, 2007, **3**(12), e220.
19. R. Holliday, *Aging: The Paradox of Life. Why We Age*, Springer, Dordrecht, 2007.
20. A. N. Khokhlov, *Biogerontology*, 2013, **14**(6), 703.
21. A. N. Khokhlov, *Russ. J. Dev. Biol.*, 2013, **44**(6), 336.
22. A. N. Khokhlov, *Adv. Gerontol.*, 2014, **4**(2), 150.
23. H. E. Swim and R. F. Parker, *Am. J. Hyg.*, 1957, **66**(2), 235.
24. L. Hayflick and P. S. Moorhead, *Exp. Cell Res.*, 1961, **25**(3), 585.
25. L. Hayflick, *Exp. Cell Res.*, 1965, **37**(3), 614.
26. L. Hayflick, *J. Invest. Dermatol.*, 1979, **73**(1), 8.
27. S. I. S. Rattan, *Biogerontology*, 2000, **1**(1), 79.
28. J. Campisi, *Curr. Opin. Genet. Dev.*, 2011, **21**(1), 107.
29. E. Sikora, T. Arendt, M. Bennett and M. Narita, *Ageing Res. Rev.*, 2011, **10**(1), 146.
30. J. Campisi, *Annu. Rev. Physiol.*, 2013, **75**, 685.
31. A. N. Khokhlov, *Moscow Univ. Biol. Sci. Bull.*, 2013, **68**(4), 158.
32. L. Hayflick, *Mech. Ageing Dev.*, 1979, **9**(5–6), 393.
33. A. N. Khokhlov, *Proliferatsiya i Starenie (Cell Proliferation and Aging), Itogi Nauki i Tekhniki VINITI AN SSSR. Ser. Obshchie Problemy Fiziko-Khimicheskoi Biologii (Advances in Science and Technology, VINITI Akad. Sci. USSR, Ser. General Problems of Physicochemical Biology)*, VINITI, Moscow, 1988 (in Russian).
34. A. N. Khokhlov, *Tsitologiia*, 2002, **44**(12), 1143.
35. A. Weismann, *Die Kontinuität des Keimplasmas als Grundlage einer Theorie der Vererbung*, G. Fisher, Jena, 1885.
36. A. Weismann, *Das Keimplasma. Eine Theorie der Vererbung*, G. Fisher, Jena, 1892.
37. L. Hayflick, *How and Why We Age*, Ballantine Books, New York, 1996.
38. T. B. Kirkwood and T. Cremer, *Hum. Genet.*, 1982, **60**(2), 101.
39. A. N. Khokhlov, in *Biomarkers of Aging: Expression and Regulation. Proceeding*, ed. F. Licastro and C. M. Caldarera, CLUEB, Bologna, 1992, p. 209.
40. A. N. Khokhlov, *Russ. J. Dev. Biol.*, 2003, **34**(5), 321.
41. E. S. Alinkina, A. K. Vorobyova, T. A. Misharina, L. D. Fatkullina, E. B. Burlakova and A. N. Khokhlov, *Moscow Univ. Biol. Sci. Bull.*, 2012, **67**(2), 52.
42. A. Carrel, *J. Exp. Med.*, 1912, **17**(1), 14.
43. A. Carrel, *J. Exp. Med.*, 1913, **18**(3), 287.
44. A. M. Olovnikov, *J. Alzheimer's Dis.*, 2007, **11**(2), 241.
45. A. M. Olovnikov, *Russ. J. Dev. Biol.*, 2007, **38**(2), 104.
46. A. Macieira-Coelho, *Biogerontology*, 2011, **12**(6), 503.
47. A. M. Olovnikov, *Dokl. Akad. Nauk SSSR*, 1971, **201**(6), 1496.
48. A. M. Olovnikov, *J. Theor. Biol.*, 1973, **41**(1), 181.
49. A. M. Olovnikov, *Exp. Gerontol.*, 1996, **31**(4), 443.
50. V. M. Mikhelson, *Mech. Ageing. Dev.*, 2001, **122**(13), 1361.
51. V. M. Mikhelson and I. A. Gamaley, *Curr. Aging Sci.*, 2012, **5**(3), 203.

52. L. Hayflick, *Mutat. Res., DNAging: Genet. Instab. Aging*, 1991, **256**(2–6), 69.
53. E. Iu. Chirkova, M. E. Golovina, T. L. Nadzharian and A. N. Khokhlov, *Dokl. Akad. Nauk SSSR*, 1984, **278**(6), 1474.
54. A. N. Khokhlov, M. E. Golovina, E. Iu. Chirkova and T. L. Nadzharian, *Tsitologiia*, 1985, **27**(8), 960.
55. E. L. Schneider and J. R. Smith, *Int. Rev. Cytol.*, 1981, **69**, 261.
56. A. N. Khokhlov, M. E. Golovina, E. Iu. Chirkova and T. L. Nadzharian, *Tsitologiia*, 1985, **27**(9), 1070.
57. A. N. Khokhlov, M. E. Golovina, E. Iu. Chirkova and T. L. Nadzharian, *Tsitologiia*, 1987, **29**(3), 353.
58. M. M. Vilenchik, A. N. Khokhlov and K. N. Grinberg, *Stud. Biophys.*, 1981, **85**(1), 53.
59. A. N. Khokhlov, *Ann. N. Y. Acad. Sci.*, 1992, **663**, 475.
60. A. N. Khokhlov, *Ann. N. Y. Acad. Sci.*, 1998, **854**, 519.
61. S. S. Akimov and A. N. Khokhlov, *Ann. N. Y. Acad. Sci.*, 1998, **854**, 520.
62. O. I. Yablonskaya, T. S. Ryndina, V. L. Voeikov and A. N. Khokhlov, *Moscow Univ. Biol. Sci. Bull.*, 2013, **68**(2), 63.
63. A. N. Khokhlov, E. Iu. Chirkova and T. L. Nadzharian, *Tsitologiia*, 1984, **26**(8), 965.
64. A. N. Khokhlov, E. Iu. Chirkova and A. N. Chebotarev, *Tsitol. Genet.*, 1985, **19**(2), 90.
65. A. N. Khokhlov, E. Yu. Chirkova and A. I. Gorin, *Bull. Exp. Biol. Med.*, 1986, **101**(4), 437.
66. A. N. Khokhlov, E. Iu. Chirkova and A. N. Chebotarev, *Tsitol. Genet.*, 1987, **21**(3), 186.
67. A. N. Khokhlov, M. D. Kirnos and B. F. Vaniushin, *Izv. Akad. Nauk SSSR, Ser. Biol.*, 1988, (3), 476.
68. L. Yu. Prokhorov, N. A. Petushkova and A. N. Khokhlov, *Age*, 1994, **17**(4), 162.
69. S. I. Shram, G. A. Shilovskii and A. N. Khokhlov, *Bull. Exp. Biol. Med.*, 2006, **141**(5), 628.
70. D. S. Esipov, T. A. Gorbacheva, G. A. Khairullina, A. A. Klebanov, T. N. T. Nguen and A. N. Khokhlov, *Adv. Gerontol.*, 2008, **21**(3), 485.
71. I. V. Vladimirova, G. A. Shilovsky, A. N. Khokhlov and S. I. Shram, in *Visualizing of Senescent Cells in Vitro and in Vivo, Programme and Abstracts, Warsaw, Poland, December 15–16, 2012*, Warsaw, 2012, p. 108.
72. T. Nyström, *Curr. Opin. Microbiol.*, 2002, **5**(6), 596.
73. P. Pletnev, I. Osterman, P. Sergiev, A. Bogdanov and O. Dontsova, *Acta Naturae*, 2015, **7**(4), 22.
74. A. N. Khokhlov, *Moscow Univ. Biol. Sci. Bull.*, 2016, **71**(1), 11.
75. V. L. Ushakov and A. N. Khokhlov, *Age*, 1992, **15**(4), 128.
76. A. N. Khokhlov, V. L. Ushakov, A. B. Kapitanov and T. L. Nadzharian, *Dokl. Akad. Nauk SSSR*, 1984, **274**(4), 930.
77. A. B. Kapitanov and M. Y. Aksenov, *Mech. Ageing Dev.*, 1990, **54**(3), 249.
78. A. N. Khokhlov, *Age*, 1994, **17**(4), 159.
79. P. Fabrizio and V. D. Longo, *Aging Cell*, 2003, **2**(2), 73.

80. *Aging research in yeast*, ed. M. Breitenbach, S. M. Jazwinski, and P. Laun, Springer, Dordrecht—Heidelberg—London—New York, 2012.
81. P. Laun, C. V. Bruschi, J. R. Dickinson, M. Rinnerthaler, G. Heeren, R. Schwimbersky, R. Rid and M. Breitenbach, *Nucleic Acids Res.*, 2007, **35**(22), 7514.
82. A. E. Roux, A. Quissac, P. Chartrand, G. Ferbeyre and L. A. Rokeach, *Aging Cell*, 2006, **5**(4), 345.
83. A. N. Khokhlov, in *Visualizing of senescent cells in vitro and in vivo. Programme and abstracts, Warsaw, Poland, December 15–16, 2012*, Warsaw, 2012, p. 49.
84. A. N. Khokhlov, A. A. Klebanov, A. F. Karmushakov, G. A. Shilovsky, M. M. Nasonov and G. V. Morgunova, *Moscow Univ. Biol. Sci. Bull.*, 2014, **69**(1), 10.
85. G. Kroemer, L. Galluzzi and P. Vandenabeele, *et al.*, *Cell Death Differ.*, 2009, **16**(1), 3.
86. *The Molecular Probes Handbook: A Guide to Fluorescent Probes and Labeling Technologies*, 11th edn, 2010 (https://www.thermofisher.com, last accessed June 2016).
87. M. V. Berridge, P. M. Herst and A. S. Tan, *Biotechnol. Annu. Rev.*, 2005, **11**, 127.
88. Y. P. Keepers, P. E. Pizao, G. J. Peters, J. van Ark-Otte, B. Winograd and H. M. Pinedo, *Eur. J. Cancer*, 1991, **27**(7), 897.
89. M. J. Stoddart, in *Mammalian Cell Viability, Methods and Protocols*, ed. M. J. Stoddart, Humana Press, 2011, p. 1.
90. A. N. Khokhlov and G. V. Morgunova, *Moscow Univ. Biol. Sci. Bull.*, 2015, **70**(2), 67.
91. A. Reddy, E. V. Caler and N. W. Andrews, *Cell*, 2001, **106**(2), 157.
92. A. N. Khokhlov, L. Yu. Prokhorov, A. S. Ivanov and A. I. Archakov, *FEBS Lett.*, 1991, **290**(1–2), 171.
93. J. R. Smith, S. Venable, T. W. Roberts, E. J. Metter, R. Monticone and E. L. Schneider, *J. Gerontol., Ser. A*, 2002, **57**(6), B239.
94. A. B. Maier, I. L. Maier, D. van Heemst and R. G. J. Westendorp, *J. Gerontol., Ser. A*, 2008, **63**(7), 655.
95. P. I. Good and J. R. Smith, *Biophys. J.*, 1974, **14**(11), 811.
96. J. R. Smith, O. Pereira-Smith and P. I. Good, *Mech. Ageing Dev.*, 1977, **6**(4), 283.
97. T. Kuczek and D. E. Axelrod, *Proc. Natl. Acad. Sci. U. S. A.*, 1987, **84**(13), 4490.
98. A. N. Khokhlov, G. V. Morgunova, T. S. Ryndina and F. Coll, *Moscow Univ. Biol. Sci. Bull.*, 2015, **70**(1), 7.
99. V. J. Cristofalo, R. G. Allen, R. J. Pignolo, B. G. Martin and J. C. Beck, *Proc. Natl. Acad. Sci. U. S. A.*, 1998, **95**(18), 10614.
100. R. W. Powers III, M. Kaeberlein, S. D. Caldwell, B. K. Kennedy and S. Fields, *Genes Dev.*, 2006, **20**(2), 174.
101. A. L. Alvers, M. S. Wood, D. Hu, A. C. Kaywell, W. A. Dunn Jr. and J. P. Aris, *Autophagy*, 2009, **5**(6), 847.

102. D. E. Harrison, R. Strong, Z. D. Sharp, J. F. Nelson, C. M. Astle, K. Flurkey, N. L. Nadon, J. E. Wilkinson, K. Frenkel, C. S. Carter, M. Pahor, M. A. Javors, E. Fernandez and R. A. Miller, *Nature*, 2009, **460**(7253), 392.
103. R. A. Miller, D. E. Harrison and C. M. Astle, *et al.*, *Aging Cell*, 2014, **13**(3), 468.
104. I. Bjedov, J. M. Toivonen, F. Kerr, C. Slack, J. Jacobson, A. Foley and L. Partridge, *Cell Metab.*, 2010, **11**(1), 35.
105. D. C. Rubinsztein, G. Mariño and G. Kroemer, *Cell*, 2011, **146**(5), 682.
106. F. Neff, D. Flores-Dominguez and D. P. Ryan, *et al.*, *J. Clin. Invest.*, 2013, **123**(8), 3272.
107. A. Alayev, S. M. Berger, M. Y. Kramer, N. S. Schwartz and M. K. Holz, *J. Cell. Biochem.*, 2015, **116**(3), 450.
108. J. C. Jeyapalan and J. M. Sedivy, *Mech. Aging Dev.*, 2008, **129**(7–8), 467.
109. T. Kuilman, C. Michaloglou, W. J. Mooi and D. S. Peeper, *Genes Dev.*, 2010, **24**(22), 2463.
110. C. Lawless, C. Wang, D. Jurk, A. Merz, T. von Zglinicki and J. F. Passos, *Exp. Gerontol.*, 2010, **45**(10), 772.
111. S. Giaimo and F. D'Adda di Fagagna, *Aging Cell*, 2012, **11**(3), 378.
112. O. Toussaint, P. Dumont, J. Remacle, J. F. Dierick, T. Pascal, C. Frippiat, J. P. Magalhaes, S. Zdanov and F. Chainiaux, *Sci. World J.*, 2002, **2**, 230.
113. D. Harman, *Biogerontology*, 2009, **10**(6), 783.
114. G. P. Dimri, X. Lee, G. Basile, M. Acosta, G. Scott, C. Roskelley, E. E. Medrano, M. Linskens, I. Rubelj, O. Pereira-Smith, M. Peacocke and J. Campisi, *Proc. Natl. Acad. Sci. U. S. A.*, 1995, **92**(20), 9363.
115. F. Debacq-Chainiaux, T. Pascal, E. Boilan, C. Bastin, E. Bauwens and O. Toussaint, *Free Radical Biol. Med.*, 2008, **44**(10), 1817.
116. Y. E. Yegorov, S. S. Akimov, R. Hass, V. Zelenina and I. A. Prudovsky, *Exp. Cell Res.*, 1998, **243**(1), 207.
117. D. R. Krishna, B. Sperker, P. Fritz and U. Klotz, *Mech. Ageing Dev.*, 1999, **109**(2), 113.
118. J. Choi, I. Shendrik, M. Peacocke, D. Peehl, R. Buttyan, E. F. Ikeguchi, A. E. Katz and M. C. Benson, *Urology*, 2000, **56**(1), 160.
119. J. Severino, R. G. Allen, S. Balin, A. Balin and V. J. Cristofalo, *Exp. Cell Res.*, 2000, **257**(1), 162.
120. G. Untergasser, R. Gander, K. Rumpold, E. Heinrich, E. Plas and P. Berger, *Exp. Gerontol.*, 2003, **38**(10), 1179.
121. H. T. Kang, C. J. Lee, E. J. Seo, Y. J. Bahn, H. J. Kim and E. S. Hwang, *Mech. Ageing Dev.*, 2004, **125**(1), 31.
122. V. J. Cristofalo, *Exp. Gerontol.*, 2005, **40**(10), 836.
123. G. V. Morgunova, A. V. Kolesnikov, A. A. Klebanov and A. N. Khokhlov, *Moscow Univ. Biol. Sci. Bull.*, 2015, **70**(4), 165.
124. B. Y. Lee, J. A. Han, J. S. Im, A. Morrone, K. Johung, E. C. Goodwin, W. J. Kleijer, D. DiMaio and E. S. Hwang, *Aging Cell*, 2006, **5**(2), 187.

CHAPTER 5

Pharmacogenomics and Epigenomics of Age-Related Neurodegenerative Disorders: Strategies for Drug Development

RAMÓN CACABELOS*[a,b], JUAN CARLOS CARRIL[a,b] AND OSCAR TEIJIDO[a,b]

[a]Chair of Genomic Medicine, Camilo José Cela University, Madrid, Spain; [b]EuroEspes Biomedical Research Center, Institute of Medical Science and Genomic Medicine, Corunna, Spain
*E-mail: rcacabelos@euroespes.com

5.1 Introduction

Cardiovascular disorders, cancer, and central nervous system (CNS) diseases are major problems of health, representing 60–80% of the morbi-mortality in developed countries.[1,2] CNS disorders are the most prevalent cause of disability and socioeconomic and family burden in our society, and among CNS disorders, Alzheimer's disease (AD) and Parkinson's disease (PD) are the most relevant neurodegenerative disorders (NDDs), compromising mental

RSC Drug Discovery Series No. 57
Anti-aging Drugs: From Basic Research to Clinical Practice
Edited by Alexander M. Vaiserman

Published by the Royal Society of Chemistry, www.rsc.org

function and psychomotor function, respectively. Genomic, epigenomic, cerebrovascular, metabolic and environmental factors are potentially involved in the pathogenesis of most NDDs.[3] The age- and sex-related syndromic profiles of NDDs reflect, at least, a tetravalent phenotype: (i) a specific neuropathological component associated with each NDD; (ii) a neurobehavioral component: cognitive deterioration, behavioral changes, functional decline; (iii) an age-related biological component (directly-, indirectly-, and un-related biochemical, hematological and metabolic phenotypes); and (iv) gender-related phenotypes.[4–6] According to this heterogeneous, complex clinical picture, therapeutic intervention in most NDDs is polymodal in order to modify the expression of all these complex phenotypes.[7]

NDDs are age-dependent processes causing premature neurodegeneration many years before the onset of the disease. In this context, post-symptomatic intervention is of poor therapeutic value and less than 30% of patients respond moderately to conventional drugs in early stages of the disease.[8] Therefore, NDDs pose two major challenges to the scientific community: (i) the characterization and validation of specific biomarkers for the early identification of people at risk in susceptible populations; and (ii) the discovery and assessment of novel compounds with preventive activity and/or pharmacological properties able to halt disease progression at a pre-symptomatic stage.[8,9]

Major determinants of therapeutic outcome in NDDs include age- and sex-related factors, pathogenic phenotype, concomitant disorders, treatment modality and polypharmacy, and pharmacogenetics. Different categories of genes are potentially involved in the pharmacogenetic network responsible for drug efficacy and safety. Pathogenic, mechanistic, metabolic, transporter, and pleiotropic genes represent the major genetic determinants of response to treatment in NDDs.[10–12] The expression of genes involved in the pharmacogenetic cascade is under the regulation of the epigenetic machinery. By-products of these genes are integrated in transcriptomic, proteomic and metabolic networks which are disrupted in NDDs and represent potential targets for therapeutic intervention (Figure 5.1).[9,11,13–16]

5.2 Age-Related Pheno-Genotypes

NDD-related polymorphic phenotypes (neuropathological, neurobehavioral, biochemical, hematological) require multifactorial interventions (combined treatments) in over 60–70% of the cases, this contributing to increasing the risk of ADRs and drug–drug interactions in these complex disorders.[17] For instance, AD patients present concomitant disorders including hypertension (20–30%), overweightness or obesity (20–40%), diabetes (20–25%), hypercholesterolemia (>40%), hypertriglyceridemia (20%); excess of urea (>80%), creatinine (6%) and uric acid (5%); alterations in transaminases (ASAT, ALAT, GGT) (>15%), alkaline phosphatase (14%), bilirubin (17%), and ions (>10%); deficits of iron (5%), ferritin (3%), folate (5%), and vitamin B_{12} (4%); thyroid dysfunction (5–7%), and reduced levels of RBC (3%), HCT (33%), and

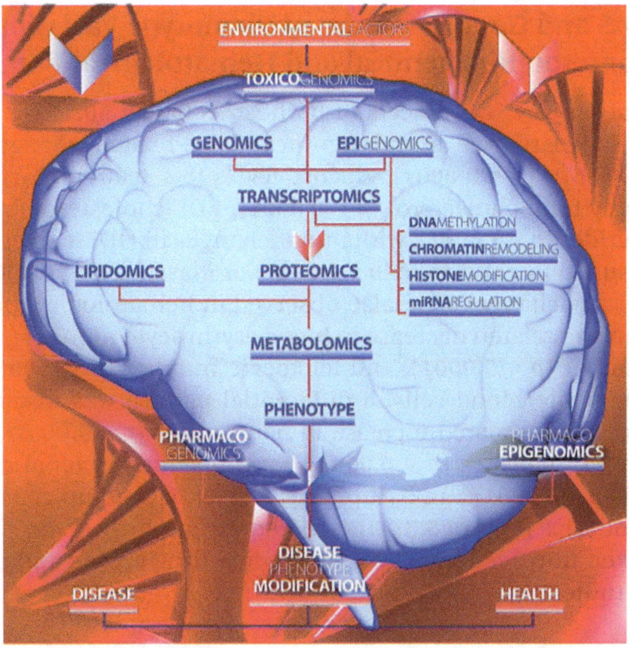

Figure 5.1 Pathogenic and therapeutic model of age-related neurodegenerative disorders.

Hb (35%).[12] Cardiovascular disorders (>40%), atherosclerosis (>60%), and different modalities of cerebrovascular damage (>60%) are also frequent among patients with AD. Most of these biochemical, hematological and metabolic anomalies exhibit gender differences and may contribute to accelerating the dementia process.

The pharmacological treatment of these concomitant pathologies adds complexity and risks to the multifactorial therapeutic intervention in patients with dementia. Of major relevance is the treatment of diabetes, hypertension, dyslipidemia, and cardiovascular, cerebrovascular and neuropsychiatric disorders. The most relevant chronic conditions among adults aged 55–64 in the USA are diabetes (18.9%), obesity (40.6%), hypercholesterolemia (50.1%), and hypertension (51.5%).[2] In the same population, the currently most-prescribed drugs are cardiovascular (45%), cholesterol-lowering (31.8%), gastric reflux (16%), analgesic (15%), antidepressant (14.4%), and antidiabetic drugs (12.9%).[2] The chronic treatment of these illnesses increases the risk of drug interactions and toxicity, aggravating the clinical condition of the demented patient. In this context, the incorporation of pharmacogenetic protocols into clinical practice is fundamental to minimize drug–drug interactions and ADRs, and to optimize the global therapeutic outcome, avoiding deleterious effects on mental function and cognition.

5.2.1 Age- and Genotype-Related Phenotype Variation in Common Biochemical and Hematological Parameters

In a population of 4747 subjects of both sexes (age = 50.55 ± 21.44 years, range: 0.3–98 years), we found significant age-related changes in body mass index (BMI) ($p < 0.0001$), glucose levels ($p < 0.001$), blood lipid levels (especially total-cholesterol ($p < 0.0001$), LDL-cholesterol ($p < 0.0001$), and triglycerides ($p < 0.0001$), with minor changes in HDL-cholesterol), and blood pressure, either systolic ($p < 0.0001$) or diastolic ($p < 0.0001$) (Figure 5.2). Interesting changes were also observed in hematological parameters, including an age-related decrease in blood erythrocyte number ($p < 0.0001$), platelet number ($p < 0.0001$), and leukocyte number ($p < 0.0001$) (Figure 5.3). Among white blood cells, a differential pattern was observed, characterized by an age-related increase in neutrophils ($p < 0.0001$), a marked decrease in lymphocytes ($p < 0.0001$), a significantly gradual decrease in eosinophils ($p < 0.0001$) and basophils ($p < 0.0001$), and no apparent variation in monocytes (Figure 5.3). Age-related variations in these phenotypes must be taken into account when monitoring drug effects and in pharmacogenetic studies.

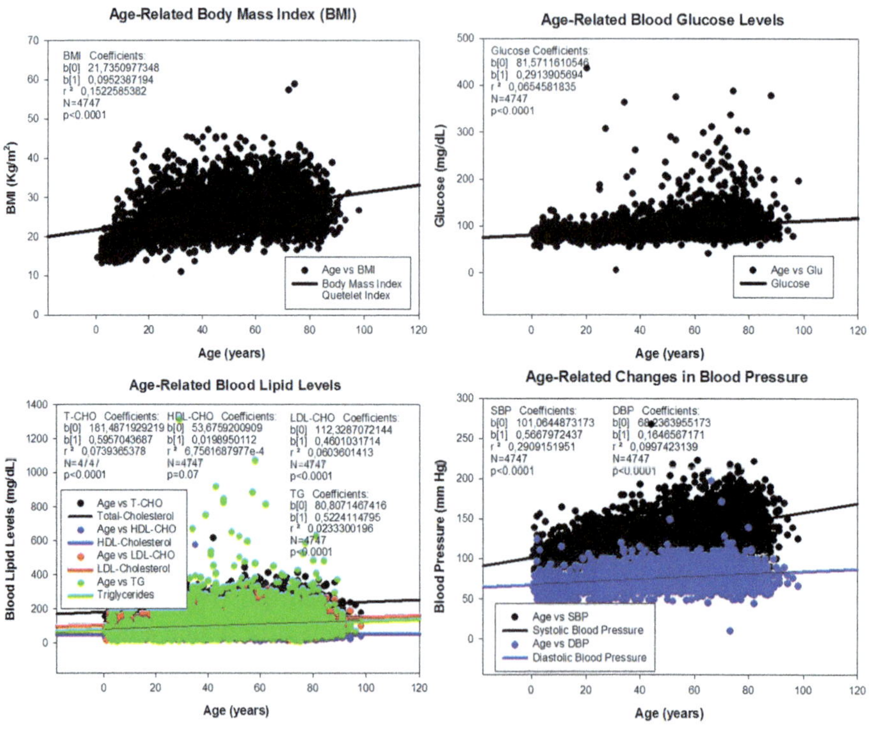

Figure 5.2 Age-related changes in body mass index (BMI), glucose levels, lipid levels and blood pressure values.

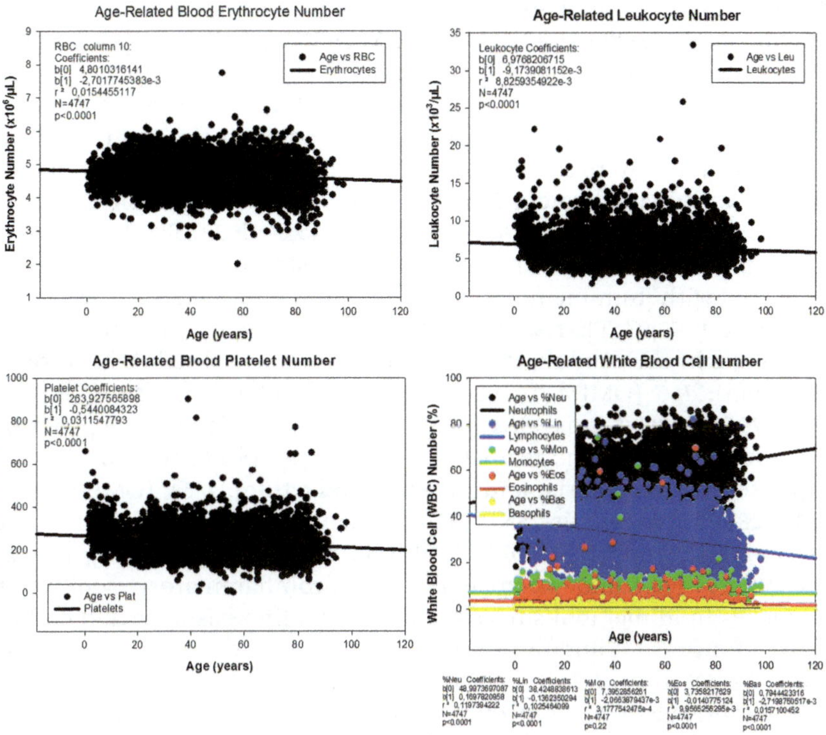

Figure 5.3 Age-related changes in erythrocyte, platelet, leukocyte and white blood cell number.

5.2.2 Common Genes with Age-Related Influence on Health Conditions in NDDs

Among hundreds of genes potentially involved in AD pathogenesis and concomitant disorders (cardiovascular and cerebrovascular disorders, hypercholesterolemia), at least 4 categories of genes deserve special attention: (i) genes associated with lipid metabolism: *APOB* (OMIM 107730; rs693 [7545C>T]; risk SNP 7545T) (participates in the atherogenic process in cooperation with *VLDL*, *IDL* and *LDL*); *APOC3* (OMIM 107720; rs5128 [3175G>C, S1/S2]; risk SNP 3175G (S2)) (associated with triglyceride levels; inhibits the activity of lipoprotein lipase and hepatic lipase); *APOE* (OMIM 107741; rs429358/rs7412 [112T>C/158T>C, *E2*, *E3*, *E4*]; risk SNP 112C/158C (*E4*)) (encodes apolipoprotein E, involved in the catabolism of triglyceride-rich lipoproteins and cholesterol homeostasis); *CETP* (OMIM 118470; rs708272 [+279G>A, B1/B2]; risk SNP +279G (B1)) (contributes to eliminate cholesterol from tissues *via* reverse cholesterol transport); and *LPL* (OMIM 609708; rs328 [1421C>G, S474X]; protective SNP 1421G) (hydrolyzes triglycerides which are part of VLDL and chylomicrons and removes lipoproteins from circulation);[19–23] (ii) genes associated with endothelial function and hypertension: *NOS3* (OMIM 163729; rs1799983 [894G>T]; risk SNP 894T) (encodes nitric oxide synthase 3, which synthesizes nitric oxide (NO) from the amino acid arginine); *ACE* (OMIM 106189; rs4332 [547C>T]; risk SNP 547T) (hydrolyzes angiotensin I to angiotensin II, a potent vasopressor and aldosterone-stimulating peptide, and inactivates bradykinin, a potent vasodilator); and *AGT* (OMIM 1906150; rs699 [9543A>G, T174M]; risk SNP 174M; rs4762 [9360G>A, M235T]; risk SNP 235T) (encodes angiotensinogen, which is converted into angiotensin I by renin);[23–27] (iii) genes associated with immune function and inflammation: *IL1B* (OMIM 147720; rs1143634 [3954C>T]; risk SNP 3954T) (encodes interleukin-1β, which is involved in the modulation of the inflammatory reaction in thrombus formation); *IL6* (OMIM 147620; rs1800795 [-174G>C]; risk SNP -174C; rs1800796 [-573G>C]; risk SNP -573C) (encodes interleukin-6, a pleiotropic cytokine involved in the regulation of the acute phase reaction, immune response, hematopoiesis, and platelet production); *IL6R* (OMIM 147880; rs8192284 [1510A>G]; risk SNP 1510C) (encodes a subunit of the IL6 receptor complex); and *TNFA* (OMIM 191160; rs1800629 [-308G>A]; risk SNP -308A) (encodes tumor necrosis factor, a proinflammatory cytokine that influences lipid metabolism, coagulation, insulin resistance and endothelial function);[23,28–38] and (iv) genes associated with thrombosis and coagulation: *F2* (OMIM 17693; rs1799983 [20210G>A]; risk SNP 2021A) (encodes Coagulation Factor 2 (Prothrombin), involved in blood clotting); *F5* (OMIM 227400; rs6025 [1691G>A]; risk SNP 1691A) (encodes Factor V Leiden, an important factor involved in blood coagulation); and *MTHFR* (OMIM 607093; rs1801133 [677C>T]; risk factor 677T; rs1801131 [1298A>C]; risk SNP 1298A) (encodes methylenetetrahydrofolate reductase, an enzyme that catalyzes the conversion of 5,10-methylenetetrahydrofolate to 5-methyltetrahydrofolate, a co-substrate for the re-methylation of homocysteine to methionine) (Figure 5.4).[23,39–43]

Although differences in genotype distribution and frequencies of all these genes between patients with Alzheimer's disease ($N = 1803$) and control subjects ($N = 1096$) are negligible, except in the case of *APOE* (Figure 5.4),[44] some of them may influence pathogenesis and the pharmacogenetic outcome in the treatment of major risk factors for dementia, such as hypercholesterolemia (Figure 5.5), cardiovascular disorders and hypertension (Figure 5.6).[44–48] Furthermore, many of these genes interact in pathogenic cascades contributing to alter brain cholesterol and Aβ metabolism, subsequently accelerating neuronal death in AD.[49–52]

In a selected group of 933 AD patients we constructed a pentagenic haplotype integrating all possible variants of the *APOE* + *APOB* + *EPOC3* + *CETP* + *LPL* genes and identified 111 haplotypes (H) (Figure 5.7) with differential basal CHO levels. About 75% of these haplotypes in the AD population had a frequency below 1%, 10% had a frequency of between 1% and 2%, 8% had a frequency of between 2% and 5%, and only 4% of the haplotypes

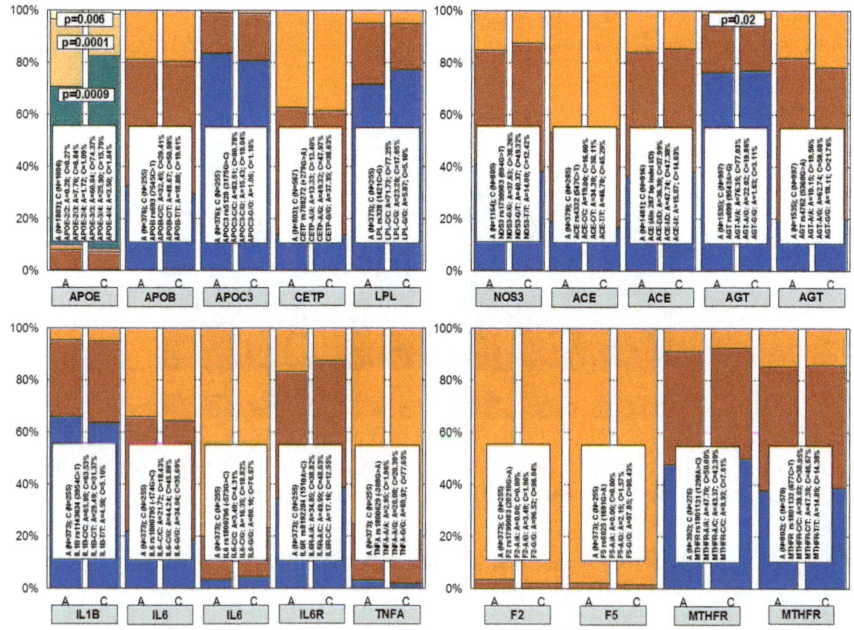

Figure 5.4 Distribution and frequency of APOE, APOB, APOC3, CETP, LPL, NOS3, ACE, AGT, IL1B, IL6, IL6R, TNFA, F2, F5, and MTHFR genes in the Spanish population.

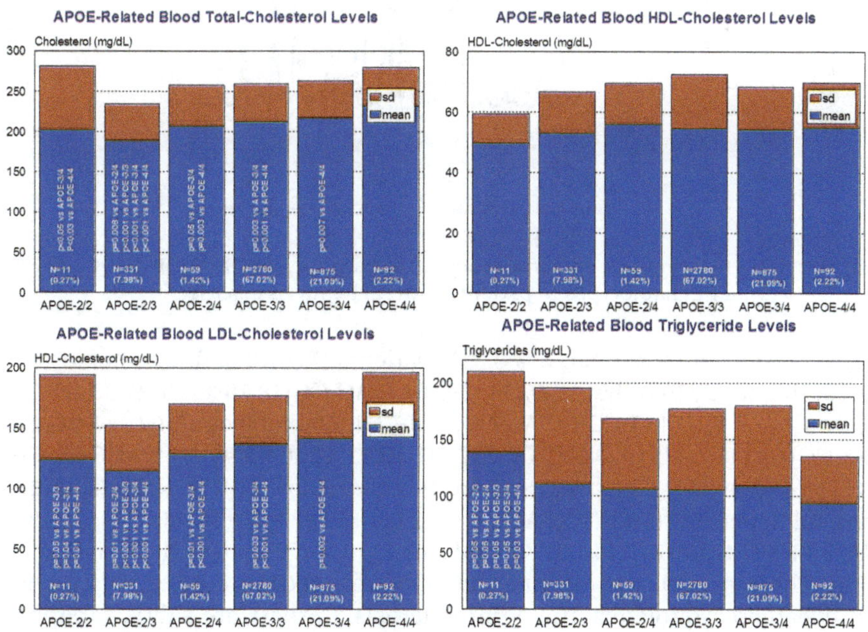

Figure 5.5 APOE-related blood lipid levels in the Spanish population.

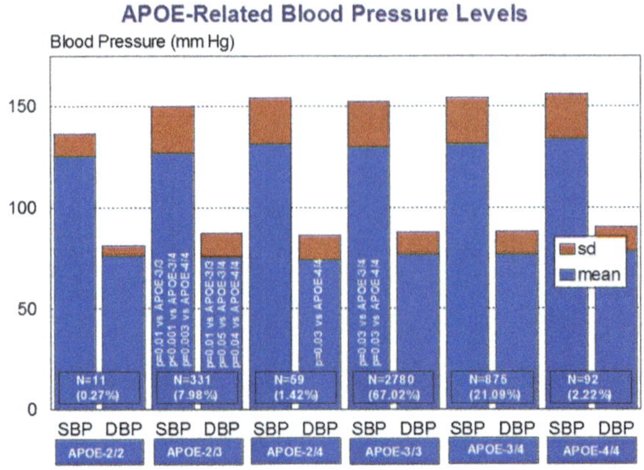

Figure 5.6 APOE-related blood pressure levels in the Spanish population. SBP: Systolic blood pressure; DBP: Diastolic blood pressure.

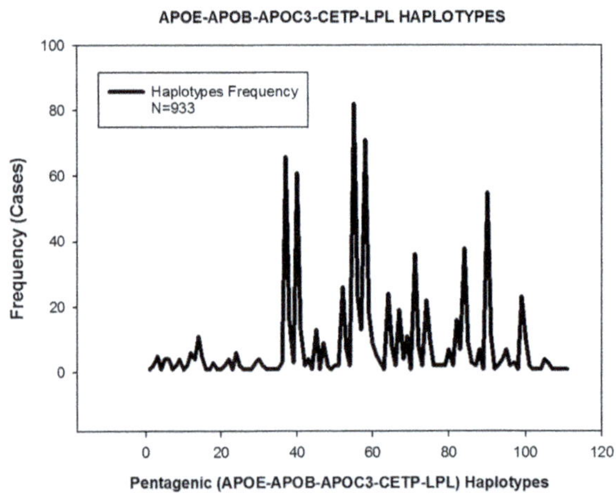

Figure 5.7 Distribution and frequency of pentagenic (APOE-APOB-APOC3-CETP-LPL) haplotypes in patients with Alzheimer's disease.

were present in more than 5% of AD patients (Figure 5.6). The haplotypes most frequently found were *H55* (33-CT-CC-AG-CC) (8.79%), *H58* (33-CT-CC-GG-CC) and *H37* (33-CC-CC-AG-CC) (7.07%). Haplotypes *H104* (44-CC-CC-AA-CC) (0.11%), *H110* (44-TT-CC-AG-CG) (0.11%) and *H98* (34-TT-CC-AA-CG) (0.11%) showed the highest CHO levels, and the lowest levels corresponded to haplotypes *H26* (23-TT-CG-AG-CC) (0.11%), *H8* (23-CC-CG-AG-CC) (0.21%), *H50* (33-CC-GG-AG-CC) (0.21%), and *H63* (33-CT-CG-AA-GG) (0.11%) (Figure 5.8). These data clearly indicate that although *APOE* is a determinant gene in hypercholesterolemia and atherosclerosis,[4,53] there are subtle variations in these haplotypes that clearly modify the cholesterolemic phenotype and associated health risks.

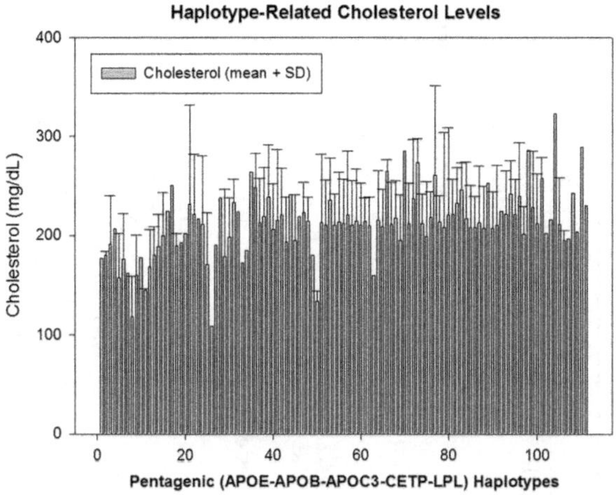

Figure 5.8 Pentagenic (APOE-APOB-APOC3-CETP-LPL) haplotype-related blood cholesterol levels in patients with Alzheimer's disease.

5.3 Pharmacogenomics

Pharmacogenomics accounts for 60–90% of the variability in pharmacokinetics and pharmacodynamics. The modest effect (and toxicity) of current AD and PD drugs (Tables 5.1 and 5.2) is in part due to their pharmacogenomic profile since over 70% of patients are deficient metabolizers.[6,11,18] The genes involved in the pharmacogenomic response to drugs in dementia fall into five major categories:

(i) Genes associated with disease pathogenesis: Mendelian mutations affect genes directly linked to AD, including >30 mutations in the amyloid beta precursor protein (*APP*) gene (21q21) (*AD1*); >160 mutations in the presenilin 1 (*PSEN1*) gene (14q24.3) (*AD3*); and >10 mutations in the presenilin 2 (*PSEN2*) gene (1q31-q42) (*AD4*).[53–57] *PSEN1* and *PSEN2* are important determinants of γ-secretase activity responsible for proteolytic cleavage of APP and NOTCH receptor proteins. Mendelian mutations are very rare in AD (1 : 1000). Mutations in exons 16 and 17 of the *APP* gene appear with a frequency of 0.30% and 0.78%, respectively, in AD patients. Likewise, *PSEN1*, *PSEN2*, and microtubule-associated protein Tau (*MAPT*) (17q21.1) mutations are present in less than 2% of the cases. Mutations in these genes confer specific phenotypic profiles to patients with dementia: amyloidogenic pathology associated with *APP*, *PSEN1* and *PSEN2* mutations and tauopathy associated with *MAPT* mutations representing the two major pathogenic hypotheses for AD.[53–59]

Multiple polymorphic risk variants can increase neuronal vulnerability to premature death. There are at least 695 genes potentially associated with AD,

Table 5.1 Pharmacogenomics of conventional anti-dementia drugs.[a]

Drug	Properties	Pharmacogenetics
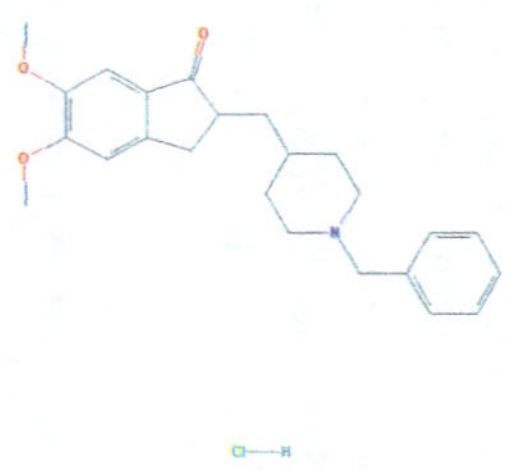	**Name**: **Donepezil hydrochloride**, Aricept, 120011-70-3, Donepezil HCl, BNAG, E-2020, E2020 **IUPAC Name**: 2-[(1-benzylpiperidin-4-yl)methyl]-5,6-dimethoxy-2,3-dihydroinden-1-one;hydrochloride **Molecular Formula**: $C_{24}H_{30}ClNO_3$ **Molecular Weight**: 415.9529 g/mol **Category**: Cholinesterase inhibitor **Mechanism**: Centrally active, reversible acetylcholinesterase inhibitor; increases the acetylcholine available for synaptic transmission in the CNS **Effect**: Nootropic agent, cholinesterase inhibitor, parasympathomimetic effect	**Pathogenic genes**: *APOE*, *CHAT* **Mechanistic genes**: *CHAT*, *ACHE*, *BCHE* **Drug metabolism-related genes**: - **Substrate**: *CYP2D6* (major), *CYP3A4* (major), *UGTs ACHE* - **Inhibitor**: *ACHE*, *BCHE* **Transporter genes**: *ABCB1*
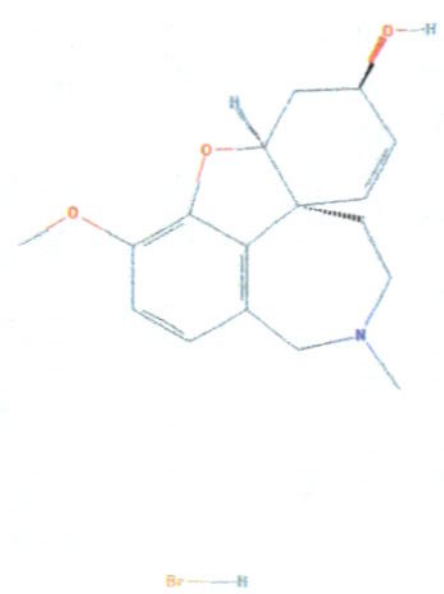	**Name**: **Galantamine hydrobromide**, Galanthamine hydrobromide, 1953-04-4, Nivalin, Razadyne, UNII-MJ4PTD2VVW, Nivaline **IUPAC Name**: (1S,12S,14R)-9-methoxy-4-methyl-11-oxa-4-azatetracyclo[8.6.1.0^{1,12}.0^{6,17}]heptadeca-6,8,10(17),15-tetraen-14-ol **Molecular Formula**: $C_{17}H_{22}BrNO_3$ **Molecular Weight**: 368.26548 g/mol **Category**: Cholinesterase inhibitor **Mechanism**: Reversible and competitive acetylcholinesterase inhibition leading to an increased concentration of acetylcholine at cholinergic synapses; modulates nicotinic acetylcholine receptor; may increase glutamate and serotonin levels **Effect**: Nootropic agent, cholinesterase inhibitor, parasympathomimetic effect	**Pathogenic genes**: *APOE, APP* **Mechanistic genes**: *ACHE, BCHE, CHRNA4, CHRNA7, CHRNB2* **Drug metabolism-related genes**: - **Substrate**: *CYP2D6* (major), *CYP3A4* (major), *UGT1A1* - **Inhibitor**: *ACHE, BCHE*

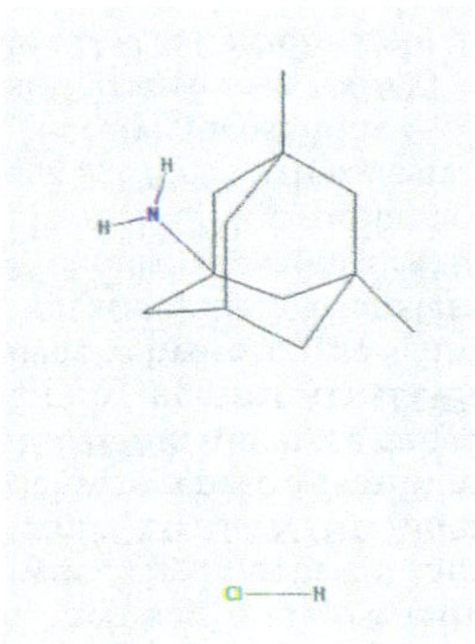	**Name: Memantine Hydrochloride**, 41100-52-1, Namenda, Memantine HCL, Axura, 3,5-Dimethyl-1-adamantanamine hydrochloride, 3,5-dimethyladamantan-1-amine hydrochloride **IUPAC Name:** 3,5-dimethyladamantan-1-amine;hydrochloride **Molecular Formula:** $C_{12}H_{22}ClN$ **Molecular Weight:** 215.76278 g/mol **Category:** N-Methyl-D-Aspartate receptor antagonist **Mechanism:** Binds preferentially to NMDA receptor-operated cation channels; may act by blocking actions of glutamate, mediated in part by NMDA receptors **Effect:** Dopamine agent, antiparkinson agent, excitatory amino acid antagonist, antidyskinetic	**Pathogenic genes:** *APOE, MAPT, PSEN1* **Mechanistic genes:** *CHRFAM7A, DLGAP1, FOS, GRIN2A, GRIN2B, GRIN3A, HOMER1, HTR3A* **Drug metabolism-related genes:** **-Inhibitor**: *CYP1A2* (weak), *CYP2A6* (weak), *CYP2B6* (strong), *CYP2C9* (weak), *CYP2C19* (weak), *CYP2D6* (strong), *CYP2E1* (weak), *CYP3A4* (weak), *NR1I2* **Transporter genes:** *NR1I2* **Pleiotropic genes:** *APOE, MAPT, MT-TK, PSEN1*
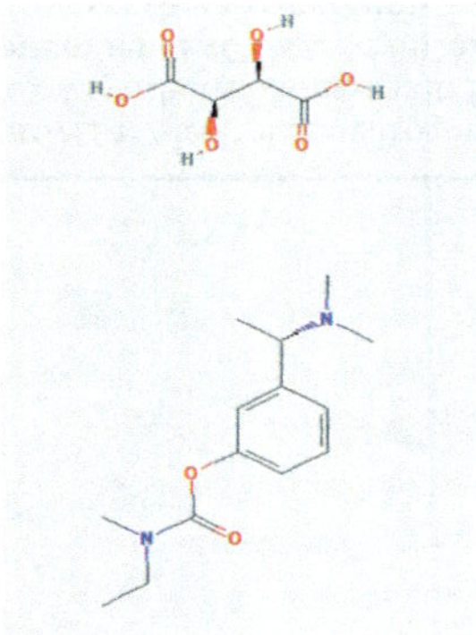	**Name: Rivastigmine tartrate**, 129101-54-8, SDZ-ENA 713, Rivastigmine hydrogentartrate, Rivastigmine Hydrogen Tartrate, ENA 713, ENA-713 **IUPAC Name:** (2R,3R)-2,3-dihydroxybutanedioic acid;[3-[(1S)-1-(dimethylamino)ethyl]phenyl] N-ethyl-N-methylcarbamate **Molecular Formula:** $C_{18}H_{28}N_2O_8$ **Molecular Weight:** 400.42352 g/mol **Category:** Cholinesterase inhibitor **Mechanism:** Increases acetylcholine in CNS through reversible inhibition of its hydrolysis by cholinesterase **Effect:** Neuroprotective agent, cholinesterase inhibitor, cholinergic agent	**Pathogenic genes:** *APOE, APP, CHAT* **Mechanistic genes:** *ACHE, BCHE, CHAT, CHRNA4, CHRNB2* **Drug metabolism-related genes:** **-Inhibitor**: *ACHE, BCHE* **Pleiotropic genes:** *APOE, MAPT*

(continued)

Table 5.1 (*continued*)

Drug	Properties	Pharmacogenetics
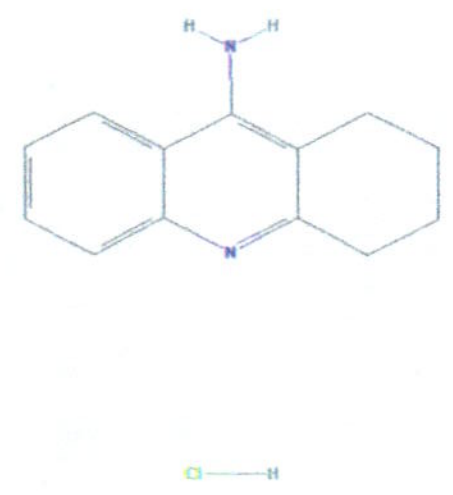	**Name**: **Tacrine Hydrochloride**, Tacrine HCl, 1684-40-8, Hydroaminacrine, tacrine.HCl, 9-AMINO-1,2,3,4-TETRAHYDROACRIDINE HYDROCHLORIDE, Tenakrin **IUPAC Name**: 1,2,3,4-tetrahydroacridin-9-amine;hydrochloride **Molecular Formula**: $C_{13}H_{15}ClN_2$ **Molecular Weight**: 234.7246 g/mol **Category**: Cholinesterase inhibitor **Mechanism**: Elevates acetylcholine in cerebral cortex by slowing degradation of acetylcholine **Effect**: Nootropic agent, cholinesterase inhibitor, Parasympathomimetic effect	**Pathogenic genes**: *APOE* **Mechanistic genes**: *ACHE, BCHE, CHRNA4, CHRNB2* **Drug metabolism-related genes**: **-Substrate**: *CYP1A2 (major), CYP2D6 (minor), CYP3A4 (major)* **-Inhibitor**: *ACHE, BCHE, CYP1A2 (weak)* **Transporter genes**: *SCN1A* **Pleiotropic genes**: *APOE, CES1, GSTM1, GSTT1, LEPR, MTHFR*

[a]***ADH1A***: Alcohol dehydrogenase 1A (class I), alpha polypeptide; ***AADAC***: Arylacetamide deacetylase; ***AANAT***: aralkylamine N-acetyltransferase; ***ACSL1***: Acyl-CoA synthetase long-chain family member 1; ***ACSL3***: Acyl-CoA synthetase long-chain family member 3; ***ACSL4***: Acyl-CoA synthetase long-chain family member 4; ***ACSM1***: Acyl-CoA synthetase medium-chain family member 1; ***ACSM2B***: Acyl-CoA synthetase medium-chain family member 2B; ***ACSM3***: Acyl-CoA synthetase medium-chain family, member 3; ***ADH1B***: Alcohol dehydrogenase 1B (class I), beta polypeptide; ***ADH1C***: Alcohol dehydrogenase 1C (class I), gamma polypeptide; ***ADH4***: Alcohol dehydrogenase 4 (class II), pi polypeptide; ***ADH5***: Alcohol dehydrogenase 5 (class III), chi polypeptide; ***ADH6***: Alcohol dehydrogenase 6 (class V); ***ADH7***: Alcohol dehydrogenase 7 (class IV), mu or sigma polypeptide; ***ADHFE1***: Alcohol dehydrogenase, iron containing, 1; ***AGXT***: Alanine-glyoxylate aminotransferase; ***AKR1A1***: Aldo-keto reductase family 1, member A1 (aldehyde reductase); ***AKR1B1***: Aldo-keto reductase family 1, member B1 (aldose reductase); ***AKR1C1***: Aldo-keto reductase family 1, member C1; ***AKR1D1***: Aldo-keto reductase family 1, member D1; ***ALDH1A1***: Aldehyde dehydrogenase 1 family, member A1; ***ALDH1A2***: Aldehyde dehydrogenase family 1, subfamily A2; ***ALDH1A3***: Aldehyde dehydrogenase family 1, subfamily A3; ***ALDH1B1***: Aldehyde dehydrogenase 1 family, member B1; ***ALDH2***: Aldehyde dehydrogenase 2 family (mitochondrial); ***ALDH3A1***: Aldehyde dehydrogenase 3 family, member A1; ***ALDH3A2***: Aldehyde dehydrogenase 3 family, member A2; ***ALDH3B1***: Aldehyde dehydrogenase 3 family, member B1; ***ALDH3B2***: Aldehyde dehydrogenase 3 family, member B2; ***ALDH4A1***: Aldehyde dehydrogenase 4 family, member A1; ***ALDH5A1***: Aldehyde dehydrogenase 5 family, member A1; ***ALDH6A1***: Aldehyde dehydrogenase 6 family, member A1; ***ALDH7A1***: Aldehyde dehydrogenase 7 family, member A1; ***ALDH8A1***: Aldehyde dehydrogenase 8 family, member A1; ***ALDH9A1***: Aldehyde dehydrogenase 9 family, member A1; ***AOX1***: Aldehyde oxidase 1; ***AS3MT***: Arsenic (+3 oxidation state) methyltransferase; ***ASMT***: Acetylserotonin O-methyltransferase; ***BAAT***: Bile acid CoA: amino acid N-acyltransferase (glycine N-choloyltransferase); ***CBR1***: Carbonyl reductase 1; ***CBR3***: Carbonyl reductase 3; ***CBR4***: Carbonyl reductase 4; ***CCBL1***: Cysteine conjugate-beta lyase, cytoplasmic; ***CDA***: Cytidine deaminase; ***CEL***: Carboxyl ester lipase; ***CES1***: Carboxylesterase 1; ***CES1P1***: Carboxylesterase 1 pseudogene 1; ***CES2***: Carboxylesterase 2; ***CES3***: Carboxylesterase 3; ***CES5A***: Carboxylesterase 5A; ***CHST1***: Carbohydrate (keratan sulfate Gal-6) sulfotransferase 1; ***CHST2***: Carbohydrate (N-acetylglucosamine-6-O) sulfotransferase 2; **CHST3**: Carbohydrate (chondroitin 6) sulfotransferase 3; ***CHST4***: Carbohydrate (N-acetylglucosamine 6-O) sulfotransferase 4; ***CHST5***: Carbohydrate (N-acetylglucosamine 6-O) sulfotransferase 5; ***CHST6***: Carbohydrate (N-acetylglucosamine 6-O) sulfotransferase 6; ***CHST7***: Carbohydrate (N-acetylglucosamine 6-O) sulfotransferase 7; ***CHST8***: Carbohydrate (N-acetylgalactosamine 4-0) sulfotransferase 8; ***CHST9***: Carbohydrate (N-acetylgalactosamine 4-0) sulfotransferase 9; ***CHST10***: Carbohydrate sulfotransferase 10; ***CHST11***: Carbohydrate (chondroitin 4) sulfotransferase 11; ***CHST12***: Carbohydrate (chondroitin 4) sulfotransferase 12; ***CHST13***: Carbohydrate (chondroitin 4) sulfotransferase 13; ***COMT***:

Catechol-O-methyltransferase; ***CYB5R3***: Cytochrome b5 reductase 3; ***CYP1A1***: Cytochrome P450, family 1, subfamily A, polypeptide 1; ***CYP1A2***: Cytochrome P450, family 1, subfamily A, polypeptide 2; ***CYP1B1***: Cytochrome P450, family 1, subfamily B, polypeptide 1; ***CYP2A6***: Cytochrome P450, family 2, subfamily A, polypeptide 6; ***CYP2A7***: Cytochrome P450, family 2, subfamily A, polypeptide 7; ***CYP2A13***: Cytochrome P450, family 2, subfamily A, polypeptide 13; ***CYP2B6***: Cytochrome P450, family 2, subfamily B, polypeptide 6; ***CYP2C8***: Cytochrome P450, family 2, subfamily C, polypeptide 8; ***CYP2C9***: Cytochrome P450, family 2, subfamily C, polypeptide 9; ***CYP2C18***: Cytochrome P450, family 2, subfamily C, polypeptide 18; ***CYP2C19***: Cytochrome P450, family 2, subfamily C, polypeptide 19; ***CYP2D6***: Cytochrome P450, family 2, subfamily D, polypeptide 6; ***CYP2D7P1***: Cytochrome P450, family 2, subfamily D, polypeptide 7 pseudogene 1; ***CYP2E1***: Cytochrome P450, family 2, subfamily E, polypeptide 1; ***CYP2F1***: Cytochrome P450, family 2, subfamily F, polypeptide 1; ***CYP2J2***: Cytochrome P450, family 2, subfamily J, polypeptide 2; ***CYP2R1***: Cytochrome P450, family 2, subfamily R, polypeptide 1; ***CYP2S1***: Cytochrome P450, family 2, subfamily S, polypeptide 1; ***CYP2W1***: Cytochrome P450, family 2, subfamily W, polypeptide 1; ***CYP3A4***: Cytochrome P450, family 3, subfamily A, polypeptide 4; ***CYP3A5***: Cytochrome P450, family 3, subfamily A, polypeptide 5; ***CYP3A7***: Cytochrome P450, family 3, subfamily A, polypeptide 7; ***CYP3A43***: Cytochrome P450, family 3, subfamily A, polypeptide 43; ***CYP4A11***: Cytochrome P450, family 4, subfamily A, polypeptide 11; ***CYP4A22***: Cytochrome P450, family 4, subfamily A, polypeptide 22; ***CYP4B1***: Cytochrome P450, family 4, subfamily B, polypeptide 1; ***CYP4F2***: Cytochrome P450, family 4, subfamily F, polypeptide 2; ***CYP4F3***: Cytochrome P450, family 4, subfamily F, polypeptide 3; ***CYP4F8***: Cytochrome P450, family 4, subfamily F, polypeptide 8; ***CYP4F11***: Cytochrome P450, family 4, subfamily F, polypeptide 11; ***CYP4F12***: Cytochrome P450, family 4, subfamily F, polypeptide 12; ***CYP4Z1***: Cytochrome P450, family 4, subfamily Z, polypeptide 1; ***CYP7A1***: Cytochrome P450, family 7, subfamily A, polypeptide 1; ***CYP7B1***: Cytochrome P450, family 7, subfamily B, polypeptide 1; ***CYP8B1***: Cytochrome P450, family 8, subfamily B, polypeptide 1; ***CYP11A1***: Cytochrome P450, family 11, subfamily A, polypeptide 1; ***CYP11B1***: Cytochrome P450, family 11, subfamily B, polypeptide 1: ***CYP11B2***: Cytochrome P450, family 11, subfamily B, polypeptide 2; ***CYP17A1***: Cytochrome P450, family 17, subfamily A, polypeptide 1; ***CYP19A1***: Cytochrome P450, family 19, subfamily A, polypeptide 1; ***CYP20A1***: Cytochrome P450, family 20, subfamily A, polypeptide 1; ***CYP21A2***: Cytochrome P450, family 21, subfamily A, polypeptide 2; ***CYP24A1***: Cytochrome P450, family 24, subfamily A, polypeptide 1; ***CYP26A1***: Cytochrome P450, family 26, subfamily A, polypeptide *1;* ***CYP26B1***: Cytochrome P450, family 26, subfamily B, polypeptide 1; ***CYP26C1***: Cytochrome P450, family 26, subfamily C, polypeptide 1; ***CYP27A1***: Cytochrome P450, family 27, subfamily A, polypeptide 1; ***CYP27B1***: Cytochrome P450, family 27, subfamily B, polypeptide 1; ***CYP39A1***: Cytochrome P450, family 39, subfamily A, polypeptide 1; ***CYP46A1***: Cytochrome P450, family 46, subfamily A, polypeptide 1; ***CYP51A1***: Cytochrome P450, family 51, subfamily A, polypeptide 1; ***DDOST***: Dolichyl-diphosphooligosaccharide--protein glycosyltransferase subunit (non-catalytic); ***DHRS1***: Dehydrogenase/reductase (SDR family) member 1; ***DHRS2***: Dehydrogenase/reductase (SDR family) member 2; ***DHRS3***: Dehydrogenase/reductase (SDR family) member 3; ***DHRS4***: Dehydrogenase/reductase (SDR family) member 4; ***DHRS7***: Dehydrogenase/reductase (SDR family) member 7; ***DHRS9***: Dehydrogenase/reductase (SDR family) member 9; ***DHRS12***: Dehydrogenase/reductase (SDR family) member 12; ***DHRS13***: Dehydrogenase/reductase (SDR family) member 13; ***DHRSX***: Dehydrogenase/reductase (SDR family) X-linked; ***DLGAP1***: discs, large (Drosophila) homolog-associated protein 1; ***DPEP1***: Dipeptidase 1 (renal); ***DPYD***: Dihydropyrimidine dehydrogenase; ***EPHX1***: Epoxide hydrolase 1, microsomal (xenobiotic); ***EPHX2***: Epoxide hydrolase 2, microsomal (xenobiotic); ***ESD***: Esterase D; ***FMO1***: Flavin containing monooxygenase 1; ***FMO2***: Flavin containing monooxygenase 2; ***FMO3***: Flavin containing monooxygenase 3; ***FMO4***: Flavin containing monooxygenase 4; ***FMO5***: Flavin containing monooxygenase 5; ***FMO6P***: Flavin containing monooxygenase 6 pseudogene; ***FOS***: FBJ murine osteosarcoma viral oncogene homolog; ***GAL3ST1***: Galactose-3-O-sulfotransferase 1; ***GAMT***: Guanidinoacetate N-methyltransferase; ***GLRX***: Glutaredoxin (thioltransferase); ***GLYAT***: Glycine-N-acyltransferase; ***GNMT***: Glycine N-methyltransferase; ***GPX1***: Glutathione peroxidase 1; ***GPX2***: Glutathione peroxidase 2 (gastrointestinal); ***GPX3***: Glutathione peroxidase 3 (plasma); ***GPX4***: Glutathione peroxidase 4; ***GPX5***: Glutathione peroxidase 5; ***GPX6***: Glutathione peroxidase 6 (olfactory); ***GPX7***: Glutathione peroxidase 7; ***GSR***: Glutathione reductase; ***GSTA1***: Glutathione S-transferase alpha 1; ***GSTA2***: Glutathione S-transferase alpha 2; ***GSTA3***: Glutathione S-transferase alpha 3; ***GSTA4***: Glutathione S-transferase alpha 4; ***GSTA5***: Glutathione S-transferase alpha 5; ***GSTCD***: Glutathione S-transferase, C-terminal domain containing; ***GSTK1***: Glutathione S-transferase kappa 1; ***GSTM1***: Glutathione S-transferase mu 1; ***GSTM2***: Glutathione S-transferase mu 2 (muscle); ***GSTM3***: Glutathione

(*continued*)

Table 5.1 (*continued*)

S-transferase mu 3 (brain); ***GSTM4***: Glutathione S-transferase mu 4; ***GSTM5***: Glutathione S-transferase mu 5; ***GSTO1***: Glutathione S-transferase omega 1; ***GSTO2***: Glutathione S-transferase omega 2; ***GSTP1***: Glutathione S-transferase pi 1; ***GSTT1***: Glutathione S-transferase theta 1; ***GSTT2***: Glutathione S-transferase theta 2; ***GSTZ1***: Glutathione S-transferase zeta 1; ***GZMA***: Granzyme A (granzyme 1, cytotoxic T-lymphocyte-associated serine esterase 3; ***GZMB***: Granzyme B (granzyme 2, cytotoxic T-lymphocyte-associated serine esterase 1); ***HNMT***: Histamine N-methyltransferase; ***HOMER1***: homer homolog 1 (Drosophila); ***HSD11B1***: Hydroxysteroid (11-beta) dehydrogenase 1; ***HSD17B10***: Hydroxysteroid (17-beta) dehydrogenase 10; ***HSD17B11***: Hydroxysteroid (17-beta) dehydrogenase 11; ***HSD17B14***: Hydroxysteroid (17-beta) dehydrogenase 14; ***INMT***: Indolethylamine N-methyltransferase; ***MAOA***: Monoamine oxidase A; ***MAOB***: monoamine oxidase B; ***METAP1***: Methionyl aminopeptidase 1; ***MGST1***: Microsomal glutathione S-transferase 1; ***MGST2***: Microsomal glutathione S-transferase 1; ***MGST3***: Microsomal glutathione S-transferase 3; ***NAA20***: N(alpha)-acetyltransferase 20, NatB catalytic subunit; ***NAT1***: N-acetyltransferase 1 (arylamine N-acetyltransferase); ***NAT2***: N-acetyltransferase 2 (arylamine N-acetyltransferase); ***NNMT***: Nicotinamide N-methyltransferase; ***NQO1***: NAD(P)H dehydrogenase, quinone 1; ***NQO2***: NAD(P)H dehydrogenase, quinone 2; ***NR1I2***: nuclear receptor subfamily 1, group I, member 2; ***PNMT***: Phenylethanolamine N-methyltransferase; ***PON1***: Paraoxonase 1; ***PON2***: Paraoxonase 2; ***PON3***: Paraoxonase 3; ***POR***: P450 (cytochrome) oxidoreductase; ***PTGES***: Prostaglandin E synthase; ***PTGS1***: Prostaglandin-endoperoxide synthase 1 (prostaglandin G/H synthase and cyclooxygenase); ***PTGS2***: Prostaglandin-endoperoxide synthase 2 (prostaglandin G/H synthase and cyclooxygenase); **SAT1**: Spermidine/spermine N1-acetyltransferase 1; ***SMOX***: Spermine oxidase; ***SOD1***: Superoxide dismutase 1, soluble; ***SOD2***: Superoxide dismutase 2, mitochondrial; ***SULT1A1***: Sulfotransferase family, cytosolic, 1A, phenol-preferring, member 1; ***SULT1A2***: Sulfotransferase family, cytosolic, 1A, phenol-preferring, member 2; ***SULT1A3***: Sulfotransferase family, cytosolic, 1A, phenol-preferring, member 3; ***SULT1B1***: Sulfotransferase family, cytosolic, 1B, member 1; ***SULT1C1***: Sulfotransferase family, cytosolic, 1C, member 1; ***SULT1C2***: Sulfotransferase family, cytosolic, 1C, member 2; ***SULT1C3***: Sulfotransferase family, cytosolic, 1C, member 3; ***SULT1C4***: Sulfotransferase family, cytosolic, 1C, member 4; ***SULT1E1***: Sulfotransferase family 1E, estrogen-preferring, member 1; ***SULT2A1***: Sulfotransferase family, cytosolic, 2A, dehydroepiandrosterone (DHEA)-preferring, member 1; ***SULT2B1***: Sulfotransferase family, cytosolic, 2B, member 1; ***SULT4A1***: Sulfotransferase family 4A, member 1; ***SULT6B1***: sulfotransferase family, cytosolic, 6B, member 1; ***TBXAS1***: Thromboxane A synthase 1 (platelet); ***TPMT***: Thiopurine S-methyltransferase; ***TST***: Thiopurine S-methyltransferase; ***UCHL1***: Ubiquitin carboxyl-terminal esterase L1 (ubiquitin thiolesterase); ***UCHL3***: Ubiquitin carboxyl-terminal esterase L3 (ubiquitin thiolesterase); ***UGT1A1***: UDP glucuronosyltransferase 1 family, polypeptide A1; ***UGT1A3***: UDP glucuronosyltransferase 1 family, polypeptide A3; ***UGT1A4***: UDP glucuronosyltransferase 1 family, polypeptide A4; ***UGT1A5***: UDP glucuronosyltransferase 1 family, polypeptide A5; ***UGT1A6***: UDP glucuronosyltransferase 1 family, polypeptide A6; ***UGT1A7***: UDP glucuronosyltransferase 1 family, polypeptide A7; ***UGT1A8***: UDP glucuronosyltransferase 1 family, polypeptide A8; ***UGT1A9***: UDP glucuronosyltransferase 1 family, polypeptide A9; ***UGT1A10***: UDP glucuronosyltransferase 1 family, polypeptide A10; ***UGT2A1***: UDP glucuronosyltransferase 2 family, polypeptide A1, complex locus; ***UGT2A3***: UDP glucuronosyltransferase 2 family, polypeptide A3; ***UGT2B10***: UDP glucuronosyltransferase 2 family, polypeptide B10; ***UGT2B11***: UDP glucuronosyltransferase 2 family, polypeptide B11; ***UGT2B15***: UDP glucuronosyltransferase 2 family, polypeptide B15; ***UGT2B17***: UDP glucuronosyltransferase 2 family, polypeptide B17; ***UGT2B28***: UDP glucuronosyltransferase 2 family, polypeptide B28; ***UGT2B4***: UDP glucuronosyltransferase 2 family, polypeptide B4; ***UGT2B7***: UDP glucuronosyltransferase 2 family, polypeptide B7; ***UGT3A1***: UDP glycosyltransferase 3 family, polypeptide A1; ***UGT8***: UDP glycosyltransferase 8; ***XDH***: Xanthine dehydrogenase.

Table 5.2 Pharmacological profile and Pharmacogenetics of selected anti-Parkinsonian drugs.[a]

Drug	Properties	Pharmacogenetics
Dopamine Precursors		
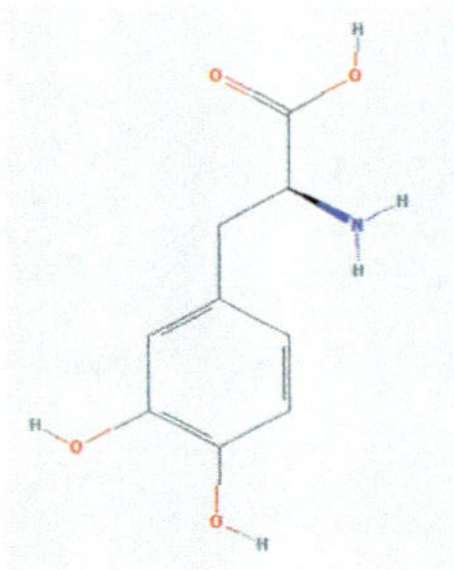	**Name: Levodopa**; 59-92-7; Levodopa; L-dopa; Dopar; Bendopa; Dopasol; 3,4-dihydroxy-L-phenylalanine; Madopar **IUPAC Name:** L-Tyrosine-3-hydroxy **Molecular Formula:** $C_9H_{11}NO_4$ **Molecular Weight:** 197.19 g/mol **Mechanism:** Levodopa circulates in the plasma to the blood-brain-barrier, where it crosses, to be converted by striatal enzymes to dopamine. Carbidopa inhibits the the peripheral plasma breakdown of levodopa by inhibiting its carboxylation, and thereby increases available levodopa at the blood-brain-barrier **Effect:** Antiparkinsonian Agents. Dopamine Precursors.	**Pathogenic genes:** *ANKK1, BDNF, LRRK2, PARK2* **Mechanistic genes:** *CCK, CCKAR, CCKBR, DRD1, DRD2, DRD3, DRD4, DRD5, GRIN2A, GRIN2B, HCRT, HOMER1, LMO3, OPRM1* **Metabolic genes** **Substrate:** *COMT, CYP1A2, CYP2B6, CYP2C19, CYP2D6, CYP3A4, CYP3A5, DBH, DDC, G6PD, MAOB, TH, UGT1A1, UGT1A9* **Transporter genes:** *SLC22A1, SLC6A3* **Pleiotropic genes:** *ACE, ACHE*
Dopaminergic Agonists		
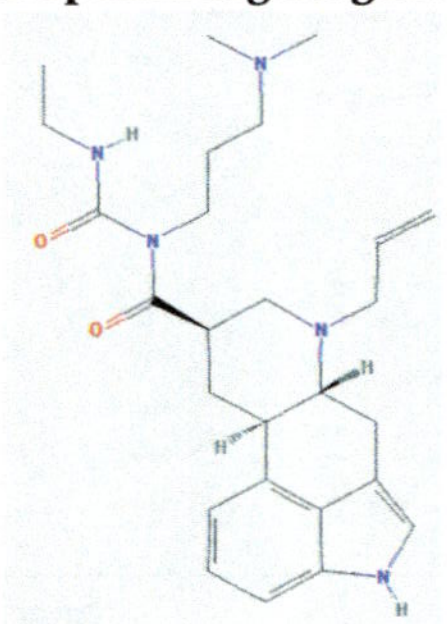	**Name: Cabergoline**; 81409-90-7; Cabergoline; Dostinex, Cabaser; Cabergolinum; Cabaseril; Cabergolina **IUPAC Name:** Ergoline-8β-carboxamide,N-[3-(dimethylamino)propyl]-N-[(ethylamino)carbonil]-6-(2-propenyl) **Molecular Formula:** $C_{26}H_{37}N_5O_2$ **Molecular Weight:** 451.60 g/mol **Mechanism:** A long-acting dopamine receptor agonist. Has high binding affinity for dopamine D_2-receptors and lesser affinity for D_1,α_1-and α_2-adrenergic, and serotonin (5-HT_1 and 5-HT_2) receptors. Reduces serum prolactin concentrations by inhibiting release of prolactin from the anterior pituitary gland (agonist activity at D_2 receptors) **Effect:** Antiparkinsonian Agents; Ergot-derivative Dopamine Receptor Agonists.	**Pathogenic genes:** *BDNF, GSK3B* **Mechanistic genes:** *ADRA2A, ADRA2B, ADRA2C, AKT1, BDNF, CNR1, DRD1, DRD2, DRD3, DRD4, DRD5, GSK3B, HTR1A, HTR1B, HTR1D, HTR2A, HTR2B, HTR2C, HTR7* **Metabolic genes** **Substrate:** *COMT, CYP1A2, CYP2B6, CYP2C19, CYP2D6, CYP3A4 (minor), CYP3A5, DDC*

(*continued*)

Table 5.2 (*continued*)

Drug	Properties	Pharmacogenetics
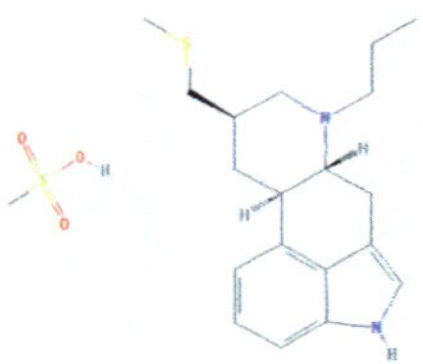	**Name: Pergolide**; 66104-22-1; Pergolide; Permax; Pergolida; Pergolidum **IUPAC Name:** Ergoline,8-[(Methylthio) methyl]-6-monomethenesulfonate **Molecular Formula:** $C_{19}H_{26}N_2SCH_4O_3S$ **Molecular Weight:** 410.59 g/mol **Mechanism:** A dopamin receptor agonist. Relieves symptoms of parkinsonism, presumably by directly stimulating postsynaptic dopamine receptors in corpus striatum. Reduces serum prolactine concentrations by inhibiting release of prolactin from anterior pituitary gland.Causes transient increass in serum somatotropin (growth hormone) concentrations and decreases in serum luteinizing hormone concentrations **Effect:** Antiparkinsonian Agents; Ergot-derivative Dopamine Receptor Agonists	**Mechanistic genes:** *ADRA1A, ADRA1B, ADRA1D, ADRA2A, ADRA2B, ADRA2C, DRD1, DRD2, DRD3, DRD4, DRD5, HTR1A, HTR1B, HTR1D, HTR2A, HTR2B, HTR2C* **Metabolic genes** **Substrate:** *COMT, CYP1A2, CY22B6, CYP2C19, CYP2D6, CYP3A4 (major), CYP3A5, DDC, UGT1A1, UGT1A9* **Transporter genes:** *SLC6A4*
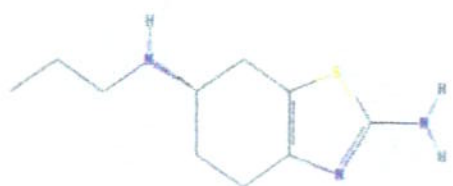	**Name: Pramipexole;** 104632-26-0; Pramipexole; Pramipexol; Parmital; Mirapex; Mirapexin; Sifrol **IUPAC Name:** 2,6-Benzothiazolediamine, 4,5,6,7-tetrahydro-N^6-propyl-, (S) **Molecular Formula:** $C_{10}H_{17}N_3S$ **Molecular Weight:** 211.33 g/mol **Mechanism:** By binding to D_2 subfamily dopamine receptor, and to D_3, and D_4 receptors, it is though that Pramipexole can stimulate dopamine activity on nerves of striatum and substantia nigra **Effect:** Antiparkinsonian Agents; Nonergot-derivative Dopamine Receptor Agonists.	**Pathogenic genes:** *ANKK1, BDNF, LRRK2* **Mechanistic genes:** *ADRA2A, ADRA2B, ADRA2C, CCK, CCKAR, CCKBR, DRD1, DRD2, DRD3, DRD4, DRD5, GRIN2A, GRIN2B, HCRT, HOMER1, HTR1A, HTR1B, HTR1D, HTR2A, HTR2B, HTR2C, LMO3, OPRM1* **Metabolic genes** **Substrate:** *COMT, CYP1A2, CY22B6, CYP2C19, CYP2D6, CYP3A4, CYP3A5, DDC, MAOB, UGT1A1, UGT1A9* **Transporter genes:** *SLC22A1, SLC6A3* **Pleiotropic genes:** *ACE, APOE*

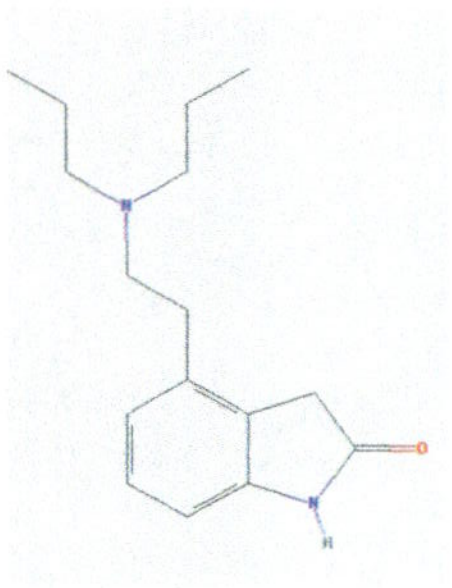	**Name: Ropinirole;** 91374-21-9; Ropinirole; ReQuip; Ropinirol; Ropinilorum; ReQuip CR **IUPAC Name:** 2-H-Indol-2-one 4-[2-(dipropylamino)ethyl]-1,3-dihydro-, monohydrochloride **Molecular Formula:** $C_{16}H_{24}N_2O$ **Molecular Weight:** 296.84 g/mol **Mechanism:** Has high relative *in vitro* specificity and full intrinsic activity at D_2 and D_3 dopamine receptor subtypes, binding with higher affinity to D_3 than to D_2 and D_4 receptor subtypes. Although precise mechanism of action unkown, it is believed to be due to stimulation of postsynaptic dopamine D_2-type receptors within caudate putamen in brain. Mechanism of Ropinirole-induced postural hypotension believed to be due to D_2-mediated blunting of noradrenergic response to standing and subsequent decrease in peripheral vascular resistance **Effect:** Antiparkinsonian Agents; Nonergot-derivative Dopamine Receptor Agonists.	**Pathogenic genes:** *ANKK1, BDNF, LRRK2* **Mechanistic genes:** *ADRA2A, ADRA2B, ADRA2C, CCK, CCKAR, CCKBR, DRD1, DRD2, DRD3, DRD4, DRD5, GRIN2A, GRIN2B, HCRT, HOMER1, HTR1A, HTR1B, HTR1D, HTR2A, HTR2B, HTR2C, LMO3, OPRM1* **Metabolic genes** **Substrate:** *COMT, CYP1A2 (major), CY22B6, CYP2C19, CYP2D6, CYP3A4 (minor), CYP3A5, DDC, MAOB, UGT1A1, UGT1A9* **Inhibitor:** *CYP1A2 (moderate), CYP2D6 (moderate), CYP3A4 (moderate)* **Transporter genes:** *SLC22A1, SLC6A3* **Pleiotropic genes:** *ACE, APOE*
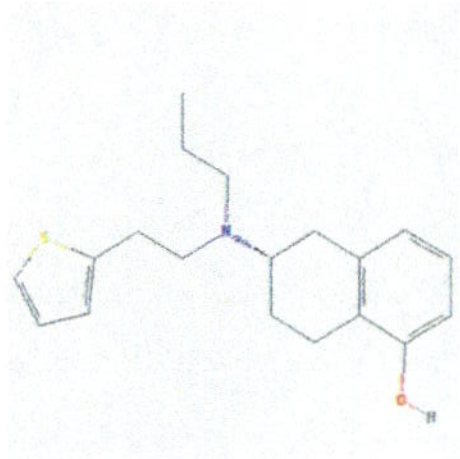	**Name: Rotigotine;** 99755-59-6; Rotigotine; Rotigotina; Neupro **IUPAC Name:** 1-Naphthalenol, 5,6,7,8-tetrahydro-6-[propyl[2-(2-thienyl)ethyl]amino]-6S **Molecular Formula:** $C_{19}H_{25}Nos$ **Molecular Weight:** 315.47 g/mol **Mechanism:** A non-ergot dopamine receptor agonist with specificity for D_3-, D_2-, and D_1-dopamine receptors. Although precise mechanism of action unkown of Rotigotine, it is believed to be due to stimulation of postsynaptic dopamine D_2-type auto receptors within substantia nigra in brain, leading to improved dopaminergic transmission in motor areas in basal ganglia, notably caudate nucleus/putamen regions **Effect:** Antiparkinsonian Agents; Nonergot-derivative Dopamine Receptor Agonists.	**Pathogenic genes:** *ANKK1, BDNF, LRRK2* **Mechanistic genes:** *CCK, CCKAR, CCKBR, DRD1, DRD2, DRD3, DRD4, DRD5, GRIN2A, GRIN2B, HCRT, HOMER1, LMO3, OPRM1* **Metabolic genes** **Substrate:** *COMT, MAOB* **Transporter genes:** *SLC22A1, SLC6A3* **Pleiotropic genes:** *ACE, APOE*

(continued)

Table 5.2 (*continued*)

Drug	Properties	Pharmacogenetics
Monoamine-Oxidase B (MOB) Inhibitors		
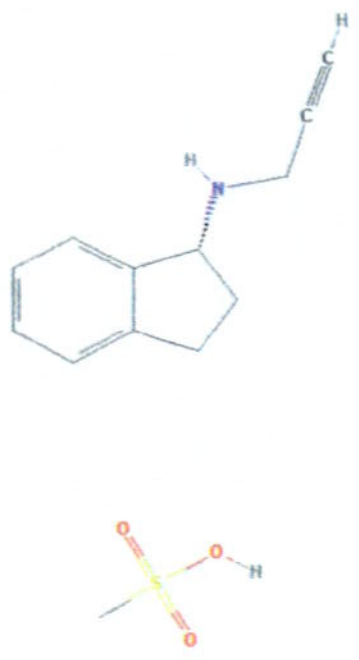	**Name: Rasagiline**; 136236-51-6; Azilet; Elbrux; Rasagilina; Raxac **IUPAC Name:** 1H-Inden-1-amine, 2,3-dihydro-N-2-propynyl-, (R)-,methanesulfonate **Molecular Formula:** $C_{12}H_{13}NCH_4O_3S$ **Molecular Weight:** 267.34 g/mol **Mechanism:** Potent, irreversible inhibitor of the monoamine oxidase (MAO) type B, which plays a major role in catabolism of dopamine. Inhibition of dopamine depletion in striatal region of brain reduces symptomatic motor deficits of Parkinson's Disease. There is also experimental evidence of Rasagiline conferring neuroprotective effects (antioxidant, antiapoptotic), which may delay onset of symptoms and progression of neuronal deterioration **Effect:** Antidepressants. Monoamine Oxidase Inhibitors. Antiparkinsonian Agents. Monoamina Oxidase B Inhibitors	**Pathogenic genes:** *ANKK1, BDNF, LRRK2, PARK2* **Mechanistic genes:** *BLC2, CCK, CCKAR, CCKBR, DRD1, DRD2, DRD3, DRD4, DRD5, GRIN2A, GRIN2B, HCRT, HOMER1, LMO3, OPRM1* **Metabolic genes** **Substrate:** *COMT, CYP1A2 (major), CYP2B6, CYP2C19, CYP2D6, CYP3A4, CYP3A5, DDC, MAOB, UGT1A1, UGT1A9* **Inhibitor:** *MAOB* **Transporter genes:** *SLC22A1, SLC6A3* **Pleiotropic genes:** *ACE, APOE*
Catecol-O-methylatransferase (COMT) Inhibitors		
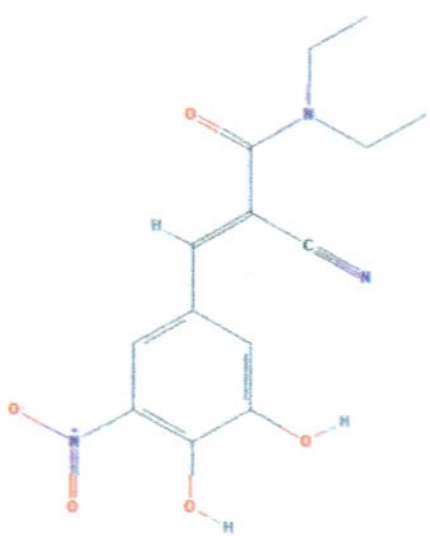	**Name: Entacapone**; 130929-57-6; Comtan; Comtess; Entacapona **IUPAC Name:** E-α-Cyano-N,N-diethyl-3,4-dihydroxy-5-nitrocinnamamida **Molecular Formula:** $C_{14}H_{15}N_3O_5$ **Molecular Weight:** 305.29 g/mol **Mechanism:** A selective and selective inhibitor of catechol-O-methylatransferase (COMT). When entacapona is taken with levodopa, the pharmacokinetics are altered, resulted in more sustained levodopa serum levels compared to levodopa taken alone **Effect:** Antiparkinsonian Agents. Catechol-O-Methylatransferase Inhibitors	**Pathogenic genes**: *ANKK1, BDNF, LRRK2, PARK2* **Mechanistic genes:** *CCK, CCKAR, CCKBR, DRD1, DRD2, DRD3, DRD4, DRD5, GRIN2A, GRIN2B, HCRT, HOMER1, LMO3, OPRM1* **Metabolic genes** **Substrate**: *COMT, CYP1A2, CYP2B6, CYP2C19, CYP2D6, CYP3A4, CYP3A5, DDC, MAOB, UGT1A1, UGT1A3, UGT1A4, UGT1A6, UGT1A9, UGT2B7, UGT2B15* **Inhibitor**: *COMT, CYP1A2 (weak), CYP2A6 (weak), CYP2C9 (weak), CYP2C19 (weak), CYP2D6 (weak), CYP2E1 (weak), CYP3A4 (weak)* **Transporter genes**: *SLC22A1, SLC6A3* **Pleiotropic genes**: *ACE, ACHE, APOE*

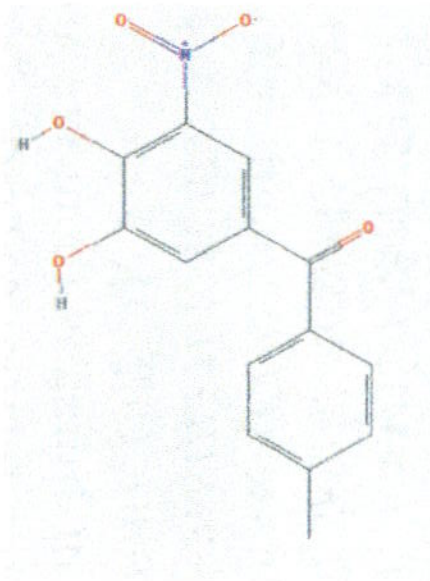

Name: Tolcapone; 134308-13-7; Tolcapona; Tasmar
IUPAC Name: Methanone, (3,4-hydroxy-5-nitrophenyl) (4-methylphenyl)
Molecular Formula: $C_{14}H_{11}NO_5$
Molecular Weight: 273.24 g/mol
Mechanism: A selective and selective inhibitor of catechol-O-methylatransferase (COMT). In the presence of a decarboxylase inhibitor (e.g. carbidopa), COMT is the major degradation pathway for levodopa. Inhibition of COMT leads to more sustained plasma levels of levodopa and enhanced central dopaminergic activity
Effect: Antiparkinsonian Agents. Catechol-O-Methylatransferase Inhibitors

Pathogenic genes: *ANKK1, BDNF, LRRK2, PARK2*
Mechanistic genes: *AKT1, CCK, CCKAR, CCKBR, CNR1, DRD1, DRD2, DRD3, DRD4, DRD5, GPT, GRIN2A, GRIN2B, GSK3B, HCRT, HOMER1, LMO3, OPRM1*
Metabolic genes
Substrate: *COMT, CYP1A2, CYP2B6, CYP2C9, CYP2C19, CYP2D6, CYP3A4, CYP3A5, DDC, MAOB, UGT1A1, UGT1A3, UGT1A4, UGT1A6, UGT1A9, UGT2B7, UGT2B15*
Transporter genes: *SLC22A1, SLC6A3*
Pleiotropic genes: *ACE, APOE*

[a]***ABCB1:*** ATP binding cassette subfamily B member 1, ***ACE:*** angiotensin I converting enzyme, ***ACHE:*** acetylcholinesterase, ***ADCY7:*** adenylate cyclase 7, ***ADRA1A:*** adrenoceptor alpha 1A, ***ADRA1B:*** adrenoceptor alpha 1B, ***ADRA1D:*** adrenoceptor alpha 1D, ***ADRA2A:*** adrenoceptor alpha 2A, ***ADRA2B:*** adrenoceptor alpha 2B, ***ADRA2C:*** adrenoceptor alpha 2C, ***AKT1:*** v-akt murine thymoma viral oncogene homolog 1, ***ANKK1:*** ankyrin repeat and kinase domain containing 1, ***APOE:*** apolipoprotein E, ***BDNF:*** brain-derived neurotrophic factor, ***BLC2:*** B-cell CLL/lymphoma 2, ***CALY:*** calcyon neuron specific vesicular protein, ***CCK:*** cholecystokinin, ***CCKAR:*** cholecystokinin A receptor, ***CCKBR:*** cholecystokinin B receptor, ***CCR5:*** C-C motif chemokine receptor 5 (gene/pseudogene), ***CHAT:*** choline O-acetyltransferase, ***CNR1:*** cannabinoid receptor 1 (brain), ***COMT:*** catechol-O-methyltransferase, ***CREB1:*** cAMP responsive element binding protein 1, ***CXCR4:*** C-X-C motif chemokine receptor 4, ***CYP1A1:*** cytochrome P450 family 1 subfamily A member 1, ***CYP1A2:*** cytochrome P450 family 1 subfamily A member 2, ***CYP1B1:*** cytochrome P450 family 1 subfamily B member 1, ***CYP2A6:*** cytochrome P450 family 2 subfamily A member 6, ***CYP2B6:*** cytochrome P450 family 2 subfamily B member 6, ***CYP2C19:*** cytochrome P450 family 2 subfamily C member 19, ***CYP2C9:*** cytochrome P450 family 2 subfamily C member 9, ***CYP2D6:*** cytochrome P450 family 2 subfamily D member 6, ***CYP2E1:*** cytochrome P450 family 2 subfamily E member 1, ***CYP3A4:*** cytochrome P450 family 3 subfamily A member 4, ***CYP3A5:*** cytochrome P450 family 3 subfamily A member 5, ***CYP19A1:*** cytochrome P450 family 19 subfamily A member 1, ***DBH:***dopamine beta-hydroxylase, ***DDC:*** dopa decarboxylase, ***DRD1:*** dopamine receptor D1, ***DRD2:*** dopamine receptor D2, ***DRD3:*** dopamine receptor D3, ***DRD4:*** dopamine receptor D4, ***DRD5:*** dopamine receptor D5, ***G6PD:*** glucose-6-phosphate dehydrogenase, ***GPT:*** glutamic-pyruvate transaminase (alanine aminotransferase), ***GRIN2A:*** glutamate ionotropic receptor NMDA type subunit 2A, ***GRIN2B:*** glutamate ionotropic receptor NMDA type subunit 2B, ***GRIN3A:*** glutamate ionotropic receptor NMDA type subunit 3A, ***GSK3B:*** glycogen synthase kinase 3 beta, ***HCRT:*** hypocretin (orexin) neuropeptide precursor, ***HOMER1:*** homer scaffolding protein 1, ***HRH1:*** histamine receptor H1, ***HTR1A:*** 5-hydroxytryptamine receptor 1A, ***HTR1B:*** 5-hydroxytryptamine receptor 1B, ***HTR1D:*** 5-hydroxytryptamine receptor 1D, ***HTR2A:*** 5-hydroxytryptamine receptor 2A, ***HTR2B:*** 5-hydroxytryptamine receptor 2B, ***HTR2C:*** 5-hydroxytryptamine receptor 2C, ***HTR7:*** 5-hydroxytryptamine receptor 7, ***LMO3:*** LIM domain only 3, ***LRRK2:*** leucine-rich repeat kinase 2, ***MAOA:*** monoamine oxidase A, ***MAOB:*** monoamine oxidase B, ***OPRM1:*** opioid receptor mu 1, ***PAH:*** phenylalanine hydroxylase, ***PARK2:*** parkin RBR E3 ubiquitin protein ligase, ***SLC22A1:*** solute carrier family 22 member 1, ***SLC6A3:*** solute carrier family 6 member 3, ***SLC6A4:*** solute carrier family 6 member 4, ***SST:*** somatostatin, ***TH:*** tyrosine hydroxylase, ***TSPO:*** translocator protein, ***UGT1A1:*** UDP glucuronosyltransferase family 1 member A1, ***UGT1A3:*** UDP glucuronosyltransferase family 1 member A3, ***UGT1A4:*** UDP glucuronosyltransferase family 1 member A4, ***UGT1A6:*** UDP glucuronosyltransferase family 1 member A6, ***UGT1A9:*** UDP glucuronosyltransferase family 1 member A9, ***UGT2B7:*** UDP glucuronosyltransferase family 2 member B7, ***UGT2B15:*** UDP glucuronosyltransferase family 2 member B15.

of which the top ten are: *APOE* (19q13.2), *BIN1* (2q14), *CLU* (8p21-p12), *ABCA7* (19p13.3), *CR1* (1q32), *PICALM* (11q14), *MS4A6A* (11q12.1), *CD33* (19q13.3), *MS4A4E* (11q12.2), and *CD2AP* (6p12).[18,57] Potentially defective genes associated with AD represent about 1.39% (35 252.69 Kb) of the human genome, which is integrated by 36 505 genes (3 095 677.41 Kb). The highest number of AD-related defective genes concentrate on chromosomes 10 (5.41%; 7337.83 Kb), 21 (4.76%; 2289.15 Kb), 7 (1.62%; 2584.26 Kb), 2 (1.56%; 3799.67 Kb), 19 (1.45%; 854.54 Kb), 9 (1.42%; 2010.62 Kb), 15 (1.23%; 1264.4 Kb), 17 (1.19%; 970.16 Kb), 12 (1.17%; 1559.9 Kb), and 6 (1.15%; 1968.22 Kb).[11] Among susceptibility genes, the apolipoprotein E (*APOE*) gene (*AD2*) is the most prevalent as a risk factor for AD, especially in those subjects harboring the *APOE-4* allele, whereas carriers of the *APOE-2* allele might be protected against dementia.[53] Polymorphic variants in other genes (GRB-associated binding protein 2 (*GAB2*), *TLR9* rs187084 variant homozygote GG, *LRRK2* R1628P variant) might also be protective.[11] Ten novel private pathogenic copy number variations (CNVs) in 10 early-onset familial Alzheimer's disease (EO-FAD) families overlapping a set of genes (*A2BP1*, *ABAT*, *CDH2*, *CRMP1*, *DMRT1*, *EPHA5*, *EPHA6*, *ERMP1*, *EVC*, *EVC2*, *FLJ35024* and *VLDLR*) have also been identified.[60]

(ii) Genes associated with the mechanism of action of drugs (enzymes, receptors, transmitters, messengers).

(iii) Genes associated with drug metabolism: (a) phase I reaction enzymes: alcohol dehydrogenases (*ADH1–7*), aldehyde dehydrogenases (*ALDH1–9*), aldo-keto reductases (*AKR1A–D*), amine oxidases (*MAOA*, *MAOB*, *SMOX*), carbonyl reductases (*CBR1–4*), cytidine deaminase (*CDA*), cytochrome P450 family (*CYP1–51*, *POR*, *TBXAS1*), cytochrome b5 reductase (*CYB5R3*), dihydropirimidine dehydrogenase (DPYD), esterases (*AADAC*, *CEL*, *CES1*, *CES1P1*, *CES2*, *CES3*, *CES5A*, *ESD*, *GZMA*, *GZMB*, *PON1*, *PON2*, *PON3*, *UCHL1*, *UCHL3*), epoxidases (*EPHX1–2*), flavin-containing monooxygenases (*FMO1–6*), glutathione reductase/peroxidases (*GPX1–7*, *GSR*), short-chain dehydrogenases/reductases (*DHRS1–13*, *DHRSX*, *HSD11B1*, *HSD17B10*, *HSD17B11*, *HSD17B14*), superoxide dismutases (*SOD1–2*), and xanthine dehydrogenase (*XDH*); and (b): phase II reaction enzymes: amino acid transferases (*AGXT*, *BAAT*, *CCBL1*), dehydrogenases (*NQO1–2*, *XDH*), esterases (*CES1–5*), glucuronosyl transferases (*UGT1–8*), glutathione transferases (*GSTA1–5*, *GSTK1*, *GSTM1–5*, *GSTO1–2*, *GSTP1*, *GSTT1–2*, *GSTZ1*, *GSTCD*, *MGST1–3*, *PTGES*), methyl transferases (*AS3MT*, *ASMT*, *COMT*, *GNMT*, *GAMT*, *HNMT*, *INMT*, *NNMT*, *PNMT*, *TPMT*), *N*-acetyl transferases (*ACSL1–4*, *ACSM1*, *ACSM2B*, *ACSM3*, *AANAT*, *GLYAT*, *NAA20*, *NAT1–2*, *SAT1*), thioltransferase (*GLRX*), and sulfotransferases (*CHST2–13*, *GAL3ST1*, *SULT1A1–3*, *SULT1B1*, *SULT1C1–4*, *SULT1E1*, *SULT2A1*, *SULT2B1*, *SULT4A1*, *SULT6B1*, *CHST1*).

(iv) Genes associated with drug transporters: in humans there are 49 ABC transporter genes and the multidrug resistance-associated proteins (*MRP1/ABCC1*, *MRP2/ABCC2*, *MRP3/ABCC3*, *MRP4/ABCC4*, *MRP5/ABCC5*, *MRP6/ABCC6*, *MRP7/ABCC10*, *MRP8/ABCC11* and *MRP9/ABCC12*), which belong to the ABCC family integrated by 13 members. Other genes encoding transporter proteins are genes of the solute carrier superfamily (SLC) and solute carrier organic (SLCO) transporter family, responsible for the transport of multiple endogenous and exogenous compounds.

(v) Pleiotropic genes involved in multifaceted cascades and metabolic reactions.[10–13]

All these genes are under the influence of the epigenetic machinery conditioning their expression and the efficiency of their drug-metabolizing products (enzymes, transporters).[13–16]

Although the *APP*, *PSEN1*, *PSEN2* and *MAPT* genes are considered major pathogenic genes for AD and classic tauopathies,[57] mutations in these genes represent less than 5% of the AD population and, consequently, their influence on AD pharmacogenetics associated with conventional anti-dementia drugs is quantitatively negligible; not so in the case of immunotherapy addressing Aβ deposition. Most anti-AD vaccines (active and passive immunization) are based on transgenic models with *APP*, *PSEN1* and *PSEN2* mutants.[61,62] In general, most pharmacogenetic studies in AD have been performed with susceptibility genes (*APOE*) and metabolic genes (CYPs).[11,18,63–65]

5.3.1 APOE-TOMM40

To date, the most influential gene in AD pharmacogenetics is the *APOE* gene.[4,5,11,13,63–66] *APOE* is a pleiotropic gene with multifaceted activities in physiological and pathological conditions, and the presence of the *APOE*-4 allele is determinant in AD pathogenesis.[53] *APOE-4* may influence AD pathology by interacting with APP metabolism and Aβ accumulation, enhancing hyperphosphorylation of tau protein and neurofibrillary tangle formation, reducing choline acetyltransferase activity, increasing oxidative processes, modifying inflammation-related neuroimmunotrophic activity and glial activation, altering lipid metabolism, lipid transport and membrane biosynthesis in sprouting and synaptic remodeling, and inducing neuronal apoptosis and premature neuronal death.[10,53] Multiple studies over the past two decades have demonstrated that *APOE* variants may affect the therapeutic response to anti-dementia drugs.[6,10,11,13,18,53,63–70] At least 20 major phenotypic features illustrate the biological disadvantage of *APOE-4* homozygotes and the potential consequences that these patients may experience when they receive pharmacological treatment for AD and/or concomitant pathologies.[10,11,23,65–67,70]

In over 100 clinical trials for dementia, *APOE* has been used as the only gene of reference for the pharmacogenomics of AD. Several studies indicate that the presence of the *APOE-4* allele differentially affects the quality and extent of drug responsiveness in AD patients treated with cholinergic enhancers, neuroprotective compounds, endogenous nucleotides, immunotrophins, neurotrophic factors, combination therapies and other drug categories;[10,11,66–72] however, controversial results are frequently found due to methodological problems, study design, and patient recruitment in clinical trials. The major conclusion in most studies is that *APOE-4* carriers are the worst responders to conventional treatments.[10,11,18] When *APOE* and *CYP2D6* genotypes are integrated in bigenic clusters and the *APOE* + *CYP2D6*-related therapeutic response to a combination therapy is analyzed in AD patients, it becomes clear that the presence of the *APOE-4/4* genotype is able to convert pure *CYP2D6*1/*1* extensive metabolizers into full poor responders to conventional treatments, indicating the existence of a powerful influence of the *APOE-4* homozygous genotype on the drug-metabolizing capacity of pure *CYP2D6* extensive metabolizers.[6] In addition, a clear accumulation of *APOE-4/4* genotypes is observed among *CYP2D6* poor and ultra-rapid metabolizers.[6]

Adjacent to the *APOE* locus (19q13.2) and in linkage disequilibrium with *APOE* is the *TOMM40* gene. A poly T repeat in an intronic polymorphism (rs10524523) (intron 6) in the *TOMM40* gene, which encodes an outer mitochondrial membrane translocase involved in the transport of Aβ and other proteins into mitochondria, has been implicated in AD.[73–86] *APOE-TOMM40* genotypes have been shown to modify disease risk and age at onset of symptoms.[74–79,87] The rs4420638 at the *TOMM40/APOE/APOC1* gene locus is associated with longevity.[88,89] The *APOE-TOMM40* genomic region is associated with cognitive aging[90] and with pathological cognitive decline.[91] There are 3 allele groups for rs10524523 ('523'), based on the number of 'T'-residues: 'Short' (S, $T \leq 19$), 'Long' (L, $20 \leq T \leq 29$) and 'Very Long' (VL, $T \geq 30$).[81] Longer lengths of rs10524523 are associated with a higher risk for late-onset AD (LOAD).[75–79] Intronic poly T (rs10524523) within this region affects expression of the *APOE* and *TOMM40* genes in the brain of patients with LOAD.[92] The 523 VL poly T shows higher expression than the S poly T, indicating that the 523 locus may contribute to LOAD susceptibility by modulating the expression of *TOMM40* and/or *APOE* transcription.[92] S/VL and VL/VL are the only *TOMM40* poly T genotypes that interact with all major *APOE* genotypes; in contrast, the *APOE-4/4-TOMM40-L/L* association is unique, representing approximately 30% of *APOE-4/4* carriers.[12,93] The first pharmacogenetic study of the *APOE-TOMM40* region in AD patients receiving a multifactorial treatment revealed that: (i) *APOE-4* carriers are the worst responders and *APOE-3* carriers are the best responders to conventional treatments; (ii) *TOMM40 poly T-S/S* carriers are the best responders, VL/VL and S/VL carriers are intermediate responders, and L/L carriers are the worst responders to treatment; (iii) patients harboring a large (L) number of poly T repeats in intron 6 of the *TOMM40* gene (L/L or S/L genotypes) in haplotypes associated with *APOE-4* are the worst responders to treatment; (iv) patients with short (S) *TOMM40* poly T variants (S/S genotype), and to a lesser extent S/VL and VL/VL carriers, in haplotypes with *APOE-3* are the best responders to treatment; and (v) in 100% of the cases, the L/L genotype is exclusively associated with the *APOE-4/4* genotype, and this haplotype (4/4-L/L) is probably responsible for early onset of the disease, a faster cognitive decline, and a poor response to different treatments.[12,93]

Other recent pharmacogenetic studies with pathogenic or mechanistic genes indicate that the response to AChEIs is associated with 2 SNPs in the intronic region of *CHAT* rs2177370 and rs3793790.[94] The *CHRNA7* T allele (rs6494223) is also associated with a better response to AChEIs and there is further confirmation that *APOE-4* carriers are the worst responders to conventional AChEIs.[95]

5.3.2 CYPs

Over 70% of AD patients are deficient metabolizers for the *CYP2D6/2C19/2C9* trigenic cluster; and for the *CYP2D6/2C19/2C9/3A4* tetragenic cluster, more than 80% of the patients exhibit a deficient metabolizer geno-phenotype.[6] These four *CYP* genes encode enzymes responsible for the metabolism of 60–80% of drugs of current use, showing ontogenic-, age-, sex-, circadian- and ethnic-related differences.[10,11,66,96] According to the database of the World Guide for Drug Use and Pharmacogenomics,[23] 982 drugs are CYP2D6-related: 371 drugs are substrates, over 300 drugs are inhibitors, and 18 drugs are CYP2D6 inducers. Over 600 drugs are *CYP2C9*-related, 311 acting as substrates (177 are major substrates, 134 are minor substrates), 375 as inhibitors (92 weak, 181 moderate, and 102 strong inhibitors), and 41 as inducers of the CYP2C9 enzyme.[23] Nearly 500 drugs are CYP2C19-related, 281 acting as substrates (151 are major substrates, 130 are minor substrates), 263 as inhibitors (72 weak, 127 moderate, and 64 strong inhibitors), and 23 as inducers of the CYP2C19 enzyme.[23] The CYP3A4/5 enzyme metabolizes over 1900 drugs, 1033 acting as substrates (897 are major substrates, 136 are minor substrates), 696 as inhibitors (118 weak, 437 moderate, and 141 strong inhibitors), and 241 as inducers of the CYP3A4 enzyme.[23]

In healthy subjects, CYP2D6 extensive metabolizers (EMs) account for 55.71% of the population, whereas intermediate metabolizers (IMs) account for 34.7%, poor metabolizers (PMs) 2.28%, and ultra-rapid metabolizers (UMs) 7.31%.[11,18] In AD, EMs, IMs, PMs, and UMs are 56.38%, 27.66%, 7.45%, and 8.51%, respectively. There is an accumulation of AD-related genes of risk in PMs and UMs. EMs and IMs are the best responders, and PMs and UMs are the worst responders to a combination therapy with AChEIs, neuroprotectants, and vasoactive substances. The pharmacogenetic response in AD appears to be dependent upon the networking activity of genes involved in drug metabolism and genes involved in AD pathogenesis.[10,11,17,53,66,67,97,98] By phenotypes, in the control population, CYP2C9-PMs represent 7.04%, IMs 32.39%, and EMs 60.56%. In AD, PMs, IMs, and EMs are 6.45%, 37.64%, and 55.91%, respectively.[11,23] The frequencies of the 3 major *CYP2C19* geno-phenotypes in the control population are: *CYP2C19-*1/*1*-EMs 68.54%, *CYP2C19-*1/*2*-IMs 30.05%, and *CYP2C19-*2/*2*-PMs 1.41%. EMs, IMs, and PMs account for 69.89%, 30.11%, and 0%, respectively, in AD.[11,23] Concerning *CYP3A4/5* polymorphisms in AD, 82.75% of the cases are EMs (*CYP3A5*3/*3*), 15.88% are IMs (*CYP3A5*1/*3*), and 1.37% are UMs (*CYP3A5*1/*1*).[11]

Most anti-dementia drugs are metabolized *via* CYP enzymes. Donepezil is a major substrate of CYP2D6, CYP3A4, ACHE, and UGTs, inhibits ACHE and BCHE, and is transported by ABCB1 (Table 5.1).[5,11,17,23,63,64,66,98–100] *CYP2D6* variants affect donepezil efficacy and safety in AD.[5,11,17,63,64,66,97,100] The common variant rs1080985 of *CYP2D6* is associated with poor response to donepezil.[101,102] A higher frequency of mutated *CYP2D6* allele **2A* was found in responder than in non-responder patients (75.38% *vs.* 43.48%).[103] In an Italian study, 67% of patients were responders and 33% were non-responders to donepezil treatment, with abnormal enzymes accumulating in responders.[104] Chinese AD patients with the mutant allele *CYP2D6*10* may respond better (58% responders) to donepezil than those with the wild allele *CYP2D6*1*.[105] In contrast, other studies revealed that CYP2D6-PMs and UMs tend to be poor responders to conventional doses of donepezil as compared to EMs and IMs.[5,11,23,63,64,66,100,106–108]

In Italian patients, no association was found between *CYP3A4* or *CYP3A5* genotypes and plasma donepezil concentrations, or between genotypes and clinical response. The most common *ABCB1* haplotypes were 1236C/2677G/3435C (46%) and 1236T/2677T/3435T (41%), and patients homozygous for the T/T/T haplotype had lower plasma donepezil concentration-to-dose ratios and better clinical response than patients with other genotypes.[109] In Brazilian patients treated with AChEIs the response rate was 27.8%, with no apparent effect of APOE and/or CYP2D6 polymorphic variants.[110]

The effects of galantamine are potentially influenced by *APOE*, *APP*, *ACHE*, *BCHE*, *CHRNA4*, *CHRNA7*, and *CHRNB2* variants. This drug is a major substrate of CYP2D6, CYP3A4, and UGT1A1, and an inhibitor of ACHE and BCHE (Table 5.1).[23,99,100,111–113] Major metabolic pathways are glucuronidation, *O*-demethylation, *N*-demethylation, *N*-oxidation, and epimerization.[114] Galantamine is extensively metabolized by the enzymes CYP2D6 and CYP3A and is a substrate of the P-gp. *CYP2D6* variants are determinant for galantamine pharmacokinetics. CYP2D6-PMs exhibit higher dose-adjusted galantamine plasma concentrations than heterozygous and homozygous CYP2D6-EMs;[115] however, these pharmacokinetic changes might not substantially affect pharmacodynamics.[116] The co-administration of galantamine with paroxetine (a CYP2D6 strong inhibitor), ketoconazole (a CYP3A4 strong inhibitor) and erythromycin increases its bioavailability.[117,118] Interaction with foods and nutritional components may alter galantamine bioavailability and therapeutic effects.[119]

APOE, *APP*, *CHAT*, *ACHE*, *BCHE*, *CHRNA4*, *CHRNB2* and *MAPT* variants may affect rivastigmine pharmacokinetics and pharmacodynamics, but CYP enzymes are not involved in the metabolism of rivastigmine.[23,99,100,117,120] UGT2B7-PMs show higher rivastigmine levels with a poor response to treatment.[121]

ACHE, *ABCB4*, *BCHE*, *CHRNA4*, *CHRNB2*, *APOE*, *MTHFR*, *CES1*, *LEPR*, *GSTM1* and *GSTT1* variants may affect the therapeutic and toxic effects of tacrine (the first AChEI introduced in 1993 and withdrawn years later due to hepatotoxicity). Tacrine is a major substrate of CYP1A2 and CYP3A4, a minor substrate of CYP2D6, and is transported *via* SCN1A and ABCB4. Tacrine is an inhibitor of ACHE, BCHE, and CYP1A2.[23] Both tacrine and some tacrine-hybrids may cause an induction of *CYP1A1*, *2B1* and *3A2* expression.[122] Tacrine is associated with transaminase elevation in up to 50% of patients. The mechanism of tacrine-induced liver damage is influenced by genetic factors. The strongest association was found between alanine aminotransferase levels and three *ABCB4* SNPs.[123]

Memantine is an *N*-methyl-D-aspartate (NMDA) receptor antagonist which binds preferentially to NMDA receptor-operated cation channels; it may act by blocking actions of glutamate, mediated in part by NMDA receptors, and is also an antagonist of GRIN2A, GRIN2B, GRIN3A, HTR3A and CHRFAM7A. Several pathogenic (*APOE*, *PSEN1*, *MAPT*) and mechanistic gene variants (*GRIN2A*, *GRIN2B*, *GRIN3A*, *HTR3A*, *CHRFAM7A*, *c-Fos*, *Homer1b* and *PSD-95*) may influence its therapeutic effects. Memantine is a strong inhibitor of CYP2B6 and CYP2D6, and a weak inhibitor of CYP1A2, CYP2A6, CYP2C9, CYP2C19, CYP2E1, and CYP3A4.[23,100,124] In human liver microsomes (HLM), memantine inhibits CYP2B6 and CYP2D6 activities, decreases CYP2A6 and CYP2C19 activities, and has no effect on CYP1A2, CYP2E1, CYP2C9, or CYP3A4 activities.[125] The co-administration of memantine with CYP2B6 substrates elicits a 65% decrease in its metabolism. In clinical studies, *NR1I2* rs1523130 was identified as the unique significant genetic covariate for memantine clearance, with carriers of the *NR1I2* rs1523130 CT/TT genotypes presenting a 16% slower memantine elimination than carriers of the CC genotype.[126]

5.3.3 Transporters

Polymorphic variants in genes encoding transporter proteins may affect drug metabolism, brain penetrance and accessibility to neuronal/glial targets, and drug resistance.[23,127–129] Of special importance in AD are the ABC and SLC family genes.[129] ABC genes (*ABCB1*, *ABCC1*, *ABCG2*), and other genes of this family encode proteins that are essential for drug metabolism and transport. Mutations in ABC transporters influence pathogenesis and therapeutics of brain disorders.[129,130] The multidrug efflux transporters (P-gp1/MDR1, multidrug-resistance associated protein 4 (MRP4), breast cancer resistance protein (BCRP)), are located on endothelial cells lining brain vasculature and play important roles in limiting the movement of substances into and enhancing their efflux from the brain.

ABCB1 is one of the most important drug transporters in the brain. Over 1270 drugs have been reported to be associated with the Abcb1 transporter protein (P-gp), of which 490 are substrates, 618 are inhibitors, 182 are inducers, and 269 additional compounds which belong to different pharmacological categories of products with potential Abcb1 interaction.[23] The *ABCB1* gene has 116 polymorphic sites in Caucasians and 127 in African–Americans, with a minor allele frequency greater than 5%. Common variants are

1236C>T, 2677G>A/T and 3435C>T, and the *ABCB1*13* haplotype involves the 1236, 2677 and 3435 (TTT) SNPs and 3 intronic SNPs (in intron 9, 13, and 14).[23] The *ABCB1* C1236T, G2677T/A and C3435T SNPs influence blood–brain barrier (BBB) *P*-glycoprotein function. AD patients with one or more T in C1236T, G2677T and C3435T have significantly higher binding potential values than patients without a T. Genetic variations in *ABCB1* might contribute to the progression of Aβ deposition in the brain[131] and some *ABCB1* SNPs (C1236T in exon 12, G2677T/A in exon 21 and C3435T in exon 26) and inferred haplotypes might represent novel biomarkers of AD.[132] ABCB1 directly transports Aβ from the brain into the blood circulation, whereas the cholesterol transporter ABCA1 neutralizes Aβ aggregation capacity in an APOE-dependent manner, facilitating subsequent Aβ elimination from the brain.[133] Some *ABCB1* variants are frequent in AD cases over 65 years of age and among females. This association of *ABCB1* 2677G>T (rs2032582) is more pronounced in *APOE4*-negative cases.[131]

Some other ABCs have shown potential association with AD.[129,134] The G allele of the *ABCA7* rs115550680 SNP is associated with AD in Europeans. The effect size for the SNP in *ABCA7* was comparable with that of the *APOE ε4*-determining SNP rs429358.[135] ABCG2 is involved in Aβ transport and is up-regulated in AD brains. The *ABCG2* gene (C421A; rs2231142) (*ABCG2* C/C genotype) is associated with AD and the *ABCG2* C/C genotype and the *APOE ε4* allele may exert an interactive effect on AD risk.[136] Also of importance for AD pharmacogenomics are transporters encoded by genes of the solute carrier superfamily (*SLC*) and solute carrier organic (*SLCO*) transporter family, responsible for the transport of multiple endogenous and exogenous compounds, including folate (*SLC19A1*), urea (*SLC14A1–2*), monoamines (*SLC29A4*, *SLC22A3*), amino-acids (*SLC1A5*, *SLC3A1*, *SLC7A3*, *SLC7A9*, *SLC38A1*, *4–5*, *7*, *SLC43A2*, *SLC45A1*), nucleotides (*SLC29A2–3*), fatty acids (*SLC27A1–6*), neurotransmitters (*SLC6A2* (noradrenaline transporter), *SLC6A3* (dopamine transporter), *SLC6A4* (serotonin transporter, SERT), *SLC6A5–6*, *9*, *11*, *12*, *14–19*), glutamate (*SLC1A6–7*), and others.[129,137] Some organic anion transporters (OAT), which belong to the solute carrier (SLC) 22A family, are also expressed at the BBB, and regulate the excretion of endogenous and exogenous organic anions and cations.[138] The transport of amino acids and di- and tripeptides is mediated by a number of different transporter families, and the bulk of oligopeptide transport is attributable to the activity of members of the SLC15A superfamily (SLC15A1-2, SLC15A2, SLC15A3-4). ABC and SLC transporters expressed at the BBB may cooperate to regulate the passage of different molecules into the brain.[11,13,18,139]

5.4 Epigenomics

Epigenomic regulation is a universal phenomenon of gene expression control during development, maturation and aging in physiological conditions. When this mechanism of control is altered by endogenous and/or exogenous factors, probably acting as an interface between the genome and the environment (nature *vs.* nurture),[140,141] then epigenomic changes become pathogenic due to the abnormal expression of genes under epigenetic control.

Harris *et al.*[142] define these metastable epialleles as mammalian genomic loci where epigenetic patterning occurs before gastrulation in a stochastic fashion, leading to systematic interindividual variation within one species. This gene expression abnormally leads to a potential reversible pathological phenotype which, in some cases, can be transferred to future generations, assuming that epigenetics refers to phenotypic changes with no apparent alterations in structural DNA. Preconceptional parental exposure to environmental stimuli may determine the offspring's phenotype *via* meiotically and mitotically heritable epigenetic mechanisms,[140] and exposure to diverse external elements (nutrition, pollutants, drugs, toxins) may condition several categories of human diseases. Classical epigenetic mechanisms, including DNA methylation, histone modifications, and regulation by microRNAs (miRNAs), are among the major regulatory elements that control metabolic pathways at the molecular level. DNA methylation/demethylation and chromatin remodeling/histone modifications regulate gene expression transcriptionally, and miRNAs suppress gene expression post-transcriptionally.[143] Mutations in the genes encoding elements of the epigenetic machinery can lead to an epigenetic Mendelian disorder.[144] Epigenetic marks contribute to natural human variation[145] and configure the emerging field of neuroepigenetics.[141] Not only nuclear DNA, but also mitochondrial DNA may be subjected to epigenetic modifications related to disease development, environmental exposure, drug treatment and aging.[146] Some epigenetic modifications are conceptually reversible and can potentially be targeted by pharmacological and dietary interventions.[13–16,147]

Age-related neuropsychiatric disorders (from neurodevelopment to aging) are complex diseases in which genomic defects, together with environmental factors and epigenetic alterations, may be involved.[17] Most of these disorders exhibit proteoepigenomic changes resulting from primary genomic traits and/or secondary epigenetic events that induce pathogenic (structural, functional, conformational) changes in key proteins.[148] Consequently, neuroepigenetic perturbations in genes involved in brain development, maturation and aging may alter gene expression and protein synthesis (and conformational protein configuration) leading to neurodevelopmental, neuropsychiatric, and neurodegenerative disorders.[149]

5.4.1 Age-Related Epigenetics

Altered DNA methylation patterns may account for phenotypic changes associated with human aging. Brain region-specific expression of genes can be epigenetically regulated by DNA methylation[150] and brain aging might be influenced by epigenetic changes in the neuronal microenvironment.[151,152]

5.4.1.1 DNA Methylation

Age- and tissue-dependent DNA hypo- and hyper-methylation has been reported.[153] It appears that global loss of DNA methylation predominates in aged cells. DNMT1, which maintains DNA methylation of CpGs, decreases with age.[154] In contrast, some loci have been found hypermethylated with age (*e.g.* estrogen receptor, interferon γ, insulin-like growth factor II, promoters of tumor-suppressor genes such as lysyl oxidase (*LOX*), *p16INK4a*, runt-related transcription factor 3 (*RUNX3*), and TPA-inducible gene 1 (*TIG1*)).[153] Xu and Taylor[155] analyzed 1006 blood DNA samples of women aged 35 to 76 from the Sister Study, and found that 7694 (28%) of the 27 578 CpGs assayed were associated with age, confirming the existence of at least 749 "high-confidence" age-related CpG (arCpGs) sites in normal blood. These age-related changes are largely concordant in a broad variety of normal tissues, and a significantly higher proportion (71–91%) than expected of increasingly-methylated arCpGs (IM-arCpGs) were over-methylated in a wide variety of tumor types. IM-arCpGs sites occurred almost exclusively at CpG islands and were disproportionately marked with the repressive H3K27me3 histone modification. These findings suggest that as cells acquire methylation at age-related sites they have a lower threshold for malignant transformation that may explain in part the increase in cancer incidence with age.

McClay *et al.*[156] performed a methylome-wide association study of aging in whole blood DNA from 718 individuals, aged 25–92 years. They sequenced the methyl-CpG-enriched genomic DNA fraction, averaging 67.3 million reads per subject, to obtain methylation measurements for the ~27 million autosomal CpGs in the human genome, and adaptively combined methylation measures for neighboring, highly correlated CpGs into 4 344 016 CpG blocks for association testing. Eleven age-associated differentially methylated regions (DMRs) passed Bonferroni correction. 42 of 70 selected DMRs showed hypomethylation and 28 showed hypermethylation with age. Hypermethylated DMRs were more likely to overlap with CpG islands and shores. Hypomethylated DMRs were more likely to be in regions associated with polycomb/regulatory proteins (EZH2) or histone modifications H3K27ac, H3K4m1, H3K4m2, H3K4m3 and H3K9ac. Among genes implicated by the top DMRs were protocadherins, homeobox genes, mitogen-activated protein kinases (MAPKs), ryanodine receptors, and genes with potential relevance for age-related disease.

The absolute levels of 5-hydroxymethylcytosine (hmC), 5-formylcytosine (fC) and 5-methylcytosine (mC) vary in human brain tissues at various ages. For hmC, an initial steady increase is observed, which levels off with age to a final steady-state value of 1.2%. This level is nearly twice as high as in mouse cerebral cortex. fC declines rapidly with age during early developmental stages. While hmC is a stable epigenetic mark, fC is more likely an intermediate of active DNA demethylation during early brain development. The trends in global cytosine modification dynamics during the lifespan are conserved between humans and mice and show similar patterns in different organs.[157]

5.4.1.2 Histone Modifications

Histone modifications are also observed with aging. Histone acetylation decreases and phosphorylation increases with age.[158] H4K20me and H3K36me3 decrease in the brain of old senescence-accelerated-prone mice (SAMP8) and H3K27m3, H3K79me, and H3K79me2 increase in these aged mouse brains.[159] The silent information regulator 2 (Sir2) in yeast and its mammalian orthologs, sirtuin 1–7 (SIRT1–7), are histone-modifying enzymes that tend to be downregulated in aging, especially SIRT1. Activation of sirtuins may extend lifespan, modulating calorie restriction mechanisms[160] and promoting healthy aging, which delays the onset of neurodegenerative processes.[161] In the epidermis, aging is associated with a limited destabilization of the epigenome at gene regulatory elements.[162] Wound treatment with sirtuin activators and class I HDAC inhibitors induces keratinocyte proliferation and enhances healing *via* a nitric oxide (NO)-dependent mechanism. Acetylation of α-tubulin and histone H3 Lysine 9 may activate cell function and gene expression to foster tissue repair. The direct activation of *P300/CBP*-associated factor (PCAF) by the histone acetylase activator pentadecylidenemalonate 1b (SPV-106) induces lysine acetylation in the wound area. An impairment of PCAF and/or other GCN5 family acetylases may delay skin repair in physiopathological conditions.[163]

5.4.1.3 Non-Coding RNAs

There is a correlation between changes in miRNA expression and aging. miRNA lin-4 regulates lifespan in *C. elegans*; several miRNAs (miRNAs-34, -669c, -709, -93, -214) were found to be upregulated with age, while others (miRNAs-103, -107,-128, -130a, -155, -24, -221, -496, -1538, -17, -19b, -20a, -106a) appeared downregulated in peripheral tissues.[164,165] 70 miRNAs were found to be upregulated in the aging brain; 27 of these miRNAs may target genes of mitochondrial complexes III, IV, and F_0F_1-ATPase involved in oxidative phosphorylation and reduced expression in aging.[166]

5.4.2 Neurodegenerative Disorders

Epigenetic dysregulation is an attractive mechanism to explain in part enigmatic areas of confusion associated with the pathogenesis of age-related neurodegenerative disorders, such as Alzheimer's disease, Parkinson's disease, and Huntington's disease (Table 5.3), where it may mediate interactions between genetic and environmental risk factors, or directly interact with disease-specific pathological factors.[13–16,153,167]

Several pathogenic genes (Table 5.3) and many other AD-related susceptibility genes with direct or indirect influence on the AD or PD phenotype (*i.e.* genes associated with vascular risk factors and lipid metabolism) (Figures 5.4 and 5.5) contain methylated CpG sites that exhibit alterations in DNA methylation.[13,153,168] Different modalities of histone aberrations are present in AD.[13–16,153,168–170] Alterations in epigenetically-regulated miRNAs may contribute to the abnormal expression of pathogenic genes in AD.[171,172] Several lncRNAs are dysregulated in AD (*Sox2OT*, *1810014B01Rik*, *BC200*,

Table 5.3 Selected genomic, epigenomic, and proteomic markers in neurodegenerative disorders.

Disease	Pathogenic gene	Locus	Promoter length (bp)	3′UTR length (bp)	Defective protein	Methylation (promoter)	Chromatin/ histone modifications	Non-coding RNAs
Alzheimer's disease (AD)	*APOE Apolipoprotein E*	19q13.32	996		APOE Apolipoprotein E	Hypomethylated	Reduced H3 acetylation	Linked to AD (miR-34a, miR-34b/c, miR-107, miR-124, miR-125b, miR-137)
	APP Amyloid beta (A4) precursor protein	21q21.3	1086	1176	A4 Amyloid beta A4 protein	Hypomethylated	Decreased SIRT1	
	MAPT Microtubule-associated protein tau	17q21.31	1094		TAU Microtubule-associated protein tau	Hypomethylated	Increased HDAC6 and HDAC2 levels	
	PSEN1 Presenilin 1	14q24.2	929	1198	PSN1 Presenilin 1	Hypomethylated		Epigenetically regulated (let-7, miR-9, miR-132/212, miR146a, miR-148a, miR-184, miR-200, miR-200c/141)
Amyotrophic lateral sclerosis (ALS)	*ALS2 Amyotrophic lateral sclerosis 2 (juvenile)*	2q33.1	1069	1394	ALS2 Alsin	Hypermethylated	Histone methylation reduces *C9orf72* gene expression	miR-9, miR-23a-b, miR132, miR-134, miR-206, miR-338-3p, miR-455
	ATXN2 Ataxin 2	12q24.12	926	699	ATX2 Ataxin 2	Hypermethylated		
	C9orf72 Chromosome 9 open reading frame 72	9p21.2		1746	CI072 Protein C9orf72	Hypermethylated		
	FUS FUS RNA binding protein	16p11.2	1087	398	FUS RNA-binding protein FUS	Arginine-methylation	HDACs overexpression induces neurodegeneration	
	SOD1 Superoxide dismutase 1, soluble	21q22.11	988	497	SODC Superoxide dismutase (Cu–Zn)	Hypomethylated		
	UBQLN2 Ubiquilin 2	Xp11.21	1089	1361	UBQLN2 Ubiquilin-2	Increase of DNMT1, DNMT3a, 5-methylcytosine		

Dementia with Lewy bodies (DLB)	*PRKAR2A Protein kinase, cAMP-dependent, regulatory, type II, alpha*	3p21.31	924	999	KAP2 cAMP-dependent protein kinase type II-alpha regulatory subunit	Hypomethylated inhibition of DNMT1	Reduced histone acetylation	Linked to *SCNA* (miR-7, miR-153)
	SEPW1 Selenoprotein W, 1	19q13.33	744	452	SELW Selenoprotein W	Hypomethylated	Reduced histone methylation	
	SNCA Synuclein, alpha (non A4 component of amyloid precursor)	4q22.1	1097	1185	SYUA Alpha-synuclein	Hypomethylated		
Frontotemporal dementia (FTD)	*C9orf72 Chromosome 9 open reading frame 72*	9p21.2		1746	CI072 Protein C9orf72	Hypermethylated	Reduced histone acetylation	miR-516a-3p, miR-571, miR-548b-5p, and miR-548c-5p, miR-922
	GRN Granulin	17q21.31	1066	356	GRN Granulin	Hypermethylated		
	MAPT Microtubule-associated protein tau	17q21.31	1094		TAU Microtubule-associated protein tau		Increased HDAC levels	
	TARDBP TAR DNA binding protein	1p36.22	982	2903	TADBP TAR DNA-binding protein 43			
Multiple sclerosis (MS)	*PADI2 Pepetidyl arginine deaminase*	1p36.13	1036	2360	PAD2 Peptidyl arginine deaminase enzyme	Hypomethylated	Increased histone acetylation	miR-18b, miR-96, miR-142-3p, miR-145, miR-146a, miR-155, miR-599
	HLA-DRB1 Major histocompatibility complex, class II, DR beta 1	6p21.32	1078	397	2B1F HLA class I histocompatibility antigen, DRB1-15 beta chain	Extensive methylation alterations		
	TNFRSF1A Tumor necrosis factor receptor superfamily, member 1A	12p13.31	997	700	TNR1A Tumor necrosis factor receptor superfamily 1A	Increase of DNMT1, DNMT3a		

(continued)

Table 5.3 (*continued*)

Disease	Pathogenic gene	Locus	Promoter length (bp)	3′UTR length (bp)	Defective protein	Methylation (promoter)	Chromatin/ histone modifications	Non-coding RNAs
Parkinson's disease (PD)	*ABCA3 ATP-binding cassette, sub-family A (ABC1), member 3*	16p13.3	2201	981	ABCA3 ATP-binding cassette sub-family A member 3	Hypomethylated	Reduced H3 acetylation	Linked to *SCNA* (miR-7, miR-153)
	ATP8A2 ATPase, aminophospholipid transporter, class I, type 8A, member 2	13q12.13	1087	1380	AT8A2 Phospholipid-transporting ATPase IB	Hypomethylated	Toxin-mediated increase H3, H4 acetylation	Changed expression in PD (miR-10a/b, miR-34b-c, miR133b, miR212, miR495)
	APBA1 Amyloid beta (A4) precursor protein-binding, family A, member	9q21.12	994	1188	APBA1 Amyloid beta A4 precursor protein-binding family A member 1	Hypomethylated	Reduced H3 methylation by PNK1	
	CNTNAP2 Contactin associated protein-like 2	7q35-q36	1038		CNTP2 Contactin-associated protein-like 2	Hypomethylated		
	CPLX2 Complexin 2	5q35.2	1050	4180	CPLX2 Complexin 2	Hypomethylated		
	FAT1 FAT atypical cadherin 1	4q35.2	991	992	FAT1 Protocadherin fat 1	Hypomethylated		
	FHIT Fragile histidine triad	3p14.2	951	382	FHIT Bis(5-adenosyl)-triphosphatase	Hypomethylated		
	GSST1 Glutathione S-transferase theta 1	22q11.2	917	379	GSTT1 Glutathione *S*-transferase theta-1	Hypomethylated		

KCNH1 Potassium channel, voltage gated eag related subfamily H, member 1	1q32.2	920		KCNH1 Potassium voltage-gated channel subfamily H member 1	Hypomethylated
MAGI2 Membrane associated guanylate kinase, WW and PDZ domain containing 2	7q21.11	1041	2375	MAGI2 Membrane-associated guanylate kinase, WW and PDZ domain-containing protein 2	Hypomethylated
SMOC2 SPARC related modular calcium binding 2	6q27	1089	1791	SMOC2 SPARC-related modular calcium-binding protein 2	Hypomethylated
SLC12A5 Solute carrier family 12 (potassium/chloride transporter), member 5	20q13.12	958	2599	S12A5 Solute carrier family 12 member 5	Hypomethylated
SNCA Synuclein, alpha (non A4 component of amyloid precursor)	4q22.1	1097	1185	SYUA Alpha-synuclein	Hypomethylated
TUBA3E Tubulin, alpha 3e	2q21.1	700		TBA3E Tubulin alpha-3E chain	Hypomethylated

BACE1-AS, *NAT-Rad18*, *17A*, *GDNFOS*).[172] Examples of miRNAs directly linked to AD pathogenesis include miR-34a (1p36.22), miR-34b/c (11q23.1), miR-107 (10q23.31), miR-124 (8p23.1/8p12.3/20q13.33), miR-125b (11q24.1/21q21.1), and miR-137 (1p21.3); and examples of epigenetically regulated miRNAs with targets linked to AD pathogenesis are let-7b (22q13.1), miR-9 (1q22/5q14.3/15q26.1), miR-132/212 (17p13.3), miR-146a (5q34), miR-148a (7p15.2), miR-184 (15q25.1), and miR-200 (miR-200b/200a/429, 1p36.33; miR-200c/141, 12p13.31).[171] AD-related SNPs interfere with miRNA gene regulation and affect AD susceptibility. The significant interactions include target SNPs present in seven genes related to AD prognosis with the miRNAs- miR-214, -23a & -23b, -486-3p, -30e*, -143, -128, -27a & -27b, -324-5p and -422a. The dysregulated miRNA network contributes to the aberrant gene expression in AD.[173–175]

5.5 Pharmacoepigenomics

Pharmacogenetics alone does not predict all phenotypic variation in drug response.[12,13] The genes involved in the pharmacogenomic network are under the regulatory control of the epigenetic machinery (DNA methylation, histone modifications, miRNA regulation), this configuring the novel pharmacoepigenomic apparatus.[12,13]

Epigenetic regulation is also responsible for the tissue-specific expression of genes involved in pharmacogenetic processes, and epigenetics plays a key role in the development of drug efficacy, safety and resistance. Epigenetic changes affect CYP expression, major transporter function, and nuclear receptor interactions.[176–179] Variable methylation patterns have been detected in genes encoding phase I–III enzymes (Table 5.4). Although this is a still poorly explored field, epigenetic regulation of genes encoding drug-metabolizing enzymes (*CYP1A1*, *1A2*, *1B1*, *1A6*, *2A13*, *2B6*, *2C8*, *2C9*, *2C18*, *2C19*, *2D6*, *2E1*, *2J2*, *2F1*, *2R1*, *2S1*, *2W1*, *3A4*, *3A5*, *3A7*, *3A43*, *UGT1*, *GSTP1*), drug transporters (*ABCB1/MDR1/P-gp*, *ABCC1/MRP1*, *ABCC11/MRP8*, *ABCG2/BCRP*, *SLC19A1*, *SLC22A8*), and nuclear receptors (*RARB2*, *ESR1*, *NR1I2*, *HNF41*) has been documented in pioneering studies of pharmacoepigenetics.[12,13,176–179]

Epigenetic modifications are also associated with drug resistance.[12,13,178,180] The acquisition of drug resistance is tightly regulated by post-transcriptional regulators such as RNA-binding proteins (RBPs) and miRNAs, which change the stability and translation of mRNA-encoding factors involved in cell survival, proliferation, epithelial–mesenchymal transition, and drug metabolism.[178] In the complex cascade of pharmacoepigenetic events, the epigenetic factory may act as a promiscuous, redundant security system in which several miRNAs target genes encoding epigenetic regulators. For example, miR-29, -29c, -370, and -450A target DNMT3A, and miR-29, -148, and -29b target DNMT3B, inducing hypomethylation and expression of tumor suppressor genes; let-7a, miR-26a, -101, -138, and -124 target EZH2, decreasing histone methylation and increasing expression of tumor suppressor genes; miR-449 and -874 target *HDAC1*, inducing growth arrest by decreasing histone

Table 5.4 Methylation patterns in genes associated with Phase I–III drugs.[a]

Category	Gene	Locus	Promoter length (bp)	Pathology	Methylation
Phase I drug metabolism genes	*ALDH1A2*	15q21.3	982	Prostate cancer	Hypermethylated
	CYP1A1	15q24.1	1200	Head and neck cancer	Hypermethylated
				Prostate cancer	Hypermethylated
				Fetal growth restriction (toxics)	Hypomethylated
				Smoking-related	Hypomethylated
	CYP1B1	2p22.2	1193	Colorectal cancer	Hypermethylated
				Prostate cancer	Hypomethylated
				Hepatoma cell lines	Hypermethylated
				Breast cancer	Hypermethylated
	CYP24A1	20q13	945	Vitamin D deficiency	Hypermethylated
				Tumor-derived endothelial cells	Hypermethylated
	CYP27B1	12q14.1	917	Breast cancer	Hypermethylated
				Choriocarcinoma	Hypermethylated
				Lymphoma and leukemia	Hypermethylated
	CYP2A13	19q13.2	928	Head and neck cancer	Hypermethylated
	CYP2C19	10q24	1048	Drug resistance	Hypermethylated
	CYP2E1	10q26.3	918	Parkinson's disease	Hypomethylated
				Toluene exposure	Hypomethylated
	CYP2R1	11p15.2	1026	Vitamin D deficiency	Hypermethylated
	CYP2W1	7p22.3	934	Colorectal cancer	Hypomethylated
				Bladder, breast, thyroid cancer	Hypomethylated
				Liver, stomach cancer	Hypomethylated
	CYP7B1	8q21.3	1052	Prostate cancer	Hypomethylated
Phase II drug metabolism genes	*GSTM1*	1p13.3	900	Head and neck cancer	Hypermethylated
	GSTP1	11q13	958	Toluene exposure	Hypomethylated
				Hepatoma cells	Hypermethylated
				Prostate cancer	Hypermethylated
				Breast cancer	Hypomethylated
	NAT1	8p22	2132	Breast cancer	Hypomethylated
	SULT1A1	16p12.1	1086	Breast cancer	Hypermethylated
	UGT3A2	5p13.2	1076	Hepatoma cells	Hypermethylated

(continued)

Table 5.4 (*continued*)

Category	Gene	Locus	Promoter length (bp)	Pathology	Methylation
Phase III transporter genes	*ABCA7*	19p13.3	967	Alzheimer's disease	Hypomethylated
	ABCB1	7q21.12	906	Breast cancer	Hypermethylated
				Resistance to chemotherapy	Hypomethylated
	ABCC6	16p13.1	975	Bladder cancer	Hypermethylated
	ABCG2	4q22	1199	T-cell acute lymphoblastic leukemia cell lines	Hypomethylated
	SLC19A1	21q22.3	1040	CNS lymphomas	Hypomethylated
	SLC22A3	6q25.3	1034	Prostate cancer	Hypermethylated
	SLC24A4	14q32.12	1029	Alzheimer's disease	Hypomethylated

[a]Phase I: *ALDH1A2*: aldehyde dehydrogenase 1 family member A2; *CYP1A1*: cytochrome P450 family 1 subfamily A member 1; *CYP1B1*: cytochrome P450 family 1 subfamily B member 1; *CYP24A1*: cytochrome P450 family 24 subfamily A member 1; *CYP27B1*: cytochrome P450 family 27 subfamily B member 1; *CYP2A13*: cytochrome P450 family 2 subfamily A member 13; *CYP2C19*: cytochrome P450 family 2 subfamily C member 19; *CYP2E1*: cytochrome P450 family 2 subfamily E member 1; *CYP2R1*: cytochrome P450 family 2 subfamily R member 1; *CYP2W1*: cytochrome P450 family 2 subfamily W member 1; *CYP7B1*: cytochrome P450 family 7 subfamily B member 1. Phase II: *GSTM1*: glutathione *S*-transferase mu 1; *GSTP1*: glutathione *S*-transferase pi 1; *NAT1*: *N*-acetyltransferase 1 (arylamine *N*-acetyltransferase); *SULT1A1*: sulfotransferase family 1A member 1; *UGT3A2*: UDP glycosyltransferase 3 family, polypeptide A2. Phase III: *ABCA7*: ATP binding cassette subfamily A member 7; *ABCB1*: ATP binding cassette subfamily B member 1; *ABCC6*: ATP binding cassette subfamily C member 6; *ABCG2*: ATP binding cassette subfamily G member 2 (Junior blood group); *SLC19A1*: solute carrier family 19 (folate transporter), member 1; *SLC22A3*: solute carrier family 22 (organic cation transporter), member 3; *SLC24A4*: solute carrier family 24 (sodium/potassium/calcium exchanger), member 4.

acetylation; miR-1 and -155 target *HDAC4*, promoting myogenesis and impairing transcriptional activity of B-cell lymphoma 6 (*BCL6*); miR-627 and -155 target JMJD1A, decreasing histone demethylation and hypoxic gene expression; miR-132 and -483-5p target *MECP2*, promoting demethylation and cell differentiation.[181] Furthermore, epigenetic drugs reverse epigenetic changes in gene expression and might open new avenues in AD therapeutics. So far, epigenetic drugs (Table 5.5) have only been approved for the treatment of neoplastic processes; most of them are not devoid of severe side effects; and concerns on their capacity to cross the blood–brain barrier and penetrate into the brain may preclude their implantation as potential drug candidates in NDDs.[15,16,170,182]

5.6 Novel Strategies

Patients with NDDs need multifactorial treatments with different drugs of diverse pharmacological profiles. AD patients may take 6–12 different drugs/day for the treatment of dementia-related symptoms, including memory deterioration (conventional anti-dementia drugs, neuroprotectants) (Table 5.1), behavioral changes (antidepressants, neuroleptics, sedatives, hypnotics), and functional decline, or for the treatment of concomitant pathologies (epilepsy, cardiovascular and cerebrovascular disorders, Parkinsonism, hypertension, dyslipidemia, anemia, arthrosis, *etc.*). Over 20% of dementia patients are current users of cardiovascular drugs. A high-throughput screening study assessed 1600 FDA-approved drugs for their ability to modulate Aβ activity; 559 of the 1600 drugs had no effect on APP processing or were toxic to neurons at the concentration tested, while 800 drugs could reduce Aβ content by over 10% in primary neurons derived from Tg2576 mice, among which, 184 drugs were able to reduce Aβ content by more than 30%; 241 drugs could potentially promote Aβ accumulation, including 26 drugs that could increase the level of Aβ by over 30%.[183] The co-administration of several drugs may cause side-effects and adverse drug reactions in over 60% of AD patients, who in 2–10% of the cases require hospitalization. The prevalence of potentially inappropriate medication (PIM) is around 50% in some European cohorts. Cerebral vasodilators are the most widely used class of PIM, accounting for 24.0% of all prescriptions, followed by atropinic drugs and long half-life benzodiazepines. Atropinic drugs were associated with cholinesterase inhibitors in 16% of patients. In over 20% of the patients, behavioral deterioration and psychomotor function can be severely altered by polypharmacy.[184] The principal causes of these iatrogenic effects are the inappropriate combination of drugs, and the genomic background of the patient, responsible for his/her pharmacogenomic outcome.

During the 2002–2012 period, 413 AD trials were performed (124 Phase 1 trials, 206 Phase 2 trials, and 83 Phase 3 trials) (78% sponsored by pharmaceutical companies). Registered trials addressed symptomatic agents (36.6%), disease-modifying small molecules (35.1%) and disease-modifying immunotherapies (18%), with a very high attrition rate (overall success

Table 5.5 Pharmacological profile and pharmacogenetics of selected epigenetic drugs.[a]

Drug	Properties	Pharmacogenetics
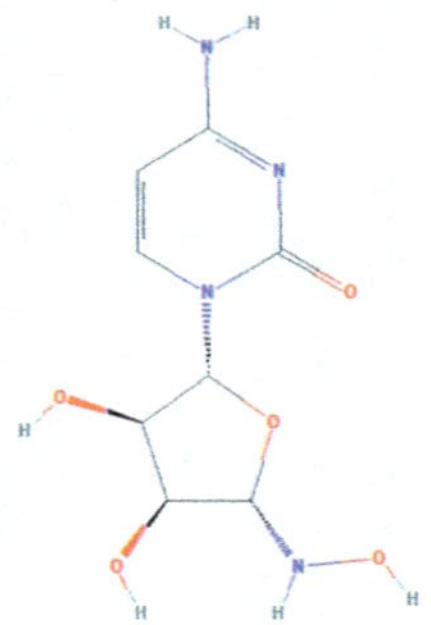	**Name: 5-Azacytidine**, Azacitidine, Azacytidine, Ladakamycin, Vidaza, Mylosar, Azacitidinum, 5-AZAC **IUPAC Name:** 4-Amino-1-[(2R,3R,4S,5R)-3,4-dihydroxy-5-(hydroxymethyl)oxolan-2-yl]-1,3,5-triazin-2-one **Molecular Formula:** $C_8H_{12}N_4O_5$ **Molecular Weight:** 244.20468 **Category:** Pyrimidine nucleoside cytidine analog **Mechanism:** DNA methyltransferase inhibitor, Telomerase inhibitor **-Target:** DNA (cytosine-5)-methyltransferase 1 (DNMT1) **-Interactions:** Cytidine deaminase **Effect:** Antineoplastic, Antimetabolite. Methylates CpG residues. Methylates hemimethylated DNA. Mediates transcriptional repression by direct binding to HDAC2	**Pathogenic genes:** *ALDH3A1, CDKN2A, MGMT, PLA2R1, RRM1, TNFRSF1B* **Mechanistic genes:** *ALDH1A1, DAPK1, DNMT1, DPYD, CDKN2A, MGMT, PLCB1* **Metabolic genes:** **Substrate:** *CDA, DCK, SLC28A1, SLC29A1, RRM1, RRM2, UCK1, UCK2* **Inhibitor:** *CYP1A2 (weak), CYP2E1 (weak), DNMT1* **Inducer:** *SULT1C2* **Transporter genes:** *SLC5A5, SLC28A1, SLC29A1* **Pleiotropic genes:** *BLK*
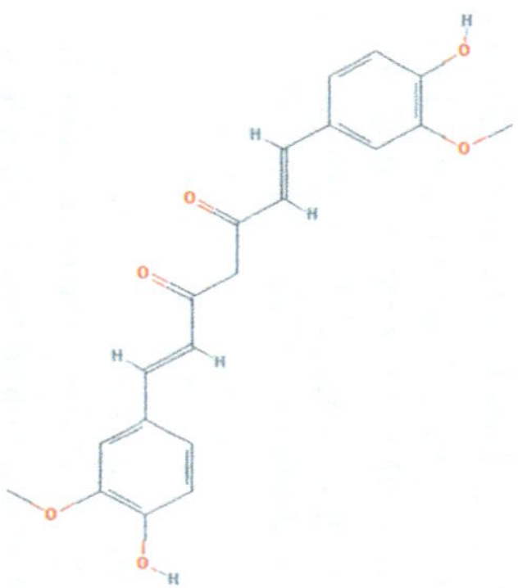	**Name: Curcumin**, Diferuloylmethane, Natural yellow 3, Turmeric yellow, Turmeric, Kacha haldi, Gelbwurz, Curcuma, Haldar, Souchet **IUPAC Name:** (1E,6E)-1,7-bis(4-hydroxy-3-methoxyphenyl)hepta-1,6-diene-3,5-dione **Molecular Formula:** $C_{21}H_{20}O_6$ **Molecular Weight:** 368.3799 **Category:** Natural product (*Curcuma longa*) **Mechanism:** Histone acetyltransferase (HAT) inhibitor **Effect:** Non-steroidal anti-inflammatory agent; Antineoplastic; Antioxidant; Cognitive enhancer; Coloring agent; Enzyme inhibitor	**Pathogenic genes:** *BACE1, CCND1, CDH1, GSK3B, IL1A, IL6, JUN, MSR1, PSEN1, PTGS2, SNCA, SREBF1, TNF* **Mechanistic genes:** *AKT1, PRKAs, BACE1, CCND1, CDH1, CDKs, CRM1, CTNNB1, EGF, GSK3B, HDACs, HIF1A, IL1A, IL6, JUN, MMPs, MSR1, NFKB1, NOS2, PDGFRs, PSEN1, PTGS2, SNCA, SOCS1, SOCS3, SREBF1, STAT3, TNF, VEGFA* **Metabolic genes:** **Inhibitor:** *CYP2C8, CYP2C9, EP300* **Inducer:** *CYP2C8, CYP2C9, CYP2D6, CYP3A4* **Transporter genes:** *ABCA1, SNCA* **Pleiotropic genes:** *CTNNB1, MSR1*

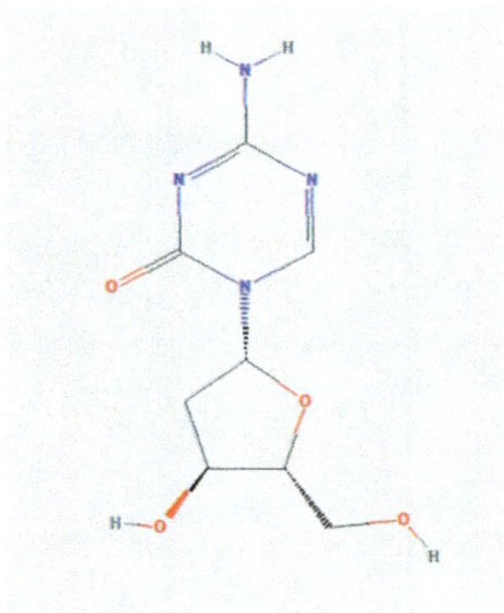	**Name:** Decitabine, 5-Aza-2'-deoxycytidine, Dacogen, Dezocitidine, 2'-Deoxy-5-azacytidine **IUPAC Name:** 4-Amino-1-[(2R,4S,5R)-4-hydroxy-5-(hydroxymethyl)oxolan-2-yl]-1,3, 5-triazin-2-one **Molecular Formula:** $C_8H_{12}N_4O_4$ **Molecular Weight:** 228.20528 **Category:** Nucleoside **Mechanism:** DNA methyltransferase inhibitor -**Target:** DNA (cytosine-5)-methyltransferase 1 (DNMT1) -**Interactions:** Deoxycytidine kinase **Effect:** Antineoplastic, Antimetabolite, Enzyme inhibitor, Teratogen	**Pathogenic genes:** *BRCA1, CDKN2B, DNMT3A, EGFR, FOS, MGMT, MLH1, MMP9, MYC, NOS3, NQO1, TP53, VHL* **Mechanistic genes:** *APAF1, BRCA1, CDKN2B, EGFR, ICAM1, MAGED1, MGMT, MLH1, MMP2, MMP9, MYC, NOS3, TIMP3, TP53, VHL, ZNF350.* **Metabolic genes:** **Substrate***: DCK, DNMT1, CDA, SLC29A1* **Inhibitor***: DNMT1, DNMT3B* **Inducer:** *DPYD* **Transporter genes:** *ABCs, SLC15s, SLC22s, SLC28A1, SLC29As* **Pleiotropic genes:** *HBG1, NQO1, NTRK2, MMP2, MSH2*
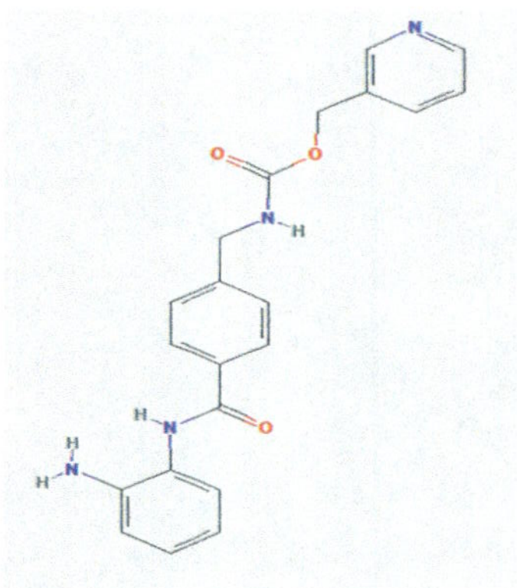	**Name: Entinostat**, ms-275, 209783-80-2, SNDX-275, MS 275, MS-27-275, SNDX 275, Histone Deacetylase Inhibitor I, S1053_Selleck, MS 27-275 **IUPAC Name:** Pyridin-3-ylmethyl N-[[4-[(2-aminophenyl)carbamoyl]phenyl]methyl]carbamate **Molecular Formula:** $C_{21}H_{20}N_4O_3$ **Molecular Weight:** 376.4085 **Category:** Benzamide **Mechanism:** Class I HDAC inhibitor (HDAC1, 2, 3) **Effect:** Antineoplastic agent; Histone deacetylase inhibitor; Memory enhancer	**Pathogenic genes:** *CDH1* **Mechanistic genes:** *CDH1, HDAC1, HDAC2, HDAC3, KLRK1* **Metabolic genes:** **Inhibitor:** *HDAC1, HDAC2, HDAC3* **Inducer:** *CYP19A1*

(*continued*)

Table 5.5 (*continued*)

Drug	Properties	Pharmacogenetics
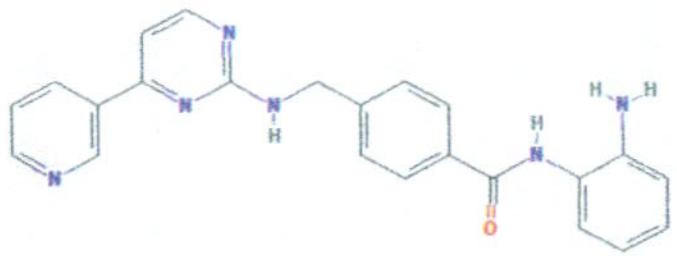	**Name:** Mocetinostat, MGCD0103, 726169-73-9, MGCD-0103, MGCD 0103, N-(2-Aminophenyl)-4-([[4-(pyridin-3-yl)pyrimidin-2 yl]amino]methyl)benzamide **IUPAC Name:** N-(2-Aminophenyl)-4-[[(4-pyridin-3-ylpyrimidin-2-yl)amino]methyl] benzamide **Molecular Formula:** $C_{23}H_{20}N_6O$ **Molecular Weight:** 396.4445 **Category:** Benzamide **Mechanism:** Class I HDAC inhibitor (HDAC1, 2, 3); Class IV HDAC inhibitor (HDAC11) **Effect:** Antineoplastic agent; Histone deacetylase inhibitor	**Pathogenic genes:** *CDKN1A, CDKN2B, TNF* **Mechanistic genes:** *CDKN1A, CDKN2B, HDAC1, HDAC2, HDAC3, HDAC11, NFKB2, TNF* **Metabolic genes:** **Inhibitor:** *HDAC1, HDAC2, HDAC3,HDAC11*
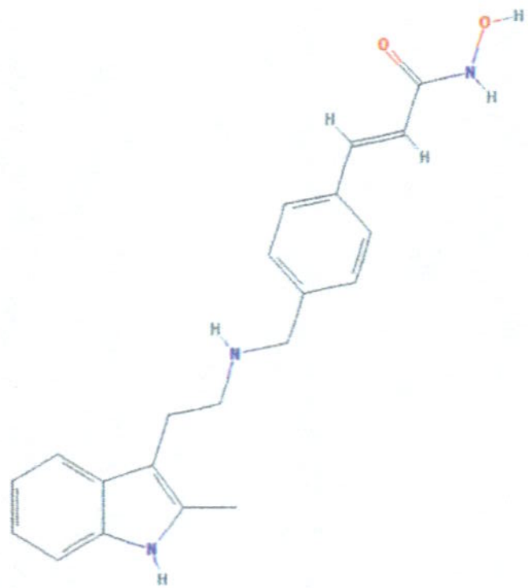	**Name: Panobinostat**, LBH-589, 404950-80-7, LBH589, Faridak, NVP-LBH589, LBH 589, S1030_Selleck, AC1OCFY8, Panobinostat (LBH589) **IUPAC Name:** (E)-N-hydroxy-3-[4-[[2-(2-methyl-1H-indol-3-yl)ethylamino]methyl]phenyl] prop-2-enamide **Molecular Formula:** $C_{21}H_{23}N_3O_2$ **Molecular Weight:** 349.42622 **Category:** Hydroxamic acid **Mechanism:** Class I HDAC inhibitor (HDAC1, 2, 3, 8); Class IIa HDAC inhibitor (HDAC4, 5, 7, 9); Class IIb HDAC inhibitor (HDAC6, 10); Class IV HDAC inhibitor (HDAC11); Pan-histone deacetylase inhibitor **Effect:** Antineoplastic agent; Histone deacetylase inhibitor	**Pathogenic genes:** *CDKN1A, EGFR, IL6, RASSF1* **Mechanistic genes:** *AKT1, CDKN1A, DAPK1, DNMT1, EGFR, HDACs, HIST3H3, HIST4H4, HSP90As, IL6, IL10, IL12, IL23A, NFKB2, RASSF1, TLR3* **Metabolic genes:** **Substrate:** *CYP2C19, CYP2D6, CYP3A4* **Inhibitor:** *AKT1, CYP19A1 (strong), HDACs* **Pleiotropic genes:** *IL10*
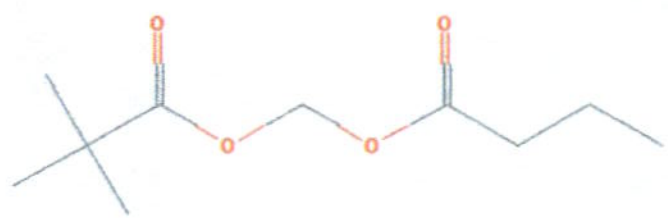	**Name: Pivanex**, AN-9, Pivalyloxymethyl butyrate, AN 9, 122110-53-6, BRN 4861411, ((2,2 Dimethylpropanoyl)oxy)methyl butanoate **IUPAC Name:** Butanoyloxymethyl 2,2-dimethylpropanoate **Molecular Formula:** $C_{10}H_{18}O_4$ **Molecular Weight:** 202.24752 **Category:** Short-chain fatty acid **Mechanism:** Class I HDAC inhibitor (HDAC1, 2, 3, 8) **Effect:** Antineoplastic agent; Histone deacetylase inhibitor	**Pathogenic genes:** *BCL2, TP53* **Mechanistic genes:** *BAX, BCL2, BCR-ABL, HDACs, TP53* **Metabolic genes:** **Inhibitor:** *ABCB1, HDACs* **Transporter genes:** *ABCB1*

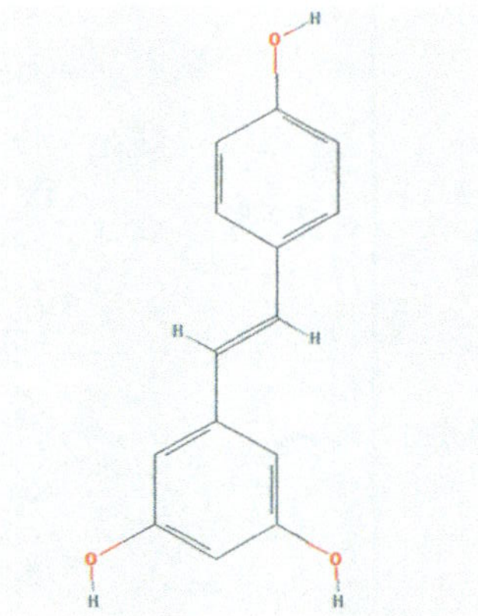	**Name: Resveratrol**, trans-resveratrol, 501-36-0, 3,4',5-Trihydroxystilbene, 3,4',5-Stilbenetriol, 3,5,4'-Trihydroxystilbene, Resvida, (E)-resveratrol **IUPAC Name:** 5-[(E)-2-(4-Hydroxyphenyl)ethenyl] benzene-1,3-diol **Molecular Formula:** $C_{14}H_{12}O_3$ **Molecular Weight:** 228.24328 **Category:** Natural polyphenol **Mechanism:** SIRT1 inducer/activator **Effect:** Non-steroidal antiinflammatory agent; Anti-carcinogenic; Antimutagenic; Antineoplastic; Antioxidant; Platelet aggregation inhibitor; Enzyme inhibitor; Lifespan extension; Memory improvement; Aβ decrease; Reduction of plaque formation	**Pathogenic genes:** *BCL2, CAV1, ESR1, ESR2, GRIN2B, NOS3, PTGS2, TNFRSF10A, TNFRSF10B* **Mechanistic genes:** *APP, ATF3, BAX, BAK1, BBC3, BCL2, BCL2L1, BCL2L11, BIRC5, CASP3, CAV1, CFTR, ESR1, ESR2, GRIN1, GRIN2B, HTR3A, NFKB1, NOS3, PMAIP1, PTGS1, PTGS2, SIRT1, SIRT3, SIRT5, SRC, TNFRSF10A, TNFRSF10B, TRPs* **Metabolic genes:** **Substrate:** *CYP1A1, CYP1A2, CYP1B1, CYP2E1, GSTP1, PTGS1, PTGS2* **Inhibitor:** *CYP1A1, CYP1B1, CYP2C9, CYP2D6, CYP3A4, NQO2* **Inducer:** *CYP1A2, SIRT1* **Transporter genes:** *ABCC1, ABCC2, ABCC3, ABCC4, ABCC8, ABCG1, ABCG2, CFTR, TRPs*
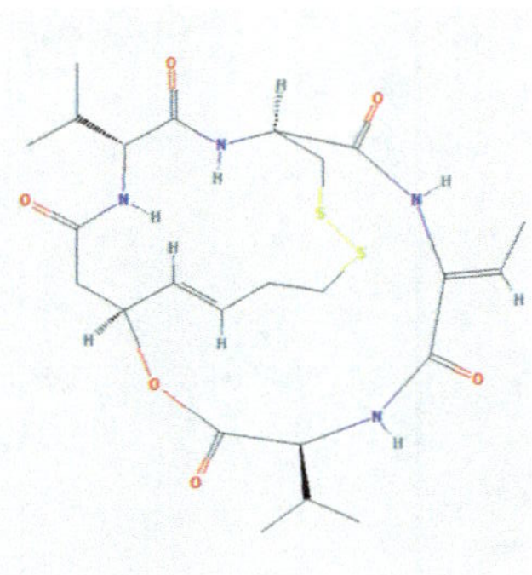	**Name: Romidepsin**, Depsipeptide, Chromadax, Istodax, Antibiotic FR 901228, FK228, FR 901228, FK-228, NSC 630176, NSC-630176 **IUPAC Name:** (1S,4S,7Z,10S,16E,21R)-7-ethylidene-4,21-di(propan-2-yl)-2-oxa-12, 13-dithia-5,8,20,23-tetrazabicyclo[8.7.6]tricos-16-ene-3,6,9,19, 22-pentone **Molecular Formula:** $C_{24}H_{36}N_4O_6S_2$ **Molecular Weight:** 540.69584 **Category:** Cyclic peptide **Mechanism:** Class I HDAC inhibitor (HDAC1, 2, 3, 8); Class IIa HDAC inhibitor (HDAC4,5,7,9); Class IIb HDAC inhibitor (HDAC6, 10); Class IV HDAC inhibitor (HDAC11) **Effect:** Antibiotic; Antineoplastic agent; Histone deacetylase inhibitor	**Pathogenic genes:** *BCL2, CCDN1, CDKN1A, MYC, NF2, RB1, ROS1, TNFSF10, VHL* **Mechanistic genes:** *BCL2, CCDN1, CDKN1A, FLT1, HDAC1, HDAC2, HDAC3, HDAC4, HSP90As, KDR, MYC, NF2, TNFSF10, VEGFs, VHL* **Metabolic genes:** **Substrate:** *ABCB1, ABCG2, CYP1A1 (minor), CYP2B6 (minor), CYP2C19 (minor), CYP3A4 (major), CYP3A5 (minor), NR1I3, SLCO1B3* **Inhibitor:** *ABCB1, HDACs* **Inducer:** *ABCG2* **Transporter genes:** *ABCB1, ABCC1, ABCG2, SLCO1B3* **Pleiotropic genes:** *CDH1, CDKN1A*

(*continued*)

Table 5.5 (*continued*)

Drug	Properties	Pharmacogenetics
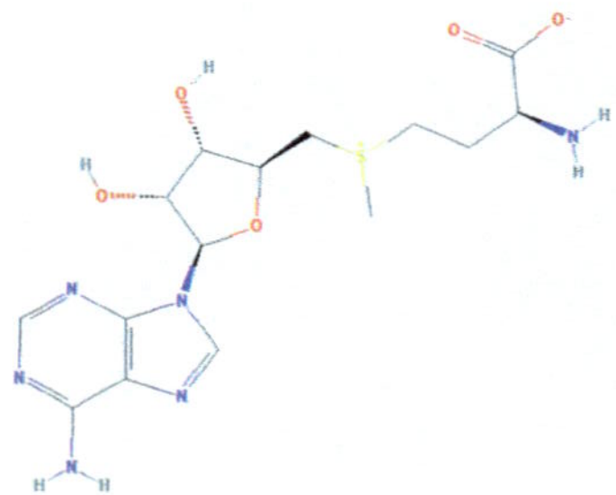	**Name: S-Adenosylmethionine**, Ademetionine, AdoMet, Donamet, S-adenosyl-L-methionine, SAMe, Methioninyladenylate, SAM-e, adenosylmethionine **IUPAC Name:** (2S)-2-Amino-4-[[(2S,3S,4R,5R)-5-(6-aminopurin-9-yl)-3,4-dihydroxyoxolan-2-yl]methyl-methylsulfonio]butanoate **Molecular Formula:** $C_{15}H_{22}N_6O_5S$ **Molecular Weight:** 398.43738 **Category:** Methyl radical donor **Mechanism:** Histone methyltransferase inhibitor **Effect:** Antineoplastic; Antiinflammatory; Memory enhancer;PSEN1 repressor	**Pathogenic genes:** *AKT1, ERK, GNMT, MAT1A, PSEN1* **Mechanistic genes:** *AMD1, CAT, CBS, GCLC, GNMT, GSS, NOS2, ROS1, STAT1, TNF* **Metabolic genes:** **Substrate:** *COMT, GNMT, TPMT, SRM* **Inhibitor:** *ABCB1, CYP2E1, NOS2* **Transporter genes:** *SLC25A26* **Pleiotropic genes:** *CAT, TNF*
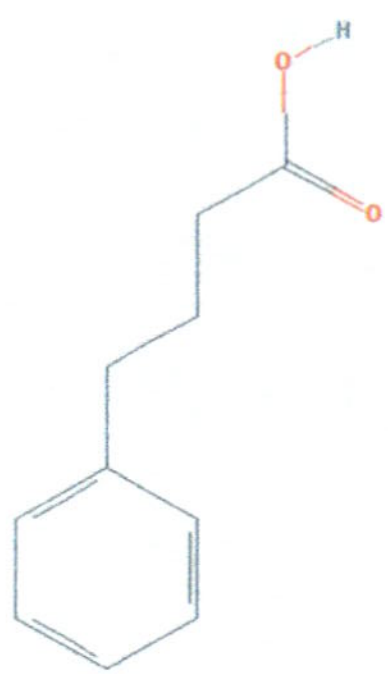	**Name: Sodium phenylbutyrate**, Buphenyl, 4-Phenylbutiric acid, 4-Phenylbutanoic acid, Benzenebutanoic acid, Benzenebutyric acid, Butyric acid, 4-phenyl-, 1821-12-1, gamma-Phenylbutyric acid, **IUPAC Name:** 4-Phenylbutanoic acid **Molecular Formula:** $C_{10}H_{12}O_2$ **Molecular Weight:** 164.20108 **Category:** Short-chain fatty acid **Mechanism:** Class I HDAC inhibitor (HDAC1, 2, 3, 8); Class IIa inhibitor (HDAC4,5,7,9); Class IIb inhibitor (HDAC6,10) **Effect:** Antineoplastic agent; Histone deacetylase inhibitor; Memory improvement; pTau decrease via GSK3β inactivation; C99 and Aβ decrease; Amyloid burden reduction	**Pathogenic genes:** *ARG1, ASS1, BCL2, CPS1, NAGS, OTC* **Mechanistic genes:** *BCL2, BDNF, EDN1, HDACs, HSPA8, ICAM1, NFKB2, NT3, VCAM1* **Metabolic genes:** **Inhibitor:** *HDACs* **Inducer:** *ARG1, CFTR, CYP2B6, NFKB2* **Transporter genes:** *CFTR* **Pleiotropic genes:** *ASL, BDNF, VCAM1*

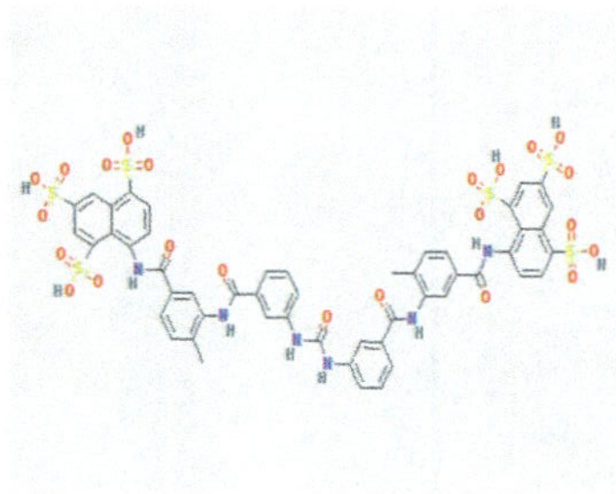	**Name: Suramin**, Naphuride, Germanin, Naganol, Belganyl, Fourneau, Farma, Antrypol, Suramine, Naganin **IUPAC Name:** 8-[[4-methyl-3-[[3-[[3-[[2-methyl-5-[(4,6, 8-trisulfonaphthalen-1-yl)carbamoyl]phenyl]carbamoyl]phenyl] carbamoylamino]benzoyl]amino]benzoyl]amino]naphthalene-1,3,5-trisulfonic acid **Molecular Formula:** $C_{51}H_{40}N_6O_{23}S_6$ **Molecular Weight:** 1297.2797 **Category:** Polyanionic compound **Mechanism:** Class III HDAC/Sirtuin inhibitor (SIRT1-3) **Effect:** Antineoplastic Agent; Trypanocidal Agent; Antiparasitic; Antinematodal (African trypanosomiasis, Onchocerca); Sirtuin inhibitor	**Mechanistic genes:** *FSHR, IL10, P2RY2, PDGFRB, RYR1, SIRT1,SIRT2, SIRT3, SIRT5* **Metabolic genes:** **Inhibitor:** *SIRT1, SIRT2, SIRT3*
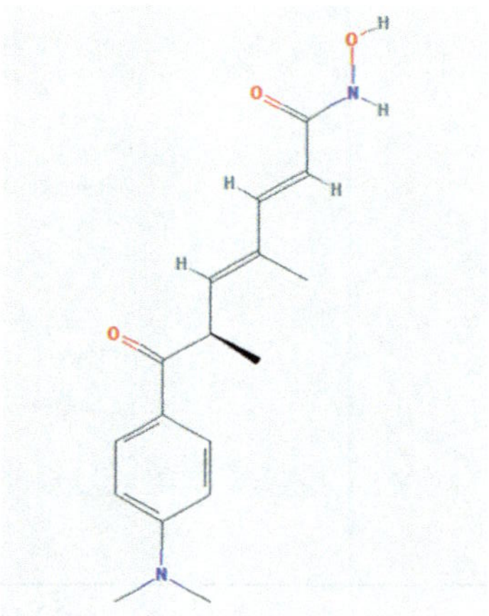	**Name: Trichostatin A,** 58880-19-6, TSA, Trichostatin A (TSA), CHEBI:46024, TSA; 2,4-Heptadienamide, 7-(4-(dimethylamino)phenyl)-N-hydroxy-4,6-dimethyl-7-oxo-7-(4-(Dimethylamino)phenyl)-N-hydroxy-4,6-dimethyl-7-oxo-2,4-heptadienamide; [R-(E,E)]-7-[4-(Dimethylamino)phenyl]-N-hydroxy-4,6-dimethyl-7-oxo-2,4-heptadienamide **IUPAC Name:** (2E,4E,6R)-7-[4-(dimethylamino) phenyl]-N-hydroxy-4,6-dimethyl-7-oxohepta-2,4-dienamide **Molecular Formula:** $C_{17}H_{22}N_2O_3$ **Molecular Weight:** 302.36818 **Category:** Hydroxamic acid **Mechanism:** Class I HDAC inhibitor (HDAC1, 2, 3); Class IIa HDAC inhibitor (HDAC4, 7, 9); Class IIb inhibitor (HDAC6) **Effect:** Antifungal agent; Antibacterial agent; Histone deacetylase inhibitor; Protein synthesis inhibitor; Antineoplastic; Memory improvement; Rescue of CA3-CA1 LTP in APP/PS1 transgenic models	**Pathogenic genes:** *BCL2* **Mechanistic genes:** *BCL2, HDACs, IL8, IL12A,IL12B, NFKB2, RARB* **Metabolic genes:** **Substrate:** *CYP3A4 (mayor)* **Inhibitor:** *HDACs* **Inducer:** *CYP1A1, CYP1B1, CYP2B6, CYP2E1, CYP7A1, SLC19A3* **Transporter genes:** *SLC19A3*

(continued)

Table 5.5 (*continued*)

Drug	Properties	Pharmacogenetics
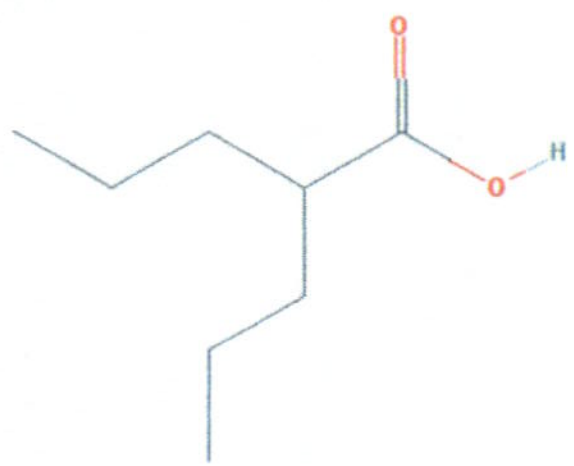	**Name: Valproic Acid**, 2-Propylpentanoic acid, Depakene, Depakine, Ergenyl, Dipropylacetic acid, Mylproin, Convulex, Myproic Acid **IUPAC Name:** 2-Propylpentanoic acid **Molecular Formula:** $C_8H_{16}O_2$ **Molecular Weight:** 144.21144 **Category:** Short-chain fatty acid **Mechanism:** Class I HDAC inhibitor (HDAC1, 2, 3, 8) **Effect:** Anticonvulsant; Mood stabilizer; Antimanic agent; Enzyme inhibitor; Histone deacetylase inhibitor; GABA modulator; Memory improvement; Aβ and pTau decrease; CDK5 inactivation	**Pathogenic genes:** *CREB1, IL6, LEP, SCN2A, TGFB1, TNF, TRNK* **Mechanistic genes:** *ABAT, CDK5, GSK3B, HDAC1, HDAC2, HDAC3, HDAC8, HDAC9, LEP, LEPR, SCNs, SMN2* **Metabolic genes:** **Substrate:** *ABCB1, CYP1A1 (minor), CYP2A6 (major), CYP2B6 (minor), CYP2C9 (major), CYP2C19 (minor), CYP2E1 (minor), CYP3A4 (minor), CYP4B1 (major), CYP4F2 (minor), UGT1A4, UGT1A6, UGT1A8, UGT1A9, UGT1A10, UGT2B7* **Inhibitor:** *ABCB1, ACADSB, AKR1A1, CYP2A6 (moderate), CYP2C9 (strong), CYP2C19 (moderate), CYP2D6 (weak), CYP3A4 (moderate), HDAC1, HDAC2, HDAC3, HDAC8, HDAC9, UGT1A9, UGT2B1, UGT2B7* **Inducer:** *ABCB1, AKR1C4, CASR, CYP2A6, CYP2B6, CYP3A4, CYP7A1, MAOA, NR1I2, SLC5A5, SLC6A2, SLC12A3, SLC22A16* **Transporter genes:** *ABCB1, ABCC2, ABCG1, ABCG2, SCNs, SLC5A5, SLC6A2, SLC12A3, SLC22A16* **Pleiotropic genes:** *ABL2, AGPAT2, ASL, ASS1, CDK4, CHRNA1, COL1A1, CPS1, CPT1A, DRD4, FMR1, FOS, HBB, HFE, HLA-A, HLA-B, ICAM1, IFNG, IL6, IL10, LEPR, NAGS, NR3C1, OTC, PTGES, STAT3, TGFB1, TNF, TP53.*

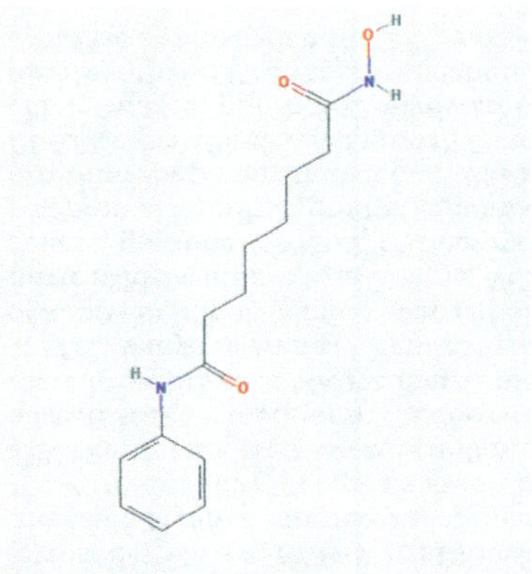	**Name: Vorinostat**, Suberoylanilide hydroxamic acid (SAHA), Zolinza, Suberanilohydroxamic acid, 149647-78-9, N-hydroxy-N'-phenyloctanediamide, SAHA cpd **IUPAC Name:** N'-Hydroxy-N-phenyloctanediamide **Molecular Formula:** $C_{14}H_{20}N_2O_3$ **Molecular Weight:** 264.3202 **Category:** Hydroxamic acid **Mechanism:** Class I HDAC inhibitor (HDAC1, 2, 3, 8) Class IIb inhibitor (HDAC6) **Effect:** Antineoplastic, Memory improvement	**Pathogenic genes:** *BIRC3, CCND1, CDKN1A, CFLAR, CYP19A1, ERBB2, ERBB3, EGFR, RB1, TP53, TNF* **Mechanistic genes:** *CDKN1A, EGFR, ERBB2, ERBB3, STATs, TYMS, VEGFs* **Metabolic genes:** **Substrate:** *CYP2A6 (minor), CYP2C9 (minor), CYP2C19 (major), CYP2D6 (minor), CYP3A4 (major)* **Inhibitor:** *HDAC1, HDAC2, HDAC3, HDAC6* **Inducer:** *CYP1A1, CYP1A2, CYP1B1* **Pleiotropic genes:** *ALPs, TNF, TYMS*

[a]***ABAT:*** 4-aminobutyrate aminotransferase; ***ABCA1:*** ATP-binding cassette, sub-family A (ABC1), member 1; ***ABCB1:*** ATP-binding cassette, sub-family B (MDR/TAP), member 1; ***ABCC1:*** ATP-binding cassette, sub-family C (CFTR/MRP), member 1; ***ABCC2:*** ATP-binding cassette, sub-family C (CFTR/MRP), member 2; ***ABCC3:*** ATP-binding cassette, sub-family C (CFTR/MRP), member 3; ***ABCC4:*** ATP-binding cassette, sub-family C (CFTR/MRP), member 4; ***ABCC8:*** ATP-binding cassette, sub-family C (CFTR/MRP), member 8; ***ABCG1:*** ATP-binding cassette, sub-family G (WHITE), member 1; ***ABCG2:*** ATP-binding cassette, sub-family G (WHITE), member 2 (Junior blood group); ***ABCs:*** ATP-binding cassette family; ***ABL2:*** ABL proto-oncogene 2, non-receptor tyrosine kinase; ***ACADSB:*** acyl-CoA dehydrogenase, short/branched chain; ***AGPAT2:*** 1-acylglycerol-3-phosphate O-acyltransferase 2; ***AKR1A1:*** aldo-keto reductase family 1, member A1 (aldehyde reductase); ***AKR1C4:*** aldo-keto reductase family 1, member C4; ***AKT1***: v-akt murine thymoma viral oncogene homolog 1; ***ALDH1A1:*** aldehyde dehydrogenase 1 family, member A1; ***ALDH3A1:*** aldehyde dehydrogenase 3 family, member A1; ***ALPs:*** alkaline phosphatases; ***AMD1:*** adenosylmethionine decarboxylase 1; ***APAF1:*** apoptotic peptidase activating factor 1; ***APP:*** amyloid beta (A4) precursor protein; ***ARG1:*** arginase 1; ***ASL:*** argininosuccinate lyase; ***ASS1:*** argininosuccinate synthase 1; ***ATF3:*** activating transcription factor 3; ***BACE1:*** beta-site APP-cleaving enzyme 1; ***BAK1:*** BCL2-antagonist/killer 1; ***BAX:*** BCL2-associated X protein; ***BBC3:*** BCL2 binding component 3; ***BCL2:*** B-cell CLL/lymphoma 2; ***BCL2L1:*** BCL2-like 1; ***BCL2L11***: BCL2-like 11 (apoptosis facilitator); ***BCR-ABL:*** *BCR-ABL* tyrosine kinase fusion; ***BDNF:*** brain-derived neurotrophic factor; ***BIRC3:*** baculoviral IAP repeat containing 3; ***BIRC5:*** baculoviral IAP repeat containing 5; ***BLK:*** BLK proto-oncogene, Src family tyrosine kinase; ***BRCA1:*** breast cancer 1, early onset; ***CASP3:*** caspase 3, apoptosis-related cysteine peptidase; ***CASR:*** calcium-sensing receptor; ***CAT:*** catalase; ***CAV1:*** caveolin 1, caveolae protein, 22kDa; ***CBS:*** cystathionine-beta-synthase; ***CCDN1:*** cyclin D1; ***CDA :*** cytidine deaminase; ***CDH1:*** cadherin 1, type 1; ***CDK4:*** cyclin-dependent kinase 4; ***CDK5:*** cyclin-dependent kinase 5; ***CDKN1A:*** cyclin-dependent kinase inhibitor 1A (p21, Cip1); ***CDKN2A:*** cyclin-dependent kinase inhibitor 2A; ***CDKN2B:*** cyclin-dependent kinase inhibitor 2B (p15, inhibits CDK4); ***CDKs***: cyclin-dependent kinases; ***CFLAR:*** CASP8 and FADD-like apoptosis regulator; ***CFTR:*** cystic fibrosis transmembrane conductance regulator (ATP-binding cassette sub-family C, member 7); ***CHRNA1:*** cholinergic receptor, nicotinic, alpha 1 (muscle); ***COL1A1:*** collagen, type I, alpha 1; ***COMT:*** catechol-O-methyltransferase; ***CPS1:*** carbamoyl-phosphate synthase 1, mitochondrial; ***CPT1A:*** carnitine palmitoyltransferase 1A (liver); ***CREB1:*** cAMP responsive element binding protein 1; ***CTNNB1:*** catenin (cadherin-associated protein), beta 1, 88kDa; ***CYP19A1:***

(*continued*)

Table 5.5 (*continued*)

cytochrome P450, family 19, subfamily A, polypeptide 1; ***CYP1A1:*** cytochrome P450, family 1, subfamily A, polypeptide 1; ***CYP1A2:*** cytochrome P450, family 1, subfamily A, polypeptide 2; ***CYP1B1:*** cytochrome P450, family 1, subfamily B, polypeptide 1; ***CYP2A6:*** cytochrome P450, family 2, subfamily A, polypeptide 6; ***CYP2C19:*** cytochrome P450, family 2, subfamily C, polypeptide 19; ***CYP2C8:*** cytochrome P450, family 2, subfamily C, polypeptide 8; ***CYP2C9:*** cytochrome P450, family 2, subfamily C, polypeptide 9; ***CYP2D6:*** cytochrome P450, family 2, subfamily D, polypeptide 6; ***CYP2E1:*** cytochrome P450, family 2, subfamily E, polypeptide 1; ***CYP3A4:*** cytochrome P450, family 3, subfamily A, polypeptide 4; ***CYP3A5:*** cytochrome P450, family 3, subfamily A, polypeptide 5; ***CYP4B1:*** cytochrome P450, family 4, subfamily B, polypeptide 1; ***CYP4F2:*** cytochrome P450, family 4, subfamily F, polypeptide 2; ***CYP7A1:*** cytochrome P450, family 7, subfamily A, polypeptide 1; ***DAPK1:*** death-associated protein kinase 1; ***DCK:*** deoxycytidine kinase; ***DNMT1:*** DNA (cytosine-5-)-methyltransferase 1; ***DNMT3A:*** DNA (cytosine-5-)-methyltransferase 3 alpha; ***DNMT3B:*** DNA (cytosine-5-)-methyltransferase 3 beta; ***DPYD:*** dihydropyrimidine dehydrogenase; ***DRD4:*** dopamine receptor D4; ***EDN1:*** endothelin 1; ***EGF:*** epidermal growth factor; ***EGFR:*** epidermal growth factor receptor; ***EP300:*** E1A binding protein p300; ***ERBB2:*** erb-b2 receptor tyrosine kinase 2; ***ERBB3:*** erb-b2 receptor tyrosine kinase 3; ***ERK:*** elk-related tyrosine kinase; ***ESR1:*** estrogen receptor 1; ***ESR2:*** estrogen receptor 2 (ER beta); ***FLT1:*** fms-related tyrosine kinase 1; ***FMR1:*** fragile X mental retardation 1; ***FOS:*** FBJ osteosarcoma oncogene; ***FSHR:*** follicle stimulating hormone receptor; ***GCLC:*** glutamate-cysteine ligase, catalytic subunit; ***GNMT:*** glycine N-methyltransferase; ***GRIN1:*** glutamate receptor, ionotropic, N-methyl D-aspartate 1; ***GRIN2B:*** glutamate receptor, ionotropic, N-methyl D-aspartate 2B; ***GSK3B:*** glycogen synthase kinase 3 beta; ***GSS:*** glutathione synthetase; ***GSTP1:*** glutathione S-transferase pi 1; ***HBB:*** hemoglobin, beta; ***HBG1:*** hemoglobin, gamma A; ***HDAC1:*** histone deacetylase 1; ***HDAC11:*** histone deacetylase 11; ***HDAC2:*** histone deacetylase 2; ***HDAC3:*** histone deacetylase 3; ***HDAC4:*** histone deacetylase 4; ***HDAC6:*** histone deacetylase 6; ***HDAC8:*** histone deacetylase 8; ***HDAC9:*** histone deacetylase 9; ***HDACs:*** histone deacetylases; ***HFE:*** hemochromatosis; ***HIF1A:*** hypoxia inducible factor 1, alpha subunit (basic helix-loop-helix transcription factor); ***HIST3H3:*** histone cluster 3, H3; ***HIST4H4:*** histone cluster 4, H4; ***HLA-A*** **:** major histocompatibility complex, class I, A; ***HLA-B:*** major histocompatibility complex, class I, B; ***HSP90As:*** heat shock protein 90kDa alpha (cytosolic), class A; ***HSPA8:*** heat shock 70kDa protein 8; ***HTR3A:*** 5-hydroxytryptamine (serotonin) receptor 3A, ionotropic; ***ICAM1:*** intercellular adhesion molecule 1; ***IFNG:*** interferon, gamma; ***IL10:*** interleukin 10; ***IL12:*** interleukin 12; ***IL12A:*** interleukin 12A; ***IL12B:*** interleukin 12B; ***IL1A:*** interleukin 1, alpha; ***IL23A:*** interleukin 23, alpha subunit p19; ***IL6:*** interleukin 6; ***IL8:*** interleukin 8; ***JUN:*** jun proto-oncogene; ***KDR:*** kinase insert domain receptor; ***KLRK1:*** killer cell lectin-like receptor subfamily K, member 1; ***LEP:*** leptin; ***LEPR:*** leptin receptor; ***MAGED1***: melanoma antigen family D1; ***MAOA:*** monoamine oxidase A; ***MAT1A:*** methionine adenosyltransferase I, alpha; ***MGMT:*** O-6-methylguanine-DNA methyltransferase; ***MLH1:*** mutL homolog 1; ***MMP2:*** matrix metallopeptidase 2; ***MMP9:*** matrix metallopeptidase 9; ***MMPs:*** matrix metallopeptidases; ***MSH2:*** mutS homolog 2; ***MSR1:*** macrophage scavenger receptor 1; ***MYC:*** v-myc avian myelocytomatosis viral oncogene homolog; ***NAGS:*** N-acetylglutamate synthase; ***NF2:*** neurofibromin 2 (merlin); ***NFKB1:*** nuclear factor of kappa light polypeptide gene enhancer in B-cells 1; ***NFKB2:*** nuclear factor of kappa light polypeptide gene enhancer in B-cells 2 (p49/p100); ***NOS2:*** nitric oxide synthase 2, inducible; ***NOS3:*** nitric oxide synthase 3 (endothelial cell); ***NQO1:*** NAD(P)H dehydrogenase, quinone 1; ***NQO2:*** NAD(P)H dehydrogenase, quinone 2; ***NR1I2:*** nuclear receptor subfamily 1, group I, member 2; ***NR1I3:*** nuclear receptor subfamily 1, group I, member 3; ***NR3C1:*** nuclear receptor subfamily 3, group C, member 1 (glucocorticoid receptor); ***NT3:*** 3'-nucleotidase; ***NTRK2:*** neurotrophic tyrosine kinase, receptor, type 2; ***OTC:*** ornithine carbamoyltransferase; ***P2RY2:*** purinergic receptor P2Y, G-protein coupled, 2; ***PDGFRB:*** platelet-derived growth factor receptor, beta polypeptide; ***PDGFRs:*** platelet-derived growth factor receptors; ***PLA2R1:*** phospholipase A2 receptor 1, 180kDa; ***PLCB1:*** phospholipase C, beta 1 (phosphoinositide-specific); ***PMAIP1:*** phorbol-12-myristate-13-acetate-induced protein 1; ***PRKAs:*** protein kinase family, AMP-activated; ***PSEN1:*** presenilin 1; ***PTGES:*** prostaglandin E synthase; ***PTGS1:*** prostaglandin-endoperoxide synthase 1 (prostaglandin G/H synthase and cyclooxygenase); ***PTGS2:*** prostaglandin-endoperoxide synthase 2 (prostaglandin G/H synthase and cyclooxygenase); ***RARB:*** retinoic acid receptor, beta; ***RASSF1:*** Ras association (RalGDS/AF-6) domain family member 1; ***RB1:*** retinoblastoma 1; ***RRM1:*** ribonucleotide reductase M1; ***ROS1:*** ROS proto-oncogene 1 , receptor tyrosine kinase; ***RRM1:*** ribonucleotide reductase M1; ***RRM2:*** ribonucleotide reductase M2; ***RYR1:*** ryanodine receptor 1 (skeletal); ***SCN2A:*** sodium channel, voltage gated, type II alpha subunit; ***SCNs:*** sodium channel family; ***SIRT1:*** sirtuin 1; ***SIRT2:*** sirtuin 2; ***SIRT3:*** sirtuin 3; ***SIRT5:*** sirtuin 5; ***SLC12A3:*** solute carrier family 12 (sodium/chloride transporter), member 3; ***SLC15s:*** solute carrier family 15; ***SLC19A3:*** solute carrier family 19 (thiamine transporter), member 3; ***SLC19A3:*** solute carrier family 19 (thiamine transporter), member 3; ***SLC22A16:*** solute carrier family 22 (organic cation/carnitine transporter), member 16; ***SLC22s:*** solute carrier family 22; ***SLC25A26:*** solute carrier family 25 (S-adenosylmethionine carrier), member 26; ***SLC28A1:*** solute carrier family 28 (concentrative nucleoside transporter), member 1; ***SLC29A1:*** solute carrier family 29 (equilibrative nucleoside transporter), member 1; ***SLC29As:*** solute carrier family 29; ***SLC5A5:*** solute carrier family 5 (sodium/iodide cotransporter), member 5; ***SLC6A2:*** solute carrier family 6 (neurotransmitter transporter), member 2; ***SLCO1B3:*** solute carrier organic anion transporter family, member 1B3; ***SMN2:*** survival of motor neuron 2, centromeric; ***SNCA:*** synuclein, alpha (non A4 component of amyloid precursor; ***SOCS1:*** suppressor of cytokine signaling 1; ***SOCS3:*** suppressor of cytokine signaling 3; ***SRC:*** SRC proto-oncogene, non-receptor tyrosine kinase; ***SREBF1:*** sterol regulatory element binding transcription factor 1; ***SRM:*** spermidine synthase; ***STAT1:*** signal transducer and activator of transcription 1, 91kDa; ***STAT3:*** signal transducer and activator of transcription 3 (acute-phase response factor); ***STATs:*** signal transducer and activator of transcription family; ***SULT1C2:*** sulfotransferase family, cytosolic, 1C, member 2; ***TGFB1:*** transforming growth factor, beta 1; ***TIMP3:*** TIMP metallopeptidase inhibitor 3; ***TLR3:*** toll-like receptor 3; ***TNF:*** tumor necrosis factor; ***TNFRSF10A:*** tumor necrosis factor receptor superfamily, member 10a; ***TNFRSF10B:*** tumor necrosis factor receptor superfamily, member 10b; ***TNFRSF1B:*** tumor necrosis factor receptor superfamily, member 1B; ***TNFSF10:*** tumor necrosis factor (ligand) superfamily, member 10; ***TP53:*** tumor protein p53; ***TPMT:*** thiopurine S-methyltransferase; ***TRNK:*** mitochondrially encoded tRNA lysine; ***TRPs:*** transient receptor potential cation channels; ***TYMS:*** thymidylate synthetase; ***UCK1:*** uridine-cytidine kinase 1; ***UCK2:*** uridine-cytidine kinase 2; ***UGT1A10:*** UDP glucuronosyltransferase 1 family, polypeptide A10; ***UGT1A4:*** UDP glucuronosyltransferase 1 family, polypeptide A4; ***UGT1A6:*** UDP glucuronosyltransferase 1 family, polypeptide A6; ***UGT1A8:*** UDP glucuronosyltransferase 1 family, polypeptide A8; ***UGT1A9:*** UDP glucuronosyltransferase 1 family, polypeptide A9; ***UGT2B1:*** UDP glucuronosyltransferase 1 family, polypeptide B1; ***UGT2B7:*** UDP glucuronosyltransferase 2 family, polypeptide B7; ***VCAM1:*** vascular cell adhesion molecule 1; ***VEGFA:*** vascular endothelial growth factor A; ***VEGFs:*** vascular endothelial growth factor family; ***VHL:*** von Hippel-Lindau tumor suppressor, E3 ubiquitin protein ligase; ***ZNF350:*** *zinc finger protein 350*.

rate: 0.4%; failure: 99.6%).[185] During the past 15 years no new drugs have been approved for the treatment of AD and the available drugs are not cost-effective.[186] Therefore, the pharmacogenetics of AD is very limited, circumscribed to cholinesterase inhibitors and memantine (Table 5.1), remaining stuck in a primitive stage of underdevelopment due to the lack of novel therapeutic options. Although many studies on the pharmacogenetics of AD have been published since the early 2000s,[64,65] many of them are redundant and contradictory, focusing mainly on the *APOE* gene and, to a lesser extent, on some *CYP* family genes and other minor genes.[63] In this context, several considerations are pertinent regarding further steps to be followed in order to achieve a more mature profile of AD pharmacogenomics: (i) a better characterization of the roles played in drug efficacy and safety by genes involved in the pharmacogenomic network is necessary; (ii) since most genes are under the influence of the epigenetic machinery, pharmacoepigenomics is becoming an attractive field that deserves special attention; (iii) drug–drug interactions represent a problematic issue in over 80% of AD patients; (iv) since the neurodegenerative process underlying AD neuropathology starts 20–30 years before the onset of the disease, novel therapeutics should be addressed to prevent premature neuronal death; (v) specific biomarkers for AD are necessary in 3 different contexts: predictive markers before disease onset, early diagnosis in initial stages, and drug monitoring (in both preventive and/or therapeutic strategies); and (vi) physicians should be aware of the usefulness of pharmacogenomics to prescribe more accurately, avoid adverse reactions and optimize the limited therapeutic resources available for the treatment of dementia.[8,187]

During the past 10 years, over 1000 different compounds have been studied as potential candidate drugs for the treatment of AD.[9,11,18,188] About 50% of these substances are novel molecules obtained from natural sources.[9,11] The candidate compounds can be classified according to their pharmacological properties and/or the AD-related pathogenic cascade to which they are addressed to halt disease progression. In addition to the FDA-approved drugs since 1993 (tacrine, donepezil, rivastigmine, galantamine, memantine) (Table 5.1), most candidate strategies fall into 6 major categories: (i) novel cholinesterase inhibitors and neurotransmitter regulators, (ii) anti-Aβ treatments (APP regulators, Aβ breakers, active and passive immunotherapy with vaccines and antibodies, β- and γ-secretase inhibitors or modulators), (iii) anti-tau treatments, (iv) pleiotropic products (most of them of natural origin), (v) epigenetic intervention, and (vi) combination therapies.[8,9,11,18]

In more global terms, prospections of diverse natural sources (vegetal, marine, animal) have allowed the identification and characterization of novel bioproducts with potential utility in the prevention and treatment of a vast array of age-related pathological phenotypes and NDDs as well. Prototypal examples of these biotechnological products are LipoFishins and Atremorine.[23]

5.6.1 LipoFishins

LipoFishins (LFs) are a new class of lipoproteins derived from the muscle of different fish species. Examples of LPs obtained from biomarine sources by means of non-denaturing biotechnological procedures include the following: E-JUR-94013 (DefenVid®), E-CAB-94011 (CabyMar®), E-Congerine-10423 (AntiGan®), E-SAR-94010 (LipoEsar®), and E-MHK-0103 (MineraXin®).[189–194] Most effects of these novel bioproducts are genotype-dependent, showing specific nutrigenomic and pharmacogenomic profiles.[10,11,195]

E-CAB-94011 is an LF obtained from the muscle of the species *Scombrus scombrus*, with anti-oxidant, anti-inflammatory, and bio-energizing properties, with potential utility in several medical conditions (anemia, debilitating disorders, alterations in growth and development, ROS generation, NDDs).[191]

E-Congerine-10423 is an LP extracted from muscular structures of the species *Conger conger*. This compound displays a powerful anti-tumoral effect in many different tumor cell-lines, with specific effects in colon cancer, ulcerative colitis, and Crohn's disease.[194]

E-MHK-0103 is an atypical LP derived from the Atlantic mollusc *Mytillus galloprovincialis* cultivated on the Atlantic coast of Galicia (Spain). This bioproduct regulates hypothalamus-pituitary hormones, influences growth and development, protects against menopause-related biological decline, and modulates bone metabolism, acting as a powerful anti-osteoporotic agent.[196]

5.6.1.1 E-SAR-94010 (LipoEsar®)

E-SAR-94010 (Sardilipin, LipoEsar®, LipoSea®) is an LP obtained from the species *Sardina pilchardus.*[189] The main chemical compounds of LipoEsar® are lipoproteins (60–80%) whose micelle structure probably mimics that of physiological lipoproteins involved in lipid metabolism. In preclinical studies, sardilipin has been shown to be effective in: (i) reducing blood cholesterol (CHO), triglyceride (TG), uric acid (UA), and glucose (Glu) levels, as well as liver alanine aminotransferase (ALT), and aspartate aminotransferase (AST) activity; (ii) enhancing immunological function by regulating both lymphocyte and microglia activity; (iii) inducing antioxidant effects mediated by superoxide dismutase activity; and (iv) improving cognitive function.[190] This LP shows a powerful effect in the regulation of lipid metabolism, especially by reducing total-cholesterol and LDL-cholesterol levels in cases of dyslipidemia or hypercholesterolemia, and also acting as an effective co-adjuvant of statins (Figure 5.9). E-SAR is effective in liver steatosis and in cases of primary or secondary transaminitis. It is also a strong anti-atherogenic agent, reducing the size of atheroma plaques in systemic atherosclerosis. E-SAR has shown cognitive-enhancing properties in hypercholesterolemic patients with AD. The therapeutic response of patients with dyslipidemia to sardilipin is *APOE*-related. The best responders are patients with *APOE-3/3*>*APOE-3/4*>*APOE-4/4*. Patients with the other *APOE* genotypes (*2/2*, *2/3*, *2/4*) do not show any hypolipemic response to this novel compound. In patients

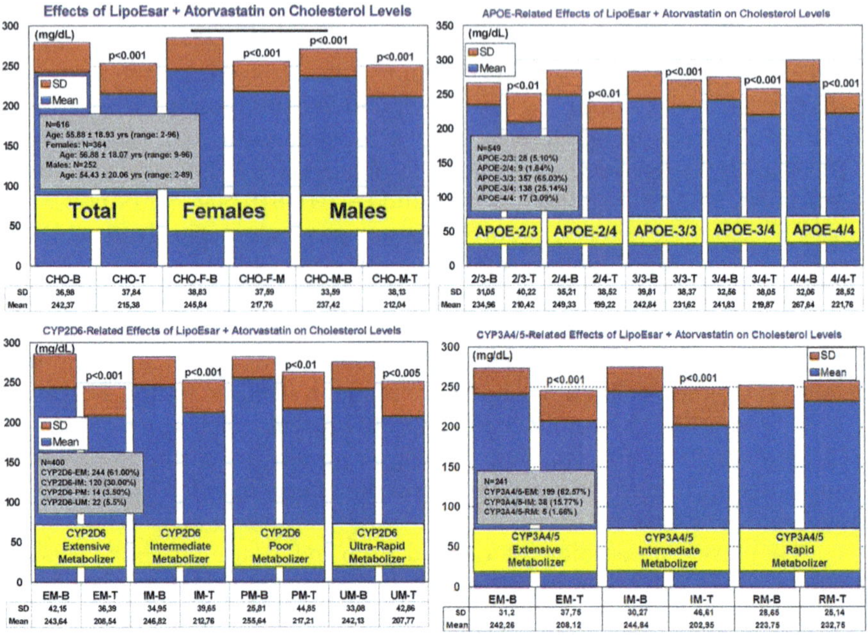

Figure 5.9 Gender-, APOE-, CYP2D6-, and CYP3A4/5-related effects of Lipo Esar + Atorvastatin on blood cholesterol levels in patients with hypercholesterolemia.

with dementia, the effects of sardilipin are very similar to those observed in patients with chronic dyslipidemia, suggesting that the lipid-lowering properties of sardilipin are *APOE*-dependent.[5,10,11,23,189,190,195]

5.6.1.2 E-JUR-94013 (DefenVid®)

E-JUR-94013 (DefenVid®) is an LF derived from the fish *Trachurus trachurus*, with anti-inflammatory activity and powerful immune-enhancing properties in cases of immunodeficiency, microbial infections and/or diseases in which there is a functional compromise of the immune system.[191–193] We investigated the effects of 1-month treatment with DefenVid (750 mg day^{-1}) in a group of 1149 patients with CNS disorders (mean age = 49.94 ± 22.06 years, range: 1–98 years; 621 females: age = 51.71 ± 20.90 years, range: 2–98 years; 528 males: age = 47.87 ± 23.19 years, range: 1–89 years). DefenVid significantly modified white blood cell (WBC) numbers in a differential fashion, decreasing neutrophils ($p < 0.02$), increasing lymphocytes ($p < 0.02$), monocytes ($p < 0.02$), and eosinophils ($p < 0.05$), and not affecting basophils (Figures 5.10 and 5.11). The effect of DefenVid is immunomodulatory due to the fact that in cases with high WBC numbers the general tendency is to reduce the excess of WBC, whereas in cases with low levels of WBC DefenVid tends to increase WBC, approaching normal levels (Figures 5.10 and 5.11).

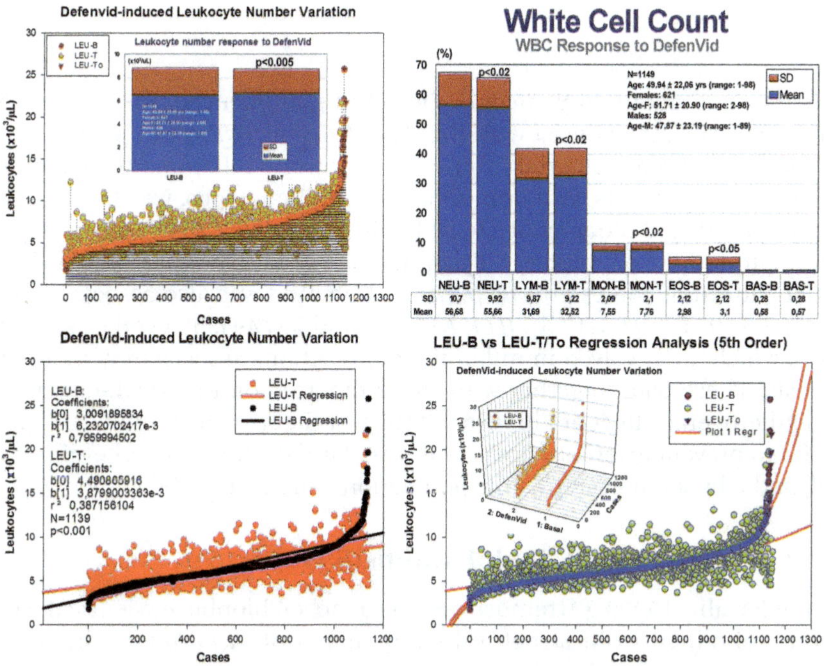

Figure 5.10 Effect of DefenVid on white blood cell number in the Spanish population. Neu: Neutrophils; LYM: Lymphocytes; Mon: Monocytes; EOS: Eosinophils; Bas: Basophils; B: Basal; T: Treatment (DefenVid: 750 mg day^{-1} for 1 month).

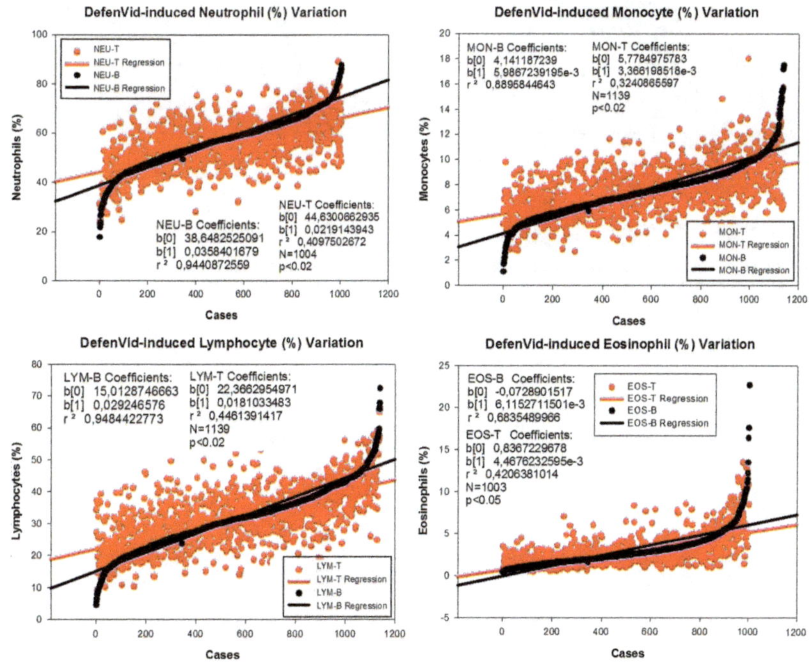

Figure 5.11 Differential effect of DefenVid on white blood cells in the Spanish population. DefenVid: 750 mg day^{-1} for 1 month.

This immunomodulatory effect of DefenVid is influenced, in part, by SNP variation associated at least with the *IL1B*, *IL6*, and *TNF* genes (Figure 5.12), classically involved in neuroimmune regulation and inflammatory reactions. In a subset of patients with immunodeficient phenotypes, we observed that DefenVid reduced blood cholesterol levels in over 60% of the cases (Figure 5.13), similarly to LipoEsar in dyslipidemic patients (Figure 5.9). A differential pattern of cholesterol response to DefenVid was also associated with the *IL1B*-T3954C, *IL6*-G174C, *IL6R*-A1510C, and *TNFA*-G308A variants (Figure 5.14), which are involved in inflammatory reactions associated with atherogenesis. These data, together with those reported on the APOE-dependent anti-atherogenic effect of LipoEsar,[195] suggest that this class of LFs might be useful to prevent arteriosclerosis and vascular risk, either peripheral or central, in the hypercholesterolemic population and in NDDs.[10,11]

5.6.2 Atremorine (E-PodoFavalin-15999)

E-PodoFavalin-15999 (Atremorine®) is a novel biopharmaceutical compound, obtained from structural components of *Vicia faba* L. by means of non-denaturing biotechnological processes, for the prevention and

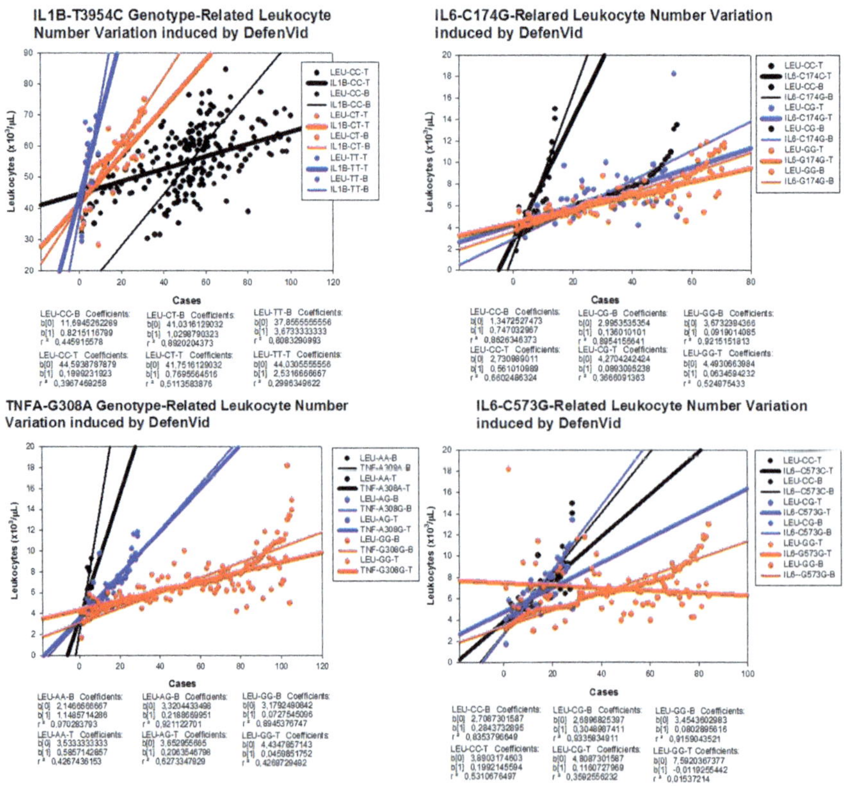

Figure 5.12 IL1B-, IL6-, and TNFA-related leukocyte variation after one-month treatment with DefenVid in the Spanish population.

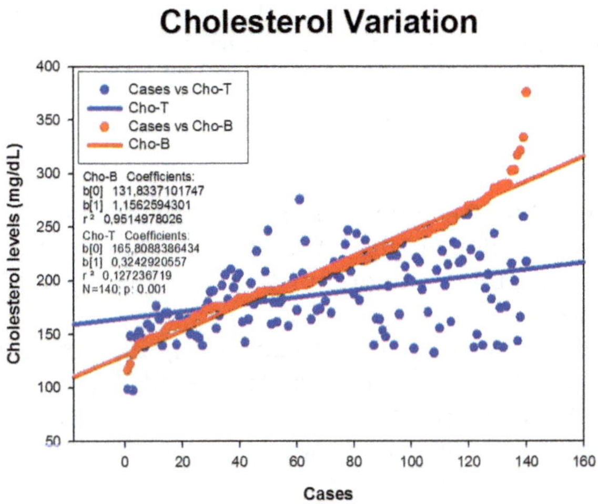

Figure 5.13 Effect of DefenVid on blood cholesterol levels in a sub-set of healthy subjects.

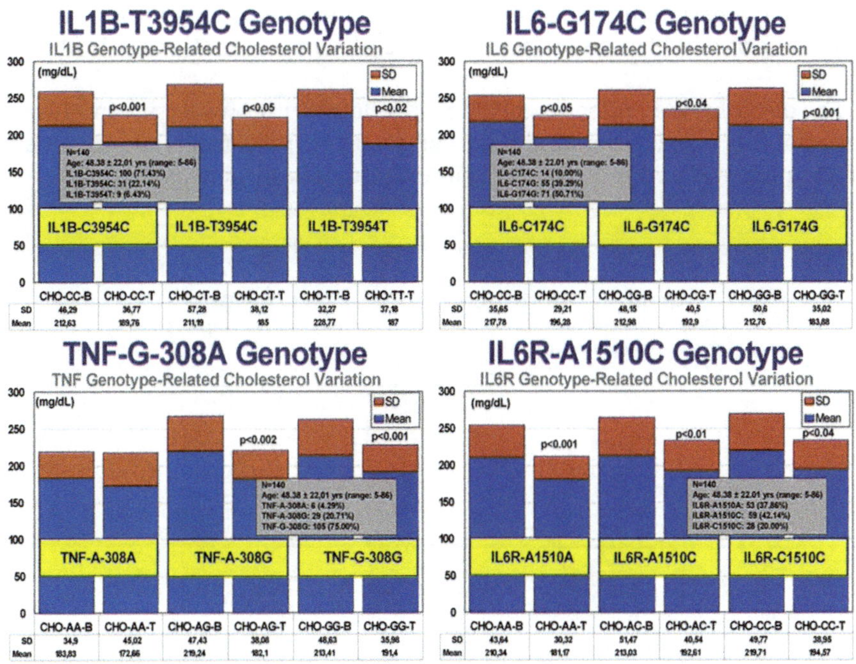

Figure 5.14 IL1B-, IL6-, IL6R-, and TNFA-related blood cholesterol variation in subjects treated with DefenVid (750 mg day^{-1}) for one month.

treatment of Parkinson's disease (PD) (Patent ID EP16382138.2). *In vitro* studies revealed that Atremorine is a powerful neuroprotectant in: (i) cell cultures of human neuroblastoma SH-SY5Y cells; (ii) hippocampal slices in conditions of oxygen and glucose deprivation; and (iii) striatal slices under conditions of neurotoxicity induced by 6-OHDA. *In vivo* studies showed that Atremorine: (i) protects against 1-methyl-4-phenyl-1,2,3,6-tetrahydropyridine (MPTP)-induced dopaminergic neurodegeneration; (ii) inhibits MPTP-induced microglia activation and neurotoxicity in substantia nigra; and (iii) improves motor function in mice with MPTP-induced neurodegeneration.

Clinical studies have been performed in 3 groups of patients: (i) NP: drug-free patients with PD; (iii) AP: Parkinsonian patients chronically treated with L-Dopa; and (iii) MX: a heterogeneous sample of patients with Parkinsonian disorders. 30–60 min. after a single dose (5 g) of Atremorine, plasma levels of dopamine increased from 16.71 ± 14.38 pg mL^{-1} to 2286 ± 4218 pg mL^{-1} ($p < 0.001$) in NP, from 4149 ± 7062 pg mL^{-1} to 13 539 ± 12 408 pg mL^{-1} ($p < 0.001$) in AP, and from 860 ± 3445 pg mL^{-1} to 4583 ± 8084 pg mL^{-1} ($p < 0.001$) in MX patients, with a parallel clinical improvement lasting for 3–6 hours (Figure 5.15). Atremorine administration also increased the plasma levels of noradrenaline in NP ($p < 0.008$) and MX ($p < 0.04$). Atremorine decreased prolactin levels in NP and MX, and GH levels in NP and MX. Pharmacogenetic studies indicate that the therapeutic response induced by Atremorine in PD is associated with the pharmacogenetic profile of each patient.

The data obtained in basic and clinical studies with Atremorine in PD suggest that this novel compound might be useful for prevention and treatment of dopamine-related movement disorders and neurodegeneration.

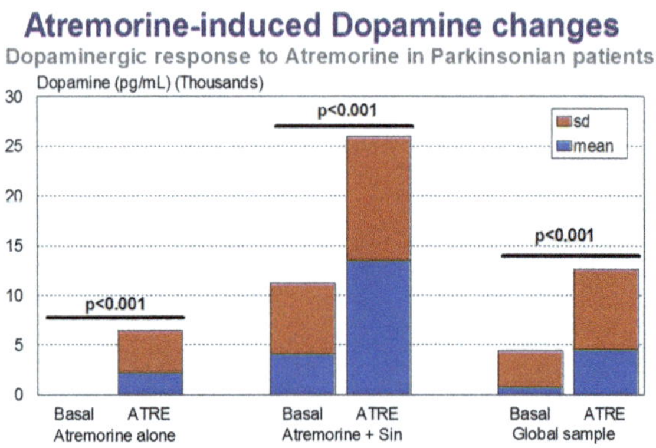

Figure 5.15 Atremorine-induced blood dopamine changes in Parkinsonian patients (i) with no previous treatment, (ii) chronically treated with L-Dopa, and (iii) and all patients pooled together. $N = 92$. Atremorine: 5 g (single dose, p.o.). Time: 30 min. after treatment.

5.7 Future Trends for the Management of Age-Related NDDs

Most CNS disorders are clinical entities which, in many instances, share some common features: (i) pathogenically, they are complex disorders in which a plethora of plural events (genomic defects, epigenetic aberrations, mitochondrial dysfunction, environmental factors) is potentially involved; (ii) many of them, especially those with a late onset, are characterized by intracellular and/or extracellular deposits of abnormal proteins; (iii) their diagnosis is difficult because they lack specific biomarkers (and their prediction is almost impossible); (iv) their treatment is symptomatic (not anti-pathogenic) and not cost-effective; and (v) the vast majority represent chronic ailments with progressive deterioration and bad prognosis.[11] The concept of epigenetics, introduced by Conrad Waddington in 1942, and its spectacular evolution, from a biotechnological perspective, has been of great help for the past 10 years in the understanding of gene regulation and expression (functional genomics), neurogenomics, and pathogenetics of CNS disorders (Figure 5.1).[13–16,141,153,197,198]

Gene expression and protein function experience profound modifications throughout the life span. It is likely that the frontier between health and disease is not only associated with specific SNP variability and epigenetic aberrations (in conjunction with environmental risks) but also with a salutary/pathogenic threshold of transformed protein accumulation in critical cells (especially in neurons). Over the past decade, progress in epigenetics and proteomics has helped to understand many aspects of pathogenic phenomena which had remained obscure or unaffordable to our technical capabilities for the assessment of genomic dysfunction, epigenetic dysregulation, and abnormal protein expression. Transcription errors represent a molecular mechanism by which cells can acquire disease phenotypes. The error rate of transcription increases as cells age, suggesting that transcription errors affect proteostasis, particularly in aging cells. Accordingly, transcription errors accelerate the aggregation of peptides and shorten the lifespan of cells.[199]

Novel methodologies have allowed us to configure new pathogenic hypotheses for a better understanding of brain disorders. In this endeavor, epigenetics and proteomics have been of great benefit. Epigenetic studies have revealed the important role that epigenetic modifications have on brain development and maturation, synaptic plasticity, brain sex differences, neurodevelopment and imprinting disorders, mental disorders, neurodegeneration, and the new field of epigenetic Mendelian disorders.[144] Structural genomic defects cannot explain in full the pathogenesis of CNS disorders. Many old concepts related to the pathogenesis of CNS disorders should be eliminated. Parkinson's disease is not the result of a single deficiency in dopamine; Alzheimer's disease is not the consequence of a cholinergic deficit; however, the basic principles for the development of the currently most-prescribed drugs for both disorders rely on a single neurotransmitter defect (enhancement of dopamine neurotransmission in Parkinson's disease, and potentiation of cholinergic transmission with acetylcholinesterase inhibitors in Alzheimer's disease). These old-fashioned pathogenic concepts are completely out-of-date, and the new conceptions on NDDs are based on the pathogenic cascade represented by genomic-epigenomic-transcriptomic-proteomic-metabolomic disturbances leading to a specific phenotype, which in the future will require a personalized therapeutic intervention (pharmacogenomics, pharmacoepigenomics) for phenotype disease modification (Figure 5.1). As pointed out by Riley *et al.*,[200] systems analysis is believed to help deconvolute complex biological responses involving hundreds or thousands of genes assayed by OMICs methods. Although systems-style approaches have been applied to CNS tissues, most studies have used simple functional overview approaches resulting in the identification of differentially expressed genes or pathways. While these approaches expanded our understanding of disease-related changes, they are not able to elucidate the complex interconnectivity of the biological and pathological processes present within diseased tissue. These approaches are "low resolution" descriptive methods with limited projection in terms of clarifying molecular pathogenesis, experimental follow-up, and clinical application.

Global protein profiling by mass spectrometry-based proteomics has evolved as a new hypothesis-free avenue to optimally unravel new candidate protein biomarkers involved in different CNS disorders. Technological developments and improvement of sensitivity, specificity and speed of different proteomic approaches have facilitated the discovery of an enormous number of biomarker candidates; however, most biomarkers have not yet been validated, which limits their application in clinical practice. The correct interpretation of thousands of data derived from proteomic and epigenomic analysis is an additional problem for the practical implementation of biomarkers in the clinical setting.[201] Novel neuroproteomic tools and powerful bioinformatic resources are needed to accelerate the incorporation of proteomic and epigenomic analysis to the diagnostic process.[202–204]

Another important field, to whose expansion epigenetics and proteomics are contributing, is drug development. Epigenetic drugs are becoming a fashion[15,16,205] and some of them have been approved by the FDA in recent years for the treatment of cancer.[206] However, most epigenetic drugs are pleiotropic and are not devoid of toxicity and biodynamic complications (*e.g.* brain penetration).[16]

The effects of drugs (pharmacokinetics and pharmacodynamics) and their therapeutic outcome in the treatment of a given disease are the result of a network of metabolomic events (genomics-epigenomics-transcriptomics-proteomics) associated with the binomial interaction of a chemical or biological molecule with a living organism. The clusters of genes currently involved in a pharmacogenomic process include pathogenic, mechanistic, metabolic, transporter, and pleiotropic genes.[11] In practice, the expression of these genes is potentially modifiable (transcriptionally and/or post-transcriptionally) by epigenetic mechanisms that may alter: (i) pathogenic events; (ii) receptor–drug interactions; (iii) drug metabolism (phase I and II enzymatic reactions); (iv) drug transport (influx–efflux across membranes and cellular barriers);

and (v) pleiotropic events leading to unexpected therapeutic outcomes. The understanding of these mechanisms is the main focus of pharmacoepigenomics, in order to optimize therapeutics and advance towards personalized medicine.[16,207]

In the coming years, important achievements must be accomplished in different areas of neuroscience: (i) brain development and maturation; (ii) toxicogenomics; (iii) functional epigenomics; (iv) proteoepigenomics; (v) pathoepigenomics; (vi) predictive proteomics; (vii) diagnostic proteomics; (viii) prognostic proteomics; (ix) pharmacoepigenomics; and (x) epitherapeutics. It is likely that systems biology will dominate the biology and medicine of the 21st century.[208] Relevant information obtained from the ENCODE Project will be incorporated into a more versatile map of clinical neuroscience and practical medicine.[209–211] Development is a dynamic process that involves interplay between genes and the environment. Postnatal environment is shaped by parent–offspring interactions that promote growth and survival and can lead to divergent developmental trajectories with implications for later-life neurobiological and behavioral characteristics.[212] The impact that nutrition, emotions, drugs and environmental toxicants during prenatal development may have on brain maturation and late CNS disorders requires urgent clarification.[213–215] Important advances related to the role of epigenetics in the pathogenesis of brain disorders will occur in the near future with reliable applications. Predictive, diagnostic, and prognostic proteomics, as well as the use of biomarkers to monitor the effects of drugs, will undergo a profound change from the present immature stage of the field to a more specific and validated area with various applications in CNS disorders.

In therapeutics, important breakthroughs will occur in some of the following areas: (i) drug discovery for different CNS disorders, age-related NDDs and cancer;[15,16,170,216,217] (ii) practical applications of pharmacogenomics[11,23] and pharmacoepigenomics[176–179,218] for the optimization and personalization of current drugs and new biopharmaceuticals; (iii) novel therapeutic approaches to decode and resolve potential resistance mechanisms in cancer, psychiatric disorders, and NDDs;[179,218–221] and (iv) targeting miRNAs in the prevention and treatment of brain disorders.[222–224]

Acknowledgements

We thank our colleagues at EuroEspes Biomedical Research Center, Institute of Medical Science and Genomic Medicine, Corunna, for their technical support in the performance of pharmacogenetic studies.

References

1. *Mortality in the United States*, 2014, NCHS Data Brief, http://www.cdc.gov/nchs/data/databriefs/db229.pdf (accessed April 2016).
2. *Health, United States*, 2014, http://www.cdc.gov/nchs/data/hus/hus14.pdf (accessed April 2016).

3. R. Cacabelos, C. Torrellas, P. Cacabelos, M. J. Villanueva, S. Piñeiro and C. Solveira, in *Preventive and Predictive Genetics: Towards Personalised Medicine*, ed. G. Grech and I. Grossman, Springer, Heidelberg, 2015, p. 173.
4. R. Cacabelos, *Pharmacogenomics*, 2003, **4**, 597.
5. R. Cacabelos, *Methods Find. Exp. Clin. Pharmacol.*, 2007, **29**, 1.
6. R. Cacabelos, *Expert Rev. Mol. Diagn.*, 2009, **9**, 567.
7. R. Cacabelos, C. Torrellas, O. Teijido and J. C. Carril, *Pharmacogenomics*, 2016, **17**, 1041.
8. R. Cacabelos, O. Teijido and J. C. Carril, *Expert Opin. Drug Discovery*, 2016, **11**, 215.
9. R. Cacabelos, C. Torrellas, I. Carrera, P. Cacabelos, L. Corzo, L. Fernández-Novoa, I. Tellado, J. C. Carril and G. Aliev, *CNS Neurol. Disord.: Drug Targets*, 2016, **15**, 141.
10. R. Cacabelos, *Methods Mol. Biol.*, 2008, **448**, 213.
11. R. Cacabelos, P. Cacabelos, C. Torrellas, I. Tellado and J. C. Carril, *Methods Mol. Biol.*, 2014, **1175**, 323.
12. R. Cacabelos, D. Goldgaber, A. D. Roses, A. Vostrov, H. Matsuki, C. Torrellas, P. Cacabelos, D. Corzo, J. C. Carril, L. Fernández-Novoa and I. Carrera, *Int. J. Mol. Genet. Gene Ther.*, 2015, **1**, DOI: 10.16966/sggt.102.
13. R. Cacabelos and C. Torrellas, *Int. J. Mol. Sci.*, 2015, **16**, 30486.
14. R. Cacabelos, C. Torrellas and F. López-Muñoz, *J. Exp. Clin. Med.*, 2014, **6**, 75.
15. R. Cacabelos and C. Torrellas, *Expert Opin. Drug Discovery*, 2014, **9**, 1059.
16. R. Cacabelos, *Drug Dev. Res.*, 2014, **75**, 348.
17. R. Cacabelos, in *The Handbook of Neuropsychiatric Biomarkers, Endophenotypes and Genes*, ed. M. S. Ritsner, Springer, Netherlands, 2011, p. 3.
18. R. Cacabelos, C. Torrellas and I. Carrera, *Future Neurol.*, 2015, **10**, 229.
19. R. H. Eckel, *N. Engl. J. Med.*, 1989, **320**, 1060.
20. G. Paré, D. Serre, D. Brisson, S. S. Anand, A. Montpetit, G. Tremblay, J. C. Engert, T. J. Hudson and D. Gaudet, *Am. J. Hum. Genet.*, 2007, **80**, 673.
21. J. Pedro-Botet, M. Sentí, X. Nogués, J. Rubiés-Prat, J. Roquer, L. D'Olhaberriaque and J. Olivé, *Stroke*, 1992, **23**, 1556.
22. J. Stocks, J. A. Thorn and D. J. Galton, *J. Lipid Res.*, 1992, **33**, 853.
23. R. Cacabelos, *World Guide for Drug Use and Pharmacogenomics*, EuroEspes Publishing, Corunna, Spain, 2012.
24. X. Jeunemaitre, F. Soubrier, Y. V. Kotelevtsev, R. P. Lifton, C. S. Williams, A. Charru, S. C. Hunt, P. N. Hopkins, R. R. Williams, J. M. Lalouel and P. Corvol, *Cell*, 1992, **71**, 7.
25. T. Katsuya, G. Koike, T. W. Yee, N. Sharpe, R. Jackson, R. Norton, M. Horiuchi, R. E. Pratt, V. J. Dzau and S. MacMahon, *Lancet*, 1995, **345**, 1600.
26. T. Rankinen, T. Rice, L. Perusse, Y. C. Chagnon, J. Gagnon, A. S. Leon, J. S. Skinner, J. H. Wilmore, D. C. Rao and C. Bouchard, *Hypertension*, 2000, **36**, 885.
27. A. Slowik, W. Turaj, T. Dziedzic, A. Haefele, J. Pera, M. T. Malecki, L. Glodzik-Sobanska, P. Szermer, D. A. Figlewicz and A. Szczudlik, *Neurology*, 2004, **63**, 359.

28. J. Danesh and P. Appleby, *Expert Opin. Invest. Drugs*, 1998, **7**, 691.
29. J. M. Fernández-Real, M. Broch, J. Vendrell, C. Richart and W. Ricart, *J. Clin. Endocrinol. Metab.*, 2000, **85**, 1334.
30. L. Iacoviello, A. Di Castelnuovo, M. Gattone, A. Pezzini, D. Assanelli, R. Lorenzet, E. Del Zotto, M. Colombo, E. Napoleone, C. Amore, A. D'Orazio, A. Padovani, G. de Gaetano, P. Giannuzzi, M. B. Donati and IGIGI Investigators, *Arterioscler., Thromb., Vasc. Biol.*, 2005, **25**, 222.
31. P. Libby, *Nature*, 2002, **420**, 868.
32. T. Luther and N. Mackman, *Trends Cardiovasc. Med.*, 2001, **11**, 307.
33. I. Rainero, L. M. Grimaldi, G. Salani, W. Valfrè, C. Rivoiro, L. Savi and L. Pinessi, *Neurology*, 2004, **62**, 141.
34. D. Reich, N. Patterson, V. Ramesh, P. L. De Jager, G. J. McDonald, A. Tandon, E. Choy, D. Hu, B. Tamraz, L. Pawlikowska, C. Wassel-Fyr, S. Huntsman, A. Waliszewska, E. Rossin, R. Li, M. Garcia, A. Reiner, R. Ferrell, S. Cummings, P. Y. Kwok, T. Harris, J. M. Zmuda, E. Ziv and Health ABC Study, *Am. J. Hum. Genet.*, 2007, **80**, 716.
35. M. Revilla, V. Obach, A. Cervera, A. Dávalos, J. Castillo and A. Chamorro, *Neurosci. Lett.*, 2002, **324**, 29.
36. R. Rosmond, M. Chagnon, C. Bouchard and P. Björntorp, *J. Clin. Endocrinol. Metab.*, 2001, **86**, 2178.
37. N. S. Rost, P. A. Wolf, C. S. Kase, M. Kelly-Hayes, H. Silbershatz, J. M. Massaro, R. B. D'Agostino, C. Franzblau and P. W. Wilson, *Stroke*, 2001, **32**, 2575.
38. Y. Yamada, N. Metoki, H. Yoshida, K. Satoh, S. Ichihara, K. Kato, T. Kameyama, K. Yokoi, H. Matsuo, T. Segawa, S. Watanabe and Y. Nozawa, *Arterioscler., Thromb., Vasc. Biol.*, 2006, **26**, 1920.
39. L. Bethke, E. Webb, A. Murray, M. Schoemaker, M. Feychting, S. Lönn, A. Ahlbom, B. Malmer, R. Henriksson, A. Auvinen, A. Kiuru, T. Salminen, C. Johansen, H. C. Christensen, K. Muir, P. McKinney, S. Hepworth, P. Dimitropoulou, A. Lophatananon, A. Swerdlow and R. Houlston, *Cancer Epidemiol., Biomarkers Prev.*, 2008, **17**, 1195.
40. J. P. Casas, A. D. Hingorani, L. E. Bautista and P. Sharma, *Arch. Neurol.*, 2004, **61**, 1652.
41. V. De Stefano, I. Martinelli, P. M. Mannucci, K. Paciaroni, P. Chiusolo, I. Casorelli, E. Rossi and G. Leone, *N. Engl. J. Med.*, 1999, **341**, 801.
42. A. Gerhardt, R. E. Scharf, M. W. Beckmann, S. Struve, H. G. Bender, M. Pillny, W. Sandmann and R. B. Zotz, *N. Engl. J. Med.*, 2000, **342**, 374.
43. D. S. Wald, M. Law and J. K. Morris, *Br. Med. J.*, 2002, **325**, 1202.
44. J. C. Carril and R. Cacabelos, *Curr. Genomics*, in press.
45. Z. Xue-Shan, P. Juan, W. Qi, R. Zhong, P. Li-Hong, T. Zhi-Han, J. Zhi-Sheng, W. Gui-Xue and L. Lu-Shan, *Clin. Chim. Acta*, 2016, **456**, 107.
46. Y. L. Chen, L. M. Wang, Y. Chen, J. Y. Gao, C. Marshall, Z. Y. Cai, G. Hu and M. Xiao, *Neuroscience*, 2016, **316**, 178.
47. H. K. Dias, C. L. Brown, M. C. Polidori, G. Y. Lip and H. R. Griffiths, *Clin. Sci.*, 2015, **129**, 1195.
48. S. Chakrabarti, V. K. Khemka, A. Banerjee, G. Chatterjee, A. Ganguly and A. Biswas, *Aging Dis.*, 2015, **6**, 282.

49. R. Cacabelos, *Pharmacogenomics*, 2004, **5**, 1049.
50. J. P. Liu, Y. Tang, S. Zhou, B. H. Toh, C. McLean and H. Li, *Mol. Cell. Neurosci.*, 2010, **43**, 33.
51. M. Maulik, D. Westaway, J. H. Jhamandas and S. Kar, *Mol. Neurobiol.*, 2013, **47**, 37.
52. J. E. Vance, *Dis. Models & Mech.*, 2012, **5**, 746.
53. R. Cacabelos, L. Fernández-Novoa, V. Lombardi, Y. Kubota and M. Takeda, *Methods Find. Exp. Clin. Pharmacol.*, 2005, **27**, 1.
54. L. Bertram, M. B. McQueen, K. Mullin, D. Blacker and R. E. Tanzi, *Nat. Genet.*, 2007, **39**, 17.
55. D. J. Selkoe and M. B. Podlisny, *Annu. Rev. Genomics Hum. Genet.*, 2002, **3**, 67.
56. Y. H. Suh and F. Checler, *Pharmacol. Rev.*, 2002, **54**, 469.
57. C. M. Karch, C. Cruchaga and A. Goate, *Neuron*, 2014, **83**, 11.
58. A. J. Larner, *J. Alzheimer's Dis.*, 2013, **37**, 653.
59. J. Hardy and D. J. Selkoe, *Science*, 2002, **297**, 353.
60. B. V. Hooli, Z. M. Kovacs-Vajna, K. Mullin, M. A. Blumenthal, M. Mattheisen, C. Zhang, C. Lange, G. Mohapatra, L. Bertram and R. E. Tanzi, *Mol. Psychiatry.*, 2014, **19**, 676.
61. D. Schenk, R. Barbour, W. Dunn, G. Gordon, H. Grajeda, T. Guido, K. Hu, J. Huang, K. Johnson-Wood, K. Khan, D. Kholodenko, M. Lee, Z. Liao, I. Lieberburg, R. Motter, L. Mutter, F. Soriano, G. Shopp, N. Vasquez, C. Vandevert, S. Walker, M. Wogulis, T. Yednock, D. Games and P. Seubert, *Nature*, 1999, **400**, 173.
62. I. Carrera, I. Etcheverría, L. Fernández-Novoa, V. Lombardi, R. Cacabelos and C. Vigo, *Int. J. Alzheimer's Dis.*, 2012, **2012**, 376138.
63. R. Cacabelos, R. Llovo, C. Fraile and L. Fernández-Novoa, *Curr. Alzheimer Res.*, 2007, **4**, 479.
64. R. Cacabelos, *Mini-Rev. Med. Chem.*, 2002, **2**, 59.
65. R. Cacabelos, X. A. Álvarez, V. Lombardi, L. Fernández-Novoa, L. Corzo, P. Pérez, M. Laredo, V. Pichel, A. Hernández, M. Varela, J. Figueroa, J. Prous Jr., M. Windisch and C. Vigo, *Drugs Today*, 2000, **36**, 415.
66. R. Cacabelos, L. Fernández-Novoa, R. Martínez-Bouza, A. McKay, J. C. Carril, V. Lombardi, L. Corzo, I. Carrera, I. Tellado, L. Nebril, M. Alcaraz, S. Rodríguez, A. Casas, V. Couceiro and A. Álvarez, *Pharmaceuticals*, 2010, **3**, 3040.
67. R. Cacabelos, *Drug Dev. Res.*, 2012, **73**, 461.
68. A. D. Roses, *Nat. Rev. Genet.*, 2004, **5**, 645.
69. A. D. Roses, *Nat. Rev. Drug Discovery*, 2008, **7**, 807.
70. A. D. Roses, A. M. Saunders, Y. Huang, J. Strum, K. H. Weisgraber and R. W. Mahley, *Pharmacogenomics J.*, 2007, **7**, 10.
71. A. D. Roses, *Neuropsychopharmacology*, 2009, **34**, 6.
72. M. E. Risner, A. M. Saunders and J. F. Altman, *et al.*, *Pharmacogenomics J.*, 2006, **6**, 246.
73. N. Takei, A. Miyashita, T. Tsukie, G. C. Ormandy, S. Craft, I. M. Foley, M. E. Zvartau-Hind, D. A. Hosford and A. D. Roses, Rosiglitazone in Alzheimer's Disease Study Group, *Genomics*, 2009, **93**, 441.

74. S. G. Potkin, G. Guffanti, A. Lakatos, J. A. Turner, F. Kruggel, J. H. Fallon, A. J. Saykin, A. Orro, S. Lupoli, E. Salvi, M. Weiner and F. Macciardi, *PLoS One*, 2009, **4**, e6501.
75. A. D. Roses, *Arch. Neurol.*, 2010, **67**, 536.
76. A. D. Roses, M. W. Lutz, H. Amrine-Madsen, A. M. Saunders, D. G. Crenshaw, S. S. Sundseth, M. J. Huentelman, K. A. Welsh-Bohmer and E. M. Reiman, *Pharmacogenomics J.*, 2010, **10**, 375.
77. A. D. Roses, A. M. Saunders, M. W. Lutz, N. Zhang, A. R. Hariri, K. E. Asin, D. G. Crenshaw, K. Budur, D. K. Burns and S. K. Brannan, *Curr. Opin. Pharmacol.*, 2014, **14**, 81.
78. A. D. Roses, M. W. Lutz, D. G. Crenshaw, I. Grossman, A. M. Saunders and W. K. Gottschalk, *Alzheimer's Dementia*, 2013, **9**, 132.
79. M. W. Lutz, D. G. Crenshaw, A. M. Saunders and A. D. Roses, *Alzheimer's Dementia*, 2010, **6**, 125.
80. L. Shen, S. Kim, S. L. Risacher, K. Nho, S. Swaminathan, J. D. West, T. Foroud, N. Pankratz, J. H. Moore, C. D. Sloan, M. J. Huentelman, D. W. Craig, B. M. Dechairo, S. G. Potkin, C. R. Jack Jr., M. W. Weiner, A. J. Saykin and Alzheimer's Disease Neuroimaging Initiative, *Neuroimage*, 2010, **53**, 1051.
81. C. Linnertz, A. M. Saunders, M. W. Lutz, D. M. Crenshaw, I. Grossman, D. K. Burns, K. E. Whitfield, M. A. Hauser, J. J. McCarthy, M. Ulmer, R. Allingham, K. A. Welsh-Bohmer, A. D. Roses and O. Chiba-Falek, *PLoS One*, 2012, **7**, e30994.
82. A. Omoumi, A. Fok, T. Greenwood, A. D. Sadovnick, H. H. Feldman and G. Y. Hsiung, *Neurobiol. Aging*, 2014, **35**, 936.
83. S. Bagnoli, I. Piaceri, A. Tedde, V. Bessi, L. Bracco, S. Sorbi and B. Nacmias, *Neurol. Sci.*, 2013, **34**, 995.
84. X. Y. Ma, J. T. Yu, W. Wang, H. F. Wang, Q. Y. Liu, W. Zhang and L. Tan, *NeuroMol. Med.*, 2013, **15**, 279.
85. V. Valant, B. T. Keenan, C. D. Anderson, J. M. Shulman, W. J. Devan, A. M. Ayres, K. Schwab, J. N. Goldstein, A. Viswanathan, S. M. Greenberg, D. A. Bennett, P. L. De Jager, J. Rosand, A. Biffi and Alzheimer's Disease Neuroimaging Initiative (ADNI), *Transl. Stroke Res.*, 2012, **3**, 102.
86. M. Vounou, E. Janousova, R. Wolz, J. L. Stein, P. M. Thompson, D. Rueckert, G. Montana and Alzheimer's Disease Neuroimaging Initiative, *Neuroimage*, 2012, **60**, 700.
87. L. Bernardi, M. Gallo, M. Anfossi, M. E. Conidi, R. Colao, G. Puccio, S. A. Curcio, F. Frangipane, A. Clodomiro, M. Mirabelli, F. Vasso, N. Smirne, R. Di Lorenzo, R. Maletta and A. C. Bruni, *J. Alzheimer's Dis.*, 2013, **37**, 285.
88. P. Sebastiani, N. Solovieff, A. T. Dewan, K. M. Walsh, A. Puca, S. W. Hartley, E. Melista, S. Andersen, D. A. Dworkis, J. B. Wilk, R. H. Myers, M. H. Steinberg, M. Montano, C. T. Baldwin, J. Hoh and T. T. Perls, *PLoS One*, 2012, **7**, e29848.
89. M. Beekman, H. Blanché, M. Perola, A. Hervonen, V. Bezrukov, E. Sikora, F. Flachsbart, L. Christiansen, A. J. De Craen, T. B. Kirkwood, I. M. Rea, M. Poulain, J. M. Robine, S. Valensin, M. A. Stazi, G. Passarino, L. Deiana, E. S. Gonos, L. Paternoster, T. I. Sørensen, Q. Tan, Q. Helmer, E. B. van den Akker, J. Deelen, F. Martella, H. J. Cordell, K. L. Ayers,

J. W. Vaupel, O. Törnwall, T. E. Johnson, S. Schreiber, M. Lathrop, A. Skytthe, R. G. Westendorp, K. Christensen, J. Gampe, A. Nebel, J. J. Houwing-Duistermaat, P. E. Slagboom, C. Franceschi and GEHA consortium, *Aging Cell*, 2013, **12**, 184.
90. G. Davies, S. E. Harris, C. A. Reynolds, A. Payton, H. M. Knight, D. C. Liewald, L. M. Lopez, M. Luciano, A. J. Gow, J. Corley, R. Henderson, C. Murray, A. Pattie, H. C. Fox, P. Redmond, M. W. Lutz, O. Chiba-Falek, C. Linnertz, S. Saith, P. Haggarty, G. McNeill, X. Ke, W. Ollier, M. Horan, A. D. Roses, C. P. Ponting, D. J. Porteous, A. Tenesa, A. Pickles, J. M. Starr, L. J. Whalley, N. L. Pedersen, N. Pendleton, P. M. Visscher and I. J. Deary, *Mol. Psychiatry*, 2014, **19**, 76.
91. K. M. Hayden, J. M. McEvoy, C. Linnertz, D. Attix, M. Kuchibhatla, A. M. Saunders, M. W. Lutz, K. A. Welsh-Bohmer, A. D. Roses and O. Chiba-Falek, *Alzheimer's Dementia*, 2012, **8**, 381.
92. C. Linnertz, L. Anderson, W. Gottschalk, D. Crenshaw, M. W. Lutz, J. Allen, S. Saith, M. Mihovilovic, J. R. Burke, K. A. Welsh-Bohmer, A. D. Roses and O. Chiba-Falek, *Alzheimer's Dementia*, 2014, **10**, 541.
93. R. Cacabelos, D. Goldgaber, A. Vostrov, H. Matsuki, C. Torrellas, D. Corzo, J. C. Carril and A. D. Roses, *J. Pharmacogenomics Pharmacoproteomics*, 2014, **5**, 135.
94. H. Yoon, W. Myung, S. W. Lim, H. S. Kang, S. Kim, H. H. Won, B. J. Carroll and D. K. Kim, *Pharmacopsychiatry*, 2015, **48**, 111.
95. I. L. Braga, P. N. Silva, T. K. Furuya, L. C. Santos, B. C. Pires, D. R. Mazzotti, P. H. Bertolucci, M. S. Cendoroglo and M. C. Smith, *Am. J. Alzheimer's Dis. Other Dementias*, 2015, **30**, 139.
96. H. G. Xie, R. B. Kim, A. J. Wood and C. M. Stein, *Annu. Rev. Pharmacol. Toxicol.*, 2001, **41**, 815.
97. R. Cacabelos, *Expert Opin. Pharmacother.*, 2005, **6**, 1967.
98. R. Cacabelos, *Neuropsychiatr. Dis. Treat.*, 2007, **3**, 303.
99. M. W. Jann, K. L. Shirley and G. W. Small, *Clin. Pharmacokinet.*, 2002, **41**, 719.
100. M. Noetzli and C. B. Eap, *Clin. Pharmacokinet.*, 2013, **52**, 225.
101. A. Pilotto, M. Franceschi, G. D'Onofrio, A. Bizzarro, F. Mangialasche, L. Cascavilla, F. Paris, M. G. Matera, A. Pilotto, A. Daniele, P. Mecocci, C. Masullo, B. Dallapiccola and D. Seripa, *Neurology*, 2009, **73**, 761.
102. D. Albani, F. Martinelli Boneschi, G. Biella, G. Giacalone, S. Lupoli, F. Clerici, L. Benussi, R. Ghidoni, D. Galimberti, R. Squitti, S. Mariani, A. Confaloni, G. Bruno, C. Mariani, E. Scarpini, G. Binetti, G. Magnani, M. Franceschi and G. Forloni, *J. Alzheimer's Dis.*, 2012, **30**, 745.
103. M. Savino, D. Seripa, A. P. Gallo, M. Garrubba, G. D'Onofrio, A. Bizzarro, G. Paroni, F. Paris, P. Mecocci, C. Masullo, A. Pilotto and S. A. Santini, *Clin. Lab.*, 2011, **57**, 887.
104. D. Seripa, A. Bizzarro, A. Pilotto, G. D'onofrio, G. Vecchione, A. P. Gallo, L. Cascavilla, F. Paris, E. Grandone, P. Mecocci, S. A. Santini, C. Masullo and A. Pilotto, *Pharmacogenet. Genomics*, 2011, **21**, 225.
105. Y. Zhong, X. Zheng, Y. Miao, L. Wan, H. Yan and B. Wang, *Am. J. Med. Sci.*, 2013, **345**, 222.

106. F. Varsaldi, G. Miglio, M. G. Scordo, M. L. Dahl, L. M. Villa, A. Biolcati and G. Lombardi, *Eur. J. Clin. Pharmacol.*, 2006, **62**, 721.
107. C. Chianella, D. Gragnaniello, P. Maisano Delser, M. F. Visentini, E. Sette, M. R. Tola, G. Barbujani and S. Fuselli, *Eur. J. Clin. Pharmacol.*, 2011, **67**, 1147.
108. B. Mannheimer, B. Wettermark, M. Lundberg, H. Pettersson, C. von Bahr and E. Eliasson, *Br. J. Clin. Pharmacol.*, 2010, **69**, 411.
109. L. Magliulo, M. L. Dahl, G. Lombardi, S. Fallarini, L. M. Villa, A. Biolcati and M. G. Scordo, *Eur. J. Clin. Pharmacol.*, 2011, **67**, 47.
110. L. F. Miranda, K. B. Gomes, J. N. Silveira, G. A. Pianetti, R. M. Byrro, P. R. Peles, F. H. Pereira, T. R. Santos, A. G. Assini, V. V. Ribeiro, P. A. Tito, R. O. Matoso, T. O. Lima, E. N. Moraes and P. Caramelli, *J. Alzheimer's Dis.*, 2015, **45**, 609.
111. S. Lilienfeld, *CNS Drug Rev.*, 2002, **8**, 159.
112. M. R. Farlow, *Clin. Pharmacokinet.*, 2003, **42**, 1383.
113. Q. Zhao, M. Brett, N. Van Osselaer, F. Huang, A. Raoult, A. Van Peer, T. Verhaeghe and R. Hust, *J. Clin. Pharmacol.*, 2002, **42**, 1002.
114. G. S. Mannens, C. A. Snel, J. Hendrickx, T. Vehaeghe, L. Le Jeune, W. Bode, L. van Beijsterveldt, K. Lavrijsen, J. Leempoels, N. Van Osselaer, A. Van Peer and W. Meuldermans, *Drug Metab. Dispos.*, 2002, **30**, 553.
115. M. Noetzli, M. Guidi, K. Ebbing, S. Eyer, S. Zumbach, P. Giannakopoulos, A. von Gunten, C. Csajka and C. B. Eap, *Ther. Drug Monit.*, 2013, **35**, 270.
116. J. A. Clarke, M. Cutler, I. Gong, U. I. Schwarz, D. Freeman and M. Dasgupta, *Am. J. Geriatr. Pharmacother.*, 2011, **9**, 224.
117. D. Bentué-Ferrer, O. Tribut, E. Polard and H. Allain, *CNS Drugs*, 2003, **17**, 947.
118. F. Huang and Y. Fu, *Curr. Clin. Pharmacol.*, 2010, **5**, 115.
119. X. J. Zhai and Y. N. Lu, *Xenobiotica*, 2012, **42**, 1151.
120. R. J. Polinsky, *Clin. Ther.*, 1998, **20**, 634.
121. N. Sonali, M. Tripathi, R. Sagar, T. Velpandian and V. Subbiah, *CNS Neurosci. Ther.*, 2013, **19**, 91.
122. A. Lupp, D. Appenroth, L. Fang, M. Decker, J. Lehmann and C. Fleck, *Arzneimittelforschung*, 2010, **60**, 229.
123. A. Alfirevic, T. Mills, D. Carr, B. J. Barratt, A. Jawaid, J. Sherwood, J. C. Smith, J. Tugwood, R. Hartkoorn, A. Owen and K. B. Park, *Pharmacogenet. Genomics*, 2007, **17**, 1091.
124. Z. Yang, X. Zhou and Q. Zhang, *J. Alzheimer's Dis.*, 2013, **36**, 445.
125. S. Micuda, L. Mundlova, E. Anzenbacherova, P. Anzenbacher, J. Chladek, L. Fuksa and J. Martinkova, *Eur. J. Clin. Pharmacol.*, 2004, **60**, 583.
126. M. Noetzli, M. Guidi, K. Ebbing, S. Eyer, L. Wilhelm, A. Michon, V. Thomazic, A. M. Alnawaqil, S. Maurer, S. Zumbach, P. Giannakopoulos, A. von Gunten, C. Csajka and C. B. Eap, *Clin. Pharmacokinet.*, 2013, **52**, 211.
127. B. Marquez and F. Van Bambeke, *Curr. Drug Targets*, 2011, **12**, 600.
128. V. Haufroid, *Curr. Drug Targets*, 2011, **12**, 631.
129. R. Cacabelos and F. López-Muñoz, *Clin. Exp. Pharmacol.*, 2014, **4**, 128.

130. G. Li, H. M. Gu and D. W. Zhang, *IUBMB Life*, 2013, **65**, 505.
131. D. M. van Assema, M. Lubberink, P. Rizzu, J. C. van Swieten, R. C. Schuit, J. Eriksson, P. Scheltens, M. Koepp, A. A. Lammertsma and B. N. van Berckel, *EJNMMI Res.*, 2012, **2**, 57.
132. K. D. Chen, P. T. Chang, Y. H. Ping, H. C. Lee, C. W. Yeh and P. N. Wang, *Neurobiol. Dis.*, 2011, **43**, 698.
133. A. Elali and S. Rivest, *Front. Physiol.*, 2013, **4**, 45.
134. I. Cascorbi, C. Fluh, C. Remmler, S. Haenisch, F. Faltraco, M. Grumbt, M. Peters, A. Brenn, D. R. Thal, R. W. Warzok and S. Vogelgesang, *Pharmacogenomics*, 2013, **14**, 485.
135. C. Reitz, G. Jun, A. Naj, R. Rajbhandary, B. N. Vardarajan, L. S. Wang, O. Valladares, C. F. Lin, E. B. Larson, N. R. Graff-Radford, D. Evans, P. L. De Jager, P. K. Crane, J. D. Buxbaum, J. R. Murrell, T. Raj, N. Ertekin-Taner, M. Logue, C. T. Baldwin, R. C. Green, L. L. Barnes, L. B. Cantwell, M. D. Fallin, R. C. Go, P. Griffith, T. O. Obisesan, J. J. Manly, K. L. Lunetta, M. I. Kamboh, O. L. Lopez, D. A. Bennett, H. Hendrie, K. S. Hall, A. M. Goate, G. S. Byrd, W. A. Kukull, T. M. Foroud, J. L. Haines, L. A. Farrer, M. A. Pericak-Vance, G. D. Schellenberg, R. Mayeux and C. Alzheimer Disease Genetics, *JAMA*, 2013, **309**, 1483.
136. A. Feher, A. Juhasz, A. Laszlo, M. Pakaski, J. Kalman and Z. Janka, *Neurosci. Lett.*, 2013, **550**, 51.
137. R. Cacabelos, *Metabolomics*, 2012, **2**, 119.
138. K. Hosoya and M. Tachikawa, *Clin. Exp. Nephrol.*, 2011, **15**, 478.
139. S. M. Carl, D. J. Lindley, D. Das, P. O. Couraud, B. B. Weksler, I. Romero, S. A. Mowery and G. T. Knipp, *Mol. Pharm.*, 2010, **7**, 1057.
140. S. A. Tammen, S. Friso and S. W. Choi, *Mol. Aspects Med.*, 2013, **34**, 753.
141. J. D. Sweatt, *Neuron*, 2013, **80**, 624.
142. R. A. Harris, D. Nagy-Szakal and R. Kellermayer, *Epigenetics*, 2013, **8**, 157.
143. K. E. Szulwach and P. Jin, *Bioessays*, 2014, **36**, 107.
144. J. A. Fahrner and H. T. Bjornsson, *Annu. Rev. Genomics Hum. Genet.*, 2014, **15**, 269.
145. H. Heyn, S. Moran, I. Hernando-Herraez, S. Sayols, A. Gomez, J. Sandoval, D. Monk, K. Hata, T. Marques-Bonet, L. Wang and M. Esteller, *Genome Res.*, 2013, **23**, 1363.
146. V. Iacobazzi, A. Castegna, V. Infantino and G. Andria, *Mol. Genet. Metab.*, 2013, **110**, 25.
147. T. Kubota, H. Takae and K. Miyake, *Pharmaceuticals*, 2012, **5**, 369.
148. M. G. Kang, K. Byun, J. H. Kim, N. H. Park, H. Heinsen, R. Ravid, H. W. Steinbusch, B. Lee and Y. M. Park, *Biochim. Biophys. Acta*, 2015, **1854**, 788.
149. P. Tognini, D. Napoli and T. Pizzorusso, *Front. Cell. Neurosci.*, 2015, **9**, 331.
150. H. Harony-Nicolas, S. Mamrut, L. Brodsky, H. Shahar-Gold, L. Barki-Harrington and S. Wagner, *Psychoneuroendocrinology*, 2014, **39**, 121.
151. H. F. Rodrigues, T. A. Souza, F. G. Ghiraldini, M. L. Mello and A. S. Moraes, *J. Cell. Biochem.*, 2014, **115**, 659.

152. Y. Q. Zhao, I. K. Jordan and V. V. Lunyak, *Neurotherapeutics*, 2013, **10**, 647.
153. J. Wang, J. T. Yu, M. S. Tan, T. Jiang and L. Tan, *Ageing Res. Rev.*, 2013, **12**, 1024.
154. K. Pan, Y. Chen, M. Roth, W. Wang, S. Wang, A. S. Yee and X. Zhang, *Mol. Cell. Biol.*, 2013, **33**, 887.
155. Z. Xu and J. A. Taylor, *Carcinogenesis*, 2014, **35**, 356.
156. J. L. McClay, K. A. Aberg, S. L. Clark, S. Nerella, G. Kumar, L. Y. Xie, A. D. Hudson, A. Harada, C. M. Hultman, P. K. Magnusson, P. F. Sullivan and E. J. Van Den Oord, *Hum. Mol. Genet.*, 2014, **23**, 1175.
157. M. Wagner, J. Steinbacher, T. F. Kraus, S. Michalakis, B. Hackner, T. Pfaffeneder, A. Perera, M. Muller, A. Giese, H. A. Kretzschmar and T. Carell, *Angew. Chem., Int. Ed. Engl.*, 2015, **54**, 12511.
158. A. Nakamura, K. Kawakami, F. Kametani, H. Nakamoto and S. Goto, *Ann. N. Y. Acad. Sci.*, 2010, **1197**, 33.
159. C. M. Wang, S. N. Tsai, T. W. Yew, Y. W. Kwan and S. M. Ngai, *Biogerontology*, 2010, **11**, 87.
160. L. Guarente and F. Picard, *Cell*, 2005, **120**, 473.
161. M. C. Haigis and D. A. Sinclair, *Annu. Rev. Pathol.*, 2010, **5**, 253.
162. G. Raddatz, S. Hagemann, D. Aran, J. Sohle, P. P. Kulkarni, L. Kaderali, A. Hellman, M. Winnefeld and F. Lyko, *Epigenet. Chromatin*, 2013, **6**, 36.
163. F. Spallotta, C. Cencioni, S. Straino, G. Sbardella, S. Castellano, M. C. Capogrossi, F. Martelli and C. Gaetano, *Commun. Integr. Biol.*, 2013, **6**, e25466.
164. N. Noren Hooten, K. Abdelmohsen, M. Gorospe, N. Ejiogu, A. B. Zonderman and M. K. Evans, *PLoS One*, 2010, **5**, e10724.
165. M. Hackl, S. Brunner, K. Fortschegger, C. Schreiner, L. Micutkova, C. Muck, G. T. Laschober, G. Lepperdinger, N. Sampson, P. Berger, D. Herndler-Brandstetter, M. Wieser, H. Kuhnel, A. Strasser, M. Rinnerthaler, M. Breitenbach, M. Mildner, L. Eckhart, E. Tschachler, A. Trost, J. W. Bauer, C. Papak, Z. Trajanoski, M. Scheideler, R. Grillari-Voglauer, B. Grubeck-Loebenstein, P. Jansen-Durr and J. Grillari, *Aging Cell*, 2010, **9**, 291.
166. N. Li, D. J. Bates, J. An, D. A. Terry and E. Wang, *Neurobiol. Aging*, 2011, **32**, 944.
167. R. Lardenoije, A. Iatrou, G. Kenis, K. Kompotis, H. W. Steinbusch, D. Mastroeni, P. Coleman, C. A. Lemere, P. R. Hof, D. L. van den Hove and B. P. Rutten, *Prog. Neurobiol.*, 2015, **131**, 21.
168. D. Mastroeni, A. Grover, E. Delvaux, C. Whiteside, P. D. Coleman and J. Rogers, *Neurobiol. Aging*, 2011, **32**, 1161.
169. J. Graff and L. H. Tsai, *Annu. Rev. Pharmacol. Toxicol.*, 2013, **53**, 311.
170. L. Adwan and N. H. Zawia, *Pharmacol. Ther.*, 2013, **139**, 41.
171. D. L. Van den Hove, K. Kompotis, R. Lardenoije, G. Kenis, J. Mill, H. W. Steinbusch, K. P. Lesch, C. P. Fitzsimons, B. De Strooper and B. P. Rutten, *Neurobiol. Aging*, 2014, **35**, 731.
172. P. Wu, X. Zuo, H. Deng, X. Liu, L. Liu and A. Ji, *Brain Res. Bull.*, 2013, **97**, 69.
173. I. A. Qureshi and M. F. Mehler, *Curr. Neurol. Neurosci. Rep*, 2011, **11**, 464.
174. A. M. Enciu, B. O. Popescu and A. Gheorghisan-Galateanu, *Mol. Biol. Rep.*, 2012, **39**, 2243.

175. B. Mallick and Z. Ghosh, *RNA Biol.*, 2011, **8**, 665.
176. R. Cacabelos, *Metabolomics*, 2015, **5**, 3.
177. X. Tang and S. Chen, *Curr. Drug Metab.*, 2015, **16**, 86.
178. H. Kang, C. Kim, H. Lee, W. Kim and E. K. Lee, *Int. J. Mol. Sci.*, 2013, **14**, 17204.
179. I. W. Kim, N. Han, G. J. Burckart and J. M. Oh, *Pharmacotherapy*, 2014, **34**, 140.
180. R. Cacabelos, *Metabolomics*, 2015, **5**, e141.
181. M. A. Varela, T. C. Roberts and M. J. Wood, *Neurotherapeutics*, 2013, **10**, 621.
182. F. Caraci, G. M. Leggio, F. Drago and S. Salomone, *Br. J. Clin. Pharmacol.*, 2013, **75**, 1154.
183. J. Wang, Z. Zhao, E. Lin, W. Zhao, X. Qian, D. Freire, A. E. Bilski, A. Cheng, P. Vempati, L. Ho, K. Ono, M. Yamada and G. M. Pasinetti, *PLoS One*, 2013, **8**, e65232.
184. F. Montastruc, V. Gardette, C. Cantet, A. Piau, M. Lapeyre-Mestre, B. Vellas, J. L. Montastruc, S. Andrieu and R. F. Group, *Eur. J. Clin. Pharmacol.*, 2013, **69**, 1589.
185. L. Jeffrey, J. L. Cummings, T. Morstorf and K. Zhong, *Alzheimer's Res. Ther.*, 2014, **6**, 37.
186. M. Bond, G. Rogers, J. Peters, R. Anderson, M. Hoyle, A. Miners, T. Moxham, S. Davis, P. Thokala, A. Wailoo, M. Jeffreys and C. Hyde, *Health Technol. Assess.*, 2012, **16**, 1.
187. Y. Otsubo, *Clin. Ther.*, 2015, **37**, 1627.
188. H. Liu, L. Wang, M. Lv, R. Pei, P. Li, Z. Pei, Y. Wang, W. Su and X. Q. Xie, *J. Chem. Inf. Model.*, 2014, **54**, 1050.
189. V. R. Lombardi and R. Cacabelos, *Drugs Future*, 1999, **24**, 167.
190. R. Cacabelos, A. I. Vallejo, V. R. Lombardi, L. Fernández-Novoa and V. Pichel, *CNS Drug Rev.*, 2004, **10**, 200.
191. V. R. Lombardi, A. Cagiao, L. Fernández-Novoa, X. A. Álvarez, M. D. Corzo, R. Zas, C. Sampedro and R. Cacabelos, *Nutr. Res.*, 2001, **21**, 1425.
192. V. R. Lombardi, L. Fernandez-Novoa, D. Corzo, R. Zas and R. Cacabelos, *Methods Find. Exp. Clin. Pharmacol.*, 2002, **24**, 573.
193. V. R. Lombardi, L. Fernandez-Novoa, I. Etcheverria, S. Seoane and R. Cacabelos, *Int. Immunopharmacol.*, 2005, **5**, 253.
194. V. R. Lombardi, J. Pereira, I. Etcheverría, L. Fernández-Novoa, S. Seoane and R. Cacabelos, *Food Agric. Immunol.*, 2006, **17**, 115.
195. R. Cacabelos and M. Takeda, *Drugs Future*, 2006, **31**, 146.
196. L. Corzo, S. Rodríguez, V. R. Lombardi, R. Alejo and R. Cacabelos, *EuroEspes J. (Gen-T)*, 2013, **9**, 31.
197. J. Landgrave-Gomez, O. Mercado-Gomez and R. Guevara-Guzman, *Front. Cell. Neurosci.*, 2015, **9**, 58.
198. J. Mill, *Neurobiol. Aging*, 2011, **32**, 1188.
199. M. Vermulst, A. S. Denney, M. J. Lang, C. W. Hung, S. Moore, M. A. Moseley, J. W. Thompson, V. Madden, J. Gauer, K. J. Wolfe, D. W. Summers, J. Schleit, G. L. Sutphin, S. Haroon, A. Holczbauer, J. Caine, J. Jorgenson, D. Cyr, M. Kaeberlein, J. N. Strathern, M. C. Duncan and D. A. Erie, *Nat. Commun.*, 2015, **6**, 8065.

200. B. E. Riley, S. J. Gardai, D. Emig-Agius, M. Bessarabova, A. E. Ivliev, B. Schule, J. Alexander, W. Wallace, G. M. Halliday, J. W. Langston, S. Braxton, T. Yednock, T. Shaler and J. A. Johnston, *PLoS One*, 2014, **9**, e102909.
201. M. Del Campo, W. Jongbloed, H. A. Twaalfhoven, R. Veerhuis, M. A. Blankenstein and C. E. Teunissen, *Front. Neurol.*, 2015, **6**, 202.
202. G. Shevchenko, A. Konzer, S. Musunuri and J. Bergquist, *Biochim. Biophys. Acta*, 2015, **1854**, 705.
203. D. S. Spellman, K. R. Wildsmith, L. A. Honigberg, M. Tuefferd, D. Baker, N. Raghavan, A. C. Nairn, P. Croteau, M. Schirm, R. Allard, J. Lamontagne, D. Chelsky, S. Hoffmann and W. Z. Potter, *Proteomics: Clin. Appl.*, 2015, **9**, 715.
204. S. Lista, S. E. O'Bryant, K. Blennow, B. Dubois, J. Hugon, H. Zetterberg and H. Hampel, *J. Alzheimer's Dis.*, 2015, **47**, 291.
205. I. F. Harrison and D. T. Dexter, *Pharmacol. Ther.*, 2013, **140**, 34.
206. A. Nebbioso, V. Carafa, R. Benedetti and L. Altucci, *Mol. Oncol.*, 2012, **6**, 657.
207. M. Rasool, A. Malik, M. I. Naseer, A. Manan, S. Ansari, I. Begum, M. H. Qazi, P. Pushparaj, A. M. Abuzenadah, M. H. Al-Qahtani, M. A. Kamal and S. Gan, *BMC Med. Genomics*, 2015, **8**(suppl. 1), S5.
208. A. A. Monte, C. Brocker, D. W. Nebert, F. J. Gonzalez, D. C. Thompson and V. Vasiliou, *Hum. Genomics*, 2014, **8**, 16.
209. L. Siggens and K. Ekwall, *J. Intern. Med.*, 2014, **276**, 201.
210. ENCODE Project Consortium, *Nature*, 2012, **489**, 57.
211. E. Pennisi, *Science*, 2012, **337**, 1159.
212. M. Kundakovic and F. A. Champagne, *Neuropsychopharmacology*, 2015, **40**, 141.
213. K. M. Bakulski, L. S. Rozek, D. C. Dolinoy, H. L. Paulson and H. Hu, *Curr. Alzheimer Res.*, 2012, **9**, 563.
214. M. J. Dauncey, *Nutrients*, 2013, **5**, 887.
215. J. Liu, S. R. Zhao and T. Reyes, *Int. J. Mol. Sci.*, 2015, **16**, 18129.
216. J. J. Day, A. J. Kennedy and J. D. Sweatt, *Annu. Rev. Pharmacol. Toxicol.*, 2015, **55**, 591.
217. G. P. Reynolds, O. O. McGowan and C. F. Dalton, *Br. J. Clin. Pharmacol.*, 2014, **77**, 654.
218. R. Cacabelos and C. Torrellas, in *Pharmacoepigenetics*, ed. T. Tollefsbol, Medical Epigenetics, Elsevier, USA, 2015.
219. U. M. Zanger and M. Schwab, *Pharmacol. Ther.*, 2013, **138**, 103.
220. U. M. Zanger, K. Klein, M. Thomas, J. K. Rieger, R. Tremmel, B. A. Kandel, M. Klein and T. Magdy, *Clin. Pharmacol. Ther.*, 2014, **95**, 258.
221. C. I. Sze, W. P. Su, M. F. Chiang, C. Y. Lu, Y. A. Chen and N. S. Chang, *Front. Oncol.*, 2013, **3**, 59.
222. S. Gupta, S. Verma, S. Mantri, N. E. Berman and R. Sandhir, *Drug Dev. Res.*, 2015, **76**, 397.
223. E. Junn and M. M. Mouradian, *Pharmacol. Ther.*, 2012, **133**, 142.
224. S. Parsi, P. Y. Smith, C. Goupil, V. Dorval and S. S. Hebert, *Mol. Ther.–Nucleic Acids*, 2015, **4**, e256.

CHAPTER 6

Nanotechnology in Anti-Aging: Nutraceutical Delivery and Related Applications

ANILA MATHEW[a], FRANCESCO MAROTTA*[b] AND D. SAKTHI KUMAR*[a]

[a]Bio-Nano Electronics Research Center, Graduate School of Interdisciplinary New Science, Toyo University, 2100 Kujirai, Kawagoe-shi, Saitama-ken 350 8585, Japan; [b]ReGenera Research Group for Aging-Intervention and Montenapoleone Clinic – Gender Healthy Aging Unit by Genomics and Biotechnology, Corso Matteotti 1/A 20121, Milano, Italy
*E-mail: fmarchimede@libero.it, sakthi@toyo.jp

6.1 Introduction

Hippocrates (460–370 BC), the Father of Western Medicine, stated "Let food be thy medicine and medicine be thy food".[1] Bioactive compounds of herbal origin have a long history of use in various traditional medical systems.[2] In the last two decades, exponential growth in the research on 'nutraceuticals' and 'functional foods' has been observed, with some studies concentrating on natural products and some concentrating on specific phytochemicals.[1] Nowadays, people are more aware and concerned about their health and well-being than they used to be earlier, and take measures to maintain good health. Nutraceuticals are foods or naturally occurring food supplements,

RSC Drug Discovery Series No. 57
Anti-aging Drugs: From Basic Research to Clinical Practice
Edited by Alexander M. Vaiserman

Published by the Royal Society of Chemistry, www.rsc.org

including vitamins, minerals and herbal products, that provide additional nutritional value and health benefits, including disease prevention.[1,3] In 1989, Dr Stephen DeFelice coined the term "nutraceutical" and it is now applied to products including herbal formulations, nutrients and dietary supplements.[4] The idea of functional food was first used in 1984 by Japanese researchers who studied the relationship between nutritional quality, sensory satisfaction and fortification of foods for advantageous physiological effects.[5] Similar to nutraceuticals, functional foods tend to improve general health, and have prophylactic and therapeutic effects.

In this chapter, we discuss the use of nanotechnology in the development of various nutraceuticals and functional foods in anti-aging medicine.

6.2 Nutraceuticals and Nanodevelopments

Nutraceuticals and functional foods encompass a large group of compounds including polyphenols, flavonoids, carotenoids, vitamins, minerals and probiotics, and all of these improve and enhance health and wellbeing.[6–8] An increasing number of bioactive compounds are now used in the food industry as supplements and the number of patents relating to the development and use of bioactive compounds has also increased. Two reviews that list recent patents on bioactive compounds used in food industry were published recently.[9,10] The identification, isolation and use of bioactives in food and related applications have increased dramatically in recent years, and one reason for this is the developments in the field of nanotechnology. Nanotechnology can help in overcoming many limitations of using bioactive compounds in food-related applications, such as solubility, color, flavor, texture, bioavailability and absorption.[9,11–14]

Many nutraceuticals are rich in antioxidants and anti-inflammatory bioactive compounds, and may be either phenolic compounds, phytosterols, carotenoids, polyunsaturated fatty acids, flavonoids or probiotics. Polyphenols are a large group of plant secondary metabolites that have high nutritive value and have recently attracted interest as a source of functional foods and nutraceuticals.[15] Unpleasant flavor, reactive nature and low bioavailability are issues that are important while considering the use of polyphenolics in food. Spray drying, entrapping in liposomes, nanoencapsulation and nanoemulsions are some methods of polyphenol delivery. Fang and Bhandari have discussed various methods by which polyphenols can be encapsulated.[16] A review discussing the unique potential of nano-antioxidants against neurodegenerative diseases was published recently.[17] Oxidative stress is a serious issue in neurodegenerative diseases and it is believed that nano-antioxidants have the capacity to offer effective preventive and therapeutic functions.[17] Liposomes are versatile carriers capable of delivering hydrophilic, hydrophobic and amphiphilic compounds, and are even capable of encapsulating multiple antioxidants. Liposomes are excellent carriers of antioxidants and facilitate prophylactic and therapeutic effects against oxidative stress.[18] Various nutraceuticals isolated from fruits, vegetables and spices can effectively and safely suppress the proinflammatory pathways. Inflammatory stress plays a key role in various age-related diseases, including cancer, neurodegenerative diseases, arthritis, and cardiovascular diseases, to cite a few. In a review that discusses the recent developments in anti-inflammatory nutraceuticals, the authors discuss their potential as targeted anti-cancerous agents.[19] Common anti-inflammatory nutraceuticals and the existing formulation strategies that help in nanodelivery of some of these nutraceuticals are also discussed in this review.[19] Anthocyanins, members of the flavonoid group of phytochemicals, are amazing compounds with broad-ranging health benefits and are used as natural colorants and nutraceuticals. It is predominant in honey, fruits like grapes, various berries, plums, vegetables like eggplant and red cabbage, kidney beans and black beans, olives and olive oil, cocoa and nuts. Although having prophylactic effect against a wide range of age-related disorders, the commercial application is limited due to the poor chemical stability.[20] Carotenoid pigments are a group of lipophilic bioactive phytochemicals that are responsible for the brilliant color in many fruits and vegetables. There are over 700 carotenoid pigments in nature, but only about 40 are absorbed and used by our body and the bioavailability from food is low. With very positive impact on health and well-being, including anti-oxidant activity and provitamin-A capacity, carotenoids are of great interest to nutritionists and food scientists.[21] Soukoulis and Bohn presented a comprehensive review on the advances in micro and nanoencapsulation technologies for improving the stability and bioavailability of carotenoids for use in food applications.[22] Factors affecting the stability of vitamin A, the available carriers and techniques for stabilizing vitamin A, and the respective formulation methods were discussed by Loveday and Singh. The authors point out that synergistic protective effects were observed when technologies were used in combination while developing vitamin A delivery systems.[23] Curcumin is another polyphenolic compound that is renowned for its various nutraceutical properties. Isolated from *Curcuma longa*, curcumin is a hydrophobic compound with low bioavailability when administered orally. Using nanotechnology, various nanoformulations of curcumin have been developed in recent years, with potential applications against cancers, neurodegenerative diseases, cardiovascular diseases and microbial infections.[24–26]

Dietary intake of essential fatty acids is very important as they are not produced in the body. These extranutritional constituents can act as vasodilators, antihypertensives, anti-inflammatory agents and anti-atherothrombotic compounds, and are proven to have beneficial effects against various diseases, including cardiovascular disease, cancer and schizophrenia.[27] Long chain omega-3 polyunsaturated fatty acids are important for normal metabolism and are acclaimed for their varied properties, including lowering of blood serum triglycerol and cholesterol.[28] With higher concentrations of eicosapentaenoic acid (EPA) and docosahexaenoic acid (DHA), fish oil is a better source of omega-3 fatty acid than plants, while plants have higher levels of α-linoleic acid (ALA), but its incorporation into edibles is challenging due to poor water solubility and oxidation.[29]

Another interesting feature affecting various functional foods is fermentation, as in the case of several Asian cuisines. For instance, fermentation may significantly modify the original active ingredient. In the case of Fermented Papaya Preparation (FPP) the ratio between complex carbohydrates and proteins increases significantly on fermentation. While papaya *per se* has significant antioxidant capacity, FPP exerts an epigenetic effect to beneficially balance the whole redox environment.[30] Similar observations open the way to applications using nanotechnology aimed at cellular targeting of functional food and bioactive compounds over the classical dose approach.

Nobel laureate Elie Metchnikoff recognized and suggested the concept of probiotics in the early 1900s while studying gut flora, and later, the term "probiotics" was coined by Lilly and Stillwell in 1965.[31] Probiotics are viable microorganisms that have a symbiotic relation with the host and play a role in modulating the mucosal and systemic immunity of the host within a wider metagenome playground. In recent years, based on Metchnikoff's perspective, the important role of microbial milieu in the oral cavity has been pointed out as a potential source of triggering factor in systemic illnesses.[32] Thus, this represents an amenable opportunity to rational interventions, but nanotechnology might be required for solving nutrient delivery strategies. Probiotics are chosen as novel potential weapons in global health strategies when considering the solid experimental and preliminary clinical literature pointing out the role of gut microbiota in virtually all chronic illnesses, from metabolic to neurodegenerative diseases, and probiotic interventions have been proved to have a pro-longevity effect on an experimental basis.[33–36] Various strains of lactic acid bacteria like lactobacilli, bifidobacteria and streptococci are considered as probiotic and the use of prebiotics like inulin, oligofructose, and galactooligosaccharides along with probiotics have been reported to support the growth of probiotics in the gut by promoting their function and/or viability *via* fermentation.[37,38] Heidarpour and coworkers reviewed the available nanocarrier systems utilizing prebiotics developed for oral delivery of bioactive compounds including vaccines, vitamin, hormones, nutraceuticals, minerals and food supplements. According to the authors, oral delivery has the best patient compliance, and prebiotic-based bioactive delivery systems can be used not only for humans, but also in veterinary therapeutic applications.[39] Microencapsulated probiotics have received a great deal of attention and have provided evidence of their higher efficacy on allergic and metabolic disease treatment, and for antimutagenic properties.[40–44] The opportunities of evolving nanotechnology related applications using probiotics hold great promise when considering the potential of probiotics that go well beyond basic gut function, nutrients and vitamin handling; probiotics are important in modulating endocrine systems, behavior and neuromodulation from birth to adult chronic and acute brain disease.[45–52] Recently, microbiome study has come up with "New Directions in Cardiovascular Disease Research, Prevention, and Treatment", as stated by the American Heart Association.[49] This is another area of paramount importance where nanotechnologies applied to bioactive compounds and specific probiotics may open up new and promising avenues to pursue.

6.3 Nanoformulations of Bioactive Compounds

Nanotechnology has revolutionized various fields of science and developed novel applications in several fields of industry, including the food industry. Nanotechnology has answered many of the quests for healthier food and related products, and shows increasing consumer acceptability. At the nanoscale, macroscale properties like sensory attributes, processability and stability of food materials or active compounds are modified. Nanotechnology can also help in designing bioactive functional food ingredients with improved physical properties, like water solubility, thermal and chemical stability, physiological performance and bioavailability, along with sensory attributes. Bioactive functional food at nanoscale is engineered by various mechanisms like encapsulation of active materials into nanoparticles or nanocapsules, formation of nanoemulsions and liposomes, incorporation into nanofibers, to list a few.[14,53–58] Many bioactive compounds that are used as nutraceuticals are highly lipophilic, which affects their bioavailability when administered orally. Excipient food matrices can help in solubilizing, transporting and controlling the release, metabolism and absorption of bioactive compounds. While the excipient ingredients do not have any bioactivity by themselves, these are able to promote the bioactivity of the co-ingested bioactives, even when administered orally. It is possible to engineer nanocolloidal systems, micro/nanoemulsions and solid lipid nanoparticles with bioactive compounds using excipient food matrices.[59] A review highlighting developments in nanodelivery systems that can overcome the challenges in incorporating lipophilic bioactives into food was published recently. Various delivery systems like emulsions, microgels and biopolymer nanoparticles can be used for this purpose (Figure 6.1).[60] The author suggests that future research should focus on developing commercially viable delivery systems.[60] While there are many reports of technologies that support the use of hydrophobic compounds, there are not many that discuss the developments in hydrophilic compounds. Development of delivery systems using hydrophilic compounds is also associated with various challenges. Liposomes, multiple emulsions, biopolymer particles and solid fat particles are some of the techniques used to develop delivery systems using hydrophilic bioactive compounds like water-soluble colors, preservatives, flavors, enzymes and vitamins.[61] Lipid based formulations are one of the most commonly used delivery system in food related application. A review on four lipid-based encapsulation systems discusses their fabrication methods, physicochemical properties and potential advantages and disadvantages when used as a delivery system.[62] Nanoemulsions, liposomes, solid lipid nanoparticles and nanostructured lipid carriers were the four lipid-based delivery system discussed in the review. The authors point out that along with the physicochemical analyses of the nanocarriers, studies on the interactions of food systems with nanoencapsulated bioactive compounds are also required.[62]

Food proteins are a versatile matrix choice for incorporating nutraceutical compounds to create a wide range of multicomponent matrices. Apart from

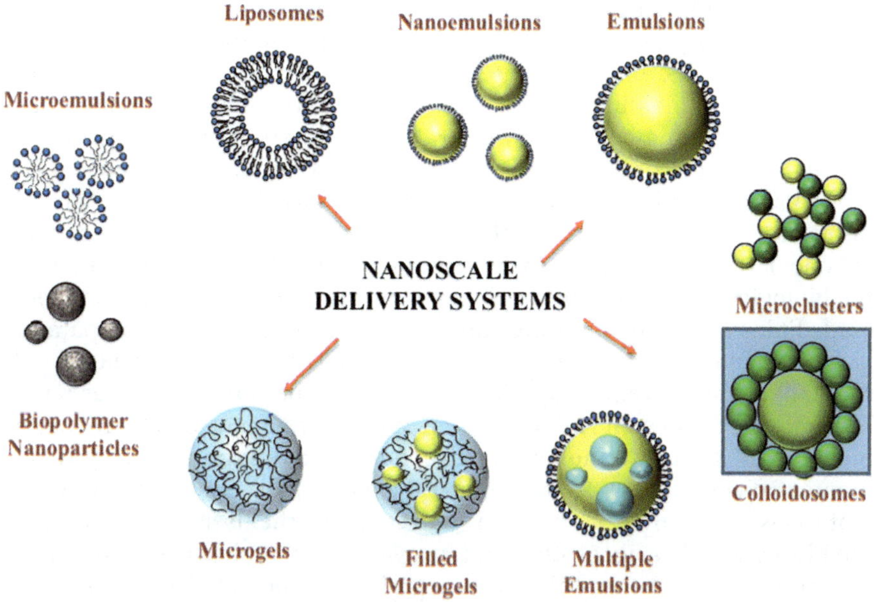

Figure 6.1 Some examples of nanoscale delivery system that can be used to deliver lipophilic bioactive compounds. Reprinted with permission from ref. 60. © 2015 Institute of Food Technologists®.

their unique physicochemical properties, high nutritional value, acceptability as natural food components, easy degradability by digestive enzymes and easy availability are some of the advantages of food protein matrices.[63] With the advent of more ecofriendly and sustainable ways to reuse industrial waste, especially in food and related industries, novel strategies are being developed to reuse them for alternative uses. In recent years, agricultural by-products have been recognized as a source of functional ingredients, including active compounds and dietary fibers. This has been achieved due to the development of technologies that help in the recovery of nutraceuticals from agro-wastes, and identification of novel approaches to reuse the recovered nutraceuticals in alternative food and related industries.[64] Of all the emerging technologies that have helped in realizing such development, nanotechnology leads the way with its versatility. Some examples of nanocarriers designed for delivering bioactive compounds are discussed below.

6.3.1 Nanoemulsions

Nanoemulsions are thermodynamically stable nanosized emulsions formed by mixing two immiscible liquids in the presence of an emulsifying agent to form a single phase. Delivery systems based on nanoemulsions can protect bioactive compounds from degradation and can be used for controlled release of the bioactive compounds with higher bioavailability. They can also be used to stabilize hydrophobic bioactive compounds in aqueous systems.[65] Nanoemulsions provide greater interfacial area due to their smaller size and this reflects in the increased interfacial reaction rate, rate of delivery and increased bioavailability. Nanoemulsions also favor increased gastrointestinal retention time by penetrating the mucus layers.[66] In a concise review by Huang and coworkers, nanotechnology-based nutraceutical delivery and bioavailability, with a focus on nanoemulsion-based delivery systems, are discussed.[54] According to the authors, nanoemulsion-based delivery systems are one of the best methods to enhance oral bioavailability while preserving the biological properties of the phytochemicals. A number of certified food-grade lipids and emulsifiers are commercially available and the preparation is easy when compared to other nanocarriers. The minimal toxicity profile also makes this system a superior nanodelivery method.[54] A similar review highlighting the design and fabrication of excipient emulsions for increased bioavailability of hydrophobic bioactive compounds was published by McClements and coworkers.[67] In this review, the authors discuss the use of excipient foods, which may not have any bioactivity by themselves, but influence the bioaccessibility, absorption and bioavailability of bioactive compounds. The authors report that this technology has the potential to create novel food products with improved bioavailability of bioactive compounds from natural products, which can be used as dressings, dips, sauces, yoghurts, *etc.* Along with the description of the advantages and applications of excipient emulsion systems, the authors do not forget to discuss the challenges of this excipient emulsion approach. According to the authors, the adverse effects and toxicity profiles of the excipient system and the bioactive compound, and the change in the activity of the bioactive compounds when administered along with excipient system, need to be studied well before adopting this system.[67] Earlier in 2011, the same author published a tutorial review on oil-in water emulsions and critiques on nanoemulsion properties and fabrication methods, along with special attention to applications in food industry.[68] The author concludes that most of the emulsion approaches that are developed so far are at lab-scale, which might not be suitable for scale-up to industrial proportion, and stresses the need for more studies on nanoemulsions.[68] The use of food-grade ingredients to create nanoemulsions to deliver lipophilic compounds has gained interest in recent years. A recently published review provided an overview of the major components used to create food-grade nanoemulsions, including examples of oil phase ingredients, aqueous phase components and stabilizers, and various nanoemulsion formulation technologies and the physicochemical properties of the nanoemulsions created are discussed in detail.[69]

Industries including the food, pharma and medical sectors require edible delivery of lipophilic functional components, such as bioactive lipids (carotenoids, phytosterols), flavors, antioxidants, antimicrobials and therapeutic moieties. One of the best methods is by creating emulsions that suit the particular compound and application. Although conventional oil-in-water emulsions are the most commonly used method of encapsulating bioactive lipids, disadvantages like breakdown of the emulsion system on exposure to environmental stress and low encapsulation efficiency are detrimental. Depending on the bioactive compound to be delivered, more complicated systems like multiple emulsions, multilayer emulsions, filled hydrogel particles and solid lipid particle emulsions might offer better advantages.[70] In a recent report, low-energy formulation techniques used for the formation of nanoemulsions are discussed in detail. Low-energy methods are simple to implement and do not need sophisticated equipment, and are suitable for applications that require relatively low levels of fats or oils, like in soft drinks and fortified waters.[71]

A review with a specific focus on nanoemulsions encapsulating, protecting and delivering omega-3 fatty acids was reported recently.[29] According to the authors, Western diets provide low levels of omega-3 fatty acids and consumers usually depend on supplements to overcome this deficiency. Along with information on omega-3 fatty acids, their sources and health benefits, this review discusses the potential of nanoemulsions in incorporating omega-3 fatty acids into edibles like beverages, dressings, sauce and dips. The review also discusses the obstacles like oxidation, physical stability, bioavailability, achieving the daily dietary requirements, flavor issues and consumer concerns while supplementing with omega fatty acids, and understands the need for carefully modulating the physical and chemical parameters for formulating the nanoemulsions to provide optimal applications in food industry.[29] With an aim to study the influence of polysaccharides on the physicochemical properties of omega-3 fatty acid emulsions, nanoemulsions were fabricated using a blend of fish oil and lemon oil.[72] Different concentrations of anionic sodium alginate, cationic chitosan and non-ionic methylcellulose were used to test the rheological properties and physical stability of the nanoemulsion, and the results indicate that sodium alginate produced both positive and negative effects on the nanoemulsion system. While the increase in viscosity with the addition of sodium alginate is useful for nanoemulsions in applications like dressings and sauces, this is undesirable in non-viscous items like beverages. The stability of the emulsion was reduced in the presence of sodium alginate. Although the physicochemical mechanism was different, both anionic and cationic polysaccharides were able to inhibit lipid oxidation, while there was no effect when a neutral polysaccharide was used. The results help in understanding the role of ionic polysaccharides in formulating nanoemulsion-based delivery system containing natural antioxidants.[72]

The development of nanoemulsions using milk proteins, with a special focus on whey proteins, is discussed in a review authored by Adjonu and coworkers.[73] Milk proteins are an important class of food ingredients, some of which are bioactive in nature and are very good emulsifiers. The authors also discuss the potential of such nanoemulsion systems for the delivery of bioactive compounds, and suggest that multifunctional peptides like those in milk proteins can act as excellent additives in the nutraceutical industry.[73] Beta-lactoglobulin is the major whey protein, which can act as a natural nanocarrier of hydrophobic molecules. A stable nanocolloid was developed by spontaneous binding of beta-lactoglobulin with DHA in the presence of pectin to form a nanocomplex, which can be potentially used for the enrichment of clear, acidic non-fat drinks. Due to the small size of the nanocomplex, no turbidity was observed and the mean size of the nanoparticle was 100 nm. While this method has been reported to have protective effects on DHA from oxidation, further tests are required to evaluate the heat stability and long term storage.[74] Folic acid, a form of vitamin B is essential for the proper development of the human body and is required for the synthesis and repair of DNA. Folic acid was encapsulated in a hydrocolloid composed of whey protein concentrate or a commercially resistant starch. The encapsulation was achieved by two different methods—electrospraying and nanospray drying methods—and the physicochemical properties of the nanoencapsulates obtained were analyzed. Electrospraying resulted in smaller particle size, and encapsulation efficiency and folic acid stability was better when whey protein concentrate was used for the nanocolloid formulation.[75]

Vitamin E is a sensitive, lipophilic vitamin that is important in maintaining health and preventing chronic diseases. The effect of glycerol on vitamin E acetate-loaded nanoemulsions synthesized by spontaneous emulsification of medium chain triglyceride oil and Tween 80 was reported.[76] The analysis of the formation, stability and properties of the nanoemulsions showed that water-soluble glycerol acts as a co-solvent and showed an appreciable effect on the particle size of the nanoemulsion, which also resulted in decreased turbidity of the nanoemulsion. Long-term stability was dependent on glycerol concentration and storage temperature. These observations are important for the development of vitamin E-enriched food and pharmaceutical applications.[76] The droplet size could be further modulated and smaller particles could be achieved by increasing the mixing temperature and stirring speed during the addition of organic phase to aqueous phase.[77] The same group of researchers reported the use of two polar co-solvents—propylene glycol and ethanol—in the development of vitamin E acetate-loaded nanoemulsions and analyzed the effect of the co-solvents on the formation, stability and physical properties of the nanoemulsion.[78] Vitamin D is a fat-soluble micronutrient essential for intestinal absorption of minerals and includes calciferol (D2) and cholecalciferol (D3). Recently, the effects of carrier oil used for nanoencapsulating vitamin D3 within an oil-in-water emulsion formulated using a natural surfactant quillaja saponin were tested. Carrier oil with distinct reactivity to lipase digestion and molecular properties such as medium chain triglycerides, long chain triglycerides like corn oil and fish oil, and indigestible oils like orange oil and mineral oil were selected for the study. The bioaccessibility of the encapsulated vitamin was studied in a simulated gastrointestinal tract, and long chain triglycerides (corn oil and fish oil) resulted in highest levels of bioaccessibility, while the rate of free fatty acid release was highest for medium chain triglycerides.[79] In 2012, a similar study was conducted to analyze the bioaccessibility of β-carotene nanoemulsions formulated using different carrier oils. The bioaccessibility of long chain triglycerides and medium chain triglycerides showed similar values, while indigestible orange oil showed almost no bioaccessibility.[80] These studies indicate that long chain triglycerides and medium chain triglycerides are most suitable for formulating nanoemulsions containing fat-soluble bioactive compounds.[79,80]

Resveratrol is a polyphenolic compound acclaimed for its anti-oxidant properties and is found in grapes, cocoa, peanuts and many berries. Studies indicate that resveratrol has prophylactic, anti-aging and disease-fighting properties, but the stability and bioavailability are very poor and it is rapidly metabolized and eliminated from the body if taken orally. To overcome these problems, Sessa and coworkers developed a nanoemulsion-based delivery system encapsulating resveratrol for potential oral administration.[65] The authors report the development of a food-grade lecithin based nanoemulsion to encapsulate resveratrol and studied the cytotoxicity, release pattern and simulated cell permeation in Caco-2 cells. Based on the confocal studies, the authors conclude that the encapsulated nanoemulsions do not alter the cell viability or the cytoskeletal structure of Caco-2 cells. The permeability of the nanoemulsion in cells was studied by measuring the transepithelial electrical resistance and the *in vitro* release of resveratrol by dialysis. The authors report that the mean droplet size of the delivery system is important in cell permeation; nanometric emulsions of subcellular size had enhanced permeability and improved bioavailability.[65] The effect of emulsifiers and the temperature of synthesis on the solubility, stability, and bioaccessibility of another polyphenolic compound, curcumin, were studied recently. Whey protein isolate, caseinate and Tween 80 were used as the emulsifier and oil-in-water emulsions were prepared with curcumin powder at either 30 °C or 100 °C. Two different temperatures were used to simulate different applications in this study: salad dressings (30 °C) and cooking sauces (100 °C). The higher temperature and surface-stabilized emulsion resulted in a higher transfer rate of curcumin into the emulsion, but on exposure to a simulated gastrointestinal tract atmosphere, the protein-stabilized emulsions had higher curcumin content. Among the emulsifiers studied, the caseinate-stabilized emulsion showed the highest absorption of curcumin. This study can be used as a base to understand the effect of excipient emulsions for developing oral bioavailability of lipophilic bioactive compounds for various applications, including in food and pharma applications.[81] An organogel-based nanoemulsion containing curcumin was developed for potential oral delivery, anticipating health-promoting benefits (Figure 6.2).[82] The fate of the curcumin nanoemulsion after oral administration was initially studied using an *in vitro* lipolysis assay and a cell permeation assay in a Caco-2 cell monolayer. This was followed by pharmacokinetic studies in mice, which confirmed the increased oral bioavailability of the curcumin in the nanoemulsion when compared to the unformulated curcumin. Other poorly soluble nutraceuticals may also be formulated in a similar way for increased bioavailability on oral delivery.[82] An excipient emulsion containing curcumin with improved solubility and bioaccessibility was prepared for use in food matrices. The excipient system was fabricated using corn oil and aqueous

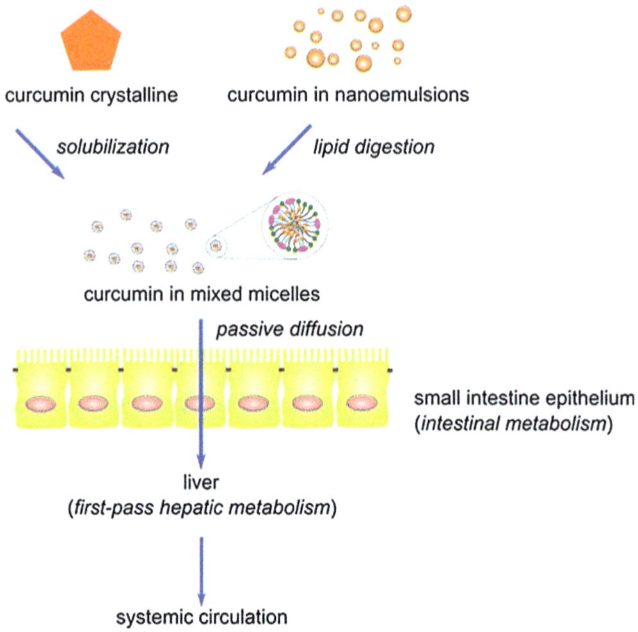

Figure 6.2 Scheme of the absorption and metabolism of unformulated (crystalline) curcumin and curcumin nanoemulsion. Reprinted with permission from ref. 82. Copyright 2012 American Chemical Society.

Tween 80, to which curcumin was added at 30 °C or 100 °C, and incubated for various intervals of time. The transfer of curcumin and bioaccessibility were higher at the higher temperature.[83] A stable nanoemulsion with two bioactive compounds, resveratrol and curcumin, was developed using soy lecithin, sugar ester and modified starch. Very fine nanoemulsions were created by high-pressure homogenization and the antioxidant activities of the encapsulated compounds were studied.[84]

In a study conducted to evaluate the influence of the size, structure and composition of droplets for use in emulsion-based delivery system, *in vitro* studies and *in vivo* experiments on a rat-feeding model were conducted.[85] Heptadecanoic acid was used as the model fatty acid and coenzyme Q10 was used as the model lipophilic nutraceutical in this emulsion-based system, which was developed for oral delivery of nutraceuticals. Coenzyme Q10 is one of the most commonly used supplements, and is an integral part of aerobic respiration. The results of the study indicate that the dimensions of the emulsion influence the bioavailability of the nutraceutical, and the rate and extent of lipid digestion. The rate of lipid digestion was increased both under *in vitro* and *in vivo* conditions when the droplet size was smaller. It was also observed that when a digestible carrier oil like corn oil was used, the rate of lipid digestion and bioavailability of coenzyme Q10 was higher than when indigestible mineral oil was used.[85] In a more recent report, the physiochemical properties and nutraceutical effects of coenzyme Q10 were studied in detail.[86] Coenzyme Q10 has the property of self-nanoemulsifying, and the authors comment that this property might be a promising property for conferring a higher nutraceutical value. The self-nanoemulsifying particles were created by spray drying a mixture of coenzyme Q10, medium-chain triglyceride, sucrose ester of fatty acid and hydroxypropyl cellulose. The improved nutraceutical value was studied by analyzing the hepatoprotective effect on carbon tetrachloride-treated rats acting as a model of acute liver injury, and was found to be better than the effects of crystalline coenzyme Q10. Storage of nanoemulsions under high humidity affected the stability, while under reduced humidity, the particles were stable for a long time.[86] Coenzyme Q10 was nanoencapsulated using octenyl succinic anhydride modified starch with rice bran oil and the stability of the resulting nanoemulsion was tested at different pH values mimicking different beverages. The particle size ranged from 200–300 nm and the freeze-dried nanoparticles were able to maintain absorbance for 3 months on storage. Authors suggest that this nanoemulsion system can be used not only in beverages, but also for supplementing coenzyme Q10 in baked foods.[87]

A nanoemulsion-based delivery system for encapsulating and delivering quercetin was designed and developed, and its activity was tested in an *in vitro* model of the gastrointestinal tract with an aim to use this system in the pharmaceutical and food industries.[88] An oil-in-water nanoemulsion of quercetin was prepared by dissolving quercetin in a medium-chain triglyceride. The multi-stage dynamic *in vitro* gastrointestinal model was created with an oral phase, gastric phase and small intestine phase. The results suggest that quercetin can be successfully encapsulated to form stable nanoemulsions, and provide improved bioaccessibility. The authors hope that this quercetin-encapsulated nanoemulsion system can be used as a successful strategy for fortifying foods with nutraceuticals.[88] A multifunctional lipophilic flavonoid, 5-demethyltangeretin, isolated from citrus plants was developed into a nanoemulsion-based delivery system.[89] Some of the properties of 5-demethyltangeretin include anti-inflammatory, anti-oxidant, anti-carcinogenic, anti-viral, anti-atherogenic and anti-thrombogenic properties. 5-Demethyltangeretin was solubilized in a medium chain triglyceride for nanoemulsion fabrication and the aqueous phase contained β-lactoglobulin. The nanoemulsion fabrication was carried out at 37 °C and emulsions of different sizes were prepared by using different homogenization conditions during fabrication. The anti-cancerous effect was studied on HCT116 human intestinal cancer cells and it was observed that when the droplet size decreased, there was increased cellular uptake and inhibition of cell viability (Figure 6.3).[89] The authors hope that this nanoemulsion system can be used as a delivery system for bioactive compounds including 5-demethyltangeretin in food and pharmaceutical industries.[89]

Three different essential oil components were developed into nanoemulsions using sunflower oil formulated with different emulsifiers, and the antimicrobial effects were studied against *Escherichia coli*, *Lactobacillus delbrueckii* and *Saccharomyces cerevisiae*.[90] Carvacrol, limonene and cinnamaldehyde were the three essential oils studied, and the emulsifiers used

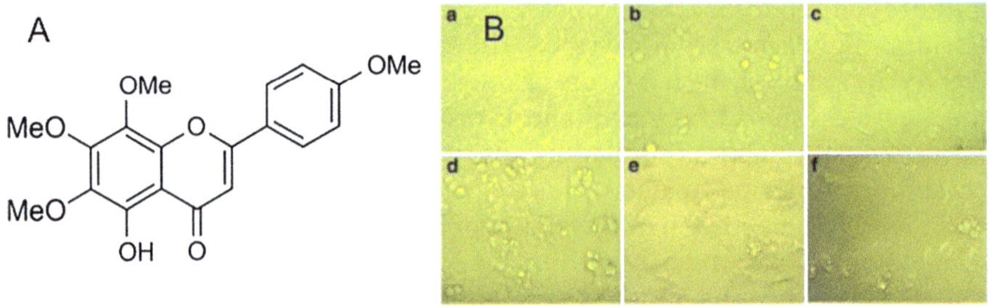

Figure 6.3 (A) Chemical structure of 5-demethyltangeretin. (B) Effect of nanoemulsion containing 5-demethyltangeretin on HCT116 cancer cells after 24 h of treatment. (a) Control cells; (b)-(f) morphological changes induced by 1.6 μM 5DT delivered in bulk water, bulk oil and emulsions with different mean droplet radii (203, 125 and 67 nm), respectively. Reprinted with permission from ref. 89, *Food Research International*, 62, Jinkai Zheng, Yan Li, Mingyue Song, Xiang Fang, Yong Cao, David Julian McClements, Hang Xiao, Improving intracellular uptake of 5-demethyltangeretin by food grade nanoemulsions, 98–103, Copyright 2014 with permission from Elsevier.

were lecithin, pea proteins, sugar ester and a combination of Tween 20 and glycerol monooleate. The anti-microbial activity was found to be dependent on the formulation method of nanoemulsions—the sugar ester-based and Tween 20 and glycerol monooleate mixture-based nanoemulsions showed antimicrobial activity for a short period of time, while the nanoemulsions based on lecithin and pea proteins showed antimicrobial activity for an extended period of time. The authors suggest that these observations can be used in a rational way to design nanoemulsion-based delivery systems for food-related applications.[90] A novel study where protein hydrogels were used as the matrix for microencapsulation of bilberry anthocyanins was reported, where whey protein was used as the matrix material. Phosphatidylcholine-depleted lecithin was used as the emulsifier. The results of this study prove that protein-based encapsulation is comparable to the well-studied polysaccharide-based systems for encapsulating active compounds.[91]

6.3.2 Nanoencapsulation/Nanoparticles

Encapsulation is the process of entrapping active agents within another carrier substance to improve the activity of the encapsulated active compound. Encapsulation can be used to create a protective covering around the active compound to stabilize it, prevent unwanted reactions and improve the delivery of the bioactive compounds. Different encapsulation technologies are available for the protection of bioactives in food.[92,93] Nedovic and coworkers explained the concept of encapsulation and described the materials used for encapsulation, techniques used in encapsulation and various examples of encapsulated food products.[92] Each bioactive food component has its own characteristic properties and the encapsulation technology used will depend on these unique properties. A review with special focus on the release of bioactive compounds, including probiotics, in the gut describes in detail the compounds that can be used for encapsulation, various methods available for encapsulation and the factors that influence the safe delivery of the bioactive compounds or probiotics.[93] Carbohydrate-based delivery systems are another group of promising vehicles for nano- and micro-encapsulation of bioactive compounds for use in food applications. Different kinds of polysaccharides including both natural and modified polysaccharides can be used for encapsulation methods. A review by Fathi and coworkers discussed various polysaccharides used for nano- and micro-encapsulation, their physicochemical properties, and the pros and cons of encapsulation. Different methods of encapsulation and the functional performance of each type of carbohydrate-based delivery system are also discussed.[94]

A highly functional and resourceful study was reported by Gökmen and coworkers, where they report the development of functional bread enriched with nanoencapsulated omega-3 fatty acids.[28] Flax seed oil was used as the source of omega-3 fatty acid, which was nanocomplexed with high amylose corn starch. The nanoparticles obtained after spray drying were incorporated into bread formulations at different concentrations. Along with the quality of bread, the authors studied the effects of encapsulation on lipid oxidation and production of thermal contaminants during baking. According to the scanning electron images of the bread, the nanoparticles remained intact in the crumb, while they were partially destroyed in the crust. Even with a significant amount of nanoencapsulated omega-3 fatty acids incorporated into the bread, no adverse sensory effect was observed. Compared to the free form of flax seed oil, nanoencapsulated flax seed oil reduced the thermoxidation of fatty acids that occured during baking.[28] Lupeol is a bioactive terpenoid with anti-inflammatory, anti-microbial, anti-protozoal and anti-cancerous activity. Recently, lupeol was encapsulated in poly(D,L-lactide-*co*-glycolide) (PLGA), and the physical properties and modulatory effects on NF-κB of Caco-2 cells were studied.[95] PLGA is a widely used copolymer that is biocompatible and approved by the FDA for use in food and therapeutic uses. The fabricated nanonutraceutical was 10% larger than empty PLGA nanoparticles and showed better a anti-inflammatory effect than pure lupeol at lower concentrations.[95] In another study, curcumin was encapsulated in PLGA for improved bioavailability and prolonged retention time. The oral bioavailability and the retention time were studied *in vivo* in a freely moving rat model. The results suggest that oral bioavailability of the curcumin nanoformulation was almost 22 times higher than that of conventional curcumin and the excretion results also support higher absorption when the curcumin nanoformulation was used.[96]

Biopolymers are versatile encapsulating agents that can provide stability and protection to the encapsulated bioactive compounds. Careful selection of food biopolymers with versatile molecular and physicochemical properties can enable the development of nanocomplexes with a wide range of functional properties.[97] Based on the nature of phase separation, biopolymers can be synthesized by associative or segregative processes, and additional stability enhancement can be brought about by internal biopolymer cross-linking. Biopolymer nanoparticles synthesized by the thermal processing and electrostatic complexation of whey protein isolate and beet pectin were used to encapsulate anthocyanins.[20] Although the encapsulation improved the heat stability of anthocyanin, the antioxidant activity was reduced due to the thermal processing. In the presence of ascorbic acid, the distinctive color of the anthocyanin was also lost. The authors suggest that this method is good for improved stability of anthocyanin, but when considering all the properties, alternative strategies are required for better applicability.[20]

Micellar nanoparticles are another group of effective carrier-delivery system that can protect and carry active compounds. Mimicking nature, casein micelles were developed to encapsulate and stabilize hydrophobic nutraceuticals, in this case, vitamin D2, for enrichment of non/low fat food products. The encapsulation efficiency of vitamin D2 and its protection from photochemical degradation inside the casein micelles were studied, and encapsulation was found to provide partial protection against UV-induced degradation. Caseinates are popular encapsulation materials and this study suggests that casein micelles are potential nano-vehicles that can be used for the entrapment, protection and delivery of bioactive nutraceuticals and functional compounds within food products.[98]

Pomegranate, also known as "the jewel of autumn", is a mine of nutrients and antioxidants. It is one of the oldest known edible fruits, and is revered as a symbol of prosperity, fertility, health and eternal life. Polyphenols of pomegranate—pomegranate extract, punicalagin or ellagic acid—were encapsulated in PLGA–poly(ethylene glycol) (PLGA–PEG) nanoparticles by double emulsion-solvent evaporation method and the anticancerous activity of the nanoparticles was tested in MCF-7 and Hs578T breast cancer cell lines. Encapsulated punicalagin nanoparticles were the most potent in inhibiting cancer cells.[99] Larger, microcapsules are also significant delivery systems. A new method was developed for pomegranate seed oil encapsulation by spray drying, which can be used in the food industry. All parts of the fruit, including the seed and peel, are nutritional, but they are discarded as waste during industrial processing of pomegranate. To utilize the nutritional properties of pomegranate seed discarded from the juice and concentrate industries, a method to extract and encapsulate pomegranate seed oil was developed. To overcome the instability and oxidative deterioration, microencapsulation by spray drying was used and skimmed milk powder was used as the encapsulating agent. The authors also report the effect of various parameters used for the encapsulation technique on the encapsulation efficiency.[100] de Conto and coworkers reported the use of commercial microencapsulated omega-3 fatty acids and rosemary extract in white pan bread. They evaluated the influence of these additives on the sensory and technological qualities of the bread. Rosemary extract from *Rosmarinus officinalis* is rich in antioxidants and is extensively used in various cuisines and the food industry.[101]

6.3.3 Liposomes

Liposomes are microscopic spherical vesicles containing at least one lipid bilayer, and can encapsulate both hydrophilic and hydrophobic materials. With industrial application in mind, a group of researchers designed liposomal formulations based on soy phosphatidylcholine, a natural lipid containing essential fatty acids like linoleic and linolenic acids. Stearic acid and calcium stearate were used to stabilize the liposomal formulation containing omega-3 and omega-6 fatty acids and vitamin E. Along with rheological properties, oxidative and thermal stability were also studied and a sensory evaluation was conducted in commercial chocolate milk. The results report the high encapsulation efficiency of folic acid and stable vitamin E after pasteurization, and conclude that the developed liposomal formulations containing bioactive compounds are suitable for food industry applications.[27] A highly disordered lipid nanomatrix containing fish oil enriched with omega-3 fatty acids that can protect and accommodate bioactive compounds was fabricated.[102] Effective encapsulation and delivery of lutein, a lipophilic bioactive pigment, was studied using this lipid nanocarrier. The nanoparticles were below 200 nm, and had high lutein entrapment efficiency and high antioxidant activity. Along with improving the solubility and stability of lutein, the presence of fish oil in the liposome increased the antioxidant capacity. Such nanolipid formulations could be successfully incorporated into food systems with enhanced nutraceutical activity.[102] The effect of nanoliposomse containing curcumin on the aggregation of amyloid fibrils associated with Alzheimer's disease was studied. Other liposomes associated with phosphatidic acid, cardiolipin and GM1 ganglioside were also compared with the anti-amyloid activity of curcumin. Nanoliposomes containing curcumin were prepared by three different approaches and the liposomes prepared by click chemistry showed the best results in inhibiting amyloid aggregation.[103] Hesperetin is a flavanone found in citrus fruits with anticancerous activity. By encapsulating in a lipid carrier, a nanoformulation of hesperetin was developed for fortifying functional foods. Nanolipid carriers have some limitations, such as rapid aggregation and burst release. The authors overcame these limitations by coating the hesperetin-loaded lipid nanostructure with various biopolymers, such as chitosan, alginate and methoxypectin, resulting in better release kinetics and higher stability. Additionally, the sensory qualities, such as taste, color and homogeneity, were also improved when tested in hesperetin-fortified milk. The authors hope that this formulation technology can be used for colon delivery of bioactive materials.[104]

6.3.4 Other Nanoformulation Strategies: Nanodisks, Nanogels, Nanofibers *etc.*

Electrospinning is a versatile method used to create nanofibers, where polymeric solutions under high electric force are spun into nanofibers. With unique physicochemical properties, electrospun nanofibers are used in many applications and recently there has been growing interest in electrospun nanofibers as delivery agents in the food industry. Recently, Ghorani and Tucker have presented a review that discusses electrospinning and its application as a delivery vehicle for bioactive compounds and probiotics.[105] The authors hope that closer interaction between academia and industry will help in overcoming the weaknesses of electrospinning technology and will inspire future developments.[105] Industrial waste management is an important issue, and newer methods of recycling and utilizing industrial wastes are an attractive alternative that can help in waste reduction. Fung and coworkers carried out a remarkable study in developing nanofibers from agrowastes for encapsulating probiotics.[106] The probiotic *Lactobacillus acidophilus* was encapsulated in electrospun nanofibers made from soluble dietary fibers obtained from the agrowastes okara (soybean solid waste) from the soybean industry and oil palm trunk and palm frond from the palm oil industry. The soluble dietary fibers were obtained after alkali treatment of the agrowastes, which was complemented with polyvinyl alcohol for synthesizing nanofibers. Surprisingly, *L. acidophilus* was able to withstand the electrospinning conditions, and later the nanofiber storage conditions. Along with probiotics, this carrier system is able to contribute to increased dietary fiber consumption. The results indicate that the nanofibers could provide thermal protection to the encapsulated probiotics. The authors expect that modification of electrospinning parameters could provide better encapsulation of viable bacteria, and controlled release of probiotics in intestines, resulting in an efficient nanoencapsulated probiotic carrier system.[106]

The visual appeal of food is very important for palatability. Addition of hydrophobic nutraceuticals to clear beverages is a challenge due to various reasons like solubility, bioavailability, stability and safety of the nutraceutical, maintenance of the clarity of the drink, and cost of production. In an attempt to overcome these problems, a study was designed based on a protein–polysaccharide conjugate formed by the Maillard reaction of casein and maltodextrin (Figure 6.4).[107] Vitamin D was used as the model nutraceutical and the physicochemical characteristics of the nanoparticles and release of vitamin D from the nanoparticles under simulated gastric digestion were analyzed. Particles encapsulating Nile red were also formulated and the fluorescent property of Nile red was used in spectroscopic analyses. A sensitive, water-soluble nutraceutical, epigallocatechin gallate, was also tested for encapsulation studies. Results suggest that this technology has the potential to encapsulate bioactives and can be used to enrich clear beverages.[107]

Nanogels are hydrogel particles in the nanometer range usually consisting of a crosslinked hydrophilic or amphiphilic polymer and can be used as carriers of therapeutics. An essential oil rich in thymol with anti-microbial activity isolated from *Lippa sidoides* was encapsulated in chitosan–cashew gum nanogel by spray drying.[108] The nanogel ranged from 335 to 558 nm and exhibited a unimodal distribution with positive zeta potential. It was observed that when the chitosan concentration was increased, larger particles were formed, and showed prolonged release compared to that from

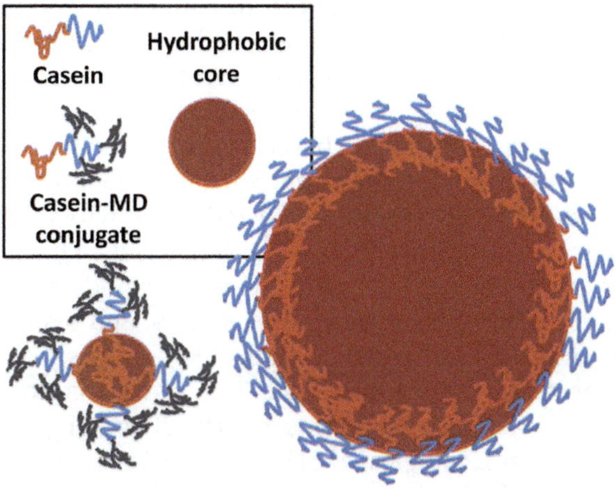

Figure 6.4 Hydrophobic compound encapsulating nanoparticle formed by protein–polysaccharide interaction between casein and maltodextrin (MD). Reprinted with permission from ref. 107 with permission from the Royal Society of Chemistry.

smaller particles. The larvicide effect of the encapsulated essential oil was tested on the third instar of *Stegomyia aegypti* larvae, and the larval mortality was related to the essential oil loading in the nanoparticle. Nanoparticles of cashew gum : chitosan concentration of 1 : 1 and 1 : 10 showed 87% and 75% larval mortality at 48 h, respectively, which further increased to over 90% at 72 h.[108] Although this is not an example of a direct food application, the anti-microbial effect of plant metabolites can be use in the food industry for ecofriendly sanitation purposes.[109] Other reports suggest the use of this essential oil in offering antimicrobial activity against oral pathogens under *in vitro* conditions, as an adjuvant in antibiotic therapy against respiratory tract infections and as an antihelmintic treatment against sheep gastrointestinal nematodes.[110–112] The development of nanoencapsulated *Lippia sidoides* essential oil is an interesting concept that has various potential uses in the food and pharma industries.

6.4 Safety and Regulatory Aspects of Nanofoods

In recent years, a number of nanofoods have been introduced into the market. According to an inventory made in 2011 by the Project on Emerging Nanotechnologies, the US leads the industry with the number of nanoproducts in the market, followed by Europe and Asia. Even with the ambiguity related to the safety and toxicological effects of nanomaterials, the number of products in the market is increasing at an exponential rate.[113] The specific use, biophysical properties and interactions, exposure, uptake and kinetics, and biological effects are some of the factors that are considered during the development and application of nanomaterials. However, the guidance and risk assessment of nanomaterials vary from country to country and a concerted "multi-perspective" approach is required for the grouping and toxicity assessment of nanomaterials.[114] A systemic and multi-tiered approach was described to assess the safety of engineered nanomaterials in direct food applications.[115] This assessment is based on the conventional risk assessment system used for novel foods and chemicals. In the initial step, both the bulk material and the nanomaterial are tested for activity in similar ways to assess the unique toxicological effects of nanomaterials. This is followed by a tiered approach, where in Tier 1, *in silico* and *in vitro* models are used to study potential engineered nanomaterial-specific effects, and a short repeat-dose *in vivo* toxicity study. In Tier 2, a 90 day repeated dosing study is conducted, and any variation from the results of the Tier 1 study is noted. Investigating the dynamics of protein interactions and the effect on gut microflora are newer dimensions of toxicological analyses.[115]

According to Boverhof and coworkers, a purely technical definition of nanomaterials based on size is not enough to efficiently evaluate their effects and may result in inconsistent identification.[116] This may give the consumers an impression that nanomaterials are not adequately evaluated and the regulations are flawed, leading to needless stigmatization of nanomaterials and their application as a whole. The authors suggest a set of characteristics, like particle size and distribution threshold, agglomeration/deagglomeration and precipitation characteristics, among others, that need careful evaluation while assessing the potential health and environmental impact.[116] An overview on the European Union's (EU) political drivers and policy processes relating to regulations in the nanotechnology sector addresses the relation between risk governance and technological innovation policy in Europe. Since 2004, the European regulations associated with nanotechnology have been undergoing significant developments, and the European Commission has formulated two horizontal nano-specific policies—the code of conduct for responsible nanosciences and nanotechnologies (N&N) research and an advocated "nanomaterials" definition that is helpful for regulatory purposes. While there are many factors that influence regulatory affairs development, most of them reflect a public interest motive, standardization of good governance principles and their practice, along with the significant role that government has in developing the innovation policies.[117] The EU including Switzerland has nano-specific regulations incorporated in existing legislation, while other countries and regions are regulated mainly based on the guidance for the specific industry. Recently, a review was published that provides an overview on the safety and regulatory measures on nanomaterials in agri- and food-related industries in EU and non-EU countries. Collaboration and understanding between countries is highly indispensable for exchanging information and ensuring safety standards.[118] An outline of the developments in food and agricultural aspects of nanotechnology with a focus on Organisation for Economic Co-operation and Development (OECD) countries was recently reported. The paper also provides an assessment on the implications of nanotechnology, which will enable the policy makers to anticipate the challenges associated with the developments in food- and agri-nanotechnology applications.[119] The Asian nanotechnology market has made great leaps in recent years and Asian countries are fervently watching the nanodevelopment scenario in other developed countries, especially with respect to regulatory affairs, mainly because of the lack of regional regulatory bodies in Asia like those that exist in Europe. An attempt was made to study the developments in nanotechnology happening in six Association of Southeast Asian Nations (ASEAN)—Singapore, Malaysia, Thailand, Indonesia, Philippine and Vietnam—and the authors suggest that issues relating to nanotechnology and safety should be discussed in ASEAN meetings. The authors hint that regulations should not only be limited to industrial development, but should also be applied at the research level and to workers who handle nanomaterials, as they are theoretically and primarily affected in case of nanotoxicity. Better budget allocation for research on risk and safety analysis is also advocated and the importance of nanosafety is highlighted as a priority.[120]

In a review authored by Sauvant and coworkers, the use of encapsulation strategies in food-related applications is urged to be applied very cautiously.[121] The authors use the example of fat-soluble nutrient Vitamin A, which has to be provided through food as it is not synthesized *de novo* in the body. Fortification of processed food needs strict regulatory guidelines. Overdose of vitamin A is dangerous for human health and, most often, there is no estimate on the intake of supplements over a period of time. Evidence suggests that subtoxicity, with no clinical or external toxicity, is a growing concern in industrialized countries. The case of vitamin A possess a dilemma, as generally, the western population does not suffer from malnutrition and does not require additional vitamin A supplements, while the situation is different in developing countries. Another aspect that requires attention is impaired vitamin A metabolism, which leads to toxicity issues. The review emphasizes the need for better understanding of the requirements of vitamin A fortification in food and its effect in the larger population across different age groups, especially in industrialized countries.[121] Currently, there are no standardized assays or analytical approaches to assess the epigenetic effects of nanoparticles and the epigenetic effects induced by exposure to nanomaterials are poorly studied. With more and more nanomaterials being used in food related applications, advanced nanotechnology testing guidelines and regulations should incorporate recommendations on epigenetic tests before proceeding with industrial manufacture of nanoproducts. A review by Smolkova and coworkers stresses the importance of epigenetic studies related to nano-exposure and discusses the advances in epigenetics-related studies in relation to nanomaterials.[122] Among environmental pollutants, heavy metals are representing an increasing "civilization" threat to global health, and are implicated in hormone-dependent cancers, obesity, type 2 diabetes, neurodegenerative diseases and cardiovascular diseases.[123–129] This has understandably generated a number of commercial proposals of "chelators", and "nano" is at times arbitrarily used to stress their supposed superiority over competitors. However, although zeolites represent an effective cation-exchange system that, once orally, ingested delivers good cations and chelates ammonium and several heavy metals, some claim that nano-clinoptilolite, which was recently marketed in Germany, may be potentially harmful, considering the reported mutagenicity of such formulations.[130–132] This example accentuates the need for expertise in nanotechnology, highly dedicated independent laboratories, vigilant and stringent regulatory bodies to analyze the claims of advantages, and tight clinical interplay when devising nano-related commercial products.

6.5 Consumer Attitude Towards Nanotechnology in Food-Related Applications

Nanotechnology has revolutionized and transformed food and related industries in the recent years, and for novel issues like those associated with nanotechnology, the media plays a crucial role in molding the consumer attitude. A study was conducted in the United States to assess the media coverage on nanotechnology, and the results show that coverage on food nanotechnology is limited and at times unpredictable. The percentage of science journalists is very small and, most often, food nanotechnology reports are written by journalists who do not have expertise in that area, which results in less thoughtful reviews and reports.[133] The commercial success of functional food depends greatly on the consumer attitude and acceptance of the product as part of the daily diet. In 2008, a study was conducted in Switzerland ($n = 249$) to examine the attitude of consumers and the results suggest that consumers preferred to buy functional foods with physiological benefits than psychological health claims. The study was based on a survey where the participants were asked to assess their willingness to buy hypothetical functional foods with various health benefits, such as cardiovascular protection, prevention of osteoporosis and cancer. While the younger consumers showed less interest in functional foods, participants who trusted the food industry were more favorable towards functional food products. The authors added that the study has its limitations; the study was based on survey and no food tasting tests were conducted. Consumers are not willing to compromise the taste of functional food over its nutritional value.[134]

Nanotechnology has attracted large-scale investments from many industries, including the food and beverage industries, and the applications of food nanotechnology are relevant in food processing, nutrient composition and packaging. Although some nanofoods and nanopackings are commercialized in various countries, the awareness towards nanoproducts is still inadequate. Frewer and coworkers analyzed the issues associated with consumer attitude and perception of nanotechnology developments in the food and agriculture sectors based on the available nanotechnology applications.[135] The authors list a number of questions that need to be answered before enabling technologies like nanotechnology are used in agri-food applications and suggest that care should be taken while analyzing the complexity of consumer acceptance, as many dynamic, complex and interdependent factors, like ethical aspects, health and environmental risks, and the perspective of stakeholders, contribute to it.[135]

Handford and coworkers conducted a study aimed at investigating the awareness and attitude of stakeholders in the agri-food industry towards nano-related applications in the agriculture and food sectors.[136] This study also reports the current nano-related applications in the agri-food industry, anticipated risks, novel advantages and opportunities of nanotechnology applications, and the hurdles and apprehensions that stop the adoption of nanoapplications. Along with competent safety assessment, the authors stress the need for increasing the awareness of nanoapplications in the agri-food sector and urge scientists and governmental bodies to actively take part in this awareness process.[136] Even with the growing demand for convenient and safe food, consumers are suspicious of food innovations and industrialized products owing to safety issues and scandals that arise. The consumers view nanoproducts skeptically and it is unclear whether the suspiciousness has its roots in a lack of information and awareness of nanotechnology and its applications. In an analysis of the sense of trust towards nanofood products, it was observed that "willingness to pay" for new products increases with trust, which is similar to the results obtained from an online survey conducted in Canada and Germany, and experimental results from Germany.[137] A similar observation based on trust was observed in an earlier study conducted in the German-speaking part of Switzerland. Naturalness of food products and trust were the significant factors that influenced the consumers, and nanofoods were viewed more suspiciously than nanofood packing strategies.[138]

6.6 Conclusion

Applications of nanotechnology can be used in various aspects of the food industry, including targeted pesticides for better crop production, anti-microbial nano-food packaging and as food-grade engineered nanoparticles.[139,140] In recent years, nanotechnology has been used for developing formulations encapsulating bioactive compounds for use as nutraceuticals and for fortifying foods. Nanoformulations have helped in protecting the bioactivity and enhancing the bioavailability of the nutraceuticals and bioactive compounds. Additionally, sensory characteristics like taste, color and texture can also be improved by using nanoformulated bioactive compounds. The safety associated with nanoformulations in food and related applications needs more detailed analyses. Increased interaction and communication with researchers, policy makers and the public are also of paramount importance. Knowledge of nanotechnology and its applications will help the public in overcoming the stigma associated with nanotechnology and recognizing the advantages associated with the technology.

References

1. J. Bernal, J. A. Mendiola, E. Ibáñez and A. Cifuentes, *J. Pharm. Biomed. Anal.*, 2011, **55**, 758.
2. S. Barnes and J. Prasain, *Curr. Opin. Plant Biol.*, 2005, **8**, 324.
3. F. Shahidi, *Trends Food Sci. Technol.*, 2009, **20**, 376.
4. V. Brower, *EMBO Rep.*, 2005, **6**, 708.
5. B. Bigliardia and F. Galati, *Trends Food Sci. Technol.*, 2013, **31**, 118.
6. L. Das, E. Bhaumik, U. Raychaudhuri and R. Chakraborty, *J. Food Sci. Technol.*, 2012, **49**, 173.
7. H. Nasri, A. Baradaran, H. Shirzad and M. Rafieian-Kopaei, *Int. J. Prev. Med.*, 2014, **5**, 1487.
8. A. Rahal, Mahima, A. K. Verma, A. Kumar, R. Tiwari, S. Kapoor, S. Chakraborty and K. Dhama, *J. Biol. Sci.*, 2014, **14**, 1.
9. E. Pérez-Esteve, A. Bernardos, R. Martínez-Máñez and J. M. Barat, *Recent Pat. Food, Nutr. Agric.*, 2013, **5**, 35.
10. J. B. Lagos, F. C. Vargas, T. G. de Oliveira, G. L. da A. Makishi and P. J. do A. Sobral, *Curr. Opin. Food Sci.*, 2015, **5**, 1.
11. P. R. Srinivas, M. Philbert, T. Q. Vu, Q. Huang, J. L. Kokini, E. Saos, H. Chen, C. M. Peterson, K. E. Friedl, C. McDade-Ngutter, V. Hubbard, P. Starke-Reed, N. Miller, J. M. Betz, J. Dwyer, J. Milner and S. A. Ross, *J. Nutr.*, 2010, **140**, 119.
12. T. Panagiotou and R. J. Fisher, *Funct. Foods Health Dis.*, 2013, **3**, 274.
13. M. Cushen, J. Kerry, M. Morris, M. Cruz-Romero and E. Cummins, *Trends Food Sci. Technol.*, 2012, **24**, 30.
14. Y. D. Livney, *Curr. Opin. Food Sci.*, 2015, **3**, 125.
15. J. C. Espín, M. T. García-Conesa and F. A. Tomás-Barberán, *Phytochemistry*, 2007, **68**, 2986.
16. Z. Fang and B. Bhandari, *Trends Food Sci. Technol.*, 2010, **21**, 510.
17. R. Sandhir, A. Yadav, A. Sunkaria and N. Singhal, *Neurochem. Int.*, 2015, **89**, 209.
18. Z. E. Suntres, *J. Toxicol.*, 2011, 152474.
19. H. B. Nair, B. Sung, V. R. Yadav, R. Kannappan, M. M. Chaturvedi and B. B. Aggarwal, *Biochem. Pharmacol.*, 2010, **80**, 1833.
20. I. J. Arroyo-Maya and D. J. McClements, *Food Res. Int.*, 2015, **69**, 1.
21. E. Fernández-García, I. Carvajal-Lérida, M. Jarén-Galán, J. Garrido-Fernández, A. Pérez-Gálvez and D. Hornero-Méndez, *Food Res. Int.*, 2012, **46**, 438.
22. C. Soukoulis and T. Bohn, *Crit. Rev. Food Sci. Nutr.*, 2015, DOI: 10.1080/10408398.2014.971353.
23. S. M. Loveday and H. Singh, *Trends Food Sci. Technol.*, 2008, **19**, 657.
24. Y. Sunagawa, Y. Katanasaka, K. Hasegawa and T. Morimoto, *PharmaNutrition*, 2015, **3**, 131.
25. A. Mathew, T. Fukuda, Y. Nagaoka, T. Hasumura, H. Morimoto, Y. Yoshida, T. Maekawa, K. Venugopal and D. S. Kumar, *PLoS One*, 2012, **7**, e32616.

26. M. M. Yallapu, M. Jaggi and S. C. Chauhan, *Drug Discovery Today*, 2012, **17**, 71.
27. M. Marsanasco, A. L. Marquez, J. R. Wagner, N. S. Chiaramoni and S. del V. Alonso, *J. Food Eng.*, 2015, **166**, 55.
28. V. Gökmen, B. A. Mogol, R. B. Lumaga, V. Fogliano, Z. Kaplun and E. Shimoni, *J. Food Eng.*, 2011, **105**, 585.
29. R. Walker, E. A. Deckera and D. J. McClements, *Food Funct.*, 2015, **6**, 42.
30. O. I. Aruoma, Y. Hayashi, F. Marotta, P. Mantello, E. Rachmilewitz and L. Montagnier, *Toxicology*, 2010, **278**, 6.
31. B. K. Bhattacharyya, *Int. J. Pharm. Sci. Nanotechnol.*, 2009, **2**(1), 383.
32. N. Illuzzi, R. Galli, A. Kushugulova, Z. Zhumadilov, O. Licciardello and F. Marotta, *Rejuvenation Res.*, 2014, **2**, 172.
33. R. Nagpal, H. Yadav and F. Marotta, *Front. Med.*, 2014, **1**, 15.
34. R. Nagpal, M. Kumar, A. K. Yadav, R. Hemalatha, H. Yadav, F. Marotta and Y. Yamashiro, *Benefic. Microbes*, 2016, **7**, 181.
35. R. Catanzaro, M. Anzalone, F. Calabrese, M. Milazzo, M. Capuana, A. Italia, S. Occhipinti and F. Marotta, *Panminerva Med.*, 2015, **57**, 127.
36. H. Nakagawa, T. Shiozaki, E. Kobatake, T. Hosoya, T. Moriya, F. Sakai, H. Taru and T. Miyazaki, *Aging Cell*, 2016, **15**, 227.
37. R. Penner, R. N. Fedorak and K. L. Madsen, *Curr. Opin. Pharmacol.*, 2005, **5**, 596.
38. K. E. Scholz-Ahrens, P. Ade, B. Marten, P. Weber, W. Timm, Y. Açil, C.-C. Glüer and J. Schrezenmeir, *J. Nutr.*, 2007, **137**, 838S.
39. F. Heidarpour, M. R. Mohammadabadi, I. S. Zaidul, B. Maherani, N. Saari, A. A. Hamid, F. Abas, M. Y. Manap and M. R. Mozafari, *Pharmazie*, 2011, **66**, 319.
40. S. Prakash, C. Tomaro-Duchesneau, S. Saha and A. Cantor, *J. Biomed. Biotechnol.*, 2011, **2011**, 981214.
41. S. Prakash, C. Tomaro-Duchesneau, S. Saha, L. Rodes, I. Kahouli and M. Malhotra, *Curr. Pharm. Des.*, 2014, **20**, 1025.
42. M. L. Jones, C. J. Martoni, M. Parent and S. Prakash, *Br. J. Nutr.*, 2012, **107**, 1505.
43. J. Bhathena, C. Martoni, A. Kulamarva, C. Tomaro-Duchesneau, M. Malhotra, A. Paul, A. M. Urbanska and S. Prakash, *PLoS One*, 2013, **8**, e58394.
44. A. M. Urbanska, J. Bhathena, S. Cherif and S. Prakash, *Artif. Cells, Nanomed., Biotechnol.*, 2016, **44**, 1.
45. J. Ochoa-Repáraz and L. H. Kasper, *Curr. Obes. Rep.*, 2016, **5**, 51.
46. A. Ghazalpour, I. Cespedes, B. J. Bennett and H. Allayee, *Curr. Opin. Lipidol.*, 2016, **27**, 141.
47. A. K. Mishra, V. Dubey and A. R. Ghosh, *Metabolism*, 2016, **65**, 48.
48. P. Forsythe, W. Kunze and J. Bienenstock, *BMC Med.*, 2016, **14**, 58.
49. J. F. Ferguson, H. Allayee, R. E. Gerszten, F. Ideraabdullah, P. M. Kris-Etherton, J. M. Ordovás, E. B. Rimm, T. J. Wang and B. J. Bennett, *Circ.: Cardiovasc. Genet.*, 2016, **9**, 291.

50. K. Keunen, R. M. van Elburg, F. van Bel and M. J. Benders, *Pediatr. Res.*, 2015, **77**, 148.
51. Y. O. Cakmak, *Mov. Disord.*, 2015, **30**, 1151.
52. C. Benakis, D. Brea, S. Caballero, G. Faraco, J. Moore, M. Murphy, G. Sita, G. Racchumi, L. Ling, E. G. Pamer, C. Iadecola and J. Anrather, *Nat. Med.*, 2016, **22**, 516.
53. N. Sozer and J. L. Kokini, *Trends Biotechnol.*, 2009, **27**, 82.
54. Q. Huang, H. Yu and Q. Ru, *J. Food Sci.*, 2010, **75**(1), R50.
55. B. S. Sekhon, *Nanotechnol., Sci. Appl.*, 2010, **3**, 1.
56. S. Sonkaria, S.-H. Ahn and V. Khare, *Recent Pat. Food, Nutr. Agric.*, 2012, **4**, 8.
57. I. J. Joye, G. Davidov-Pardo and D. J. McClements, *Trends Food Sci. Technol.*, 2014, **40**, 168.
58. F. Gallocchio, S. Belluco and A. Riccia, *Procedia Food Sci.*, 2015, **5**, 85.
59. D. J. McClements and H. Xiao, *Food Funct.*, 2014, **5**, 1320.
60. D. J. McClements, *J. Food Sci.*, 2015, **80**, N1602.
61. D. J. McClements, *Adv. Colloid Interface Sci.*, 2015, **219**, 27.
62. M. Fathi, M. R. Mozafari and M. Mohebbi, *Trends Food Sci. Technol.*, 2012, **23**, 13.
63. L. Chen, G. E. Remondetto and M. Subirade, *Trends Food Sci. Technol.*, 2006, **17**, 272.
64. C. M. Galanakis, *Food Bioprod. Process.*, 2013, **91**, 575.
65. M. Sessa, M. L. Balestrieri, G. Ferrari, L. Servillo, D. Castaldo, N. D'Onofrio, F. Donsì and R. Tsao, *Food Chem.*, 2014, **147**, 42.
66. M. Douaire and I. T. Norton, *J. Sci. Food Agric.*, 2013, **93**, 3147.
67. D. J. McClements, L. Saliva-Trujillo, R. Zhang, Z. Zhang, L. Zou, M. Yao and H. Xiao, *Food Res. Int.*, 2016, **88A**, 140.
68. D. J. McClements, *Soft Matter*, 2011, 7, 2297.
69. D. J. McClements and J. Rao, *Crit. Rev. Food Sci. Nutr.*, 2011, **51**, 285.
70. D. J. Mc Clements, E. A. Decker and J. Weiss, *J. Food Sci.*, 2007, **72**, R109.
71. J. S. Komaiko and D. J. McClements, *Compr. Rev. Food Sci. Food Saf.*, 2016, **15**, 331.
72. L. Salvia-Trujillo, E. A. Decker and D. J. McClements, *Food Hydrocolloids*, 2016, **52**, 690.
73. R. Adjonu, G. Doran, P. Torley and S. Agboola, *J. Food Eng.*, 2014, **122**, 15.
74. P. Zimet and Y. D. Livney, *Food Hydrocolloids*, 2009, **23**, 1120.
75. R. Pérez-Masiá, R. López-Nicolás, M. J. Periago, G. Ros, J. M. Lagaron and A. López-Rubio, *Food Chem.*, 2015, **168**, 124.
76. A. H. Saberi, Y. Fang and D. J. McClements, *J. Colloid Interface Sci.*, 2013, **411**, 105.
77. A. H. Saberi, Y. Fang and D. J. McClements, *J. Colloid Interface Sci.*, 2013, **391**, 95.
78. A. H. Saberi, Y. Fang and D. J. McClements, *Food Res. Int.*, 2013, **54**, 812.
79. B. Ozturk, S. Argin, M. Ozilgen and D. J. McClements, *Food Chem.*, 2015, **187**, 499.

80. C. Qian, E. A. Decker, H. Xiao and D. J. McClements, *Food Chem.*, 2012, **135**, 1440.
81. L. Zou, W. Liu, C. Liu, H. Xiao and D. J. McClements, *Food Funct.*, 2015, **6**, 2475.
82. H. Yu and Q. Huang, *J. Agric. Food Chem.*, 2012, **60**, 5373.
83. L. Zou, W. Liu, C. Liu, H. Xiao and D. J. McClements, *J. Agric. Food Chem.*, 2015, **63**, 2052.
84. F. Donsì, M. Sessa, H. Mediouni, A. Mgaidi and G. Ferrari, *Procedia Food Sci.*, 2011, **1**, 1666.
85. H. T. Cho, L. Salvia-Trujillo, J. Kim, Y. Park, H. Xiao and D. J. McClements, *Food Chem.*, 2014, **156**, 117.
86. S. Onoue, A. Uchida, T. Nakamura, K. Kuriyama, J. Hatanaka, T. Tanaka, H. Miyoshi, Y. Seto and S. Yamada, *PharmaNutrition*, 2015, **3**, 153.
87. S. Y. Cheuk, F. F. Shih, E. T. Champagne, K. W. Daigle, J. A. Patindol, C. P. Mattison and S. M. Boue, *Food Chem.*, 2015, **174**, 585.
88. H. Pool, S. Mendoza, H. Xiaob and D. J. McClements, *Food Funct.*, 2013, **4**, 162.
89. J. Zheng, Y. Li, M. Song, X. Fang, Y. Cao, D. J. McClements and H. Xiao, *Food Res. Int.*, 2014, **62**, 98.
90. F. Donsì, M. Annunziata, M. Vincensi and G. Ferrari, *J. Biotechnol.*, 2012, **159**, 342.
91. M. Betz and U. Kulozika, *Procedia Food Sci.*, 2011, **1**, 2047.
92. V. Nedovic, A. Kalusevic, V. Manojlovic, S. Levic and B. Bugarski, *Procedia Food Sci.*, 2011, **1**, 1806.
93. P. de Vos, M. M. Faas, M. Spasojevic and J. Sikkema, *Int. Dairy J.*, 2010, **20**, 292.
94. M. Fathi, Á. Martín and D. J. McClements, *Trends Food Sci. Technol.*, 2014, **39**, 18.
95. M. H. Cháirez-Ramírez, J. A. Sánchez-Burgos, C. Gomes, M. R. Moreno-Jiménez, R. F. González-Laredo, M. J. Bernad-Bernad, L. Medina-Torres, M. V. Ramírez-Mares, J. A. Gallegos-Infante and N. E. Rocha-Guzmán, *Food Chem. Toxicol.*, 2015, **85**, 2.
96. Y.-M. Tsai, W.-C. Jan, C.-F. Chien, W.-C. Lee, L.-C. Lin and T.-H. Tsai, *Food Chem.*, 2011, **127**, 918.
97. I. J. Joye and D. J. McClements, *Curr. Opin. Colloid Interface Sci.*, 2014, **19**, 417.
98. E. Semo, E. Kesselman, D. Danino and Y. D. Livney, *Food Hydrocolloids*, 2007, **21**, 936.
99. A. B. Shirode, D. J. Bharali, S. Nallanthighal, J. K. Coon, S. A. Mousa and R. Reliene, *Int. J. Nanomed.*, 2015, **10**, 475.
100. A. M. Goula and K. G. Adamopoulos, *Food Bioprod. Process.*, 2012, **90**, 639.
101. L. C. de Conto, R. S. P. Oliveira, L. G. P. Martin, Y. K. Chang and C. J. Steel, *LWT–Food Sci. Technol.*, 2012, **45**, 103.
102. I. Lacatusu, E. Mitrea, N. Badea, R. Stan, O. Oprea and A. Meghea, *J. Funct. Foods*, 2013, **5**, 1260.

103. M. Taylor, S. Moore, S. Mourtas, A. Niarakis, F. Re, C. Zona, B. L. Ferla, F. Nicotra, M. Masserini, S. G. Antimisiaris, M. Gregori and D. Allsop, *Nanomedicine*, 2011, **7**, 541.
104. M. Fathi and J. Varshosaz, *J. Funct. Foods*, 2013, **5**, 1382.
105. B. Ghorani and N. Tucker, *Food Hydrocolloids*, 2015, **51**, 227.
106. W.-Y. Fung, K.-H. Yuen and M.-T. Liong, *J. Agric. Food Chem.*, 2011, **59**, 8140.
107. G. Markman and Y. D. Livney, *Food Funct.*, 2012, **3**, 262.
108. F. O. M. S. Abreu, E. F. Oliveira, H. C. B. Paula and R. C. M. de Paula, *Carbohydr. Polym.*, 2012, **89**, 1277.
109. C. E. Castro, J. M. Ribeiro, T. T. Diniz, A. C. Almeida, L. C. Ferreira, E. R. Martins and E. R. Duarte, *Rev. Bras. Plant. Med.*, 2011, **13**, 293.
110. M. A. Botelho, N. A. P. Nogueira, G. M. Bastos, S. G. C. Fonseca, T. L. G. Lemos, F. J. A. Matos, D. Montenegro, J. Heukelbach, V. S. Rao and G. A. C. Brito, *Braz. J. Med. Biol. Res.*, 2007, **40**, 349.
111. H. N. H. Veras, F. F. G. Rodrigues, A. V. Colares, I. R. A. Menezes, H. D. M. Coutinho, M. A. Botelho and J. G. M. Costa, *Fitoterapia*, 2012, **83**, 508.
112. A. L. F. Camurça-Vasconcelos, C. M. L. Bevilaqua, S. M. Morais, M. V. Maciel, C. T. C. Costa, I. T. F. Macedo, L. M. B. Oliveira, R. R. Braga, R. A. Silva, L. S. Vieira and A. M. C. Navarro, *Vet. Parasitol.*, 2008, **154**, 167.
113. T. Mennini, *Nutrafoods*, 2013, **12**, 64.
114. J. H. E. Arts, M. Hadi, A. M. Keene, R. Kreiling, D. Lyon, M. Maier, K. Michel, T. Petry, U. G. Sauer, D. Warheit, K. Wiench and R. Landsiedel, *Regul. Toxicol. Pharmacol.*, 2014, **70**, 492.
115. A. Cockburn, R. Bradford, N. Buck, A. Constable, G. Edwards, B. Haber, P. Hepburn, J. Howlett, F. Kampers, C. Klein, M. Radomski, H. Stamm, S. Wijnhovenm and T. Wildemann, *Food Chem. Toxicol.*, 2012, **50**, 2224.
116. D. R. Boverhof, C. M. Bramante, J. H. Butala, S. F. Clancy, M. Lafranconi, J. West and S. C. Gordon, *Regul. Toxicol. Pharmacol.*, 2015, **73**, 137.
117. R. Justo-Hanani and T. Dayan, *Res. Policy*, 2015, **44**, 1527.
118. V. Amenta, K. Aschberger, M. Arena, H. Bouwmeester, F. B. Moniz, P. Brandhoff, S. Gottardo, H. J. P. Marvin, A. Mech, L. Q. Pesudo, H. Rauscher, R. Schoonjans, M. V. Vettori, S. Weigel and R. J. Peters, *Regul. Toxicol. Pharmacol.*, 2015, **73**, 463.
119. G. P. Gruère, *Food Policy*, 2012, **37**, 191.
120. E. Karim, A. B. Munir, A. W. Reza, F. Muhammad-Sukki, S. H. M. Yasin, S. H. Abu-Bakar and R. A. Rahim, *Technol. Forecast. Soc. Change*, 2015, **92**, 168.
121. P. Sauvant, M. Cansell, A. H. Sassi and C. Atgié, *Food Res. Int.*, 2012, **46**, 469.
122. B. Smolkova, N. E. Yamani, A. R. Collins, A. C. Gutleb and M. Dusinska, *Food Chem. Toxicol.*, 2015, **77**, 64.
123. K. H. Chang, M. Y. Chang, C. H. Muo, T. N. Wu, B. F. Hwang, C. Y. Chen, T. H. Lin and C. H. Kao, *Medicine*, 2015, **94**, e733.
124. A. M. Soto and C. Sonnenschein, *Nat. Rev. Endocrinol.*, 2010, **6**, 363.

125. A. Nadal, P. Alonso-Magdalena, S. Soriano, I. Quesada and A. B. Ropero, *Mol. Cell. Endocrinol.*, 2009, **304**, 63.
126. F. Ranciére, J. G. Lyons, V. H. Loh, J. Botton, T. Galloway, T. Wang, J. E. Shaw and D. J. Magliano, *Environ. Health*, 2015, **31**, 14.
127. F. Grün and B. Blumberg, *Mol. Cell. Endocrinol.*, 2009, **304**, 19.
128. S. Bae and Y. C. Hong, *Hypertension*, 2015, **65**, 313.
129. F. Cavaleri, *Med. Hypotheses*, 2015, **85**, 779.
130. K. Bhattacharya, P. C. Naha, I. Naydenova, S. Mintova and H. J. Byrne, *Toxicol. Lett.*, 2012, **215**, 151.
131. L. C. Thomassen, D. Napierska, D. Dinsdale, N. Lievens, J. Jammaer, D. Lison, C. E. Kirschhock, P. H. Hoet and J. A. Martens, *Nanotoxicology*, 2012, **6**, 472.
132. T. Kihara, Y. Zhang, Y. Hu, Q. Mao, Y. Tang and J. Miyake, *J. Biosci. Bioeng.*, 2011, **11**, 725.
133. A. Dudo, D.-H. Choi and D. A. Scheufele, *Appetite*, 2011, **56**, 78.
134. M. Siegrist, N. Stampfli and H. Kastenholz, *Appetite*, 2008, **51**, 526.
135. L. J. Frewer, N. Gupta, S. George, A. R. H. Fischer, E. L. Giles and D. Coles, *Trends Food Sci. Technol.*, 2014, **40**, 211.
136. C. E. Handford, M. Dean, M. Spence, M. Henchion, C. T. Elliott and K. Campbell, *Food Control*, 2015, **57**, 24.
137. J. Roosen, A. Bieberstein, S. Blanchemanche, E. Goddard, S. Marette and F. Vandermoere, *Food Policy*, 2015, **52**, 75.
138. M. Siegrist, N. Stampfli, H. Kastenholz and C. Keller, *Appetite*, 2008, **51**, 283.
139. J. C. Hannon, J. Kerry, M. Cruz-Romero, M. Morris and E. Cummins, *Trends Food Sci. Technol.*, 2015, **43**, 43.
140. C. E. Handford, M. Dean, M. Henchionb, M. Spence, C. T. Elliott and K. Campbell, *Trends Food Sci. Technol.*, 2014, **40**, 226.

CHAPTER 7

Hormetins as Drugs for Healthy Aging

SURESH I. S. RATTAN[a]

[a]Laboratory of Cellular Ageing, Department of Molecular Biology and Genetics, Aarhus University, Gustav Wieds Vej 10C, DK8000 Aarhus – C, Denmark
*E-mail: rattan@mbg.au.dk

7.1 Introduction

A promising strategy for anti-aging interventions is that of mild stress-induced hormesis. The consequences of stress can be either harmful or beneficial depending on the intensity, duration and frequency of the stress. Biological stress response (SR) is not linear with respect to the dose of the stressor. SR is characterized by a nonlinear biphasic relationship. Meta-analyses performed on a large number of papers published within toxicology, pharmacology, medicine, and radiation biology have revealed that the dose–response curve is neither linear nor threshold-based linear.[1] Instead, the shape of the dose response curve is U- or inverted U-shaped, depending on the endpoint being measured. This phenomenon of biphasic dose response was termed as hormesis in the early 1940s.[2] The terminology for hormesis has been expanded to specify the nature of the hormetic responses, such as physiological hormesis, pre-conditioning hormesis, and post-exposure conditioning hormesis.[3] Therefore, any condition that is potentially hormetic in terms of providing biological benefits through

RSC Drug Discovery Series No. 57
Anti-aging Drugs: From Basic Research to Clinical Practice
Edited by Alexander M. Vaiserman

Published by the Royal Society of Chemistry, www.rsc.org

the process of hormesis is termed as a hormetin.[4,5] Exercise is one of the best-known examples of a hormetin.

The aim of this chapter is to discuss the role of hormetins in the maintenance and enhancement of health, and in the prevention and treatment of age-related diseases. However, in order to appreciate the rationale behind the possible use of hormetins as drugs for healthy aging, it will be useful to review first the present understanding of the biological basis of aging and age-related diseases. This will then be followed by a discussion of the phenomenon of hormesis, molecular basis of hormesis and the strategies for the discovery and development of hormetins as drugs for healthy aging.

7.2 Aging in a Nutshell

Biogerontological research has developed a comprehensive scientific understanding about the evolutionary and mechanistic explanations for aging, longevity and age-related diseases. It is now generally accepted that progressive aging and senescence at the biological level occur primarily during the life time beyond the natural lifespan of a species, termed as the essential lifespan (ELS).[6,7] This view has developed from the basic biological understanding that survival of an organism is a dynamic tug between the occurrence of damage and the processes of maintenance and repair systems (MARS). The main MARS that comprise the ELS-assurance processes are listed in Table 7.1. These are also known as the longevity assurance pathways and involve hundreds of longevity assurance genes (LAG).

"Homeodynamic space" is another way of conceptualizing MARS. Homeodynamic space may also be considered as the "survival ability" of a biological

Table 7.1 Main maintenance and repair pathways (MARS) in biological systems arranged from molecular to whole body level.

Level of MARS	Pathway
Molecular	Nuclear and mitochondrial DNA repair
	Anti-oxidative enzymes and free radical scavengers
	Degradation of damaged DNA and RNA
	Protein repair
	Degradation of damaged proteins
Cellular	Degradation of damaged organelles
	Programmed cell death—apoptosis
	Intra-cellular stress responses
Systemic	Detoxification of harmful chemicals and metabolites
	Immune responses
	Wound healing and tissue regeneration
Whole-organism	Thermal regulation
	Neuro-endocrine balance
	Daily rhythms
	Other higher order defenses, including psychological mechanisms

system.[8,9] The term "homeodynamics"[10] means "the same dynamics", and is modified from the classical term "homeostasis", which means "the same state". This is because the notion of homeostasis ignores the reality of the constantly changing, interacting, and adapting dynamics of living systems. Biological systems are never static. There is no fixed and rigid static state for intracellular molecules, for inter-cellular interactions, and for higher order organization and functioning. Therefore, it is more precise and accurate to use the term homeodynamics for living systems. It is in this context that the term homeodynamic space refers to the survival ability of a system. Three main characteristics of homeodynamic space are: (1) stress response; (2) damage control; and (3) constant remodeling and adaptation in dynamic interactions.[8,9] A large number of molecular, cellular and physiological pathways and their interconnected networks, including MARS mentioned above, determine the nature and extent of the homeodynamic space.

At the species level, biological evolutionary processes have assured ELS by optimizing for homeodynamic space. However, the period of survival beyond ELS is accompanied by the progressive shrinkage of the homeodynamic space owing mainly to the accumulation of molecular damage.[7,11] This shrinkage of the homeodynamic space manifests as altered stress response, reduced ability to adapt, and increased probability of the emergence of chronic diseases and eventual death. Thus, although the process of aging in itself is a continuum of life, aging is a condition that allows the emergence of one or more diseases. Therefore, the issues of aging, quality of life and longevity need to be approached with holistic health-oriented paradigms. Hormesis is one such holistic approach.

7.3 Hormesis and Stress Response

The mechanistic bases of hormesis lie in the primary event of disruption of homeodynamics through a stressor, which then leads to the stress response (SR) to counteract the disruption. The molecular and physiological processes initiated by SR are not strictly limited to matching the level of disruption, and almost always lead to modest overcompensation.[1,12] A successful SR not only results in the re-establishment of homeodynamics, but also strengthens the homeodynamic space.[13]

A list of the main molecular pathways of SR in human and other mammalian systems is given in Table 7.2. Each of these pathways is well understood in terms of its molecular biology, and several excellent review articles can be found in the published literature and online. For example, heat shock response (HSR) is a universal and primordial stress response achieved by the activation of the HS transcription factor(s), followed by the preferential synthesis of several heat shock proteins (Hsp).[14–16] Similarly, accumulation of misfolded proteins in the endoplasmic reticulum (ER) leads to the so-called ER stress response, also known as the UPR stress response, resulting in the synthesis, activation and translocation of several chaperones.[17–21] In addition, there is a mitochondrial-specific stress response in mammalian cells, which involves the

Table 7.2 Major molecular pathways of stress response in human cells.

Stress response	Common stressors	Sensors and effectors
Heat shock response (HSR)	Heat, exercise, heavy metals, natural and synthetic small molecules, antibiotics	Heat shock transcription factors (HSF), heat shock proteins (Hsp), proteasome
Unfolded protein response (UPR) in endoplasmic reticulum (ER) and mitochondria (mt)	Unfolded and misfolded proteins, cytokines	ER-chaperones, mt-chaperonins, Hsp, proteasome
Autophagic response	Nutritional limitation, hypoxia, damaged organelles	Autophagosomes, lysosomes
Oxidative stress response	Oxidants, free radicals, reactive oxygen species	Transcription factors (Nrf2, FOXO), heme-oxygenase, antioxidative enzymes (SOD, catalase, glutathione)
DNA damage response (DDR)	Radiation, reactive oxygen species	DNA damage sensors (ATM, ATR), p53, DNA repair proteins
Inflammatory response	Pathogens, allergens, damaged macromolecules	Nuclear factor—κB transcription factors, cytokines, nitric oxide synthase (NOS)
Sirtuin-mediated response	Energy depletion, metabolic imbalance	Sirtuins

induction and activation of various chaperones, including mortalin, chaperonin-10 (Cpn10/Hsp10), chaperonin-60 (Cpn60/Hsp60), and mortalin.[22,23]

Autophagy response is the lysosome-mediated and chaperone-mediated sequestering of damaged membranes and organelles, which is a stress response induced during nutritional limitation, starvation, and hypoxia.[24–26] Sirtuin SR is the activator of sirtuins, which cause the deacetylation of histones and other proteins in response to reduced levels of metabolic energy.[27,28]

Another widely studied SR by which cells respond to oxidative stress is through the regulation of transcription of antioxidant genes. The main regulator of this specific antioxidant phenotype is the nuclear factor-erythroid-2 (Nrf2) transcription factor, which regulates the basal and inducible expression of numerous detoxifying and antioxidant genes.[29,30] Under normal conditions, Nrf2 is held in the cytoplasm by the specific inhibitory protein KEAP1. Oxidative modification of cysteine residues of KEAP1 induces conformational changes and a loss of Nrf-2 binding, allowing Nrf2 to translocate to the nucleus where it heterodimerizes with specific co-factors, and leads to the transcription of various genes through the regulatory regions of antioxidant response elements (AREs).[29,31] Some of the genes activated by stress-induced activation of Nrf2 are heme oxygenase1 (HO-1), NAD(P)H-quinone oxidoreductase-1 (NQO1), and glutathione *S*-transferases (GSTs).

Activation of DNA repair enzymes in response to DNA damage is another SR, which is essential for the maintenance of genomic stability.[32–34] Both acute and chronic inflammations are protective SR mechanisms in the wake of cell and tissue injury.[35–38] Similarly, one of the main inflammatory mediators is the transcription factor NF-κB, which together with other mediators, such as TNF-α, IL-6, NOS, and prostaglandins, has important homeodynamic functions, including repair of tissues, control of metabolism, and regulation of the hypothalamus–pituitary axis.[37]

Although the exact nature of the initial molecular damage caused by a condition may not be easily identified, activation of one or more SR pathways is a good indicator of the potential occurrence of hormesis. However, an induction of a specific SR pathway as the first response (immediate response) does not rule out the induction of one or more other SR pathways later on (delayed response). A complete and successful SR for effective homeodynamics and for the maintenance of the homeodynamic space includes both immediate and delayed SR. As argued previously,[13] it is important that all SR pathways are analyzed simultaneously and a complete stress response profile (SRP) is established under a given condition, such as age, health and disease status, and during and after exposure to single or multiple stressors. Determining SRP is essential for establishing the nature and extent of the homeodynamic space of cells, tissues and organisms. Furthermore, being able to map the kinetics and amplitude of different SR, and their effects on each other, can be the basis to evaluate the health status of an individual and to develop effective means of aging modulators and maintainers of homeodynamic space.

7.4 Hormetins for Health and Longevity

A condition that can bring about health beneficial effects through mild stress-induced hormesis is termed as a hormetin.[39–41] Sometimes the term "adaptogens" is also used for such agents.[42] Hormetins are further categorized as: (1) physical hormetins, for example exercise, heat, gravity and irradiation; (2) psychological or mental hormetins, such as intense brain activity and focused attention or meditation; and (3) nutritional hormetins, such as flavonoids, polyphenols and other chemicals, including micronutrients in spices and other food sources. Chronic calorie restriction (CR) and intermittent fasting are also considered as nutritional hormetins because these conditions also induce SR and lead to physiological hormesis.[43–45]

A very important observation in studies of hormesis is that a single hormetin can strengthen the overall homeodynamics of cells by initiating a cascade of processes resulting in a biological amplification of the beneficial effects.[46,47] Moderate and repeated physical exercise as a hormetin is the best example of holistic stress-induced hormesis. Exercise initially increases the production of free radicals, acids and aldehydes, which leads to the activation of a series of SR pathways, and eventual health beneficial effects are achieved.[48–50] Most importantly, exercise-associated health benefits are not limited to the site of exercise, but spread to the whole body level, including overall physical and mental well-being.

Various hormetins that have been reported to modulate aging and longevity in cells and model organisms include heat shock, irradiation, heavy metals, pro-oxidants, acetaldehyde, alcohols, hypergravity, exercise, mechanical stretching, electromagnetic field, food restriction and mental challenge.[39,51,52] Nutritional hormetins, especially those derived from plant sources, have generated much scientific interest for their health beneficial effects. This is because of the realization that not all chemicals found in plants are beneficial in a simple and straight-forward manner. Instead, these non-nutritional food components cause molecular damage by virtue of their electrochemical properties and have a typical biphasic hormetic dose response. Some examples of nutritional hormetins are those containing phenolic acids, polyphenols, flavanoids, ferulic acid, geranylgeranyl, rosmarinic acid, resveratrol, kinetin, zinc, and the extracts of tea, dark chocolate, saffron and spinach.[5,53] Chronic CR, intermittent fasting and CR-mimetics, including rapamycin and its analogues, are other examples of nutritional hormetins as drugs for healthy aging and longevity.[54–56]

7.5 Discovering Novel Hormetins

Putting test materials through a screening process for their ability to induce one or more SR pathways in cells and organisms is a promising strategy for discovering novel hormetins.[5] A general scheme for screening natural and synthetic single compounds or complex extracts as hormetins for human beings involves initial testing by using normal diploid human cells in culture. The use of normal diploid cells is very important for such studies, since immortal cell lines usually have one or more genetic and metabolic deviations, which are rarely comparable to normal cells. An important aspect of normal diploid cells is the Hayflick phenomenon of limited proliferative capacity and replicative senescence, which is a model of aging *in vitro*.[57]

Determining dose-dependent, time-dependent and age-dependent SR profiles is the first step in discovering novel hormetins.[13,40,41] Since most of the early SR markers are transcription factors (see Table 7.2), which undergo post-translational modifications and translocate from the cytoplasm to the nucleus, immunofluorescence microscopy showing this cytoplasm-to-nuclear shift may be sufficient at this stage. However, for identifying the late SR effectors, such as induced synthesis of Hsp, chaperones, cytokines, sirtuins and other antioxidative enzymes, both gene-array expression analysis for mRNA levels and proteomic analysis for protein levels will be required.

The initial screening of test materials as potential hormetins by determining their effects on early and late SR markers must be followed by performing cell type-specific functional assays. Furthermore, the cell type to be used for such a screening will depend on the biological end-point that one expects to improve by hormetin treatment. Some of the cell type-specific assays that can be used for testing novel hormetins are: cellular motility and wound healing assay for fibroblasts, induction of differentiation for stem cells and keratinocytes, blood vessel formation by endothelial cells, osteocalcin and mineralized matrix formation by osteoblasts, and muscle fiber formation by muscle cells. Similarly, other cell type-specific markers for other cells, such as neurons, hepatocytes, immune system cells and others, should be employed. Only after this step, one could take the tested material for its further testing as a prospective hormetin at the tissue, organ and organismal level.

A recent example of a successful hormetin product development following the strategy outlined above is a skin care cosmetic.[5] This was achieved by analyzing the stress-inducing effects of active ingredients extracted from the roots of the Chinese herb Sanchi (*Panax notoginseng*). Ginsenosides extracted from Sanchi induced the transcription of stress genes and increased the synthesis of stress proteins, especially HSP1A1 in normal human keratinocytes and dermal fibroblasts.[5] Once recognized as a potential hormetin based on the criteria described above, Sanchi extract was then further tested for its effects on human skin by following the established clinical protocols used in skin care research.[53]

7.6 Drugs for Health and Longevity

A drug, as defined by the World Health Organisation (WHO) in 1994, refers to *"...any substance with a potential to prevent or cure disease or enhance physical or mental welfare..., and that alters the biochemical or physiological processes..."* (http://www.who.int/substance_abuse/terminology/who_ladt/en/). Thus a drug can be both AGAINST a disease or FOR health and well-being. Hormetins are drugs for health by strengthening the homeodynamic space.

An important issue here is how to define and measure health. Health is often described either in the context of the absence of one or more diseases or as a vague concept of well-being, without having any objective measures for that. For example, the WHO's definition of health as *"a state of complete physical, mental and social well-being and not merely the absence of disease or infirmity"* (http://www.who.int/topics/mental_health/en/), does not clarify what is meant by well-being in definitive terms. Furthermore, this is an idealized state, which perhaps no one can have.

Health has also been defined as the ability to adapt and self manage.[58] Although this definition includes certain aspects of functionality, it still implies a kind of an idealized state of adaptation and self-management. A pragmatic and realistic definition of health has been put forward taking into consideration the functionality of the living system as a crucial phenotype. Employing the concept of activities of daily living (ADL), health is defined as having "adequate" physical and mental independence in ADL.[9] This state of adequacy can vary widely depending on biological factors, such as genetics and age, and on psycho-social factors, such as personal temperament, cultural values and peer pressure. Yet, it can be possible to establish "adequate health" objectively by measuring a series of functional markers, including basic characteristics of the homeodynamic space.

Finally, the notion of applying single or multiple hormetins as drugs for health and longevity is based in the fundamental understanding of aging as a complex, holistic and individualistic phenomenon. Of course there are several important issues yet to be resolved with respect to the dose, frequency and the combination of hormetins. This will be surely achieved by ongoing systematic research in the field of hormetics, using a wide variety of experimental systems, including cultured cells, simple organisms, mammals and healthy human beings.[59]

References

1. E. J. Calabrese, *Environ. Pollut.*, 2013, **182**, 452.
2. C. M. Southam and J. Ehrlich, *Phytopathology*, 1943, **33**, 517.
3. E. J. Calabrese, K. A. Bachmann, A. J. Bailer, P. M. Bolger, J. Borak, L. Cai, N. Cedergreen, M. G. Cherian, C. C. Chiueh, T. W. Clarkson, R. R. Cook, D. M. Diamond, D. J. Doolittle, M. A. Dorato, S. O. Duke, L. Feinendegen, D. E. Gardner, R. W. Hart, K. L. Hastings, A. W. Hayes, G. R. Hoffmann, J. A. Ives, Z. Jaworowski, T. E. Johnson, W. B. Jonas, N. E. Kaminski, J. G. Keller, J. E. Klaunig, T. B. Knudsen, W. J. Kozumbo, T. Lettieri, S. Z. Liu, A. Maisseu, K. I. Maynard, E. J. Masoro, R. O. McClellan, H. M. Mehendale, C. Mothersill, D. B. Newlin, H. N. Nigg, F. W. Oehme, R. F. Phalen, M. A. Philbert, S. I. S. Rattan, J. E. Riviere, J. Rodricks, R. M. Sapolsky, B. R. Scott, C. Seymour, D. A. Sinclair, J. Smith-Sonneborn, E. T. Snow, L. Spear, D. E. Stevenson, Y. Thomas, M. Tubiana, G. M. Williams and M. P. Mattson, *Toxicol. Appl. Pharmacol.*, 2007, **222**, 122.
4. R. E. Ali and S. I. S. Rattan, *Ann. N. Y. Acad. Sci.*, 2006, **1067**, 394.
5. S. I. S. Rattan, *Expert Opin. Drug Discovery*, 2012, **7**, 439.
6. S. I. S. Rattan, *Ann. N. Y. Acad. Sci.*, 2000, **908**, 282.
7. S. I. S. Rattan, *Free Radical Res.*, 2006, **40**, 1230.
8. S. I. S. Rattan, in *Encyclopedia of Gerontology*, ed. J. Birren, Elsevier Inc., UK, 2nd edn, 2007, pp. 696–699.
9. S. I. S. Rattan, *Biogerontology*, 2013, **14**, 673.
10. F. E. Yates, *Math. Comput. Model.*, 1994, **19**, 49.
11. S. I. S. Rattan, *Biol. Chem.*, 2008, **389**, 267.
12. E. J. Calabrese, I. Iavicoli and V. Calabrese, *Biogerontology*, 2012, **13**, 215.
13. D. Demirovic and S. I. S. Rattan, *Exp. Gerontol.*, 2013, **48**, 94.
14. P. Verbeke, J. Fonager, B. F. C. Clark and S. I. S. Rattan, *Cell Biol. Int.*, 2001, **25**, 845.
15. D. R. Ciocca and S. K. Caderwood, *Cell Stress Chaperones*, 2005, **10**, 86.
16. R. K. Leak, *J. Cell Commun. Signal.*, 2014, **8**, 293.
17. A. Lee, *Trends Biochem. Sci.*, 2001, **26**, 504.
18. J. H. Lin, H. Li, D. Yasumura, H. R. Cohen, C. Zhang, B. Panning, K. M. Shokat, M. M. LaVail and P. Walter, *Science*, 2007, **318**, 944.
19. M. Ni and A. S. Lee, *FEBS Lett.*, 2007, **581**, 3641.
20. H. Yoshida, *FEBS J.*, 2007, **274**, 630.
21. G. Banhegyi, P. Baumeister, A. Benedetti, D. Dong, Y. Fu, A. S. Lee, J. li, C. Mao, E. Margittai, M. Ni, W. Paschen, S. Piccirella, S. Sensei, R. Sitia, M. Wang and W. Yang, *Ann. N. Y. Acad. Sci.*, 2007, **1113**, 58.

22. Q. Zhao, J. Wang, I. V. Levichkin, S. Stasinopoulos, M. T. Ryan and N. J. Hoogenraad, *EMBO J.*, 2002, **21**, 4411.
23. S. C. Kaul, C. C. Deocaris and R. Wadhwa, *Exp. Gerontol.*, 2007, **42**, 263.
24. S. W. Ryter and A. M. K. Choi, *J. Biochem. Pharmacol. Res.*, 2013, **1**, 176.
25. G. Filomeni, D. De Zio and F. Cecconi, *Cell Death Differ.*, 2015, **22**, 377.
26. J. Zhang, *Redox Biol.*, 2015, **4**, 242.
27. W. Giblin, M. E. Skinner and D. B. Lombard, *Trends Genet.*, 2014, **30**, 271.
28. A. Efeyan, W. C. Comb and D. M. Sabatini, *Nature*, 2015, **517**, 302.
29. H. Motohashi and M. Yamamoto, *Trends Mol. Med.*, 2004, **10**, 549.
30. S. Singh, S. Vrishni, B. K. Singh, I. Rahman and P. Kakkar, *Free Radical Res.*, 2010, **44**, 1267.
31. T. Yamamoto, K. Yoh, A. Kobayashi, Y. Ishii, S. Kure, A. Koyama, T. Sakamoto, K. Sekizawa, H. Motohashi and M. Yamamoto, *Biochem. Biophys. Res. Commun.*, 2004, **321**, 72.
32. R. Hakem, *Embo J.*, 2008, **27**, 589.
33. J. Vijg, *Mech. Ageing Dev.*, 2008, **129**, 498.
34. A. A. Moskalev, M. V. Shaposhnikov, E. N. Plyusnina, A. Zhavoronkov, A. Budovsky, H. Yanai and V. E. Fraifeld, *Ageing Res. Rev.*, 2012, **11**, 51.
35. J. M. Kyriakis and J. Avruch, *J. Biol. Chem.*, 1996, **271**, 24313.
36. R. Njemini, C. Demanet and T. Mets, *Biogerontology*, 2004, **5**, 31.
37. R. Medzhitov, *Nature*, 2008, **454**, 428.
38. S. Vasto, G. Carruba, D. Lio, G. Colonna-Romano, D. Di Bona, G. Candore and C. Caruso, *Mech. Ageing Dev.*, 2009, **130**, 40.
39. S. I. S. Rattan and D. Demirovic, in *Hormesis: A Revolution in Biology, Toxicology and Medicine*, ed. M. P. Mattson and E. Calabrese, Springer, New York, 2009, pp. 153–175.
40. S. I. S. Rattan and D. Demirovic, *Dose-Response*, 2010, **8**, 58.
41. S. I. S. Rattan and D. Demirovic, in *Calorie Restriction, Aging and Longevity*, ed. A. V. Everitte, S. I. S. Rattan, D. G. Le Couteur and R. de Cabo, Springer, Dordrecht, 2010, pp. 233–245.
42. F. A. Wiegant, S. Surinova, E. Ytsma, M. Langelaar-Makkinje, G. Wikman and J. A. Post, *Biogerontology*, 2009, **10**, 27.
43. *Calorie Restriction, Aging and Longevity*, ed. A. V. Everitt, S. I. S. Rattan, D. G. Le Couteur and C. de Cabo, Springer, New York, 2010.
44. B. D. Horne, J. B. Muhlestein and J. L. Anderson, *Am. J. Clin. Nutr.*, 2015, **102**, 464.
45. M. P. Mattson, *Dose-Response*, 2014, **12**, 600.
46. *Hormesis – A Revolution in Biology, Toxicology and Medicine*, ed. M. P. Mattson and E. Calabrese, Springer, New York, 2010.
47. S. I. S. Rattan, R. A. Fernandes, D. Demirovic, B. Dymek and C. F. Lima, *Dose-Response*, 2009, **7**, 93.
48. *Handbook of Oxidants and Antioxidants in Exercise*, ed. C. K. Sen, L. Packer and O. Hänninen, Elsevier, Amsterdam, The Netherlands, 1st edn, 2000.
49. Z. Radak, H. Y. Chung and S. Goto, *Biogerontology*, 2005, **6**, 71.
50. J. Williamson and M. Pahor, *Arch. Intern. Med.*, 2010, **170**, 124.

51. *Mild Stress and Healthy Aging: Applying Hormesis in Aging Research and Interventions*, ed. E. Le Bourg and S. I. S. Rattan, Springer, Dordrecht, The Netherlands, 2008.
52. *Hormesis in Health and Disease*, ed. S. I. S. Rattan and E. Le Bourg, CRC Press, Boca Raton, 2014.
53. S. I. S. Rattan, *Biogerontology*, 2012, **13**, 83.
54. D. K. Ingram and G. S. Roth, *Ageing Res. Rev.*, 2015, **20C**, 46.
55. J. I. Castillo-Quan, K. J. Kinghorn and I. Bjedov, *Adv. Genet.*, 2015, **90**, 1.
56. H. E. Walters, S. Deneka-Hannemann and L. S. Cox, *Aging*, 2016, **8**, 1.
57. *Cellular Ageing and Replicative Senescence*, ed. S. I. S. Rattan and L. Hayflick, Springer, Dordrecht, 2016.
58. M. Huber, J. A. Knottnerus, L. Green, H. van der Horst, A. R. Jadad, D. Kromhout, B. Leonard, K. Lorig, M. I. Loureiro, J. W. van der Meer, P. Schnabel, R. Smith, C. van Weel and H. Smid, *BMJ*, 2011, **343**, d4163.
59. W. Luyten, P. Antal, B. P. Braeckman, J. Bundy, F. Cirulli, C. Fang-Yen, G. Fuellen, A. Leroi, Q. A. Liu, P. Martorell, A. Metspalu, M. Perola, M. Ristow, N. Saul, L. Schoofs, K. Siems, L. Temmerman, T. Smets, A. Wolk and S. I. S. Rattan, *Biogerontology*, 2016, **17**, 771.

Section III

Antioxidants

CHAPTER 8

Antioxidant Therapy of Aging: From Free Radical Chemistry to Systems Theory of Reliability

V. K. KOLTOVER[a]

[a]Institute of Problems of Chemical Physics, Russian Academy of Sciences, Chernogolovka, Moscow Region, 142432, Russia
*E-mail: koltover@icp.ac.ru

8.1 Introduction: Historical Synopsis

Aging is a universal process to which all organisms, both multi-cellular and unicellular, are subjected. Some researchers consider aging as the last stage of a genetic program of ontogenesis and believe that there exist special "aging genes" that regulate aging and death. Other researchers believe that aging is a stochastic process occurring through the progressive accumulation of damage in macromolecules, including DNA, proteins and lipids.[1,2] In this research direction, the free radical theory of aging, which was put forward in the 1950s, has determined the most heuristic lines of investigations.

The vigorous research of free radical processes and antioxidants in biology and medicine started in the mid-20th century. This was stimulated by awarding the Nobel Prize in Chemistry to Cyril Hinshelwood and Nikolay Semenov in 1956 for the studies of free radical mechanisms of chain radical reactions. By the mid-1950s, it was already known that free radicals, specifically the

RSC Drug Discovery Series No. 57
Anti-aging Drugs: From Basic Research to Clinical Practice
Edited by Alexander M. Vaiserman

Published by the Royal Society of Chemistry, www.rsc.org

hydroxyl radical ($OH^•$), arise in water under the action of ionizing radiation and that toxic products of free radical-induced lipid oxidation appear in animal tissues on exposure to ionizing radiation.[3,4]

The idea that free radicals of oxygen are responsible for the oxygen toxicity was stated by Gershman and Gilbert in 1954 (quoted from ref. 4). At the same time, the free radical hypothesis of aging was put forward by Harman.[5–7] In his works, inhibitors of free radical chain oxidation, the so-called antioxidants, were proposed for prolongation of life of humans and animals. Harman was the first to test this hypothesis in experiments with laboratory animals. It was discovered that the antioxidant, radiation-protector 2-mercaptoethylamine, prolongs life spans of C3H female mice (26%) and AKR male mice (29.2%).[6,7] Since then, the beneficial effects of antioxidant therapy were experimentally proved over and over again (see ref. 7 and 8). Currently, however, the mechanism of action of antioxidants in living systems (*in vivo*) does not seem as unambiguous as half a century ago.

At the end of the 20th century, a new approach to the problems of aging was developed on the basis of theory of reliability.[9–19] This review is designed to show that the systems theory of reliability allows the programmed and stochastic events to be integrated into a single united theory. In addition, this review presents the data that antioxidants provide preventive protection from free radicals *via* the beneficial effects of the antioxidants on the system of neuro-hormonal regulation along with their systems reliability effects on microbiota cells.

8.2 Aging *Versus* Reliability

8.2.1 Theory of Reliability: Basic Ideas

Biological objects are constructs, *i.e.* all of them are designed according to special genetic programs with the aim to perform predetermined functions. Similarly to technical devices, biological constructs are not perfectly reliable in operation: malfunctions happen alternating with the normal operation function acts. In engineering, reliability is defined as the ability of an object to perform its function for a given time under given conditions. The same intuitive definition of reliability fits biological systems.[10–19] Conferences that were initiated by Ukrainian Academy of Sciences in Kiev, Ukraine, starting from 1975,[9] spurred the studies on reliability ("robustness") of biological systems, confirming the old saying of the Middle Ages that "Teaching comes from Kiev".

In engineering, the longevity of a device is programmed inasmuch as structures of functional elements, their quality and quantity, interconnections between them, the means of replacing failed elements and so on are predetermined, *i.e.* "preset". At the same time, a device is subjected to the influence of random factors of internal and external origin: load fluctuations, noises in the elements, random disturbances in interconnections, *etc.* Thus, both the constructional particularities of the device (program) and the random factors (stochastics), should be taken into account. Hence, the theory of reliability naturally suggests methodology and mathematical apparatus for building a unified theory of aging of living systems that can encompass both programmed and stochastic events.

The important line of supporting the systems reliability is redundancy, when redundant components of the same type are introduced to fulfill one and the same function in the device. Indeed, all essential biomolecular constructions in cells are present in superfluous amounts. The redundant amounts of mitochondria and other organelles represent the examples of structural reservation. The elimination of hydrogen peroxide (H_2O_2) and organic peroxides from cells by two different enzyme systems, catalase and glutathione peroxidase, is an example of functional reservation.

The main line of creating reliable devices from unreliable functional components is, however, in the timely supplying of repair and replacement of unreliable functional elements. Repair and renewal processes proceed in all complex biological systems starting from the level of cells. In order to provide a failure rate as low as possible, components are to be replaced for novel ones before the phase of their wear-out begins. The failure rate of the system may become intolerably high if components are replaced only as they have been damaged or worn-out. Hence, the preventive replacement of functional elements that follows the pattern preset in the cell genome, the so-called metabolic turnover, seems to be the main line of providing the high reliability of biological systems.

8.2.2 Preset Reliability Prescribes Lifespan

It is generally known that any organism is a hierarchical structure in which a relatively small number of key elements that manage a large number of executive elements can be distinguished. The template principle of organization of living systems implies that information DNA structures are of the first operation importance in the cell hierarchy. A multi-cellular organism is governed by the genes of a special anatomically isolated group of cells. In animals, for instance, these are the specialized neurons of the hypothalamus. Furthermore, from the mathematical theory of reliability, it is known that the effectiveness of the operation of a complex system is determined mainly by the reliability of its governing elements, "the power structure".[19]

Following this line, the systems reliability approach to the problems of aging was developed in our papers.[10–19] This approach is based on simple general principles. The first one is the template principle of organization of living systems implying that information structures rank first in cell hierarchy. An organism works like a system of biomolecular constructions designed in accordance with the genetic program (*information plan*) in order to perform the preset programmed functions (*purpose*). The second one is that all biomolecular constructions operate with limited reliability, namely, for each and every biological device or bionanoreactor, starting from enzymes, normal operation acts alternate with accidental malfunctions. The third principle states that preventive maintenance replacement of functional elements in cells and tissues is the main line of assuring the high systems reliability. Following the preset genome pattern, unreliable elements should be timely replaced for novel ones before the phase of their wear-out begins. It is the so-called metabolic turnover. The fourth principle states that there is a finite number of critical elements that perform the supervisory functions over the organism's repair and renewal processes, *i.e.* over the metabolic turnover. Since these critical elements of the highest hierarchic level exert the control over the systems reliability, they can be called "longevity-assurance structures" (LAS). Inasmuch as all reliability facilities—among them preventive maintenance, repair, and redundancy of functional elements—are genetically limited, stochastic damages in LAS accumulate up to the preset threshold dysfunction levels. As a result, each organism has a limited life-span.[18,19] Indeed, it is common knowledge that there are neither mice nor rats exceeding 3–4 years of age, and that a human life-span does not exceed ≈120 years provided we take reliable data into account, not sensational press reports or legends. The limited lifetime of diploid cell strains is also a well-known phenomenon. For example, human fibroblasts *in vitro* die or mutate into cancer cells after performing about 50 doublings. American biologist Hayflick discovered this effect in 1961 (see ref. 1 and 2). Russian biologist Olovnikov explained Hayflick's limit suggesting the mechanism of the incomplete copying of telomere ends of DNA. According to Olovnikov's theory of marginotomy, every cell division is accompanied by the reduction of the telomere ends of cell chromosomes.[20] In essence, it means that the cell division stops as soon as the telomere circumcision runs up to the limit fatal level.

Following the reliability-theory approach, the simple mathematical model of aging was suggested first in our papers.[10–12] It was taken that LAS accumulate stochastic flaws resulting in disarray of their functions. Account was also taken of another widespread peculiarity of living systems, *i.e.* the existence of threshold values for the most important functional parameters. The organism has been assumed to perish the moment that any of LAS develops a threshold dysfunction (a limit, m_c). As a matter of fact, the life-span of the organism is determined by the threshold dysfunction of the worst LAS, *i.e.* by the weakest link's longevity. If N is the number of LAS, then the survival function is given by the smallest value of the random sample of size N with the following approximation for mortality rate:

$$h(t) = \Delta n(t)/n(t)\Delta t = h_0 \exp(\gamma t),\ h_0 = \gamma N/[\exp(\gamma T) - 1] \qquad (8.1)$$

where $n(t)$ is the number of live persons of the age t, Δn is the number of those who died during the time interval Δt, parameters h_0 and γ are independent of time.[10–19]

The exponential growth of mortality rate with time has long been known in quantitative gerontology and demography as the so-called "Gompertz law" of mortality. It has been confirmed for people (of age approximately from 35 to 90 years), other mammals, flies, and mollusks.[1,2] Moreover, it was shown that aging of prokaryote cells, *Acholeplasma laidlawii*, in the stationary phase of growth—namely loss of viability measured as their ability to form macro-colonies—follows the same kinetic pattern.[21] It is noteworthy that the cited work[21] was the first one where it was demonstrated that the cell viability in cell cultures declines accordingly to the Gompertz law. Thus, the universal feature of aging, the Gompertz law of mortality, gets its explanation in the context of the reliability-theory approach stated above.

The limit life-span T in eqn (8.1) has appeared as the direct result of existence of the limit dysfunction, m_c, for LAS. This limit T is the life-span of an "ideal" organism with no initial flaws at $t = 0$. If we take from the review[1] that the maximum life-span for human populations, on average, is about 95 years, the magnitude of γ varies from 0.0612 to 0.119 years^{-1} and the magnitude of h_0 varies from 0.820×10^{-3} to 0.022×10^{-3} years^{-1}, then, using the expression for h_0, we find that $N \approx 5$–15. The values of N for dogs, mice and mares calculated by the relevant values of parameters (T, h_0, γ) fall in the same order of magnitude too.

An analytical transition from the abstract "longevity-assurance structures" to real biomolecular structures seems to be not easier than similar transitions from the "generalized co-ordinates" in theoretical physics. It is worthy, however, to note that this estimation corresponds, by the order of magnitude, to the number of the so-called "longevity-assurance genes" which have been recently discovered in nematodes, yeasts, drosophilae, mice, and other organisms (see in ref. 2 and 22). In humans and animals, these "longevity-assurance genes" are believed to be located in the special neurons of the suprachiasmatic nucleus of the hypothalamus.

8.3 Free-Radical Failures

8.3.1 Free-Radical Malfunctions of Electron-Transport Nanoreactors

The oxygen radical anion ($O_2^{\bullet-}$), the most important source of chemically reactive "toxic" oxygen species, is produced in cells and tissues of all aerobic organisms.[23–27] The main bulk of $O_2^{\bullet-}$ is formed as by-product of electron transport in cell mitochondria, the organelles that use up to 99% of all oxygen consumed by cells for ATP synthesis. Normal functioning of electron-transport nanoreactors (ETN) of mitochondria lies in the transport of electrons from the oxidation substrates, NADH and succinate, to cytochrome oxidase and then to oxygen with reduction of oxygen molecules to water and synthesis of ATP (see ref. 24). However, the reliability characteristics of mitochondrial nanoreactors are not perfect. As a result, normal elementary acts of electron transfers alternate with accidental malfunctions, which result in the formation of $O_2^{\bullet-}$. From the reliability point of view, the fact that this radical appears is to be considered as the random malfunction of ETN, similarly to "recurrent failures" in engineering.[10–19] Among other possible generators of

$O_2^{\bullet-}$ in cells and tissues, there are NADPH-cytochrome-C-reductase and cytochrome P-450 of endoplasmic reticulum, xanthine oxidase, catecholamine and other biogenic amines, mono- and di-amine oxidases, aldehyde oxidases, oxidases of D-aminoacids, D-galactosidase, lipoxigenase, nitric oxide synthase, leukoflavines, hemoglobin and myoglobin, ascorbate, NADPH-oxidase of phagocytes and other so-called NOX enzymes (NADPH oxidases).[23,26]

From chemistry, it is known that hydrogen peroxide (H_2O_2) is formed as the product of the reaction of dismutation of the $O_2^{\bullet-}$ radicals:

$$O_2^{\bullet-} + O_2^{\bullet-} + 2H^+ \Rightarrow H_2O_2 + O_2 \tag{8.2a}$$

Next, $O_2^{\bullet-}$ reacts with H_2O_2 with the formation of the $OH^{\bullet}$ radical, which is known as a strong oxidant, the so-called Haber–Weiss reaction:

$$O_2^{\bullet-} + H_2O_2 \Rightarrow OH + OH^{\bullet} + O_2 \tag{8.2b}$$

Moreover, this reaction is accelerated in the presence of ions of variable valence, like iron (Fenton reaction):

$$O_2^{\bullet-} + Fe^{3+} \Rightarrow Fe^{2+} + O_2 \tag{8.2c}$$

$$Fe^{2+} + H_2O_2 \Rightarrow Fe^{3+} + HO^{\bullet} + HO^{-} \tag{8.2d}$$

Besides, there are reasons to believe that $O_2^{\bullet-}$ can react with the nitric oxide ($NO^{\bullet}$) radical with the formation of peroxynitrite ($ONOO^-$):

$$O_2^{\bullet-} + NO^{\bullet} + H^+ \Rightarrow O{=}N{-}O{-}OH \tag{8.3a}$$

$$O{=}N{-}O{-}OH \Rightarrow O{=}NO^{\bullet} + HO^{\bullet} \tag{8.3b}$$

$$O{=}N{-}O{-}OH \Rightarrow ONOO^- + H^+ \Rightarrow NO_3^- + H^+ \tag{8.3c}$$

Peroxynitrite is considered as a strong oxidant and nitrating species that mediates the biological effects of the superoxide and the nitric oxide.[25,27]

To protect cell structures from $O_2^{\bullet-}$ and its toxic chemical products, there is a special defense enzyme, superoxide dismutase (SOD), which catalyzes the reaction of dismutation of $O_2^{\bullet-}$ into H_2O_2 and oxygen. There are three kinds of SOD, mitochondrial Mn-SOD, cytosolic Cu,Zn-SOD and periplasmatic Fe-SOD. SOD enzymes work in cooperation with other antioxidant enzymes, catalase and glutathione peroxidase, which catalyze decomposition of H_2O_2.[23–26]

The reliability theory approach to the problems of free-radicals and aging was first proposed in.[10–12] Elementary acts of occurrence of $O_2^{\bullet-}$, as well as elementary acts of disappearance of $O_2^{\bullet-}$ in the dismutation reaction, are stochastic processes. Inasmuch as SOD, like all other enzymes, operates with limited reliability, the $O_2^{\bullet-}$ radicals can slip through the SOD defense system.

We have analyzed the stochastic dynamics of this system by the mathematical "Birth and Death" model often used in mathematical reliability theory. The calculations, based on the experimental data from the available literature, show that the probability of the slipping of $O_2^{\bullet-}$ through the mtSOD defense is about 1.9×10^{-5}, *i.e.*, about 2 radicals from every hundred thousand may penetrate the defense system.[10–12] It is noteworthy that $O_2^{\bullet-}$ can penetrate through lipid membranes.[28] Hence, with the intense electron transport fluxes in mitochondria and outside, the probability of $O_2^{\bullet-}$-induced free-radical damages in cells can be high enough.

8.3.2 Free-Radical Redox-Timer of Aging

In the case of functional damages in LAS caused by the $O_2^{\bullet-}$ radicals, the following equation was derived for the maximum lifespan:[11–14]

$$T = mc/[(qV/E)u + D] \tag{8.4}$$

In this equation, V is respiration rate, q is probability of the malfunction in electron transport nanoreactors leading to occurrence of $O_2^{\bullet-}$, E is the activity of SOD in LAS, and u is the probability of the free-radical failures to provoke functional violations. This parameter, u, is to take into account the next lines of defense, from special enzymes of reparation of DNA to other repair and renewal processes in cells and tissues. The higher the reliability of the defense, the less the value of this parameter is. In essence, the u parameter takes into account that deleterious effects of $O_2^{\bullet-}$, which slipped through the SOD defense, are of no concern if the preventive replacement of the damaged biological constructions is properly maintained. Lastly, D is the index to incorporate other damage factors that are not associated with oxygen free radicals. As a matter of fact, eqn (8.4) is the reliability explanation of the well-known "Rubner scaling relation", that there is an inverse correlation between the species-specific resting metabolism and the maximum life spans of species.

The data on SOD activity in tissues of brain, liver and heart of men and animals of thirteen species were published by Cutler's group in ref. 29. These data and the data on resting specific metabolic rates in the tissues of the species were used to plot the graphs of the reciprocal of the maximum life-spans ($1/T$) as a function of the ratio V/E values, according to eqn (8.4).

Figure 8.1 represents the graph for the brain tissues.

In line with the prediction of our model, the linear correlations were obtained for brain, liver and heart:

$$1/T = (0.0132 \pm 0.0002)(V/E) + (0.004 \pm 0.002),\ r = 0.997\ \text{(brain)} \tag{8.5a}$$

$$1/T = (0.0144 \pm 0.0003)(V/E) + (0.005 \pm 0.002),\ r = 0.997\ \text{(liver)} \tag{8.5b}$$

$$1/T = (0.0110 \pm 0.0009)(V/E) + (0.011 \pm 0.006),\ r = 0.981\ \text{(heart)} \tag{8.5c}$$

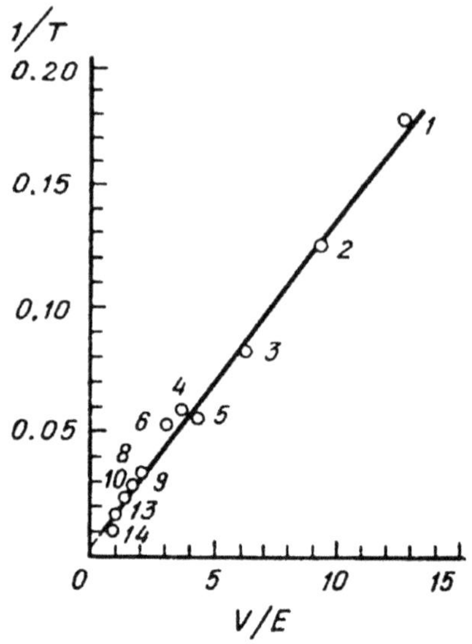

Figure 8.1 The correlation between the reciprocal maximum life-span (*1*/*T*) and the ratio of specific metabolic rate to SOD (*V*/*E*) in brain for mammalian species: 1: house mouse; 2: deer mouse; 3: common tree shrew; 4: squirrel monkey; 5: bush baby; 6: moustache tamarin; 7: lemur; 8: African green monkey; 9: Rhesus monkey; 10: olive baboon; 11: gorilla; 12: chimpanzee; 13: orangutan; 14: man (data compiled from ref. 13).

By using the free coefficient *D*, it was estimated that the longevity of the human brain could reach 250 years should the reliability of the antioxidant SOD defense be absolutely perfect. By using the free coefficient *D* of the relevant equations for the heart and liver, the limit longevity values were estimated to be 100 and 200 years, respectively.[11,12] Although these estimations are illustrative, they do emphasize the large role of ROS in the pathogenesis of the cardiovascular system.

It has long been known that free radical oxidation damages of DNA, proteins and lipids take place in cells and tissues (see ref. 24 and 26). However, biochemical mechanisms of toxicity of $O_2^{\bullet-}$ are not quite clear. The oxygen radical anion $O_2^{\bullet-}$ is not so much an oxidant but, on the contrary, it is a rather powerful reductant. It is noteworthy that tissues like the liver, kidney and adrenals with a high amount of reduced glutathione (GSH) are also characterized with a high level of activity of SOD. This positive correlation between the GSH amount and the SOD activity, first noted half a century ago,[10,12] has been proved experimentally.[30] It suggests that the GSH pool is tracking the intensity of the $O_2^{\bullet-}$ fluxes, thereby regulating the SOD activity in cells and tissues.

Since then, a new protein family of NAD^+-dependent protein deacetylases or the so-called sirtuins (Silent Information Regulators) has been discovered. The first from this protein family, Sir2, was discovered in yeast cells. In human and animal organisms, seven sirtuins (Sirt1–7) orthological to yeast cell Sir2 have now been found.[31–37] They are localized in cell nuclei (Sirt1,

Sirt2, Sirt6, Sirt7), cytoplasm (Sirt1, Sirt2), and mitochondria (Sirt3, Sirt4, Sirt5). The sirtuins serve as key regulators of many important cell processes, including metabolism of glucose, secretion of insulin and adaptation to oxidative stress and hypoxia. For example, Sirt1 produces structural changes in chromatin that activate the synthesis of protective proteins, including antioxidant enzymes, and increase the number of mitochondria in the cells.[33] Some data indicate that Sirt1 and, possibly, other sirtuins regulate the amplitude and duration of the expression of circadian genes (metabolic clock).[34] All these proteins are NAD^+-dependent deacetylases/ADP-monoribosyltransferases. Moreover, it has been shown that expression of the sirtuin genes depends on the redox state of their intracellular environment.[35–37]

It stands to reason that $O_2^{\bullet-}$, as a powerful reducing agent, would significantly affect the redox ratio [NADH]/[NAD^+] and, thus, provoke undesirable changes in expression of the sirtuin genes in the cells that perform the supervisory functions over the organism's repair and renewal processes, be it the suprachiasmatic nucleus of the hypothalamus or another kind of "longevity-assurance structures". The evident consequence of this will be accumulation of free-radical products and other metabolic slag in peripheral cells and tissues with the resulting impetus to autophagic or apoptotic cell death and age-associated clinical disorders. As a matter of fact, the radicals are targeted onto the [NADH]/[NAD^+]-dependent sirtuin system that performs, in its turn, the function of the biological amplifier of $O_2^{\bullet-}$.

8.4 Extension of Lifespan by Antioxidants

8.4.1 Antioxidants: Radical Chemistry Standpoint

In chemistry, antioxidants are compounds, synthetic and natural, capable of terminating branching chain oxidation. These are mainly derivatives of phenols, secondary aromatic amines, organic phosphites and sulfides whose valence-saturated molecules containing an active hydrogen atom (InH) react with an active free radical $R^\bullet$ or $RO_2^\bullet$, that runs the oxidative chain, to give radical ($In^\bullet$) of the antioxidant:

$$R^\bullet + InH \Rightarrow RH + In^\bullet \qquad (8.6a)$$

$$RO_2^\bullet + InH \Rightarrow ROOH + In^\bullet \qquad (8.6b)$$

The relatively unreactive free radical $In^\bullet$, thus formed, cannot participate in chain propagation reactions and is destroyed upon collision with another radical or the vessel wall. The most common antioxidants are phenolic derivatives in which the OH group is shielded, the so-called "sterically hindered phenols" (see ref. 38–40).

Some of the synthetic antioxidants are depicted in Scheme 8.1. The antioxidant 2,6-di-*tert*-butyl-4-methylphenol, called butylated hydroxytoluene (BHT) in English-language literature or ionol or dibunol in Russian

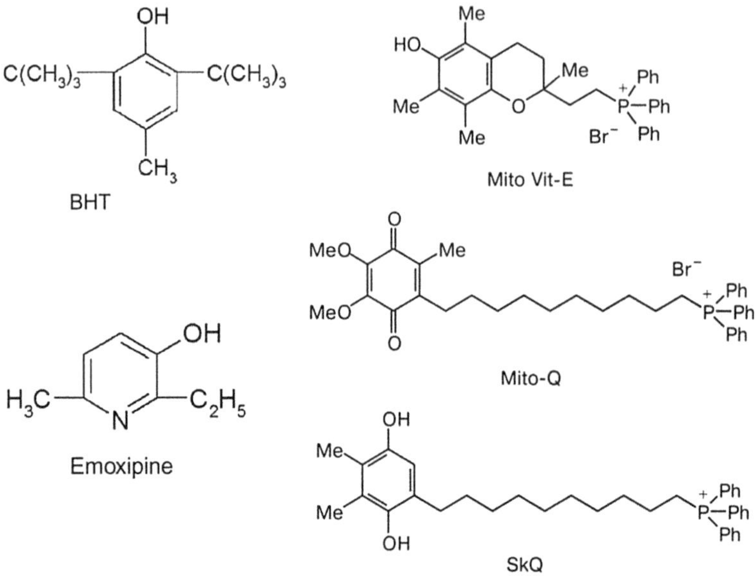

Scheme 8.1 Synthetic antioxidants BHT (2,6-di-*tert*-butyl-4-methylphenol), Emoxipine (2-ethyl-3-hydroxy-6-methylpyridine hydrochloride), MitoVit E, Mito-Q, and SkQ. Me = CH_3-group.

literature, first found wide use as stabilizer for industrial oils and edible fats. The antioxidants based on alkyl-substituted hydroxypyridine, such as 2-ethyl-3-hydroxy-6-methylpyridine hydrochloride (Emoxipine), are water-soluble, unlike BHT.[8,38,40] The antioxidant derivatives of vitamin E and ubiquinone, the so-called MitoVit E and Mito-Q, also contain in addition to the hydrophobic antioxidant group a lipophilic cationic group.[41] The mitochondria-targeted plastoquinone compounds with rhodamine and a triphenylphosphonium as cations (SkQ) were also synthesized.[42] Owing to the electric charge of the cation, these molecules are presumed to use the mitochondrial transmembrane potential to pass through the mitochondrial membranes and get into mitochondria, just those organelles that generate the highest amount of ROS.[41,42]

There are plenty of natural antioxidants starting from "classical" α-tocopherol (vitamin E) and ascorbic acid (vitamin C). Most natural antioxidants are substituted phenols or polyphenolic compounds, which, owing to their hydroxyl groups, are capable of inhibiting free-radical chain oxidation reactions in model systems (*in vitro*), for example, oxidation of linolenic acid. Among antioxidants, even melatonin and other human and animal hormones and even amino acids and peptides are mentioned sometimes.[43] Most functionally substituted phenols and polyphenols are of plant origin. These are querticin, flavones and other flavonoids, simple catechols, which are present in large amounts in green tea, and catechol oligomers present in high concentrations in grapes, cocoa beans, *etc.*, carotenoids, tannins, anthocyanins, coumarins, hydroxycinnamic acid derivatives, and resveratrol, which is

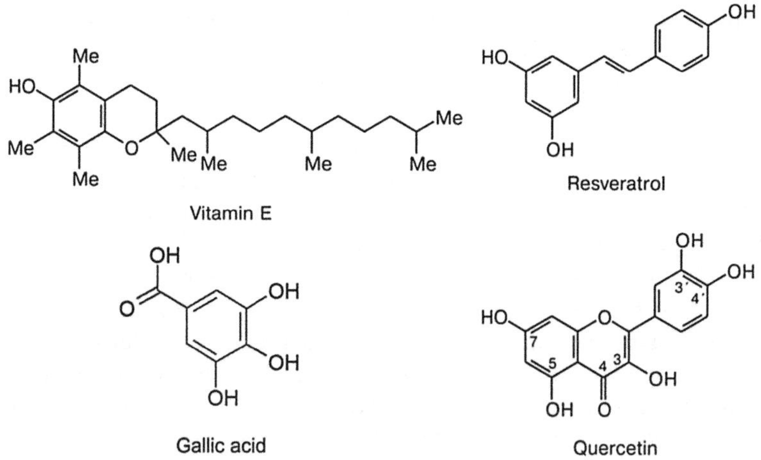

Scheme 8.2 Natural antioxidants α-tocopherol (vitamin E), resveratrol (3,5,4′-trihydroxy-stilbene), garlic acid, quercetin.

especially abundant in grapes and red wine. Some of the natural antioxidants are depicted in Scheme 8.2.

With appropriate dosage, natural and synthetic antioxidants exert various favorable therapeutic effects and therefore they have been long used with success in medical practice. For example, BHT is used to treat burns, gunshot wounds, trophic ulcers, and bladder cancer. Mexidol (2-ethyl-3-hydroxy-6-methylpyridine succinate) is used to treat brain circulation disorders, and Emoxipine (hydrochloride of the same pyridine derivative) is used in ophthalmology (see ref. 38). The same synthetic antioxidants have proved to be effective geroprotectors, *i.e.*, compounds that extend the life span of laboratory animals when added to food or drinking water on a regular basis. By adding BHT to food, it was possible in some cases to extend the lives of some lines of mice and fruit flies by 25 and 30%.[44,45] Emoxipine when added to drinking water extends the average life spans of fruit flies and mice by 24 and 38%, respectively (see ref. 8 and 40). The mitochondria-targeted Mito-Vit E and Mito-Q had beneficial effects for treating sepsis.[41] Similar compounds based on plastoquinone (SkQ) were applied profitably against age pathologies in animals, in particular, for treating a number of cardiovascular and ocular diseases and even as geroprotectors in experiments on mice or other animal species.[42] The remarkable geroprotective effects were found for resveratrol (3,5,4′-trihydroxy-stilbene) in experiments with *Drosophila* flies, mice, yeasts, nematodes and fish. The extension of the average life span by 30% in experiments with mice fed on a fatty diet and the extension of the maximum life span by 59% in experiments on fishes were revealed.[46–48] However, the classical natural antioxidant, α-tocopherol, appears to be of comparatively low efficiency in analogous biomedical testing.[49]

The results of analysis of the rate constants and actual concentrations of antioxidants also raise doubts in the fact that antioxidants operate *in vivo* in as simple way as *in vitro*, *i.e.*, as free-radical inhibitors. The oxygen radical anion $O_2^{\bullet-}$ seems to be the main source of ROS in aerobic organisms. However, there are specific enzymes, superoxide dismutases (SOD), in cells and tissues and the enzyme reacts with the $O_2^{\bullet-}$ with a rate constant of about 2×10^9 L mol^{-1} s^{-1}. Meanwhile, the rate constants for the reactions of ascorbic acid and 5,7,8-trimethyltocol (water-soluble α-tocopherol derivative) with the $O_2^{\bullet-}$ radical do not exceed 10^5 L mol^{-1} s^{-1} while those for hydroxypyridine antioxidants are no more than 10^2 L mol^{-1} s^{-1} (see ref. 38–40). For Mito-Q (ubiquinone-based antioxidant), the rate constant for the reaction with $O_2^{\bullet-}$ in water can be as high as 10^8 L mol^{-1} s^{-1} according to pulse radiolysis data.[41] However, in this case, too, it remains an order of magnitude lower than for SOD. In principle, mitochondria-targeted antioxidants can be accumulated in mitochondria.[41,42] However, they can hardly be accumulated up to a concentration comparable with the amount of SOD (about 10^{-5} mol L^{-1}) without considerable disturbance of the operation of mitochondrial bionanoreactors.

Yet, as was mentioned above, the reliability of the SOD protection is limited so that there exists a finite probability that the $O_2^{\bullet-}$ would penetrate the SOD defense, about 2 radicals from every hundred thousand.[10–12] The radicals that penetrate the defense system can react with H_2O_2 to give the hydroxyl $OH^{\bullet}$ radical. However, it is also known that the enzymes catalase and glutathione peroxidase, which catalyze hydrogen peroxide decomposition to water and oxygen, always occur near SOD. The rate constant for the reaction of the antioxidant α-tocopherol with the $OH^{\bullet}$ radical can be as high as 8×10^{10} L mol^{-1} s^{-1} (see ref. 38–40). The $OH^{\bullet}$ radical is known to react, however, with any organic molecules as a strong oxidant with rate constants close to the diffusion limit ($>10^{10}$–10^{11} L mol^{-1} s^{-1}).[38–40] Therefore, none of antioxidants can compete for hydroxyl radicals *in vivo* with other organic molecules that are obviously always present around this radical in considerably greater numbers than the molecules of any antioxidant. Of course, the peroxyl radicals $RO_2^{\bullet}$ can appear in reactions of $OH^{\bullet}$ radicals with lipids. In addition, $OH^{\bullet}$ radicals initiate oxidation of proteins, oxidative degradation of DNA and so on (see, for example, ref. 50–52). *In vivo*, however, $RO_2^{\bullet}$ and other products of peroxidation arise mainly as secondary products in the reactions that accompany cell death on apoptosis and autophagocytosis during utilization of the cellular waste by lyzosomes and peroxysomes (see, for example, a review in ref. 53). The rate constants for the reactions of synthetic and natural antioxidants with $RO_2^{\bullet}$ in model reactions may range up to about 10^6 L mol^{-1} s^{-1}. However, the antioxidants are unlikely to be highly necessary for scavenging the active radicals in the catabolism. Besides, the reports on *in vivo* yields of the DNA oxidation products for both mitochondrial and nuclear DNA are overestimated due to various artifacts.[50] Thus, manifold effects of antioxidants *in vivo* can hardly be interpreted on the basis of simple chemical analogy with the action of the same antioxidants as radical scavengers *in vitro*.

8.4.2 Antioxidants: Reliability-Theory Standpoint

The most efficient way to increase the systems reliability of complex systems is well-timed prevention of malfunctions (failures) of functional elements.[13,15,18,19] Following this reliability-theory guide-line, it was proposed[11,12] that antioxidants provide preventive protections from free radicals *in vivo*. In this regard, the particular protection mechanisms may be different for antioxidants of different types. For BHT, it was found that this antioxidant prevents generation of $O_2^{\bullet-}$ radicals as by-products of electron transport in mitochondria. In a study of low-temperature ESR (electron spin resonance) signals of rat tissues, we found that BHT increases the myocardium oxygenation.[54] It is known that hypoxia results in structural damage in mitochondrial membranes resulting in considerable decrease in the reliability of electron transport, so that the mitochondria become generators of intense $O_2^{\bullet-}$ fluxes.[55] It stands to reason that BHT prevents the development of hypoxia by increasing the degree of myocardium oxygenation that prevents the transformation of mitochondria into $O_2^{\bullet-}$ generators.

Furthermore, BHT produces the dramatic hormonal changes in the animal's blood. Figure 8.2 demonstrates the increase of corticotropin and

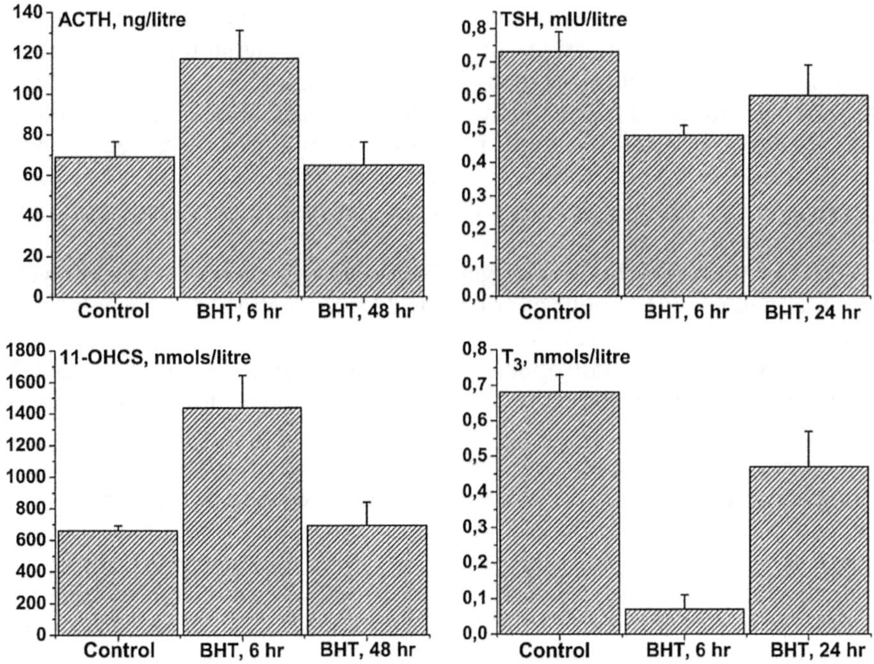

Figure 8.2 Concentrations of corticotropin (ACTH), thyrotropin (TSH), 11-oxycorticosteroids (11-OHCS) and L-3,3′,5-triiodothyronine (T_3) in blood plasma of rats (adult, 4–6 months, male Wistar) in control and after injection of antioxidant BHT (data compiled from ref. 57).

corticosteroids along with the decrease of thyrotropin and L-3,3′,5-triiodothyronine in blood plasma of rats after the BHT administration.

Hence, BHT induces the substantial shift in the activity of adenohypophysis gland, which is the source of corticotropin and thyrotropin hormones, and this is accompanied by the relevant shifts in the activity of peripheral endocrine glands, the adrenal cortex (the source of corticosteroids) and the thyroid gland (the source of triiodothyronine).[56,57] It is common knowledge that the release of corticotropin into blood followed by an increase in the synthesis of corticosteroids and a decrease in the synthesis of thyroid hormones is a significant phase of the system's adaptation to stress. It seems that, with regular introduction into animals' food, BHT as a mild stress factor 'trains' the neuro-hormonal system and, thus, increases the systems reliability, *i.e.* adaptive capabilities of the organism.[56–59] Hence, BHT is actually able to decrease the level of active oxygen species in myocardial cells and, probably, other cells too. However, the beneficial effect of this antioxidant is manifested not through direct radical elimination (scavenging), but in a preventive manner, *i.e.*, upon a decrease in the probability of their generation. Besides, it was shown by the spin probe technique that serum albumin,[54] a blood protein among the functions of which is transport of the hormone aldosterone, sorbs the hydrophobic BHT molecules. The hormone transport proteins can presumably serve as the molecular targets of the antioxidant. In addition, the BHT injections gave rise to the ESR signal from the nitrosyl complex of hemoglobin (NO-Hb) in the animal's blood.[58] It is generally known that nitric oxide serves as the signal molecule that causes, in particular, the relaxation of arterial smooth muscles that enhances the oxygen supply in myocardium.[60] Thus, there are the reasons to believe that BHT performs preventive maintenance against $O_2^{\bullet-}$ and its reactive products *via* hormonal/NO regulation.

The so-called mitochondria-targeted antioxidants can also act in a preventive manner. As the phenolic compounds, MitoVit-E and SkQ[41,42] have weakly acidic properties and, as such, they can serve as protonophore uncouplers, like, for example, 2,4-dinitrophenol, uncoupling electron transport and ATP synthesis in mitochondria.[24] In addition, hydrophobic cations can transfer counter-ions (anions) through a mitochondrial lipid membrane, thereby decreasing the transmembrane potential. Again, it should produce the uncoupling effect, according to the Mitchell theory of oxidative phosphorylation.[24] Indeed, molecules combining a hydrophobic part and a cationic group, in particular, a triphenylphosphonium group, serve as efficient uncouplers of oxidative phosphorylation in mitochondria. Actually, they were synthesized for this purpose, as uncouplers, about 40 years ago.[61] It is also known that the electron transport in mitochondria experiences a "back pressure" from the transmembrane potential.[24] Therefore, oxidative phosphorylation uncouplers, in particular, transmembrane transfer agents of protons and anions, decrease the transmembrane potential and thus decrease the generation of $O_2^{\bullet-}$ and other ROS in mitochondria. This provides grounds for believing that mitochondria-targeted antioxidants not so much scavenge directly the

$O_2^{\bullet -}$ (or its protonated form $HO_2^{\bullet}$) but prevent the formation of these radicals in mitochondria. Besides, it is noteworthy that SkQ synthesized from plant plastoquinone[42] proved to be a substantially more efficient therapeutic agent in biomedical investigations than similar mitochondria-targeted antioxidants MitoVit-E and MitoQ synthesized from vitamin E and animal ubiquinone.[41]

Another synthetic antioxidant 2-ethyl-3-hydroxy-6-methylpyridine increases the reliability of the electron transporting bionanoreactors as well, but the mechanisms of antioxidant prophylaxis in this case differ from those for BHT. The antioxidants based on hydroxypyridines are analogs of pyridoxine and pyridoxal phosphate, which are group B_6 vitamins. Meanwhile, pyridoxal phosphate is a cofactor of glutamate aspartate aminotransferase, RNA polymerase, and some other enzymes of biosynthesis of nitrogen-containing compounds.[24] This implies that hydroxypyridine antioxidants are the anti-metabolites of vitamin B_6 and, as such, they inhibit the key enzymes of synthesis of amino acids and nucleotides. In particular, this allows understanding of the efficiency of 2-ethyl-3-hydroxy-6-methylpyridine as a radioprotector in yeast cell experiments. Inhibition of biosynthesis retards cell division and thus provides the cells with additional time for restoring the genetic structures damaged by ionizing radiation.[62]

Flavonoids can provide preventive protection from oxygen radicals by induction of specific antioxidant enzymes. For example, the induction of synthesis of Cu,Zn-SOD and catalase was detected in blood erythrocytes of humans who received the food additive *Protandim* (extracts of five medical plants).[63] It was concluded that modest induction of the antioxidant enzymes SOD and catalase may be a much more effective approach than supplementation with antioxidants "that can, at best, stoichiometrically scavenge a very small fraction of total oxidant production".[63] It is noteworthy that resveratrol increases the expression and activity of mitochondrial Mn-SOD *in vivo*.[64] Moreover, resveratrol was found to activate the expression of the sirtuin proteins.[46–48]

Inasmuch as expression of SOD and other antioxidant enzymes in humans and animals is under hormonal control, flavonoids also seem to make their preventive maintenance defense through hormonal regulation mechanisms. Indeed, in experiments with *Macaca mulatta* monkeys it was found that the diurnal changes (circadian rhythms) in the SOD activity in erythrocytes tightly and positively correlate with the diurnal changes in the levels of cortisol and dehydroepiandrosterone sulfate (DHEAS) in blood plasma.[65] For young animals, the values of correlation coefficient were 0.92 ± 0.09 (cortisol *versus* SOD) and 0.99 ± 0.02 (DHEAS *versus* SOD). With aging, the circadian rhythms of SOD, cortisol and DHEAS are smoothed out, although the correlation between the diurnal changes in cortisol and in SOD still maintains, even for old animals. These results, like the above-mentioned experiments with BHT, testify that corticosteroid hormones play an essential role in regulation of SOD activity and that reliability of the hormonal regulation decreases with aging.[65]

Furthermore, the monkeys were subjected to psycho-emotional stress, *i.e.* two-hour immobilization.[39,66] The stress hormones cortisol and DHEAS in the plasma of the animals' peripheral blood were measured before the stress (basal conditions) and after the stress. In parallel, the levels of activity of SOD, glutathione peroxidase and glutathione reductase were measured in the erythrocytes. It is well known that stress is accompanied by intensification of respiration. This, in turn, leads to enhanced production of $O_2^{\bullet-}$ as a by-product of respiration. Therefore, the SOD induction should have been anticipated. As was expected, the plasma levels of cortisol and DHEAS sharply increased under the stress. However, the SOD activity in erythrocytes remained invariable during the stress and even for several hours after the stress release (see Figure 8.3).

Meanwhile, in the same experiments the glutathione reductase activity demonstrated significant growth, especially in the case of old animals, which obviously reflects the increased consumption of GSH due to the increased production of ROS. Thus, an acute stress is accompanied by kinetic deficiency of the antioxidant enzymes. Under the stress conditions, this deficiency seems to be a critical factor in the age-associated clinical disorders. Accordingly, a timely, *i.e.* before the stress, introduction of antioxidants, for example, flavonoids that induce SOD synthesis, increases the systems reliability, providing the preventive maintenance against ROS.

Quite a lot of data have now been accumulated demonstrating that even vitamin E (α-tocopherol), a "key antioxidant", can hardly serve as a free radical inhibitor *in vivo* and, hence, this issue in handbooks should be revised.[67] There are four tocopherol isomers: α-, β-, γ-, and δ-tocopherol. All

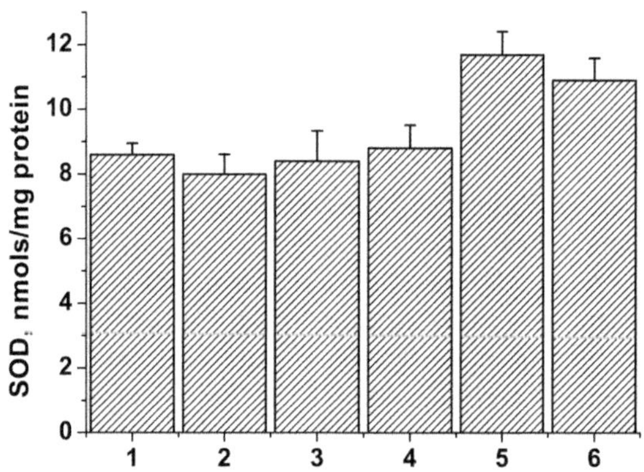

Figure 8.3 Activity of superoxide dismutase (SOD) in erythrocytes of *Macaca mulatta* monkeys before psycho-emotional stress (1), during the stress (2, 3, 4) and after the stress (5, 6). Seven adult (6–8 years) healthy male rhesus monkeys underwent the stress procedure, *i.e.* moderate immobilization restraint in a metabolic cage for 2 hours. Blood samples were taken before the stress (1) and at 0.5 h (2), 1 h (3) and 2 h (4) during immobilization. Besides, the blood samples were taken at 2 h (5) and 22 h (6) after termination of immobilization, *i.e.* 4 and 24 h after the onset of the stressor (data compiled from ref. 39).

four isomers react *in vitro* with the $RO_2^{\bullet}$ radical with approximately the same rate constants of about 10^6 L mol^{-1} s^{-1}. In living nature, however, mainly α-tocopherol is encountered. As shown in experiments on cell cultures and isolated enzymes, this form of vitamin E inhibits a key regulator enzyme of biosynthesis, protein kinase C, inhibits 5-lipoxygenase and phospholipase A_2 and also activates protein phosphatase 2A and diacylglycerol kinase. It was proved that α-tocopherol modulates the expression of genes encoding the synthesis of a number of protective proteins, including α-TTP, α-tropomyosin, and collagenase. Moreover, α-tocophenyl phosphate, rather than the antioxidant phenolic form of vitamin E, serves as the bioregulator. It has been suggested that α-tocopherol acts as a ligand for yet unidentified specific proteins, membrane receptors or transcription factors, capable of regulating signal transduction and gene expression.[67]

Furthermore, there are more and more data indicating that the therapeutic effects of many pharmaceutical drugs are due to their beneficial action not only on the cells and tissues of the host organism but also on gastric and intestinal microbiota. The number of microbiota cells in the gastrointestinal tract, on the skin, and in some other organs and tissues nearly exceeds the number of cells of the host organism.[68–70] Of even greater importance is that the microbial cells produce physiologically active substances that markedly affect all organs and tissues, including the immune system.[71–74] Moreover, there are the data that show that the microbial metabolites promote metabolic benefits in the brain cells *via* gut–brain neural circuits.[69,74,75] As a matter of fact, a new synthetic biomedical concept has emerged that the human microbiota is a source of therapeutic drug targets.[76–78]

Meanwhile, most polyphenol compounds, including flavonoids, which are traditionally regarded as "natural antioxidants", refer to the extensive class of physiologically active compounds long known as phytoalexins. Moreover, phytoalexins are synthesized in plant tissues for fighting against bacterial and fungal infections and for acting like antibiotics as inhibitors of transcription and translation of particular proteins in the cells of the infecting organism.[79–81] In view of the advances of systems biology, one can suggest that the so-called antioxidants, both natural and synthetic ones, attack the organism's microbial population. In high doses, these substances are toxic, as implied, because of their deleterious effects on the microbiota. In low doses, however, the same compounds produce favorable effects on the organism's microbiota and, thereby, increase the system reliability and lifespan of the organism. One can further assume that the so-called "mitochondria-targeted" compounds like MitoVit-E and SkQ affect actually the microbiotic cells. Thus, in this century, which is the century of systems biology, the theory that was put forward in the early 20th century by Metchnikoff about the considerable effect of the microbial population on the body health and aging is actually revived.[82] One can say with reasonable confidence that "Metchnikoff arises".

It has rather long been questioned whether the synthetic and natural antioxidant molecules work *in vivo* in the same way as *in vitro*, *i.e.* as simple chemical scavengers of $OH^{\bullet}$ and other active radicals.[12,39,40,56–59] Indeed, over the years, more and more experimental results indicate that the true mechanisms of the "antioxidant prophylaxis" are to be studied using the ways of systems biology instead of free-radical chemistry. Last years, such terms as "polyphenols", instead of "antioxidants", and "redox regulation/redox signaling pathways", instead of "oxidative stress", came into use.[83–85] Moreover, the Society for Free Radical Biology and Medicine has recently been renamed to the Society for Redox Biology and Medicine. As said in ref. 86, "it is harder to overcome old ideas, rather than create the new ones."

8.5 Conclusions

The systems approach, based on the engineering theory of reliability, integrates the concept of the aging program and the free-radical theory of aging in a unified pattern. The universal features of aging, such as the exponential growth of mortality rate with time and the correlation of longevity with the species-specific resting metabolism, are naturally explained on this basis. From the systems reliability standpoint, aging is a stochastic consequence of the genetically preprogrammed limits of bioreliability at all functional levels, from biomolecular nanoreactors to the organism as a whole. The stochastic malfunctions of the mitochondrial electron transport nanoreactors, which produce the oxygen anion-radicals ("superoxide radicals") as by-products of respiration, seem to be of first importance. The free-radical redox-timer, presumably located in the specialized neurons of the central nervous system, serves as the effective stochastic mechanism of realization of the preset deficiency in bioreliability. As a consequence, the oxidative-stress products and other metabolic slag accumulate with the resulting impetus to autophagic or apoptotic cell death accompanied with age-associated clinical disorders. Some antioxidants, both natural and synthetic ones, extend the life span of animals when added to food or drinking water. However, the antioxidant power of such compounds is negligible *in vivo* because their rate constants and concentrations are too small to compete with the specialized antioxidant enzymes for the reactive oxygen species. The so-called antioxidants provide a preventive protection against ROS, *i.e.* the prophylactic reliability maintenance operating *via* their beneficial effects on the organism's neuro-hormonal system and/or microbial cells of the body. Thus, the systems reliability approach serves as a heuristic methodology in searching for realistic mechanisms of aging and anti-aging therapy.

Acknowledgements

The motives that predetermined this work have been my lectures and presentations at the international symposia and seminars at the Institute of Gerontology, National Academy of Medical Science of Ukraine, in Kiev. Support from the organizers of these symposia, colleagues and friends of mine is greatly acknowledged, as well as the financial support from the Russian Foundation for Fundamental Research.

References

1. G. A. Sacher, in *Handbook of the Biology of Aging*, ed. C. E. Finch and L. Hayflick, Van Nostrand Reinhold, New York, 1977, pp. 582–638.
2. V. N. Anisimov, *Molecular and Physiological Mechanisms of Aging*, Nauka, St. Petersburg, 2nd edn, 2008.
3. O. Risse, *Strahlentherapie*, 1929, **34**, 578.
4. *Oxygen and Living Processes: An Interdisciplinary Approach*, ed. D. L. Gilbert, Springer-Verlag, New York, 1981.
5. D. Harman, *J. Gerontol.*, 1956, **11**, 298.
6. D. Harman, *J. Gerontol.*, 1957, **12**, 257.
7. D. Harman, in *Free Radicals and Aging*, ed. I. Emerit and B. Chance, Birkhauser, Basel, 1992, pp. 1–10.
8. L. K. Obukhova, *Adv. Gerontol.*, 1999, **3**, 27.
9. *Systems of Reliability of Cell (Proceedings of the 1st All-Union Conference Systems of Reliability of Cell, 1975)*, ed. D. M. Grodzinsky, Naukova Dumka, Kiev (Ukraine, former SU), 1977, (Russian).
10. V. K. Koltover, *Dokl. Biophys. (Engl. Transl.)*, 1981, **256**, 3.
11. (a) V. K. Koltover, *Theory of Reliability, Superoxide Radicals and Aging*, ICPH USSR Acad. Sci, Chernogolovka (Moscow Region), 1983, p. 42; (b) V. K. Koltover, *Achiev. Life Sci. (Uspekhi Sovremennoj Biologii)*, 1983, **96**, 85, (Russian).
12. V. K. Koltover, *Reliability of Electron–Transport Membranes and the Role of Oxygen Radical Anions in Aging (Doctoral Dissertation in Biology)*, A. Bogomolets Institute of Physiology, Acad. Sci. of Ukraine SSR, Kiev; N. Semenov Institute of Chemical Physics, Acad. Sci. of USSR, 1987, (Russian).
13. D. M. Grodzinsky, V. P. Vojtenko, Y. A. Kutlakhmedov and V. K. Koltover, *Reliability and Aging of Biological Systems*, Naukova Dumka, Kiev (Ukraine, former SU), 1987, (Russian).
14. V. K. Koltover, in *Free Radicals and Aging*, ed. I. Emerit and B. Chance, Birkhauser, Basel, 1992, pp. 11–19.
15. V. K. Koltover, *J. Theor. Biol.*, 1997, **184**, 157.
16. V. K. Koltover, in *Vitality, Mortality and Aging*, ed. A. Viidik and G. Hofecker, Facultas, Vienna, 1996, pp. 173–180.
17. V. K. Koltover, *Biophysics*, 2011, **56**, 125.
18. V. K. Koltover, in *Recent Advances in Systems Biology Research*, ed. A. X. Valente, A. Sarkar and Y. Gao, Nova Science Publishers Inc., New York, 2014, pp. 109–130.
19. V. K. Koltover, in *Proceedings of the Second International Symposium on Stochastic Models in Reliability Engineering, Life Science and Operations Management (SMRLO'16)*, ed. I. Frenkel and A. Lisnianski, IEEE CPS (The Institute of Electrical and Electronics Engineers Inc., Conference Publishing Services), 2016, pp. 386–392.
20. A. M. Olovnikov, *Exp. Gerontol.*, 1996, **31**, 443.
21. A. B. Kapitanov, M. Y. Aksenov, O. S. Tatishchev and V. K. Koltover, *Dokl. Biophys. (Engl. Transl.)*, 1985, **281**, 186.

22. G. A. Garinis, G. T. J. van der Horst, J. Vijg and J. H. J. Hoeijmakers, *Nat. Cell Biol.*, 2008, **10**, 1241.
23. I. Fridovich, *Annu. Rev. Biochem.*, 1995, **64**, 97.
24. D. L. Nelson and M. M. Cox, *Lehninger Principles of Biochemistry*, Freeman, New York, 2008.
25. F. Murad and R. Barber, *Free Radicals Biol. Med.*, 2009, **47**, 1325.
26. G. J. Maghzal, K.-H. Krause, R. Stocker and V. Jaquet, *Free Radicals Biol. Med.*, 2012, **53**, 1903.
27. S. Shiva, *Redox. Biol.*, 2013, **1**, 40.
28. R. I. Gus'kova, I. I. Ivanov, V. K. Koltover, V. V. Akhobadze and A. B. Rubin, *Biochim. Biophys. Acta*, 1984, **778**, 579.
29. J. Tolmasoff, T. Ono and R. G. Cutler, *Proc. Natl. Acad. Sci. U. S. A.*, 1980, **77**, 2777.
30. B. Morgan, D. Ezeria, T. N. E. Amoako, J. Riemer, M. Seedorf and T. P. Dick, *Nat. Chem. Biol.*, 2013, **9**, 119.
31. J. A. Baur and D. A. Sinclair, *Nat. Rev. Drug Discovery*, 2006, **5**, 493.
32. T. Finkel, C.-X. Deng and R. Mostoslavsky, *Nature*, 2009, **460**, 587.
33. E. M. Dioum, R. Chen, M. S. Alexander, Q. Zhang, R. T. Hogg, R. D. Gerard and J. A. Garcia, *Science*, 2009, **324**, 1289.
34. S. Imai, *Biochim. Biophys. Acta*, 2010, **1804**, 1584.
35. A. A. Sauve and D. Y. Youn, *Curr. Opin. Chem. Biol.*, 2012, **16**, 535.
36. R. H. Houtkooper, E. Pirinen and J. Auwerx, *Nat. Rev. Mol. Cell Biol.*, 2012, **13**, 225.
37. L. Naia and A. C. Rego, *Biochim. Biophys. Acta*, 2015, **1852**, 2183.
38. E. B. Burlakova, *Mendeleev Chem. J.*, 2007, **51**, 3.
39. V. K. Koltover, *Toxicol. Ind. Health*, 2009, **25**, 295.
40. V. K. Koltover, *Russ. Chem. Bull.*, 2010, **59**, 37.
41. (a) M. P. Murphy, *Biochim. Biophys. Acta*, 2008, **1777**, 1028; (b) A. Maroz, R. F. Anderson, R. A. J. Smith and M. P. Murphy, *Free Radicals Biol. Med.*, 2009, **46**, 105.
42. V. P. Skulachev, V. N. Anisimov, Y. N. Antonenko, L. E. Bakeeva, B. V. Chernyak, V. P. Erichev, O. F. Filenko, N. I. Kalinina, V. I. Kapelko, N. G. Kolosova, B. P. Kopnin, G. A. Korshunova, M. R. Lichinitser, L. A. Obukhova, E. G. Pasyukova, O. I. Pisarenko, V. A. Roginsky, E. K. Ruuge, I. I. Senin, I. I. Severina, M. V. Skulachev, I. M. Spivak, V. N. Tashlitsky, V. A. Tkachuk, M. Yu. Vyssokikh, L. S. Yaguzhinsky and D. B. Zorov, *Biochim. Biophys. Acta*, 2009, **1787**, 437.
43. L. S. Kozina, A. V. Arutjunyan and V. Kh. Khavinson, *Arch. Gerontol. Geriatr.*, 2007, **44**, 213.
44. N. K. Clapp, L. C. Satterfield and N. D. Bowles, *J. Gerontol.*, 1979, **34**, 497.
45. S. P. Sharma and R. Wadhwa, *Mech. Ageing Dev.*, 1983, **23**, 67.
46. J. A. Baur, K. J. Pearson, N. L. Price, H. A. Jamieson, C. Lerin, A. Kalra, V. V. Prabhu, J. S. Allard, G. Lopez-Lluch, K. Lewis, P. J. Pistell, S. Poosala, K. G. Becker, O. Boss, D. Gwinn, M. Wang, S. Ramaswamy, K. W. Fishbein, R. G. Spencer, E. G. Lakatta, D. Le Couteur, R. J. Shaw, P. Navas, P. Puigserver, D. K. Ingram, R. de Cabo and D. A. Sinclair, *Nature*, 2006, **444**, 337.

47. D. R. Valenzano, E. Terzibasi, T. Genade, A. Cattaneo, L. Domenici and A. Cellerino, *Curr. Biol.*, 2006, **16**(3), 296.
48. L. Guarente, *Nat. Med.*, 2012, **18**, 500.
49. N. Sumien, M. J. Forster and R. S. Sohal, *Exp. Gerontol.*, 2003, **38**, 699.
50. A. F. Cadet, T. Douki and J. L. Ravanat, *Acc. Chem. Res.*, 2008, **41**, 1075.
51. T. Schewe, Y. Steffen and H. Sies, *Arch. Biochem. Biophys.*, 2008, **476**, 102.
52. C. C. Winterbourn, *Nat. Chem. Biol.*, 2008, **4**, 278.
53. T. Kurz, A. Terman, B. Gustafsson and U. T. Brunk, *Biochim. Biophys. Acta*, 2008, **1780**, 1291.
54. V. K. Koltover, E. N. Gorban and P. S. Maior, *Dokl. Biophys. (Engl. Transl.)*, 1984, **277**, 130.
55. H. Nohl, V. Koltover and K. Stolze, *Free Radical Res. Commun.*, 1993, **18**, 127.
56. V. V. Frol'kis, E. N. Gorban and V. K. Koltover, *Dokl. Biophys. (Engl. Transl.)*, 1985, **284**, 210.
57. V. V. Frol'kis, E. N. Gorban and V. K. Koltover, *Age*, 1990, **13**, 5.
58. V. K. Koltover, *Age*, 1995, **18**, 85.
59. (a) V. K. Koltover, in *Abstracts of the 17th World Congress of the Inter. Ass. of Gerontology, Vancouver, Canada, July 1-6, 2001*, 2001, p. 13; (b) V. K. Koltover, *Free. Radicals Biol. Med.*, 2002, **33**, S173.
60. A. F. Vanin, *Nitric Oxide*, 2016, **54**, 15.
61. E. A. Liberman, V. P. Topaly, L. M. Tsofina, A. A. Jasaitis and V. P. Skulachev, *Nature*, 1969, **222**, 1076.
62. V. K. Koltover, V. G. Korolev and Y. A. Kutlakhmedov, in *Ionizing Radiation: Applications, Sources and Biological Effects*, ed. E. Belotserkovsky and Z. Ostaltsov, Nova Science Publishers Inc., New York, 2012, pp.117–128.
63. S. K. Nelson, S. K. Bose, G. K. Grunwald, P. Myhill and J. M. McCord, *Free Radicals Biol. Med.*, 2006, **40**, 341.
64. E. L. Robb, L. Winkelmolen, N. Visanji, J. Brotchie and J. A. Stuart, *Biochem. Biophys. Res. Commun.*, 2008, **372**, 254.
65. N. D. Goncharova, A. V. Shmaliy, T. N. Bogatyrenko and V. K. Koltover, *Exp. Gerontol.*, 2006, **41**, 778.
66. V. K. Koltover, T. N. Bogatyrenko, A. V. Shmaliy and N. D. Goncharova, *Free. Radicals Biol. Med.*, 2006, **41**, S164.
67. A. Azzi, S. N. Meydani, M. Meydani and J. M. Zingg, *Arch. Biochem. Biophys.*, 2015, **595**, 100.
68. J. K. Nicholson, E. Holmes, J. Kinross, R. Burcelin, G. Gibson, W. Jia and S. Pettersson, *Science*, 2012, **336**, 1262.
69. J. F. Cryan and T. G. Dinan, *Nat. Rev. Neurosci.*, 2012, **13**, 701.
70. C. Heintz and W. Mair, *Cell*, 2014, **156**, 408.
71. N. Kamada, S.-U. Seo, G. Y. Chen and G. Núñez, *Nat. Rev. Immunol.*, 2013, **13**, 321.
72. Y. Belkaid and T. W. Hand, *Cell*, 2014, **157**, 121.
73. D. Zhang, G. Chen, D. Manwani, A. Mortha, C. Xu, J. J. Faith, R. D. Burk, Y. Kunisaki, J.-E. Jang, C. Cheiermann, M. Meras and P. S. Frenette, *Nature*, 2015, **525**, 528.

74. F. De Vadder, P. Kovatcheva-Datchary, D. Goncalves, J. Vinera, C. Zitoun, A. Duchampt, F. Bäckhed and G. Mithieux, *Cell*, 2014, **156**, 84.
75. P. A. Smith, *Nature*, 2015, **526**, 312.
76. B. D. Wallace and M. R. Redinbo, *Curr. Opin. Chem. Biol.*, 2013, **17**, 379.
77. B. Olle, *Nat. Biotechnol.*, 2013, **31**, 309.
78. T. J. Ford and P. A. Silver, *Curr. Opin. Chem. Biol.*, 2015, **28**, 20.
79. *Phytoalexins*, ed. J. A. Bailey and J. W. Mansfield, Blackie, Glasgow-London, 1982.
80. M. Daglia, *Curr. Opin. Biotechnol.*, 2012, **23**, 174.
81. D. S. Guttman, A. C. McHardy and P. Schulze-Lefert, *Nat. Rev. Genet.*, 2014, **15**, 797.
82. E. Metchnikoff, *The Prolongation of Life. Optimistic Studies*, Heinemann, London, 1907.
83. M. Schieber and N. S. Chandel, *Curr. Biol.*, 2014, **24**, R453.
84. A.-L. Levonen, B. G. Hill, E. Kansanen, J. Zhang and V. M. Darley-Usmar, *Free Radical Biol. Med.*, 2014, **71**, 196.
85. K. J. A. Davies, *Arch. Biochem. Biophys.*, 2016, **595**, 28.
86. J. M. Keynes, *The General Theory of Employment, Interest and Money*, Palgrave Macmillan, London, 2006.

CHAPTER 9

Mitochondria-Targeted Rechargeable Antioxidants as Potential Anti-Aging Drugs

ELENA G. PASYUKOVA[a], BORIS A. FENIOUK[b] AND VLADIMIR P. SKULACHEV*[b,c]

[a]Institute of Molecular Genetics of Russian Academy of Sciences, Moscow, 123182, Russia; [b]Faculty of Bioengineering and Bioinformatics, Lomonosov Moscow State University, Moscow, Russia; [c]A. N. Belozersky Institute of Physico-Chemical Biology, Lomonosov Moscow State University, Moscow, Russia
*E-mail: skulach@belozersky.msu.ru

9.1 Introduction

Advances in medical care and improvements in living conditions during the last century led to a marked increase in human health span and lifespan. Nevertheless, age remains the most important risk factor for many disabling diseases and conditions, including cardiovascular disease, stroke, diabetes, neurodegeneration, and cancer. Moreover, in the human population health span increased at a slower pace compared to lifespan in the recent half century.[1] As a consequence, the number of people suffering from age-related diseases is anticipated to almost double over the next two decades,[2,3] creating a social and economic burden[4] that urgently needs appropriate interventions.

RSC Drug Discovery Series No. 57
Anti-aging Drugs: From Basic Research to Clinical Practice
Edited by Alexander M. Vaiserman

Published by the Royal Society of Chemistry, www.rsc.org

Targeting aging itself, rather than combating each age-related pathology individually, seems to be a promising new strategy.[5–7] The question of whether we should consider aging as a disease is now discussed.[8–10] Growing evidence also supports the hypothesis stating that aging of living organisms can be considered as a particular case of programmed death of an organism (slow phenoptosis),[7,11] and that switching off this deleterious program may slow down or even abolish aging.[11]

Despite the fact that aging is a highly diverse phenomenon across the variety of living organisms, there are a few cellular processes linked to aging that are conserved over a broad evolutionary distance, from yeast to humans: mitochondrial function, nutrient signaling, proteostasis,[12,13] and autophagy.[14] In this review we focus mainly on one aspect of mitochondrial function: the generation of reactive oxygen species (ROS), and application of the mitochondria-targeted rechargeable antioxidants as a tool to suppress this generation. Experiments on several animal models are reviewed, with a special focus on the data obtained on the fruit fly *Drosophila melanogaster*.

9.2 Mitochondria Malfunction and Aging

Mitochondrial malfunction is a feature inherent in the aging of living creatures. It is documented for yeast,[15,16] invertebrates[17–19] and mammals[20,21] (see ref. 22 and 23 for recent reviews). The stability of mitochondria during purification procedures also seems to decrease with age, as demonstrated by experiments with rat skeletal muscle.[24] Mitochondrial genetic diseases, including those caused by defects in mitochondrial DNA polymerase gamma that lead to frequent mutations in mitochondrial DNA (mutator mouse[25,26]), result in phenotypes that resemble premature aging.

Deletions and mutations in mitochondrial DNA accumulate with age and clonally expand in tissues, accompanied by the decline in the respiratory function. A causative role for mtDNA mutations in mammalian aging was suggested (evidence for this hypothesis is reviewed in ref. 27), but quantitative analysis of the data obtained on the mutator mouse does not support this suggestion. In homozygous mutator mice with mtDNA mutation load much higher than that detected in aged animals or elderly humans, the lifespan is shortened, but heterozygous animals have normal phenotype and lifespan, despite having an mtDNA mutation burden at birth 30 times higher than that of aged wild-type mice.[28] It is therefore likely that mtDNA mutations increase is just one of the manifestations of damage accumulation that accompany normal, physiological aging rather than the cause of the latter.[22]

In mammals, the consequences of age-dependent mitochondrial malfunction involve: (i) oxidative stress due to excess generation of mitochondrial reactive oxygen species (mROS); (ii) proteotoxicity caused by impaired mitochondrial unfolded protein response (UPR^{mt});[12,13] and (iii) inflammation.[14,29] In invertebrates, mitochondria functions also decline with aging. Surprisingly, there are examples when mitochondrial dysfunction results in life extension rather than shortening. For example, certain Mit mutants of the nematode *Caenorhabditis elegans* are long-lived. The first Mit mutant discovered was clk-1(e2519) with longer embryonic and postembryonic development, and increased lifespan, as well as the periods of the defecation, swimming and pumping cycles. Disrupted timing of several developmental and behavioral processes in the mutant led to a hypothesis that the clk-1 gene is a component of some biological clock.[30] In fact, the clk-1 gene encodes an enzyme involved in the synthesis of ubiquinone (coenzyme Q, CoQ), a mitochondrial electron carrier molecule necessary for respiration. It is noteworthy that transgenic expression of mouse clk-1 homologue in *C. elegans* completely rescued the slowed rhythmic behaviors of clk-1 nematode mutants and reverted their extended lifespan to a level comparable with that of the wild-type control.[31]

Since this initial seminal discovery, it has been found that disruption of several components of the mitochondrial respiratory chain or its assembly factors, as well as low doses of mitochondrial toxins (such as rotenone and antimycin A) can increase the lifespan in *C. elegans* (see ref. 17 for a review). In *D. melanogaster*, alterations in mitochondrial functions caused by overexpression of mtDNA polymerase and uncoupling protein 3 or knockdown of cytochrome c oxidase[32–34] decreased lifespan, whereas elevated expression of heat shock protein 22 and uncoupling proteins 1 and 2[35–37] increased lifespan. In mice, knockouts of the cytochrome c oxidase assembly factor Surf1[38] or of clk-1 orthologues,[39,40] as well as the reduced expression of mitochondrial ribosomal protein S5[41], have all been reported to prolong life.

Overall, at present, the accepted consensus is that a moderate lowering of functioning of mitochondria and, consequently, a moderate decrease in mROS production is the major factor contributing to the lifespan increase in animals, whereas its strong lowering shortens lifespan (see ref. 42 for a review). The observations of increased lifespan caused by disruption of mitochondrial functions can be explained in terms of mitohormesis: mildly increased mROS generation may cause an adaptive response that triggers a stress resistance increase that eventually causes long-term lowering of oxidative stress.[43] Another (and simpler) explanation of the same paradox is an assumption that mROS are intermediates of the aging program.[44,45]

9.3 The Link Between Oxidative Stress and Aging

The key role of free radicals in aging was first proposed by Harman in 1956;[46] he also suggested that free radicals arise as by-products of respiration,[46] and that mROS-generating mitochondria might serve as a biologic clock that determines the rate of aging[47,48] (see also ref. 49–51 for excellent reviews on the role of mROS in the pathophysiology of aging).

The modern version of free radical theory of aging suggests that senescence is not only caused by direct ROS-induced damage to DNA, proteins, lipids and other cellular components, but, mainly, is a consequence of the imbalance in cellular ROS signaling (see ref. 52–55 for recent reviews). In particular, mROS play a central role in the regulation of programmed cell death and other vital processes in organisms ranging from single-cell eukaryotes to humans.

The link between mROS production and degenerative processes leading to disease and aging in animals is supported by an overwhelming body of experimental evidence (see ref. 52,54–58 for reviews). This link is also supported by the correlation of mROS production and longevity in animals.[59,60] It is worth noting that this correlation is not universal: the longest-living rodent, the naked mole rat, exhibits strong ROS production.[60] This amazing animal also has high levels of oxidative stress markers in multiple tissues at quite an early age,[61–63] and can tolerate significant oxidative damage while having a lifespan of over 30 years.[64] Apparently they produce more ROS but overcome the oxidative stress due to an excellent defense against these ROS. ROS generation in mitochondria is an inevitable side-effect of respiration (see ref. 57,65–68 for reviews). However, the extent of mROS production can vary significantly, and while a modest, "normal" level of mROS is essential for cellular signaling; excessive mROS induce regulatory imbalance.

One of the primary targets of mROS attack is the mitochondrial membrane. Polyunsaturated fatty acids are particularly vulnerable to lipid peroxidation—a free radical chain reaction of oxidative degradation of such fatty acids. The reaction is initiated by a free radical attack at a double bond-forming carbon atom resulting in fatty acid radical formation. The latter readily reacts with molecular oxygen, producing a peroxyl-fatty acid radical that in turn attacks another unsaturated fatty acid, propagating the chain reaction. In this way a single ROS radical can "burn" a large fraction of unsaturated lipids in the mitochondrial membrane, disrupting the membrane organization. At the same time, the products of lipid peroxidation have been shown to act as mediators signaling that oxidative stress occurs.[69]

Cardiolipin (CL), a tetra-acylated anionic phospholipid localized normally in the inner mitochondrial membrane, is a particularly likely target for lipid peroxidation. Unlike other phospholipids, CL has four fatty acid residues instead of two; all four are polyunsaturated. In some protein complexes of the inner mitochondrial membrane (*e.g.*, complex III of the respiratory chain), CL forms dimers, so that 8 unsaturated fatty acids rich with double bonds are present in close vicinity to each other. Such dimers, especially when localized near respiratory chain proteins that generate mROS, are very probable starting points for the chain reaction of lipid peroxidation.

Another important side of pathological consequences of excessive ROS generation in mammalian mitochondria is inflammatory response. Oxidative stress is interlinked with inflammation *via* several feedback loops. ROS serve as secondary messengers in the inflammatory response[70] that can induce the activation of leukocytes and stimulate the expression of other mediators, such as cytokines, chemokines and adhesion molecules. And *vice versa*, inflammatory agents can also induce oxidative stress.[71] According to the oxidation-inflammation theory of ageing (see ref. 72 for a review), the age-related changes in the organism are caused by a chronic oxidative and inflammatory stress that results in the damage of cell components. It is still a matter of debate whether chronic inflammation is responsible for the development of age-related degenerative chronic diseases, or whether these chronic pathologies cause the inflammatory state observed in aging. But regardless of the cause–effect relationship between age-related diseases and inflammation, oxidative stress has been recognized to play a major role in determining and maintaining the low-grade inflammation observed in aging and age-associated diseases,[72] a process called inflamm-aging.[73]

It was also found in several studies (see ref. 74 for a review) that DNA damage by excess ROS may cause telomere erosion. Experiments in mice suggest that an increase in ROS-mediated DNA damage might enhance telomere dysfunction and thus accelerate accumulation of senescent cells. In turn, cell senescence stimulates chronic inflammation, limits tissue regeneration and accelerates ageing.[75]

In invertebrates, mROS also seem to be linked to aging. It was found that mitochondria of the mud clam *Arctica islandica*, one of the longest-living metazoan species (maximum reported longevity = 507 years), produced significantly less H_2O_2 than those of the two short-lived species.[76] The susceptibility of membrane lipids to peroxidation was also lower in *A. islandica* compared to that in short-lived bivalve mollusks.[77] (A similar situation is inherent in the long-living rodent naked mole rat).

Experiments in annelids *Aeolosoma viride* revealed that oxidative stress status in these worms significantly depended on age, following a Gaussian function centered at nearly half-life.[78] These small limnetic freshwater worms age rather quickly (average survival is 69 days) and share many metabolic processes with nematodes and vertebrates, including some related to the aging process. The radical scavenger bis(1-hydroxy-2,2,6,6-tetramethyl-4-piperidinyl)-decandioate (IAC), which effectively quenches ROS (including peroxyl radicals and superoxide radical-anion) and is able to attenuate several pathologies associated with oxidative stress,[79,80] was shown to prolong the mean lifespan of *A. viride*. IAC added to the cultured medium to a final concentration of 1.25 μM increased the resistance of *A. viride* to oxygen-derived damage without affecting mitochondrial respiration or reproductive activity, and extended the mean lifespan by 170%. Another antioxidant, super-oxide dismutase (SOD)-mimetic EUK134, also extended *A. viride*'s lifespan, although by mere 50%,[78] a figure very close to the 44% increase in the mean lifespan observed previously in the EUK134-treated *C. elegans*.[81]

9.4 Mitochondria-Targeted Rechargeable Antioxidants

Small molecules targeted to mitochondria recently became a powerful new tool for drug delivery, as well as for fundamental studies of mitochondrial functions, including the role of mitochondria in aging. Several comprehensive reviews (ref. 44,82–87) published in the last 10 years cover this topic in great detail.

In 1970, it was suggested that membrane-permeable cations could be used as locomotives to deliver various components specifically to mitochondria.[88] Release of protons from the mitochondrial matrix to the intermembrane space by the respiratory chain proteins results in a large transmembrane electrical potential difference (≈180 mV, negative inside mitochondria). Cations that are able to cross lipid membranes are driven into mitochondria by $\Delta\psi$. According to the Nernst equation, the concentration of permeable monovalent cations increases ~1000-fold inside mitochondria compared to that in the cytoplasm (approximately one order of magnitude for each 60 mV). Bearing in mind that the plasma membrane of eukaryotic cells is also charged (60 mV, cell interior negative), one can expect a further concentration increase by a factor of 10. In total this yields about a 10,000-fold higher concentration of membrane-permeable cations inside respiring mitochondria compared with that in the extracellular space. Moreover, the equilibrium distribution of some synthetic penetrating cations between the aqueous phase and the membrane is greatly shifted towards the latter, so their concentration in the inner mitochondrial membrane is further increased. Therefore, antioxidants conjugated with penetrating membranophilic cations can be used in extremely low concentrations because of their ability to selectively accumulate in the inner mitochondrial membrane, where the respiratory chain enzymes are located. For potential drug candidates, this property greatly diminishes the risks of unwanted side effects.

The first mitochondria-targeted antioxidants, lipophilic cations bound to thiobutyl,[89] vitamin E,[90] and ubiquinone,[91] were synthesized by Murphy's group in the end of 1990s. The ubiquinone derivative, known as MitoQ, had an important advantage: its oxidized form could accept electrons from the respiratory chain, rendering the reduced form of MitoQ a rechargeable mitochondrial antioxidant.[82] Later studies performed in our group indicated that plastoquinone derivatives (SkQ compounds, see Figure 9.1) are more promising antioxidants than ubiquinone conjugates. The "antioxidant concentration window", *i.e.* the difference between anti- and pro-oxidant concentrations, was much larger for SkQ than for MitoQ.[44,92,93] Recently, we determined that a conjugate of lipophilic cations to 2-demethylplastoquinone (a component of black cumin seeds) has an antioxidant concentration window even larger than SkQ1. The novel compounds, SkQT1 and SkQTR1, were also readily reduced by the respiratory chain and strongly inhibited the H_2O_2-induced apoptosis at pico- and nano-molar concentrations in cell cultures.[94]

The rechargeable mitochondrially targeted antioxidants have been extensively studied in the last 20 years, both *in vitro* and *in vivo*. Physicochemical properties of a broad spectrum of these compounds were determined; the compounds were tested in model systems and in animals, and clinical trials were carried out (covered in detail in ref. 45,85,87,95). Below we provide information on the results obtained in the animal models, with special focus on *D. melanogaster* studies.

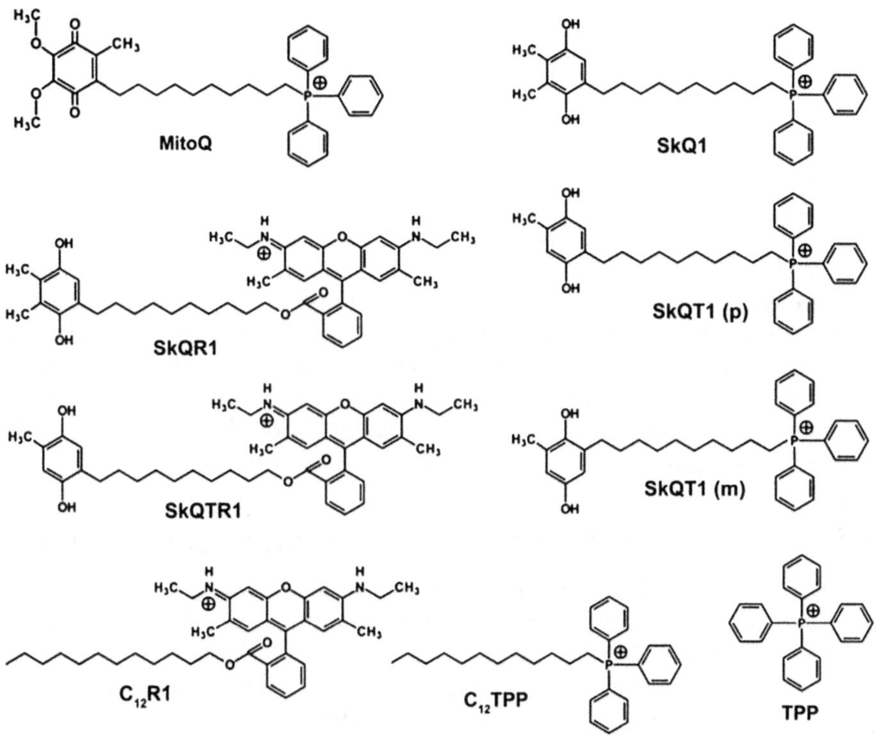

Figure 9.1 Some of the cationic mitochondria-targeted antioxidants and their analogs lacking a quinol residue.

9.4.1 Mitochondria-Targeted Antioxidants in Invertebrate Models

Identifying compounds that slow mammalian aging is clearly more relevant for human drug development compared to invertebrate animal models studies. However, the prohibitive cost of aging studies on mammals, as well as the substantial amount of time necessary for such studies, make invertebrate model organisms an attractive choice for anti-aging drug candidates screening (see ref. 96 for an extensive review of the progress being made in identifying compounds that extend the lifespan of invertebrates). The next step is to elucidate the genetic pathways that are targeted by the compounds found. Finally, it is possible to check the role of these pathways in mammalian aging.

The two most popular invertebrate models for such screenings are the nematode *C. elegans* and the fruit fly *D. melanogaster*. MitoQ was shown to extend lifespan and to protect cardiolipin from oxidation in *C. elegans* overexpressing human amyloid β (Alzheimer's disease model). It is worth noting that MitoQ failed to protect the mitochondrial DNA from oxidative damage, indicating that the protective effects are limited by the mitochondrial membrane.[97]

SkQ1 failed to increase the lifespan of the wild-type *C. elegans* (A. P. Grigorenko, unpublished observation), but curcumin extended the lifespan of *C. elegans* and reduced lipofuscin levels during aging. The effect was attributed to ROS quenching and to the antioxidant activity of curcumin, but not to its antimicrobial properties.[98] It is therefore possible that mitochondrially targeted derivatives of 2-demethylplastoquinone (a compound abundant in black cumin) that have a significantly larger window between anti- and pro-oxidant concentrations compared to SkQ1 [94] might be more efficient.

D. melanogaster is a recognized and well-established model system for gerontological studies. Flies are easily maintained on a controlled food under established temperature and light regimens, they have a short lifespan compared to vertebrates, and numerous genetically well characterized lines are readily available for research purposes. Experiments with *D. melanogaster* were used to investigate the consequences of diet supplementation with SkQ1 on aging, lifespan and correlated life-history traits.

9.4.2 SkQ1 Affects Early Survival and Aging in Unmated Flies

Genetically identical unmated flies of an isogenic line marked by w^{1118} mutation were selected for experiments. SkQ1 prolonged lifespan in both males and females when 100 μl of 20 pM to 20 nM SkQ1 solution was applied to the food surface once a week throughout life. 20 μM solution decreased lifespan.[99] The average SkQ1 effect was approximately 10% of the mean lifespan, and it was more prominent at early ages, increasing the survival of juveniles, mostly in females. Indeed, a significant positive effect on female survival was observed in the first 25% of the population, whereas the longevity of the longest-living 10% of the population was not affected.[100] Moreover, the survival of flies receiving SkQ1 during the first week of life was the same as the survival of flies receiving the drug lifelong.[99] Comparison of the Gompertz function parameters showed that in SkQ1-treated females, the initial level of mortality was substantially lower; this effect was less pronounced in males. Reducing early mortality led to an increase in the mean and median lifespan, but this had almost no effect on maximum lifespan.[100] These observations indicated that the drug, like many other antioxidants,[101,102] acted mainly on the short-lived part of the population, and its geroprotective effect was directed primarily at improving the quality of life, not its maximal extension. In addition, the variance of lifespan in flies treated with SkQ1 was smaller than that in control flies, confirming that SkQ1 affected the quality of life.[100] At the same time, analysis of a correlation between the parameters of a Gompertz function in normal physiological conditions (the Strehler–Mildvan correlation, which reflects the rate of loss of "vitality" in aging organisms[103]) allowed us to suggest that SkQ1 reduced the rate of the age-related decrease in fly vitality and, consequently, slightly slowed aging both in males and in females.[100]

To confirm these results, we assessed the effects of SkQ1 on general locomotor activity, which is often considered a marker of vitality and age (see ref. 104 for a review). Unmated females fed SkQ1 were characterized by an increase in locomotion as early as the second day post treatment; the difference reached its maximum by the age of 10 days and was maintained lifelong. No difference in locomotion was revealed between 10 day-old SkQ1-treated and control males; however, the difference was detected later, by the age of 20 days.[105] These observations indicate that SkQ1 effects on locomotion parallel SkQ1 effects on lifespan, thus confirming an important property of the drug to improve vitality at some ages. At an old age, a decline in locomotion in both SkQ1-treated and control flies was observed, as expected.[104] However, the difference in locomotion between the SkQ1-treated and control flies remained constant throughout life, indicating that SkQ1 did not affect the age-dependent rate of locomotion decline.[105] This result failed to confirm the effect of SkQ1 on aging, probably because this effect is small. There is, of course, a possibility that SkQ1 directly stimulates locomotion independently of effects on lifespan. Even in this case, however, an improvement in locomotor activity can be regarded as an indicator of the health-beneficial effects of SkQ1, which is an important property of any therapeutic.

Our results demonstrate slight sex-specificity of SkQ1 effects on lifespan and locomotion. In *D. melanogaster*, sex-specificity of lifespan control was reported earlier,[106–108] and was supposed to be associated with changes in protein homeostasis,[109] insulin[110–113] and steroid[114] signaling, and changes in expression of sexual differentiation pathway genes in adults.[115] Furthermore, it was shown that several groups of genes involved in reproductive physiology, amino acid utilization, sensory perception, immune response, and growth control are regulated in a sex-biased manner under stress conditions.[116] Given that often genes are involved in both stress resistance and lifespan control, their sex-biased expression might account for the sex-specific patterns of aging. The fundamental evolutionarily conserved systemic regulation of aging by the reproductive system may also be responsible for the sex-specificity of lifespan.[112] It is important to stress, however, that the sex-specificity of SkQ1 effects on lifespan does not violate the life prolonging properties of the drug.

As for any anti-aging drug, it was of special interest to know whether SkQ1 is able to prolong lifespan when applied late in life. A short-term SkQ1 treatment proved to be ineffective when started on the 30th day. However, constant administration of SkQ1 from the 30th day up to the end of life was quite effective, and the survival curve of treated females in this case was significantly different from the control curve.[99]

Another important notion is that SkQ1 treatment did not affect the rate of feeding in flies,[105] and therefore its effects on lifespan can hardly be attributed to caloric restriction.

9.4.3 SkQ1 Affects Reproduction in Mated Flies

An SkQ1 solution of the same 20 pM concentration did not increase the lifespan of mated *D. melanogaster* females and males of an isogenic line marked by w^{1118} mutation, and the effect on the early survival was not observed in mated females.[117] Not surprisingly, SkQ1 also did not affect locomotion of mated males and females: no difference was observed between treated and control flies, both males and females, at 10 and 20 days of age.[105] It should be stressed that in mated flies, SkQ1 had no negative effects on lifespan and locomotion.

Early fertility and the total number of adult progeny were elevated in flies reared on the SkQ1-supplemented diet. A significantly higher number of progeny was registered for 10 day-old treated parents compared to controls; this difference became insignificant at 20 days and disappeared completely later in life. However, the early effect of SkQ1 on reproduction was sufficient to provide a significant increase in the total number of progeny produced by SkQ1-treated flies.[117] The increase in reproduction ability observed in young mated females instead of the increased survival typical in young virgin females may illustrate the widely discussed complicated relationship between lifespan and reproduction (see ref. 118 for a review).

Hypothetically, the increase in reproduction stimulated by SkQ1 could be due to elevated mating activity, fecundity and/or improved viability of offspring. Experimental assessment of these traits showed that viability was not affected by SkQ1. No significant difference was found in the egg-to-pupa or egg-to-adult viability of the progeny from SkQ1-treated *versus* control young, 1 to 3 day-old and 10 day-old females. At the same time, a slight but reproducible increase in fecundity and mating activity was observed in young 10 day-old flies fed SkQ1. The accumulation of these small effects could be responsible for the increase in overall reproductive ability observed in SkQ1-treated flies.[117]

One may speculate that the general activity of flies is raised due to SkQ1 treatment, with both elevated locomotion in unmated flies and increased mating frequency in mated flies being just a part of this general effect. It was suggested that SkQ1 treatment may lead to an increase in energy supply based on the feedback mechanisms between ROS and energy production in mitochondria.[119] This would reasonably explain the observed trade-off between lifespan/locomotion and reproduction. In this case, an increase in reproduction due to increased mating frequency is expected because, of all the traits, frequency of mating is significantly associated with the extent of the female survival cost of mating.[120] Another possibility is that the effect of SkQ1 on reproduction is more specific and based on an interaction with the metabolism of sex peptides. These male seminal fluid proteins can considerably affect female gene expression and physiology, including egg production and frequency of mating (see ref. 121 for a review).

9.4.4 SkQ1 Acts as a Mitochondria-Targeted Antioxidant Combating ROS in *D. melanogaster*

A mixture of TPP (tetraphenylphosphonium) and PQ (decyl plastoquinone), the two constituents of SkQ1, applied at the same concentrations and under the same regiment as SkQ1 was completely ineffective,[99] indicating that a combination of the mitochondria-targeted cation and the oxidant in the same molecule is essential for prolonging *D. melanogaster* lifespan. Most probably, the direct antioxidant effect of SkQ1 carried out by its quinol residue should be related to those mROS that are generated by Complex I inside mitochondria.[122]

A solution of C_{12}TPP, a compound containing two additional methylene groups instead of the plastoquinone in SkQ1, was found to increase lifespan; however, unlike SkQ1, C_{12}TPP did not affect early survival but rather was effective later in life.[99] Hydrophobic cations such as SkQs and C_{12}TPP can operate as carriers of fatty acid anions, in this way mediating uncoupling of oxidative phosphorylation by these acids.[123] This, in turn, should decrease the amount of mROS produced not only by Complex I but also by Complex III,[124] which might explain the difference in the life-prolonging effects of SkQ1 and C_{12}TPP. These data allowed us to speculate that mROS increasing the early and the late mortality risks are generated by different mechanisms.[99]

ROS production can be affected by mutations in genes encoding enzymes of the respiratory electron transport chain (ETC). As it was mentioned above, it is believed that a moderate decrease and an increase in mROS amounts are associated with lifespan extension and reduction, respectively. Homozygous lethal mutations in genes encoding components of the ETC most likely reduce their function and, consequently, ROS production by the ETC to a level that is incompatible with life. However, in flies, both males and females, heterozygous for lethal mutations in genes encoding components of the complexes I, II and IV of the ETC the lifespan was increased. In both males and females heterozygous for these mutations, increased lifespan was not further increased by SkQ1 treatment.[105] This is in agreement with the hypothesis that mutations and SkQ1 affect the same pathway: in mutants the amount of mROS is already reduced, and therefore SkQ1 is not effective. Interestingly, the other mutation in the same complex II gene, *SdhB*, which reduced transcription of the gene, was associated with an increased level of mitochondrial hydrogen peroxide production and the decreased lifespan of the mutant flies.[125] It is worth noting that mutation in the gene encoding one of the subunits of cytochrome c-oxidase (complex IV of the ETC that is not supposed to generate mROS[67,126]) had the same effect on lifespan, both in the presence or absence of SkQ1, as mutations in genes encoding enzymes of complexes I and II generating mROS. This result indicates either that not mROS production, but other characteristics of mitochondria are altered in mutants and cured by SkQ1, or complex IV of the ETC is able to affect mROS production indirectly (*e.g.*, *via* changing the reduction degree of the CoQ in complex III).

We have also demonstrated that in both males and females heterozygous for a mutation in *Sod2*, a gene encoding mitochondrial superoxide dismutase, lifespan was decreased compared to controls, presumably, due to decreased detoxification of superoxide anion in mitochondria. SkQ1 treatment restored the decreased lifespan in both males and females.[105] This result agrees with the hypothesis that SkQ1 can combat ROS produced by mitochondria directly in the mitochondria and thus compensate for the impaired function of mitochondrial superoxide dismutase.

Overall, we speculate that *in vivo* SkQ1 acts as a mitochondria-targeted antioxidant capable of alleviating the detrimental effects of increased production of mROS on lifespan but is not effective when mROS production is already decreased by other means. At the same time, current evidence suggests that the mitochondrial role in lifespan control is largely associated with mitochondrial biogenesis and turnover, energy sensing, apoptosis, and calcium dynamics,[127] and SkQ1 may interfere with any of these processes.

9.4.5 SkQ1 Effects are Stable Under Different Experimental Scenarios and Across Different Wild-Type Genotypes

An important property of any geroprotector is the stability of its effects in an uncontrolled and changing environment, under different administration methods, to individuals with genetic constitutions that vary within the normal range.

Survival curves of females and males of an isogenic line marked by w^{1118} mutation were analyzed in 11 and 7 experiments, respectively, conducted over six years, with six experiments in the autumn, three in the spring, one in the winter, and one in the summer. In the course of the study, we attempted to increase the effect of SkQ1 on lifespan. To achieve this, we gave the drug starting at an earlier stage of development, used freshly prepared solutions, and increased the frequency of feeding. However, in all variants of the experiments the effect of the drug was almost identical.[100] Thus, the effect of the chemical was resistant to a variety of changes in the protocol of its administration and to chronological time, which is a very valuable quality for potential practical applications. Of course, slight variations in the effect of the drug were observed in different experiments.

Over time, the mean and median lifespan of w^{1118} females decreased in both the control and experimental cohorts. Changes in the mean lifespan of flies of a particular line during long-term observations have been previously reported.[128,129] However, the change in the mean and median lifespan did not affect the magnitude of the SkQ1 effect.[100] This result was inconsistent with the widespread view that the impact of various factors on lifespan is always greater if the original lifespan is smaller.[101]

To examine again if the effect of SkQ1 depends on the baseline lifespan and to determine whether SkQ1 affects lifespan of other, genetically unrelated lines of flies, three wild type lines, R340 (extremely low lifespan[130]), Canton S (medium lifespan) and Oregon RC (long lifespan), which have no genetic similarity to the w^{1118} line, were used in the experiments. A significant increase in female and male survival was observed in experiments with lines R340[100] and Canton S (Tsybul'ko, Pasyukova, unpublished results), the effect of about 10% was similar to that observed previously for the w^{1118} line. Thus, the relative increase in the fly lifespan under SkQ1 treatment did not depend on initial lifespan and was very similar for three genetically different lines. No significant increase in female and male survival was observed in experiments with the Oregon RC line.[100] In this case, the baseline lifespan of both females and males was quite high, and higher than in the other three lines used. It is possible that there is a certain baseline threshold lifespan, and if this is exceeded, SkQ1 fails to increase lifespan. An alternative explanation for the lack of effect of SkQ1 on the lifespan of Oregon RC flies is the genetic characteristics of the line.

We also assessed the effectiveness of SkQ1 treatment when severely changing environmental factors: temperature (18 °C and 8 °C) and diet (starvation, 12.5% and 25% of regular food supply). Physiologically, the possible temperature range for fruit flies varies from 5 °C to 31 °C; the environmentally optimal temperature for flies is considered to range from 17 °C to 25 °C;[131] lowering the temperature below 11 °C induces a state of diapauses.[132] Hence, lowering the temperature to 18 °C places flies on the border of the optimal environment, and we can view these conditions as moderately stressful. Under such a moderate stress, SkQ1 increased the lifespan of males by 27% and of females by 12%, *i.e.*, slightly more effectively than in a standard environment.[105] This result supports the idea that the impact of antioxidants is more effective in weak organisms or under suboptimal conditions[133] and is in agreement with the fact that the effect of SkQ1 on lifespan was more pronounced in mice kept in a vivarium with unfavorable conditions[134] (see Section 9.5, "Mitochondria-targeted Antioxidants in Rodents"). Thus, SkQ1 is effective not only in healthy organisms but also, and even more so, when some factors weaken the organism's status. Lowering the temperature to 8 °C places flies in a strong stressful condition, as does starvation. In our experiments, in both cases of severe stress, SkQ1 failed to affect lifespan of either males or females.[105] Thus, SkQ1 was not effective when the environment severely impaired the organism's status.

The ambient temperature affects the lifespan and the rate of aging of fruit flies.[135] Generally, low temperatures are associated with a longer lifespan in both wild populations and in laboratory conditions (see ref. 136 for a review). In line with this conclusion, we observed that lifespan was considerably increased at low temperatures: in males, the effect was more pronounced at 8 °C and in females the effect was similar at both temperatures. It is always of interest to see if a drug is able to increase a lifespan that is already quite long. Our data showed that SkQ1 substantially increased the lifespan of long-living flies under certain conditions (Figure 9.2). In addition, under starvation conditions the lifespan was extremely low, whereas at 8 °C it was extremely high, and in both cases SkQ1 failed to increase survival under severe stress conditions. All these data confirmed that the SkQ1 effect does not depend on the mean lifespan.

Thus, SkQ1 positively affected the lifespan of individuals with different wild type genotypes living in a variety of environments; it demonstrated properties of a promising life-prolonging drug unsusceptible to fluctuations in the mean lifespan of recipients, methods of preparation and administration of the drug, seasons, or calendar years.

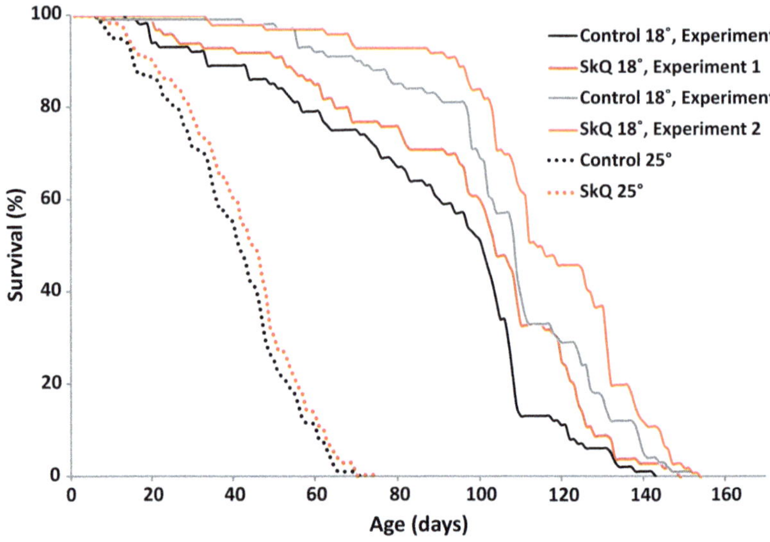

Figure 9.2 Effects of SkQ1 on the lifespan of *D. melanogaster* females at different temperatures. Original data were published in ref. 100 and 105.

9.5 Mitochondria-Targeted Antioxidants in Rodents

Lifespan extension by mitochondria-targeted rechargeable antioxidants was also documented for rodents.[134] SkQ1 was shown to extend the lifespan of female outbred mice kept under conditions close to natural in a conventional vivarium. In such a case the mice died mainly of various infections, the mortality being age-dependent due to the gradual lowering of immunity. The median lifespan in the control group was about 300 days, and was doubled by very low doses of SkQ1 (5 nmol SkQ1 kg^{-1} per day), which greatly decreased the infection-related mortality. In the SkQ1-treated female mice mammary carcinomas, rather than infections, became the primary cause of death. Moreover, SkQ1 prevented the age-dependent disappearance of estrous cycles. The latter effect was observed for outbred mice in both the low-pathogen and the conventional vivaria. Mole-voles and dwarf hamsters kept under conditions close to natural also lived longer if treated with SkQl.[134]

Besides the lifespan extension, mitochondria-targeted rechargeable antioxidants were efficient in health span prolongation. SkQ1 was found to diminish the age-dependent fertility decline in spontaneously hypertensive rats (SHR). In particular, we found that in the SkQ1-treated group, 9 females of 10 became pregnant, but only 5 out of 10 animals were pregnant in the control group (N. A. Medvedeva, V. P. Skulachev, unpublished observation). Since both males and females obtained SkQ1 in these experiments, it is difficult to distinguish if the beneficial effect was caused by sperm quality improvement or by a direct influence of SkQ1 on the female organism. However, other experiments in mice revealed that SkQ1 ameliorates the age-dependent disturbances of estrous function, as well as some other manifestations of aging,[137] making the direct action of SkQ1 on the female organism a more probable explanation.

We have recently demonstrated that life-long treatment with SkQ1 retarded the progression of age-related cardiac dysfunction (cardiomyopathy, cardiac hypertrophy, and diffuse myocardial fibrosis) in mice, presumably *via* a reduction in age-related inflammation.[138] SkQ1 also accelerated the resolution of the inflammatory phase, formation of granulation tissue, vascularization and epithelization of cutaneous wounds in aged mice.[139] Electron microscopy study of rat mitochondrial ultrastructure revealed that SkQ1 treatment prevented the development of age-dependent destructive changes (skeletal muscle sarcopenia) in both the control Wistar animals and OXYS rats suffering from excessive oxidative stress and accelerated aging.[140] This finding is in good agreement with earlier experiments demonstrating that SkQ1 at nanomolar concentrations slows down the cerebral dysfunctions in OXYS rats and decreases the pathological accumulation of AbetaPP, Abeta, and hyperphosphorylation of tau-protein in OXYS rats, as well as age-dependent changes in healthy Wistar rats.[141]

The positive effect of mitochondria-targeted rechargeable antioxidants on the lifespan of rodents is likely due to the anti-inflammatory action of these compounds. In murine endothelial cell culture, SkQl attenuated the TNF-induced increase in adhesion molecule ICAM1, VCAM, and E-selectin expression and secretion of IL-6 and IL-8, and prevented neutrophil adhesion to the endothelial monolayer. Although treatment with SkQl did not prevent age-related elevation of the major proinflammatory cytokines TNF and IL-6 in serum, it completely abolished the increase in ICAM1 expression in aortas of 24 month-old mice.[142] It was also demonstrated that both classic and mitochondria-targeted antioxidants inhibited the TNF induced NFκB-dependent activation of endothelium.[143]

SkQ1 might also suppress the inflammatory response by a partial lowering of the transmembrane electrical potential difference ($\Delta\psi$) on the inner mitochondrial membrane, *i.e.* acting as a mild uncoupler. We have recently found that the inflammatory activation of endothelial cells can be suppressed by low doses of classic mitochondrial uncouplers 2,4-dinitrophenol and 4,5,6,7-tetrachloro-2-trifluoromethylbenzimidazole, as well as by the mitochondria-targeted cationic uncoupler dodecyltriphenylphosphonium (C_{12}TPP). Mild uncoupling lowered the expression of E-selectin as well as adhesion molecules ICAM1 and VCAM1, and suppressed the adhesion of neutrophils to endothelium induced by tumor necrosis factor (TNF). Such an anti-inflammatory effect can be explained by inhibition of NFκB activation. These results suggest that the anti-inflammatory effect of mild uncoupling might be explained by decreased mROS production and by reduction of oxidative stress.[144] This hypothesis is in good agreement with the well-known correlation between high $\Delta\psi$ and excessive mROS production.[145] The beneficial action of mitochondrially targeted rechargeable antioxidants could also be mediated by the protective effect on cardiolipin

(CL) oxidation. It is experimentally established that under oxidative stress cytochrome c, an electron carrier in the mitochondrial respiratory chain, exerts peroxidase activity and oxidizes CL.[146] Addition of H_2O_2 + cytochrome c to CL-containing liposomes induced membrane permeabilization for molecules up to 3 kDa. The requirement of unsaturated CL for the permeabilization suggests that cardiolipin oxidation plays a critical role in the formation of membrane defects induced by H_2O_2 + cytochrome c.[147] Besides membrane permeabilization, cardiolipin oxidation leads to respiratory chain enzymes inactivation, cellular dysfunction and eventually apoptotic cell death.[148] Micro- and submicro-molar concentrations of mitochondrially targeted rechargeable antioxidants fully protected CL from peroxidation in liposomes.[149] Previously, a similar effect was shown in isolated mitochondria.[44]

9.6 Conclusion

Mitochondria-targeted rechargeable antioxidants represent a novel class of prospective anti-aging drug candidates. The important advantages of these compounds include: (1) extremely low effective concentrations due to selective accumulation in mitochondria (over 10,000-fold concentration); (2) ability to quench excess mitochondrial reactive oxygen species and thereby prevent oxidative stress and pathologies related to it, including chronic inflammation; (3) ability to regenerate the antioxidant (reduced) form of their quinone residue by reduction of the oxidized form by the mitochondrial respiratory chain; and (4) to decrease the $\Delta\psi$ and thereby reduce the mROS production (mild uncoupling).

The experiments in animal models suggest that mitochondria-targeted rechargeable antioxidants might be a tool to increase health span and lifespan in humans by lowering chronic inflammation and quenching oxidative damage during acute oxidative stress.

Acknowledgements

This work was supported by the Russian Science Foundation [project No. 14-50-00029 – review of the role of ROS in aging (B.A.F.); project No. 14-24-00107 – review of the mitochondria-targeted rechargeable antioxidants effects in animal models (V.P.S.)].

References

1. W. W. Hung, J. S. Ross, K. S. Boockvar and A. L. Siu, *BMC Geriatr.*, 2011, **11**, 47.
2. I. Akushevich, J. Kravchenko, S. Ukraintseva, K. Arbeev and A. I. Yashin, *Age Ageing*, 2013, **42**, 494–500.
3. K. Christensen, G. Doblhammer, R. Rau and J. W. Vaupel, *Lancet*, 2009, **374**, 1196–1208.

4. K. E. Thorpe and M. Philyaw, *Annu. Rev. Public Health*, 2012, **33**, 409–423.
5. A. M. Vaiserman and F. Marotta, *Trends Pharmacol. Sci.*, 2016, **37**, 331–333.
6. D. P. Goldman, D. Cutler, J. W. Rowe, P. C. Michaud, J. Sullivan, D. Peneva and S. J. Olshansky, *Health Aff. (Millwood)*, 2013, **32**, 1698–1705.
7. M. V. Skulachev, F. F. Severin and V. P. Skulachev, *Curr. Aging Sci.*, 2015, **8**, 95–109.
8. S. Bulterijs, R. S. Hull, V. C. Bjork and A. G. Roy, *Front. Genet.*, 2015, **6**, 205.
9. D. Gems, *Front. Genet.*, 2015, **6**, 212.
10. R. G. Faragher, *Front Genet*, 2015, **6**, 171.
11. M. V. Skulachev and V. P. Skulachev, *Biochemistry (Mosc)*, 2014, **79**, 977–993.
12. V. Jovaisaite, L. Mouchiroud and J. Auwerx, *J. Exp. Biol.*, 2014, **217**, 137–143.
13. A. M. Schulz and C. M. Haynes, *Biochim. Biophys. Acta*, 2015, **1847**, 1448–1456.
14. D. R. Green, L. Galluzzi and G. Kroemer, *Science*, 2011, **333**, 1109–1112.
15. S. M. Jazwinski, *Prog. Mol. Biol. Transl. Sci.*, 2014, **127**, 133–154.
16. L. Guarente, *Cell*, 2008, **132**, 171–176.
17. E. Munkacsy and S. L. Rea, *Exp. Gerontol.*, 2014, **56**, 221–233.
18. A. A. Moskalev, A. M. Aliper, Z. Smit-McBride, A. Buzdin and A. Zhavoronkov, *Cell Cycle*, 2014, **13**, 1063–1077.
19. C. E. Finch and J. Tower, *Cell*, 2014, **156**, 398–399.
20. D. C. Wallace, *Environ. Mol. Mutagen.*, 2010, **51**, 440–450.
21. F. Sevini, C. Giuliani, D. Vianello, E. Giampieri, A. Santoro, F. Biondi, P. Garagnani, G. Passarino, D. Luiselli, M. Capri, C. Franceschi and S. Salvioli, *Exp. Gerontol.*, 2014, **56**, 234–244.
22. Y. Wang and S. Hekimi, *Science*, 2015, **350**, 1204–1207.
23. N. Sun, R. J. Youle and T. Finkel, *Mol. Cell*, 2016, **61**, 654–666.
24. M. Picard, D. Ritchie, K. J. Wright, C. Romestaing, M. M. Thomas, S. L. Rowan, T. Taivassalo and R. T. Hepple, *Aging Cell*, 2010, **9**, 1032–1046.
25. G. C. Kujoth, A. Hiona, T. D. Pugh, S. Someya, K. Panzer, S. E. Wohlgemuth, T. Hofer, A. Y. Seo, R. Sullivan, W. A. Jobling, J. D. Morrow and H. Van Remmen, *et al.*, *Science*, 2005, **309**, 481–484.
26. A. Trifunovic, A. Wredenberg, M. Falkenberg, J. N. Spelbrink, A. T. Rovio, C. E. Bruder, Y. M. Bohlooly, S. Gidlof, A. Oldfors, R. Wibom, J. Tornell and H. T. Jacobs, *et al.*, *Nature*, 2004, **429**, 417–423.
27. G. C. Kujoth, P. C. Bradshaw, S. Haroon and T. A. Prolla, *PLoS Genet.*, 2007, **3**, e24.
28. M. Vermulst, J. H. Bielas, G. C. Kujoth, W. C. Ladiges, P. S. Rabinovitch, T. A. Prolla and L. A. Loeb, *Nat. Genet.*, 2007, **39**, 540–543.
29. A. Currais, *Ageing Res. Rev.*, 2015, **21**, 30–42.
30. A. Wong, P. Boutis and S. Hekimi, *Genetics*, 1995, **139**, 1247–1259.
31. M. Takahashi, S. Asaumi, S. Honda, Y. Suzuki, D. Nakai, H. Kuroyanagi, T. Shimizu, Y. Honda and T. Shirasawa, *Biochem. Biophys. Res. Commun.*, 2001, **286**, 534–540.

32. F. Martinez-Azorin, M. Calleja, R. Hernandez-Sierra, C. L. Farr, L. S. Kaguni and R. Garesse, *J. Neurochem.*, 2008, **105**, 165–176.
33. D. M. Humphrey, J. M. Toivonen, M. Giannakou, L. Partridge and M. D. Brand, *Exp. Gerontol.*, 2009, **44**, 316–327.
34. V. Klichko, B. H. Sohal, S. N. Radyuk, W. C. Orr and R. S. Sohal, *Biochem. J.*, 2014, **459**, 127–135.
35. G. Morrow, M. Samson, S. Michaud and R. M. Tanguay, *FASEB J.*, 2004, **18**, 598–599.
36. Y. W. Fridell, A. Sanchez-Blanco, B. A. Silvia and S. L. Helfand, *Cell Metab.*, 2005, **1**, 145–152.
37. Y. W. Fridell, M. Hoh, O. Kreneisz, S. Hosier, C. Chang, D. Scantling, D. K. Mulkey and S. L. Helfand, *Aging (Albany NY)*, 2009, **1**, 699–713.
38. C. Dell'agnello, S. Leo, A. Agostino, G. Szabadkai, C. Tiveron, A. Zulian, A. Prelle, P. Roubertoux, R. Rizzuto and M. Zeviani, *Hum. Mol. Genet.*, 2007, **16**, 431–444.
39. K. Takahashi, Y. Noda, I. Ohsawa, T. Shirasawa and M. Takahashi, *Exp. Gerontol.*, 2014, **58**, 146–153.
40. X. Liu, N. Jiang, B. Hughes, E. Bigras, E. Shoubridge and S. Hekimi, *Genes Dev.*, 2005, **19**, 2424–2434.
41. R. H. Houtkooper, L. Mouchiroud, D. Ryu, N. Moullan, E. Katsyuba, G. Knott, R. W. Williams and J. Auwerx, *Nature*, 2013, **497**, 451–457.
42. W. C. Orr, S. N. Radyuk and R. S. Sohal, *Antioxid. Redox Signaling*, 2013, **19**, 788–803.
43. M. Ristow and K. Zarse, *Exp. Gerontol.*, 2010, **45**, 410–418.
44. V. P. Skulachev, V. N. Anisimov, Y. N. Antonenko, L. E. Bakeeva, B. V. Chernyak, V. P. Erichev, O. F. Filenko, N. I. Kalinina, V. I. Kapelko, N. G. Kolosova, B. P. Kopnin and G. A. Korshunova, *et al.*, *Biochim. Biophys. Acta, Bioenerg.*, 2009, **1787**, 437–461.
45. M. V. Skulachev, Y. N. Antonenko, V. N. Anisimov, B. V. Chernyak, D. A. Cherepanov, V. A. Chistyakov, M. V. Egorov, N. G. Kolosova, G. A. Korshunova, K. G. Lyamzaev, E. Y. Plotnikov and V. A. Roginsky, *et al.*, *Curr. Drug Targets*, 2011, **12**, 800–826.
46. D. Harman, *J. Gerontol.*, 1956, **11**, 298–300.
47. D. Harman, *J. Am. Geriatr. Soc.*, 1972, **20**, 145–147.
48. D. Harman, *AGE*, 1983, **6**, 86–94.
49. L. A. Sena and N. S. Chandel, *Mol. Cell*, 2012, **48**, 158–167.
50. S. I. Liochev, *Free Radical Biol. Med.*, 2013, **60**, 1–4.
51. N. B. Metcalfe and C. Alonso-Alvarez, *Funct. Ecol.*, 2010, **24**, 984–996.
52. H. C. Lee and Y. H. Wei, *Exp. Biol. Med. (Maywood)*, 2007, **232**, 592–606.
53. S. Hekimi, J. Lapointe and Y. Wen, *Trends Cell Biol.*, 2011, **21**, 569–576.
54. A. Raffaello and R. Rizzuto, *Biochim. Biophys. Acta*, 2011, **1813**, 260–268.
55. A. Bratic and N. G. Larsson, *J. Clin. Invest.*, 2013, **123**, 951–957.
56. G. Barja, *Prog. Mol. Biol. Transl. Sci.*, 2014, **127**, 1–27.
57. M. P. Murphy, *Biochem. J.*, 2009, **417**, 1–13.
58. A. Bokov, A. Chaudhuri and A. Richardson, *Mech. Ageing Dev.*, 2004, **125**, 811–826.

59. H. H. Ku, U. T. Brunk and R. S. Sohal, *Free Radical Biol. Med.*, 1993, **15**, 621–627.
60. A. J. Lambert, H. M. Boysen, J. A. Buckingham, T. Yang, A. Podlutsky, S. N. Austad, T. H. Kunz, R. Buffenstein and M. D. Brand, *Aging Cell*, 2007, **6**, 607–618.
61. B. Andziak and R. Buffenstein, *Aging Cell*, 2006, **5**, 525–532.
62. B. Andziak, T. P. O'Connor, W. Qi, E. M. DeWaal, A. Pierce, A. R. Chaudhuri, H. Van Remmen and R. Buffenstein, *Aging Cell*, 2006, **5**, 463–471.
63. V. I. Perez, R. Buffenstein, V. Masamsetti, S. Leonard, A. B. Salmon, J. Mele, B. Andziak, T. Yang, Y. Edrey, B. Friguet, W. Ward and A. Richardson, *et al.*, *Proc. Natl. Acad. Sci. U. S. A.*, 2009, **106**, 3059–3064.
64. K. N. Lewis, B. Andziak, T. Yang and R. Buffenstein, *Antioxid. Redox Signaling*, 2013, **19**, 1388–1399.
65. A. Y. Andreyev, Y. E. Kushnareva and A. A. Starkov, *Biochemistry (Mosc)*, 2005, **70**, 200–214.
66. A. J. Kowaltowski, N. C. de Souza-Pinto, R. F. Castilho and A. E. Vercesi, *Free Radical Biol. Med.*, 2009, **47**, 333–343.
67. M. D. Brand, *Exp. Gerontol.*, 2010, **45**, 466–472.
68. S. Drose and U. Brandt, *Adv. Exp. Med. Biol.*, 2012, **748**, 145–169.
69. E. Niki, *Free Radical Biol. Med.*, 2009, **47**, 469–484.
70. H. J. Forman, M. Maiorino and F. Ursini, *Biochemistry*, 2010, **49**, 835–842.
71. M. E. Bauer and M. Fuente, *Mech. Ageing Dev.*, 2016, **158**, 27–37.
72. E. S. Cannizzo, C. C. Clement, R. Sahu, C. Follo and L. Santambrogio, *J. Proteomics*, 2011, **74**, 2313–2323.
73. C. Franceschi, M. Bonafe, S. Valensin, F. Olivieri, M. De Luca, E. Ottaviani and G. De Benedictis, *Ann. N. Y. Acad. Sci.*, 2000, **908**, 244–254.
74. A. K. Koliada, D. S. Krasnenkov and A. M. Vaiserman, *Front. Genet.*, 2015, **6**, 82.
75. D. Jurk, C. Wilson, J. F. Passos, F. Oakley, C. Correia-Melo, L. Greaves, G. Saretzki, C. Fox, C. Lawless, R. Anderson, G. Hewitt and S. L. Pender, *et al.*, *Nat. Commun.*, 2014, **2**, 4172.
76. D. Munro, N. Pichaud, F. Paquin, V. Kemeid and P. U. Blier, *Aging Cell*, 2013, **12**, 584–592.
77. D. Munro and P. U. Blier, *Aging Cell*, 2012, **11**, 845–855.
78. D. Canistro, C. Boccia, R. Falconi, B. Bonamassa, L. Valgimigli, F. Vivarelli, A. Soleti, M. L. Genova, G. Lenaz, A. Sapone, F. Zaccanti and S. Z. Abdel-Rahman, *et al.*, *J. Gerontol., Ser. A*, 2015, **70**, 936–943.
79. A. Nurmi, T. K. Miettinen, J. Puolivali, R. Pussinen, A. Soleti, K. Bagate, F. Riccardino, R. I. Grundy, J. Yrjanheikki, D. Canistro and M. Paolini, *Brain Res.*, 2008, **1207**, 174–181.
80. M. Novelli, V. D'Aleo, R. Lupi, M. Paolini, A. Soleti, P. Marchetti and P. Masiello, *Pancreas*, 2007, **35**, e10–e17.
81. S. Melov, J. Ravenscroft, S. Malik, M. S. Gill, D. W. Walker, P. E. Clayton, D. C. Wallace, B. Malfroy, S. R. Doctrow and G. J. Lithgow, *Science*, 2000, **289**, 1567–1569.

82. M. P. Murphy and R. A. J. Smith, *Annu. Rev. Pharmacol.*, 2007, **47**, 629–656.
83. A. T. Hoye, J. E. Davoren, P. Wipf, M. P. Fink and V. E. Kagan, *Acc. Chem. Res.*, 2008, **41**, 87–97.
84. S. Fulda, L. Galluzzi and G. Kroemer, *Nat. Rev. Drug Discovery*, 2010, **9**, 447–464.
85. V. P. Skulachev, *J. Alzheimer's Dis.*, 2012, **28**, 283–289.
86. L. F. Yousif, K. M. Stewart and S. O. Kelley, *ChemBioChem*, 2009, **10**, 1939–1950.
87. R. A. Smith, R. C. Hartley and M. P. Murphy, *Antioxid. Redox Signaling*, 2011, **15**, 3021–3038.
88. S. E. Severin, V. P. Skulachev and L. S. Yaguzhinsky, *Biokhimiia*, 1970, **35**, 1250–1252.
89. R. J. Burns, R. A. Smith and M. P. Murphy, *Arch. Biochem. Biophys.*, 1995, **322**, 60–68.
90. R. A. Smith, C. M. Porteous, C. V. Coulter and M. P. Murphy, *Eur. J. Biochem.*, 1999, **263**, 709–716.
91. G. F. Kelso, C. M. Porteous, C. V. Coulter, G. Hughes, W. K. Porteous, E. C. Ledgerwood, R. A. Smith and M. P. Murphy, *J. Biol. Chem.*, 2001, **276**, 4588–4596.
92. Y. N. Antonenko, A. V. Avetisyan, L. E. Bakeeva, B. V. Chernyak, V. A. Chertkov, L. V. Domnina, O. Y. Ivanova, D. S. Izyumov, L. S. Khailova, S. S. Klishin, G. A. Korshunova and K. G. Lyamzaev, *et al.*, *Biochemistry (Mosc.)*, 2008, **73**, 1273–1287.
93. L. E. Bakeeva, I. V. Barskov, M. V. Egorov, N. K. Isaev, V. I. Kapelko, A. V. Kazachenko, V. I. Kirpatovsky, S. V. Kozlovsky, V. L. Lakomkin, S. B. Levina, O. I. Pisarenko and E. Y. Plotnikov, *et al.*, *Biochemistry (Mosc.)*, 2008, **73**, 1288–1299.
94. I. I. Severina, F. F. Severin, G. A. Korshunova, N. V. Sumbatyan, T. M. Ilyasova, R. A. Simonyan, A. G. Rogov, T. A. Trendeleva, R. A. Zvyagilskaya, V. B. Dugina, L. V. Domnina and E. K. Fetisova, *et al.*, *FEBS Lett.*, 2013, **587**, 2018–2024.
95. A. N. Lukashev, M. V. Skulachev, V. Ostapenko, A. Y. Savchenko, V. V. Pavshintsev and V. P. Skulachev, *Prog. Mol. Biol. Transl. Sci.*, 2014, **127**, 251–265.
96. M. Lucanic, G. J. Lithgow and S. Alavez, *Ageing Res. Rev.*, 2013, **12**, 445–458.
97. L. F. Ng, J. Gruber, I. K. Cheah, C. K. Goo, W. F. Cheong, G. H. Shui, K. P. Sit, M. R. Wenk and B. Halliwell, *Free Radical Biol. Med.*, 2014, **71**, 390–401.
98. V. H. C. Liao, C. W. Yu, Y. J. Chu, W. H. Li, Y. C. Hsieh and T. T. Wang, *Mech. Ageing Dev.*, 2011, **132**, 480–487.
99. V. N. Anisimov, L. E. Bakeeva, P. A. Egormin, O. F. Filenko, E. F. Isakova, V. N. Manskikh, V. M. Mikhelson, A. A. Panteleeva, E. G. Pasyukova, D. I. Pilipenko, T. S. Piskunova and I. G. Popovich, *et al.*, *Biochemistry (Mosc.)*, 2008, **73**, 1329–1342.
100. A. V. Krementsova, N. V. Roshina, E. A. Tsybul'ko, O. Y. Rybina, A. V. Symonenko and E. G. Pasyukova, *Biogerontology*, 2012, **13**, 595–607.

101. W. C. Orr, S. N. Radyuk, L. Prabhudesai, D. Toroser, J. J. Benes, J. M. Luchak, R. J. Mockett, I. Rebrin, J. G. Hubbard and R. S. Sohal, *J. Biol. Chem.*, 2005, **280**, 37331–37338.
102. V. N. Erokhin, A. V. Krementsova, V. A. Semenov and E. B. Burlakova, *Izv. Akad. Nauk. Ser. Biol.*, 2007, 583–590.
103. B. L. Strehler and A. S. Mildvan, *Science*, 1960, **132**, 14–21.
104. A. L. Ridgel and R. E. Ritzmann, *Ageing Res. Rev.*, 2005, **4**, 23–39.
105. E. A. Tsybul'ko, A. V. Symonenko, O. Y. Rybina, N. V. Roshina and E. G. Pasyukova, *J. Gerontol., Ser. A*, 2016, DOI: 10.1039/gerona/glw084.
106. S. V. Nuzhdin, E. G. Pasyukova, C. L. Dilda, Z. B. Zeng and T. F. Mackay, *Proc. Natl. Acad. Sci. U. S. A.*, 1997, **94**, 9734–9739.
107. M. M. Magwire, A. Yamamoto, M. A. Carbone, N. V. Roshina, A. V. Symonenko, E. G. Pasyukova, T. V. Morozova and T. F. Mackay, *PLoS Genet.*, 2010, **6**, e1001037.
108. N. V. Roshina, A. V. Symonenko, A. V. Krementsova, M. V. Trostnikov and E. G. Pasyukova, *Aging*, 2014, **6**, 1076–1093.
109. M. Ruiz, D. Sanchez, I. Canal, A. Acebes and M. D. Ganfornina, *Exp. Gerontol.*, 2011, **46**, 579–589.
110. D. J. Clancy, D. Gems, L. G. Harshman, S. Oldham, H. Stocker, E. Hafen, S. J. Leevers and L. Partridge, *Science*, 2001, **292**, 104–106.
111. M. Tatar, A. Kopelman, D. Epstein, M. P. Tu, C. M. Yin and R. S. Garofalo, *Science*, 2001, **292**, 107–110.
112. T. Flatt, K. J. Min, C. D'Alterio, E. Villa-Cuesta, J. Cumbers, R. Lehmann, D. L. Jones and M. Tatar, *Proc. Natl. Acad. Sci. U. S. A.*, 2008, **105**, 6368–6373.
113. S. E. Schriner, S. Kuramada, T. E. Lopez, S. Truong, A. Pham and M. Jafari, *Exp. Gerontol.*, 2014, **60**, 220–230.
114. H. Tricoire, V. Battisti, S. Trannoy, C. Lasbleiz, A. M. Pret and V. Monnier, *Mech. Ageing Dev.*, 2009, **130**, 547–552.
115. J. Shen, D. Ford, G. N. Landis and J. Tower, *BMC Geriatr.*, 2009, **9**, 56.
116. M. Bauer, J. D. Katzenberger, A. C. Hamm, M. Bonaus, I. Zinke, J. Jaekel and M. J. Pankratz, *Physiol. Genomics*, 2006, **25**, 393–404.
117. E. A. Tsybul'ko, N. V. Roshina, O. Rybina and E. G. Pasyukova, *Biochemistry (Mosc.)*, 2010, **75**, 265–268.
118. T. Flatt, *Exp. Gerontol.*, 2011, **46**, 369–375.
119. J. Gruber, S. Fong, C. B. Chen, S. Yoong, G. Pastorin, S. Schaffer, I. Cheah and B. Halliwell, *Biotechnol. Adv.*, 2013, **31**, 563–592.
120. A. I. Barnes, S. Wigby, J. M. Boone, L. Partridge and T. Chapman, *Proc. Biol. Sci.*, 2008, **275**, 1675–1683.
121. E. Kubli, *Cell. Mol. Life Sci.*, 2003, **60**, 1689–1704.
122. V. P. Skulachev, *Biochemistry (Mosc)*, 2007, **72**, 1385–1396.
123. F. F. Severin, I. I. Severina, Y. N. Antonenko, T. I. Rokitskaya, D. A. Cherepanov, E. N. Mokhova, M. Y. Vyssokikh, A. V. Pustovidko, O. V. Markova, L. S. Yaguzhinsky, G. A. Korshunova and N. V. Sumbatyan, *et al.*, *Proc. Natl. Acad. Sci. U. S. A.*, 2010, **107**, 663–668.
124. V. P. Skulachev, *Q. Rev. Biophys.*, 1996, **29**, 169–202.
125. D. W. Walker, P. Hajek, J. Muffat, D. Knoepfle, S. Cornelison, G. Attardi and S. Benzer, *Proc. Natl. Acad. Sci. U. S. A.*, 2006, **103**, 16382–16387.

126. J. F. Turrens, *J. Physiol.*, 2003, **552**, 335–344.
127. M. Gonzalez-Freire, R. de Cabo, M. Bernier, S. J. Sollott, E. Fabbri, P. Navas and L. Ferrucci, *J. Gerontol., Ser. A*, 2015, **70**, 1334–1342.
128. F. A. Lints, C. V. Lints, P. Bullens, M. Bourgois and J. Delince, *Exp. Gerontol.*, 1989, **24**, 265–271.
129. D. M. Izmailov and L. K. Obukhova, *Adv. Gerontol.*, 2004, **15**, 30–35.
130. M. De Luca, N. V. Roshina, G. L. Geiger-Thornsberry, R. F. Lyman, E. G. Pasyukova and T. F. Mackay, *Nat. Genet.*, 2003, **34**, 429–433.
131. J. R. David, R. Allemand, J. van Herrewege and Y. Cohet, in *The Genetics and Biology of Drosophila*, ed. M. Ashburner, H. L. Carson and J. N Thompson, Academic Press, New York, 1983, pp. 105–170.
132. O. I. Kubrak, L. Kucerova, U. Theopold and D. R. Nassel, *PLoS One*, 2014, **9**(11):e113051.
133. R. S. Sohal, R. J. Mockett and W. C. Orr, *Free Radical Biol. Med.*, 2002, **33**, 575–586.
134. V. N. Anisimov, M. V. Egorov, M. S. Krasilshchikova, K. G. Lyamzaev, V. N. Manskikh, M. P. Moshkin, E. A. Novikov, I. G. Popovich, K. A. Rogovin, I. G. Shabalina, O. N. Shekarova and M. V. Skulachev, *et al.*, *Aging (Albany NY)*, 2011, **3**, 1110–1119.
135. J. Miquel, P. R. Lundgren, K. G. Bensch and H. Atlan, *Mech. Ageing Dev.*, 1976, **5**, 347–370.
136. G. Keil, E. Cummings and J. P. de Magalhaes, *Biogerontology*, 2015, **16**, 383–397.
137. M. H. Iurova, M. A. Zabezhinskii, T. S. Piskunova, M. L. Tyndyk, I. G. Popovich and V. N. anisimov, *Adv. Gerontol.*, 2010, **23**, 430–441.
138. V. N. Manskikh, O. S. Gancharova, A. I. Nikiforova, M. S. Krasilshchikova, I. G. Shabalina, M. V. Egorov, E. M. Karger, G. E. Milanovsky, I. I. Galkin, V. P. Skulachev and R. A. Zinovkin, *Histol. Histopathol.*, 2015, **30**, 353–360.
139. I. A. Demyanenko, E. N. Popova, V. V. Zakharova, O. P. Ilyinskaya, T. V. Vasilieva, V. P. Romashchenko, A. V. Fedorov, V. N. Manskikh, M. V. Skulachev, R. A. Zinovkin, O. Y. Pletjushkina and V. P. Skulachev, *et al.*, *Aging (Albany NY)*, 2015, **7**, 475–485.
140. V. B. Vays, C. M. Eldarov, I. M. Vangely, N. G. Kolosova, L. E. Bakeeva and V. P. Skulachev, *Aging (Albany NY)*, 2014, **6**, 140–148.
141. N. A. Stefanova, N. A. Muraleva, V. P. Skulachev and N. G. Kolosova, *J. Alzheimers Dis.*, 2014, **38**, 681–694.
142. R. A. Zinovkin, V. P. Romaschenko, I. I. Galkin, V. V. Zakharova, O. Y. Pletjushkina, B. V. Chernyak and E. N. Popova, *Aging*, 2014, **6**, 661–674.
143. I. I. Galkin, O. Y. Pletjushkina, R. A. Zinovkin, V. V. Zakharova, I. S. Birjukov, B. V. Chernyak and E. N. Popova, *Biochemistry (Mosc)*, 2014, **79**, 124–130.
144. V. P. Romaschenko, R. A. Zinovkin, I. I. Galkin, V. V. Zakharova, A. A. Panteleeva, A. V. Tokarchuk, K. G. Lyamzaev, O. Y. Pletjushkina, B. V. Chernyak and E. N. Popova, *Biochemistry (Mosc)*, 2015, **80**, 610–619.

145. S. S. Korshunov, V. P. Skulachev and A. A. Starkov, *FEBS Lett.*, 1997, **416**, 15–18.
146. I. Wiswedel, A. Gardemann, A. Storch, D. Peter and L. Schild, *Free Radical Res.*, 2010, **44**, 135–145.
147. A. M. Firsov, E. A. Kotova, E. A. Korepanova, A. N. Osipov and Y. N. Antonenko, *Biochim. Biophys. Acta*, 2015, **1848**, 767–774.
148. A. J. Chicco and G. C. Sparagna, *Am. J. Physiol.: Cell Physiol.*, 2007, **292**, C33–C44.
149. A. V. Lokhmatikov, N. E. Voskoboynikova, D. A. Cherepanov, N. V. Sumbatyan, G. A. Korshunova, M. V. Skulachev, H. J. Steinhoff, V. P. Skulachev and A. Y. Mulkidjanian, *Biochemistry (Mosc)*, 2014, **79**, 1081–1100.

Section IV

Mimicking Caloric Restriction

CHAPTER 10

Mimetics of Caloric Restriction

OLEH LUSHCHAK[a]* AND DMYTRO GOSPODARYOV[a]

[a]Department of Biochemistry and Biotechnology, Vasyl Stefanyk Precarpathian National University, 57 Shevchenko Str., Ivano-Frankivsk, 76018, Ukraine
*E-mail: olehl@pu.if.ua

10.1 Introduction

Caloric restriction or calorie restriction (CR) is a relatively old concept, initially based on the data of McCay and colleagues.[1] It was found that dietary-restricted rats lived longer than the corresponding controls.[1] It was generally accepted for a long time that total amount of calories in consumed food did influence the life span of tested animals.[2,3] After a while, life span extension by CR was observed for many model organisms, including fruit flies and nematodes.[3] However, it was difficult to draw a distinct line between malnutrition and life span-prolonging CR. The mechanism of life span extension by limitation of ingested food was obscure. Hence, it is not distinguished whether food restriction improves some health parameters or overeating worsens them. In other words, an optimal caloric state was not set. On the one hand, a calorically restricted diet can be optimal for an organism while increasing the amount of food will worsen health and shorten life span.[2,4] On the other hand, a "zero" point for amount of food consumed can be located between a supposedly "calorically restricted" diet and a diet promoting over-nutrition. It was found in the 2000s (Figure 10.1) that not only calories but particular nutrients, namely proteins, may play a role in

RSC Drug Discovery Series No. 57
Anti-aging Drugs: From Basic Research to Clinical Practice
Edited by Alexander M. Vaiserman

Published by the Royal Society of Chemistry, www.rsc.org

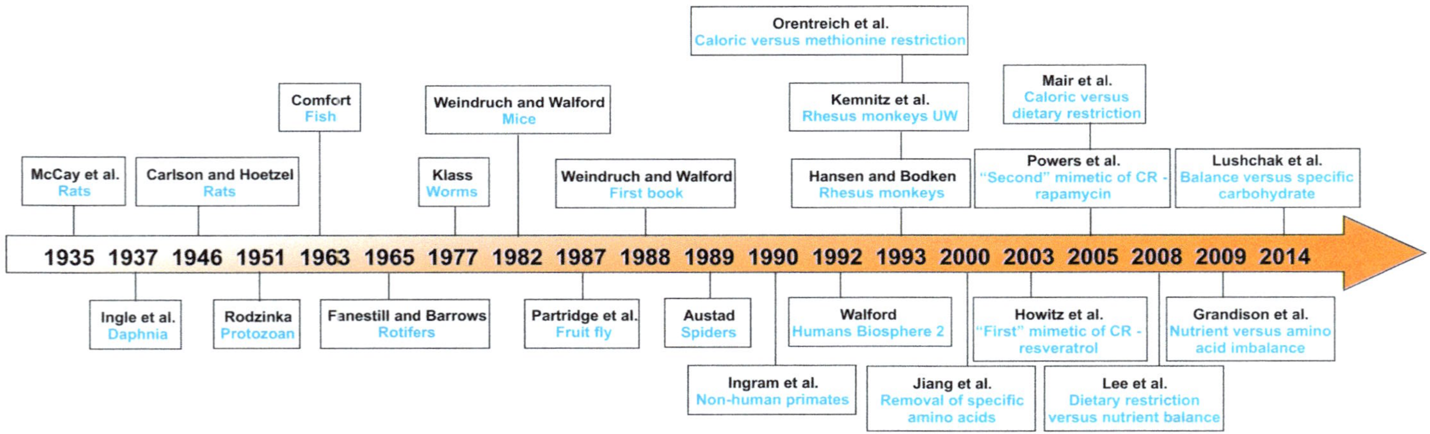

Figure 10.1 Development of CR concept and related findings. Data were obtained from Masoro.[19] The figure is modified from Benjamin *et al.*[20] Briefly, the first book on CR collected data on the effects of CR on organismal life span and was published in 1988. Until the mid-1990s, the life-prolonging effects of CR had been proven for main classes of the *Animalia* kingdom, including humans. Since the mid-1990s, several other concepts have been overlapping with the CR concept (*e.g.*, the hypothesis of methionine restriction). The first compound that prolongs life span by mimicking CR, resveratrol, was discovered in 2003. In each box: in black font: surname(s) of the investigator(s); in blue font: either model organism tested, or experiment mode, or key finding.

determining life span.[3,5–9] So, isocaloric diets with different amounts of proteins may result in different life spans of model organisms.[3,9] At nearly the same time, the life-prolonging effects of resveratrol, a low molecular phenolic compound, were discovered.[10] It was first noticed that resveratrol mimics CR.[11,12] The molecular mechanisms of resveratrol were found to be mediated by NAD^+-dependent histone deacetylases, sirtuins.[11,13–15] Parallels between the life-extending effect and mimicking physiological and molecular signs of CR were found for many anti-aging drugs, including glitazones[16] and rapamycin.[17] Simultaneously, a number of compounds that block catabolism of macronutrients or even ingestion were found to provide life span prolongation.[18] For instance, acarbose and 2-deoxyglucose, which block glycogen breakdown and glycolysis, respectively, are among these compounds.[18] It is suggested that some CR mimetics are not only inhibitors of catabolism but are signaling molecules shamming energy and/or nutrient sensors of cells. Indeed, a way of blockage of nutrient delivery to cells may not matter. The result is always either a lack of energy in the form of adenosine triphosphate (ATP) or deficiency of some building blocks for synthesis of proteins, polysaccharides and lipids. The question about the superiority of energy over plastic material or *vice versa* is still unresolved. In addition, it is still not clear whether we should take into account the number of calories in the diet, or dietary composition, or macronutrient balance, or even certain dietary components, which when taken away from the diet would provide life span extension. In this review we will try to provide a comprehensive picture for development of the CR concept, the tissue-specific and intracellular consequences of CR, a description of drugs that mimic CR outcomes, and suggest a mechanism explaining why a decrease in either energy or specific nutrients may lead to an increased life expectancy in particular biological species.

10.2 Aging and CR

The first evidence on the ability of CR to extend life span came from experiments performed in rats by McCay and colleagues.[1] Further studies showed that CR extended the life span by shifting mortality factors, such as diseases and tumors, to older age. Using different protocols, researchers revealed the life span extension by CR in many organisms like yeast, rotifers, spiders, nematodes, flies, fishes and mammals, including non-human primates (Figure 10.1).

10.2.1 CR in Yeast: *Saccharomyces cerevisiae*

Replicative and chronological life spans are described for unicellular yeast. The replicative life span is measured by the amount of formed daughter cells produced by a single cell. Reduction of the glucose concentration from 2% to 0.5% was the first CR intervention in yeast and this extended the replicative life span two-fold. The same study showed the importance of certain genes (*SIR2*, *FOB1*, *CYT1*) for mediating this phenotype. In addition, gene expression profiles of mutants with overexpression of *HXK2* and *HAP4* showed significant overlap with CR-treated cells. A similar protocol was used in later studies to show the importance of the TOR pathway and transcription factors Msn2 and Msn4 in mediating life span extension by CR. In addition, many genes and factors were identified in yeast as important regulators of life span. The life-extending properties of many drugs, remedies and natural extracts were later discovered using this method.[21–23]

Chronological life span is also described for unicellular organisms such as yeast *Saccharomyces cerevisiae*. Determination of chronological life span requires a measurement of the number of alive cells able to produce colonies while transferred to fresh media over different time points.[24] Cells are quiescent in the G_0 phase of the cell cycle but are metabolically active thus having some similarities to those of postmitotic tissues in adult multicellular organisms. Taking into account the similarities of basic biochemical pathways in yeast and other organisms, the model of yeast replicative and chronological life spans can be successfully used for studies involving the effects of metabolism on the life span of postmitotic cells.[25,26] A few pathways mediate chronological life span and act as mediators of its extension under CR. Well-known yeast signaling pathways, namely Tor/Sch9 and Ras/AC/PKA, were shown to be sensitive to nutrient availability and activate transcription factors Msn2, Msn4, and Gis1, which are responsible for metabolic reprogramming during starvation *via* protein kinase Rim15. The Tor/Sch9 pathway also regulates the respiration and membrane potential of mitochondria. CR inhibits this pathway to decrease ROS production and increase the stress response of the cell. However, many intermediate components of the aforementioned signaling pathways are still missing to link CR effects on apoptosis, protein aggregation, genome stability and epigenetic machinery of gene regulation.[27–30]

10.2.2 CR in Worms: *Caenorhabditis elegans*

C. elegans round worms are extensively used in aging research. The standard diet for worms consists of attenuated *E. coli* bacteria placed on solid plates. There are several methods to induce CR in these organisms: dilution of bacteria in liquid cultures; dilution of peptone, which reduces bacterial growth; using axenic medium or chemically defined liquid medium; serial bacterial dilution or total absence of bacteria in plates. Klass observed life span extension by bacterial restriction for the first time in 1977 (Figure 10.1). He did a 10-fold dilution of the initial bacterial culture and in these conditions worms lived about 52% longer.[31] A similar protocol was used in other studies that showed longer life spans in animals fed diluted bacterial culture.[32–34] Removal of peptone from plates with nematode growth medium (NGM) agar extended the life span of worms by 30% but reduced reproduction.[35] Significant extension of worm life span was observed on axenic medium with killed *E. coli*, soy protein, yeast extract and hemoglobin.[36] Similarly, a longer life span was observed for worms fed chemically defined *C. elegans* maintenance medium (CeMM) in comparison to the NGM one.[37] CR created by serial dilution of bacterial culture has been widely used in studies of worm longevity.[38] The advantage of this method is the possibility of testing life span in a broad range of bacterial concentrations. Finally, the complete absence of bacteria on plates also extended the life span of nematodes.[39,40] Interestingly, in these conditions worms lived 50% longer but restriction of bacteria by 90% increased life span by 20%.[39]

10.2.3 CR in Fruit Flies

Studies on various arthropod species show the possibility of extending the life span by CR in these animals. The most useful model is fruit fly *Drosophila melanogaster*. This model has many advantages like short generation time, life span of about 2–3 months, availability of many mutants and the possibility to manipulate genes of interest in the whole organism or very specific cells. In addition, flies can be fed on a diet with only sucrose and yeast. The combination of sucrose and yeast in the diet makes it possible to produce a diet with different caloricities or ratios of macronutrients, such as protein and carbohydrate. A few types of dietary manipulations can be distinguished for fruit flies. The first one is food dilution, which is usually achieved by simple reduction of sucrose and yeast within the diet. Dilution decreases the caloric value of the diet. The only problem is that flies eat more to compensate for the lower caloric value of the diet by increasing amount of volume eaten when given full access to food.[41] Second, dietary manipulation is usually called dietary protein restriction and is experimentally realized by using different concentrations of yeast (or protein in some cases). In principle, this means that researchers use diets with different protein-to-carbohydrate ratios ($P{:}C$). The power of using fruit flies in studies involving dietary effects on life span was most fully used by evolvement of the geometric framework (GF) firstly introduced by Simpson and colleagues.[7] Later GF was used on other models, including mice. GF suggests using an array of diets with varied ratios between macronutrients with different total diet caloricity. Since the caloric value of yeast and sucrose is about 4 cal gram^{-1}, an isocaloric diet can be simply prepared by adding these compounds in the same concentrations.

CR achieved by food dilution had a beneficial effect on fly life span. It was shown that restriction of the initial diet by 60% and 40% caused life extension in males and females, respectively.[42] It was discovered that dietary restriction (DR), as well as lowering the amount of only one or a few components of the diet, affects life span by decreasing the mortality rate. This decrease can be observed in two days after changing the diet from the control to the restricted one.[43] Conversely, the mortality rate returned to the control values when flies were switched from a dietary restricted to an *ad libitum* diet. Taken together, these results show that DR can be implicated at any age to extend the life span. Many studies were performed to study the interaction of specific genes and diet in regulation of life span. It was shown that Sir2 mediates longevity in the same way as CR. Ubiquitous expression of dSir2 in whole flies or specifically in the nervous system extended fly life span.[44] That effect was not observed when flies were tested in CR conditions. Another study showed that flies mutant for the histone deacetylase *RPD3* gene live longer and this life extension involved mechanisms similar to CR since the life span of these flies was not different when fed the restricted diet.[45] In *Drosophila*, mutation in gene *Chico* (insulin receptor substrate) produced a phenotype similar to those in Ames dwarf mice. *Chico*[1] mutant flies were shown to live longer on calorie-rich diets. However, under restricted nutritional conditions they had a shorter life span due to starvation.[46] Finally, some genetic interventions that extend life span are fully independent of dietary composition. A longer life span was observed in Or83b mutant flies that cannot sense smell in all diets tested.[47] A study in which GF was applied to flies showed that the $P:C$ ratio rather than caloricity influences the life span.[7] In that study, the life span was maximized at a $P:C$ of about 1:8 while fecundity was observed at 1:2. Probably, when the $P:C$ had been changed from 1:8 to 1:2, the life span became shorter because of increased reproduction, so, a trade-off between these two parameters can be observed. Other studies in flies also pointed out the importance of $P:C$ in the regulation of fly longevity.[8,48,49] Thus, it can be assumed that the $P:C$ ratio rather than calorie intake *per se* can explain the life span extension by DR in *Drosophila*.[5]

10.2.4 CR in Mammals

The influence of CR on life span was tested in mammals including rodents, dogs and nonhuman primates. In many studies CR was induced by food reduction by 20–40% from the *ad libitum* amount and was mostly called dietary restriction (DR). Furthermore, the effects of CR induced in different ways were studied in human volunteers. CR extended mean and maximum life spans in brown rats (*Rattus norvegicus*)[1,50,51] and in most laboratory strains of mice (*Mus musculus*). Meta-analysis of laboratory experiments since 1934 showed an increased median life span by 14–45% in DR animals. In mice, the effects were much weaker than those observed in rats: the difference in life span was about 4–27%. The magnitude of extension was significantly lower among inbred mouse strains. In the inbred DBA/2 strain, DR did not affect life span at all. In addition, the lack of effects of DR on life span or even its shortening was shown in experiments with the ILSXISS recombinant inbred panel.[52] There were some strains with significantly shortened life spans under DR. In addition, DR was not beneficial in offspring derived from wild-delivered mice. A possible explanation of these diverse effects might be the difference in genetic background, so DR protocols have to be different to extend the life span in animals with different genomes. Recent exciting studies performed at Sydney University under the supervision of Prof. Stephen Simpson demonstrate the possibility and power of GF to evaluate interactive effects of dietary energy, protein, fat, and carbohydrate on life span and other traits, such as food intake, metabolism, and reproduction. The authors concluded that longevity and health are optimized when protein is replaced with carbohydrate to limit compensatory feeding for protein and suppress protein intake. These consequences were in part associated with the activation of mTOR and mitochondrial function in the liver to affect branched-chain amino acids and glucose in blood. In addition, CR achieved by high-protein diets or dietary dilution had no beneficial effects on life span, suggesting that longevity can be extended by manipulating the ratio of macronutrients to inhibit mTOR activation.[3,53,54]

A 15 year-long study was performed to study the effect of CR in domesticated dogs, namely Labrador Retrievers.[55] Restricted dogs were fed 25% less food and the median and maximum life spans were 16 and 9% increased, respectively. Furthermore, calorically restricted animals had lower weights and fat contents as well as reduced serum triglycerides, triiodothyronine, insulin, and glucose concentrations. In addition, the onset of clinical signs of chronic diseases was delayed for diet-restricted dogs.

Rhesus monkeys (*Macaca mulatta*), with great similarities to humans in terms of genetics, endocrinology, physiology, neuroanatomy, cognitive function, and features of aging,[56] are commonly used in biomedical research. Few trials have studied the life span of these animals in respect to CR. The properly designed studies started in 1987 and 1989 at the National Institute on Aging (NIA) and the University of Wisconsin–Madison (UW), respectively (Figure 10.1). These studies identified that the median life expectancy of rhesus monkeys is about 26 years, 10% of animals survive up to 35 years, and the maximum life span was approximately 40 years.[56] Two different strategies were chosen to deal with chronic conditions and diseases. In the NIA study, animals with chronic disease were euthanized.[57] In contrast, a strategy to treat sick animals in a similar way to human clinical medicine to prolong life was used in the UW study. The most important message from both studies is that in non-human primates (NHP), CR increases life span.[58] It was shown that long-term CR improves heath in part by affecting signs related to metabolic syndrome. Restricted animals had decreased body weight,[56,60–63] fat mass[56,59–63] and amount of triglycerides.[64,65] CR decreased the basal level of glucose and insulin[66–69] and increased cholesterol in terms of high density lipoproteins (HDL)[64,65] and sensitivity to insulin. Benefits of CR in NHP suggest that it may be beneficial in humans as well.

Even if there are no strong data on how CR affects longevity in humans, many studies were performed to understand the effects of CR. Some epidemiological observations make it possible to suggest that an inverse relationship exists between caloric intake and aging.[70,71] Similar results to the NHP results were obtained in experiments performed at the Biosphere 2 research facility, where 8 subjects were calorically restricted for about two years. The participants had decreased weight and fat content that was accompanied by decreased levels of basal glucose and insulin, blood pressure and increased insulin sensitivity.[72–74] Similar changes were observed in a more recent study where the effect of 25% restriction was tested in 150 healthy people. Briefly, CR individuals had decreased body weight and fat,[75–77] triglycerides,[75,78] basal insulin and glucose levels,[75,78,79] and blood pressure.[75] Increased insulin sensitivity[80] and high density lipoprotein (HDL) levels[75,81] were observed under CR. Finally we have to agree that the capacity of CR to extend life span in humans is still unknown. However, similarities between CR effects on animal models and in humans suggest that CR may be useful to extend healthspan.

10.3 Beneficial Effects of CR

Many studies have shown that CR induces various beneficial changes both in short and long term perspectives for the extension of the life span and prolongation of healthspan. These changes can be used as early markers during the studies of chemicals or drugs with suggested action mechanisms similar to CR. The biological effects of CR include, but are not limited to: modification of important regulatory pathways *via* the expression and activity of key enzymes involved in metabolism; reduction of damage to macromolecules like proteins and nucleic acids and intensification of their clearance if damaged and cannot be repaired; reduction of chronic inflammation and decreasing of inflammatory markers; modulation of apoptosis and action of chaperone molecules; prevention of glucose and insulin intolerance; specific alteration of processes controlling cell repair or death.

10.3.1 Cardiovascular System

A body of evidence predominantly from animal studies and from some limited human trials indicates that CR has beneficial effects on the cardiovascular system. It reduces blood pressure and improves vascular function by decreasing oxidative stress[82] and increasing the availability of nitric oxide (NO) by activation of endothelial nitric oxide synthase (eNOS) and affecting the histone deacetylase Sirt1.[82–85] Increased levels of adiponectin and activation of AMP-activated protein kinase (AMPK) by CR prevented hypertension and cardiac hypertrophy in spontaneously hypertensive rats.[86] Reduced atherosclerosis and improved insulin sensitivity, as well as prevention of oxidative damage in cells of arterial walls, are achieved by decreased blood lipids (triglycerides, cholesterol) and glucose.[58,67,75,87,88] Additionally, CR decreases inflammation markers like TNF-α and IL-6.[75,89,90] Increased myocardial oxidation and ATP production,[91] Sirt1 activity,[92] mitochondrial function and biogenesis[93–95] as well as activation of pro-survival kinases Akt and extracellular signal-regulated kinases (ERKs) under CR conditions reduce the myocardial injury caused by ischemia-reperfusion.[91] Reduction of ventricular hypertrophy and improved diastolic function are possible by reduction of blood pressure, oxidative stress and myocardial fibrosis in parallel with activation of AMPK[86] and increased expression of sarco/endoplasmic reticulum calcium ATPase, SERCA2.[96]

10.3.2 Brain Function

An interesting interrelationship occurs between brain functioning and CR. On the one hand, CR affects the brain by enhancing cognitive function and preventing age-related changes and neurodegeneration.[97] On the other hand, the brain plays an important role as a mediator of the response to CR by activating nutrient-sensitive hypothalamic circuitries.[98] The beneficial effects of CR on the brain mostly include impact on neurogenesis, synaptic plasticity, and neuroprotection, and are mediated by induction of mild stress.[99] This mechanism, called hormesis, modulates key pathways regulating neuronal activity and cell resistance in response to stronger stress.[100] CR affects neurogenesis by maintaining neuronal stem cells.[101] These cells can proliferate and differentiate into either neuronal or glial cells to recover tissue after damage.[102] The ability to produce new cells is also important for learning and memory consolidation.[103]

Synaptic plasticity is the ability of synapses to transmit stronger or weaker signals between cells. Both regulation of neurotransmitter release from presynaptic cells and the amount of receptors in postsynaptic cells define the strengths of a signal. Production, release and reabsorption of neurotransmitters require energy for synthesis and active transport. Thus, an increased mitochondrial biogenesis to support energetic needs could be among the beneficial effects of CR.[104] Additionally, it was shown that more mitochondria are accumulated in the synapses[105] and their damage impairs learning and memory consolidation.[106] Mitochondrial biogenesis is partially activated by a higher concentration of NO, which is also involved in the formation of synapses in hippocampus.[107] This fact is also supported by results obtained with Sirt1-deficient mice, which have weaker synaptic plasticity, impaired memory and upregulation of eNOS.[108,109] CR affects the expression levels of many genes encoding receptors by preventing age-dependent decrease in expression.[110,111] Neurotrophins, Trk-B, NR1 and NR2B subunits of the *N*-methyl-D-aspartate-sensitive receptor are among them. This mechanism prevents the loss of synaptic plasticity and increases cognitive functions that rely on hippocampus-dependent memory tasks.

CR induces pro-longevity and anti-aging mechanisms in various cells, including neuronal ones. Altered mitochondrial biogenesis[112] and an increased level of free radical species were observed in aged brain and in several models for neurodegeneration.[113,114] Respiration capacity and biogenesis of mitochondria are regulated primarily by ***p***eroxisome proliferator-activated nuclear receptor ***g***amma ***c***oactivators (PGC-1α and PGC-1β), regulators of gene transcription whose activation is mediated by NO.[94,115,116] Furthermore, the activity of PGC-1α is regulated by Sirt1-driven deacetylation.[117] Being an important regulator of brain health and mediator of response to CR, Sirt1 targets the transcription factors such as FOXO1 and NF-κB, which, in turn, regulate metabolism, stress resistance and inflammation.[118–120] Thus, brain Sirt1-deficient mice did not properly respond to CR by increasing locomotor activity and insulin sensitivity.[118] The key molecular pathways affected by CR in cells belonging to cardiovascular and neural system are summarized in Figure 10.2.

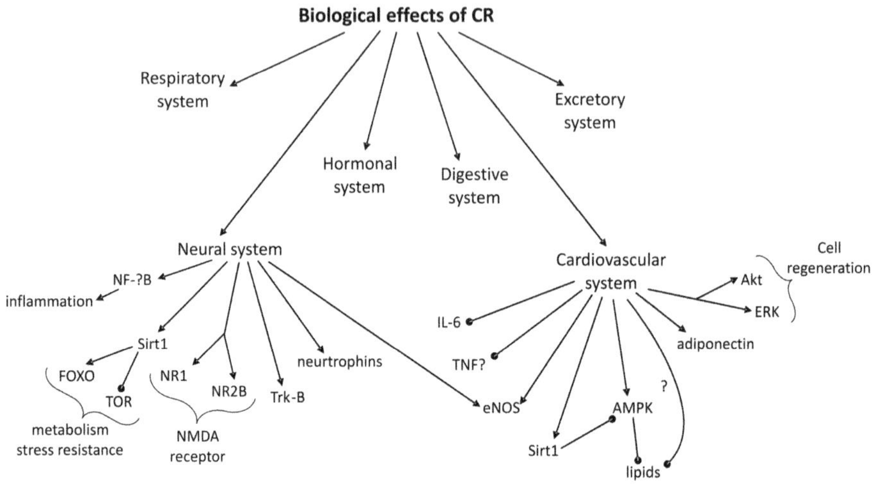

Figure 10.2 Key molecular pathways affected by CR in the cardiovascular and neural systems. Transcription regulators: NF-κB: nuclear factor κB; FOXO: forkhead box O. Modifying enzymes: TOR: target-of-rapamycin kinase, AMPK: AMP-activated protein kinase; ERK: extracellular signal-regulated kinase; Trk-B: tropomyosin receptor kinase B; Akt: pro-carcinogenic protein kinase B; Sirt1: sirtuin, histone/protein deacetylase. Metabolic enzymes: eNOS: endothelial nitric oxide synthase. Receptors/ligands: NMDA: *N*-methyl-D-aspartate; NR1 and NR2B: subunits of NMDA receptor; TNFα: tumor necrosis factor α; IL-6: interleukin 6.

10.3.3 Hormonal Regulation

Aging significantly affects the endocrine system. Both decreased hormone secretion and responsiveness of endocrine tissues to stimuli are observed in aging.[121] Aging causes decreases in the levels of estrogen (menopause), testosterone (andropause), growth hormone and insulin-like growth factor-1 (IGF-1) (somatopause), and dehydroepiandrosterone sulfate (adrenopause). In addition, the response of targeted tissues may induce development of pathologies and diseases. For example, the level of fasting insulin increases because of the development of resistance to it in peripheral tissues due to increased adiposity, decreased physical activity, and loss of muscle mass.[122] Thus, insulin, IGF-1, and growth hormone are key mediators of life span extension by CR.

The insulin concentration in plasma may serve as a biomarker for aging since it increases with age. Lower levels of insulin were observed in calorie restricted rodents and monkeys.[69] Increased longevity was observed in people with lower than median plasma insulin levels.[123] In CRONies (*CR* with *O*ptimal *N*utrition) study, restriction for about 6 years decreased serum insulin and glucose concentrations.[75] A randomized controlled trial for 12 months where aged individuals were subjected to 20% CR showed decreased concentrations of fasting insulin but increased sensitivity to this hormone.[124]

A decrease of fasting insulin by about 30% was observed under the 6 month CR.[79] Additionally, decreased acute insulin response to glucose supports improvement of cell responsiveness.[80] This study also showed that CR decreases fat content in adipose tissue and that the size of fat cells explains how the organism became leaner. However, the link between insulin and CR is not so obvious. Decreased insulin levels in circulation cause lower exposure of tissues to this hormone. In addition, increased sensitivity to insulin has been observed in tissues that require insulin for glucose uptake, such as muscle. Glucose uptake by these tissues can be virtually the same in control and CR conditions. In tissues that do not require insulin for effective glucose uptake, such as the nervous system, where insulin signaling at CR conditions is decreased, CR can both increase and decrease insulin signaling in a tissue-specific manner.

The importance of decreased insulin signaling in the nervous system is supported by experiments with long-living mutants that have impaired signaling in this tissue.[125] A similar situation is observed in IGF signaling. The level of this peptide in plasma is decreased in long-term CR-treated animals,[126] but the expression of its receptors is increased by 1.5–2.0-fold in the liver, muscles and heart.[127] These facts support the importance of tissue-specific increases or decreases of IGF signaling for life span extension.[128]

10.4 Intracellular Consequences of CR

Taking together all the knowledge described above and in previous chapters of this book about CR, we see that CR triggers many pro-longevity processes. The most exciting discussion is about the balance between anabolic and catabolic processes. Logically, to live long the organism should be set in a way to avoid damage or incorrect operation of any essential component, and to repair this damage quickly and effectively in case it happens. There are many strategies for protecting an organism from loss of functionality. The most simple of them is slowing down metabolism[129–132] and thus preventing deterioration of cellular components, cells, tissues and organs. The other strategy is keeping effective regeneration of all worn-out components along with slow wearing out of the repair machinery.[132] Food consumption, anabolism and accumulation of storage metabolites force organisms to operate constantly and wear out. CR promotes life-extending processes, such as mitochondrial biogenesis, ATP production, and autophagy. In some cases, apoptosis of outworn cells followed by cell division and tissue regeneration may lead to life span extension. These pro-longevity events are partially controlled by AMPK, which promotes mitochondrial biogenesis. The latter process is controlled by several transcription factors and their co-activators, particularly PGC-1α, ***p***eroxisome proliferator-activated receptor ***g***amma ***c***o-activator ***1α***. It was demonstrated that both sirtuins and AMPK are involved in activation of PGC-1α.[133–135] Inhibition of mTOR by AMPK results in activation of autophagy. Autophagy allows cells to clear impaired or age-worn proteins and/or organelles. Surely, the autophagy should be followed by biosynthetic processes and expenditure of storage metabolites.[136–138] In addition, extensive autophagy may promote biosynthetic processes, resulting in wearing out of biosynthetic machinery.

10.4.1 Autophagy

Autophagy is a "self-eating" process involved in elimination of cytoplasmic macromolecules, organelles and their parts. Autophagy enables reuse of simpler compounds derived from damaged molecules and cytoplasmic organelles for energetic and biosynthetic needs. At the cellular level, it is involved in maintenance of nutrient fluctuations, disposition of dangerous protein aggregates and removal of dysfunctional or damaged organelles. It also has benefits at the organismal level due to reducing oncogenesis and inflammation, maintaining neuronal function, improving lipid mobilization, and cleaning of apoptotic and necrotic cells. Autophagy mediates protective effects in rodent models of damage in organs like the liver,[139] heart,[140] nervous system[141] or kidney.[142]

Many studies indicate that expression of genes encoding proteins related to autophagy is downregulated in aged organisms. Protein kinase ULK1, adaptor proteins and autophagosome components, such as Beclin 1, LC3, Atg5, and Atg7, are among these proteins. This fact suggests an important role of autophagy in regulation of longevity and this was experimentally shown in various models. Screening in yeast *S. cerevisiae* identified that mutations in 10 ATG genes have shortened chronological aging.[143] The short-lived phenotype was observed in nematodes with loss-of-function mutations in Atg1, 7 and 18 and Beclin 1.[144] Atg1, Atg8 and Sestrin1 ablation shortened the life span by induction of mitochondrial dysfunction, triglyceride accumulation and muscle degeneration in flies.[145,146] Tissue-specific knockout of ATG genes in mice resulted in accumulation of ubiquitinylated proteins and lipofuscin, disorganized mitochondria, and protein carbonylation, carboxymethylation, or nitrosylation.

CR is the strongest physiological inducer of autophagy,[147] and inhibition of autophagy prevents the anti-aging effects of CR in all species investigated in this respect. CR may induce autophagy by TOR inhibition either by activation of AMPK[148,149] or inhibition of insulin/insulin-like growth (IGF) factor signaling.[150] CR is not able to increase life span if TOR signaling is inhibited in yeast, worms, or flies.[151] Pharmacological inhibition of TOR by rapamycin extends the life span in various organisms. However, life span extension by rapamycin in *C. elegans* was abolished by loss-of-functions in Atg1 and Atg 8.[152] In *Drosophila*, the life span extension by TOR inhibition was mediated by activities of its phosphorylation targets, such as S6K and 4E-BP.[153] Impaired expression of S6K extended the life span of *C. elegans*, *D. melanogaster*, and mice.[154,155] Rapamycin failed to extend the life span of flies that overexpress a constitutively active form of S6K.[153]

10.4.2 Metabolism of Reactive Oxygen Species

Oxidation of cell components by reactive oxygen species (ROS) is one of the factors that hasten cell deterioration. Now, it has become clear that endogenous production of ROS is well regulated by several signaling pathways. This fine control of the ROS level may even represent a certain kind of biological clock. The control of ROS production is mediated by transcription factor Nrf2 (***N***uclear factor erythroid 2-***r***elated ***f***actor ***2***) and partially by FOXO (***fo***rkhead bo***x*** ***O***), which induce expression of antioxidant enzymes.[156] Indeed, increased ROS production is observed in aged animals.[157,158] This increase is caused by mitochondrial dysfunction, leading to ROS-induced damage. CR was shown to decrease production of ROS by mitochondria in various tissues. Also, decreased amounts of damaged macromolecules represent the benefits of CR. However, organism still has to generate energy even if energetic substrates are limited. This problem could be resolved by increase of respiration under CR conditions. However, this logic is rather ambiguous. For instance, it is accepted that mitochondria are ones of the main ROS sources. So, an increase in mitochondrial biogenesis attributable to CR may potentially lead to an increase in ROS production. The steady state level of oxidative damage (including oxidized proteins, lipid peroxides, and modified nitrogen bases) is indeed higher in long-lived naked mole rats compared with mice that have a shorter life span.[159–161]

Oxidative stress is developed when the balance of free radical production and detoxification is changed. ROS, such as superoxide anion, hydrogen peroxide and hydroxyl radical, are produced as side products of energy production by the electron transport chain within mitochondria. In addition, some amount of them can be produced by catalytic action of specific enzymes (*e.g.*, xanthine oxidase). ROS are active molecules that can damage cellular macromolecules if not detoxified by antioxidant enzymes or exogenous antioxidants. Accumulation of oxidized lipids, proteins or nucleic acids during aging has been suggested to affect the life span and supports the free radical theory of aging (FRTA). However, many recent pieces of evidence suggest that FRTA is not correct. Naked mole rats are probably the most exciting evidence against FRTA. Short-, middle- and long-time CR decreases ROS production by mitochondria in many species. The longer life span of CR animals can support FRTA to some extent. However, even if FRTA cannot fully explain aging, ROS definitely play an important role in the regulation of longevity.

It was shown that chemical or environmental stresses of moderate intensity might extend the life span by activation of pro-survival pathways. This phenomenon is called 'hormesis'. It was shown that CR increases organism mobility and exercise to increase ROS production and mitochondrial metabolism.[162–164] This observation may suggest hormesis as a mechanism of CR action. Here we have a contradiction between two processes—decreased ROS production under CR and increased ROS production due to more exercise. Partial explanation of this contradiction came recently. New data reveal more complex links between aging, anabolism, catabolism, autophagy, and oxidation by ROS. It was recently found that a relatively slight increase in the ROS level may account for life span extension in fruit flies and nematodes.[165,166] ROS production by respiratory complex I increased because of over-reduction of CoQ and thus leaking of electrons. In fruit flies, expression of non-proton-pumping rotenone-insensitive NADH dehydrogenase increased ROS production and extended the life span.[166] This effect was not observed when other complexes were affected. An explanation may come from the regulatory function of ROS as signaling molecules from mitochondria to other cellular compartments. ROS can affect the activity of specific MAP-kinase cascades and redox-sensitive transcription factors to increase antioxidant defense and capacity. Interestingly, co-treatment with antioxidants to inhibit ROS production reduces signal transduction and abolished life span extension under CR.[167,168] Moreover, the hypothesis of mitohormesis assumes a potentially beneficial influence of moderate levels of mitochondrial ROS on organismal healthspan.[169] Furthermore, the oxidatively modified molecules may not cause death themselves.[132,160,170]

An organism would deteriorate in case of oxidation of essential proteins, key metabolic enzymes, metabolite transporters, components of signaling machinery, *etc.* It becomes more evident that oxidation of relatively unessential proteins may instead protect essential ones.[171,172] The oxidation and clearance of incorrectly translated proteins may also protect organisms.[173,174] Oxidation of special susceptible amino acids, particularly methionine, may protect organisms from death.[175] There is also evidence that protein aggregation is necessary for homeostasis.[137] In some cases, oxidation of susceptible proteins may allow the re-direction of metabolism in order to withstand stressful conditions.[176] An increase in mitochondrial biogenesis may also not lead to a corresponding increase in ROS production: several works show that mitochondria are sinks for ROS rather than ROS generators.[177–179]

Finally, one can conclude that chemical compounds that would activate the same pro-longevity processes as CR, particularly mitochondrial biogenesis, autophagy and antioxidant response, may partially or completely mimic CR and extend life span.

10.5 Ways to Achieve CR

10.5.1 Decreased Food Consumption

Food consumption is regulated by many factors, both external and internal. Among external ones, food quality, visual and smell input signals are probably the most important. Internal factors primarily involve neuronal and hormonal regulation. Decreased food consumption is the easiest way to achieve CR. It is quite simple to decrease food consumption in mice, rats or humans by giving less food, but not in some other animals. For example, in budding yeast *S. cerevisiae*, the only way to give less food is to decrease the amount of carbohydrate within the initial medium. For the round worm *C. elegans* the dilution of the initial culture makes it possible to decrease the amount of bacteria given to a certain amount of organisms. Interestingly, worms with mutation in the *eat2* gene have significantly reduced food uptake and thus live longer due to CR.[180,181] In the fruit fly *D. melanogaster*, there is no way to give less food so far and thus food dilution is used. There are some important factors to be pointed out. Firstly, the restriction has to be designed to avoid malnutrition by important exogenous factors like vitamins and microelements. Secondly, it is important to measure *ad libitum* food uptake in order to ensure proper reduction of given food. Finally, the restricted protocol should not induce starvation.

10.5.2 Dietary Composition

Every diet consists of carbohydrates, fats, proteins, water, vitamins, and minerals. Total caloric value or food energy is the amount of chemical energy the organism can get. Carbohydrates, fats and proteins are macronutrients that give about 95% of food energy. However, even with the same energetic value, one diet can be calorie restricted in comparison to other. For instance, sucrose and yeast are used in studies with fruit flies. Sucrose is a pure carbohydrate whilst yeast consists of protein and simple and complex carbohydrates, lipids and indigestible fibers. The caloric value of these components is virtually the same as 4 kcal g^{-1} of sucrose and 4.02 kcal g^{-1} of yeast. The diet with 65 g of yeast and 150 g of sucrose per liter has a caloricity of 861 kcal l^{-1}. The diet with 150 g of yeast and 65 g of sucrose per liter has 862 kcal l^{-1}. Experiments showed that flies of the first group are long-lived in comparison with those of the second group. This experiment shows that the caloric value of the diet is not the primary factor to regulate life span. However, these results show the possibility of using an isocaloric diet to extend the life span.

The way to extend the life span by using isocaloric diets is partial replacement of components that can be easily used to produce energy with those that cannot be used for this purpose. The easiest example is fiber. For example, the human body cannot digest cellulose, a polysaccharide composed of glucose monomers. Additionally, consumption of soluble dietary fibers before meals slows the absorption of carbohydrates by blunting the postprandial insulin spike.[182,183] Thus, addition of this compound to the diet will keep the same food energy density but the organism will produce less energy in the form of ATP during its metabolism. CR can be also achieved by addition of non-metabolizable or "zero-calorie" sugars like sucralose or L-glucose.

10.5.3 Inhibition of Food Digestion and Absorption

Affecting both food digestion and absorption processes may be a way to create CR conditions. Multiple enzymes are involved in digestion of complex nutrients to more simple forms that can be absorbed. To decrease the uptake of calories derived from complex carbohydrates, inhibitors of enzymes such as amylases, glycosidases and disaccharidases can be used. Amylases and glycosidases convert starch, glycogen and other molecules to simpler disaccharides. The latter undergo further digestion on the surface of enterocytes in the gut. For instance, *Streptomyces tendae* produces tendamistat, a 74 amino acid inhibitor that targets a wide range of mammalian alpha-amylases by steric blockage of the active site of the enzyme. Disaccharidase inhibitors, such as acarbose,[184,185] miglitol or L-arabinose,[186] effectively decrease the breakdown of disaccharides like maltose and sucrose to monosaccharides in the intestinal epithelium. Inhibition of carbohydrate breakdown was shown for white beans[187,188] and some commercial formulations, such as InSea[189] or Irvingia.[190,191] Interestingly, many inhibitors may be effectively used under diabetic complications or obesity.

Lipase enzymes in the stomach and small intestines break down dietary fats. Orlistat, the lipase-inhibiting drug, was found to be able to reduce the dietary fat absorption by 30%.[192] The polyphenols from green and black tea are able to inhibit lipase activity enabling to eat more calories without absorbing all of the fats.[193,194] Along with lower caloric absorption from fat, green or black tea polyphenols substantially reduce blood glucose, triglycerides, cholesterol, and other vascular risk factors.[195–199] Similar changes happen when calorie intake is reduced,[81,200] supporting the fact that blocking fat digestive enzymes prevents excess calories from being absorbed.

10.5.4 Decrease in Appetite and Satiety

Appetite is regulated by interaction between digestive tract, hormonal and neuronal systems. Appetite serves to regulate food intake to support organisms with energy and is closely related with satiety. The hypothalamus is the part of the brain responsible for regulation of the appetite. Its neurons sense the concentration of hormones such as ghrelin, leptin, PYY 3–36, orexin and cholecystokinin produced in other parts of the organism. Thus, modification of the hypothalamic sensitivity or the concentration of hormones can be useful for decreasing energy intake. Interestingly, pinolenic acid found in pine nuts can suppress appetite and thus reduce food intake by 36%.[201] It stimulates the secretion of the hunger-suppressing hormone cholecystokinin[202] and glucagon-like peptide-1.

10.5.5 Mimetics of CR

10.5.5.1 Biguanides

Biguanide metformin, which has been used for a long time as an anti-diabetic drug, has been found to prolong life span in mice.[203–205] A similar compound, phenformin, demonstrated a stronger effect, but in relatively low concentrations.[206–208] In high concentrations, phenformin was shown to be toxic.[208] It is widely accepted that the effect of biguanides such as metformin, phenformin, buformin, galegine, and some other guanidine-containing compounds is mediated by the activation of AMPK.[209–214] However, AMPK is believed to be a secondary target of biguanides while the primary target and exact mechanism of AMPK activation are not understood completely. Metformin was shown to inhibit complex I (proton-pumping NADH:ubiquinone oxidoreductase) of the mitochondrial respiratory chain. Logically, this inhibition may hinder the formation of mitochondrial membrane potential by mitochondrial respiratory complexes. The decrease in membrane potential may result in a lowering of proton flux through ATP synthase and, finally, a decrease in ATP production.[215–218] The latter may lead to a change of energy charge and an increase in the concentration of AMP, which, in turn, activates AMPK. However, the exact binding site of biguanides on complex I and/or mechanism of inhibition has not been found yet.[216,218] It was suggested that the inhibitory action of metformin is connected with its metal-chelating properties, especially its ability to bind copper.[218,219] However, it remains unclear how copper sequestration may affect complex I, which does not contain this metal, and not affect copper-containing cytochrome c oxidase (complex IV). Indeed, it was demonstrated that metformin affects only complex I without an influence on other complexes of the mitochondrial respiratory chain.[212,216,218] An alternative mechanism of AMPK activation by biguanides, particularly metformin, suggests a direct interaction between the γ-subunit of AMPK and the drug.[220] Other authors showed that metformin helps the formation of a heterotrimeric complex of α-, β- and γ-subunits.[221] Recent findings have demonstrated that metformin inhibits mitochondrial glycerophosphate dehydrogenase.[213,217] This inhibition impairs recycling of the cytosolic NAD^+ pool by means of the mitochondrial respiratory chain. In the latter case, cytosolic NADH would tend to be oxidized *via* the lactate dehydrogenase reaction.[217] Lactic acidosis was declared to be a common side-effect for biguanides. Moreover, inhibition of glycolysis by accumulation of lactic acid and NADH may account for the anti-cancer effects of metformin. The dependence of cancer cells on the glycolytic way of energy production has been well documented.[222]

10.5.5.2 Natural Phenols

It was shown in a number of studies that the life span of model animals can be prolonged by administration of flavonoids and natural plant phenols as wells as alkaloids or multicomponent plant extracts.[223–228] The most well known examples are curcumin,[145,229–231] quercetin[225] and epigallocatechine gallate.[135,226] Life span-extending properties were also attributed to resveratrol,[13,14,224,231–234] which was first studied for its life extension properties in a yeast model.[10] Resveratrol was first found to mimic effects of CR.[11,12] Possible mechanisms of this mimicking are believed to be connected with nutrient sensing. The early studies on resveratrol demonstrated its ability to activate sirtuins.[11,13–15,235] These deacetylases, in turn, modulate the activity of many proteins, including protein kinases of signaling pathways, transcription factors and their co-activators. Proteins involved in life span regulation, such as AMPK and transcription factor FOXO, are counted among those regulated by sirtuin-catalyzed deacetylation. Naturally, sirtuins respond to the cellular energy state, using oxidized nicotinamide adenine dinucleotide (NAD^+) as the substrate and being inhibited by nicotinamide.[235] An increase in NAD^+ occurs during intensive mitochondrial respiration, indirectly indicating about the rate of ATP production by oxidative phosphorylation.[236] Some studies have shown the direct interaction of resveratrol with sirtuins.[134,234] Other studies suggest a more global effect of resveratrol based on its ability to activate mitochondrial complex I.[236] However, this action of resveratrol contradicts the supposed life span-prolonging mode of metformin mentioned above. It can also be possible that cellular longevity can be promoted by either inhibition of complex I or its activation. At the same time, undisturbed operation of complex I may be attributable to a "healthy" cell senescence.

Recently, a few more explanations of life span extension by resveratrol have been proposed. The transcription factor Nrf2 (***N****uclear factor erythroid 2-****r****elated* ***f****actor* ***2***), which regulates genes of antioxidant response and xenobiotic detoxification, has been found to be crucial for cellular longevity.[237–240] This factor is, in turn, regulated by redox sensing mechanism: thiol-containing adaptor protein Keap1 binds Nrf2 when reduced and targets this transcription factor for ubiquitination and subsequent proteolytic breakdown by proteasome. Under oxidative stress, thiol groups of Keap1 are oxidized and form disulfide bonds, leading to conformational changes and, eventually, the inability to bind Nrf2.[239,241] As a result, Nrf2 is directed to the nucleus, where it activates expression of target genes. Nrf2 target genes are those encoding antioxidant enzymes like superoxide dismutase and catalase,[239,240,242] NADPH-producing enzymes—glucose-6-phosphate dehydrogenase and malic enzyme,[243,244] cytosolic NAD(P)H-quinone dehydrogenase, and thioredoxins.[239,243,245] It was shown that resveratrol and natural plant phenols (*e.g.*, curcumin) are able to bind and inactivate Keap1, inducing Nrf2.[239,240]

10.5.5.3 Rapamycin

Rapamycin (also known as sirolimus) is a macrolide immunosuppressant drug. Its ability to prolong life span in yeast was first observed in 2006.[246] Soon, it was found that the target of rapamycin in animals is peptidylprolyl isomerase FKBP12 (FK-binding protein 12).[247–249] The complex of FKBP12 with rapamycin inhibits downstream kinase, literally called mammalian (or mechanistic) target-of-rapamycin (mTOR) kinase. This kinase was shown to be involved in regulation of protein synthesis and autophagy. Particularly, mTOR kinase phosphorylates downstream P70 S6 kinase, which promotes protein biosynthesis and inhibits autophagy. The final result of this regulation is fostering of cell division and tissue growth, and accumulation of storage metabolites like fat.[136,250] The processes such as tissue growth and accumulation of reserve metabolites rely on calorie and nutrient intake, and are considered to be pro-aging ones.[136,251] The anti-aging properties of rapamycin have been confirmed on many model organisms, including mammals.[251,252] It suggests that mTOR signaling is a universal pathway that regulates the impact of nutrients on senescence in eukaryotic organisms. It is proven that the mTOR pathway is activated by nutrients, especially amino acids. The activation of mTOR by amino acids is mediated by small GTPases called Rag, which get the signal by directly binding amino acids like leucine.[253–256] In this case, rapamycin may mimic CR effects by preventing utilization of ingested nutrients for tissue proliferation. However, there is no direct connection of mTOR with cell energy state, *i.e.* ATP level. An indirect connection is possible *via* AMPK. Particularly, AMPK activates adaptor tuberous sclerosis proteins, hamartin (also known as TSC1) and tuberin (TSC2), which, in turn, indirectly inhibit TOR by means of small GTPase Rheb.[247,248,254] Thus, mTOR seems to be a common final target for biguanides and rapamycin.

Despite beneficial effects on longevity, rapamycin also confers negative side effects, including its immunosuppressant action and, as a consequence, an increased probability of infectious diseases.[257,258] Derivatives of rapamycin (rapalogs), temsirolimus, everolimus, zotarolimus, zidaforolimus and a few others were designed in order to treat cancer *via* inhibition of the mTOR pathway. The rapalogs exhibited only moderate anti-cancer properties in clinical trials.[248] Nevertheless, it does not exclude the use of rapalogs as anti-aging drugs. A possible way to decrease the side effects of rapalogs is to elaborate a chemical inhibiting the mTOR pathway by direct binding of mTOR kinase complexes or even their downstream target P70 S6 kinase. The deletion of ribosomal P70 S6 kinase 1 was shown to be sufficient to prolong life span in mice.[155,259]

10.6 Intracellular Targets of CR

10.6.1 Sensors of Nutrient and Energy State

In previous subsections we have shown that a number of compounds that suppress catabolism led eventually to accumulation of certain crucial metabolites, namely NAD^+ and acetyl-coenzyme A (hereinafter, acetyl-CoA). Of note, these two compounds appear to be on the crossroads of metabolic pathways. For example, acetyl-CoA can be formed as a product of fatty acid β-oxidation, while dietary polysaccharides are converted into glucose and then into pyruvate, which, in turn, yields acetyl-CoA, being oxidized by the pyruvate dehydrogenase complex. Proteins are broken into amino acids, some of which are turned into ketoacids by transamination or deamination. Many ketoacids can easily be converted into acetyl-CoA.[138,230,260,261] Thus, we expect that acetyl-CoA is the compound that may signal about a lack of specific nutrients.

True signaling suggests the presence of a specific receptor and a signaling pathway that regulates cell processes mainly by post-translational modification of specific enzymes and transcription factors. The modification of transcription factors results in deep changes to cell metabolism at the level of gene expression. Indeed, this way of nutrient and energy signaling may take place in some cases we describe below. However, more evidence has been obtained, supporting the idea that post-translational modifications can be induced directly by acetyl-CoA and NAD^+.[235,262] Particularly, both acetyl-CoA and NAD^+ are involved in post-translational modifications of proteins, such as acetylation and deacetylation, respectively. Moreover, NAD^+ is a source of ADP-ribose, a modifying molecule for ADP-ribosylation and poly-ADP-ribosylation. The latter two processes were found to be closely connected with stress resistance and extension of life span.[235,262] The levels of NAD^+ and acetyl-CoA grow during a lack of nutrients.[230,260] This likely happens because the organism restricted in nutrients tries to use all possible sources of energy for ATP production. They include fatty acids from fat stores and amino acids provided by autophagy, and particularly *via* hydrolysis of proteins. We can see that some of above-mentioned mimetics of CR may mimic the effects of NAD^+ and acetyl-CoA. For instance, resveratrol is supposed to activate sirtuins for which NAD^+ is a natural co-substrate and activator. Activators of AMP-activated protein kinase promote mitochondrial biogenesis and, as a consequence, utilization of acetyl-CoA by citrate synthase, a tricarboxylic cycle enzyme. Thus, agents decreasing the NADH/NAD^+ ratio and those depleting acetyl-CoA are considered to be potent CR mimetics.[230,260] A decrease in the NADH/NAD^+ ratio was found to prolong life span.[263,264] Yet, ectopic expression of rotenone-insensitive non-proton-pumping NADH dehydrogenase, which converts NADH to NAD^+, extended life span in model organisms in virtually all cases.[263,265–267]

It seems that metformin and other biguanides drop from this logic since they were shown to inhibit NADH:ubiquinone oxidoreductase (also known as complex I), thus they may increase the NADH/NAD^+ ratio.[215] However, it was recently found that metformin may indeed decrease the NADH/NAD^+ ratio as an inhibitor of mitochondrial sn-glycerol-3-phosphate dehydrogenase[217] or affects neither NADH/NAD^+ ratio[213] nor ATP levels.[268]

10.6.2 Signaling Pathways

Aging is proven to be closely connected with an organism's growth and reproduction.[136,251] Relatively long-lived organisms have either a high regenerative capacity, are able to re-program their cells, or slow down their metabolism.[129–132] There are a few examples of life span extension due to metabolic slowdown: (1) nematode, *Caenorhabditis elegans*, is able to live longer in the dormancy dauer stage;[269] (2) fruit flies live longer at temperatures moderately lower than the standard cultivation temperature;[270,271] and (3) naked mole rats (*Heterocephalus glaber*), long-lived rodents, have a relatively low metabolic rate.[272] CR was shown to specifically down-regulate cell signaling pathways responsible for regulation of anabolic processes: protein, glycogen, and lipid synthesis, and accumulation. Conversely, CR itself and its mimetics up-regulate pro-longevity catabolic processes like autophagy, oxidation of monosaccharides and organic acids, including amino and fatty acids, by means of mitochondria.

10.6.2.1 Insulin Signaling Pathway

Numerous studies on centenarians show that long-lived humans bear single nucleotide polymorphisms that lead to downregulation of the insulin signaling pathway.[273–275] This pathway regulates processes of tissue growth and, in case of insulin as a first messenger, results in accumulation of glycogen and lipid stores in cells.[276] It starts from receptor tyrosine kinase and through a series of downstream kinases leads to phosphorylation of effectors: transcription factor FOXO (***F**orkhead b**ox O***) and glycogen synthase kinase 3β, which become inactivated, and proteins involved in glycogen and lipid synthesis, which become activated.[276,277] Many components of this pathway, namely phosphatidyl inositol-3 kinase (PI3K) and protein kinase B (PKB; also called Akt), are pro-carcinogenic proteins.[278] Protein kinase B is at the crossroads of several signaling pathways regulating the mTOR pathway.[20,249,279,280] The insulin signaling pathway is switched on in animals when the blood glucose level is increased. Conversely, food restriction may turn off this pathway by multiple mechanisms: a decrease in insulin release, activation of AMPK (*via* an increase in the AMP level), which in turn phosphorylates and inactivates Akt, and activation of sirtuins, which in turn activate FOXO.

There are several compounds that prolong life span by affecting the insulin signaling pathway. Wortmannin and LY-294002, well-known inhibitors of PI3K, were found to extend median and maximum life span in *Drosophila* and mice.[281–283] The effect of these drugs was, however, not so pronounced as that of rapamycin.[153,252,283,284] A greater effect, though denied by some studies,[285–287] was found regarding life span extension by resveratrol, which is believed to activate sirtuins.[10–12,288–290] A few types of sirtuins deacetylate FOXO, thus enabling its re-localization from the cytoplasm to the cell nucleus.[277,291] In turn, FOXO controls expression of pro-longevity genes coding for antioxidant enzymes (catalase and mitochondrial superoxide dismutase), DNA repair proteins (Gadd45 and DDB1), inhibitors of cell cycle ($p27^{Kip1}$),[277] heat shock proteins,[292] and so forth. Life extending properties of the medicinal herb *Rhodiola rosea* may also be accounted for by its action on the insulin signaling pathway. Particularly, it was shown that Daf-16, a FOXO homolog in nematode *Caenorhabditis elegans*, re-localizes from the cytoplasm to the nucleus after the consumption of *R. rosea* extract by worms.[223] In addition, the *R. rosea* extract induced expression of heat shock proteins.[223,293] Interestingly, genes that code for both orexigenic (promoting food consumption) and anorexigenic proteins are among FOXO targets.[277,294,295] In humans, these are Agouti-related protein and neuropeptide Y, which foster appetite, and proopiomelanocortin, which suppresses food consumption. Similar proteins were found in other organisms.[296,297] It was found that fruit flies consume less food when it is mixed with *Rhodiola rosea* extract.[227]

Generally, the CR mimetics that supposedly affect the insulin signaling pathway confer relatively slight and sometimes controversial life span extensions, as in the case with resveratrol. Insulin signaling pathway is activated by carbohydrates and regulates carbohydrate and lipid metabolism. Nevertheless, some components of this pathway, like PI3K and PKB, are at the crossroads of many signaling pathways that regulate processes of cell growth and division. The relatively weak effects of insulin signaling inhibition on life span suggest a moderate role of carbohydrates as nutrients for CR. Indeed, many studies reveal that restriction in the protein component of the diet plays a more important role in providing longevity.[3,5,7,8,227] The pathway responsible for protein metabolism is governed by the above-mentioned mTOR kinase.

10.6.2.2 Mechanistic Target-of-Rapamycin (mTOR) Kinase Pathway

Down-regulation of this pathway often provides a more profound life span extension than that of the insulin signaling pathway. Mimetics of CR that affect this pathway include rapamycin and metformin. The suppression of the mTOR pathway responds with pro-longevity processes, such as inhibition of protein synthesis and activation of autophagy. As was already mentioned, rapamycin binds immunophilin FKBP12 and this complex is, in turn, attached to a particular domain of mTOR kinase, suppressing its enzymatic activity.[247–249]

Metformin's influence is mediated by AMPK, which inhibits mTOR *via* phosphorylation of tuberous sclerosis complex proteins. Phosphorylation activates these proteins, whereas they, in turn, inhibit mTOR. There are several reports about the gerosuppressant activity of aspirin, another well-known activator of AMPK.[298–302] Moreover, the life-extending effect of the anticonvulsant valproic acid can be mediated by activation of AMPK.[303,304] Another anticonvulsant, ethosuximide, can prolong life span in *Caenorhabditis elegans*, likely by activation of insulin signaling components such as Daf-16, a homolog of FOXO in this nematode.[305–309] The other anticonvulsant, trimethadione, was shown to enhance the effects of valproic acid on *C. elegans* life span.[303,309] The latter findings suggest a possible link between AMPK, FOXO and mTOR. Indeed, mTOR activity is partly regulated by the insulin signaling pathway *via* Akt/PKB.[20,249,279,280] While AMPK activates TSC, Akt was shown to inhibit it by phosphorylation.[279] Overall, mTOR signaling is intrinsically linked with epilepsy,[310] thus anticonvulsants may represent a novel class of direct or indirect modulators of mTOR activity, simultaneously mimicking CR.

The mechanism of AMPK-mediated mTOR inhibition can also be suggested for resveratrol. It is widely accepted that this natural phenol activates FOXO, a pro-longevity transcription factor suppressed by insulin signaling, *via* deacetylation by sirtuins.[234,311–314] However, it has recently been found that human sirtuin Sirt1 can activate AMPK *via* upstream kinase LKB1 (*l*iver *k*inase *B1*; also known as STK11, *s*erine/*t*hreonine *k*inase *11*).[315] Thus, sirtuin activation may link both mTOR and insulin signaling pathways of life span extension. The strength of the effect and thus the difference between the life-prolonging effects of CR mimetics, which act through the sirtuin-AMPK-mTOR axis, may depend on probable side effects and the degree of sirtuin activation. Extract of *Rhodiola rosea* and its main component, salidroside, which we already mentioned as possible activators of FOXO, were also found to affect mTOR signaling in cultured cells.[316] Many recent studies show that AMPK can be activated by a vast number of natural chemicals, including curcumin, berberine, quercetin, epigallocatechin-3-gallate, ginsenoside, hispidulin, caffeine, and others,[317] for some of which a life-prolonging effect was also found. Further studies are necessary to reveal a possible interplay between insulin and mTOR signaling pathways, along with the AMPK pathway.

Interestingly, p70 S6 kinase, a downstream, positively regulated target of mTOR, was shown to suppress food consumption itself *via* down-regulation of neuropeptide expression.[318,319] Thus, regulation of feeding behavior by mTOR signaling closes the negative feedback regulatory loop between mTOR and food consumption.

10.6.2.3 Nrf2/Keap1 Signaling Pathway

A growing body of evidence suggests a role of the Nrf2/Keap1 signaling pathway in prolonging life span by natural and synthetic geroprotectants, including CR mimetics.[161,239–241] The pathway is conserved among mammals, fruit flies and nematodes. In fruit flies, a homolog of Nrf2 called CncC (*c*ap-***n***-*c*ollar isoform ***C***) was found to be involved in life span extension.[320] A similar role in prolonging life span was declared for SKN-1, a homolog of Nrf2 in *C. elegans*.[321–323] It was also revealed that transcription factor Nrf2 is activated by many CR mimetics, such as resveratrol, curcumin[241] metformin,[239] epigallocatechin gallate,[324,325] and others. Nrf2 is activated by reactive oxygen species (ROS) *via* Keap1 protein. It is also activated by xenobiotic compounds with electrophilic properties, including plant phenols and alkaloids. Many xenobiotics undergo oxidation by P-450 type cytochromes, the enzymes that produce ROS as by-products.[326,327] Then, in phase II of detoxification, oxidized xenobiotics are conjugated with glutathione. After this conjugation, xenobiotics are decomposed by cellular metabolic systems and/or excreted from cells. Activation of Nrf2 by xenobiotics directly or indirectly *via* ROS produced by cytochromes P-450 leads to increased synthesis of low molecular weight antioxidants, glutathione and NADPH,[240,327] as well as antioxidant and related enzymes, such as superoxide dismutase, catalase, thioredoxin reductase, glutathione reductase, glutathione peroxidase, and glutathione-*S*-transferase.[161,239] Thus, Nrf2 governs potential pro-longevity processes, saving cell components from oxidation. However, it turns out that naked mole rats, long-lived rodents, have both up-regulated components of Nrf2 signaling pathway and a higher level of oxidized proteins.[159,161,328] The phenomenon is actually explained by the higher rate of oxidized protein clearance despite a higher basal level of oxidized proteins.[161,329] Nevertheless, we have demonstrated recently that accumulation of oxidized molecules may not be a life-shortening event itself.[158] Instead, the onset of death may depend on the functional roles of oxidized proteins and lipids, rather than on their net amount. Recent studies more often link Nrf2 with CR. Indeed, it has also been shown that a lack of nutrients may lead to activation of antioxidant enzymes, accompanied by a simultaneous increase in the amount of oxidized molecules.[330,331] However, it would be more important to know whether activation of Nrf2 under CR depends on insulin, AMPK and mTOR signaling. There are reasons to suppose that Nrf2 activation under CR is additional to those pro-longevity processes that are governed by AMPK and sirtuins.

Intracellular molecular pathways affected by CR and CR mimetics are summarized in Figures 10.3 and 10.4.

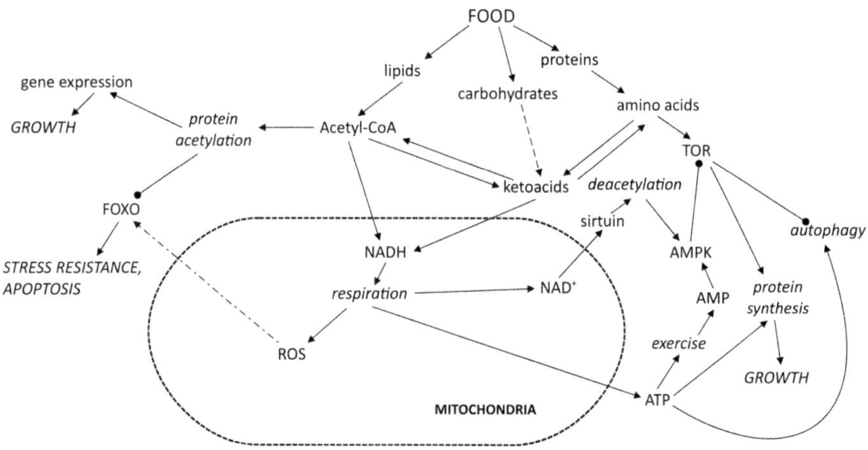

Figure 10.3 The interaction between the main cellular signaling pathways affected by CR, the processes governed by these pathways and the mechanisms of pathway inhibition by food. Food biopolymers are digested to monomers, namely monosaccharides and amino acids. Amino acids can directly activate pro-carcinogenic kinase TOR. Carbohydrates are converted to ketoacids *via* glycolysis and the tricarboxylic acid cycle, giving rise to NADH. In turn, NADH is oxidized by the mitochondrial respiratory chain, some parts of which are the sites of ROS generation. Acetyl coenzyme A (acetyl-CoA) is obtained *via* fatty acid β-oxidation and then oxidized by the respiratory chain. NADH, NAD^+ and acetyl-CoA are nutrient sensors that can themselves allosterically or covalently activate or inhibit life span-prolonging and life span-shortening pathways (*e.g.* acetyl-CoA is a substrate for acetylation, NAD^+ is a co-substrate and allosteric activator of life span-extending deacetylation, *etc.*).

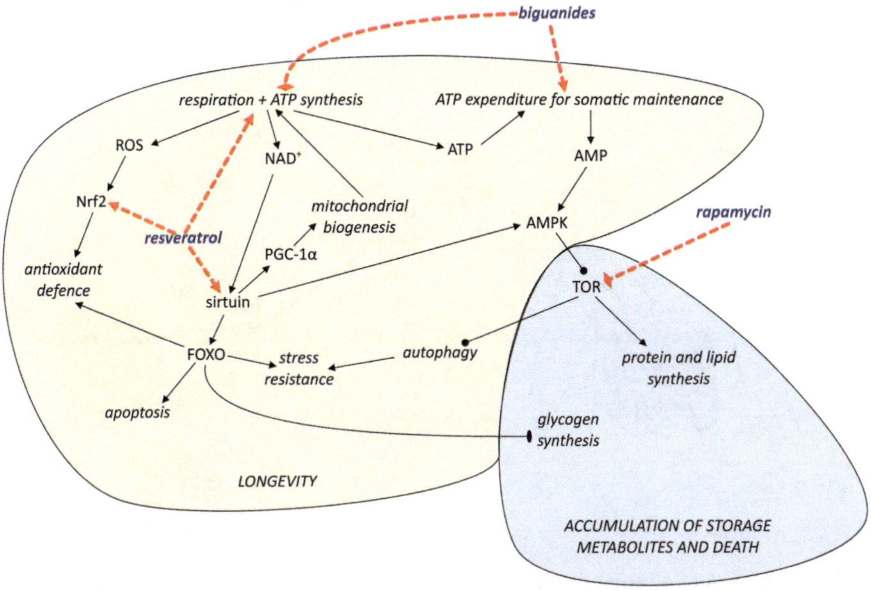

Figure 10.4 Summary of molecular mechanisms of action of the most prominent CR mimetics that act on the cellular level.

10.7 Conclusion

We would classify CR mimetics into "inhibitors of food consumption", "inhibitors of primary catabolism" (acarbose, 2-deoxyglucose) and "true mimetics" (resveratrol and rapamycin), which likely affect energy and nutrient sensing pathways (Figure 10.5).

However, it is not always easy to delineate the main mechanisms of action for a CR mimetic. Metabolic and signaling pathways in cells and organs are tightly tangled. Thus, known CR mimetics likely act *via* different pathways simultaneously. Even those that are supposed to block food consumption may also influence metabolism, affecting signaling molecules, and *vice versa*. It has been revealed that resveratrol and preparations of *Rhodiola rosea* suppress appetite.[227] It is difficult to conclude whether these preparations prolong life span by their anorectic activity and subsequent induction of CR or by their interaction with signaling proteins, which then lead to diminution of appetite. In previous sections we described the role of important signaling pathways, FOXO and mTOR, in the regulation of the foraging behavior of animals. Activation of FOXO or inhibition of mTOR can both boost appetite.[294,295,297,332] However, it was found that over-expression of FOXO in *D. melanogaster* can suppress appetite.[333]

One can notice that the CR mimetics described in this chapter are all secondary metabolites produced by plants, fungi or bacteria. They also affect many targets in cells and likely act *via* multiple pathways. These observations suggest co-evolution of heterotrophic multicellular organisms with plants,

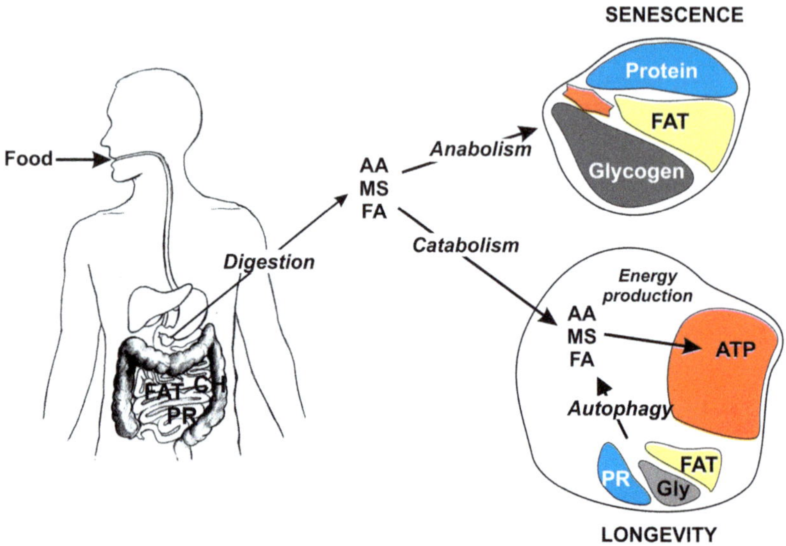

Figure 10.5 A generalized scheme for longevity-modulating effects of CR and targets of all groups of CR mimetics. AA: amino acids, MS: monosaccharides, FA: fatty acids, PR: proteins, Gly: glycogen.

fungi or bacteria, which serve as food sources. On the one hand, most of these secondary metabolites are toxic in high concentrations. Natural selection in this case could be directed against the most voracious herbivorous eaters, which are able to consume much of the toxicants along with carbohydrates. On the other hand, moderate eaters could be beneficial for maintenance of certain plant populations, keeping them away from overgrowth and resource exhaustion. In this case, plant secondary metabolites would prolong the life span of moderate eaters while being toxic for voracious ones. This understanding might help in the search for novel CR mimetics among endemic plant species, especially those growing in stressful environments.

In addition to the search for novel CR mimetics, the prospective goal is to create a mixture of CR mimetics that would act by different cellular pathways.[334] The mTOR pathway seems to be the most effective one among others. Moreover, affecting only S6 kinase is enough for life span extension.[155] In addition, activators of Nrf2, such as plant-derived isothiocyanates, can substantially prolong life span seemingly by other paths than mTOR or S6 kinase inhibitors.[335] Activation of aforementioned PGC-1α, which regulates mitochondrial biogenesis, may provide another independent line for promoting longevity. Recent studies show that fibrates, compounds that activate PGC-1α, indeed prolong life span.[336] So far, the properties of a cocktail containing S6 kinase inhibitors and Nrf2, and PGC-1α activators have not been studied, so it can be a goal for future studies. It remains a hard task to create a non-toxic combination of CR mimetics with a true synergistic effect on life span. However, such an effect was shown for genetic blocking of multiple anti-aging signaling pathways,[337] which gives hope of getting the same result for the chemical inhibition.

Finally, most well-known CR mimetics are natural compounds produced in biosynthetic pathways, which had passed through a long way of complex evolution. Elucidation of mechanisms underlying life span-modulating effects and development of novel life-extending chemicals is likely to be the biggest challenge for contemporary biogerontology and pharmacology. Some solutions for this have already been found and are described in the next chapters of this book.

Acknowledgements

We thank Ouliana Djaman for improvements to the quality of language and Ihor Yurkevych for help in manuscript preparation.

References

1. C. M. McCay, M. F. Crowell and L. A. Maynard, *J. Nutr.*, 1935, **10**, 63.
2. S. J. Simpson and D. Raubenheimer, *J. Gerontol., Ser. A*, 2007, **62**, 707.
3. S. M. Solon-Biet, S. J. Mitchell, R. de Cabo, D. Raubenheimer, D. G. Le Couteur and S. Simpson, *J. Endocrinol.*, 2015, **226**, R17.
4. D. Raubenheimer, K. P. Lee and S. J. Simpson, *Proc. R. Soc. B*, 2005, **272**, 2429.
5. W. Mair, M. D. W. Piper and L. Partridge, *PLoS Biol.*, 2005, **3**, e223.
6. K. Min, T. Flatt, I. Kulaots and M. Tatar, *Exp. Gerontol.*, 2007, **42**, 247.
7. K. P. Lee, S. J. Simpson, F. J. Clissold, R. Brooks, J. W. O. Ballard, P. W. Taylor, N. Soran and D. Raubenheimer, *Proc. Natl. Acad. Sci. U. S. A.*, 2008, **105**, 2498.
8. K. D. Bruce, S. Hoxha, G. B. Carvalho, R. Yamada, H.-D. Wang, P. Karayan, S. He, T. Brummel, P. Kapahi and W. W. Ja, *Exp. Gerontol.*, 2013, **48**, 1129.
9. L. Fontana and L. Partridge, *Cell*, 2015, **161**, 106.
10. K. T. Howitz, K. J. Bitterman, H. Y. Cohen, D. W. Lamming, S. Lavu, J. G. Wood, R. E. Zipkin, P. Chung, A. Kisielewski, L.-L. Zhang, B. Scherer and D. A. Sinclair, *Nature*, 2003, **425**, 191.
11. J. G. Wood, B. Rogina, S. Lavu, K. Howitz, S. L. Helfand, M. Tatar and D. Sinclair, *Nature*, 2004, **430**, 686.
12. K. J. Pearson, J. A. Baur, K. N. Lewis, L. Peshkin, N. L. Price, N. Labinskyy, W. R. Swindell, D. Kamara, R. K. Minor, E. Perez, H. A. Jamieson, Y. Zhang, S. R. Dunn, K. Sharma, N. Pleshko, L. A. Woollett, A. Csiszar, Y. Ikeno, D. Le Couteur, P. J. Elliott, K. G. Becker, P. Navas, D. K. Ingram, N. S. Wolf, Z. Ungvari, D. A. Sinclair and R. de Cabo, *Cell Metab.*, 2008, **8**, 157.
13. J. A. Baur, *Mech. Ageing Dev.*, 2010, **131**, 261.
14. S. Timmers, J. Auwerx and P. Schrauwen, *Aging*, 2012, **4**, 146.
15. L. Guarente, *Genes Dev.*, 2013, **27**, 2072.

16. M. Jafari, B. Khodayari, J. Felgner, I. I. Bussel, M. R. Rose and L. D. Mueller, *Biogerontology*, 2007, **8**, 639.
17. T. Peng, T. R. Golub and D. M. Sabatini, *Mol. Cell. Biol.*, 2002, **22**, 5575.
18. D. K. Ingram and G. S. Roth, *Ageing Res. Rev.*, 2015, **20**, 46.
19. E. J. Masoro, in *Calorie Restriction, Aging and Longevity*, ed. A. V. Everitt, S. I. S. Rattan, D. G. le Couteur and R. de Cabo, Springer, Netherlands, 2010, p. 3.
20. D. Benjamin, M. Colombi, C. Moroni and M. N. Hall, *Nat. Rev. Drug Discovery*, 2011, **10**, 868.
21. S.-J. Lin, M. Kaeberlein, A. A. Andalis, L. A. Sturtz, P.-A. Defossez, V. C. Culotta, G. R. Fink and L. Guarente, *Nature*, 2002, **418**, 344.
22. S.-J. Lin, E. Ford, M. Haigis, G. Liszt and L. Guarente, *Genes Dev.*, 2004, **18**, 12.
23. O. Medvedik, D. W. Lamming, K. D. Kim and D. A. Sinclair, *PLoS Biol.*, 2007, **5**, e261.
24. P. Fabrizio and V. D. Longo, *Aging Cell*, 2003, **2**, 73.
25. M. MacLean, N. Harris and P. W. Piper, *Yeast*, 2001, **18**, 499.
26. Q. Chen, Q. Ding and J. N. Keller, *Biogerontology*, 2005, **6**, 1.
27. V. D. Longo, G. S. Shadel, M. Kaeberlein and B. Kennedy, *Cell Metab.*, 2012, **16**, 18.
28. P. Fabrizio and V. D. Longo, *Methods Mol. Biol.*, 2007, **371**, 89.
29. M. Kaeberlein and R. W. Powers, *Ageing Res. Rev.*, 2007, **6**, 128.
30. S. Michan and D. Sinclair, *Biochem. J.*, 2007, **404**, 1.
31. M. R. Klass, *Mech. Ageing Dev.*, 1977, **6**, 413.
32. K. Houthoofd, B. P. Braeckman, T. E. Johnson and J. R. Vanfleteren, *Exp. Gerontol.*, 2003, **38**, 947.
33. N. A. Bishop and L. Guarente, *Nature*, 2007, **447**, 545.
34. S. H. Panowski, S. Wolff, H. Aguilaniu, J. Durieux and A. Dillin, *Nature*, 2007, **447**, 550.
35. R. Hosono, S. Nishimoto and S. Kuno, *Exp. Gerontol.*, 1989, **24**, 251.
36. K. Houthoofd, B. P. Braeckman, I. Lenaerts, K. Brys, A. De Vreese, S. Van Eygen and J. R. Vanfleteren, *Exp. Gerontol.*, 2002, **37**, 1371.
37. N. J. Szewczyk, I. A. Udranszky, E. Kozak, J. Sunga, S. K. Kim, L. A. Jacobson and C. A. Conley, *J. Exp. Biol.*, 2006, **209**, 4129.
38. E. L. Greer, D. Dowlatshahi, M. R. Banko, J. Villen, K. Hoang, D. Blanchard, S. P. Gygi and A. Brunet, *Curr. Biol.*, 2007, **17**, 1646.
39. T. L. Kaeberlein, E. D. Smith, M. Tsuchiya, K. L. Welton, J. H. Thomas, S. Fields, B. K. Kennedy and M. Kaeberlein, *Aging Cell*, 2006, **5**, 487.
40. G. D. Lee, M. A. Wilson, M. Zhu, C. A. Wolkow, R. de Cabo, D. K. Ingram and S. Zou, *Aging Cell*, 2006, **5**, 515.
41. G. B. Carvalho, P. Kapahi and S. Benzer, *Nat. Methods*, 2005, **2**, 813.
42. T. Magwere, T. Chapman and L. Partridge, *J. Gerontol., Ser. A*, 2004, **59**, 3.
43. W. Mair, P. Goymer, S. D. Pletcher and L. Partridge, *Science*, 2003, **301**, 1731.
44. B. Rogina and S. L. Helfand, *Proc. Natl. Acad. Sci. U. S. A.*, 2004, **101**, 15998.

45. B. Rogina, S. L. Helfand and S. Frankel, *Science*, 2002, **298**, 1745.
46. D. J. Clancy, D. Gems, E. Hafen, S. J. Leevers and L. Partridge, *Science*, 2002, **296**, 319.
47. S. Libert, J. Zwiener, X. Chu, W. Vanvoorhies, G. Roman and S. D. Pletcher, *Science*, 2007, **315**, 1133.
48. O. V. Lushchak, D. V. Gospodaryov, B. M. Rovenko, A. D. Glovyak, I. S. Yurkevych, V. P. Klyuba, M. V. Shcherbij and V. I. Lushchak, *J. Gerontol., Ser. A*, 2012, **67**, 118.
49. K. P. Lee, *J. Insect Physiol.*, 2015, **75**, 12.
50. B. P. Yu, E. J. Masoro, I. Murata, H. A. Bertrand and F. T. Lynd, *J. Gerontol., Ser. A*, 1982, **37**, 130.
51. A. Turturro, W. W. Witt, S. Lewis, B. S. Hass, R. D. Lipman and R. W. Hart, *J. Gerontol., Ser. A*, 1999, **54**, B492.
52. W. R. Swindell, *Ageing Res. Rev.*, 2012, **11**, 254.
53. S. M. Solon-Biet, A. C. McMahon, J. W. O. Ballard, K. Ruohonen, L. E. Wu, V. C. Cogger, A. Warren, X. Huang, N. Pichaud, R. G. Melvin, R. Gokarn, M. Khalil, N. Turner, G. J. Cooney, D. A. Sinclair, D. Raubenheimer, D. G. Le Couteur and S. J. Simpson, *Cell Metab.*, 2014, **19**, 418.
54. D. G. Le Couteur, S. Solon-Biet, V. C. Cogger, S. J. Mitchell, A. Senior, R. de Cabo, D. Raubenheimer and S. J. Simpson, *Cell. Mol. Life Sci.*, 2016, **73**, 1237.
55. R. D. Kealy, D. F. Lawler, J. M. Ballam, S. L. Mantz, D. N. Biery, E. H. Greeley, G. Lust, M. Segre, G. K. Smith and H. D. Stowe, *J. Am. Vet. Med. Assoc.*, 2002, **220**, 1315.
56. R. J. Colman and J. W. Kemnitz, in *Methods in Aging Research*, ed. B. P. Yu, CRC Press, Boca Raton, FL, 1998.
57. J. A. Mattison, M. A. Ottinger, D. Powell, D. L. Longo and D. K. Ingram, *J. Med. Primatol.*, 2007, **36**, 391.
58. R. J. Colman, R. M. Anderson, S. C. Johnson, E. K. Kastman, K. J. Kosmatka, T. M. Beasley, D. B. Allison, C. Cruzen, H. A. Simmons, J. W. Kemnitz and R. Weindruch, *Science*, 2009, **325**, 201.
59. R. J. Colman, J. J. Ramsey, E. B. Roecker, T. Havighurst, J. C. Hudson and J. W. Kemnitz, *J. Gerontol., Ser. A*, 1999, **54**, B283.
60. M. A. Lane, D. K. Ingram, R. G. Cutler, J. J. Knapka, D. E. Barnard and G. S. Roth, *Ann. N. Y. Acad. Sci.*, 1992, **673**, 36.
61. B. C. Hansen and N. L. Bodkin, *Diabetes*, 1993, **42**, 1809.
62. M. A. Lane, D. K. Ingram and G. S. Roth, *Toxicol. Sci.*, 1999, **52**, 41.
63. W. T. Cefalu, Z. Q. Wang, A. D. Bell-Farrow, J. Collins, T. Morgan and J. D. Wagner, *J. Gerontol., Ser. A*, 2004, **59**, 1007.
64. I. J. Edwards, L. L. Rudel, J. G. Terry, J. W. Kemnitz, R. Weindruch and W. T. Cefalu, *J. Gerontol., Ser. A*, 1998, **53**, B443.
65. R. B. Verdery, D. K. Ingram, G. S. Roth and M. A. Lane, *Am. J. Physiol.: Endocrinol. Metab.*, 1997, **273**, E714.
66. T. A. Gresl, R. J. Colman, T. C. Havighurst, D. B. Allison, D. A. Schoeller and J. W. Kemnitz, *J. Gerontol., Ser. A*, 2003, **58**, 598.

67. T. A. Gresl, R. J. Colman, E. B. Roecker, T. C. Havighurst, Z. Huang, D. B. Allison, R. N. Bergman and J. W. Kemnitz, *Am. J. Physiol.: Endocrinol. Metab.*, 2001, **281**, E757.
68. J. W. Kemnitz, E. B. Roecker, R. Weindruch, D. F. Elson, S. T. Baum and R. N. Bergman, *Am. J. Physiol.*, 1994, **266**, E540.
69. M. A. Lane, S. S. Ball, D. K. Ingram, R. G. Cutler, J. Engel, V. Read and G. S. Roth, *Am. J. Physiol.: Endocrinol. Metab.*, 1995, **268**, E941.
70. I. M. Lee, S. N. Blair, D. B. Allison, A. R. Folsom, T. B. Harris, J. E. Manson and R. R. Wing, *J. Gerontol., Ser. A*, 2001, **56**(1), 7.
71. R. Weindruch and R. S. Sohal, *N. Engl. J. Med.*, 1997, **337**, 986.
72. R. L. Walford, S. B. Harris and M. W. Gunion, *Proc. Natl. Acad. Sci. U. S. A.*, 1992, **89**, 11533.
73. R. L. Walford, D. Mock, T. MacCallum and J. L. Laseter, *Toxicol. Sci.*, 1999, **52**, 61.
74. R. L. Walford, D. Mock, R. Verdery and T. MacCallum, *J. Gerontol., Ser. A*, 2002, **57**, B211.
75. L. Fontana, T. E. Meyer, S. Klein and J. O. Holloszy, *Proc. Natl. Acad. Sci. U. S. A.*, 2004, **101**, 6659.
76. S. B. Racette, E. P. Weiss, D. T. Villareal, H. Arif, K. Steger-May, K. B. Schechtman, L. Fontana, S. Klein, J. O. Holloszy and Washington University School of Medicine CALERIE Group, *J. Gerontol., Ser. A*, 2006, **61**, 943.
77. L. M. Redman, L. K. Heilbronn, C. K. Martin, A. Alfonso, S. R. Smith, E. Ravussin and Pennington CALERIE Team, *J. Clin. Endocrinol. Metab.*, 2007, **92**, 865.
78. L. Fontana, D. T. Villareal, E. P. Weiss, S. B. Racette, K. Steger-May, S. Klein, J. O. Holloszy and Washington University School of Medicine CALERIE Group, *Am. J. Physiol.: Endocrinol. Metab.*, 2007, **293**, E197.
79. L. K. Heilbronn, L. de Jonge, M. I. Frisard, J. P. DeLany, D. E. Larson-Meyer, J. Rood, T. Nguyen, C. K. Martin, J. Volaufova, M. M. Most, F. L. Greenway, S. R. Smith, W. A. Deutsch, D. A. Williamson, E. Ravussin and Pennington CALERIE Team, *JAMA*, 2006, **295**, 1539.
80. D. E. Larson-Meyer, L. K. Heilbronn, L. M. Redman, B. R. Newcomer, M. I. Frisard, S. Anton, S. R. Smith, A. Alfonso and E. Ravussin, *Diabetes Care*, 2006, **29**, 1337.
81. M. Lefevre, L. M. Redman, L. K. Heilbronn, J. V. Smith, C. K. Martin, J. C. Rood, F. L. Greenway, D. A. Williamson, S. R. Smith, E. Ravussin and Pennington CALERIE team, *Atherosclerosis*, 2009, **203**, 206.
82. C. Rippe, L. Lesniewski, M. Connell, T. LaRocca, A. Donato and D. Seals, *Aging Cell*, 2010, **9**, 304.
83. I. Mattagajasingh, C.-S. Kim, A. Naqvi, T. Yamamori, T. A. Hoffman, S.-B. Jung, J. DeRicco, K. Kasuno and K. Irani, *Proc. Natl. Acad. Sci. U. S. A.*, 2007, **104**, 14855.
84. M. Zanetti, R. Barazzoni, M. Vadori, M. Stebel, G. Biolo and G. Guarnieri, *Atherosclerosis*, 2004, **175**, 253.
85. A. M. Sharifi, S. Mohseni, S. Nekoparvar, B. Larijani, H. Fakhrzadeh and S. Oryan, *Acta Physiol. Hung.*, 2008, **95**, 55.

86. V. W. Dolinsky, J. S. Morton, T. Oka, I. Robillard-Frayne, M. Bagdan, G. D. Lopaschuk, C. Des Rosiers, K. Walsh, S. T. Davidge and J. R. B. Dyck, *Hypertension*, 2010, **56**, 412.
87. Z. Guo, F. Mitchell-Raymundo, H. Yang, Y. Ikeno, J. Nelson, V. Diaz, A. Richardson and R. Reddick, *Mech. Ageing Dev.*, 2002, **123**, 1121.
88. K. Shinmura, K. Tamaki, K. Saito, Y. Nakano, T. Tobe and R. Bolli, *Circulation*, 2007, **116**, 2809.
89. C. C. Spaulding, R. L. Walford and R. B. Effros, *Mech. Ageing Dev.*, 1997, **93**, 87.
90. J. Matsuzaki, M. Kuwamura, R. Yamaji, H. Inui and Y. Nakano, *J. Nutr.*, 2001, **131**, 2139.
91. M. M. Y. Sung, C.-L. M. Soltys, G. Masson, J. J. Boisvenue and J. R. B. Dyck, *J. Mol. Med.*, 2011, **89**, 291.
92. K. Shinmura, K. Tamaki and R. Bolli, *Am. J. Physiol.: Heart Circ. Physiol.*, 2008, **295**, H2348.
93. R. Gredilla, A. Sanz, M. Lopez-Torres and G. Barja, *FASEB J.*, 2001, **15**, 1589.
94. E. Nisoli, C. Tonello, A. Cardile, V. Cozzi, R. Bracale, L. Tedesco, S. Falcone, A. Valerio, O. Cantoni, E. Clementi, S. Moncada and M. O. Carruba, *Science*, 2005, **310**, 314.
95. K. Shinmura, K. Tamaki, M. Sano, M. Murata, H. Yamakawa, H. Ishida and K. Fukuda, *J. Mol. Cell. Cardiol.*, 2011, **50**, 117.
96. K. Shinmura, K. Tamaki, M. Sano, N. Nakashima-Kamimura, A. M. Wolf, T. Amo, S. Ohta, Y. Katsumata, K. Fukuda, K. Ishiwata, M. Suematsu and T. Adachi, *Circ. Res.*, 2011, **109**, 396.
97. J. Guo, V. Bakshi and A.-L. Lin, *Front. Aging Neurosci.*, 2015, **7**, 213.
98. P. A. Dacks, C. L. Moreno, E. S. Kim, B. K. Marcellino and C. V. Mobbs, *Front. Neuroendocrinol.*, 2013, **34**, 95.
99. M. Maalouf, J. M. Rho and M. P. Mattson, *Brain Res. Rev.*, 2009, **59**, 293.
100. V. Calabrese, C. Cornelius, A. T. Dinkova-Kostova, E. J. Calabrese and M. P. Mattson, *Antioxid. Redox Signaling*, 2010, **13**, 1763.
101. J. Lee, K. B. Seroogy and M. P. Mattson, *J. Neurochem.*, 2002, **80**, 539.
102. P. K. Dash, S. A. Mach and A. N. Moore, *J. Neurosci. Res.*, 2001, **63**, 313.
103. J. S. Snyder, N. S. Hong, R. J. McDonald and J. M. Wojtowicz, *Neuroscience*, 2005, **130**, 843.
104. O. Kann and R. Kovács, *Am. J. Physiol.: Cell Physiol.*, 2007, **292**, C641.
105. J. J. Tong, *Biol. Bull.*, 2007, **212**, 169.
106. M. Levy, G. C. Faas, P. Saggau, W. J. Craigen and J. D. Sweatt, *J. Biol. Chem.*, 2003, **278**, 17727.
107. J. L. Dinerman, T. M. Dawson, M. J. Schell, A. Snowman and S. H. Snyder, *Proc. Natl. Acad. Sci. U. S. A.*, 1994, **91**, 4214.
108. S. Michán, Y. Li, M. M. H. Chou, E. Parrella, H. Ge, J. M. Long, J. S. Allard, K. Lewis, M. Miller, W. Xu, R. F. Mervis, J. Chen, K. I. Guerin, L. E. H. Smith, M. W. McBurney, D. A. Sinclair, M. Baudry, R. de Cabo and V. D. Longo, *J. Neurosci.*, 2010, **30**, 9695.

109. S. Fusco, C. Ripoli, M. V. Podda, S. C. Ranieri, L. Leone, G. Toietta, M. W. McBurney, G. Schütz, A. Riccio, C. Grassi, T. Galeotti and G. Pani, *Proc. Natl. Acad. Sci. U. S. A.*, 2012, **109**, 621.
110. M. P. Mattson, W. Duan and Z. Guo, *J. Neurochem.*, 2003, **84**, 417.
111. M. M. Adams, L. Shi, M. C. Linville, M. E. Forbes, A. B. Long, C. Bennett, I. G. Newton, C. S. Carter, W. E. Sonntag, D. R. Riddle and J. K. Brunso-Bechtold, *Exp. Neurol.*, 2008, **211**, 141.
112. A. Martin-Montalvo and R. de Cabo, *Antioxid. Redox Signaling*, 2013, **19**, 310.
113. R. Stefanatos, A. Sriram, E. Kiviranta, A. Mohan, V. Ayala, H. T. Jacobs, R. Pamplona and A. Sanz, *Cell Cycle*, 2012, **11**, 3876.
114. T. Nakamura, D.-H. Cho and S. A. Lipton, *Exp. Neurol.*, 2012, **238**, 12.
115. G. Lopez-Lluch, N. Hunt, B. Jones, M. Zhu, H. Jamieson, S. Hilmer, M. V. Cascajo, J. Allard, D. K. Ingram, P. Navas and R. De Cabo, *Proc. Natl. Acad. Sci. U. S. A.*, 2006, **103**, 1768.
116. F. M. Cerqueira, F. M. Cunha, F. R. M. Laurindo and A. J. Kowaltowski, *Free Radical Biol. Med.*, 2012, **52**, 1236.
117. K. Higashida, S. H. Kim, S. R. Jung, M. Asaka, J. O. Holloszy and D.-H. Han, *PLoS Biol.*, 2013, **11**, e1001603.
118. D. E. Cohen, A. M. Supinski, M. S. Bonkowski, G. Donmez and L. P. Guarente, *Genes Dev.*, 2009, **23**, 2812.
119. M. Jiang, J. Wang, J. Fu, L. Du, H. Jeong, T. West, L. Xiang, Q. Peng, Z. Hou, H. Cai, T. Seredenina, N. Arbez, S. Zhu, K. Sommers, J. Qian, J. Zhang, S. Mori, X. W. Yang, K. L. K. Tamashiro, S. Aja, T. H. Moran, R. Luthi-Carter, B. Martin, S. Maudsley, M. P. Mattson, R. H. Cichewicz, C. A. Ross, D. M. Holtzman, D. Krainc and W. Duan. *Nat. Med.*, 2011, **18**, 153.
120. H. Jeong, D. E. Cohen, L. Cui, A. Supinski, J. N. Savas, J. R. Mazzulli, J. R. Yates, L. Bordone, L. Guarente and D. Krainc, *Nat. Med.*, 2012, **18**, 159.
121. H. S. Chahal and W. M. Drake, *J. Pathol.*, 2007, **211**, 173.
122. N. Møller, J. Gjedsted, L. Gormsen, J. Fuglsang and C. Djurhuus, *Growth Horm. IGF Res.*, 2003, **13**(suppl. A), S18.
123. G. S. Roth, M. A. Lane, D. K. Ingram, J. A. Mattison, D. Elahi, J. D. Tobin, D. Muller and E. J. Metter, *Science*, 2002, **297**, 811.
124. E. P. Weiss, S. B. Racette, D. T. Villareal, L. Fontana, K. Steger-May, K. B. Schechtman, S. Klein, J. O. Holloszy and Washington University School of Medicine CALERIE Group, *Am. J. Clin. Nutr.*, 2006, **84**, 1033.
125. S. Broughton and L. Partridge, *Biochem. J.*, 2009, **418**, 1.
126. M. Holzenberger, J. Dupont, B. Ducos, P. Leneuve, A. Géloën, P. C. Even, P. Cervera and Y. Le Bouc, *Nature*, 2003, **421**, 182.
127. P. D'Costa, J. E. Lenham, R. L. Ingram and W. E. Sonntag, *Mech. Ageing Dev.*, 1993, **71**, 59.
128. M. S. Bonkowski, J. S. Rocha, M. M. Masternak, K. A. Al Regaiey and A. Bartke, *Proc. Natl. Acad. Sci. U. S. A.*, 2006, **103**, 7901.
129. *Handbook of the biology of aging*, ed. E. J. Masoro and S. N. Austad, Elsevier Academic Press, Amsterdam, Boston, 6th edn, 2006.

130. A. J. Hulbert, R. Pamplona, R. Buffenstein and W. A. Buttemer, *Physiol. Rev.*, 2007, **87**, 1175.
131. R. S. Petralia, M. P. Mattson and P. J. Yao, *Ageing Res. Rev.*, 2014, **16**, 66.
132. T. Bilinski, A. Bylak and R. Zadrag-Tecza, *Aging*, 2016, **8**, 589.
133. C. Cantó and J. Auwerx, *Curr. Opin. Lipidol.*, 2009, **20**, 98.
134. B. P. Hubbard, A. P. Gomes, H. Dai, J. Li, A. W. Case, T. Considine, T. V. Riera, J. E. Lee, S. Y. E, D. W. Lamming, B. L. Pentelute, E. R. Schuman, L. A. Stevens, A. J. Y. Ling, S. M. Armour, S. Michan, H. Zhao, Y. Jiang, S. M. Sweitzer, C. A. Blum, J. S. Disch, P. Y. Ng, K. T. Howitz, A. P. Rolo, Y. Hamuro, J. Moss, R. B. Perni, J. L. Ellis, G. P. Vlasuk and D. A. Sinclair, *Science*, 2013, **339**, 1216.
135. A. E. Wagner, S. Piegholdt, D. Rabe, N. Baenas, A. Schloesser, M. Eggersdorfer, A. Stocker and G. Rimbach, *Oncotarget*, 2015, **6**, 30568.
136. M. V. Blagosklonny and M. N. Hall, *Aging*, 2009, **1**, 357.
137. G. M. Fimia, G. Kroemer and M. Piacentini, *Cell Death Differ.*, 2013, **20**, 1.
138. G. Filomeni, D. De Zio and F. Cecconi, *Cell Death Differ.*, 2015, **22**, 377.
139. C. Zhang and A. M. Cuervo, *Nat. Med.*, 2008, **14**, 959.
140. R. A. Gottlieb and R. M. Mentzer Jr, *Annu. Rev. Physiol.*, 2010, **72**, 45.
141. B. Ravikumar, S. Sarkar, J. E. Davies, M. Futter, M. Garcia-Arencibia, Z. W. Green-Thompson, M. Jimenez-Sanchez, V. I. Korolchuk, M. Lichtenberg, S. Luo, D. C. O. Massey, F. M. Menzies, K. Moreau, U. Narayanan, M. Renna, F. H. Siddiqi, B. R. Underwood, A. R. Winslow and D. C. Rubinsztein, *Physiol. Rev.*, 2010, **90**, 1383.
142. M. Jiang, K. Liu, J. Luo and Z. Dong, *Am. J. Pathol.*, 2010, **176**, 1181.
143. M. Matecic, D. L. Smith Jr, X. Pan, N. Maqani, S. Bekiranov, J. D. Boeke and J. S. Smith, *PLoS Genet.*, 2010, **6**, e1000921.
144. M. L. Tóth, T. Sigmond, E. Borsos, J. Barna, P. Erdélyi, K. Takács-Vellai, L. Orosz, A. L. Kovács, G. Csikós, M. Sass and T. Vellai, *Autophagy*, 2008, **4**, 330.
145. K.-S. Lee, B.-S. Lee, S. Semnani, A. Avanesian, C.-Y. Um, H.-J. Jeon, K.-M. Seong, K. Yu, K.-J. Min and M. Jafari, *Rejuvenation Res.*, 2010, **13**, 561.
146. A. Simonsen, R. C. Cumming, A. Brech, P. Isakson, D. R. Schubert and K. D. Finley, *Autophagy*, 2008, **4**, 176.
147. B. Levine and G. Kroemer, *Cell*, 2008, **132**, 27.
148. D. Egan, J. Kim, R. J. Shaw and K. L. Guan, *Autophagy*, 2011, **7**, 643.
149. J. Kim, M. Kundu, B. Viollet and K. L. Guan, *Nat. Cell Biol.*, 2011, **13**, 132.
150. C. J. Kenyon, *Nature*, 2010, **464**, 504.
151. L. Fontana, L. Partridge and V. D. Longo, *Science*, 2010, **328**, 321.
152. A. L. Alvers, M. S. Wood, D. Hu, A. C. Kaywell, W. A. Dunn, Jr. and J. P. Aris, *Autophagy*, 2009, **5**, 847.
153. I. Bjedov, J. M. Toivonen, F. Kerr, C. Slack, J. Jacobson, A. Foley and L. Partridge, *Cell Metab.*, 2010, **11**, 35.
154. S. Hands, C. Proud and A. Wyttenbach, *Aging*, 2009, **1**, 586.

155. C. Selman, J. M. A. Tullet, D. Wieser, E. Irvine, S. J. Lingard, A. I. Choudhury, M. Claret, H. Al-Qassab, D. Carmignac, F. Ramadani, A. Woods, I. C. A. Robinson, E. Schuster, R. L. Batterham, S. C. Kozma, G. Thomas, D. Carling, K. Okkenhaug, J. M. Thornton, L. Partridge, D. Gems and D. J. Withers, *Science*, 2009, **326**, 140.
156. C. Gorrini, I. S. Harris and T. W. Mak, *Nat. Rev. Drug Discovery*, 2013, **12**, 931.
157. R. L. Levine, *Free Radical Biol. Med.*, 2002, **32**, 790.
158. O. V. Lushchak, D. V. Gospodaryov, I. S. Yurkevych and K. B. Storey, *Arch. Insect Biochem. Physiol.*, 2016, **91**, 52.
159. B. Andziak, T. P. O'Connor, W. Qi, E. M. DeWaal, A. Pierce, A. R. Chaudhuri, H. Van Remmen and R. Buffenstein, *Aging Cell*, 2006, **5**, 463.
160. K. N. Lewis, B. Andziak, T. Yang and R. Buffenstein, *Antioxid. Redox Signaling*, 2013, **19**, 1388.
161. D. R. Bruns, J. C. Drake, L. M. Biela, F. F. Peelor, B. F. Miller and K. L. Hamilton, *Oxid. Med. Cell. Longevity*, 2015, **2015**, 732596.
162. K. J. Davies, A. T. Quintanilha, G. A. Brooks and L. Packer, *Biochem. Biophys. Res. Commun.*, 1982, **107**, 1198.
163. S. Chevion, D. S. Moran, Y. Heled, Y. Shani, G. Regev, B. Abbou, E. Berenshtein, E. R. Stadtman and Y. Epstein, *Proc. Natl. Acad. Sci. U. S. A.*, 2003, **100**, 5119.
164. S. K. Powers and M. J. Jackson, *Physiol. Rev.*, 2008, **88**, 1243.
165. C. E. Schaar, D. J. Dues, K. K. Spielbauer, E. Machiela, J. F. Cooper, M. Senchuk, S. Hekimi and J. M. Van Raamsdonk, *PLoS Genet.*, 2015, **11**, e1004972.
166. F. Scialò, A. Sriram, D. Fernández-Ayala, N. Gubina, M. Lõhmus, G. Nelson, A. Logan, H. M. Cooper, P. Navas, J. A. Enríquez, M. P. Murphy and A. Sanz, *Cell Metab.*, 2016, **23**, 725.
167. M.-C. Gomez-Cabrera, E. Domenech, M. Romagnoli, A. Arduini, C. Borras, F. V. Pallardo, J. Sastre and J. Viña, *Am. J. Clin. Nutr.*, 2008, **87**, 142.
168. M. Ristow, K. Zarse, A. Oberbach, N. Klöting, M. Birringer, M. Kiehntopf, M. Stumvoll, C. R. Kahn and M. Blüher, *Proc. Natl. Acad. Sci. U. S. A.*, 2009, **106**, 8665.
169. J. Yun and T. Finkel, *Cell Metab.*, 2014, **19**, 757.
170. K. A. Rodriguez, E. Wywial, V. I. Perez, A. J. Lambert, Y. H. Edrey, K. N. Lewis, K. Grimes, M. L. Lindsey, M. D. Brand and R. Buffenstein, *Curr. Pharm. Des.*, 2011, **17**, 2290.
171. S.-C. Seehuus, K. Norberg, U. Gimsa, T. Krekling and G. V. Amdam, *Proc. Natl. Acad. Sci. U. S. A.*, 2006, **103**, 962.
172. H. Havukainen, D. Munch, A. Baumann, S. Zhong, O. Halskau, M. Krogsgaard and G. V. Amdam, *J. Biol. Chem.*, 2013, **288**, 28369.
173. S. Dukan, A. Farewell, M. Ballesteros, F. Taddei, M. Radman and T. Nyström, *Proc. Natl. Acad. Sci. U. S. A.*, 2000, **97**, 5746.
174. T. Jung and T. Grune, *IUBMB Life*, 2008, **60**, 743.
175. S. Luo and R. L. Levine, *FASEB J.*, 2008, **23**, 464.
176. C. M. Grant, *J. Biol.*, 2008, **7**, 1.

177. G. C. Brown and V. Borutaite, *Mitochondrion*, 2012, **12**, 1.
178. A. A. Starkov, A. Y. Andreyev, S. F. Zhang, N. N. Starkova, M. Korneeva, M. Syromyatnikov and V. N. Popov, *J. Bioenerg. Biomembr.*, 2014, **46**, 471.
179. J. R. Treberg, D. Munro, S. Banh, P. Zacharias and E. Sotiri, *Redox Biol.*, 2015, **5**, 216.
180. L. Avery, *Genetics*, 1993, **133**, 897.
181. B. Lakowski and S. Hekimi, *Proc. Natl. Acad. Sci. U. S. A.*, 1998, **95**, 13091.
182. M. F. McCarty, *Med. Hypotheses*, 2002, **58**, 487.
183. S. D. Poppitt, J. D. E. van Drunen, A.-T. McGill, T. B. Mulvey and F. E. Leahy, *Asia Pac. J. Clin. Nutr.*, 2007, **16**, 16.
184. T. Oyama, A. Saiki, K. Endoh, N. Ban, D. Nagayama, M. Ohhira, N. Koide, Y. Miyashita and K. Shirai, *J. Atheroscler. Thromb.*, 2008, **15**, 154.
185. M. Hanefeld, J. L. Chiasson, C. Koehler, E. Henkel, F. Schaper and T. Temelkova-Kurktschiev, *Stroke*, 2004, **35**, 1073.
186. S. Osaki, T. Kimura, T. Sugimoto, S. Hizukuri and N. Iritani, *J. Nutr.*, 2001, **131**, 796.
187. J. Udani, M. Hardy and D. C. Madsen, *Altern. Med. Rev.*, 2004, **9**, 63.
188. J. Udani and B. B. Singh, *Altern. Ther. Health Med.*, 2007, **13**, 32.
189. M. Lamela, J. Anca, R. Villar, J. Otero and J. M. Calleja, *J. Ethnopharmacol.*, 1989, **27**, 35.
190. J. L. Ngondi, B. C. Etoundi, C. B. Nyangono, C. M. Mbofung and J. E. Oben, *Lipids Health Dis.*, 2009, **8**, 7.
191. J. E. Oben, J. L. Ngondi and K. Blum, *Lipids Health Dis.*, 2008, **7**, 44.
192. S. Henness and C. M. Perry, *Drugs*, 2006, **66**, 1625.
193. C. Juhel, M. Armand, Y. Pafumi, C. Rosier, J. Vandermander and D. Lairon, *J. Nutr. Biochem.*, 2000, **11**, 45.
194. M. Kobayashi, M. Ichitani, Y. Suzuki, T. Unno, T. Sugawara, T. Yamahira, M. Kato, T. Takihara, Y. Sagesaka, T. Kakuda and I. Ikeda, *J. Agric. Food Chem.*, 2009, **57**, 7131.
195. K. Hosoda, M.-F. Wang, M.-L. Liao, C.-K. Chuang, M. Iha, B. Clevidence and S. Yamamoto, *Diabetes Care*, 2003, **26**, 1714.
196. Y. Fukino, A. Ikeda, K. Maruyama, N. Aoki, T. Okubo and H. Iso, *Eur. J. Clin. Nutr.*, 2008, **62**, 953.
197. C.-L. Lin, H.-C. Huang and J.-K. Lin, *J. Lipid Res.*, 2007, **48**, 2334.
198. T. Nagao, T. Hase and I. Tokimitsu, *Obesity*, 2007, **15**, 1473.
199. L. M. Yung, F. P. Leung, W. T. Wong, X. Y. Tian, L. H. Yung, Z. Y. Chen, X. Q. Yao and Y. Huang, *Inflammopharmacology*, 2008, **16**, 230.
200. C. Martins, L. M. Morgan and M. D. Robertson, *Physiol. Behav.*, 2009, **96**, 703.
201. W. J. Pasman, J. Heimerikx, C. M. Rubingh, R. van den Berg, M. O'Shea, L. Gambelli, H. F. J. Hendriks, A. W. C. Einerhand, C. Scott, H. G. Keizer and L. I. Mennen, *Lipids Health Dis.*, 2008, **7**, 10.
202. T. J. Little, M. Horowitz and C. Feinle-Bisset, *Obes. Rev.*, 2005, **6**, 297.
203. S. R. Spindler, *Aging Cell*, 2006, **5**, 39.
204. V. N. Anisimov, L. M. Berstein, I. G. Popovich, M. A. Zabezhinski, P. A. Egormin, T. S. Piskunova, A. V. Semenchenko, M. L. Tyndyk, M. N. Yurova, I. G. Kovalenko and T. E. Poroshina, *Aging*, 2011, **3**, 148.

205. A. Martin-Montalvo, E. M. Mercken, S. J. Mitchell, H. H. Palacios, P. L. Mote, M. Scheibye-Knudsen, A. P. Gomes, T. M. Ward, R. K. Minor, M.-J. Blouin, M. Schwab, M. Pollak, Y. Zhang, Y. Yu, K. G. Becker, V. A. Bohr, D. K. Ingram, D. A. Sinclair, N. S. Wolf, S. R. Spindler, M. Bernier and R. de Cabo, *Nat. Commun.*, 2013, **4**, 2192.
206. R. A. Miller, Q. Chu, J. Xie, M. Foretz, B. Viollet and M. J. Birnbaum, *Nature*, 2013, **494**, 256.
207. A. Janzer, N. J. German, K. N. Gonzalez-Herrera, J. M. Asara, M. C. Haigis and K. Struhl, *Proc. Natl. Acad. Sci. U. S. A.*, 2014, **111**, 10574.
208. L. He and F. E. Wondisford, *Cell Metab.*, 2015, **21**, 159.
209. G. Zhou, R. Myers, Y. Li, Y. Chen, X. Shen, J. Fenyk-Melody, M. Wu, J. Ventre, T. Doebber, N. Fujii, N. Musi, M. F. Hirshman, L. J. Goodyear and D. E. Moller, *J. Clin. Invest.*, 2001, **108**, 1167.
210. S. A. Hawley, A. E. Gadalla, G. S. Olsen and D. G. Hardie, *Diabetes*, 2002, **51**, 2420.
211. M. H. Mooney, S. Fogarty, C. Stevenson, A. M. Gallagher, P. Palit, S. A. Hawley, D. G. Hardie, G. D. Coxon, R. D. Waigh, R. J. Tate, A. L. Harvey and B. L. Furman, *Br. J. Pharmacol.*, 2008, **153**, 1669.
212. X. Stephenne, M. Foretz, N. Taleux, G. C. van der Zon, E. Sokal, L. Hue, B. Viollet and B. Guigas, *Diabetologia*, 2011, **54**, 3101.
213. A. K. Madiraju, D. M. Erion, Y. Rahimi, X.-M. Zhang, D. T. Braddock, R. A. Albright, B. J. Prigaro, J. L. Wood, S. Bhanot, M. J. MacDonald, M. J. Jurczak, J.-P. Camporez, H.-Y. Lee, G. W. Cline, V. T. Samuel, R. G. Kibbey and G. I. Shulman, *Nature*, 2014, **510**, 542.
214. M. Foretz, B. Guigas, L. Bertrand, M. Pollak and B. Viollet, *Cell Metab.*, 2014, **20**, 953.
215. M.-Y. El-Mir, V. Nogueira, E. Fontaine, N. Averet, M. Rigoulet and X. Leverve, *J. Biol. Chem.*, 2000, **275**, 223.
216. B. Viollet, B. Guigas, N. S. Garcia, J. Leclerc, M. Foretz and F. Andreelli, *Clin. Sci.*, 2012, **122**, 253.
217. J. A. Baur and M. J. Birnbaum, *Cell Metab.*, 2014, **20**, 197.
218. R. Pryor and F. Cabreiro, *Biochem. J.*, 2015, **471**, 307.
219. L. Logie, J. Harthill, K. Patel, S. Bacon, D. L. Hamilton, K. Macrae, G. McDougall, H.-H. Wang, L. Xue, H. Jiang, K. Sakamoto, A. R. Prescott and G. Rena, *Diabetes*, 2012, **61**, 1423.
220. Y. Zhang, Y. Wang, C. Bao, Y. Xu, H. Shen, J. Chen, J. Yan and Y. Chen, *Mol. Cell. Biochem.*, 2012, **368**, 69.
221. S. Meng, J. Cao, Q. He, L. Xiong, E. Chang, S. Radovick, F. E. Wondisford and L. He, *J. Biol. Chem.*, 2015, **290**, 3793.
222. R. A. Gatenby and R. J. Gillies, *Nat. Rev. Cancer*, 2004, **4**, 891.
223. F. A. C. Wiegant, S. Surinova, E. Ytsma, M. Langelaar-Makkinje, G. Wikman and J. A. Post, *Biogerontology*, 2009, **10**, 27.
224. V. Calabrese, C. Cornelius, A. T. Dinkova-Kostova, I. Iavicoli, R. Di Paola, A. Koverech, S. Cuzzocrea, E. Rizzarelli and E. J. Calabrese, *Biochim. Biophys. Acta, Mol. Basis Dis.*, 2012, **1822**, 753.

225. J. A. Menendez, J. Joven, G. Aragonès, E. Barrajón-Catalán, R. Beltrán-Debón, I. Borrás-Linares, J. Camps, B. Corominas-Faja, S. Cufí, S. Fernández-Arroyo, A. Garcia-Heredia, A. Hernández-Aguilera, M. Herranz-López, C. Jiménez-Sánchez, E. López-Bonet, J. Lozano-Sánchez, F. Luciano-Mateo, B. Martin-Castillo, V. Martin-Paredero, A. Pérez-Sánchez, C. Oliveras-Ferraros, M. Riera-Borrull, E. Rodríguez-Gallego, R. Quirantes-Piné, A. Rull, L. Tomás-Menor, A. Vazquez-Martin, C. Alonso-Villaverde, V. Micol and A. Segura-Carretero, *Cell Cycle*, 2013, **12**, 555.
226. Y. Niu, L. Na, R. Feng, L. Gong, Y. Zhao, Q. Li, Y. Li and C. Sun, *Aging Cell*, 2013, **12**, 1041.
227. D. V. Gospodaryov, I. S. Yurkevych, M. Jafari, V. I. Lushchak and O. V. Lushchak, *Longevity Healthspan*, 2013, **2**, 5.
228. M. Jafari, *Rejuvenation Res.*, 2015, **18**, 573.
229. V. H.-C. Liao, C.-W. Yu, Y.-J. Chu, W.-H. Li, Y.-C. Hsieh and T.-T. Wang, *Mech. Ageing Dev.*, 2011, **132**, 480.
230. G. Mariño, F. Pietrocola, F. Madeo and G. Kroemer, *Autophagy*, 2014, **10**, 1879.
231. A. C. Carvalho, A. C. Gomes, C. Pereira-Wilson and C. F. Lima, *Mol Basis Nutr Aging*, Elsevier, 2016, p. 491.
232. M. Kaeberlein, *BioEssays*, 2010, **32**, 96.
233. J. H. Chung, V. Manganiello and J. R. B. Dyck, *Trends Cell Biol.*, 2012, **22**, 546.
234. K. S. Bhullar and B. P. Hubbard, *Biochim. Biophys. Acta, Mol. Basis Dis.*, 2015, **1852**, 1209.
235. C. Canto and J. Auwerx, *Pharmacol. Rev.*, 2012, **64**, 166.
236. V. Desquiret-Dumas, N. Gueguen, G. Leman, S. Baron, V. Nivet-Antoine, S. Chupin, A. Chevrollier, E. Vessieres, A. Ayer, M. Ferre, D. Bonneau, D. Henrion, P. Reynier and V. Procaccio, *J. Biol. Chem.*, 2013, **288**, 36662.
237. C.-Y. Chen, J.-H. Jang, M.-H. Li and Y.-J. Surh, *Biochem. Biophys. Res. Commun.*, 2005, **331**, 993.
238. T. Hsieh, X. Lu, Z. Wang and J. M. Wu, *Med. Chem.*, 2006, **2**, 275.
239. A. Martín-Montalvo, J. M. Villalba, P. Navas and R. de Cabo, *Oncogene*, 2011, **30**, 505.
240. C. M. Hine and J. R. Mitchell, *J. Clin. Exp. Pathol.*, 2012, **S4**, 7329.
241. K. N. Lewis, J. Mele, J. D. Hayes and R. Buffenstein, *Integr. Comp. Biol.*, 2010, **50**, 829.
242. E. P. Weiss and L. Fontana, *Am. J. Physiol. Heart Circ. Physiol.*, 2011, **301**, H1205.
243. C. Gorrini, I. S. Harris and T. W. Mak, *Nat. Rev. Drug Discovery*, 2013, **12**, 931.
244. D. V. Gospodaryov, *J. Vasyl Stefanyk Precarp. Nat. Univ.*, 2015, **2**, 26.
245. V. Bocci and G. Valacchi, *Front. Chem.*, 2015, **3**, 4.
246. R. W. Powers, *Genes Dev.*, 2006, **20**, 174.
247. L. M. Ballou and R. Z. Lin, *J. Chem. Biol.*, 2008, **1**, 27.

248. D. W. Lamming, L. Ye, D. M. Sabatini and J. A. Baur, *J. Clin. Invest.*, 2013, **123**, 980.
249. M. Shimobayashi and M. N. Hall, *Nat. Rev. Mol. Cell Biol.*, 2014, **15**, 155.
250. A. A. Soukas, E. A. Kane, C. E. Carr, J. A. Melo and G. Ruvkun, *Genes Dev.*, 2009, **23**, 496.
251. M. V. Blagosklonny, *Cell Death Dis.*, 2014, **5**, e1552.
252. Y. Zhang, A. Bokov, J. Gelfond, V. Soto, Y. Ikeno, G. Hubbard, V. Diaz, L. Sloane, K. Maslin, S. Treaster, S. Rendon, H. van Remmen, W. Ward, M. Javors, A. Richardson, S. N. Austad and K. Fischer, *J. Gerontol., Ser. A*, 2014, **69A**, 119.
253. G. Bonfils, M. Jaquenoud, S. Bontron, C. Ostrowicz, C. Ungermann and C. De Virgilio, *Mol. Cell*, 2012, **46**, 105.
254. B. Carroll, D. Maetzel, O. D. Maddocks, G. Otten, M. Ratcliff, G. R. Smith, E. A. Dunlop, J. F. Passos, O. R. Davies, R. Jaenisch, A. R. Tee, S. Sarkar and V. I. Korolchuk, *eLife*, 2016, **5**, e11058.
255. J. Kim and E. Kim, *Amino Acids*, 2016, **48**, 915.
256. K. Powis and C. De Virgilio, *Cell Disc.*, 2016, **2**, 15049.
257. M. D. Kaymakcalan, Y. Je, G. Sonpavde, M. Galsky, P. L. Nguyen, D. Y. C. Heng, C. J. Richards and T. K. Choueiri, *Br. J. Cancer*, 2013, **108**, 2478.
258. S. Rafii, D. Roda, E. Geuna, B. Jimenez, K. Rihawi, M. Capelan, T. A. Yap, L. R. Molife, S. B. Kaye, J. S. de Bono and U. Banerji, *Clin. Cancer Res.*, 2015, **21**, 1869.
259. S. H. Um, F. Frigerio, M. Watanabe, F. Picard, M. Joaquin, M. Sticker, S. Fumagalli, P. R. Allegrini, S. C. Kozma, J. Auwerx and G. Thomas, *Nature*, 2004, **431**, 200.
260. F. Madeo, F. Pietrocola, T. Eisenberg and G. Kroemer, *Nat. Rev. Drug Discovery*, 2014, **13**, 727.
261. J. C. Newman and E. Verdin, *Trends Endocrinol. Metab.*, 2014, **25**, 42.
262. L. Mouchiroud, R. H. Houtkooper and J. Auwerx, *Crit. Rev. Biochem. Mol. Biol.*, 2013, **48**, 397.
263. A. Sanz, M. Soikkeli, M. Portero-Otín, A. Wilson, E. Kemppainen, G. McIlroy, S. Ellilä, K. K. Kemppainen, T. Tuomela, M. Lakanmaa, E. Kiviranta, R. Stefanatos, E. Dufour, B. Hutz, A. Naudí, M. Jové, A. Zeb, S. Vartiainen, A. Matsuno-Yagi, T. Yagi, P. Rustin, R. Pamplona and H. T. Jacobs, *Proc. Natl. Acad. Sci. U. S. A.*, 2010, **107**, 9105.
264. R. Stefanatos and A. Sanz, *Cell Cycle*, 2011, **10**, 1528.
265. S. Bahadorani, J. Cho, T. Lo, H. Contreras, H. O. Lawal, D. E. Krantz, T. J. Bradley and D. W. Walker, *Aging Cell*, 2010, **9**, 191.
266. J. H. Hur, S. Bahadorani, J. Graniel, C. L. Koehler, M. Ulgherait, M. Rera, D. L. Jones and D. W. Walker, *Aging (Albany NY)*, 2013, **5**, 662.
267. D. V. Gospodaryov, O. V. Lushchak, B. M. Rovenko, N. V. Perkhulyn, M. Gerards, T. Tuomela and H. T. Jacobs, *Biochim. Biophys. Acta, Bioenerg.*, 2014, **1837**, 1861.
268. J. Cao, S. Meng, E. Chang, K. Beckwith-Fickas, L. Xiong, R. N. Cole, S. Radovick, F. E. Wondisford and L. He, *J. Biol. Chem.*, 2014, **289**, 20435.
269. K. B. Storey, *J. Exp. Biol.*, 2015, **218**, 150.

270. J. Miquel, P. R. Lundgren, K. G. Bensch and H. Atlan, *Mech. Ageing Dev.*, 1976, **5**, 347.
271. F. Scialò, A. Sriram, A. Naudí, V. Ayala, M. Jové, R. Pamplona and A. Sanz, *Cell Cycle*, 2015, **14**, 2949.
272. R. Buffenstein, *J. Gerontol., Ser. A*, 2005, **60**, 1369.
273. M. Capri, A. Santoro, P. Garagnani, M. G. Bacalini, C. Pirazzini, F. Olivieri, A. Procopio, S. Salvioli and C. Franceschi, *Curr. Vasc. Pharmacol.*, 2014, **12**, 707.
274. L. Bucci, R. Ostan, E. Cevenini, E. Pini, M. Scurti, G. Vitale, D. Mari, C. Caruso, P. Sansoni, F. Fanelli, R. Pasquali, P. Gueresi, C. Franceschi and D. Monti, *Aging*, 2016, **8**, 510.
275. S. Milman and N. Barzilai, *Cold Spring Harbor Perspect. Med.*, 2016, **6**, a025098.
276. *Functional metabolism: regulation and adaptation*, Ed. K. B. Storey, John Wiley & Sons, Hoboken, NJ, 2004.
277. D. R. Calnan and A. Brunet, *Oncogene*, 2008, **27**, 2276.
278. X. Zhang, N. Tang, T. J. Hadden and A. K. Rishi, *Biochim. Biophys. Acta, Mol. Cell Res.*, 2011, **1813**, 1978.
279. N. Hay, *Biochim. Biophys. Acta, Mol. Cell Res.*, 2011, **1813**, 1965.
280. C. C. Dibble and L. C. Cantley, *Trends Cell Biol.*, 2015, **25**, 545.
281. A. A. Moskalev and M. V. Shaposhnikov, *Rejuvenation Res.*, 2010, **13**, 246.
282. K. B. MacKay, J. D. Lowenson and S. G. Clarke, *PLoS One*, 2012, **7**, e46719.
283. A. Danilov, M. Shaposhnikov, E. Plyusnina, V. Kogan, P. Fedichev and A. Moskalev, *Oncotarget*, 2013, **4**, 1507.
284. S. Emran, M. Yang, X. He, J. Zandveld and M. D. W. Piper, *Aging*, 2014, **6**, 390.
285. T. M. Bass, D. Weinkove, K. Houthoofd, D. Gems and L. Partridge, *Mech. Ageing Dev.*, 2007, **128**, 546.
286. C. Wang, C. T. Wheeler, T. Alberico, X. Sun, J. Seeberger, M. Laslo, E. Spangler, B. Kern, R. de Cabo and S. Zou, *Age*, 2013, **35**, 69.
287. R. Strong, R. A. Miller, C. M. Astle, J. A. Baur, R. de Cabo, E. Fernandez, W. Guo, M. Javors, J. L. Kirkland, J. F. Nelson, D. A. Sinclair, B. Teter, D. Williams, N. Zaveri, N. L. Nadon and D. E. Harrison, *J. Gerontol., Ser. A*, 2013, **68**, 6.
288. M. Viswanathan, S. K. Kim, A. Berdichevsky and L. Guarente, *Dev. Cell*, 2005, **9**, 605.
289. J. A. Baur, K. J. Pearson, N. L. Price, H. A. Jamieson, C. Lerin, A. Kalra, V. V. Prabhu, J. S. Allard, G. Lopez-Lluch, K. Lewis, P. J. Pistell, S. Poosala, K. G. Becker, O. Boss, D. Gwinn, M. Wang, S. Ramaswamy, K. W. Fishbein, R. G. Spencer, E. G. Lakatta, D. Le Couteur, R. J. Shaw, P. Navas, P. Puigserver, D. K. Ingram, R. de Cabo and D. A. Sinclair, *Nature*, 2006, **444**, 337.
290. D. R. Valenzano, E. Terzibasi, T. Genade, A. Cattaneo, L. Domenici and A. Cellerino, *Curr. Biol.*, 2006, **16**, 296.
291. L.-O. Klotz, C. Sánchez-Ramos, I. Prieto-Arroyo, P. Urbánek, H. Steinbrenner and M. Monsalve, *Redox Biol.*, 2015, **6**, 51.

292. G. Morrow and R. M. Tanguay, *Front Genet.*, 2015, **6**, 103.
293. A. Hernández-Santana, V. Pérez-López, J. M. Zubeldia and M. Jiménez-del-Rio, *Phytother. Res.*, 2014, **28**, 623.
294. M.-S. Kim, Y. K. Pak, P.-G. Jang, C. Namkoong, Y.-S. Choi, J.-C. Won, K.-S. Kim, S.-W. Kim, H.-S. Kim, J.-Y. Park, Y.-B. Kim and K.-U. Lee, *Nat. Neurosci.*, 2006, **9**, 901.
295. L. Varela and T. L. Horvath, *EMBO Rep.*, 2012, **13**, 1079.
296. G. W. Millington, *Nutr. Metab.*, 2007, **4**, 18.
297. K. R. Watterson, D. Bestow, J. Gallagher, D. L. Hamilton, F. B. Ashford, P. J. Meakin and M. L. J. Ashford, *Neurosignals*, 2013, **21**, 28.
298. T. Phillips and C. Leeuwenburgh, *Rejuvenation Res.*, 2004, **7**, 243.
299. R. Strong, R. A. Miller, C. M. Astle, R. A. Floyd, K. Flurkey, K. L. Hensley, M. A. Javors, C. Leeuwenburgh, J. F. Nelson, E. Ongini, N. L. Nadon, H. R. Warner and D. E. Harrison, *Aging Cell*, 2008, **7**, 641.
300. S. Ayyadevara, P. Bharill, A. Dandapat, C. Hu, M. Khaidakov, S. Mitra, R. J. Shmookler Reis and J. L. Mehta, *Antioxid. Redox Signaling*, 2013, **18**, 481.
301. Q.-L. Wan, S.-Q. Zheng, G.-S. Wu and H.-R. Luo, *Exp. Gerontol.*, 2013, **48**, 499.
302. H. Hans, A. Lone, V. Aksenov and C. D. Rollo, *Age*, 2015, **37**, 31.
303. K. Evason, J. J. Collins, C. Huang, S. Hughes and K. Kornfeld, *Aging Cell*, 2008, **7**, 305.
304. L. B. Avery and N. N. Bumpus, *Mol. Pharmacol.*, 2013, **85**, 1.
305. K. Kornfeld and K. Evason, *Arch. Neurol.*, 2006, **63**, 491.
306. J. J. Collins, K. Evason, C. L. Pickett, D. L. Schneider and K. Kornfeld, *PLoS Genet.*, 2008, **4**, e1000230.
307. H. Choi, H. Schneider, S. Klum, D. Chandler-Brown, M. Kaeberlein and L. Shamieh, *PLoS One*, 2013, **8**, e82543.
308. X. Chen, H. V. McCue, S. Q. Wong, S. S. Kashyap, B. C. Kraemer, J. W. Barclay, R. D. Burgoyne and A. Morgan, *Mol. Neurodegener.*, 2015, **10**, 51.
309. K. Evason, C. Huang, I. Yamben, D. F. Covey and K. Kornfeld, *Science*, 2005, **307**, 258.
310. M. Wong, *Epilepsia*, 2010, **51**, 27.
311. Q. Chen, S. Ganapathy, K. P. Singh, S. Shankar and R. K. Srivastava, *PLoS One*, 2010, **5**, e15288.
312. L. Xiang, Y. Nakamura, Y.-M. Lim, Y. Yamasaki, Y. Kurokawa-Nose, W. Maruyama, T. Osawa, A. Matsuura, N. Motoyama and L. Tsuda, *Aging*, 2011, **3**, 1098.
313. L. Wu, Y. Zhang, X. Ma, N. Zhang and G. Qin, *Mol. Biol. Rep.*, 2012, **39**, 9085.
314. T. K. Sin, B. Y. Yung and P. M. Siu, *Cell. Physiol. Biochem.*, 2015, **35**, 541.
315. F. Lan, J. M. Cacicedo, N. Ruderman and Y. Ido, *J. Biol. Chem.*, 2008, **283**, 27628.
316. Z. Liu, X. Li, A. R. Simoneau, M. Jafari and X. Zi, *Mol. Carcinog.*, 2012, **51**, 257.
317. D. G. Hardie, *Diabetes*, 2013, **62**, 2164.

318. Q. Wu, Y. Zhang, J. Xu and P. Shen, *Proc. Natl. Acad. Sci. U. S. A.*, 2005, **102**, 13289.
319. D. Stevanovic, V. Trajkovic, S. Müller-Lühlhoff, E. Brandt, W. Abplanalp, C. Bumke-Vogt, B. Liehl, P. Wiedmer, K. Janjetovic, V. Starcevic, A. F. H. Pfeiffer, H. Al-Hasani, M. H. Tschöp and T. R. Castañeda, *Mol. Cell. Endocrinol.*, 2013, **381**, 280.
320. G. P. Sykiotis and D. Bohmann, *Dev. Cell*, 2008, **14**, 76.
321. T. K. Blackwell, M. J. Steinbaugh, J. M. Hourihan, C. Y. Ewald and M. Isik, *Free Radical Biol. Med.*, 2015, **88**, 290.
322. C. Edwards, J. Canfield, N. Copes, A. Brito, M. Rehan, D. Lipps, J. Brunquell, S. D. Westerheide and P. C. Bradshaw, *BMC Genet.*, 2015, **16**, 8.
323. T. Ogawa, Y. Kodera, D. Hirata, T. K. Blackwell and M. Mizunuma, *Sci. Rep.*, 2016, **6**, 21611.
324. W.-S. Jeong, M. Jun and A.-N. T. Kong, *Antioxid. Redox Signaling*, 2006, **8**, 99.
325. K. Sahin, M. Tuzcu, H. Gencoglu, A. Dogukan, M. Timurkan, N. Sahin, A. Aslan and O. Kucuk, *Life Sci.*, 2010, **87**, 240.
326. Y. Lu, X. H. Zhang and A. I. Cederbaum, *Toxicol. Sci.*, 2012, **128**, 427.
327. V. I. Lushchak, *J. Amino Acids*, 2012, **2012**, 736837.
328. K. N. Lewis, E. Wason, Y. H. Edrey, D. M. Kristan, E. Nevo and R. Buffenstein, *Proc. Natl. Acad. Sci. U. S. A.*, 2015, **112**, 3722.
329. K. A. Rodriguez, P. A. Osmulski, A. Pierce, S. T. Weintraub, M. Gaczynska and R. Buffenstein, *Biochim. Biophys. Acta, Mol. Basis Dis.*, 2014, **1842**, 2060.
330. B. M. Rovenko, O. I. Kubrak, D. V. Gospodaryov, N. V. Perkhulyn, I. S. Yurkevych, A. Sanz, O. V. Lushchak and V. I. Lushchak, *J. Insect Physiol.*, 2015, **79**, 42.
331. B. M. Rovenko, O. I. Kubrak, D. V. Gospodaryov, I. S. Yurkevych, A. Sanz, O. V. Lushchak and V. I. Lushchak, *Comp. Biochem. Physiol., Part A: Mol. Integr. Physiol.*, 2015, **187**, 27.
332. D. Cota, K. Proulx, K. A. B. Smith, S. C. Kozma, G. Thomas, S. C. Woods and R. J. Seeley, *Science*, 2006, **312**, 927.
333. J. M. Kramer, J. T. Davidge, J. M. Lockyer and B. E. Staveley, *BMC Dev. Biol.*, 2003, **3**, 5.
334. D. K. Ingram and G. S. Roth, *Exp. Gerontol.*, 2011, **46**, 148.
335. S. Grünwald, J. Stellzig, I. V. Adam, K. Weber, S. Binger, M. Boll, E. Knorr, R. M. Twyman, A. Vilcinskas and U. Wenzel, *Genes Nutr.*, 2013, **8**, 439.
336. S. Brandstädt, K. Schmeisser, K. Zarse and M. Ristow, *Aging*, 2013, **5**, 270.
337. D. Chen, P. W.-L. Li, B. A. Goldstein, W. Cai, E. L. Thomas, F. Chen, A. E. Hubbard, S. Melov and P. Kapahi, *Cell Rep.*, 2013, **5**, 1600.

CHAPTER 11

Allosteric SIRT1 Activators as Putative Anti-Aging Drugs

BASIL P. HUBBARD[a]

[a]Department of Pharmacology, University Alberta, Edmonton, Alberta, Canada
*E-mail: bphubbard@ualberta.ca

11.1 Introduction

The underlying basis for biological aging is one of the great mysteries of the life sciences. Two criteria have been established to define the phenomenon of aging.[1] First, the probability of death of an organism at a given moment must increase with its chronological age.[1] Second, characteristic changes in the phenotype of the organism must occur over time due to limiting processes.[1] The current consensus among evolutionary biologists is that aging likely occurs due to a lack of selective pressure in the post-reproductive phase of life,[2,3] but that genes capable of extending lifespan during periods of delayed reproduction might undergo selection.[1] Recent advances in molecular biology have allowed researchers to begin identifying these putative longevity genes.[4]

While it was originally thought that biological aging was a fixed condition,[4] experimental work in the 1960s on the replicative life span of the yeast *Saccharomyces cerevisiae*,[5] followed by later work in *Caenorhabditis elegans* and *Drosophila melanogaster* in the 1970s and 1980s, demonstrated that acute mutagenesis and selection could be used to extend the lifespan of

RSC Drug Discovery Series No. 57
Anti-aging Drugs: From Basic Research to Clinical Practice
Edited by Alexander M. Vaiserman

Published by the Royal Society of Chemistry, www.rsc.org

small model organisms, transforming this view.[6–8] Subsequently, studies in *C. elegans* identified the insulin/IGF-1/FOXO cascade as the first pathway influencing aging.[9] Mutations that cause inactivation of the insulin/IGF-1 signaling pathway (IIS) extend lifespan and delay a number of age-related phenotypes in *C. elegans*.[10] For example, point mutations in the daf-2 gene (the insulin/IGF-1 receptor homolog) increase both mean and maximal lifespan in worms,[10,11] and mutations in the age-1 gene, a gene acting downstream of daf-2, also extend lifespan.[7,12] Furthermore, IIS has been shown to influence aging in mammals such as mice.[10] At least three other conserved longevity pathways have since been identified using similar genetic approaches in small model organisms: TSC/mTOR,[13] AMPK,[14] and Sir2/SIRT1.[15,16]

In addition to the discovery of longevity genes, a number of therapeutic interventions that effectively counteract aging have recently been formulated.[15] Caloric restriction (CR), the first intervention strategy reported to extend the lifespan of mammals,[17] consists of a defined reduction in caloric intake without malnutrition.[17] Studies have demonstrated that CR not only increases lifespan in mammals, but also delays the onset of a number of age-related diseases, including cancer, diabetes, and others.[18,19] Moreover, the effects of CR on lifespan are conserved in a wide variety of organisms ranging from yeast to mammals.[19,20] Resveratrol, a naturally occurring Sir2/SIRT1 activator (STAC), was the first small molecule shown to extend the lifespan of a model organism.[21] Subsequent work led to the identification of other anti-aging drugs, including metformin (activates AMPK),[22] rapamycin (inhibits TOR),[13] spermidine (regulates autophagy),[23] as well more potent second and third generation synthetic STACs.[24–26] However, amongst these, STACs have arguably received the greatest amount of attention due their apparent lack of toxicity and their protective effects against a number of age-related disorders.[15,27] In this chapter, the sirtuin longevity pathway, the discovery of allosteric STACs, and the effects of STACs on aging and age-related disease will be summarized. In addition, the difficulties faced in translating experimental findings on STACs to human trials will be critically examined.

11.2 The Sirtuin Longevity Pathway

The silent information regulator (SIR) genes, or sirtuins, have become the focus of much research over the past decade due to their ability to regulate numerous critical cell processes and to modulate lifespan across diverse species.[15] SIR1–4 were originally identified in screens for mutations causing sterility in *S. cerevisiae*.[28–32] Later, it was shown that these genes play important roles in repression of the silent mating type loci and in genomic silencing at telomeres.[33,34] While mutations in SIR2, SIR3, and SIR4 shorten lifespan in yeast,[1] overexpression of SIR2 increases replicative lifespan by up to 30% *via* suppression of extrachromosomal rDNA circle formation, a cause of aging in yeast.[15,35] While somewhat controversial,[36] several studies have shown that overexpression of Sir2 homologues in worms and flies also extends lifespan.[37,38] Finally, overexpression of the mammalian Sir2 homolog SIRT1 in the brain extends mouse lifespan,[39] and whole body overexpression of SIRT6 (another mammalian sirtuin homolog) extends the lifespan of male mice.[40]

Sirtuins are categorized as class III, or β-nicotinamide adenine dinucleotide (β-NAD)-dependent, histone deacetylases.[41,42] Using β-NAD as a co-substrate, sirtuins catalyze the conversion of acetylated lysine residues on histones and other proteins into deacetylated substrates, nicotinamide (NAM), and *O*-acetyl adenosine diphosphate ribose (*O*-AcADPR) *via* a sequential 2-step mechanism (Figure 11.1).[43] In the first step, the acetyllysine substrate initiates a nucleophilic attack on β-NAD resulting in the formation of an alkylamidate intermediate and the release of free NAM.[43–47] Interestingly, re-binding of NAM can inhibit the sirtuin reaction, making it an endogenous inhibitor.[42] In the second step, an intramolecular attack from the 2′ hydroxyl group on the ribose ring results in cyclization of the alkylamidate; this 1′2′-cyclic intermediate[45,48] is subsequently cleaved by a water molecule to release the final products.[45,49] This mechanism of catalysis appears to be multifaceted, as several sirtuins also exhibit ADP-ribosyltransferase activity,[50] as well as a broad range of other deacylase (desuccinylase, demalonylase, *etc.*) activities.[51–53]

In mammals, seven sirtuin homologs (SIRT1-7) exist, with varied intracellular localization, activity, and function.[16] SIRT1 is localized to the nucleus and acts both as a chromatin-associated histone deacetylase as well as a general nuclear protein deacetylase.[15] SIRT2 is found mainly in the cytoplasm where it deacetylates tubulin[54] and other cytoplasmic targets.[16] SIRT3, SIRT4, and SIRT5 all reside in the mitochondrion.[16] While SIRT3 is the primary mitochondrial protein deacetylase enzyme,[55] SIRT4 appears to

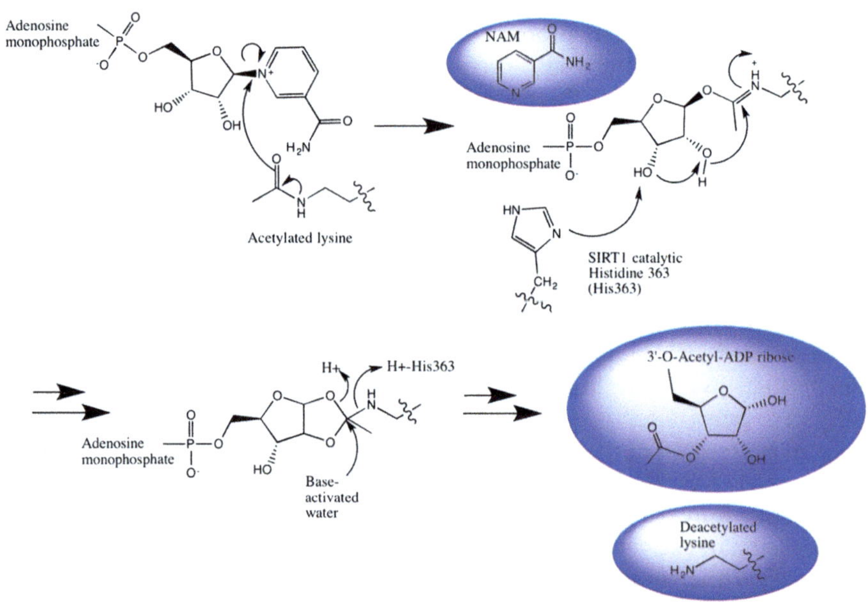

Figure 11.1 Catalytic mechanism of sirtuin-mediated lysine deacetylation.

possess lipoamidase activity towards the pyruvate dehydrogenase (PDH) complex,[56] and SIRT5 removes succinyl,[57] malony,[57] and glutaryl,[58] but not acetyl modifications from proteins.[52] Like SIRT1, SIRT6 displays nuclear localization, but it is predominately associated with chromatin.[59] SIRT6 deacetylates several key residues on histone proteins such as H3K56(ac)[60] and also catalyzes fatty long-chain deacylation of proteins.[61] While comparatively little work has been done on SIRT7, recent studies suggest that it localizes to the nucleus where it performs histone deacetylation.[62]

Of the seven mammalian sirtuin genes, SIRT1, which bears the closest phylogenetic relationship to the yeast longevity gene Sir2,[15] has been the most extensively studied. Through deacetylation of a myriad of nuclear targets including histones, transcription factors, such as p53,[63] FOXO,[64] NFκB,[65] PGC1α,[66] and other proteins, such as Ku70[67] and BMAL,[68] SIRT1 regulates multiple critical cell processes including RNA transcription, apoptosis, DNA damage response, regulation of muscle and fat differentiation, neurogenesis, mitochondrial biogenesis, insulin signaling, and circadian rhythms.[15,16] Moreover, SIRT1 appears to play an important role in a number of age-related disease states.[15,16] For example, overexpression of SIRT1 is protective in mouse models of Alzheimer's disease,[69] type II diabetes,[70] colon cancer,[71] prostate cancer,[72] and thymic-induced lymphomas.[73]

Interestingly, there is growing evidence that suggests SIRT1 underlies many of the beneficial effects of CR.[15] For instance, Sir2 is required for CR-mediated lifespan extension in *S. cerevisiae*[74,75] and in *D. melanogaster*.[76] CR-mediated lifespan extension in worms is also partially dependent on Sir2.1, albeit in a diet-specific manner.[77] Two mechanisms for how SIRT1 might be regulated during CR in these organisms have been proposed. One model suggests that CR activates a nicotinamidase enzyme that lowers levels of NAM,[78] resulting in SIRT1 activation. A second model proposes that an increase in the ratio of NAD/NADH resulting from low caloric intake activates SIRT1.[79] In mammals, SIRT1 protein levels seem to be increased in numerous tissues following caloric restriction,[67] revealing yet another possible link between SIRT1 activity and CR. Consistent with this hypothesis, deletion of SIRT1 in mice abrogates the metabolic benefits CR and blocks lifespan extension.[80] Furthermore, SIRT1 transgenic overexpressing mice display metabolic phenotypes that resemble CR.[81] The notion that overexpression or activation of SIRT1 could mimic the effects of CR has been a driving force in the search for small-molecule activators of SIRT1.

11.3 Small-Molecule SIRT1 Activators

A number of pharmacological approaches to activate SIRT1 have been tested.[15] Based on research showing that physiological levels of NAM are sufficient to inhibit sirtuin activity, Sauve *et al.* showed that a transient competitive inhibitor of NAM binding, isonicotinamide (iNAM) (Figure 11.2), could be used be used as a pan-sirtuin activator in yeast cells.[82] The disadvantages of this approach are that iNAM can interfere with other enzymes using NAM,

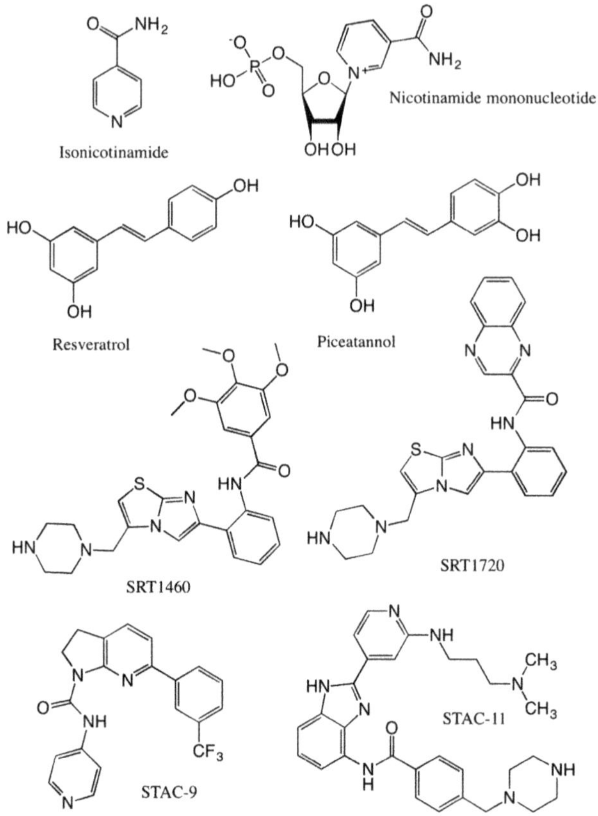

Figure 11.2 Structures of SIRT1 activating compounds.

and that intracellular levels of iNAM must reach millimolar concentrations in order for sirtuin activation to occur.[82] More recent reports have examined the possibility of boosting SIRT1 activity by increasing endogenous levels of β-NAD, similar to what is thought to occur during CR.[83] A number of NAD+ precursor molecules, including nicotinamide mononucleotide (NMN)[83] and the more cell-permeable nicotinamide riboside (NR),[84] have been shown to increase intracellular β-NAD reserves and to activate SIRT1 in cells and in mice. Remarkably, dietary supplementation with NR increases mouse lifespan.[85] However, since β-NAD is a ubiquitous co-factor that is utilized by many enzymes,[86] further work will be needed to dissect if the health effects of NMN and NR are due solely to activation of SIRT1. For this reason, most pharmacological research on SIRT1 has focused on the development of specific, allosteric SIRT1-activators.[15]

The first putative allosteric SIRT1 activators were identified using a high-throughput small-molecule screen in 2003.[21] This screen employed a fluorometric assay (BIOMOL Fleur de Lys™) in which a peptide substrate bearing an aminomethylcoumarin (AMC) moiety directly adjacent to the acetyl-lysine is deacetylated by SIRT1 in the presence of β-NAD.[21] In the second step of the assay, deacetylated peptides are cleaved by trypsin to liberate free AMC, which produces a fluorescent signal.[15,21] This screen identified a wide-array of plant-derived polyphenols, including flavones, stilbenes, isoflavones, catechins (flavan-3-ols), chalcones, and anthocyanidins, with the ability to activate SIRT1 through a proposed allosteric mechanism involving a lowering of the peptide substrate K_m.[21] The stilbenes resveratrol (3,5,4′-trihydroxy-*trans*-stilbene) and piceatannol (3′,4′,3,5-tetrahydroxy-*trans*-stilbene) were the two most potent activators identified in the screen.[21] In the fluorometric assay, resveratrol increased SIRT1 activity by up to 8-fold (EC_{50} ~ 50 μM), concomitant with a 35-fold decrease in peptide K_m (a slight reduction in the K_m of β-NAD was also observed).[21] Consistent with these *in vitro* results, resveratrol was shown to extend the lifespan of yeast in a Sir2-dependent manner.[21] Previous to its discovery as a SIRT1 activator, resveratrol had been characterized as a phytoalexin with the ability to impart many health-promoting effects in mammals including protection from cancer and heart disease.[27] While most of these effects were originally attributed to its antioxidant potential, they have been re-evaluated in the context of SIRT1 activation.[15,27]

The discovery of naturally occurring SIRT1 activators stimulated research into the development of more potent and more specific synthetic molecules.[15] Using a high-throughput fluorescence polarization screen employing a carboxytetramethylrhodamine (TAMRA)-tagged p53 acetyl-lysine peptide, Sirtris Pharmaceuticals reported the discovery of second generation STACs in 2007.[24] SRT1460, SRT1720, and SRT2183, which are derivatives of an imidazothiazole scaffold structurally distinct from resveratrol, were shown to have EC_{50} values for SIRT1 nearly 1000-fold lower than resveratrol.[24] These compounds were shown to induce effects resembling SIRT1 overexpression in cells, and to ameliorate insulin resistance in obese mice.[24] Later, third generation STACs based on saturated urea (STAC-9) and benzimidazole (STAC-11) scaffolds were reported, and these also displayed phenotypes in cells consistent with SIRT1 activation.[25,26] Importantly, all of these molecules have been shown to activate SIRT1 *in vitro via* a peptide K_m-lowering mechanism similar to resveratrol, indicating a common mechanism of activation.[25,26]

Controversy over the mechanism of action of STACs erupted when several reports were published claiming that although resveratrol and synthetic STACs activated SIRT1 deacetylation of fluorophore-tagged peptides *in vitro*, no activation was observed when the corresponding non-tagged peptides were used.[87–89] Furthermore, it was proposed that the SIRT1-dependent effects of resveratrol in cells could be due to increased β-NAD levels resulting from inhibition of phosphodiesterase (PDE).[90] These reports questioned the validity of STACs as direct SIRT1 activators, and led to speculation that the effects of STACs in cells and in mice could be due to coincidental off-target effects on other proteins.[15]

A study published in 2013 shed light on the mechanism of STAC-mediated SIRT1 activation, and helped to partially resolve the controversy.[26] This study demonstrated that STACs can directly bind to SIRT1 and that activation is substrate specific.[26] More precisely, it was shown that while STACs were ineffective at enhancing SIRT1 deacetylation of most natural amino acid peptides, certain peptides containing hydrophobic amino acids (*e.g.* tryptophan or phenylalanine) at the +1 and +6 positions relative to the acetyl-lysine, could support STAC-mediated activation.[26] Over 500 naturally occurring SIRT1 substrates bearing these structural features were identified computationally, including acetyl lysine sites on the well characterized SIRT1 substrates PGC1α and FOXO3a.[26] In addition, this report also identified Glutamate 230 (E230) as a critical residue mediating activation.[26] Specifically, mutation of E230 to lysine (E230K) was shown to block STAC-mediated SIRT1 activation on all types of substrates (natural, fluorophore-tagged), using all classes of STACs (natural, synthetic) without affecting basal enzyme activity.[26] Furthermore, SIRT1 knockout cells reconstituted with SIRT1-E230K were shown to be nonresponsive to STAC treatment, suggesting that the effects of these compounds in cells are likely due to a direct effect on SIRT1.[26]

Based on these findings, two models of SIRT1 activation have been proposed.[26] Steegborn and colleagues proposed that STACs interact with the enzyme-bound substrate and allow it to dock more efficiently.[91] In contrast, based on the discovery of E230K and the identification of a surrounding N-terminal “activation domain”, Sinclair and others proposed that SIRT1 activation occurs through an assisted-allosteric activation (AAA) mechanism.[26] Briefly, this model posits that binding of certain peptide substrates to SIRT1 induces an exosite, which allows STACs to bind and subsequently stabilize the docked substrate, thus lowering its K_M.[26] Recent crystallographic data on SIRT1 has supported this hypothesis. For example, Cao *et al.* confirmed the presence of an ordered N-terminal activation domain on SIRT1,[92] and Dai and colleagues have crystallized a SIRT1–STAC complex and shown that interactions between E230 and the catalytic core govern allosteric binding of STACs.[93]

11.4 STACs in Aging and Age-Related Disease

11.4.1 Lifespan Extension

Resveratrol was the first small molecule shown to extend the lifespan of a laboratory model organism.[21] In 2003 it was demonstrated that supplementing yeast growth media with 10 μM resveratrol could extend the mean replicative lifespan of *S. cerevisiae* by up to 70%.[21] Subsequent work showed that this effect was conserved in other small model organisms.[27] For example, resveratrol dosed at 100 μM was shown to extend the mean lifespan of worms and flies by up to 30% and 15%, respectively, in a Sir-2-dependent manner.[37] Furthermore, dietary supplementation with 130 μM resveratrol was shown to extend the mean lifespan of *Apis mellifera* (the common honeybee) by ~33%, likely through a caloric restriction-like mechanism.[94] Finally, resveratrol extends both the mean and maximum lifespans of the short-lived fish species *Nothobranchius furzeri* and *Nothobranchius guentheri*, and delays age-dependent decline of locomotor activity in these animals.[95,96]

It is important to note that the reported effects of resveratrol on organismal lifespan have been controversial.[15] One study claimed that resveratrol only marginally increased the lifespan of worms and flies, and that any observable effect was independent of Sir2.[97] This claim was bolstered by a subsequent report claiming that Sir2 overexpression has no effect on worm lifespan.[36] However, recent findings have supported the original results and highlighted several experimental factors that could account for the dramatically different findings.[15] In worms, the effect of Sir2 on lifespan appears to be diet-dependent,[98] and in flies the effect of resveratrol on lifespan appears to be both diet- and sex-dependent.[99] Importantly, resveratrol does not appear to extend the lifespan of the crustacean *Daphnia pulex*[100] or the mosquito *Anopheles stephensi*.[101] Moreover, at least 3 studies have failed to show a lifespan extending effect for resveratrol in mice fed a normal chow diet,[102–104] despite a report showing that mice eating a regular diet supplemented with resveratrol display a transcriptional profile resembling that of mice on CR.[102] Nonetheless, resveratrol does extend the lifespan of mice fed a high-fat diet (HFD)[105] (Figure 11.3), and it has been suggested that resveratrol may only extend average lifespan in populations suffering from specific causes of death.[15]

In contrast to resveratrol, promising data exist on the ability of potent synthetic SIRT1 activators to extend the lifespan of healthy mammals.[15] It was recently shown that dietary supplementation of mice fed a standard diet with the STAC SRT1720 delayed the onset of age-related metabolic diseases, improved general health, and extended mean lifespan by ~9%.[106] This study also confirmed previous results showing that supplementation with SRT1720 extends the mean lifespan of mice fed a high-fat diet by >20%, and that it reverses many of the metabolic abnormalities caused by this calorie-rich diet.[106,107] Most recently, SRT2104, an imidazothiazole-based STAC similar to

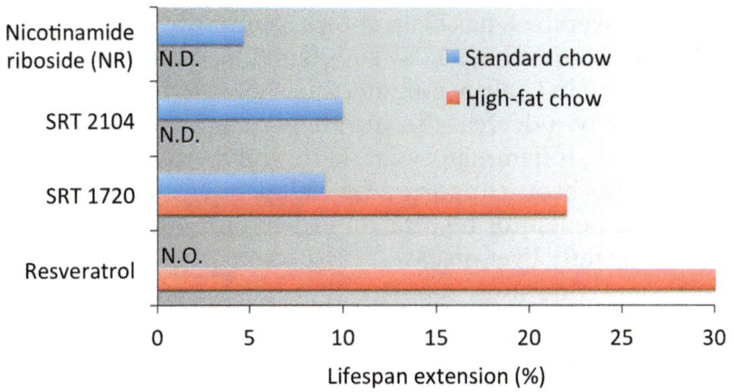

Figure 11.3 Mean lifespan extension by SIRT1 activators in mice fed either a standard chow diet or a high-fat chow diet. N.D.; Not Determined, N.O.; Not Observed.

SRT1720, was shown to extend both mean and maximal lifespan of mice fed a standard diet.[108] The lifespan extension caused by SRT2104 was accompanied by enhanced motor coordination and performance, bone mineral density, insulin sensitivity, and decreased inflammation in the treated mice.[108] Further studies will need to be carried out on synthetic STACs to see if their effects are conserved in higher organisms such as non-human primates.

11.4.2 Obesity, Metabolism, and Type II Diabetes

One of the most well studied effects of STACs on healthspan is their ability to regulate metabolic parameters and type II diabetes.[15] In mice fed a high-fat diet, dietary supplementation with resveratrol not only reduces the risk by death by 31% but also protects against obesity, fatty liver disease, and insulin resistance.[15,109] In addition, resveratrol reverses transcriptional changes associated with HFD feeding, and improves mitochondrial function by increasing the activity of AMPK and PGC1α, two key regulators of mitochondrial biogenesis.[105] The beneficial effects of resveratrol treatment in the HFD model appear to depend on both SIRT1 and AMPK, a kinase that is activated by resveratrol in a dose-dependent manner.[105] Studies in conditional SIRT1 knockout mice have demonstrated that a low dose of resveratrol (~24 mg kg^{-1}) stimulates AMPK *via* SIRT1-dependent deacetylation of LKB1, while at higher doses (~240 mg kg^{-1}) AMPK is activated in a SIRT1-independent manner.[110]

The effects of resveratrol on metabolism and obesity have also been examined in rhesus monkeys and humans.[27] In monkeys fed a high fat (HF)/high sucrose (HS) diet that mimics the typical Western diet, supplementation with resveratrol was shown to improve adipose insulin signaling and decrease inflammation in the adipose tissue.[111] Subsequent studies have demonstrated that resveratrol also decreases arterial wall inflammation and stiffening,[112] improves muscle function,[113] and confers neuroprotection in monkeys fed a HF/HS diet.[114] In humans, SRT501, a proprietary formulation of resveratrol, has been shown to improve glucose tolerance in type II diabetes.[115] Moreover, 30 day supplementation with resveratrol (150 mg day^{-1} resVida™) in healthy humans was shown to induce a CR-like phenotype consisting of a decrease in intrahepatic lipids, circulating glucose levels, triglycerides, inflammatory markers, and systolic blood pressure.[116] These results have been supported by independent studies demonstrating a therapeutic benefit of resveratrol in patients with type II diabetes[115] and nonalcoholic fatty liver disease.[117] However, other studies have shown conflicting results. One study reported that 12 week supplementation with 75 mg day^{-1} of resveratrol in non-obese healthy women did not improve metabolic performance,[118] and a more recent clinical trial reported that resveratrol dosed twice daily at 500 mg for 5 weeks had no effect on glucagon-like peptide (GLP-1) secretion, or on glycemic control in type II diabetic patients.[119] Further studies will need to be performed in order to resolve the factors underlying these discrepancies.

In addition to resveratrol, three synthetic STACs, SRT1720, SRT2104, and SRT3025, may help prevent metabolic decline and type II diabetes.[15] In both diet- and genetically-induced models of obesity, SRT1720 has been shown to improve insulin sensitivity, lower plasma glucose levels, and increase mitochondrial capacity.[24] Moreover, SRT1720 extends the lifespan of mice fed a high-fat diet and prevents liver steatosis and insulin resistance.[107] Another STAC, SRT2104, also appears to improve insulin sensitivity and boost mitochondrial performance in mice.[108] However, the effectiveness of SRT2104 in humans remains less clear. While dietary supplementation with 2 g day^{-1} of SRT2014 in otherwise healthy cigarette smokers was shown to improve their lipid profile,[120] a recent phase II clinical trial in patients with type II diabetes failed to yield evidence of improved insulin sensitivity or glucose tolerance, ostensibly due to poor drug pharmacokinetics.[121] SRT3025 has been shown to reduce hyperglycemia and promote beta cell expansion in a mouse model of diet-induced type II diabetes, but human studies using this compound have not been attempted.[122]

11.4.3 Cancer

Even before its characterization as a SIRT1 activator, resveratrol had been investigated as a potential anti-tumorigenic agent. One of the first studies to test the effects of resveratrol on cancer, published in 1997, examined its topical application in a model of skin cancer.[123] This study demonstrated that resveratrol shows chemopreventive activity in three major stages of carcinogenesis, including anti-initiation activity, anti-promotion activity, and anti-progression acitivty.[123] Subsequent studies have demonstrated that systemic administration of resveratrol also prevents tumor growth in mouse and rat models of colon cancer,[124] prostate cancer,[125] gastric cancer,[126] and gamma-radiation-induced thymic lymphoma,[73] and that at least some of these effects are SIRT1-dependent.[73,127] However, resveratrol does not appear to protect against all cancers, and at least one study has suggested that it could in fact promote growth of certain tumor types.[27] For example, resveratrol was shown to be ineffective in treating breast cancer[109] and cancer-related deaths in old mice,[102] and to worsen survival in certain models of prostate cancer.[128]

Several classes of synthetic STAC appear to inhibit tumor growth and enhance tumor cell apoptosis. For example, both resveratrol and SRT2183 were recently shown to induce growth arrest and apoptosis of malignant lymphoid cells.[129] In addition, SRT1460, SRT1720, and SRT3025 block tumor growth and chemosensitize pancreatic cancer cells through a mechanism involving SIRT1-dependent lysosomal-mediated cell death.[130] SRT1720, which is the most well-studied synthetic STAC in the context of cancer, induces lysosomal-dependent cell death in breast cancer cells[131] and attenuates tumor growth in a mouse model of multiple myeloma.[132] However, like resveratrol, synthetic STACs may not always be beneficial for cancer treatment. SRT1720 has been shown to promote cell migration of breast cancer tumor cells to the lung[133] and to attenuate the antitumor activity of melatonin, a potent suppressor of osteosarcoma, in bone tumors.[134] Thus the effects of STACs in cancer appear to be tumor- and context-specific, and will need to be carefully evaluated in the future prior to their use in the clinic.

11.4.4 Neurodegenerative Disease

The pharmacokinetic properties of resveratrol allow it to cross the blood–brain barrier where it displays a broad range of neuroprotective effects.[15] For example, intraventricular injection of resveratrol for one week was shown to improve learning and memory in aged mice, and to prevent cognitive decline.[135] These effects also appear to be conserved in nonhuman primates, as dietary supplementation with resveratrol was recently shown to confer neuroprotection in cortical brain tissue of monkeys fed a high-fat/high-sucrose diet.[114] It also protects against the damaging effects of ischemia in the brain *via* a SIRT1-dependent mechanism.[136] In addition to these protective effects, resveratrol appears to be well suited for the treatment of a number of age-related brain disorders. First, resveratrol has been shown to prevent accumulation of beta-amyloid peptide[137] and to reduce plaque formation,[138] two processes that are involved in Alzheimer's disease. It also extends the lifespan of the senescence-accelerated mouse (SAMP), a model of Alzheimer's disease in which the amyloid precursor protein (APP) is overexpressed in the brain.[139] Second, resveratrol slows down the progression of Parkinson's disease.[140] Finally, two reports have shown that resveratrol improves motor neuron function and extends the shortened lifespan of the SOD1(G93A) mouse model of amyotrophic lateral sclerosis (ALS).[141,142]

While several STACs, including SRT1720, SRT2104, and SRT3025, have been shown to penetrate into the brain, few studies have examined their efficacy for the treatment of age-related brain diseases.[15] Nonetheless, one study demonstrated that SRT3025 protects against neurodegeneration and mimics the effects of CR on the brain.[143] Other studies have focused on using synthetic STACs to treat specific diseases not necessarily associated with age. For example, it was shown that SRT1720 is protective in a mouse model of multiple sclerosis,[144] and that SRT2104 can attenuate brain atrophy, improve motor function, and extend survival in the N171-82Q mouse model of Huntington's disease.[145] Given that synthetic STACs can cross the blood–brain barrier and appear to be well tolerated, further studies examining their neuroprotective properties and effects in Alzheimer's disease are warranted.

11.4.5 Cardiovascular Disease

Due to its many protective effects on the cardiovascular system,[146] ingestion of resveratrol has been proposed to account for the "French Paradox", the fact that certain European populations with high wine consumption have a low risk of heart disease despite consuming a fat-rich diet.[109] Resveratrol acts to prevent cardiovascular disease in at least four ways, not all of which are mediated through SIRT1.[109] First, resveratrol prevents the oxidation of low-density lipoprotein (LDL) particles, a contributing factor to heart disease, by scavenging free radicals and chelating copper.[109,146] Second, resveratrol thins the blood and prevents excessive platelet aggregation, which can result in thrombus formation leading to ischemia, infarction, or stroke.[109,146] This property is thought to be due to its ability to inhibit COX1.[147] Third, resveratrol increases expression levels of both endothelial and inducible nitric oxide synthetase (eNOS and iNOS), and acts as a vasodilator to relax arteries and decrease blood pressure.[148] Another mechanism by which it appears to protect endothelial cells is *via* SIRT1-mediated activation of the transcription factor KLF2.[149] Finally, resveratrol has been shown to reduce the formation of atherosclerotic plaques in rabbits fed a high-cholesterol diet,[150] and it may lower serum cholesterol and triglyceride levels *via* SIRT1-independent regulation of the bile acid transporter ASBT.[151]

Synthetic STACs that are structurally unrelated to resveratrol will prove to be a useful tool in dissecting which of the cardioprotective effects of resveratrol are due to activation of SIRT1 and which are caused by ancillary effects on other enzymes. While synthetic STACs have not been extensively studied in the context of cardiovascular disease, a few reports have suggested that they might impart protective effects. For example, administration of SRT1720 reduces myocardial infarction in mice[152] and reverses vascular endothelial dysfunction in aging mice.[153] Moreover, SRT1720 has been shown to reduce vascular inflammation in an angiotensin II apoE-knockout mouse model of atherosclerosis,[154] and SRT3025 also provides protection in this model.[155] In humans, SRT2104 has been shown to improve the serum lipid profile of otherwise healthy cigarette smokers by reducing LDL, cholesterol, and triglyceride levels,[156] in a manner similar to resveratrol studies in mice.[151] However, analysis of these patients also revealed that SRT2104 had no effect on vascular or platelet function,[156] or arterial stiffness.[120] Thus, the effectiveness of STACs in improving a patient's cardiovascular health could depend on a number of factors, including diet, age, and pre-existing medical conditions.

11.4.6 Inflammation and Immunity

Many diseases of aging, as well as autoimmune diseases, are influenced by inflammation. In general, STACs are thought to control inflammation through activation of SIRT1 and subsequent deacetylation of NF-kB, a transcription factor that acts as a central mediator for the immune response.[15] However, in the case of resveratrol, its effects on immunity are likely multifaceted due its antioxidant and antiviral properties, and its ability to regulate a number of other enzymes involved in inflammation and immunity (*e.g.* COX1, COX2).[15,109] Resveratrol decreases inflammation in autoimmune models of Crohn's disease,[157] psoriasis,[158] and inflammatory arthiritis[159] in rats, mice, and rabbits, respectively. In addition to acting as a natural antibacterial and antiviral agent (especially against HSV1 and HSV2),[109,160] resveratrol decreases inflammation in response to pathogen infection by *Listeria monocytogenes*[161] and reduces levels of inflammatory cytokines in

LPS-stimulated macrophages by inhibiting the proteasome.[162] Finally, resveratrol decreases inflammation in models of respiratory disorders, including chronic obstructive pulmonary diseases (COPD)[163] and asthma.[164] Given its pervasive anti-inflammatory activity and apparent safety, resveratrol may prove to be a practical treatment for many low-grade chronic inflammatory disorders.

In addition to its anti-inflammatory effects on the cardiovascular system,[154] SRT1720 has demonstrated promising effects in mouse models of chronic inflammation,[165] ashma[166] and COPD/emphysema.[167] Importantly, its effects in COPD occur *via* a FOXO3-mediated reduction in premature lung cell senescence, an effect that was shown to be SIRT1-dependent.[167] In addition, SRT1720 ameliorates colitis in mice through a mechanism involving PGC1α.[168] In humans, clinical trials on inflammation using synthetic STACs have yielded mixed results. For example, SRT2104 has been reported to attenuate LPS-induced inflammation in human patients.[169] In addition, a randomized, placebo-controlled study demonstrated that SRT2104 may be a viable treatment for patients with moderate to severe psoriasis.[170] However, another study assessing the safety and clinical activity of SRT2104 in patients with mild to moderate ulcerative colitis failed to show any effect.[171] This result is quite unexpected since SIRT1 has been shown to play an important role in colitis,[172] but it could be the result of poor drug pharmacokinetics or pharmacodynamics.

11.4.7 Fertility and Development

Since fertility decreases in both males and females with age, it may be considered another marker for aging. Because resveratrol has been shown to have weak estrogenic activity (*via* activation of ERα),[173] initially there was concern that supplementation could alter male fertility and possibly influence the development of offspring in mammals.[109] The results of numerous studies, however, have led to dismissal of these concerns.[109] Treatment of pregnant mothers with resveratrol had virtually no effect on their offspring,[174] and even very high doses of resveratrol up to 300 mg kg^{-1} do not appear to result in any observable toxic effects on fertility or development.[175] In fact, resveratrol appears to protect against age-associated infertility in female mice. In one study, young mice fed a diet containing resveratrol for 12 months retained the capacity to reproduce while their age-matched counterparts on the control diet did not.[176] This phenotype was associated with an increase in the number and quality of oocytes.[176] In male rodents, resveratrol protects sperm against a number of chemical and environmental insults.[177] While researchers have only started to investigate the use of synthetic STACs for the treatment of age-related infertility, one report has demonstrated that SRT1720 improves follicle reserve and prolongs the ovarian lifespan of obese female mice *via* a SIRT1-dependent mechanism.[178] Thus, recent work is supporting the notion that an extension of lifespan and healthy aging could also lead to an extension of the reproductive phase of life. Figure 11.4 summarizes the beneficial effects of STACs on age-related diseases and physiological declines.

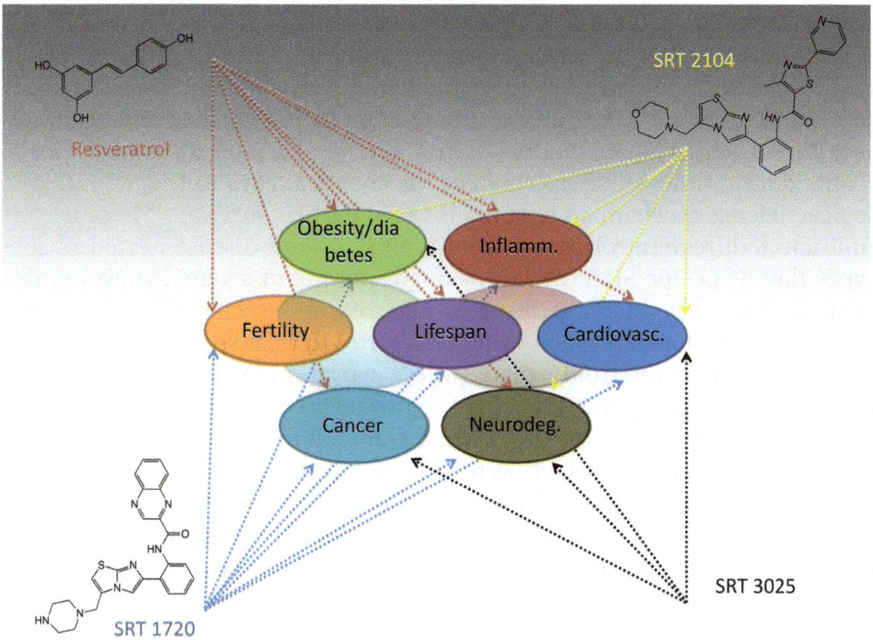

Figure 11.4 Documented effects of SIRT1 activators on lifespan and the treatment and prevention of various age-related diseases in mammals.

11.5 Clinical Challenges with STACs

11.5.1 Pharmacology

To date, resveratrol and synthetic STACs have been tested in over 125 phase I and phase II clinical trials (http://www.clinicaltrials.gov). Some of the most promising results from these trials include data showing that resveratrol supplementation can improve insulin sensitivity in patients suffering from metabolic syndrome,[179] and that it can reduce inflammation in patients with ulcerative colitis.[180] In addition, SRT2104 showed promising effects in the treatment of psoriasis in phase I trials.[170] Unfortunately, very few phase III trials have been performed using these molecules, and even fewer are in the pipeline (http://www.clinicaltrials.gov). Pharmaceutical companies may be discouraged from pursuing large-scale studies using resveratrol because it is a non-patentable natural compound that displays relatively poor pharmacological properties, including molecular promiscuitiy[181] and low bioavailabilty.[115] In the case of synthetic STACs, which are patented new chemical entities, much more work is still needed to adequately characterize their molecular pharmacodynamics and toxicity.[15]

Resveratrol exhibits complex pharmacodynamics.[15] In addition to activating SIRT1 and modulating cyclooxygenase enzymes,[109,182] PDE,[90] and the estrogen receptor,[173] resveratrol also targets cytochrome P450 enzymes (CYPs) involved in phase I drug metabolism,[183,184] the aryl hydrocarbon receptor (AHR),[185] quinone reductase II (QRII),[186] and S6 kinase (S6K),[187] amongst others.[109] Yet, the cumulative effect of these many interactions on health appears to be generally positive,[188] and resveratrol demonstrates an apparent lack of toxicity in animal models.[109] This observation has given rise to the 'xenohormesis' hypothesis, which states that resveratrol and similar molecules produced by plants during times of stress could act as chemical cues to trigger a physiological defense mechanism in animals to aid in their survival.[188] Because it is multifaceted, resveratrol may protect against several disease-related phenotypes that a specific SIRT1 activator might not guard against. However, this property could also increase the occurrence of unforeseen and possibly detrimental off-target effects when using resveratrol to treat specific diseases. For example, while resveratrol-mediated SIRT1 activation might help treat type II diabetes,[189] its off-target effects could aggravate other pre-existing conditions in these patients. In this regard, second generation synthetic STACs such as SRT1720 and SRT1460, which were designed to specifically activate SIRT1, may be more appropriate.[24] However, while these molecules also appear to be well-tolerated,[190] they too display off-target effects.[191] No work has yet been published on the molecular specificity or toxicity of third generation STACs.

Poor bioavailability and pharmacokinetics are additional clinical challenges facing STACs.[27] Studies have shown that while resveratrol absorption in humans is dose-dependent,[115] low plasma levels are achieved due to poor bioavailability and rapid metabolism.[27] For example, one study showed that supplementation of a 5 g dose of resveratrol in humans resulted in a mean plasma concentration of only ~52 μg L^{-1} over a one day period.[192] The maximum plasma concentration (C_{max}) reached during this period was 540 μg L^{-1}, which occurred 1.5 hours after administration.[192] It is thought that human gut microbiota could limit the bioavailability of resveratrol through conversion into non-absorbable metabolites such as 3,4′-dihydroxy-*trans*-stilbene and 3,4′-dihydroxybibenzyl.[193] Time of intake (morning or night), prior food intake, and fat content in food are other factors that appear to affect absorption.[194] Once in the blood stream, resveratrol is rapidly metabolized into sulfate, disulfate, and glucuronide derivatives that are quickly excreted from the body,[109,195,196] resulting in a short half-life of only 8–14 minutes for the primary molecule.[196] To circumvent these issues, a proprietary micronized formulation of resveratrol, SRT501, with improved bioavailability has been developed.[197] In clinical trials, SRT501 has achieved blood levels 5–8 times higher than standard resveratrol, suggesting that this formulation may represent a promising pharmaceutical.[198] Likewise, SRT2104 shows moderate bioavailability of roughly 14%, and a mean clearance of ~400 mL min^{-1} in human patients.[199]

11.5.2 Regulatory Paradigms

Regulatory agencies do not typically consider aging a disease given its widespread prevalence. Furthermore, funding for research promoting health is far scarcer than funding for disease-related research. The cost of testing STACs in clinical trials for healthy aging would be enormous given the length of time and logistical complications associated with completing such a study.

For these reasons, it is unlikely that the effects of STACs on human lifespan and healthspan will be assessed directly. Instead, it has been speculated that the first insights into the effects of STACs on human aging and disease prevention will be inferred from longer-term disease-specific trials, many of which are currently ongoing.[15] In addition, it has been estimated that over two thirds of Americans who consume multiple dietary supplements take resveratrol.[115] This large uncontrolled public drug trial could contribute important data in the form of individual medical reports and case studies about how STACs affect human health.

While research into aging and anti-aging drugs was once looked upon with skepticism, there are now clear indications that this perception is changing. For example, in 1974 the US National Institute for Aging (NIA) was founded with the specific directive of funding research focusing on the biological, social, and economic implications of aging (http://www.nia.nih.gov). More recently, the FDA has revised its policies and recommendations regarding clinical trials for lifespan extending therapeutics and given approval for the first anti-aging drug trial. The aim of the Targeting Aging with Metformin (TAME) study is to treat 3000 volunteers aged 70–80 years for 5–7 years with metformin and subsequently observe if age-related disease is delayed.[4] Metformin, an AMPK activator, is currently used for the treatment of type II diabetes and has previously been shown to extend the lifespan of mice.[22] If successful, this study could revolutionize medicine by demonstrating the possibility of treating multiple age-related diseases using a single compound. Moreover, positive results with metformin could help advocate approval for similar studies using STACs.

11.6 Conclusion

Over the past century, developments in medicine relating to vaccination, disinfectants, and antibiotics have led to a dramatic decrease in deaths due to infectious diseases, an increase in lifespan, and an increase in the elderly demographic.[200] In fact, it has been predicted that by the year 2050, 20% of the world's population will be over the age of 60.[200] Unfortunately, this increase in life expectancy is not always coincident with health, as individuals over the age of 65 suffer from 1–3 chronic age-related diseases on average.[200] Therefore, now more than ever, there is a need to develop treatments for aging and age-related disease. STACs are one of several new classes of molecules recently discovered that could potentially be used to treat multiple age-related diseases in humans. Whether STACs fulfill their promise as human anti-aging drugs or not, it is becoming increasingly clear that such compounds do exist, and that the medical and societal impact of the dissemination of these pharmaceuticals will be immense.

References

1. F. B. Johnson, D. A. Sinclair and L. Guarente, *Cell*, 1999, **96**, 291.
2. T. B. Kirkwood, *Nature*, 1977, **270**, 301.
3. P. B. Medawar, *An Unsolved Problem of Biology*, London, H.K. Lewis, 1952.

4. A. M. Vaiserman and F. Marotta, *Trends Pharmacol. Sci.*, 2016, **37**, 331.
5. J. Sun, S. P. Kale, A. M. Childress, C. Pinswasdi and S. M. Jazwinski, *J. Biol. Chem.*, 1994, **269**, 18638.
6. V. D. Longo, G. S. Shadel, M. Kaeberlein and B. Kennedy, *Cell Metab.*, 2012, **16**, 18.
7. D. B. Friedman and T. E. Johnson, *J. Gerontol.*, 1988, **43**, B102.
8. M. R. Klass, *Mech. Ageing Dev.*, 1983, **22**, 279.
9. C. Kenyon, *Philos. Trans. R. Soc., B*, 2011, **366**, 9.
10. J. M. Tullet, *Biogerontology*, 2015, **16**, 221.
11. C. Kenyon, J. Chang, E. Gensch, A. Rudner and R. Tabtiang, *Nature*, 1993, **366**, 461.
12. S. Ayyadevara, R. Alla, J. J. Thaden and R. J. Shmookler Reis, *Aging cell*, 2008, **7**, 13.
13. D. E. Harrison, R. Strong, Z. D. Sharp, J. F. Nelson, C. M. Astle, K. Flurkey, N. L. Nadon, J. E. Wilkinson, K. Frenkel, C. S. Carter, M. Pahor, M. A. Javors, E. Fernandez and R. A. Miller, *Nature*, 2009, **460**, 392.
14. W. Mair, I. Morantte, A. P. Rodrigues, G. Manning, M. Montminy, R. J. Shaw and A. Dillin, *Nature*, 2011, **470**, 404.
15. B. P. Hubbard and D. A. Sinclair, *Trends Pharmacol. Sci.*, 2014, **35**, 146.
16. B. J. Morris, *Free Radical Biol. Med.*, 2013, **56**, 133.
17. C. M. McCay and L. A. Maynard, *J. Nutr.*, 1935, **1935**, 63.
18. *Caloric Restriction: A Key to Understanding and Modulating Aging*, ed. E. J. Masoro, Elsevier, Amsterdam, 2002.
19. R. Weindruch, P. H. Naylor, A. L. Goldstein and R. L. Walford, *J. Gerontol.*, 1988, **43**, B40.
20. *Dietary Restriction and Life Extension, Biological Mechanisms*, ed. C. Barrows and G. C. Kokkonen, CRC Press Inc, Boca Raton, FL, 1982.
21. K. T. Howitz, K. J. Bitterman, H. Y. Cohen, D. W. Lamming, S. Lavu, J. G. Wood, R. E. Zipkin, P. Chung, A. Kisielewski, L. L. Zhang, B. Scherer and D. A. Sinclair, *Nature*, 2003, **425**, 191.
22. A. Martin-Montalvo, E. M. Mercken, S. J. Mitchell, H. H. Palacios, P. L. Mote, M. Scheibye-Knudsen, A. P. Gomes, T. M. Ward, R. K. Minor, M. J. Blouin, M. Schwab, M. Pollak, Y. Zhang, Y. Yu, K. G. Becker, V. A. Bohr, D. K. Ingram, D. A. Sinclair, N. S. Wolf, S. R. Spindler, M. Bernier and R. de Cabo, *Nat. Commun.*, 2013, **4**, 2192.
23. E. Morselli, L. Galluzzi, O. Kepp, A. Criollo, M. C. Maiuri, N. Tavernarakis, F. Madeo and G. Kroemer, *Aging*, 2009, **1**, 961.
24. J. C. Milne, P. D. Lambert, S. Schenk, D. P. Carney, J. J. Smith, D. J. Gagne, L. Jin, O. Boss, R. B. Perni, C. B. Vu, J. E. Bemis, R. Xie, J. S. Disch, P. Y. Ng, J. J. Nunes, A. V. Lynch, H. Yang, H. Galonek, K. Israelian, W. Choy, A. Iffland, S. Lavu, O. Medvedik, D. A. Sinclair, J. M. Olefsky, M. R. Jirousek, P. J. Elliott and C. H. Westphal, *Nature*, 2007, **450**, 712.
25. H. Dai, L. Kustigian, D. Carney, A. Case, T. Considine, B. P. Hubbard, R. B. Perni, T. V. Riera, B. Szczepankiewicz, G. P. Vlasuk and R. L. Stein, *J. Biol. Chem.*, 2010, **285**, 32695.

26. B. P. Hubbard, A. P. Gomes, H. Dai, J. Li, A. W. Case, T. Considine, T. V. Riera, J. E. Lee, S. Y. E, D. W. Lamming, B. L. Pentelute, E. R. Schuman, L. A. Stevens, A. J. Ling, S. M. Armour, S. Michan, H. Zhao, Y. Jiang, S. M. Sweitzer, C. A. Blum, J. S. Disch, P. Y. Ng, K. T. Howitz, A. P. Rolo, Y. Hamuro, J. Moss, R. B. Perni, J. L. Ellis, G. P. Vlasuk and D. A. Sinclair, *Science*, 2013, **339**, 1216.
27. K. S. Bhullar and B. P. Hubbard, *Biochim. Biophys. Acta*, 2015, **1852**(6), 1209–1218.
28. J. E. Haber and J. P. George, *Genetics*, 1979, **93**, 13.
29. A. J. Klar, S. Fogel and K. Macleod, *Genetics*, 1979, **93**, 37.
30. J. Rine, J. N. Strathern, J. B. Hicks and I. Herskowitz, *Genetics*, 1979, **93**, 877.
31. J. Rine and I. Herskowitz, *Genetics*, 1987, **116**, 9.
32. D. Shore, M. Squire and K. A. Nasmyth, *EMBO J.*, 1984, **3**, 2817.
33. B. K. Kennedy, N. R. Austriaco Jr, J. Zhang and L. Guarente, *Cell*, 1995, **80**, 485.
34. B. K. Kennedy, M. Gotta, D. A. Sinclair, K. Mills, D. S. McNabb, M. Murthy, S. M. Pak, T. Laroche, S. M. Gasser and L. Guarente, *Cell*, 1997, **89**, 381.
35. M. Kaeberlein, M. McVey and L. Guarente, *Genes Dev.*, 1999, **13**, 2570.
36. C. Burnett, S. Valentini, F. Cabreiro, M. Goss, M. Somogyvari, M. D. Piper, M. Hoddinott, G. L. Sutphin, V. Leko, J. J. McElwee, R. P. Vazquez-Manrique, A. M. Orfila, D. Ackerman, C. Au, G. Vinti, M. Riesen, K. Howard, C. Neri, A. Bedalov, M. Kaeberlein, C. Soti, L. Partridge and D. Gems, *Nature*, 2011, **477**, 482.
37. J. G. Wood, B. Rogina, S. Lavu, K. Howitz, S. L. Helfand, M. Tatar and D. Sinclair, *Nature*, 2004, **430**, 686.
38. H. A. Tissenbaum and L. Guarente, *Nature*, 2001, **410**, 227.
39. A. Satoh, C. S. Brace, N. Rensing, P. Cliften, D. F. Wozniak, E. D. Herzog, K. A. Yamada and S. Imai, *Cell Metab.*, 2013, **18**, 416.
40. Y. Kanfi, S. Naiman, G. Amir, V. Peshti, G. Zinman, L. Nahum, Z. Bar-Joseph and H. Y. Cohen, *Nature*, 2012, **483**, 218.
41. R. A. Frye, *Biochem. Biophys. Res. Commun.*, 1999, **260**, 273.
42. K. J. Bitterman, R. M. Anderson, H. Y. Cohen, M. Latorre-Esteves and D. A. Sinclair, *J. Biol. Chem.*, 2002, **277**, 45099.
43. B. C. Smith and J. M. Denu, *Biochemistry*, 2006, **45**, 272.
44. K. E. Dittenhafer-Reed, J. L. Feldman and J. M. Denu, *ChemBioChem*, 2011, **12**, 281.
45. A. A. Sauve, I. Celic, J. Avalos, H. Deng, J. D. Boeke and V. L. Schramm, *Biochemistry*, 2001, **40**, 15456.
46. M. D. Jackson, M. T. Schmidt, N. J. Oppenheimer and J. M. Denu, *J. Biol. Chem.*, 2003, **278**, 50985.
47. B. C. Smith and J. M. Denu, *Biochemistry*, 2007, **46**, 14478.
48. A. A. Sauve, C. Wolberger, V. L. Schramm and J. D. Boeke, *Annu. Rev. Biochem.*, 2006, **75**, 435.

49. M. D. Jackson and J. M. Denu, *J. Biol. Chem.*, 2002, **277**, 18535.
50. P. W. Pan, J. L. Feldman, M. K. Devries, A. Dong, A. M. Edwards and J. M. Denu, *J. Biol. Chem.*, 2011, **286**, 14575.
51. H. Dai, J. L. Ellis, D. A. Sinclair and B. P. Hubbard, *Methods Enzymol.*, 2016, **574**, 213–244.
52. F. Fischer, M. Gertz, B. Suenkel, M. Lakshminarasimhan, M. Schutkowski and C. Steegborn, *PLoS One*, 2012, **7**, e45098.
53. B. P. Hubbard and D. A. Sinclair, *Methods Mol. Biol.*, 2013, **1077**, 167.
54. B. J. North, B. L. Marshall, M. T. Borra, J. M. Denu and E. Verdin, *Mol. Cell.*, 2003, **11**, 437.
55. Y. Yang, B. P. Hubbard, D. A. Sinclair and Q. Tong, *J. Cell. Biochem.*, 2010, **111**, 1051.
56. R. A. Mathias, T. M. Greco, A. Oberstein, H. G. Budayeva, R. Chakrabarti, E. A. Rowland, Y. Kang, T. Shenk and I. M. Cristea, *Cell*, 2014, **159**, 1615.
57. J. Du, Y. Zhou, X. Su, J. J. Yu, S. Khan, H. Jiang, J. Kim, J. Woo, J. H. Kim, B. H. Choi, B. He, W. Chen, S. Zhang, R. A. Cerione, J. Auwerx, Q. Hao and H. Lin, *Science*, 2011, **334**, 806.
58. L. Zhou, F. Wang, R. Sun, X. Chen, M. Zhang, Q. Xu, Y. Wang, S. Wang, Y. Xiong, K. L. Guan, P. Yang, H. Yu and D. Ye, *EMBO Rep.*, 2016, **17**, 811.
59. J. E. Dominy Jr, Y. Lee, M. P. Jedrychowski, H. Chim, M. J. Jurczak, J. P. Camporez, H. B. Ruan, J. Feldman, K. Pierce, R. Mostoslavsky, J. M. Denu, C. B. Clish, X. Yang, G. I. Shulman, S. P. Gygi and P. Puigserver, *Mol. Cell*, 2012, **48**, 900.
60. B. Yang, B. M. Zwaans, M. Eckersdorff and D. B. Lombard, *Cell Cycle*, 2009, **8**, 2662.
61. J. L. Feldman, J. Baeza and J. M. Denu, *J. Biol. Chem.*, 2013, **288**, 31350.
62. M. F. Barber, E. Michishita-Kioi, Y. Xi, L. Tasselli, M. Kioi, Z. Moqtaderi, R. I. Tennen, S. Paredes, N. L. Young, K. Chen, K. Struhl, B. A. Garcia, O. Gozani, W. Li and K. F. Chua, *Nature*, 2012, **487**, 114.
63. H. Vaziri, S. K. Dessain, E. Ng Eaton, S. I. Imai, R. A. Frye, T. K. Pandita, L. Guarente and R. A. Weinberg, *Cell*, 2001, **107**, 149.
64. A. Brunet, L. B. Sweeney, J. F. Sturgill, K. F. Chua, P. L. Greer, Y. Lin, H. Tran, S. E. Ross, R. Mostoslavsky, H. Y. Cohen, L. S. Hu, H. L. Cheng, M. P. Jedrychowski, S. P. Gygi, D. A. Sinclair, F. W. Alt and M. E. Greenberg, *Sciencc*, 2004, **303**, 2011.
65. F. Yeung, J. E. Hoberg, C. S. Ramsey, M. D. Keller, D. R. Jones, R. A. Frye and M. W. Mayo, *EMBO J.*, 2004, **23**, 2369.
66. J. T. Rodgers, C. Lerin, W. Haas, S. P. Gygi, B. M. Spiegelman and P. Puigserver, *Nature*, 2005, **434**, 113.
67. H. Y. Cohen, C. Miller, K. J. Bitterman, N. R. Wall, B. Hekking, B. Kessler, K. T. Howitz, M. Gorospe, R. de Cabo and D. A. Sinclair, *Science*, 2004, **305**, 390.
68. G. Asher, D. Gatfield, M. Stratmann, H. Reinke, C. Dibner, F. Kreppel, R. Mostoslavsky, F. W. Alt and U. Schibler, *Cell*, 2008, **134**, 317.
69. G. Donmez, D. Wang, D. E. Cohen and L. Guarente, *Cell*, 2010, **142**, 320.
70. A. S. Banks, N. Kon, C. Knight, M. Matsumoto, R. Gutierrez-Juarez, L. Rossetti, W. Gu and D. Accili, *Cell Metab.*, 2008, **8**, 333.

71. R. Firestein, G. Blander, S. Michan, P. Oberdoerffer, S. Ogino, J. Campbell, A. Bhimavarapu, S. Luikenhuis, R. de Cabo, C. Fuchs, W. C. Hahn, L. P. Guarente and D. A. Sinclair, *PLoS One*, 2008, **3**, e2020.
72. B. Jung-Hynes and N. Ahmad, *Cell Cycle*, 2009, **8**, 1478.
73. P. Oberdoerffer, S. Michan, M. McVay, R. Mostoslavsky, J. Vann, S. K. Park, A. Hartlerode, J. Stegmuller, A. Hafner, P. Loerch, S. M. Wright, K. D. Mills, A. Bonni, B. A. Yankner, R. Scully, T. A. Prolla, F. W. Alt and D. A. Sinclair, *Cell*, 2008, **135**, 907.
74. R. M. Anderson, K. J. Bitterman, J. G. Wood, O. Medvedik and D. A. Sinclair, *Nature*, 2003, **423**, 181.
75. S. J. Lin, P. A. Defossez and L. Guarente, *Science*, 2000, **289**, 2126.
76. B. Rogina and S. L. Helfand, *Proc. Natl. Acad. Sci. U. S. A.*, 2004, **101**, 15998.
77. E. L. Greer and A. Brunet, *Aging Cell*, 2009, **8**, 113.
78. C. M. Gallo, D. L. Smith Jr and J. S. Smith, *Mol. Cell. Biol.*, 2004, **24**, 1301.
79. S. J. Lin, E. Ford, M. Haigis, G. Liszt and L. Guarente, *Genes Dev.*, 2004, **18**, 12.
80. G. Boily, E. L. Seifert, L. Bevilacqua, X. H. He, G. Sabourin, C. Estey, C. Moffat, S. Crawford, S. Saliba, K. Jardine, J. Xuan, M. Evans, M. E. Harper and M. W. McBurney, *PLoS One*, 2008, **3**, e1759.
81. L. Bordone, D. Cohen, A. Robinson, M. C. Motta, E. van Veen, A. Czopik, A. D. Steele, H. Crowe, S. Marmor, J. Luo, W. Gu and L. Guarente, *Aging Cell*, 2007, **6**, 759.
82. A. A. Sauve, R. D. Moir, V. L. Schramm and I. M. Willis, *Mol. Cell.*, 2005, **17**, 595.
83. J. Yoshino, K. F. Mills, M. J. Yoon and S. Imai, *Cell Metab.*, 2011, **14**, 528.
84. R. Cerutti, E. Pirinen, C. Lamperti, S. Marchet, A. A. Sauve, W. Li, V. Leoni, E. A. Schon, F. Dantzer, J. Auwerx, C. Viscomi and M. Zeviani, *Cell Metab.*, 2014, **19**, 1042.
85. H. Zhang, D. Ryu, Y. Wu, K. Gariani, X. Wang, P. Luan, D. D'Amico, E. R. Ropelle, M. P. Lutolf, R. Aebersold, K. Schoonjans, K. J. Menzies and J. Auwerx, *Science*, 2016, **352**, 1436.
86. S. Wang, X. Yang, Y. Lin, X. Qiu, H. Li, X. Zhao, L. Cao, X. Liu, Y. Pang, X. Wang and Z. Chi, *Brain Res.*, 2013, **1535**, 14.
87. M. Kaeberlein, T. McDonagh, B. Heltweg, J. Hixon, E. A. Westman, S. D. Caldwell, A. Napper, R. Curtis, P. S. DiStefano, S. Fields, A. Bedalov and B. K. Kennedy, *J. Biol. Chem.*, 2005, **280**, 17038.
88. M. T. Borra, B. C. Smith and J. M. Denu, *J. Biol. Chem.*, 2005, **280**, 17187.
89. M. Pacholec, J. E. Bleasdale, B. Chrunyk, D. Cunningham, D. Flynn, R. S. Garofalo, D. Griffith, M. Griffor, P. Loulakis, B. Pabst, X. Qiu, B. Stockman, V. Thanabal, A. Varghese, J. Ward, J. Withka and K. Ahn, *J. Biol. Chem.*, 2010, **285**, 8340.
90. S. J. Park, F. Ahmad, A. Philp, K. Baar, T. Williams, H. Luo, H. Ke, H. Rehmann, R. Taussig, A. L. Brown, M. K. Kim, M. A. Beaven, A. B. Burgin, V. Manganiello and J. H. Chung, *Cell*, 2012, **148**, 421.
91. M. Lakshminarasimhan, D. Rauh, M. Schutkowski and C. Steegborn, *Aging*, 2013, **5**, 151.

92. D. Cao, M. Wang, X. Qiu, D. Liu, H. Jiang, N. Yang and R. M. Xu, *Genes Dev.*, 2015, **29**, 1316.
93. H. Dai, A. W. Case, T. V. Riera, T. Considine, J. E. Lee, Y. Hamuro, H. Zhao, Y. Jiang, S. M. Sweitzer, B. Pietrak, B. Schwartz, C. A. Blum, J. S. Disch, R. Caldwell, B. Szczepankiewicz, C. Oalmann, P. Yee Ng, B. H. White, R. Casaubon, R. Narayan, K. Koppetsch, F. Bourbonais, B. Wu, J. Wang, D. Qian, F. Jiang, C. Mao, M. Wang, E. Hu, J. C. Wu, R. B. Perni, G. P. Vlasuk and J. L. Ellis, *Nat. Commun.*, 2015, **6**, 7645.
94. B. Rascon, B. P. Hubbard, D. A. Sinclair and G. V. Amdam, *Aging*, 2012, **4**, 499.
95. X. Yu and G. Li, *Exp. Gerontol.*, 2012, **47**, 940.
96. D. R. Valenzano, E. Terzibasi, T. Genade, A. Cattaneo, L. Domenici and A. Cellerino, *Curr. Biol.*, 2006, **16**, 296.
97. T. M. Bass, D. Weinkove, K. Houthoofd, D. Gems and L. Partridge, *Mech. Ageing Dev.*, 2007, **128**, 546.
98. M. Viswanathan and L. Guarente, *Nature*, 2011, **477**, E1.
99. C. Wang, C. T. Wheeler, T. Alberico, X. Sun, J. Seeberger, M. Laslo, E. Spangler, B. Kern, R. de Cabo and S. Zou, *Age*, 2013, **35**, 69.
100. E. Kim, C. M. Ansell and J. L. Dudycha, *J. Exp. Zool., Part A*, 2014, **321**, 48.
101. A. A. Johnson and M. A. Riehle, *Rejuvenation Res.*, 2015, **18**, 473.
102. K. J. Pearson, J. A. Baur, K. N. Lewis, L. Peshkin, N. L. Price, N. Labinskyy, W. R. Swindell, D. Kamara, R. K. Minor, E. Perez, H. A. Jamieson, Y. Zhang, S. R. Dunn, K. Sharma, N. Pleshko, L. A. Woollett, A. Csiszar, Y. Ikeno, D. Le Couteur, P. J. Elliott, K. G. Becker, P. Navas, D. K. Ingram, N. S. Wolf, Z. Ungvari, D. A. Sinclair and R. de Cabo, *Cell Metab.*, 2008, **8**, 157.
103. R. A. Miller, D. E. Harrison, C. M. Astle, J. A. Baur, A. R. Boyd, R. de Cabo, E. Fernandez, K. Flurkey, M. A. Javors, J. F. Nelson, C. J. Orihuela, S. Pletcher, Z. D. Sharp, D. Sinclair, J. W. Starnes, J. E. Wilkinson, N. L. Nadon and R. Strong, *J. Gerontol., Ser. A*, 2011, **66**, 191.
104. R. Strong, R. A. Miller, C. M. Astle, J. A. Baur, R. de Cabo, E. Fernandez, W. Guo, M. Javors, J. L. Kirkland, J. F. Nelson, D. A. Sinclair, B. Teter, D. Williams, N. Zaveri, N. L. Nadon and D. E. Harrison, *J. Gerontol., Ser. A*, 2013, **68**, 6.
105. J. A. Baur, K. J. Pearson, N. L. Price, H. A. Jamieson, C. Lerin, A. Kalra, V. V. Prabhu, J. S. Allard, G. Lopez-Lluch, K. Lewis, P. J. Pistell, S. Poosala, K. G. Becker, O. Boss, D. Gwinn, M. Wang, S. Ramaswamy, K. W. Fishbein, R. G. Spencer, E. G. Lakatta, D. Le Couteur, R. J. Shaw, P. Navas, P. Puigserver, D. K. Ingram, R. de Cabo and D. A. Sinclair, *Nature*, 2006, **444**, 337.
106. S. J. Mitchell, A. Martin-Montalvo, E. M. Mercken, H. H. Palacios, T. M. Ward, G. Abulwerdi, R. K. Minor, G. P. Vlasuk, J. L. Ellis, D. A. Sinclair, J. Dawson, D. B. Allison, Y. Zhang, K. G. Becker, M. Bernier and R. de Cabo, *Cell Rep.*, 2014, **6**, 836.
107. R. K. Minor, J. A. Baur, A. P. Gomes, T. M. Ward, A. Csiszar, E. M. Mercken, K. Abdelmohsen, Y. K. Shin, C. Canto, M. Scheibye-Knudsen, M.

Krawczyk, P. M. Irusta, A. Martin-Montalvo, B. P. Hubbard, Y. Zhang, E. Lehrmann, A. A. White, N. L. Price, W. R. Swindell, K. J. Pearson, K. G. Becker, V. A. Bohr, M. Gorospe, J. M. Egan, M. I. Talan, J. Auwerx, C. H. Westphal, J. L. Ellis, Z. Ungvari, G. P. Vlasuk, P. J. Elliott, D. A. Sinclair and R. de Cabo, *Sci. Rep.*, 2011, **1**, 70.

108. E. M. Mercken, S. J. Mitchell, A. Martin-Montalvo, R. K. Minor, M. Almeida, A. P. Gomes, M. Scheibye-Knudsen, H. H. Palacios, J. J. Licata, Y. Zhang, K. G. Becker, H. Khraiwesh, J. A. Gonzalez-Reyes, J. M. Villalba, J. A. Baur, P. Elliott, C. Westphal, G. P. Vlasuk, J. L. Ellis, D. A. Sinclair, M. Bernier and R. de Cabo, *Aging Cell*, 2014, **13**, 787.
109. J. A. Baur and D. A. Sinclair, *Nat. Rev. Drug Discovery*, 2006, **5**, 493.
110. N. L. Price, A. P. Gomes, A. J. Ling, F. V. Duarte, A. Martin-Montalvo, B. J. North, B. Agarwal, L. Ye, G. Ramadori, J. S. Teodoro, B. P. Hubbard, A. T. Varela, J. G. Davis, B. Varamini, A. Hafner, R. Moaddel, A. P. Rolo, R. Coppari, C. M. Palmeira, R. de Cabo, J. A. Baur and D. A. Sinclair, *Cell Metab.*, 2012, **15**, 675.
111. Y. Jimenez-Gomez, J. A. Mattison, K. J. Pearson, A. Martin-Montalvo, H. H. Palacios, A. M. Sossong, T. M. Ward, C. M. Younts, K. Lewis, J. S. Allard, D. L. Longo, J. P. Belman, M. M. Malagon, P. Navas, M. Sanghvi, R. Moaddel, E. M. Tilmont, R. L. Herbert, C. H. Morrell, J. M. Egan, J. A. Baur, L. Ferrucci, J. S. Bogan, M. Bernier and R. de Cabo, *Cell Metab.*, 2013, **18**, 533.
112. J. A. Mattison, M. Wang, M. Bernier, J. Zhang, S. S. Park, S. Maudsley, S. S. An, L. Santhanam, B. Martin, S. Faulkner, C. Morrell, J. A. Baur, L. Peshkin, D. Sosnowska, A. Csiszar, R. L. Herbert, E. M. Tilmont, Z. Ungvari, K. J. Pearson, E. G. Lakatta and R. de Cabo, *Cell Metab.*, 2014, **20**, 183.
113. J. P. Hyatt, L. Nguyen, A. E. Hall, A. M. Huber, J. C. Kocan, J. A. Mattison, R. de Cabo, J. R. LaRocque and R. J. Talmadge, *Front. Physiol.*, 2016, **7**, 77.
114. M. Bernier, D. Wahl, A. Ali, J. Allard, S. Faulkner, A. Wnorowski, M. Sanghvi, R. Moaddel, I. Alfaras, J. A. Mattison, S. Tarantini, Z. Tucsek, Z. Ungvari, A. Csiszar, K. J. Pearson and R. de Cabo, *Aging*, 2016, **8**, 899.
115. J. M. Smoliga, J. A. Baur and H. A. Hausenblas, *Mol. Nutr. Food Res.*, 2011, **55**, 1129.
116. S. Timmers, E. Konings, L. Bilet, R. H. Houtkooper, T. van de Weijer, G. H. Goossens, J. Hoeks, S. van der Krieken, D. Ryu, S. Kersten, E. Moonen-Kornips, M. K. Hesselink, I. Kunz, V. B. Schrauwen-Hinderling, E. E. Blaak, J. Auwerx and P. Schrauwen, *Cell Metab.*, 2011, **14**, 612.
117. F. Faghihzadeh, P. Adibi, R. Rafiei and A. Hekmatdoost, *Nutr. Res.*, 2014, **34**, 837.
118. J. Yoshino, C. Conte, L. Fontana, B. Mittendorfer, S. Imai, K. B. Schechtman, C. Gu, I. Kunz, F. Rossi Fanelli, B. W. Patterson and S. Klein, *Cell Metab.*, 2012, **16**, 658.
119. S. S. Thazhath, T. Wu, M. J. Bound, H. L. Checklin, S. Standfield, K. L. Jones, M. Horowitz and C. K. Rayner, *Am. J. Clin. Nutr.*, 2016, **103**, 66.

120. S. Venkatasubramanian, R. M. Noh, S. Daga, J. P. Langrish, N. L. Mills, B. R. Waterhouse, E. Hoffmann, E. W. Jacobson, N. N. Lang, B. M. Frier and D. E. Newby, *Open Heart*, 2016, **3**, e000402.
121. A. Baksi, O. Kraydashenko, A. Zalevkaya, R. Stets, P. Elliott, J. Haddad, E. Hoffmann, G. P. Vlasuk and E. W. Jacobson, *Br. J. Clin. Pharmacol.*, 2014, **78**, 69.
122. R. E. Gilbert, K. Thai, S. L. Advani, C. L. Cummins, D. M. Kepecs, S. A. Schroer, M. Woo and Y. Zhang, *Diabetologia*, 2015, **58**, 819.
123. M. Jang, L. Cai, G. O. Udeani, K. V. Slowing, C. F. Thomas, C. W. Beecher, H. H. Fong, N. R. Farnsworth, A. D. Kinghorn, R. G. Mehta, R. C. Moon and J. M. Pezzuto, *Science*, 1997, **275**, 218.
124. M. Sengottuvelan, P. Viswanathan and N. Nalini, *Carcinogenesis*, 2006, **27**, 1038.
125. L. G. Carter, J. A. D'Orazio and K. J. Pearson, *Endocr.-Relat. Cancer*, 2014, **21**, R209.
126. Q. Gao, Y. Yuan, H. Z. Gan and Q. Peng, *Oncol. Lett.*, 2015, **9**, 2381.
127. G. Boily, X. H. He, B. Pearce, K. Jardine and M. W. McBurney, *Oncogene*, 2009, **28**, 2882.
128. J. C. Klink, A. K. Tewari, E. M. Masko, J. Antonelli, P. G. Febbo, P. Cohen, M. W. Dewhirst, S. V. Pizzo and S. J. Freedland, *Prostate*, 2013, **73**, 754.
129. A. Scuto, M. Kirschbaum, R. Buettner, M. Kujawski, J. M. Cermak, P. Atadja and R. Jove, *Cell Death Dis.*, 2013, **4**, e635.
130. C. C. Chini, J. M. Espindola-Netto, G. Mondal, A. M. Guerrico, V. Nin, C. Escande, M. Sola-Penna, J. S. Zhang, D. D. Billadeau and E. N. Chini, *Clin. Cancer Res.*, 2016, **22**, 2496.
131. T. J. Lahusen and C. X. Deng, *Mol. Cancer Ther.*, 2015, **14**, 183.
132. D. Chauhan, M. Bandi, A. V. Singh, A. Ray, N. Raje, P. Richardson and K. C. Anderson, *Br. J. Haematol.*, 2011, **155**, 588.
133. K. Suzuki, R. Hayashi, T. Ichikawa, S. Imanishi, T. Yamada, M. Inomata, T. Miwa, S. Matsui, I. Usui, M. Urakaze, Y. Matsuya, H. Ogawa, H. Sakurai, I. Saiki and K. Tobe, *Oncol. Rep.*, 2012, **27**, 1726.
134. Y. Cheng, L. Cai, P. Jiang, J. Wang, C. Gao, H. Feng, C. Wang, H. Pan and Y. Yang, *Eur. J. Pharmacol.*, 2013, **715**, 219.
135. Y. N. Zhao, W. F. Li, F. Li, Z. Zhang, Y. D. Dai, A. L. Xu, C. Qi, J. M. Gao and J. Gao, *Biochem. Biophys. Res. Commun.*, 2013, **135**, 597.
136. Z. Meng, J. Li, H. Zhao, H. Liu, G. Zhang, L. Wang, H. E. Hu, D. I. Li, M. Liu, F. Bi, X. Wang, G. Tian, Q. Liu and B. Buren, *Exp. Ther. Med.*, 2015, **10**, 525.
137. P. Marambaud, H. Zhao and P. Davies, *J. Biol. Chem.*, 2005, **280**, 37377.
138. S. S. Karuppagounder, J. T. Pinto, H. Xu, H. L. Chen, M. F. Beal and G. E. Gibson, *Neurochem. Int.*, 2009, **54**, 111.
139. D. Porquet, G. Casadesus, S. Bayod, A. Vicente, A. M. Canudas, J. Vilaplana, C. Pelegri, C. Sanfeliu, A. Camins, M. Pallas and J. del Valle, *Age*, 2013, **35**, 1851.
140. M. M. Khan, A. Ahmad, T. Ishrat, M. B. Khan, M. N. Hoda, G. Khuwaja, S. S. Raza, A. Khan, H. Javed, K. Vaibhav and F. Islam, *Brain Res.*, 2010, **1328**, 139.

141. R. Mancuso, J. del Valle, L. Modol, A. Martinez, A. B. Granado-Serrano, O. Ramirez-Nunez, M. Pallas, M. Portero-Otin, R. Osta and X. Navarro, *Neurotherapeutics*, 2014, **11**, 419.
142. L. Song, L. Chen, X. Zhang, J. Li and W. Le, *BioMed Res. Int.*, 2014, **2014**, 483501.
143. J. Graff, M. Kahn, A. Samiei, J. Gao, K. T. Ota, D. Rei and L. H. Tsai, *J. Neurosci.*, 2013, **33**, 8951.
144. K. S. Shindler, E. Ventura, M. Dutt, P. Elliott, D. C. Fitzgerald and A. Rostami, *J Neuroophthalmol.*, 2010, **30**, 328.
145. M. Jiang, J. Zheng, Q. Peng, Z. Hou, J. Zhang, S. Mori, J. L. Ellis, G. P. Vlasuk, H. Fries, V. Suri and W. Duan, *Ann. Clin. Transl. Neurol.*, 2014, **1**, 1047.
146. J. M. Wu and T. C. Hsieh, *Ann. N. Y. Acad. Sci.*, 2011, **1215**, 16.
147. M. Jang and J. M. Pezzuto, *Drugs Exp. Clin. Res.*, 1999, **25**, 65.
148. S. Das, V. K. Alagappan, D. Bagchi, H. S. Sharma, N. Maulik and D. K. Das, *Vasc. Pharmacol.*, 2005, **42**, 281.
149. J. Gracia-Sancho, G. Villarreal Jr, Y. Zhang and G. Garcia-Cardena, *Cardiovasc. Res.*, 2010, **85**, 514.
150. Z. Wang, J. Zou, K. Cao, T. C. Hsieh, Y. Huang and J. M. Wu, *Int. J. Mol. Med.*, 2005, **16**, 533.
151. P. P. Chothe and P. W. Swaan, *Biochem. J.*, 2014, **459**, 301.
152. C. Tong, A. Morrison, S. Mattison, S. Qian, M. Bryniarski, B. Rankin, J. Wang, D. P. Thomas and J. Li, *FASEB J.*, 2013, **27**, 4332.
153. L. B. Gano, A. J. Donato, H. M. Pasha, C. M. Hearon Jr, A. L. Sindler and D. R. Seals, *Am. J. Physiol.: Heart Circ. Physiol.*, 2014, **307**, H1754.
154. Y. X. Chen, M. Zhang, Y. Cai, Q. Zhao and W. Dai, *Biochem. Biophys. Res. Commun.*, 2015, **465**, 732.
155. M. X. Miranda, L. J. van Tits, C. Lohmann, T. Arsiwala, S. Winnik, A. Tailleux, S. Stein, A. P. Gomes, V. Suri, J. L. Ellis, T. A. Lutz, M. O. Hottiger, D. A. Sinclair, J. Auwerx, K. Schoonjans, B. Staels, T. F. Luscher and C. M. Matter, *Eur. Heart J.*, 2015, **36**, 51.
156. S. Venkatasubramanian, R. M. Noh, S. Daga, J. P. Langrish, N. V. Joshi, N. L. Mills, E. Hoffmann, E. W. Jacobson, G. P. Vlasuk, B. R. Waterhouse, N. N. Lang and D. E. Newby, *J. Am. Heart Assoc.*, 2013, **2**, e000042.
157. K. Rahal, P. Schmiedlin-Ren, J. Adler, M. Dhanani, V. Sultani, A. C. Rittershaus, L. Reingold, J. Zhu, B. J. McKenna, G. M. Christman and E. M. Zimmermann, *Inflammatory Bowel Dis.*, 2012, **18**, 613.
158. T. N. Kjaer, K. Thorsen, N. Jessen, K. Stenderup and S. B. Pedersen, *PLoS One*, 2015, **10**, e0126599.
159. N. Elmali, O. Baysal, A. Harma, I. Esenkaya and B. Mizrak, *Inflammation*, 2007, **30**, 1.
160. J. J. Docherty, M. M. Fu, B. S. Stiffler, R. J. Limperos, C. M. Pokabla and A. L. DeLucia, *Antiviral Res.*, 1999, **43**, 145.
161. D. W. Park, J. S. Kim, B. R. Chin and S. H. Baek, *J. Med. Food*, 2012, **15**, 788.
162. A. A. Qureshi, X. Q. Guan, J. C. Reis, C. J. Papasian, S. Jabre, D. C. Morrison and N. Qureshi, *Lipids Health Dis.*, 2012, **11**, 76.

163. J. Knobloch, C. Wahl, M. Feldmann, D. Jungck, J. Strauch, E. Stoelben and A. Koch, *Basic Clin. Pharmacol. Toxicol.*, 2014, **114**, 202.
164. G. Chen, J. Tang, Z. Ni, Q. Chen, Z. Li, W. Yang, J. Din, X. Luo and X. Wang, *Biol. Pharm. Bull.*, 2015, **38**, 507.
165. W. R. Swindell, X. Xing, Y. Fritz, D. Diaconu, D. I. Simon, N. L. Ward and J. E. Gudjonsson, *Oncotarget*, 2016, **7**(24), 35535–35551.
166. T. Ichikawa, R. Hayashi, K. Suzuki, S. Imanishi, K. Kambara, S. Okazawa, M. Inomata, T. Yamada, Y. Yamazaki, Y. Koshimizu, T. Miwa, S. Matsui, I. Usui, M. Urakaze, Y. Matsuya, M. Sasahara and K. Tobe, *Respirology*, 2013, **18**, 332.
167. H. Yao, S. Chung, J. W. Hwang, S. Rajendrasozhan, I. K. Sundar, D. A. Dean, M. W. McBurney, L. Guarente, W. Gu, M. Ronty, V. L. Kinnula and I. Rahman, *J. Clin. Invest.*, 2012, **122**, 2032.
168. K. E. Cunningham, G. Vincent, C. P. Sodhi, E. A. Novak, S. Ranganathan, C. E. Egan, D. B. Stolz, M. B. Rogers, B. Firek, M. J. Morowitz, G. K. Gittes, B. S. Zuckerbraun, D. J. Hackam and K. P. Mollen, *J. Biol. Chem.*, 2016, **291**, 10184.
169. A. J. van der Meer, B. P. Scicluna, P. D. Moerland, J. Lin, E. W. Jacobson, G. P. Vlasuk and T. van der Poll, *Crit. Care Med.*, 2015, **43**, e199.
170. J. G. Krueger, M. Suarez-Farinas, I. Cueto, A. Khacherian, R. Matheson, L. C. Parish, C. Leonardi, D. Shortino, A. Gupta, J. Haddad, G. P. Vlasuk and E. W. Jacobson, *PLoS One*, 2015, **10**, e0142081.
171. B. E. Sands, S. Joshi, J. Haddad, J. M. Freudenberg, D. E. Oommen, E. Hoffmann, S. W. McCallum and E. Jacobson, *Inflammatory Bowel Dis.*, 2016, **22**, 607.
172. R. Caruso, I. Marafini, E. Franze, C. Stolfi, F. Zorzi, I. Monteleone, F. Caprioli, A. Colantoni, M. Sarra, S. Sedda, L. Biancone, P. Sileri, G. S. Sica, T. T. MacDonald, F. Pallone and G. Monteleone, *Mucosal Immunol.*, 2014, **7**, 1467.
173. A. M. el-Mowafy, L. A. Abou-Zeid and I. Edafiogho, *Med. Princ. Pract.*, 2002, **11**, 86.
174. K. Kubo, O. Arai, M. Omura, R. Watanabe, R. Ogata and S. Aou, *Neurosci. Res.*, 2003, **45**, 345.
175. J. A. Crowell, P. J. Korytko, R. L. Morrissey, T. D. Booth and B. S. Levine, *Toxicol. Sci.*, 2004, **82**, 614.
176. M. Liu, Y. Yin, X. Ye, M. Zeng, Q. Zhao, D. L. Keefe and L. Liu, *Hum. Reprod.*, 2013, **28**, 707.
177. A. L. Katen, S. J. Stanger, A. L. Anderson, B. Nixon and S. D. Roman, *Reprod. Toxicol.*, 2016, **63**, 1.
178. X. L. Zhou, J. J. Xu, Y. H. Ni, X. C. Chen, H. X. Zhang, X. M. Zhang, W. J. Liu, L. L. Luo and Y. C. Fu, *J. Ovarian Res.*, 2014, **7**, 97.
179. M. Mendez-del Villar, M. Gonzalez-Ortiz, E. Martinez-Abundis, K. G. Perez-Rubio and R. Lizarraga-Valdez, *Metab. Syndr. Relat. Disord.*, 2014, **12**, 497.
180. M. Samsami-Kor, N. E. Daryani, P. R. Asl and A. Hekmatdoost, *Arch. Med. Res.*, 2015, **46**, 280.

181. E. Barrajon-Catalan, M. Herranz-Lopez, J. Joven, A. Segura-Carretero, C. Alonso-Villaverde, J. A. Menendez and V. Micol, *Adv. Exp. Med. Biol.*, 2014, **824**, 141.
182. M. Murias, N. Handler, T. Erker, K. Pleban, G. Ecker, P. Saiko, T. Szekeres and W. Jager, *Bioorg. Med. Chem.*, 2004, **12**, 5571.
183. C. Yu, Y. G. Shin, J. W. Kosmeder, J. M. Pezzuto and R. B. van Breemen, *Rapid Commun. Mass Spectrom.*, 2003, **17**, 307.
184. B. Piver, F. Berthou, Y. Dreano and D. Lucas, *Toxicol. Lett.*, 2001, **125**, 83.
185. G. H. Perdew, B. D. Hollingshead, B. C. Dinatale, J. L. Morales, M. P. Labrecque, M. K. Takhar, K. J. Tam and T. V. Beischlag, *J. Pharmacol. Exp. Ther.*, 2010, **335**, 273.
186. J. B. Cai, Z. H. Zhang, D. J. Xu, Z. Y. Qian, Z. R. Wang, Y. Z. Huang, J. G. Zou and K. J. Cao, *Clin. Exp. Pharmacol. Physiol.*, 2008, **35**, 1419.
187. M. V. Blagosklonny, *Aging*, 2009, **1**, 511.
188. K. T. Howitz and D. A. Sinclair, *Cell*, 2008, **133**, 387.
189. N. B. Ruderman, X. J. Xu, L. Nelson, J. M. Cacicedo, A. K. Saha, F. Lan and Y. Ido, *Am. J. Physiol.: Endocrinol. Metab.*, 2010, **298**, E751.
190. V. Libri, A. P. Brown, G. Gambarota, J. Haddad, G. S. Shields, H. Dawes, D. J. Pinato, E. Hoffman, P. J. Elliot, G. P. Vlasuk, E. Jacobson, M. R. Wilkins and P. M. Matthews, *PLoS One*, 2012, **7**, e51395.
191. J. L. Huber, M. W. McBurney, P. S. Distefano and T. McDonagh, *Future Med. Chem.*, 2010, **2**, 1751.
192. D. J. Boocock, G. E. Faust, K. R. Patel, A. M. Schinas, V. A. Brown, M. P. Ducharme, T. D. Booth, J. A. Crowell, M. Perloff, A. J. Gescher, W. P. Steward and D. E. Brenner, *Cancer Epidemiol., Biomarkers Prev.*, 2007, **16**, 1246.
193. L. M. Bode, D. Bunzel, M. Huch, G. S. Cho, D. Ruhland, M. Bunzel, A. Bub, C. M. Franz and S. E. Kulling, *Am. J. Clin. Nutr.*, 2013, **97**, 295.
194. M. Vaz-da-Silva, A. I. Loureiro, A. Falcao, T. Nunes, J. F. Rocha, C. Fernandes-Lopes, E. Soares, L. Wright, L. Almeida and P. Soares-da-Silva, *Int. J. Clin. Pharmacol. Ther.*, 2008, **46**, 564.
195. H. H. Chow, L. L. Garland, C. H. Hsu, D. R. Vining, W. M. Chew, J. A. Miller, M. Perloff, J. A. Crowell and D. S. Alberts, *Cancer Prev. Res.*, 2010, **3**, 1168.
196. J. F. Marier, P. Vachon, A. Gritsas, J. Zhang, J. P. Moreau and M. P. Ducharme, *J. Pharmacol. Exp. Ther.*, 2002, **302**, 369.
197. P. J. Elliott and M. Jirousek, *Curr. Opin. Invest. Drugs*, 2008, **9**, 371.
198. L. M. Howells, D. P. Berry, P. J. Elliott, E. W. Jacobson, E. Hoffmann, B. Hegarty, K. Brown, W. P. Steward and A. J. Gescher, *Cancer Prev. Res.*, 2011, **4**, 1419.
199. E. Hoffmann, J. Wald, S. Lavu, J. Roberts, C. Beaumont, J. Haddad, P. Elliott, C. Westphal and E. Jacobson, *Br. J. Clin. Pharmacol.*, 2013, **75**, 186.
200. B. K. Kennedy and J. K. Pennypacker, *Transl. Res.*, 2014, **163**, 456.

CHAPTER 12

Therapeutic Potential of Sirtuin Inhibitors in Cancer

FRANCISCO JAVIER ALCAIN*[a,b], CONSUELO M. NIEVA[a], LUCÍA FERNÁNDEZ DEL RÍO[c], RAQUEL SANTIAGO-MORA[a,b], ELENA GUTIÉRREZ-CASADO[c], MARIO DURAN-PRADO[a,b] AND JOSE M. VILLALBA[c]

[a]Department of Medical Sciences, Faculty of Medicine, University of Castilla-La Mancha, Ciudad Real, Spain; [b]Oxidative Stress and Neurodegeneration Group, Regional Centre for Biomedical Research, Ciudad Real, Spain; [c]Department of Cell Biology, Physiology and Immunology, Faculty of Sciences, University of Córdoba, Agrifood Campus of International Excellence ceiA3, Córdoba, Spain
*E-mail: franciscoj.alcain@uclm.es

12.1 Introduction

Acetylation in the ε-amino group of lysine residues results in a gain-of-function or loss-of-function for many proteins, such as histones, transcription factors and proteins that are involved in DNA repair and replication, metabolism, cytoskeletal dynamics, apoptosis, protein folding and cellular signaling. *N*-ε-acetylation is reversible and is dynamically controlled by histone lysine deacetylases (HDACs).[1] So far, four different classes of HDACs have been identified in humans, and sirtuins belong to the third class of deacetylase enzymes. Mammalian sirtuins comprise seven forms (SIRT1–7) and,

RSC Drug Discovery Series No. 57
Anti-aging Drugs: From Basic Research to Clinical Practice
Edited by Alexander M. Vaiserman

Published by the Royal Society of Chemistry, www.rsc.org

although deacetylase activity has not been reported for all members, all sirtuins contain a conserved catalytic core domain of 275 amino acids and require nicotinamide adenine dinucleotide (NAD^+) as an essential co-factor.[2,3] Sirtuins also differ in subcellular locations. SIRT1, SRT6 and SIRT7 are nuclear proteins with distinct subnuclear localizations. For example, SIRT1 is detected in the nuclei but not the nucleoli, whereas SIRT7 is a widely expressed nucleolar protein associated with active rRNA. SIRT2 is predominantly a cytoplasmic protein, but it can also be located in the nucleus. SIRT3, SIRT4 and SIRT5 are mitochondrial proteins, although SIRT3 has also been reported to move from the nucleus to mitochondria during cellular stress.[4] Sirtuins are controlling the activity of many proteins related to cell survival, inflammation, lifespan regulation, metabolism, cell cycle regulation, apoptosis and senescence, DNA repair and genome stability. Moreover, sirtuins can modulate lifespan in worms and flies, and their activities may underlie the beneficial effects of caloric restriction and moderate exercise, which are the best known experimental interventions leading to healthy aging and lifespan extension.[4,5] Thus, sirtuin activation may help in preventing the aging-related decline in heart function and neuronal loss, as well as tumorigenesis.

Cancer can be considered as a collection of related diseases where its key feature is the uncontrolled rate of cell division. In the evolution from a normal to a neoplastic state, cells acquire different hallmarks that make them tumorigenic and ultimately malignant. These hallmarks are implicated in sustaining proliferative signaling, evading growth suppressors, avoiding replicative senescence, reprogramming energy metabolism, inducing angiogenesis, activating invasion by regulation of the cell motility and epithelial-to-mesenchymal transition, and increasing genome instability.[6] Different sirtuins are key regulators of this wide variety of cellular and physiological processes, and can act either as tumor suppressors or as tumor promoters, depending of the cellular context. In this sense, sirtuins can be recognized as multiple functional proteins acting solely as tumor suppressor, as described for SIRT4, having a role as tumor suppressor or oncogene, as described for SIRT1, SIRT2, SIRT3 and SIRT6, or showing only oncogene-like activity, as SIRT5 and SIRT7.[7] In those models where sirtuin activation favors cell transformation, pharmacological inhibition of sirtuins can be a valuable tool to aid in the fight against cancer. In this chapter we review the potential use of sirtuin inhibitors in the treatment of cancer.

12.2 Expression of Sirtuins in Cancer Cells

SIRT1, which is an orthologue of the yeast Sir2 protein,[8] is the most studied among sirtuins, which have attracted great interest as a result of reports that calorie restriction (CR) could extend lifespan in mammals by inducing SIRT1 expression, promoting the long-term survival of irreplaceable cells and reducing cancer incidence.[9] SIRT1 is a multiple functional protein that deacetylates various substrates, including histones, transcription factors, DNA repair factors and signaling proteins, thereby modulating their activity and presenting a dual role in cancer.[7] SIRT1 inhibits NF-kB transcription, sensitizing cells to undergo apoptosis in response to TNFα, but also inhibits p53 activity, the first discovered non-histone target of SIRT1, suggesting that SIRT1 can play a central role in tumorigenesis and senescence.[10] In genetic mouse models it was shown to act as a tumor suppressor,[7,11–13] as was also the case for several human cancers where its expression was found reduced in comparison with in normal tissues. In this way, 30% of colorectal adenocarcinomas displayed lower than normal SIRT1 expression that was gradually decreased during carcinogenesis and tumor progression.[14,15] In addition, both mRNA and protein levels were downregulated as compared with the corresponding non-neoplastic tissue in gastric cancer.[16] Moreover, SIRT1 expression is associated with good prognosis for head and neck squamous cell carcinoma (HNSCC) patients.[17] Human breast cancers, hepatocellular carcinoma (HCC), esophageal adenocarcinomas, gastric adenocarcinomas, and HNSCC exhibited reduced SIRT2 levels compared with normal tissues.[12,18] The expression levels of both SIRT2 and SIRT3 were significantly reduced in tumor tissues from basal cell carcinoma.[19] Gastric cancer patients with SIRT3 expression have better prognosis than those without,[20] and low levels of SIRT3 and SIRT7 expression correlated with more aggressive tumor phenotypes and poorer outcome, as measured by disease-free and disease-specific survival time 12 months post-diagnosis.[21] Finally, protein levels of SIRT6 were decreased in human non-small cell lung cancer (NSCLC) tissues.[22]

However, there is also strong evidence that SIRT1, SIRT3 and SIRT6 can behave as tumor promoters in human tumorigenesis because their expression is significantly increased in many tumors. In accordance, SIRT1 was consistently overexpressed in multiple hematopoietic malignant diseases, including acute myeloid leukemia (AML), samples compared with all controls,[23] in diffuse large B-cell lymphoma where it is associated with poor prognosis,[24] and in chronic myelogenous leukemia (CML) where it is required for efficient BCR-ABL transformation of hematopoietic progenitor cells.[25] In a solid tumor, such as human prostate cancer, it was shown that cancer cells had greater SIRT1 expression than uninvolved cells.[26] High levels of SIRT1 expression correlated with advanced stages and poor prognosis of colorectal carcinoma (CRC),[27,28] gastric cancer,[29] pancreatic ductal adenocarcinoma,[30] hepatocellular carcinoma (HCC) and cholangiocarcinoma,[31] NSCLC,[32,33] breast cancer,[34] endometrial cancer,[35] ovarian carcinoma,[36] Barret's esophagus carcinoma,[37] synovial sarcomas,[38] and osteosarcoma.[39] How SIRT1 is overexpressed in many cancers is not fully understood, but it has been proposed that it is due to evasion of SIRT1 mRNA from repression by a group of SIRT1-targeting microRNAs that might be robustly silenced in cancer.[40] In gastric cancer, the activating transcription factor 4 (ATF4), a stress response gene involved in homeostasis and cellular protection, binds directly to the SIRT1 promoter resulting in SIRT1 upregulation.[41] ATF4 overexpression has been also correlated with multiple malignant characteristics and indicates poor prognosis in esophageal squamous cell carcinoma (ESCC) patients.[42]

Moreover, it promoted cell proliferation, migration and lung metastasis in osteosarcoma cell lines and patient clinical samples, as compared to matched non-tumor tissue.[43]

High SIRT3 expression in the cytoplasm significantly correlates with high tumor grades in positive lymph node status, and poor prognosis in colon cancer patients.[44] Again highlighting the double face of sirtuins in cancer, low levels of SIRT3 have been associated with poor outcome in breast cancer[45] but high levels of SIRT3 expression also predicted a poor prognosis in patients, increasing lymph node metastasis, pathological grade and tumor size for breast cancer.[46,47] Only SIRT3 and, to a lesser extent, SIRT7 were overexpressed in three cell lines of oral squamous cell carcinomas (OSCC) compared with primary keratinocytes.[48]

SIRT5 and SIRT7 have been described acting only as oncogenes in relation to tumorigenesis. SIRT5 was significantly increased in human NSCLC tissues at both protein and mRNA levels compared with adjacent normal lung tissues; it was associated with large tumor volume, metastasis and high disease stage, and predicted poor overall and disease-free survival.[49] Meanwhile, SIRT7 overexpression has been detected in hepatocarcinoma, thyroid and breast cancers.[12,46,50] In addition to SIRT1, microRNAs are direct suppressors of SIRT7 and may function as tumor suppressors by controlling aberrant expression of SIRT7 in HCC tumorigenesis.[51]

12.3 Sirtuins and the Hallmarks of Cancer

The main feature during tumorigenesis is the increase in the number of cells. This increment can be produced by different mechanisms: by increasing the rate of cell proliferation, by avoiding apoptosis, or by a combination of both. The first non-histone substrate found for SIRT1 was p53, where the acetylation in lysine residues is indispensable for p53 activation. The tumor suppressor p53 inhibits the formation of tumors by controlling the cell cycle, apoptosis and DNA repair in response to various forms of genotoxic stress.[52] Luo *et al.* and Vaziri *et al.* found that SIRT1-mediated deacetylation, with a specificity for its C-terminal Lys382 residue, antagonized p53-dependent transcriptional activation and apoptosis. SIRT1 activation avoided the increase in the levels of acetylated p53 upon exposure of cells to ionizing radiation. Conversely, levels of acetylated p53 were enhanced when the cells were treated with nicotinamide, a general competitive inhibitor of sirtuin activity.[10,53] SIRT2 is also able to deacetylate and downregulate the transcriptional activity of p53 in HEK293 cells,[54] and recently it has been reported that p53 levels are also regulated by SIRT7 in NIH3T3 and U2OS cells.[55] SIRT7 overexpressing cells resisted both senescent and apoptotic effects of doxorubicin relocalizing SIRT7 from the nucleolus to the nucleoplasm, which resulted in decreased accumulation of p53 as well as its transcriptional targets, the well-known tumor suppressor p21.[55] Epigenetic suppression of p21 has been also reported in patients with HCC where SIRT7 gene expression was significantly upregulated and mediated mitotic stimulation of cells.[51]

One of the cancer hallmarks is the evasion of growth suppressors, and p53 is a key regulator that governs the decisions of cells to activate senescence and apoptotic programs.[6] Similarly to p53, E2F1, which is activated and stabilized by DNA damage, plays an important role regulating cell growth and apoptosis. E2F1 transcriptional activity is enhanced by acetylation and increased E2F1 levels were associated with increased SIRT1 expression, which reduces the acetylation level of E2F1, thus forming a negative loop.[56] The inhibition of the apoptotic program induced by SIRT1 is also mediated by inhibition of FOXO3, another nuclear factor.[57] Furthermore, SIRT1 does not regulate all FOXO target genes in the same manner but has a dual effect on FOXO3 function. FOXO3 can be acetylated at five different lysine residues in response to oxidative stress, and SIRT1 deacetylated FOXO3, increasing its ability to inhibit cell death by diminishing the expression of proapoptotic FOXO targets (*Fas ligand* and *BIM*) but inducing cell cycle arrest and resistance to oxidative stress.[57] This pathway may be very important for regulating survival of cancer cells, which are characterized by high levels of oxidative stress.[58]

Another tumor suppressor regulated by SIRT1 is p27, a potent stoichiometric inhibitor of all G1 cyclin-dependent kinase (CDK) complexes with roles in cell proliferation, senescence, differentiation, migration, and invasion. Experimental drugs targeting CDKs have been shown to exhibit profound anti-tumor activity.[59] It has been reported that SIRT1 negatively regulates p27 expression. SIRT1 regulates p27 stability through the ubiquitin proteolysis pathway and p27 downregulation is consistently associated with poor prognosis in NSCLC.[60]

Alterations in the Wnt/FZD ligands could be tumorigenic. The Wnt/β-catenin signaling pathway transmits signals through specific Frizzled (FZD) receptor that are connected through Dishevelled (Dvl) proteins to the canonical β-catenin-dependent pathway. In the absence of Wnts, β-catenin is sequestered in a complex that promotes its ubiquitination and subsequent degradation.[61] SIRT1 loss of function leads to a significant decrease in the levels of all three Dvl proteins, which have been reported to be overexpressed in colon tumors and invasive ductal breast carcinomas, and to contribute to pancreatic cancer malignancy. Moreover, SIRT1 promotes constitutive Wnt signaling and Wnt-induced cell migration.[62]

The universal feature of cancer cells is their ability to sustain chronic proliferation deregulating growth-promoting pathways, and the c-Myc oncogene, another substrate for SIRT1, is a "master regulator" that controls cellular growth regulation. Although it was first described as a negative loop showing that c-Myc binding to the SIRT1 promoter enhanced SIRT1 levels leading to increased rates of SIRT1-mediated deacetylation of c-Myc, which ultimately triggered its destruction,[63] it was later reported that SIRT1 gene transcription is upregulated by N-Myc in neuroblastoma cell lines and N-Myc protein is stabilized by upregulation of ERK protein phosphorylation, which in turn phosphorylates N-Myc protein at S62 and blocks its degradation. N-Myc and SIRT1 then repressed mitogen-activated protein kinase phosphatase 3 (MKP3) gene transcription by forming a transcriptional repressor complex at the Sp1-binding sites of the MKP3 gene promoter. When SIRT1 activity was inhibited by cambinol before tumor initiation in TH-MYCN transgenic mice, N-Myc-induced neuroblastoma initiation was suppressed.[64] Similarly to SIRT1, N-Myc and c-Myc also upregulated SIRT2 expression respectively in neuroblastoma cells and pancreatic cancer cells, which stabilized Myc oncoproteins because SIRT2 represses gene transcription of the ubiquitin-protein ligase NEDD4 that mediates the ubiquitination and degradation of Myc oncoproteins.[65] Amplification of the proto-oncogene N-Myc occurs in 25% of neuroblastoma tumors and is the best characterized genetic-risk factor for high-risk chemotherapy-refractory disease,[66] and deregulation of c-Myc protein is a common feature in pancreatic cancer that may be involved in early neoplastic development and progression.[67] Furthermore, the Fms-like tyrosine kinase (FLT3) receptor is present in 25–30% of acute myeloid leukemia patients, constituting the most commonly observed mutation in this disease, in which SIRT1 is selectively overexpressed by Myc-mediated induction. FLT3-ITD is associated with reduced length of remission and survival, consistent with a lack of elimination of leukemic stem cells.[68] In addition, SIRT1 is overexpressed in cancerous neural stem cells and has a critical role in the maintenance of cellular growth potential as well as neural stemness, and is also critical for the oncogenic transformation of cancerous neural stem cells.[69]

When a tumor grows and forms large masses of tumor cells, the core of that tumor become hypoxic and cells have to adapt to this stressing low oxygen condition. The hypoxia inducible factor-1α (HIF-1α), a key transcription factor, is quickly degraded under normal oxygen tension due to its continuous posttranslational modification by propyl-hydroxylases that targets HIF-1α to proteosomal degradation. However, low oxygen concentration induces an increase in the level of cellular ROS and inhibits these enzymes, leading to HIF-1α stabilization.[70] HIF-1α and HIF-2α are the best characterized isoforms of the HIF family that interact physically with sirtuins. SIRT3 has been shown to decrease HIF-1α stability *via* regulation of ROS and oxygen levels,[71] while SIRT2 decreases the levels of HIF-1α by deacetylating Lys-709 of HIF-1α and stimulating its binding to prolyl-hydroxylase, which leads to the subsequent hydroxylation and ubiquitination of the protein.[72] However, how SIRT1 regulates HIF-1α stability is controversial. SIRT1 binds to HIF-1α and deacetylates it at Lys-674, repressing HIF-1α target genes in HEK293T cells in normoxia, whereas in hypoxic conditions SIRT1 is downregulated due to decreased NAD^+ levels.[73] However, conflicting data have been reported showing that high levels of SIRT1 are necessary for HIF-1α protein accumulation and activation of HIF-1α target genes under hypoxic conditions in the HCC cell line. Moreover, treatment with the SIRT1 inhibitor sirtinol added at the onset of hypoxia inhibited the accumulation of HIF-1α protein.[74] Confirming these results, Joo *et al.* recently reported that SIRT1 stabilizes HIF-1α *via* direct binding and deacetylation during hypoxia leading to increased expression of HIF-1α target genes, including VEGF, GLUT1 and MMP2, and the ultimate promotion of cancer cell invasion.[75]

In tumor cells, HIF-1α is stabilized leading to the enhancement of VEGF expression, increasing capillary density but also triggering the expression of numerous genes involved in cell physiology and survival, including those related to a shift in cell metabolism.[76] Angiogenesis and the shift to aerobic glycolytic metabolism (Warburg effect) are two of the hallmarks of cancer mediated in part by HIF-1α stabilization, so inhibitors of this process have potential use as antitumor drugs.[6] As glucose is degraded by aerobic glycolysis, pyruvate is not used in the mitochondria but it is converted into lactate, so the amount of ATP synthesis produced per glucose molecule drops, and HIF-1α increases the expression of transporters, like GLUT1, necessary for the entry of glucose into the cell[77] and the mitochondrial metabolism is shifted in tumor cells. p53, which is also deacetylated by SIRT3 in addition to SIRT1,[78] and c-Myc are master regulators of metabolism. When p53 is triggered by metabolic stress the GLUT expression is suppressed, inhibiting the aerobic glycolysis in cancer cells,[79] so the inhibition of p53 activity by SIRT1 facilitates the aerobic glycolytic phenotype during tumorigenesis. On the other hand, c-Myc is frequently downregulated in cancer and was found to directly target and induce GLUT1 gene expression and increase glucose uptake.[80] Furthermore, in cancer cells, lactate could drive cell migration and radioresistance. Lactate dehydrogenase-A (LDH-A) is acetylated at Lys-5 and its substitution for Arg dramatically decreased the LDH-A acetylation by approximately 70%, and the mutated protein displayed only 18% of the wild-type activity. SIRT2, but not SIRT1, decreased LDH-A acetylation and increased LDH-A protein in both 293T and a pancreatic cancer cell line. The acetylation-induced decrease of LDH-A is independent of proteasome but involves chaperone-mediated autophagy-dependent degradation of LDH-A.[81]

In normal cells the activity of many tricarboxylic acid cycle enzymes and electron transport chain proteins implicated in the aerobic metabolism is regulated by SIRT3[82] and the overexpression of SIRT3 makes aerobic metabolism more efficient, reducing the levels of ROS. For these reasons the SIRT3 gene is considered as a tumor suppressor and consequently its expression has been found to be reduced in many human tumors.[44–47] However, the first substrate identified for SIRT3 was acetyl-CoA synthetase 2 (AceCS2), an enzyme that is important in converting acetate to acetyl-CoA in the presence of ATP and CoA, contributing to cancer cell growth under low oxygen conditions.[83] In mammalian cells, there are two types of ACeCS that are regulated by reversible acetylation: cytosolic ACeCS1 and mitochondrial ACeCS2.[84,85] Human AceCS2 is a mitochondrial matrix protein that was completely inactivated upon acetylation at Lys-642 in the active site whereas deacetylation of this residue by SIRT3 resulted in the activation of the enzyme.[83,86] Acetylation of mouse cytosolic ACeCS1 on Lys-661 inhibited its activity and deacetylation restored its ability to synthetize acetyl-CoA from acetate.[86] Induced acetate/acetyl CoA metabolism is a notable feature that is related to fatty acid synthesis, which is essential for tumor growth.[87] In accordance, ACeCS2 is overexpressed in human breast tumors and its expression correlates with disease progression.[88]

Furthermore, isocitrate dehydrogenase 2 (IDH2), another substrate for SIRT3, is a mitochondrial enzyme that converts isocitrate to α keto-glutarate (α-KG) and can be activated by SIRT3 thorough deacetylation of Lys-413, an evolutionarily invariant residue. Acetylated IDH2 displays a 44-fold loss in activity.[89] IDH2 activities are a major factor in cancer because they convert glutamine to α-KG, which can enter the tricarboxylic acid cycle as energy fuel in cancer cells to sustain cell proliferation in hypoxic conditions.[90] In this context, SIRT3 is acting as an oncogene.

Most of the cancers are of epithelial origin. One of the main characteristics of these kind of cells is the formation of strong cell junctions to other epithelial cells and to the basal membrane, so they do not have the ability to move from their place. In their places, epithelial cancer cells can grow to form primary tumors that are responsible for only about 10% of deaths from cancer. The rest of patients die by cancerous cells growing at sites far from these locations in their bodies where their primary tumors first arose. The ability to invade other organs depends on a series of complex biological steps. The first of the many necessary steps leading to metastasis is to undergo the epithelial–mesenchymal transition (EMT), losing the epithelial cell phenotype and acquiring mesenchymal characteristics. A role of SIRT2 as a tumor promoter in HCC by promoting EMT has been reported.[91] SIRT2 was highly expressed in HCC tissues and its depletion induced the expression of the epithelial markers *E*-cadherin and α-catenin, which was accompanied by a concomitant reduction of mesenchymal marker *N*-cadherin and α-SMA. SIRT2 expression led to the accumulation and nuclear import of β-catenin by regulating the Akt/GSK-3β/β-catenin-signaling axis, deacetylating Akt and regulating its activity.[91] Furthermore, overexpression of SIRT1 promoted EMT and enhanced the invasive and metastatic potential in HCC cell lines downregulating *E*-cadherin levels, whereas vimentin, Snail, and Twist were upregulated. Inhibiting SIRT1 activity with nicotinamide yielded the opposite result, suggesting the potential of reversal of the EMT. Furthermore, SIRT1 expression was significantly linked to poor prognosis after surgical resection in patients with HCC.[92] Moreover, SIRT2 worked synergistically with the deacetylase HDAC6 to promote cell migration and invasion in bladder cancer by targeting cortactin, a protein located in regions of cells undergoing membrane remodeling that is frequently overexpressed in several types of cancers.[93] Cortactin was identified as a novel substrate for SIRT1 in breast tumor tissues that expressed SIRT1 and cortactin more abundantly than normal surrounding tissues, and deacetylation of cortactin was associated with high levels of SIRT1 and tumorigenesis, suggesting a possible role for SIRT1 in cell motility through deacetylation of cortactin.[94] SIRT1 promotes cell migration because sir2α$^{-/-}$ mouse embryonic fibroblasts were less motile than those of sir2α$^{+/+}$ and showed higher levels of acetylated cortactin.[94]

In addition to SIRT1 and SIRT2, SIRT7 also participate in the EMT process. SIRT7 is a nucleolar sirtuin that acts only as an oncogene in tumorigenesis, which is highly expressed in several epithelial cancers as well as in sarcomas.[95] SIRT7 regulates the expression of invasion-related genes and

EMT markers and promotes cell migration and invasiveness in epithelial and mesenchymal cancer cells. SIRT7 cooperates with SIRT1 to repress *E*-cadherin expression interacting physically with SIRT1, with the deacetylase activity of SIRT1 not being required for *E*-cadherin repression, but only the physical interaction between SIRT1 and SIRT7.[95] In accordance with these results, over-expression of SIRT7 also increased the motility of ovarian cancer cells.[96]

High levels of SIRT1 expression have also been associated with the regulation of EMT in other human cancers. SIRT1 expression has been correlated with poor prognosis of colorectal cancer and co-localized with the stem marker CD133. SIRT1 decreased the level of p53 allowing the expression of several genes associated with the stemness, whereas the SIRT1 inhibitor nicotinamide significantly decreased the percentage of $CD133^+$ cells.[27] However, the role of SIRT1 in tumorigenesis is, once again, controversial because over-expression of SIRT1 inhibited migration of OSCC cells *in vitro*, as well as their metastasis to the lung *in vivo*, increasing the expression of *E*-cadherin and decreasing the expression of mesenchymal markers, involving the deacetylation of Smad4, which can influence MMP7 expression, cell migration, invasion, and tumor metastasis in OSCCs.[97]

Finally, unlike other sirtuins, SIRT7 selectively binds to promoter regions of target genes and deacetylates Lys-18 of histone H3, stabilizing the transformed state of cancer cells, including anchorage independent growth and escape from contact inhibition, two important hallmarks of transformed cells.[98] Among the genes regulated by SIRT7, some microRNAs have been reported, with microR-34a downregulation being markedly correlated with tumor size, metastasis, disease stage and prognosis, which played a pivotal role in SIRT7-mediated effects on gastric cancer.[99]

12.4 Sirtuin Inhibitors as Anticancer Agents

In view of their varied functions in cells, sirtuins are a druggable class of enzymes that could have beneficial effects on a number of human diseases when selectively activated or inhibited by different molecules. Mammalian sirtuins are characterized by N- and C-terminal sequences of variable length and a 275 amino acid catalytic core region that consists of a large domain with a Rossmann fold, a small domain containing a three-stranded zinc ribbon motif and a cleft between the domains that form the binding sites for both substrates: NAD^+ and the acetylated Lys residue of a protein substrate.[100,101] This structural similarity means that a given molecule could inhibit different sirtuins simultaneously. Compared with sirtuin activators, which have been mainly developed for SIRT1, more studies have been performed for inhibitors against different sirtuins, especially as anticancer drugs.

12.4.1 Nicotidamine and Its Analogues

In the sirtuin-operated deacetylation of a protein, the enzyme binds acetylated proteins and NAD^+, releasing a molecule of nicotinamide by each acetyl-Lys hydrolyzed.[3] Nicotinamide (Figure 12.1A) has been demonstrated to be

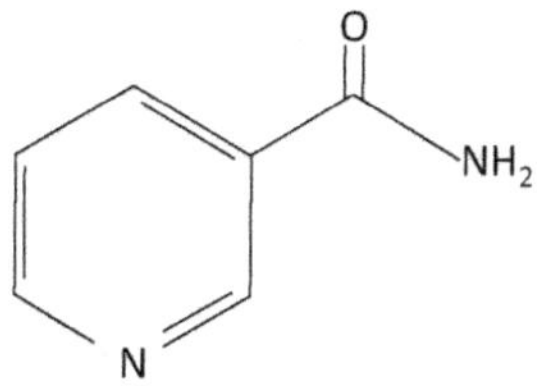

(A) Nicotinamide

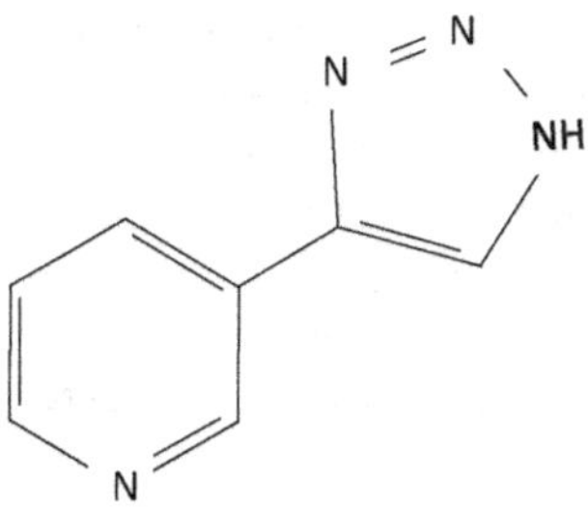

(B) Compound 2

Figure 12.1 (A) Nicotinamide and (B) compound 2.

a potent physiological inhibitor of SIRT1 enzyme by a reaction mechanism in which base exchange and deacetylation are competitive chemical processes.[102] Nicotinamide blocked proliferation and promoted apoptosis selectively in chronic lymphocytic leukemia cells that expressed p53 by means of a dual mechanism: blocking the enzymatic activity and increasing the miR-34a levels through a pathway involving p53.[103] SIRT1 was also significantly higher in poorly differentiated carcinomas and is an independent prognosticator of poor survival in pancreatic ductal adenocarcinoma. Inhibition of SIRT1 by increasing concentrations of nicotinamide led to a dose-dependent decrease of viability in MiaPaCa-2 and PANC-1 cells, and a combinatory treatment with gefitinib, an EGFR tyrosine kinase inhibitor, plus nicotinamide showed a synergistic effect on cell viability.[30] These observations suggest that SIRT1 inhibition in combination with other anti-cancer therapies may be a future approach that could enhance the sensitivity of tumor cells and impair their escape mechanisms. In addition, nicotinamide (as well as sirtinol, another sirtuin inhibitor) inhibited SIRT3 and induced apoptosis in OSCC cells that overexpressed SIRT3 and SIRT7, diminishing cell growth and proliferation in these cells in comparison with the untreated controls.[48] Galli *et al.* synthesized and tested several analogues of nicotinamide and found that compound 2 (Figure 12.1B) was more selective for SIRT3 over SIRT1 and SIRT2. Whereas nicotinamide inhibited SIRT3 activity with an IC_{50} of about 377 ± 52 μM, compound 2 showed an IC_{50} of 38 ± 5 μM and inhibited the cell growth of some tumor cell lines by more than 50%. However, these effects could be due to the inhibition of several off-targets.[104]

Cell migration is involved in tumorigenesis and deacetylation of cortactin is associated with high levels of mobility. Although SIRT1 is a protein mainly located in the nucleus, SIRT1 can also be found in the cytoplasm, as well as in the nucleus in ovarian cancer specimens. Inhibition of SIRT1 induced greater amounts of acetylated cortactin in nicotinamide-treated C13 and A2780cp cells, and migration was significantly slower than that of vehicle-treated cells.[94]

A clinical trial to determine the maximal tolerated dose and dose-limiting toxicity of vorinostat and nicotinamide in combination enrolled 25 patients with different types of lymphoma. The treatment was well tolerated, and the most significant toxicity was related directly to nicotinamide. The most common toxicities included fatigue (84%), nausea (80%), diarrhea (72%), and anorexia (56%). 24% of patients with relapsed or refractory lymphoma attained a response to vorinostat and nicotinamide, and 57% experienced disease stabilization.[105] Currently, a clinical trial is recruiting patients to determine whether nicotinamide is effective in the treatment of human lung cancer (http://www.ClinicalTrials.gov Identifier: NCT02416739).

12.4.2 Splitomicin and Its Derivatives

In a cell-based screen for inhibitors of the yeast Sir2p, Bedalov *et al.* found that splitomicin (Figure 12.2A) inhibited the deacetylase activity with an IC_{50} of 60 μM.[106] However, splitomicin showed rather weak inhibition on human enzymes and to clarify the spatial orientations that the splitomicins adopt within the SIRT2 binding pocket, a series of splitomicins derivatives was synthesized. Among the different compounds tested, a β-phenylsplitomicin (compound HR73, Figure 12.2B) increased enzyme inhibition and showed antiproliferative properties and tubulin hyperacetylation in MCF-7 breast cancer cells.[107]

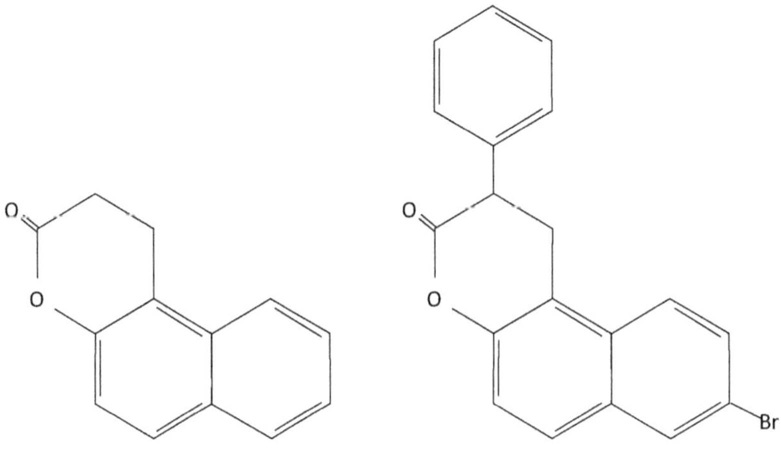

(A) Splitomicin (B) HR73

Figure 12.2 (A) Splitomicin and (B) the analogue HR 73.

12.4.3 Sirtinol

Sirtinol (2-[(2-hydroxy-naphthalen-1-ylmethylene)-amino]-*N*-(1-phenyl-ethyl)-benzamide) (Figure 12.3) was found to be an efficient inhibitor of yeast Sir2 and human SIRT2 in a screening of more than 1500 compounds from two chemical libraries, with an IC_{50} of 68 and 38 μM, respectively. The 2-hydroxyl-1-napthol moiety was sufficient for inhibition because it makes important contacts with the enzyme active site.[108]

Sirtinol induced senescence-like growth arrest in human breast cancer MCF-7 cells and lung cancer H1299 cells by impairing activation of mitogen activated protein kinase (MAPK) pathways, but neither expression nor acetylation of p53 were found to be upregulated by sirtinol.[109] However, later studies using MCF-7 cell lines confirmed that p53 is essential for sirtinol-induced apoptosis. Sirtinol was found to be a potent inhibitor of both SIRT1 and SIRT2 and combined targeting of both sirtuins was required to significantly increase the acetylation of p53 and induce cell death.[110,111] MCF-7 is an estrogen receptor (ER)-positive human breast cancer cell line expressing higher levels of SIRT1 than ERα-negative tumors and cancer cell lines. ERα and SIRT1 physically interact, but catalytically active SIRT1 was required for the interaction between ERα and SIRT1, increasing levels of antioxidative enzymes, especially Mn-SOD and glutathione peroxidase, and protecting tumor cells from ROS-induced cell death. In addition, ERα only bound directly to p53 and repressed its transcriptional activity when interacting with SIRT1. A combined treatment with tamoxifen (an antiestrogen) and sirtinol produced a greater magnitude of apoptosis in these cells than the individual treatment, suggesting that interaction at the molecular level could be used for the treatment of breast cancer.[112] Sirtinol also impaired cell

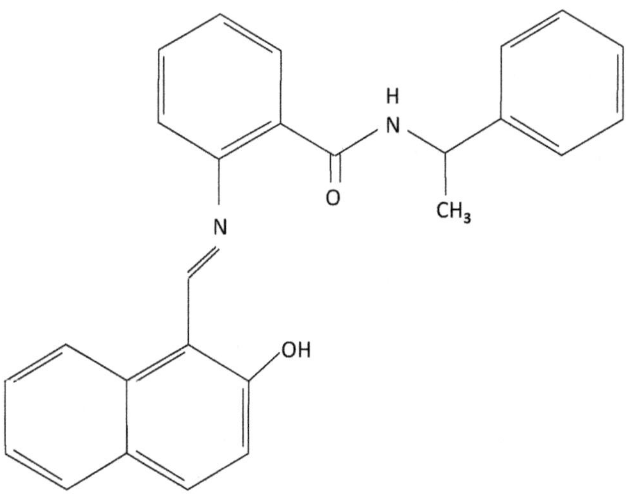

Figure 12.3 Sirtinol.

growth and increased ROS production and apoptosis in primary chronic lymphocytic leukemia cells and cell lines,[113] adult T-cell leukemia-lymphoma,[114] melanoma cells[115] and NSCLC cells.[116]

Furthermore, SIRT1 expression was upregulated in the androgen refractory PC3 and DU145 human prostate cancer cells and cell growth was dependent on the SIRT1 expression. Treatment of PC3 cells with sirtinol at 50 μM resulted in a significant inhibition of cell growth and also attenuated the chemoresistance that these cells presented against camptothecin and cisplatin by a mechanism that involved inhibition of SIRT1 activity.[117] The exposure of the human prostate cancer cell line LNCaP to sirtinol had a pleiotropic effect leading to G_1/G_0 arrest and the inhibition of cell growth through p53-dependent pathways. Furthermore, the treatment inhibited the expression of markers for both androgen and IGF-1 pathways, which is of great importance since high circulating level of IGF-1 correlated with increased prostate cancer risk.[118] Furthermore, in the BxPC-3 pancreatic cancer xenogeneic mice model a combined treatment of sirtinol and gemcitabine, the standard chemotherapeutic and first line drug for patients suffering from pancreatic cancer, improved the efficacy and survival time compared with either single inhibition of SIRT1 or single gemcitabine therapy,[119] suggesting that high levels of SIRT1 may play an important role in promoting chemoresistance in these cancer cells.

SIRT1 also plays an important role in angiogenesis, modulating the stability of HIF-1α. Although SIRT1 was overexpressed in HCC and many liver cancer cell lines, HIF-1α protein was not detected in cells cultured at 21% O_2 but was stabilized in cells cultured at 1% O_2.[74] In this oxygen tension, there was a dose-dependent repression of HIF-1α transcriptional activity in cells treated with sirtinol, but also a dose-dependent repression of HIF-1α protein accumulation due to a decrease of newly stabilized HIF-1α protein, rather than enhanced degradation of mature HIF-1α.[74]

12.4.4 Cambinol

Cambinol (Figure 12.4), a β-naphthol compound, was discovered in a screening of the National Cancer Institute repository of drugs as an inhibitor of human SIRT1 and SIRT2 activity *in vitro* with IC_{50} values of 56 and 59 μmol L^{-1}, respectively. The substitution of β-naphthol in cambinol with phenol led to loss of inhibitory activity.[120] Cambinol treatment induced apoptosis in Burkitt lymphoma cells by hyperacetylation of p53, even in the absence of any DNA-damaging agent, but was also less toxic to most of the carcinoma and primary cells. Cambinol-induced sirtuin inhibition also increased the acetylation and reduced the activity of BCL6, an oncoprotein that functions as a transcriptional repressor downregulating essential tumor suppressors like p53.[121] In addition, inhibition of SIRT1 and SIRT2 by cambinol was effective, arresting the growth of ER-positive human breast cancer MCF-7 cells. Moreover, inhibiting these sirtuins by treatment with cambinol decreased

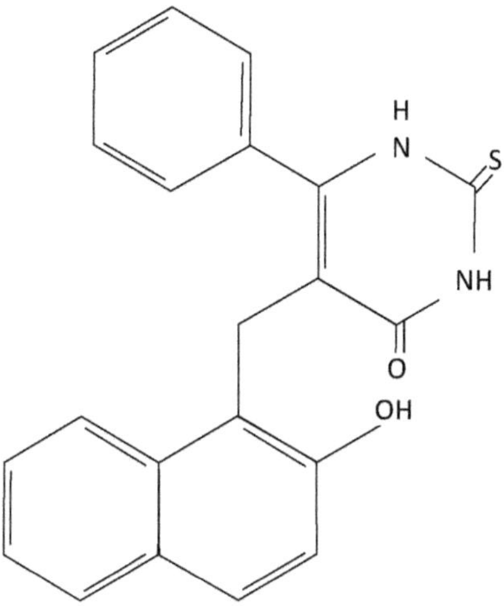

Figure 12.4 Cambinol.

the levels of aromatase, an enzyme that converts androgen to estrogen, thus contributing significantly to the malignancy of most breast cancers.[122]

Cambinol was also active in a mouse xenograft model reducing tumor growth relative to mice treated with vehicle alone, and was well tolerated at a dose of 100 mg kg^{-1}.[120] SIRT1 is overexpressed in HCC and liver cancer cell lines and knocking down of SIRT1 reduced tumor formation in an orthotopic xenograft model showing an impairment of tumor angiogenesis due to the inhibition of hypoxia-induced VEGF expression in SIRT1 knockdown cells. The pharmacologic intervention of this orthotopic xenograft HCC tumors with cambinol resulted in overall smaller tumors than vehicle-treated controls and the suppression of tumor growth in 3 of 4 animals or even in the reversion of tumor growth in the remaining animal after day 23. Cambinol treatment was safe and mice treated daily with 50 or 100 mg kg^{-1} cambinol for 2 weeks did not show any hepatotoxic effects and the treatments did not impair the regenerative capacity of normal liver.[74,123] Inhibition of SIRT1 in HepG2 cells also reduced colony formation *in vitro* in a dose-dependent manner but only when cells were expressing p53wt.[123] *In vivo*, cambinol also decreased the HIF-1α transcriptional activity. In mice exposed to 6% oxygen, HIF-1α protein accumulated in various tissues and activated HIF target genes but the treatment with 100 mg kg^{-1} cambinol decreased mRNA levels of the HIF target gene and pro-angiogenesis factor, VEGF. Furthermore, tumors of cambinol-treated animals showed less vascular density and intratumoral hemorrhage.[74]

Similar to nicotinamide, a combination of cambinol and gefitinib led to a synergistic inhibitory cell growth effect in pancreatic cancer MiaPaCa-2 and PANC-1 cell lines overexpressing SIRT1.[30]

Since cambinol shows no selectivity among SIRT1, SIRT2 and SIRT3, Mahajan *et al.* designed different cambinol analogues to modulate sirtuin activity for their potential use as chemotherapeutics. Cambinol analogue compound 17 (Figure 12.5A) showed selectivity for SIRT1, compound 8 (Figure 12.5C) for SIRT2 and compound 24 (Figure 12.5B) for SIRT3 *versus* SIRT1 and SIRT2. Cell viability assays using these cambinol analogues suggested that SIRT2 may be primarily responsible for the observed antilymphoma activity of these compound against B-cell lymphoma cell lines.[124]

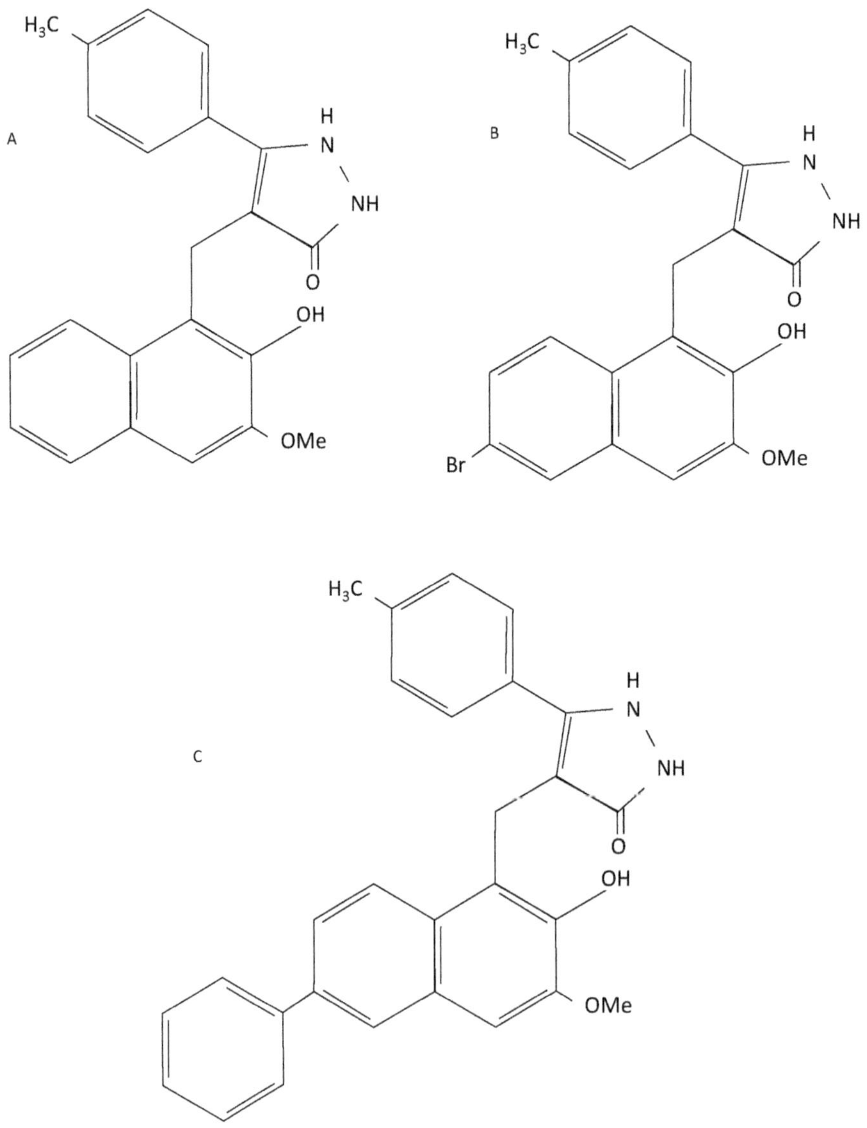

Figure 12.5 Cambinol analogues. (A) Compound 17; (B) compound 24; and (C) compound 8.

12.4.5 Salermide

Lara *et al.* (2009) synthesized salermide (*N*-phenyl-propionamide) (Figure 12.6), which exhibits higher selectivity against SIRT2 than against SIRT1.[125] Similar to sirtinol, which inhibited both SIRT1 and SIRT2, salermide significantly impaired MCF-7 cell proliferation through a p53-dependent mechanism. This dual inhibition of SIRT1 and SIRT2 was essential to achieve MCF-7 cell death.[110] Salermide also upregulated the expression of death receptor 5 and induced apoptosis in NSCLC, BE(2)-C and MiaPaca-2 pancreatic cells, leukemia MOLT4 cells, and MCF-7 cells.[65,126–128] The effect of salermide was more potent in cancerous cells, but not in non-tumorigenic cell lines with low level expression of SIRT1,[126] indicating that SIRT1 is a cancer-related gene that inhibits p53 function.

12.4.6 Indole Derivatives

Napper *et al.* discovered a series of indoles as potent inhibitors that were selective for SIRT1. The most active compound was named EX-527 (Figure 12.7), which inhibited SIRT1 in the nanomolar range with a 500-fold

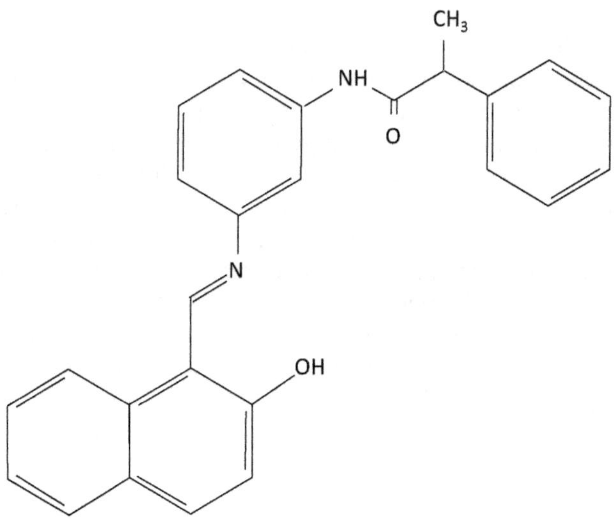

Figure 12.6 Salermide.

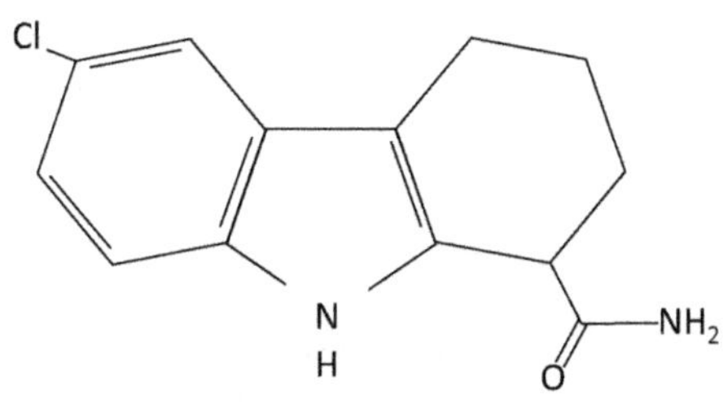

Figure 12.7 Ex-527.

improvement over previously reported SIRT inhibitors. Furthermore, EX-527 showed a higher degree of selectivity for SIRT1 over two other sirtuins: SIRT2 and SIRT3. This compound entered into the cells and inhibited the deacetylation of p53 at a concentration of 1 μM.[129]

Deacetylation of cortactin is associated with high levels of SIRT1 and tumorigenesis. EX-527 inhibition of SIRT1 induced greater amounts of acetylated cortactin in C13 and A2780cp cells.[94] SIRT1 is involved in tumorigenesis and drug resistance. Gemcitabine, which is used as a first-line therapy in pancreatic cancer patients, induced SIRT1 expression and potentially SIRT1-mediated pathways in the PANC-1 cell line. In this cell line, EX-527 inhibited or reduced proliferation PANC-1 cells *in vitro* and enhanced their sensitivity to gemcitabine treatment through increased apoptosis and the augmentation of caspase 3/7 activity, but had no effect on EMT.[130] Moreover, SIRT1 was upregulated in gastric cancer and ESCC with SIRT1 being required for the ATF4-induced multidrug resistance (MDR) effect in these cancers.[41,131] The inhibition of SIRT1 with EX-527 could partly reverse the gastric cancer MDR phenotype mediated by ATF4 in a dose–response manner.[41] Furthermore, the growth of endometrial carcinoma cells was also mediated by SIRT1. This sirtuin was significantly higher in endometrial carcinoma than in normal endometria and its overexpression was associated with a shorter survival and significantly enhanced the resistance for cisplatin and paclitaxel. EX-527 significantly suppressed the proliferation and cisplatin resistance of three endometrial carcinoma cell lines and also markedly inhibited tumor growth in a mouse xenograft model of endometrial carcinoma cell lines, regardless of the p53 mutational status.[132]

A significant increase, not only in SIRT1 expression, but also in SIRT2 and SIRT7, was noted during different stages of cervical cancer progression. Similar to ovarian cancer, SIRT1 expression was noted in both the cytoplasm and nucleus of the preneoplastic lesions. The treatment of cancer cell lines with EX-527 and AGK2, which specifically inhibit SIRT1 and SIRT2, respectively, also inhibited cell growth.[133] Furthermore, it has recently been reported that EX-527, like sirtinol, impaired cell growth and increased ROS production and apoptosis in primary chronic lymphocytic leukemia cells.[113]

The first Phase I pharmacokinetics studies has been carried out in seven cohorts of eight subjects that received a single dose of EX-527 (selisistat) at levels of 5, 25, 75, 150, 300 and 600 mg, and four cohorts of eight subjects that were administered 100, 200 and 300 mg once daily for 7 days. EX-527 was rapidly absorbed in proportion to the dose in the 5–300 mg range and a plateau in plasma was achieved within 4 days of repeated dosing. No serious adverse events were reported and EX-527 was considered safe and well tolerated by healthy male and female subjects after single doses up to 600 mg and multiple doses up to 300 mg per day.[134]

12.4.7 Tenovin

Using a cell-based screen for small molecules able to activate p53 and decrease tumor growth, Lain *et al.* (2008) found two SIRT1 inhibitors, tenovin-1 and its more water-soluble analog tenovin-6 (Figure 12.8), which

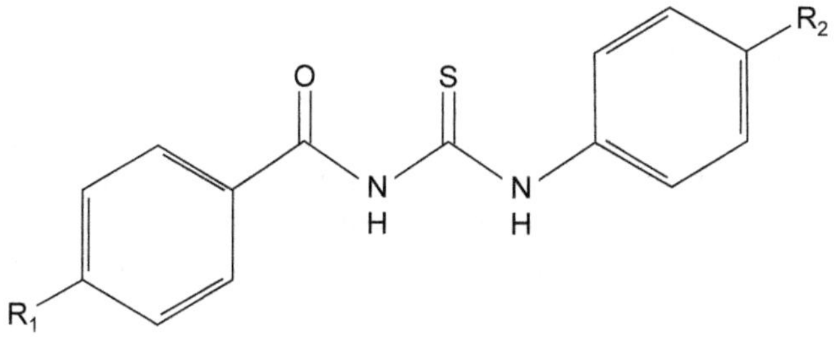

	R_1	R_2
Tenovin-1	**^{t}Bu**	**$NHCOCH_3$**
Tenovin-6	**^{t}Bu**	**$NHCO(CH_2)_4NMe_2$.HCL**

Figure 12.8 Tenovins.

decreased tumor growth *in vitro* at one-digit micromolar concentrations and delayed tumor growth *in vivo* as single agents.[135]

SIRT1 plays a critical role in chronic myelogenous leukemia (CML), an age-dependent malignancy, and further increases in the advanced phases of CML, and this was correlated with increasing BCR-ABL expression. SIRT1 activation promoted CML cell survival and proliferation associated with deacetylation of multiple SIRT1 substrates, including FOXO1, p53, and Ku70. Tenovin-6 administered at 50 mg kg^{-1} day^{-1} showed no hematologic toxicity or body weight loss. SIRT1 inhibition with tenovin-6 further sensitized CML cells to imatinib-induced apoptosis. However, the combination of SIRT1 inhibition with imatinib for treatment of mouse CML-like disease did not increase survival advantage compared with any of the individual drug treatments.[25]

Tenovin-6 also arrested cell growth and induced apoptosis in the acute promyelocytic leukemia NB4 cell line and promoted granulocytic differentiation of these cells when tenovin-6 and BML-266 were added in combination by a mechanism that involved SIRT2 inhibition, whereas the acetylation status of p53 was unchanged at 3 mM tenovin-6.[136] On the other hand, in chronic lymphocytic leukemia, tenovin-6 neither induced cellular apoptosis or p53-pathway activity but caused non-genotoxic cytotoxicity deregulating protective autophagy pathways by increases in the autophagy-regulatory proteins LC3 (LC3-II) and p62/Sequestosome.[137,138] Although p53 seems to not be involved in the inhibitory mechanism mediated by tenovin-6 in T-cell lymphoma that strongly expressed SIRT1, tenovin-1 reduced SIRT1 enzymatic activity and SIRT1 expression and led to increased apoptosis accompanied by increased acetylated p53.[139] Also in acute lymphoblastic leukemia (ALL), primary ALL cells from patients expressed higher levels of SIRT1 and SIRT2 than peripheral blood mononuclear cells from healthy individuals, and tenovin-6 treatment increased the level of hyperacetylated p53 protein. p53 potently inhibited the growth of pre-B ALL cells and primary ALL cells with IC_{50} values of 0.36 μM and 2.5 μM, respectively, sensitized ALL cells to the conventional chemotherapeutic agents etoposide and cytarabine, and inhibited the Wnt/β-catenin signaling pathway,[140] probably because SIRT1 loss of function led to a significant decrease in the levels of Dvl proteins and β-catenin was sequestered in a complex that promoted its degradation.[61]

In melanoma, a solid tumor where SIRT1 was upregulated, tenovin-induced SIRT1 inhibition arrested cell proliferation and clonogenic survival of melanoma cells, possibly *via* the activation of p53,[141,142] which is not mutated in the majority of melanomas. Some SIRT1 downstream targets were identified in melanoma cells by a proteomic approach, and tenovin-1-mediated SIRT1 inhibition affected apoptotic signaling, one of the major hallmarks of cancer cells. The protein network analysis highlighted p53 as a central hub relating to the response to stresses and DNA damage.[141] Recently, a positive loop where Myc protein upregulated SIRT1 and SIRT2 expression, which in turn stabilized Myc oncoproteins, has been described.[64,65] Inhibition of melanoma cells with tenovin-1 also revealed new downstream targets of SIRT1, decreasing the levels of BUB3, BUB1 and BUBR1, which are spindle assembly checkpoint proteins that were connected to p53 in the protein network.[141] The effect of tenovin-1 on six NSCLC cell lines was also p53-dependent. Tenovin-1 significantly decreased the growth of all NSCLC cell lines tested, inhibited colony formation, and the anchorage-independent cell growth in soft agar was also deeply restrained.[33] The effect of tenovin-6 on different uveal melanoma (UM) cell lines was also p53-dependent and was mediated in part by an increment in the level of ROS. Tenovin-6 suppressed the growth of UM cells inducing a massive apoptotic cell death, inhibited the clonogenicity in a concentration-dependent fashion and attenuated the migration of UM cells, decreasing the secretion of matrix metalloproteinase (MMP) 2 and MMP9. Furthermore, the combination of tenovin-6 with vinblastine, a conventional chemotherapeutic agent used for systemic therapy of UM patients, synergistically inhibited the viability of UM cells.[143]

As shown for sirtuins, the consequences of activated NOTCH signaling are cell type-specific and may have either tumor-suppressive or oncogenic effect. In Erwing sarcoma, the NOTCH pathway acts a tumor suppressor leading to strong transcriptional induction of p53. Moreover, SIRT1 was highly expressed in Ewing sarcoma and was associated with metastasis and poor prognosis.[144] The amplitude and duration of the NOTCH response were regulated by acetylation of NICD on specific Lys residues, and SIRT1 can modulate NICD activity by deacetylation.[145] The SIRT1/2 inhibitor tenovin-6 killed Ewing sarcoma cells *in vitro*. The silencing of SIRT1 but not of SIRT2 in TC252 cells induced p53 acetylation and impaired Ewing sarcoma growth and migration *in vivo* in an established xenograft model in zebrafish.[144]

However, the antiproliferative effect of tenovin-6 in different gastric cancer cell lines was p53-independent because a similar inhibitory effect was achieved with cell lines containing wild-type p53 and with cells endowed with mutant-type or null versions of p53 protein.[146] The inhibitory effect of tenovin-6 was linked to upregulation of the death receptor 5, a member of the tumor necrosis factor receptor family. Tenovin-6 showed a slight to moderate synergistic effect in treatment with chemotherapeutic agents, including docetaxel, SN-38, cisplatin, and 5-FU, in gastric cancer cell lines.[146]

In synovial sarcoma tumors and soft tissue sarcoma cell lines, SIRT1 expression was higher than in normal mesenchymal cells, but this difference was not statistically significant. Tenovin-6 treatment inhibited cell proliferation and induced p21 in all sarcoma cell lines tested independently of p53 status, without affecting the viability of primary mesenchymal stem cells, and antitumor growth effect of tenovin-6 was enhanced in starving nutrient-deprived conditions. Furthermore, treatment with tenovin-6 had an inhibitory effect on the growth of rhabdomyosarcoma xenografts.[38] Similarly to chronic lymphocytic leukemia,[137,138] tenovin-6 induced a time-dependent accumulation of LC3-II in all cell lines.[38]

12.4.8 Other Inhibitors of Human Sirtuins

Recently, new inhibitors for SIRT1, SIRT2 and SIRT3 have been designed and their ability as anticancer drugs tested in cell culture. An SIRT2 inhibitor, named AC-93253 (Figure 12.9), was identified by Zhang *et al.* in 2009 and exhibited selective inhibition of SIRT2 with an IC_{50} value of 6.0 μM, compared with related sirtuins for SIRT1 and SIRT3 for which IC_{50} values of 45.3 μM and 24.6 μM, respectively, were reported.[147] AC-93253 arrested cell growth in four different cancer cell lines derived from the prostate (DU145), pancreas (MiaPaCa2), and lung (A549 and NCI-H460), but was dramatically less active against non-transformed cell lines. AC-93253 toxicity was mediated by a significant induction of apoptotic cell death with few necrotic cells being observed.[147] In addition, AC-93253 exerted a negative effect on the expression of a set of genes involved in the progression and chemoresistance in melanoma and HeLa cells. The compound decreased expression of ABC transporters that mediate doxorubicin resistance in melanomas, sensitizing

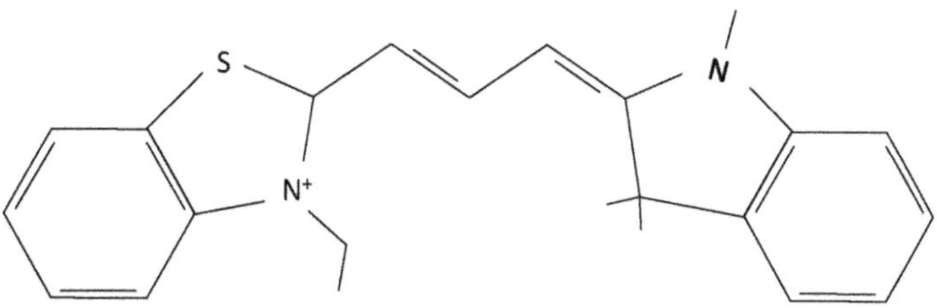

Figure 12.9 AC-93253.

the melanoma cell line to doxorubicin, and also impaired the migration of MDA-MB-435S melanoma cells.[148]

A novel benzimidazole derivative that showed SIRT1/SIRT2 inhibition activity with micromolar IC_{50} values has been described. This compound was able to inhibit MCF-7 breast cancer, as well as the triple-negative breast cancer cells (MDA-MB-468), although target therapies currently do not exist.[149]

SIRT3 was overexpressed in 3 cell lines of oral squamous cell carcinomas that have a poor 5 year survival rate.[48] A novel SIRT3-specific inhibitor, named LC-0296 (Figure 12.10), showed 10-fold greater inhibition towards SIRT3 enzymatic activity in comparison with SIRT1 and SIRT2. LC-0296 inhibited cell growth and proliferation and promoted apoptosis of HNSCC cells, in part by increasing ROS levels, and enhanced the sensitivity of these cells to both radiation and chemotherapeutic drugs without affecting normal human oral keratinocytes at any of the inhibitor doses.[150]

Finally, a number of compounds have been reported to inhibit SIRT1 expression instead of affecting its enzymatic activity. Divalproex sodium, a drug used in the treatment of epilepsy, downregulated SIRT1 expression in K562 cells enhancing the antileukemic effects of imatinib.[151] In addition, alisertib, an Aurora kinase A inhibitor, suppressed the expression of Sirt1 in human pancreatic cancer cells and exerted a potent cell growth inhibitory effect on two human pancreatic cancer cell lines.[152]

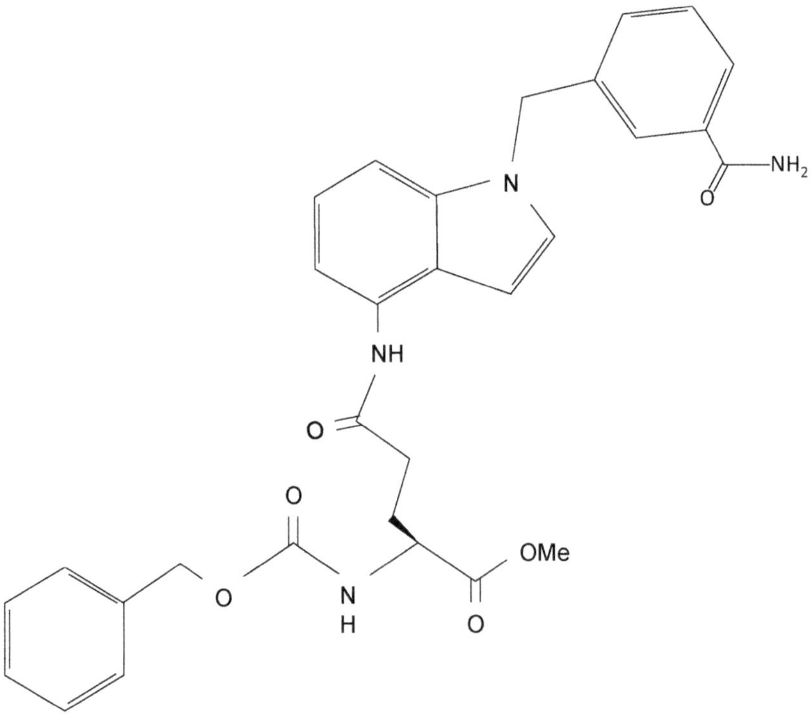

Figure 12.10 LC-0296.

12.5 Concluding Remarks

The seven mammalian sirtuins have been shown to be implicated in processes as diverse as gene expression, cell survival, angiogenesis, motility and the regulation of energy metabolism. The expression of different sirtuins is deregulated in many cancers and, depending of the cellular context, sirtuins may present a dual face, acting as tumor suppressors, as reported for SIRT1, SIRT2, SIRT3 SIRT4, and SIRT6, whose expression was found to be reduced in several cancer cells when compared with normal tissues, but also as tumor promoters in human tumorigenesis, as reported for SIRT1, SIRT2, SIRT3 and SIRT6, which were significantly increased in many tumors. In this context, the majority of studies suggest that inhibition of sirtuins is a promising strategy in the fight against cancer. However, most sirtuin inhibitors target SIRT1 and SIRT2 with similar affinity and SIRT3 with less affinity. A current effort is focused towards the development of inhibitors showing high specificity against a particular sirtuin. However, SIRT1 and SIRT2 also share substrates, so the redundant inhibition of both of them could be better in cancer treatment than inhibiting only one sirtuin.

Some studies on the role of SIRT1 in cancer are suggesting the great importance of this sirtuin in chemoresistance and radioresistance, decreasing drug penetration and increasing cells' ability to repair DNA damage or tolerate stress conditions. For that, the use of inhibitors targeting SIRT1 in combination with other cancer therapies to enhance sensitivity and to impair tumor cell escape mechanisms may be a future approach.

Acknowledgements

The authors' work is supported by grants: Instituto de Salud Carlos III grant PI13/00907 (to FJA), Junta de Comunidades de Castilla La-Mancha grant PPII-2014-010-P (to FJA), University of Castilla La-Mancha grant GI20163390 (to FJA), NIH 1R01AG028125-01A1 (to JMV), *Ministerio de Economía y Competitividad* BFU2011-23578 and BFU2015-64630-R (to JMV), *Junta de Andalucía Proyectos de Excelencia* grant P09-CVI-4887 (to JMV), *Junta de Andalucía Proyectos Internacionales* grant (to JMV), and BIO-276 (Junta de Andalucía and the University of Córdoba, to JMV). LFdR and EGC were funded by a FPU predoctoral fellowship of the Spanish *Ministerio de Educación, Cultura y Deporte*.

References

1. X. J. Yang and E. Seto, *Nat. Rev. Mol. Cell Biol.*, 2008, **9**, 206.
2. J. Landry, A. Sutton, S. T. Tafrov, R. C. Heller, J. Stebbins, L. Pillus and R. Sternglanz, *Proc. Natl. Acad. Sci. U. S. A.*, 2000, **97**, 5807.
3. J. M. Denu, *Curr. Opin. Chem. Biol.*, 2005, **9**, 431.
4. S. Michan and D. Sinclair, *Biochem. J.*, 2007, **404**, 1.
5. A. R. Konopka and K. Sreekumaran Nair, *Mol. Cell. Endocrinol.*, 2013, **379**, 19.
6. D. Hanahan and R. A. Weinberg, *Cell*, 2011, **144**, 646.

7. A. Chalkiadaki and L. Guarente, *Nat. Rev. Cancer*, 2015, **15**, 608.
8. S. G. Gray and T. J. Ekstrom, *Exp. Cell Res.*, 2001, **262**, 75.
9. H. Y. Cohen, C. Miller, K. J. Bitterman, N. R. Wall, B. Hekking, B. Kessler, K. T. Howitz, M. Gorospe, R. de Cabo and D. A. Sinclair, *Science*, 2004, **305**, 390.
10. H. Vaziri, S. K. Dessain, E. Ng Eaton, S. I. Imai, R. A. Frye, T. K. Pandita, L. Guarente and R. A. Weinberg, *Cell*, 2001, **107**, 149.
11. H. S. Kim, K. Patel, K. Muldoon-Jacobs, K. S. Bisht, N. Aykin-Burns, J. D. Pennington, R. van der Meer, P. Nguyen, J. Savage, K. M. Owens, A. Vassilopoulos, O. Ozden, S. H. Park, K. K. Singh, S. A. Abdulkadir, D. R. Spitz, C. X. Deng and D. Gius, *Cancer Cell*, 2010, **17**, 41.
12. H. S. Kim, A. Vassilopoulos, R. H. Wang, T. Lahusen, Z. Xiao, X. Xu, C. Li, T. D. Veenstra, B. Li, H. Yu, J. Ji, X. W. Wang, S. H. Park, Y. I. Cha, D. Gius and C. X. Deng, *Cancer Cell*, 2011, **20**, 487.
13. R. H. Wang, K. Sengupta, C. Li, H. S. Kim, L. Cao, C. Xiao, S. Kim, X. Xu, Y. Zheng, B. Chilton, R. Jia, Z. M. Zheng, E. Appella, X. W. Wang, T. Ried and C. X. Deng, *Cancer Cell*, 2008, **14**, 312.
14. S. H. Jang, K. W. Min, S. S. Paik and K. S. Jang, *J. Clin. Pathol.*, 2012, **65**, 735.
15. N. Kabra, Z. Li, L. Chen, B. Li, X. Zhang, C. Wang, T. Yeatman, D. Coppola and J. Chen, *J. Biol. Chem.*, 2009, **284**, 18210.
16. Q. Yang, B. Wang, W. Gao, S. Huang, Z. Liu, W. Li and J. Jia, *Mol. Cancer Res.*, 2013, **11**, 1497.
17. A. Noguchi, X. Li, A. Kubota, K. Kikuchi, Y. Kameda, H. Zheng, Y. Miyagi, I. Aoki and Y. Takano, *Oral Surg., Oral Med., Oral Pathol., Oral Radiol.*, 2013, **115**, 385.
18. C. J. Peters, J. R. Rees, R. H. Hardwick, J. S. Hardwick, S. L. Vowler, C. A. Ong, C. Zhang, V. Save, M. O'Donovan, D. Rassl, D. Alderson, C. Caldas and R. C. Fitzgerald, *Gastroenterology*, 2010, **139**, 1995–2004 e1915.
19. M. Temel, M. N. Koc, S. Ulutas and B. Gogebakan, *Tumour Biol.*, 2016, **37**, 6429.
20. K. H. Huang, C. C. Hsu, W. L. Fang, C. W. Chi, M. T. Sung, H. L. Kao, A. F. Li, P. H. Yin, M. H. Yang and H. C. Lee, *World J. Surg.*, 2014, **38**, 910.
21. L. M. McGlynn, S. McCluney, N. B. Jamieson, J. Thomson, A. I. MacDonald, K. Oien, E. J. Dickson, C. R. Carter, C. J. McKay and P. G. Shiels, *PLoS One*, 2015, **10**, e0131344.
22. Z. Han, L. Liu, Y. Liu and S. Li, *Int. J. Clin. Exp. Pathol.*, 2014, **7**, 4774.
23. C. A. Bradbury, F. L. Khanim, R. Hayden, C. M. Bunce, D. A. White, M. T. Drayson, C. Craddock and B. M. Turner, *Leukemia*, 2005, **19**, 1751.
24. K. Y. Jang, S. H. Hwang, K. S. Kwon, K. R. Kim, H. N. Choi, N. R. Lee, J. Y. Kwak, B. H. Park, H. S. Park, M. J. Chung, M. J. Kang, D. G. Lee, H. S. Kim, H. Shim and W. S. Moon, *Am. J. Surg. Pathol.*, 2008, **32**, 1523.
25. H. Yuan, Z. Wang, L. Li, H. Zhang, H. Modi, D. Horne, R. Bhatia and W. Chen, *Blood*, 2012, **119**, 1904.
26. D. M. Huffman, W. E. Grizzle, M. M. Bamman, J. S. Kim, I. A. Eltoum, A. Elgavish and T. R. Nagy, *Cancer Res.*, 2007, **67**, 6612.

27. X. Chen, K. Sun, S. Jiao, N. Cai, X. Zhao, H. Zou, Y. Xie, Z. Wang, M. Zhong and L. Wei, *Sci. Rep.*, 2014, **4**, 7481.
28. L. Lv, Z. Shen, J. Zhang, H. Zhang, J. Dong, Y. Yan, F. Liu, K. Jiang, Y. Ye and S. Wang, *Med. Oncol.*, 2014, **31**, 965.
29. A. Noguchi, K. Kikuchi, H. Zheng, H. Takahashi, Y. Miyagi, I. Aoki and Y. Takano, *Cancer Med.*, 2014, **3**, 1553.
30. A. Stenzinger, V. Endris, F. Klauschen, B. Sinn, K. Lorenz, A. Warth, B. Goeppert, V. Ehemann, A. Muckenhuber, C. Kamphues, M. Bahra, P. Neuhaus and W. Weichert, *BMC Cancer*, 2013, **13**, 450.
31. R. Al-Bahrani, D. Tuertcher, S. Zailaie, Y. Abuetabh, S. Nagamori, N. Zetouni, W. Bahitham and C. Sergi, *Ann. Clin. Lab. Sci.*, 2015, **45**, 3.
32. T. Zhang, N. Rong, J. Chen, C. Zou, H. Jing, X. Zhu and W. Zhang, *PLoS One*, 2013, **8**, e79162.
33. I. Grbesa, M. J. Pajares, E. Martinez-Terroba, J. Agorreta, A. M. Mikecin, M. Larrayoz, M. A. Idoate, K. Gall-Troselj, R. Pio and L. M. Montuenga, *PLoS One*, 2015, **10**, e0124670.
34. Y. W. Cao, W. Q. Li, G. X. Wan, Y. X. Li, X. M. Du, Y. C. Li and F. Li, *J. Exp. Clin. Cancer Res.*, 2014, **33**, 97.
35. L. Lin, X. Zheng, C. Qiu, S. Dongol, Q. Lv, J. Jiang, B. Kong and C. Wang, *Oncol. Rep.*, 2014, **32**, 2831.
36. D. H. Mvunta, T. Miyamoto, R. Asaka, Y. Yamada, H. Ando, S. Higuchi, K. Ida, H. Kashima and T. Shiozawa, *Appl. Immunohistochem. Mol. Morphol.*, 2016, DOI: 10.1097/PAI.0000000000000316.
37. S. Zhang and X. I. Wang, *Ann. Clin. Lab. Sci.*, 2013, **43**, 373.
38. L. Ma, W. Maruwge, A. Strambi, P. D'Arcy, P. Pellegrini, L. Kis, A. de Milito, S. Lain and B. Brodin, *Cell Death Dis.*, 2014, **5**, e1483.
39. H. Feng, P. Guo, J. Wang, J. Xu, C. Xie and F. Gao, *Pathol., Res. Pract.*, 2016, **212**, 319.
40. K. Kiga, Y. Fukuda-Yuzawa, M. Tanabe, S. Tsuji, C. Sasakawa and T. Fukao, *RNA Biol.*, 2014, **11**, 1347.
41. H. Zhu, L. Xia, Y. Zhang, H. Wang, W. Xu, H. Hu, J. Wang, J. Xin, Y. Gang, S. Sha, B. Xu, D. Fan, Y. Nie and K. Wu, *PLoS One*, 2012, **7**, e31431.
42. H. Zhu, X. Chen, B. Chen, W. Song, D. Sun and Y. Zhao, *PLoS One*, 2014, **9**, e103882.
43. H. Zeng, J. M. Zhang, Y. Du, J. Wang, Y. Ren, M. Li, H. Li, Z. Cai, Q. Chu and C. Yang, *Oncotarget*, 2016, **7**, 7329.
44. C. Liu, Z. Huang, H. Jiang and F. Shi, *BioMed Res. Int.*, 2014, **2014**, 871263.
45. M. M. Desouki, I. Doubinskaia, D. Gius and S. A. Abdulkadir, *Hum. Pathol.*, 2014, **45**, 1071.
46. N. Ashraf, S. Zino, A. Macintyre, D. Kingsmore, A. P. Payne, W. D. George and P. G. Shiels, *Br. J. Cancer*, 2006, **95**, 1056.
47. S. He, C. He, H. Yuan, S. Xiong, Z. Xiao and L. Chen, *Cell. Physiol. Biochem.*, 2014, **34**, 2061.
48. T. Y. Alhazzazi, P. Kamarajan, N. Joo, J. Y. Huang, E. Verdin, N. J. D'Silva and Y. L. Kapila, *Cancer*, 2011, **117**, 1670.

49. W. Lu, Y. Zuo, Y. Feng and M. Zhang, *Tumour Biol.*, 2014, **35**, 10699.
50. F. de Nigris, J. Cerutti, C. Morelli, D. Califano, L. Chiariotti, G. Viglietto, G. Santelli and A. Fusco, *Br. J. Cancer*, 2002, **86**, 917.
51. J. K. Kim, J. H. Noh, K. H. Jung, J. W. Eun, H. J. Bae, M. G. Kim, Y. G. Chang, Q. Shen, W. S. Park, J. Y. Lee, J. Borlak and S. W. Nam, *Hepatology*, 2013, **57**, 1055.
52. Y. Tang, W. Zhao, Y. Chen, Y. Zhao and W. Gu, *Cell*, 2008, **133**, 612.
53. J. Luo, A. Y. Nikolaev, S. Imai, D. Chen, F. Su, A. Shiloh, L. Guarente and W. Gu, *Cell*, 2001, **107**, 137.
54. Y. H. Jin, Y. J. Kim, D. W. Kim, K. H. Baek, B. Y. Kang, C. Y. Yeo and K. Y. Lee, *Biochem. Biophys. Res. Commun.*, 2008, **368**, 690.
55. S. Kiran, T. Anwar, M. Kiran and G. Ramakrishna, *Cell. Signalling*, 2015, **27**, 673.
56. C. Wang, L. Chen, X. Hou, Z. Li, N. Kabra, Y. Ma, S. Nemoto, T. Finkel, W. Gu, W. D. Cress and J. Chen, *Nat. Cell Biol.*, 2006, **8**, 1025.
57. A. Brunet, L. B. Sweeney, J. F. Sturgill, K. F. Chua, P. L. Greer, Y. Lin, H. Tran, S. E. Ross, R. Mostoslavsky, H. Y. Cohen, L. S. Hu, H. L. Cheng, M. P. Jedrychowski, S. P. Gygi, D. A. Sinclair, F. W. Alt and M. E. Greenberg, *Science*, 2004, **303**, 2011.
58. C. Gorrini, I. S. Harris and T. W. Mak, *Nat. Rev. Drug Discovery*, 2013, **12**, 931.
59. A. Roy and S. Banerjee, *J. Cell. Physiol.*, 2015, **230**, 504–509.
60. L. Zhu, C. Y. Chiao, K. G. Enzer, A. J. Stankiewicz, D. V. Faller and Y. Dai, *Mol. Cancer Res.*, 2015, **13**, 41.
61. T. D. King, M. J. Suto and Y. Li, *J. Cell. Biochem.*, 2012, **113**, 13.
62. K. R. Holloway, T. N. Calhoun, M. Saxena, C. F. Metoyer, E. F. Kandler, C. A. Rivera and K. Pruitt, *Proc. Natl. Acad. Sci. U. S. A.*, 2010, **107**, 9216.
63. J. Yuan, K. Minter-Dykhouse and Z. Lou, *J. Cell Biol.*, 2009, **185**, 203.
64. G. M. Marshall, P. Y. Liu, S. Gherardi, C. J. Scarlett, A. Bedalov, N. Xu, N. Iraci, E. Valli, D. Ling, W. Thomas, M. van Bekkum, E. Sekyere, K. Jankowski, T. Trahair, K. L. Mackenzie, M. Haber, M. D. Norris, A. V. Biankin, G. Perini and T. Liu, *PLoS Genet.*, 2011, **7**, e1002135.
65. P. Y. Liu, N. Xu, A. Malyukova, C. J. Scarlett, Y. T. Sun, X. D. Zhang, D. Ling, S. P. Su, C. Nelson, D. K. Chang, J. Koach, A. E. Tee, M. Haber, M. D. Norris, C. Toon, I. Rooman, C. Xue, B. B. Cheung, S. Kumar, G. M. Marshall, A. V. Biankin and T. Liu, *Cell Death Differ.*, 2013, **20**, 503.
66. W. C. Gustafson and W. A. Weiss, *Oncogene*, 2010, **29**, 1249.
67. C. Schleger, C. Verbeke, R. Hildenbrand, H. Zentgraf and U. Bleyl, *Mod. Pathol.*, 2002, **15**, 462.
68. L. Li, T. Osdal, Y. Ho, S. Chun, T. McDonald, P. Agarwal, A. Lin, S. Chu, J. Qi, Y. T. Hsieh, C. Dos Santos, H. Yuan, T. Q. Ha, M. Popa, R. Hovland, O. Bruserud, B. T. Gjertsen, Y. H. Kuo, W. Chen, S. Lain, E. McCormack and R. Bhatia, *Cell Stem Cell*, 2014, **15**, 431.
69. J. S. Lee, J. R. Park, O. S. Kwon, T. H. Lee, I. Nakano, H. Miyoshi, K. H. Chun, M. J. Park, H. J. Lee, S. U. Kim and H. J. Cha, *Neuro-Oncology*, 2015, **17**, 95.

70. A. J. Majmundar, W. J. Wong and M. C. Simon, *Mol. Cell*, 2010, **40**, 294.
71. E. L. Bell, B. M. Emerling, S. J. Ricoult and L. Guarente, *Oncogene*, 2011, **30**, 2986.
72. K. S. Seo, J. H. Park, J. Y. Heo, K. Jing, J. Han, K. N. Min, C. Kim, G. Y. Koh, K. Lim, G. Y. Kang, J. Uee Lee, Y. H. Yim, M. Shong, T. H. Kwak and G. R. Kweon, *Oncogene*, 2015, **34**, 1354.
73. J. H. Lim, Y. M. Lee, Y. S. Chun, J. Chen, J. E. Kim and J. W. Park, *Mol. Cell*, 2010, **38**, 864.
74. A. Laemmle, A. Lechleiter, V. Roh, C. Schwarz, S. Portmann, C. Furer, A. Keogh, M. P. Tschan, D. Candinas, S. A. Vorburger and D. Stroka, *PLoS One*, 2012, **7**, e33433.
75. H. Y. Joo, M. Yun, J. Jeong, E. R. Park, H. J. Shin, S. R. Woo, J. K. Jung, Y. M. Kim, J. J. Park, J. Kim and K. H. Lee, *Biochem. Biophys. Res. Commun.*, 2015, **462**, 294.
76. G. N. Masoud and W. Li, *Acta Pharm. Sin. B*, 2015, **5**, 378.
77. N. C. Denko, *Nat. Rev. Cancer*, 2008, **8**, 705.
78. S. Li, M. Banck, S. Mujtaba, M. M. Zhou, M. M. Sugrue and M. J. Walsh, *PLoS One*, 2010, **5**, e10486.
79. C. C. Barron, P. J. Bilan, T. Tsakiridis and E. Tsiani, *Metab., Clin. Exp.*, 2016, **65**, 124.
80. R. C. Osthus, H. Shim, S. Kim, Q. Li, R. Reddy, M. Mukherjee, Y. Xu, D. Wonsey, L. A. Lee and C. V. Dang, *J. Biol. Chem.*, 2000, **275**, 21797.
81. D. Zhao, S. W. Zou, Y. Liu, X. Zhou, Y. Mo, P. Wang, Y. H. Xu, B. Dong, Y. Xiong, Q. Y. Lei and K. L. Guan, *Cancer Cell*, 2013, **23**, 464.
82. C. Sebastian and R. Mostoslavsky, *Semin. Cell Dev. Biol.*, 2015, **43**, 33.
83. B. Schwer, J. Bunkenborg, R. O. Verdin, J. S. Andersen and E. Verdin, *Proc. Natl. Acad. Sci. U. S. A.*, 2006, **103**, 10224.
84. T. Fujino, J. Kondo, M. Ishikawa, K. Morikawa and T. T. Yamamoto, *J. Biol. Chem.*, 2001, **276**, 11420.
85. M. Yun, S. H. Bang, J. W. Kim, J. Y. Park, K. S. Kim and J. D. Lee, *J. Nucl. Med.*, 2009, **50**, 1222.
86. W. C. Hallows, S. Lee and J. M. Denu, *Proc. Natl. Acad. Sci. U. S. A.*, 2006, **103**, 10230.
87. Y. Yoshii, T. Furukawa, T. Saga and Y. Fujibayashi, *Cancer Lett.*, 2015, **356**, 211.
88. Z. T. Schug, B. Peck, D. T. Jones, Q. Zhang, S. Grosskurth, I. S. Alam, L. M. Goodwin, E. Smethurst, S. Mason, K. Blyth, L. McGarry, D. James, E. Shanks, G. Kalna, R. E. Saunders, M. Jiang, M. Howell, F. Lassailly, M. Z. Thin, B. Spencer-Dene, G. Stamp, N. J. van den Broek, G. Mackay, V. Bulusu, J. J. Kamphorst, S. Tardito, D. Strachan, A. L. Harris, E. O. Aboagye, S. E. Critchlow, M. J. Wakelam, A. Schulze and E. Gottlieb, *Cancer Cell*, 2015, **27**, 57.
89. W. Yu, K. E. Dittenhafer-Reed and J. M. Denu, *J. Biol. Chem.*, 2012, **287**, 14078.

90. D. R. Wise, P. S. Ward, J. E. Shay, J. R. Cross, J. J. Gruber, U. M. Sachdeva, J. M. Platt, R. G. DeMatteo, M. C. Simon and C. B. Thompson, *Proc. Natl. Acad. Sci. U. S. A.*, 2011, **108**, 19611.
91. J. Chen, A. W. Chan, K. F. To, W. Chen, Z. Zhang, J. Ren, C. Song, Y. S. Cheung, P. B. Lai, S. H. Cheng, M. H. Ng, A. Huang and B. C. Ko, *Hepatology*, 2013, **57**, 2287.
92. C. Hao, P. X. Zhu, X. Yang, Z. P. Han, J. H. Jiang, C. Zong, X. G. Zhang, W. T. Liu, Q. D. Zhao, T. T. Fan, L. Zhang and L. X. Wei, *BMC Cancer*, 2014, **14**, 978.
93. Q. Zuo, W. Wu, X. Li, L. Zhao and W. Chen, *Oncol. Rep.*, 2012, **27**, 819.
94. Y. Zhang, M. Zhang, H. Dong, S. Yong, X. Li, N. Olashaw, P. A. Kruk, J. Q. Cheng, W. Bai, J. Chen, S. V. Nicosia and X. Zhang, *Oncogene*, 2009, **28**, 445.
95. S. Malik, L. Villanova, S. Tanaka, M. Aonuma, N. Roy, E. Berber, J. R. Pollack, E. Michishita-Kioi and K. F. Chua, *Sci. Rep.*, 2015, **5**, 9841.
96. H. L. Wang, R. Q. Lu, S. H. Xie, H. Zheng, X. M. Wen, X. Gao and L. Guo, *Asian Pac J Cancer Prev.*, 2015, **16**, 3573.
97. I. C. Chen, W. F. Chiang, H. H. Huang, P. F. Chen, Y. Y. Shen and H. C. Chiang, *Mol. Cancer*, 2014, **13**, 254.
98. M. F. Barber, E. Michishita-Kioi, Y. Xi, L. Tasselli, M. Kioi, Z. Moqtaderi, R. I. Tennen, S. Paredes, N. L. Young, K. Chen, K. Struhl, B. A. Garcia, O. Gozani, W. Li and K. F. Chua, *Nature*, 2012, **487**, 114.
99. S. Zhang, P. Chen, Z. Huang, X. Hu, M. Chen, S. Hu, Y. Hu and T. Cai, *Sci. Rep.*, 2015, **5**, 9787.
100. M. S. Finnin, J. R. Donigian and N. P. Pavletich, *Nat. Struct. Biol.*, 2001, **8**, 621.
101. J. Min, J. Landry, R. Sternglanz and R. M. Xu, *Cell*, 2001, **105**, 269.
102. A. A. Sauve and V. L. Schramm, *Biochemistry*, 2003, **42**, 9249.
103. V. Audrito, T. Vaisitti, D. Rossi, D. Gottardi, G. D'Arena, L. Laurenti, G. Gaidano, F. Malavasi and S. Deaglio, *Cancer Res.*, 2011, **71**, 4473.
104. U. Galli, O. Mesenzani, C. Coppo, G. Sorba, P. L. Canonico, G. C. Tron and A. A. Genazzani, *Eur. J. Med. Chem.*, 2012, **55**, 58.
105. J. E. Amengual, S. Clark-Garvey, M. Kalac, L. Scotto, E. Marchi, E. Neylon, P. Johannet, Y. Wei, J. Zain and O. A. O'Connor, *Blood*, 2013, **122**, 2104.
106. A. Bedalov, T. Gatbonton, W. P. Irvine, D. E. Gottschling and J. A. Simon, *Proc. Natl. Acad. Sci. U. S. A.*, 2001, **98**, 15113.
107. R. C. Neugebauer, U. Uchiechowska, R. Meier, H. Hruby, V. Valkov, E. Verdin, W. Sippl and M. Jung, *J. Med. Chem.*, 2008, **51**, 1203.
108. C. M. Grozinger, E. D. Chao, H. E. Blackwell, D. Moazed and S. L. Schreiber, *J. Biol. Chem.*, 2001, **276**, 38837.
109. H. Ota, E. Tokunaga, K. Chang, M. Hikasa, K. Iijima, M. Eto, K. Kozaki, M. Akishita, Y. Ouchi and M. Kaneki, *Oncogene*, 2006, **25**, 176.
110. B. Peck, C. Y. Chen, K. K. Ho, P. Di Fruscia, S. S. Myatt, R. C. Coombes, M. J. Fuchter, C. D. Hsiao and E. W. Lam, *Mol. Cancer Ther.*, 2010, **9**, 844.

111. J. Wang, T. H. Kim, M. Y. Ahn, J. Lee, J. H. Jung, W. S. Choi, B. M. Lee, K. S. Yoon, S. Yoon and H. S. Kim, *Int. J. Oncol.*, 2012, **41**, 1101.
112. S. Elangovan, S. Ramachandran, N. Venkatesan, S. Ananth, J. P. Gnana-Prakasam, P. M. Martin, D. D. Browning, P. V. Schoenlein, P. D. Prasad, V. Ganapathy and M. Thangaraju, *Cancer Res.*, 2011, **71**, 6654.
113. S. Bhalla and L. I. Gordon, *Cancer Biol. Ther.*, 2016, **17**, 300.
114. T. Kozako, A. Aikawa, T. Shoji, T. Fujimoto, M. Yoshimitsu, S. Shirasawa, H. Tanaka, S. Honda, H. Shimeno, N. Arima and S. Soeda, *Int. J. Cancer*, 2012, **131**, 2044.
115. M. J. Wilking, C. K. Singh, M. Nihal, M. A. Ndiaye and N. Ahmad, *Cell Cycle*, 2014, **13**, 2821.
116. Y. Fong, Y. C. Lin, C. Y. Wu, H. M. Wang, L. L. Lin, H. L. Chou, Y. N. Teng, S. S. Yuan and C. C. Chiu, *Sci. World J.*, 2014, **2014**, 937051.
117. K. Kojima, R. Ohhashi, Y. Fujita, N. Hamada, Y. Akao, Y. Nozawa, T. Deguchi and M. Ito, *Biochem. Biophys. Res. Commun.*, 2008, **373**, 423.
118. T. T. Wang, N. W. Schoene, E. K. Kim and Y. S. Kim, *Mol. Carcinog.*, 2013, **52**, 676.
119. D. J. Gong, J. M. Zhang, M. Yu, B. Zhuang and Q. Q. Guo, *Clin. Interventions Aging*, 2013, **8**, 889.
120. B. Heltweg, T. Gatbonton, A. D. Schuler, J. Posakony, H. Li, S. Goehle, R. Kollipara, R. A. Depinho, Y. Gu, J. A. Simon and A. Bedalov, *Cancer Res.*, 2006, **66**, 4368–4377.
121. R. T. Phan and R. Dalla-Favera, *Nature*, 2004, **432**, 635.
122. K. R. Holloway, A. Barbieri, S. Malyarchuk, M. Saxena, A. Nedeljkovic-Kurepa, M. Cameron Mehl, A. Wang, X. Gu and K. Pruitt, *Mol. Endocrinol.*, 2013, **27**, 480.
123. S. Portmann, R. Fahrner, A. Lechleiter, A. Keogh, S. Overney, A. Laemmle, K. Mikami, M. Montani, M. P. Tschan, D. Candinas and D. Stroka, *Mol. Cancer Ther.*, 2013, **12**, 499.
124. S. S. Mahajan, M. Scian, S. Sripathy, J. Posakony, U. Lao, T. K. Loe, V. Leko, A. Thalhofer, A. D. Schuler, A. Bedalov and J. A. Simon, *J. Med. Chem.*, 2014, **57**, 3283.
125. E. Lara, A. Mai, V. Calvanese, L. Altucci, P. Lopez-Nieva, M. L. Martinez-Chantar, M. Varela-Rey, D. Rotili, A. Nebbioso, S. Ropero, G. Montoya, J. Oyarzabal, S. Velasco, M. Serrano, M. Witt, A. Villar-Garea, A. Imhof, J. M. Mato, M. Esteller and M. F. Fraga, *Oncogene*, 2009, **28**, 781.
126. M. N. Dastjerdi, M. R. Salahshoor, M. Mardani, M. Rabbani, B. Hashemibeni, M. Gharagozloo, M. Kazemi, N. Esmaeil, S. Roshankhah, R. Golmohammadi and M. Mobarakian, *Res. Pharm. Sci.*, 2013, **8**, 79.
127. G. Liu, L. Su, X. Hao, N. Zhong, D. Zhong, S. Singhal and X. Liu, *J. Cell. Mol. Med.*, 2012, **16**, 1618.
128. D. Rotili, D. Tarantino, A. Nebbioso, C. Paolini, C. Huidobro, E. Lara, P. Mellini, A. Lenoci, R. Pezzi, G. Botta, M. Lahtela-Kakkonen, A. Poso, C. Steinkuhler, P. Gallinari, R. De Maria, M. Fraga, M. Esteller, L. Altucci and A. Mai, *J. Med. Chem.*, 2012, **55**, 10937.

129. A. D. Napper, J. Hixon, T. McDonagh, K. Keavey, J. F. Pons, J. Barker, W. T. Yau, P. Amouzegh, A. Flegg, E. Hamelin, R. J. Thomas, M. Kates, S. Jones, M. A. Navia, J. O. Saunders, P. S. DiStefano and R. Curtis, *J. Med. Chem.*, 2005, **48**, 8045.
130. C. E. Oon, C. Strell, K. Y. Yeong, A. Ostman and J. Prakash, *Eur. J. Pharmacol.*, 2015, **757**, 59.
131. H. Zhu, X. Chen, B. Chen, J. Fan, W. Song, Z. Xie, D. Jiang, Q. Li, M. Zhou, D. Sun and Y. Zhao, *Cancer Lett.*, 2014, **354**, 142.
132. R. Asaka, T. Miyamoto, Y. Yamada, H. Ando, D. H. Mvunta, H. Kobara and T. Shiozawa, *Lab. Invest.*, 2015, **95**, 1363.
133. S. Singh, P. U. Kumar, S. Thakur, S. Kiran, B. Sen, S. Sharma, V. V. Rao, A. R. Poongothai and G. Ramakrishna, *Tumour Biol.*, 2015, **36**, 6159.
134. G. Westerberg, J. A. Chiesa, C. A. Andersen, D. Diamanti, L. Magnoni, G. Pollio, B. Darpo and M. Zhou, *Br. J. Clin. Pharmacol.*, 2015, **79**, 477.
135. S. Lain, J. J. Hollick, J. Campbell, O. D. Staples, M. Higgins, M. Aoubala, A. McCarthy, V. Appleyard, K. E. Murray, L. Baker, A. Thompson, J. Mathers, S. J. Holland, M. J. Stark, G. Pass, J. Woods, D. P. Lane and N. J. Westwood, *Cancer Cell*, 2008, **13**, 454.
136. Y. Sunami, M. Araki, Y. Hironaka, S. Morishita, M. Kobayashi, E. L. Liew, Y. Edahiro, M. Tsutsui, A. Ohsaka and N. Komatsu, *PLoS One*, 2013, **8**, e57633.
137. M. J. Groves, C. E. Johnson, J. James, A. R. Prescott, J. Cunningham, S. Haydock, C. Pepper, C. Fegan, L. Pirrie, N. J. Westwood, P. J. Coates, I. G. Ganley and S. Tauro, *Br. J. Cancer*, 2013, **109**, 2434.
138. S. F. MacCallum, M. J. Groves, J. James, K. Murray, V. Appleyard, A. R. Prescott, A. A. Drbal, A. Nicolaou, J. Cunningham, S. Haydock, I. G. Ganley, N. J. Westwood, P. J. Coates, S. Lain and S. Tauro, *Sci. Rep.*, 2013, **3**, 1275.
139. M. Nihal, N. Ahmad and G. S. Wood, *Cell Cycle*, 2014, **13**, 632.
140. Y. Jin, Q. Cao, C. Chen, X. Du, B. Jin and J. Pan, *BMC Cancer*, 2015, **15**, 226.
141. C. K. Singh, J. George, M. Nihal, G. Sabat, R. Kumar and N. Ahmad, *Oncotarget*, 2014, **5**, 1987.
142. M. J. Wilking, C. Singh, M. Nihal, W. Zhong and N. Ahmad, *Arch. Biochem. Biophys.*, 2014, **563**, 94.
143. W. Dai, J. Zhou, B. Jin and J. Pan, *Sci. Rep.*, 2016, **6**, 22622.
144. J. Ban, D. N. Aryee, A. Fourtouna, W. van der Ent, M. Kauer, S. Niedan, I. Machado, C. Rodriguez-Galindo, O. M. Tirado, R. Schwentner, P. Picci, A. M. Flanagan, V. Berg, S. J. Strauss, K. Scotlandi, E. R. Lawlor, E. Snaar-Jagalska, A. Llombart-Bosch and H. Kovar, *Cancer Res.*, 2014, **74**, 6578.
145. V. Guarani, G. Deflorian, C. A. Franco, M. Kruger, L. K. Phng, K. Bentley, L. Toussaint, F. Dequiedt, R. Mostoslavsky, M. H. Schmidt, B. Zimmermann, R. P. Brandes, M. Mione, C. H. Westphal, T. Braun, A. M. Zeiher, H. Gerhardt, S. Dimmeler and M. Potente, *Nature*, 2011, **473**, 234.
146. S. Hirai, S. Endo, R. Saito, M. Hirose, T. Ueno, H. Suzuki, K. Yamato, M. Abei and I. Hyodo, *PLoS One*, 2014, **9**, e102831.

147. Y. Zhang, Q. Au, M. Zhang, J. R. Barber, S. C. Ng and B. Zhang, *Biochem. Biophys. Res. Commun.*, 2009, **386**, 729.
148. I. Karwaciak, M. Gorzkiewicz, K. Ryba, J. Dastych, L. Pulaski and M. Ratajewski, *Chem.-Biol. Interact.*, 2015, **236**, 9.
149. Y. K. Yoon, M. A. Ali, A. C. Wei, T. S. Choon, H. Osman, K. Parang and A. N. Shirazi, *Bioorg. Med. Chem.*, 2014, **22**, 703.
150. T. Y. Alhazzazi, P. Kamarajan, Y. Xu, T. Ai, L. Chen, E. Verdin and Y. L. Kapila, *Anticancer Res.*, 2016, **36**, 49.
151. W. Wang, J. Zhang, Y. Li, X. Yang, Y. He, T. Li, F. Ren and R. Lin, *Cancer Lett.*, 2015, **356**, 791–799.
152. F. Wang, H. Li, X. G. Yan, Z. W. Zhou, Z. G. Yi, Z. X. He, S. T. Pan, Y. X. Yang, Z. Z. Wang, X. Zhang, T. Yang, J. X. Qiu and S. F. Zhou, *Drug Des., Dev. Ther.*, 2015, **9**, 575.

CHAPTER 13

Lifespan-Extending Effect of Resveratrol and Other Phytochemicals

SHIN-HAE LEE[a] AND KYUNG-JIN MIN*[a]

[a]Department of Biological Sciences, Inha University, Incheon, Korea
*E-mail: minkj@inha.ac.kr

13.1 Introduction

Today, anti-aging methods and ways to prolong life have become the focus of people's attention. As such, researchers throughout the world are searching for ways to live a healthier and longer life. Many research studies have shown that controlling diet leads to a healthier life that ultimately results in longevity. Implementation of restricted diet, known as caloric/dietary restriction, allows people to live longer; in addition, the intake of natural plants has been shown to have beneficial effects on age-related diseases and longevity.[1,2] Such natural plant products used in a therapeutic approach are called nutraceuticals, and their active compounds are phytochemicals.

Phytochemicals are secondary metabolites synthesized by plants—including fruits, vegetables, cereals, nuts, and cacao—to assist in their survival and protect from microbial infection and environmental pollutants. Phytochemicals include several groups of compounds and the phenolic compounds, called polyphenols, are the most well-known group.[1] A lot of phytochemicals

RSC Drug Discovery Series No. 57
Anti-aging Drugs: From Basic Research to Clinical Practice
Edited by Alexander M. Vaiserman

Published by the Royal Society of Chemistry, www.rsc.org

have anti-oxidative and anti-inflammatory characteristics, which contribute to the improvement of health and prevention of age-related diseases, such as cancer, diabetes, cardiovascular disease, and neurodegenerative disorders.[3] The beneficial effects of phytochemicals on the prevention and treatment of age-related diseases are well-documented in the literature.[4–7] In addition to these benefits, some phytochemicals have attracted attention from the general public due to their potential in retarding the aging process and for their previous results associated with extending the lifespan of various organisms including yeast, worms, flies, fishes, and rodents.[1,2] The most representative phytochemical associated with the extension of lifespan is resveratrol, a compound rich in red wine with a lot of biological and physiological activities.[8] In this chapter, we summarize and review the longevity benefits and the putative underlying mechanisms of these phytochemicals, with particular attention to resveratrol.

13.2 Resveratrol

Resveratrol (3,4′,5-trihydroxy stilbene) is the most extensively investigated phytochemical to retard aging, extend lifespan, and improve health. Resveratrol is a phytoalexin and polyphenolic compound found in various plants, including berries and peanuts, especially in the skin of red grapes, produced in response to stresses such as fungal infection and ultraviolet irradiation.[8] Although resveratrol has been shown to exist naturally in both isomeric *trans* and *cis* forms, *trans*-resveratrol has been the focus of past investigations, mostly for its beneficial effects. Resveratrol was first isolated from white hellebore (*Veratrum grandiflorum*) in 1940, and was identified to have cancer-chemopreventive activity in mice.[9] Since then, resveratrol has been well established to have various biological effects, such as anti-oxidant and anti-inflammation effects.[10,11] Moreover, it has been shown to be beneficial against various age-related diseases, such as cardiovascular disease, cancer, neural disease, and metabolic disease, including diabetes.[2,12–16]

13.2.1 Lifespan-Extending Effect of Resveratrol in Invertebrates: Yeasts, Worms and Flies

The longevity benefit of resveratrol has been highlighted since the study by Howitz and his colleagues in 2003. They screened for activators of sirtuin and showed the lifespan extension in *Saccharomyces cerevisiae* by resveratrol.[18] In this initial screening, resveratrol was found to be the most potent Sirt1 activator mimicking the effects of calorie restriction; resveratrol administration increased the lifespan of yeasts by 70%.[17] After this finding, several studies showed that resveratrol increased the lifespan of evolutionarily distant species, including worms, flies, fishes, and rodents.[17–21]

Several studies tested the lifespan-extending effect of resveratrol in various concentrations and in several mutant strains using worms. Wood *et al.*

reported that 100 μM resveratrol activated the deacetylase activity of sirtuin and increased the lifespan by up to 14% in the N2 wild type, but not in the *sir2.1* mutant, indicating the sir2-dependency of resveratrol.[18] Viswanathan *et al.* showed that 1 mM resveratrol extended worm lifespan in the wild type and *daf16* mutants—daf16 is a homologue to mammalian Forkhead box O (FOXO) transcription factor—but not in mutants of *sir2.1*, indicating that the lifespan-extending effect of resveratrol is sir2-dependent, but FOXO-independent.[22] Additionally, Gruber *et al.* demonstrated that 50 μM resveratrol extended the mean and maximum lifespan of *Caenorhabditis elegans* N2 strain by 64% and 30%, respectively, and increased the survival rate under oxidative stress condition by 24.6%.[23] Greer *et al.* reported that 100 μM resveratrol extended the lifespan of N2 and *daf-16* worms by 14.2% and 13.7%, respectively,[24] but resveratrol did not extend the lifespan of AMPK/aak2 mutant worms,[24] indicating that resveratrol extends the lifespan of worms in an AMPK-dependent, but FOXO-independent manner. In addition, Morselli *et al.* also showed that 100 μg ml^{-1} resveratrol extended the lifespan of N2 worms by 10.9%, but not in *bec-1* mutants, one of the essential genes for autophagy.[25]

The prolongevity effect of resveratrol was also tested in *Drosophila melanogaster*. Wood *et al.* showed that 100 μM resveratrol increased lifespan by up to 29% in Canton S wild-type strain without reducing the fecundity and food uptake.[18] Bauer *et al.* found that 200 μM resveratrol extended lifespan by 9% in short lived flies newly developed for rapid screening of lifespan.[26] In addition to the healthy wild type flies, resveratrol also has a beneficial effect in diseased flies. Supplementation of 200 μM resveratrol was reported to extend the mean lifespan of flies with Parkinson's disease.[27]

Although a lot of studies have demonstrated the prolongevity effect of resveratrol, several laboratories failed to reproduce such effects. In 2007, Bass *et al.* showed that resveratrol ranging from 1 to 1000 μM did not increase the lifespan of Canton-S and Dahomey strains of flies.[28] Moreover, the administration of 100 μM resveratrol slightly changed the lifespan of worms in only one trial out of four independent experiments using the N2 wild type strain.[28] Such conflicting results of resveratrol administration on lifespan may be due to the differences in macronutrient composition in the diets. In 2009, Zou *et al.* reported that thc prolongevity effect of resveratrol was dependent on dietary composition in tephritid fruit flies (*Anastrepha ludens*).[29] In brief, the prolongevity effect of resveratrol was tested in 24 diets in a combination of 4 different sugar:yeast ratios (1:0, 24:1: 9:1, 3:1), three food dilution levels (1×, 0.5×, 0.25×), and with or without 100 μM resveratrol. The lifespan of male tephritid flies was unresponsive to the supplementation of resveratrol. Conversely, female tephritid flies extended their lifespan in a dietary composition-dependent manner. Resveratrol increased the mean lifespan of females from 71 to 82 days, when the flies were fed with the 0.25 × 9:1 diet. The nutrient composition-dependent effect of resveratrol was also tested in *D. melanogaster*.[30] Supplementation of resveratrol at 200 μM but not 100 μM extended the mean lifespan of females fed with a low sugar/high protein diet by approximately 15%. Supplementation of resveratrol at either concentration did not extend the lifespan of males fed with a high-fat diet. Similarly, supplementation of resveratrol at 400 μM increased the mean lifespan of females fed with a high-fat diet by about 10% without changes in daily food intake.[30] These reports suggest that the composition of dietary nutrients should be considered when testing the longevity effects of nutraceuticals like resveratrol.

In addition to *C. elegans* and *D. melanogaster*, the effect of resveratrol on longevity was tested in other lower organisms. Supplementation of 30 and 130 μM resveratrol extended the median and maximum lifespan of honey bees (*Apis melllifera*) by 38% and 33%, respectively, under normal oxygen condition, but not in hyperoxic conditions.[31] However, resveratrol failed to extend the lifespan in mosquitoes (*Anopheles stephensi*) and the crustacean *Daphnia*.[32,33]

13.2.2 Lifespan-Extending Effects of Resveratrol in Vertebrate: Fishes and Rodents

The longevity benefits of resveratrol were also investigated in higher model organisms. Three independent studies documented the effects of lifespan extension by resveratrol in the annual fish *Nothobranchius*. In 2006, Valenzano *et al.* reported that the lifespan of the Gonarezhous strain of *Nothobranchius furzeri* was increased by food supplementation of 24–600 μg g^{-1} resveratrol in a dose-dependent manner without any loss of fertility.[20] The maximum lifespan of this seasonal fish was extended by 59% with 600 μg g^{-1} resveratrol supplementation. In 2012, another *Nothobranchius* strain was also tested for the effects of resveratrol.[34] Supplementation of 200 μg g^{-1} resveratrol in food extended the maximum lifespan by 28%, from 64 to 82 weeks in *Nothobranchius guentheri*.[34] Similarly, in 2013, Genade *et al.* reported that 12 μg per fish per day of resveratrol extended the median and maximum lifespan of *Nothobranchius guentheri* by 42.9% and 17%, respectively.[35]

Resveratrol also has been shown to have a prolongevity effect in mammals. In 2006, Baur *et al.* fed 1 year-old male C57BL/6NIA mice with a high-calorie diet containing 22.4 mg kg^{-1} day^{-1} of resveratrol for 110 weeks.[19] Supplementation of resveratrol changed the physiology of mice fed with the high-fat diet towards those with a standard diet, reducing the risk of death in mice fed with the high-fat diet by 31%.[19] Unlike the prolongevity effect of resveratrol on lifespan in those with a high-fat diet, resveratrol failed to extend the lifespan of healthy mice supplemented with a standard diet. In 2008, Pearson *et al.* supplemented 1 year-old C57BL/6NIA mice with 100 or 400 mg kg^{-1} resveratrol.[36] In this study, they showed that resveratrol had no effect on the lifespan of mice.[36] In 2010 and 2013, the National Institute on Aging's Intervention Testing Program also reported that supplementation of 300 or 1200 mg kg^{-1} resveratrol to 1 year-old or 4 month-old genetically heterogeneous mice did not extend their lifespan, both in males and females.[37,38] Similarly, supplementation of 4 mg kg^{-1} resveratrol did not extend the lifespan of 12 month-old Wistar rats.[39]

Resveratrol treatment has been documented to attenuate symptoms and prolong the lifespan of mice with age-related diseases. In 2011, supplementation of 25 mg kg^{-1} resveratrol to *KtrA2* knock-out mice exhibiting Parkinsonian phenotype increased the lifespan by 30%, delaying the deterioration of motor activity.[40] Furthermore, supplementation of 1 g kg^{-1} resveratrol to age-accelerated mice (SAMP8)—a model with Alzheimer's disease—reduced cognitive impairment, amyloid accumulation in the brain, and levels of phosphorylated tau, a marker of Alzheimer's disease severity.[41] In addition, resveratrol treatment increased the levels of Sirt1 and pAMPK, and extended the lifespan of SAMP8 mice by 33%.[41] In an experiment using an amyotrophic lateral sclerosis model of $SOD1^{G93A}$ transgenic mice, 25 mg kg^{-1} day^{-1} of resveratrol reduced the disease onset, increased lifespan, and attenuated the loss of motor neuron and atrophy of mitochondria.[42] Similarly, it was independently reported that resveratrol supplementation at 160 mg kg^{-1} day^{-1} extended the lifespan by 10% and improved the motor function in $SOD1^{G93A}$ mice.[43] Resveratrol was also effective in rodents with hypertension. Supplementation of resveratrol at 18 mg kg^{-1} day^{-1} to Dahl salt-sensitive rats extended the lifespan by 64%, and improved mitochondrial respiration and biogenesis.[44] However, supplementation of 50 mg kg^{-1} day^{-1} resveratrol decreased the survival of immunodeficient mice with prostate cancer xenografts.[45] These conflicting effects of resveratrol on lifespan between the healthy and diseased animals can be explained by the requirement of metabolic stresses to achieve lifespan-extending properties of resveratrol.

13.2.3 Clinical Trials of Resveratrol in Human Subjects

Although there have been no trials investigating the longevity effect of resveratrol in primates, including human subjects, an inference can be made from previously published reports with respect to the beneficial effects of resveratrol on human health. Recently, two studies reported by the National Institute on Aging showed that the supplementation of resveratrol improved insulin sensitivity in adipose and ameliorated arterial wall inflammation in rhesus monkeys fed with a high-fat/high-sucrose diet.[46,47] Several clinical trials of resveratrol have been attempted, especially in patients with metabolic disease, cardiovascular disease, and cancer, but the results have been controversial thus far. Daily supplementation of 500–600 mg resveratrol for 12 weeks to patients with non-alcoholic fatty liver reduced alanine aminotransferase, hepatic steatosis, tumor necrosis factor-α, and low-density lipoprotein cholesterol, but did not affect insulin sensitivity, lipid profile, and blood pressure.[48,49] Higher doses of resveratrol were also attempted to treat patients with non-alcoholic fatty liver. The supplementation of 1.5 g of resveratrol for 6 months or 3 g of resveratrol for 8 weeks was reported to decrease liver lipid contents, but did not change the histological features of the liver and insulin sensitivity.[50,51] In addition, daily supplementation of 3 g of resveratrol for 12 weeks significantly increased the expression of SIRT1 and the phosphorylation of AMPK in type 2 diabetes.[52] Inconsistently with the data from rodents, daily supplementation of 1–2 g resveratrol for 2 weeks or 500 mg for 4 weeks to obese men did not affect the blood pressure, resting energy expenditure, visceral fat contents, and inflammatory biomarkers.[53,54] In addition, non-obese and postmenopausal women with normal glucose tolerance did not respond to daily administration of 75 mg of resveratrol for 12 weeks.[55] A 1 year supplementation of resveratrol-containing (8 mg) grape-extract in coronary artery disease patients increased the level of anti-inflammatory adiponectin and decreased the levels of high-sensitivity C-reactive protein, tumor necrosis factor α, and thrombogenic plasminogen activator inhibitor type 1.[56,57] Moreover, daily supplementation of 10 mg of resveratrol capsule for 3 months to patients with coronary artery disease after myocardial infarction showed an improvement in the ventricle function and lowered the LDL-cholesterol level.[58] The effect of resveratrol on the host metabolism and clinical trials related to metabolic diseases were well reviewed by Timmers *et al.*[59]

According to the data from these reports, the optimal concentration and duration of resveratrol treatment seem to be distinct for the specific disease. Thus, large-scale and long-term studies are required to further evaluate the clinical values of resveratrol.

13.2.4 Putative Target Molecules for Lifespan-Extending Effect of Resveratrol

The first identified molecular target of resveratrol was sirtuin, an NAD+ dependent deacetylase. However, sirtuins are not the only target of resveratrol, and resveratrol is reported to have numerous molecular targets, including AMP-activated protein kinase (AMPK), cyclooxygenases, lipoxygenases, adenylyl cyclase, DNA polymerase, ribonucleotide reductase, quinone reductase 2, aryl hydrocarbon receptors, cytochrome P450 enzymes, F1-ATPase, and phosphodiesterases.[8,12,60–65] In addition, the metabolic effects of resveratrol were recently reported to be mediated by inhibiting cAMP phosphodiesterases (PDEs), particularly PDE4.[64] The underlying mechanism of lifespan-extending effects by phytochemicals is well reviewed by Leonov *et al.*[1] In this section, we focus on the putative mechanisms of lifespan-extending effects by resveratrol.

13.2.4.1 Caloric Restriction Mimetics

Caloric restriction is defined as the reduction of calorie uptake without malnutrition, and is well established to extend the lifespan of almost all species, including non-human primates. Several studies have shown that resveratrol did not further increase the lifespan under caloric restriction conditions, indicating that caloric restriction and resveratrol share similar anti-aging mechanisms. In 2003, Howitz *et al.* showed that resveratrol did not further extend the lifespan of *S. cerevisiae* under glucose-restricted conditions.[17]

Similarly, Wood *et al.* showed that resveratrol did not further extend the lifespan of flies under a restricted diet.[18] In this regard, resveratrol was considered as a calorie restriction mimetic—a compound producing the beneficial effects of caloric restriction without the actual restriction of energy intake. The concept of resveratrol having caloric restriction mimetic properties was supported by the fact that resveratrol shifts the physiology of mice fed with excess calories towards that of mice fed with a standard diet.[19] Supplementation of resveratrol make the mice fed with high calories healthier as indicated by survival, motor function, insulin sensitivity, organ pathology, and mitochondrial number.[19]

To find the link between resveratrol and caloric restriction, several studies compared the transcriptional profile of resveratrol treatment and caloric restriction implementation in rodents, and found that the transcriptional response to resveratrol resembles that by caloric restriction.[36,66,67] The gene expression alterations by low-dose resveratrol (4.9 mg kg^{-1}) or caloric restriction occurred in the same direction in the heart, skeletal muscle, and brain.[66] A strong repression of age-related transcriptional alterations by resveratrol and caloric restriction was shown in the heart.[66] Additionally, it was reported that the transcriptional effects of calorie restriction exhibited a variable degree of overlap with resveratrol in the liver of mice.[67] However, the transcriptional changes of resveratrol and caloric restriction were reported to be largely independent of the increase in SIRT1 activity.[66]

Although many studies indicate the possible role of resveratrol as a caloric restriction mimetic, not all studies support this notion. In a report published by Zou *et al.*, resveratrol supplementation still extended the lifespan of the tephritid fruit fly under caloric restriction.[29] Furthermore, caloric restriction reduced the circulating IGF-1 level but not resveratrol,[66] and resveratrol failed to slow down the heart rate, to decrease the core body temperature, and to extend the lifespan in non-obese animals unrelated to caloric restriction.[36,68,69]

13.2.4.2 NAD^+-Dependent Deacetylase Sirtuin

The NAD^+-dependent deacetylase sirtuins, including silent information regulator 1 (SIRT1), play a role in DNA damage response, metabolism, longevity, and carcinogenesis. SIRT1 regulates cellular processes such as proliferation, differentiation, and apoptosis through deacetylation of important regulatory proteins such as p53, transcription factor forkhead box O (FOXO), nuclear factor κB (NF-κB) subunit p65, and peroxisome proliferator-activated receptorγcoactivator-1α (PGC-1α). In addition, histones H1, H3, and H4 are also reported as substrates of SIRT1.[70] Much experimental evidence indicates that sirtuin is the major mediator of the health-improving and lifespan-extending effects of caloric restriction. Resveratrol was initially selected in the process of sirtuin activator screening in yeast, and it was revealed that resveratrol significantly increased SIRT1 activity through an allosteric interaction, resulting in an increase of SIRT1 affinity for both NAD^+ and acetylated substrate.[17,70]

Moreover, several studies have shown that the lifespan-extending effect of resveratrol is mediated by sirtuin. Both in worms and flies, resveratrol failed to extend the lifespan due to the lack of a functional Sir2—a major member of sirtuin associated with longevity and anti-aging effects.[17,18,22]

Although it has been attested that sirtuin activation is involved in the longevity benefit of resveratrol *in vivo*, whether resveratrol directly binds to and activates sirtuin is controversial since the allosteric interaction of resveratrol with SIRT1 has been challenged. In the first report that showed the longevity effect of resveratrol in yeast, authors used a fluorescence-conjugated peptide substrate, Fluor-de-Lys, to show the binding and activation of sirtuin by resveratrol.[17] However, such an outcome was not reproducible in subsequent studies that used other peptides. Resveratrol failed to activate SIRT1 when *in vitro* full-length endogenous substrates or short, fluorescence-unconjugated peptide substrates were used.[63,71,72] Therefore, a possibility of other potential pathways has been suggested, such as the activation of AMPK or other unknown pathways.[8,73] Conversely, several studies confirmed the initial finding. In 2013, Hubbard *et al.* identified the allosteric binding site of SIRT1 with resveratrol.[74] They showed that when glutamate at position 230 was substituted for lysine or alanine, the activation of SIRT1 by resveratrol was attenuated regardless of the substrates used.[74] In addition, the substrate sequence was reported to affect the activation of SIRT1 by resveratrol.[75] In summary, although whether resveratrol directly binds to and activates SIRT1 remains to be clarified, the requirement of sirtuin activation for the lifespan-extending effects of resveratrol seems to be obvious.

13.2.4.3 Autophagy

Autophagy is an evolutionarily conserved lysosomal degradation process for old, supernumerary, and damaged cytoplasmic components, including proteins, lipids, organelles, and membranes, contributing eukaryotic cellular homeostasis. Moreover, several pieces of evidence show that autophagy is involved in organismal survival. The induction of autophagy has been reported to improve lifespan and reduce age-associated mortality,[75–77] and the lifespan-extending effect of rapamycin and spermidine has been shown to be mediated by autophagy in *C. elegans* and *D. melanogaster*.[76,78] In addition, the autophagy processes within the central nervous system have been shown to be particularly important for determining longevity, since brain-specific overexpression of *Atg8* extends the longevity of *Drosophila* by 50%.[79,80]

Recently, autophagy and its related signaling pathways, including the mTOR pathway, were reported to contribute to the beneficial effects of resveratrol.[81] Induction of autophagy by resveratrol has been well established thus far, although controversial evidence showed that autophagy induced by nutrient starvation or rapamycin treatment was suppressed by resveratrol.[82] 100 μM resveratrol induced autophagy indicated by the accumulation of LC3B-II and LC3 puncta through the inhibition of mTOR signaling in GFP-LC3-expressing HeLa Cells.[83] In addition, resveratrol can induce autophagy in yeast, human cancer cells, and even in mice.[84–86] Furthermore, resveratrol prolonged the lifespan of autophagy-proficient nematodes, whereas this longevity benefit was abolished by the knockdown of essential autophagy modulator beclin-1.[25] The autophagy induced by resveratrol seems to be mediated by sirtuin. The knockdown, knockout, or pharmacological inhibition of SIRT1 prevented the induction of autophagy by resveratrol.[25,77] In addition, the lifespan-extension effect by *sir2.1* overexpression in nematodes was diminished when the beclin-1 was depleted by RNAi, indicating that the longevity benefits by resveratrol is dependent on autophagy.[25,87]

13.2.4.4 Other Molecular Targets

Several evidences indicate that AMPK is a key mediator of metabolic effect by resveratrol. AMPK is an enzyme involved in cellular energy homeostasis, which can be activated by physical exercise, ischemia, glucose deprivation, and caloric restriction.[88–90] Resveratrol 10–50 μM is reported to increase AMPK phosphorylation in HepG2 cells,[91] and 100–300 μM resveratrol is known to decrease the intracellular ATP levels through the activation of AMPK.[92] Although SIRT1 can activate AMPK through deacetylation of upstream kinase LKB1,[65] resveratrol is known to activate AMPK independent of SIRT1.[93] In addition, AMPK is known to increase the NAD^+ level, which promotes the deacetylation of SIRT1 substrates.[94–97] These results—together with the conflicting results regarding resveratrol binding with SIRT1—suggest that AMPK is proposed to be an alternative target of resveratrol. However, to date, there is a lack of experimental evidence about the requirement for AMPK in the longevity effect by resveratrol.

Similar to caloric restriction, resveratrol increased the mitochondrial content in several tissues, including the liver and skeletal muscle, and increased PGC-1α expression, which is a transcriptional cofactor in the regulation of mitochondrial biogenesis, respiration, and glucose homeostasis.[98–100] Furthermore, SIRT1 has been recently shown to function together with PGC-1α in glucose hormeostasis.[101] Moreover, an induction of PGC-1α in the intestine was reported to increase the lifespan of *Drosophila*.[102] Given the above mentioned results, PGC-1α may be the target of lifespan-extending effect of resveratrol, but direct evidence is currently lacking.

13.2.5 Uncertainty of Resveratrol as a Clinical Drug

Despite the obvious benefits of resveratrol on health and lack of apparent toxicity at high doses, use of resveratrol as a clinical drug remains questioned. The concentration of resveratrol is around 50–100 μg g^{-1} in fresh grapes, and 1.5–3 mg l^{-1} in red wine.[61] However, the bioavailability of resveratrol is too low since resveratrol in plasma is quickly taken up by enterocytes of the intestine, and metabolized into glucuronide/sulfate conjugates or dihydroresveratrol. This rapid clearance from plasma leads to poor bioavailability of resveratrol.[103–106] Several *in vitro* studies showed that 10–200 μM resveratrol is necessary in plasma for therapeutic activity, but *in vivo* studies indicated that a high concentration of resveratrol in plasma may not be achieved. In models involving rabbit, rat, and mice, the highest concentration of resveratrol—reached in the first 5 min after oral administration of 20 mg kg^{-1} resveratrol—was less than 3 μM in the plasma.[107] In addition, the level of resveratrol dropped to half after only 14.4 min of intravenous administration of 20 mg kg^{-1} resveratrol in rabbits.[107] Furthermore, the concentration of resveratrol in the plasma of rats generated a peak value not greater than 6.6 μM in 5–10 min after oral administration of 50 mg kg^{-1} resveratrol, and it dropped to 50 nM in 2 h.[108,109] Similarly, neither resveratrol nor resveratrol conjugates were detected in the plasma after 24 h fasting in rats fed for 8 weeks with a diet containing 50 mg kg^{-1} resveratrol.[110] Although the conjugates of resveratrol seem to reach much higher plasma levels than resveratrol,[111] the peak plasma levels of resveratrol were 2 μM after an oral ingestion of 25 mg resveratrol with a half-life of 9.2 ± 0.6 h.[104,112] Human plasma concentrations as high as 0.5 μM for resveratrol and 2 to 10 μM for its 4′-*O*-sulfate conjugate have been reported after oral administration of pure resveratrol.[111] Resveratrol absorbed and modified in the intestine is secreted back into the intestine, where it may be de-conjugated, reabsorbed, or excreted in the feces.[104] Interestingly, a recent report demonstrated that resveratrol sulfate conjugates can be taken up by several tissues, and that subsequent processing can regenerate free unmodified resveratrol inside the cells.[113] In addition, human gut microbiota may also limit the bioavailability of resveratrol through conversion into metabolites, such as 3,4′-dihydroxy-trans-stilbene and 3,4′-dihydroxybibenzyl.[114] To improve the low bioavailability of resveratrol, several theoretical solutions have been tried, such as a combination with additional phytochemicals and nano materials. In the report using rats, piperine—a polyphenol found in black pepper—is reported to increase the maximal plasma concentration of resveratrol.[115] Moreover, the load of resveratrol on lipid-core nanocapsules was reported to improve the bioavailability of resveratrol in rats and mice.[116–118] The issue of low bioavailability of resveratrol and its solutions are well reviewed by Smoliga *et al.*[119]

Despite the uncertainty of resveratrol as a clinical drug, the investigation of resveratrol is sufficiently valuable. As resveratrol has low bioavailability and interacts with multiple molecular targets, the development of new molecules with better bioavailability and higher affinity with sirtuin is a promising direction for the field of medicinal chemistry.

13.3 Other Phytochemicals with Lifespan-Extending Effects

13.3.1 Curcumin

Curcumin [1,7-bis-(4-hydroxy-3-methoxyphenyl)-1,6-heptadiene-3,5-dione] is a non-flavonoid polyphenolic yellow pigment, extracted from the rhizome of the plant *Curcuma longa* (turmeric). It has been widely used as a spice, food additive, dye, and herbal medicine in Asia, and its biological activity—anti-oxidative, anti-inflammatory, anti-cancer, anti-neurodegenerative, anti-diabetic, anti-allergic activity, and prolongevity effects—has been actively investigated.

The first report to show the lifespan-extending effect of curcumin was performed by Suckow and Suckow using *D. melanogaster* in 2006.[120] They showed that supplementation of 1.0 mg g^{-1} curcumin in media extended the lifespan of wild type male flies by 20%. However, the population size used in the study was rather small, and the secondary confounding factors, such as changes in food consumption and fecundity, were not examined. Additionally, several studies investigated the prolongevity effect of curcumin in various animal models. In *D. melanogaster*, 250 μM curcumin extended the lifespan and retarded the age-related decline of movement in Canton-S and Ives wild type flies,[121] and 1.0 mg g^{-1} curcumin in media extended the lifespan and increased superoxide dismutase activity in the Oregon-R wild type strain.[122] Furthermore, 100 μM curcumin extended the survival of *Drosophila* with Parkinson's disease.[123] Supplementation of curcumin at 20 or 100 μM extended the lifespan of *C. elegans* by 45%.[124,125] In addition to the feeding of curcumin during the adult stage, supplementation of curcumin during the developmental stage was also examined. 10 μM curcumin at the larvae stage increased the lifespan of adult flies,[126] and pretreatment with 100 μM curcumin recovered the shortened lifespan of flies irradiated with 10 Gy (0.8 Gy min^{-1}) radiation.[127] Furthermore, Ra strain flies supplemented with 10 mM curcumin during the larval stage were reported to have a long adult lifespan.[128] The authors also provided evidence of stage-specific lifespan-extension effect from curcumin supplementation. Curcumin supplementation during early-adulthood (days 5–27) extended the median lifespan by 49%, while curcumin supplementation throughout the adult life or late-adult stage (days 38–89) decreased the median lifespan by 30% or 4%, respectively.[128] These results suggest that the longevity benefit of curcumin is strain- and stage-specific. In addition to curcumin, the lifespan-extension effects of tetrahydrocurcumin, a metabolite of curcumin, were also investigated. Supplementation of 0.2% tetrahydrocurcumin with mouse pellets that began at the age of 13 months extended the lifespan of male C57BL/6JHsd mice with a reduction of body weight.[129] In addition, 50 μM tetrahydrocurcumin extended the lifespan of Oregon-R or *yw* flies by 20%.[130]

Several studies have suggested the molecular target of the longevity effect by curcumin. Soh *et al.* showed that supplementation of curcumin during the larval stage did not further extend the lifespan of flies under adult dietary restriction.[128] This suggests curcumin as a putative caloric restriction mimetic. In addition, Xiang *et al.* showed that tetrahydrocurcumin extended the lifespan for wild type flies, but not for *foxo* and *sir2* mutants.[130] These reports suggest that curcumin extends the lifespan *via* a similar pathway with dietary restriction.

Despite the beneficial effect of curcumin on longevity and health, curcumin also has an issue of low bioavailability. The problems and promises of the bioavailability of curcumin are well described in a review by Anand *et al.*[131] The low bioavailability of curcumin is contributed by curcumin's poor absorption from the intestine, rapid metabolism, and rapid systemic elimination. For example, a 650 mg capsule of curcuminoids was not detected in the serum for 6 h in healthy volunteers, and oral administration of 12 g of curcumin yielded only nanogram concentrations in serum in other human trials.[131–133] To solve this problem, numerous methods have been tried. For example, curcumin analogues and highly stabilized curcumin nanoparticles have been developed, and their effects have been investigated.[134,135] Despite its low bioavailability, curcumin is safe in animals and humans, even at high doses.[136–138]

13.3.2 Quercetin

Quercetin (3,3′,4′,5,7-pentahydroxyflavon) is a flavonoid found in herbal edibles like onions, apples, and broccolis, as well as in red wine, tea, and extracts of Ginkgo biloba. Quercetin has been well established to possess neuroprotective, cardioprotective, and chemopreventive properties.

The major longevity effect of quercetin was determined using *C. elegans*. In 2007, Kampkotter *et al.* reported, for the first time, the longevity effect of quercetin.[139] They showed that 100 μM quercetin extended the median lifespan of N2 wild-type *C. elegans* by 19%. Subsequently, several reports followed, showing that 100–200 μM quercetin extended the lifespan of *C. elegans*.[140–142] Moreover, quercetin 3-*O*-β-ᴅ-glucophyranoside—a quercetin derivate from onions—was reported to have stronger activity on the extension of the lifespan of worms.[143] Metabolites of quercetin—quercetin-3′-*O*-methylether or quercetin-4′-*O*-methylether—were also investigated using *C. elegans* for potential longevity effects.[144] In addition, quercetin-3-*O*-glucoside was absorbed by worms to a greater degree than quercetin.[142] Furthermore, 100 mg kg^{-1} quercetin administration decreased the number of tumor cells but did not affect the survival of Swiss albino mice bearing Ehrlich ascites tumor cells.[145] Using an intricate meta-analysis technique, quercetin was reported to have a lifespan-extending effect through the following pathways: TGF-β signaling, insulin-like signaling, and the p38 MAPK pathways.[146] In addition, quercetin was reported to extend the lifespan independent of daf16, sir2.1, and caloric restriction, but dependent on daf2, sek1, and unc43.[140,141]

13.3.3 Catechin

Catechin is a flavonol-type flavonoid (flavan-3-ol) possessing antioxidant, cardioprotective, anti-atherogenic, and anti-carcinogenic effects. Catechin and its derivatives, such as epicatechins (EC), epicatechins-3-gallate (ECG), epigallocatechin (EGC), epigallocatechin-3-gallate (EGCG), and gallocatecin

(GC), were known to increase the expression and activity of antioxidant enzymes.[147,148] Commonly, catechin is found in foods and edible plants, such as green tea, cacao, and red wine. Interestingly, the levels of catechins and epicatechins are 15 times higher than that of resveratrol in red wine; green and black teas are known to be rich in catechin.

The supplementation of green or black tea extract was reported to extend the lifespan of flies.[149,150] Similarly, Kitani *et al.* reported that green tea water extract, at a concentration of 80 mg l^{-1}, extended the lifespan of C57L/6JNIA male mice by 6.4% without any changes to body weight.[151] Li *et al.* showed that 10 mg ml^{-1} catechins extracted from green tea extended the lifespan of Oregon-R wild type *D. melanogaster* by 15.7% and increased resistance to oxidative stress.[152] They also showed that green tea catechin increased the activity of SOD1, SOD2, and catalase in *D. melanogaster*, and did not extend the lifespan of *SOD* mutant flies, indicating that the longevity effect of catechin is mediated by the activation of antioxidant enzymes.[152] The longevity benefit of catechin was also investigated in *C. elegans*. Supplementation of 200 μM catechin increased the survival of N2 wild type worms, independent of internal bacterial growth, but dependent on akt2, daf2, and mev1.[153] In addition, EGCG—a main active ingredient of green tea—extended the lifespan of worker honeybees (*Apismellifera scutellata*)[154] and healthy Wistar rats.[155] Catechin also showed a longevity benefit in animals with a high-fat diet. Supplementation of 10 mg ml^{-1} green tea catechin extended the mean lifespan of flies fed with 10% lard fatty acid.[156] Moreover, supplementation of 0.25% epicatechins in drinking water for 15 weeks to obese diabetic mice (db/db) significantly reduced the mortality rate from 50% to 8.3%.[157] However, not all studies were able to show a lifespan-extending effect from catechin. In particular, Zhang *et al.* showed that EGCG improved the survival time under stressed conditions, and they were unable to show any influence of lifespan-extending effect on *C. elegans* under normal conditions.[147]

13.3.4 Others

Other phytochemicals have been actively investigated for their anti-aging properties, including vitamins, lipoic acid, carotenoids, anthocyanins, saponin, and morphine.

Alpha-tocopherol, also known as vitamin E, is a well-known potent antioxidant. The lifespan-extending effect of α-tocopherol has been demonstrated in rotifer, *S. cerevisiae*, worms, flies, and rodents. The optimal concentration for the longevity effect of α-tocopherol is different in each species. For example, in single-cell organisms, such as rotifer and yeast, the longevity effect of α-tocopherol was shown in the concentration range of 0.05–10 000 μg ml^{-1}, and 80–200 μg ml^{-1} in *C. elegans*, 5–10 μg ml^{-1} in *D. melanogaster*, and 250–5000 μg g^{-1} in rodents.[158] However, not all reports confirmed the longevity benefits of α-tocopherol. In 2003, supplementation of 20 μg ml^{-1} α-tocopherol was reported to extend the lifespan of Canton-S wild type *D. melanogaster* by 16%.[159] However, Zou *et al.* failed to show the longevity benefit of α-tocopherol in both *C. elegans* and *D. melanogaster*.[160] Bahadorani *et al.* also showed that supplementation of 0.005–25 I.U ml^{-1} α-tocopherol did not extend the lifespan of *rosy*$^{+5}$ wild type flies in normoxic conditions.[161] In rodents, several studies have shown that supplementation of α-tocopherol with a concentration range of 20–4000 μg g^{-1} did not extend the mean and maximum lifespan of rats.[162–166] Moreover, supplementation of vitamin E on short-tailed field voles (*Microtous agrestis*) was reported to significantly shorten the lifespan.[167] However, 250–500 mg g^{-1} α-tocopherol was recently reported to extend the lifespan of C57BL/6, CD1/UCadiz, and MRL/lymproliferative mice.[168–170] The reports regarding the longevity effect of α-tocopherol in various model organisms are well summarized in a review by Ernst *et al.*[158]

Carotenoids—lipid-soluble pigments synthesized by plants, bacteria, and algae—are also well established to be potent antioxidants and immunostimulants.[171] Carotenoids, including β-carotene, lutein, zeaxanthin, and fucoxanthin, contain pigments responsible for yellow, red, and orange colors in food. Although the beneficial effects of carotenoids, like antioxidant, anti-mutagenic, anti-inflammatory, and anti-tumor properties, have been well investigated, studies on the longevity effect of carotenoids have only recently begun. Beta-carotene, an oxygen-lacking form of carotenoid, extended the lifespan of Canton-S wild type *D. melanogaster* at 0.3–1 μM,[172] and extended the lifespan and ameliorated the damage by γ-ray irradiation in Wistar rats.[173] In addition, supplementation of lutein, an oxycarotenoid was reported to extend the mean lifespan of Oregon-R wild type *D. melanogaster* at a concentration of 0.1 mg ml^{-1} by 11.35%.[174] Furthermore, fucoxanthin—a marine carotenoid—showed a beneficial effect on cancer, improvement of the plasma lipid profile,[175,176] and a lifespan-extending effect on flies.[172] Supplementation of 1 μM fucoxanthin increased the median lifespan of Canton-S wild type *D. melanogaster* by 33%.

Lipoic acid is a potent antioxidant found in spinach, broccoli, tomatoes, and rice. Lipoic acid is known to have beneficial effects for several diseases, including diabetes, cardiovascular disease, cognitive decline, and dementia. Supplementation of 24 μM α-lipoic acid was reported to extend the mean and maximum lifespan of *C. elegans* by 24% and 14%, respectively.[177] However, supplementation of 600 mg kg^{-1} α-lipoic acid had no effect on the lifespan of C57BL/6C3F1 mice,[178] and administration of 100 mg kg^{-1} α-lipoic acid decreased the median lifespan of SAMP8 mice.[179]

Morphine—an analgesic extracted from the opium poppy (*Papaver somniferum*)—is also reported to possess longevity benefits in mice and flies. Supplementation of 10 mg kg^{-1} morphine hydrochloride once a week extended the residual lifespan of 28 month-old CBA mice, and 0.25 mg ml^{-1} morphine hydrochloride extended the residual lifespan of Oregon-R wild type flies when supplied once a week since 5 days or 54 days.[180] Nolinospiroside F—a steroidal saponin extracted from *Ophiopogon japonicas*—was reported to increase the replicative lifespan of K6001 *S. cerevisiae* at concentrations of 1, 3, and 10 μM.[181] Caffeic acid (300 μM) and rosmarinic acid (200 μM) extended the lifespan of N2 *C. elegans* fed with heat-killed *E. coli*.[182] There is an infinite variety of plants in the world, and the possibility to discover better phytochemicals, superior to resveratrol, is also inexhaustible.

13.4 Conclusion

In this chapter, the lifespan-extending effects of resveratrol and other phytochemicals, such as curcumin, quercetin, and catechins, were reviewed. Phytochemicals are better therapeutic candidates than synthetic drugs with low toxicity and safety issues as they have been extensively used for centuries. Moreover, there is obvious evidence of longevity benefits from phytochemicals. However, the issues of low bioavailability and uncertainty in the molecular targets of phytochemicals need be addressed. More methodical and profound approaches to explore the mechanisms underlying the lifespan-extending effects of these phytochemicals and probe for solutions to address the aforementioned issue of low bioavailability are required for clinical application.

References

1. A. Leonov, A. Arlia-Ciommo, A. Piano, V. Svistkova, V. Lutchman, Y. Medkour and V. I. Titorenko, *Molecules*, 2015, **20**, 6544.
2. L. Fontana and L. Partridge, *Cell*, 2015, **161**, 106.
3. A. P. Gollucke and D. A. Ribeiro, *Recent Pat. Food, Nutr. Agric.*, 2012, **4**, 26.
4. C. I. Gamboa-Gomez, N. E. Rocha-Guzman, J. A. Gallegos-Infante, M. R. Moreno-Jimenez, B. D. Vazquez-Cabral and R. F. Gonzalez-Laredo, *EXCLI J.*, 2015, **14**, 809.
5. Y. J. Zhang, R. Y. Gan, S. Li, Y. Zhou, A. N. Li, D. P. Xu and H. B. Li, *Molecules*, 2015, **20**, 21138.
6. B. Pagliaro, C. Santolamazza, F. Simonelli and S. Rubattu, *BioMed Res. Int.*, 2015, **2015**, 918069.
7. Y. H. Li, Y. B. Niu, Y. Sun, F. Zhang, C. X. Liu, L. Fan and Q. B. Mei, *World J. Gastroenterol.*, 2015, **21**, 9262.
8. J. A. Baur, *Mech. Ageing Dev.*, 2010, **131**, 261.
9. M. Jang, L. Cai, G. O. Udeani, K. V. Slowing, C. F. Thomas, C. W. Beecher, H. H. Fong, N. R. Farnsworth, A. D. Kinghorn, R. G. Mehta, R. C. Moon and J. M. Pezzuto, *Science (N. Y.)*, 1997, **275**, 218.
10. K. Sinha, G. Chaudhary and Y. K. Gupta, *Life Sci.*, 2002, **71**, 655.
11. R. Falchetti, M. P. Fuggetta, G. Lanzilli, M. Tricarico and G. Ravagnan, *Life Sci.*, 2001, **70**, 81.
12. J. A. Baur and D. A. Sinclair, *Nat. Rev.*, 2006, **5**, 493.
13. K. Mizutani, K. Ikeda and Y. Yamori, *Biochem. Biophys. Res. Commun.*, 2000, **274**, 61.
14. T. C. Hsieh, P. Burfeind, K. Laud, J. M. Backer, F. Traganos, Z. Darzynkiewicz and J. M. Wu, *Int. J. Oncol.*, 1999, **15**, 245.
15. M. Nielsen, R. J. Ruch and O. Vang, *Biochem. Biophys. Res. Commun.*, 2000, **275**, 804.

16. B. P. Hubbard and D. A. Sinclair, *Trends Pharmacol. Sci.*, 2014, **35**, 146.
17. K. T. Howitz, K. J. Bitterman, H. Y. Cohen, D. W. Lamming, S. Lavu, J. G. Wood, R. E. Zipkin, P. Chung, A. Kisielewski, L. L. Zhang, B. Scherer and D. A. Sinclair, *Nature*, 2003, **425**, 191.
18. J. G. Wood, B. Rogina, S. Lavu, K. Howitz, S. L. Helfand, M. Tatar and D. Sinclair, *Nature*, 2004, **430**, 686.
19. J. A. Baur, K. J. Pearson, N. L. Price, H. A. Jamieson, C. Lerin, A. Kalra, V. V. Prabhu, J. S. Allard, G. Lopez-Lluch, K. Lewis, P. J. Pistell, S. Poosala, K. G. Becker, O. Boss, D. Gwinn, M. Wang, S. Ramaswamy, K. W. Fishbein, R. G. Spencer, E. G. Lakatta, D. Le Couteur, R. J. Shaw, P. Navas, P. Puigserver, D. K. Ingram, R. de Cabo and D. A. Sinclair, *Nature*, 2006, **444**, 337.
20. D. R. Valenzano, E. Terzibasi, T. Genade, A. Cattaneo, L. Domenici and A. Cellerino, *Curr. Biol.*, 2006, **16**, 296.
21. J. Marchal, F. Pifferi and F. Aujard, *Ann. N. Y. Acad. Sci.*, 2013, **1290**, 67.
22. M. Viswanathan, S. K. Kim, A. Berdichevsky and L. Guarente, *Dev. Cell*, 2005, **9**, 605.
23. J. Gruber, S. Y. Tang and B. Halliwell, *Ann. N. Y. Acad. Sci.*, 2007, **1100**, 530.
24. E. L. Greer and A. Brunet, *Aging Cell*, 2009, **8**, 113.
25. E. Morselli, M. C. Maiuri, M. Markaki, E. Megalou, A. Pasparaki, K. Palikaras, A. Criollo, L. Galluzzi, S. A. Malik, I. Vitale, M. Michaud, F. Madeo, N. Tavernarakis and G. Kroemer, *Cell Death Discovery*, 2010, **1**, e10.
26. J. H. Bauer, S. Goupil, G. B. Garber and S. L. Helfand, *Proc. Natl. Acad. Sci. U. S. A.*, 2004, **101**, 12980.
27. J. Long, H. Gao, L. Sun, J. Liu and X. Zhao-Wilson, *Rejuvenation Res.*, 2009, **12**, 321.
28. T. M. Bass, D. Weinkove, K. Houthoofd, D. Gems and L. Partridge, *Mech. Ageing Dev.*, 2007, **128**, 546.
29. S. Zou, J. R. Carey, P. Liedo, D. K. Ingram, H. G. Muller, J. L. Wang, F. Yao, B. Yu and A. Zhou, *Exp. Gerontol.*, 2009, **44**, 472.
30. C. Wang, C. T. Wheeler, T. Alberico, X. Sun, J. Seeberger, M. Laslo, E. Spangler, B. Kern, R. de Cabo and S. Zou, *Age (Dordrecht, Neth.)*, 2013, **35**, 69.
31. B. Rascon, B. P. Hubbard, D. A. Sinclair and G. V. Amdam, *Aging*, 2012, **4**, 499.
32. E. Kim, C. M. Ansell and J. L. Dudycha, *J. Exp. Zool.*, 2014, **321**, 48.
33. A. A. Johnson and M. A. Riehle, *Rejuvenation Res.*, 2015, **18**, 473.
34. X. Yu and G. Li, *Exp. Gerontol.*, 2012, **47**, 940.
35. T. Genade and D. M. Lang, *Exp. Gerontol.*, 2013, **48**, 202.
36. K. J. Pearson, J. A. Baur, K. N. Lewis, L. Peshkin, N. L. Price, N. Labinskyy, W. R. Swindell, D. Kamara, R. K. Minor, E. Perez, H. A. Jamieson, Y. Zhang, S. R. Dunn, K. Sharma, N. Pleshko, L. A. Woollett, A. Csiszar, Y. Ikeno, D. Le Couteur, P. J. Elliott, K. G. Becker, P. Navas, D. K. Ingram, N. S. Wolf, Z. Ungvari, D. A. Sinclair and R. de Cabo, *Cell Metab.*, 2008, **8**, 157.

37. R. A. Miller, D. E. Harrison, C. M. Astle, J. A. Baur, A. R. Boyd, R. de Cabo, E. Fernandez, K. Flurkey, M. A. Javors, J. F. Nelson, C. J. Orihuela, S. Pletcher, Z. D. Sharp, D. Sinclair, J. W. Starnes, J. E. Wilkinson, N. L. Nadon and R. Strong, *J. Gerontol.*, 2010, **66**, 191.
38. R. Strong, R. A. Miller, C. M. Astle, J. A. Baur, R. de Cabo, E. Fernandez, W. Guo, M. Javors, J. L. Kirkland, J. F. Nelson, D. A. Sinclair, B. Teter, D. Williams, N. Zaveri, N. L. Nadon and D. E. Harrison, *J. Gerontol.*, 2013, **68**, 6.
39. P. L. da Luz, L. Tanaka, P. C. Brum, P. M. Dourado, D. Favarato, J. E. Krieger and F. R. Laurindo, *Atherosclerosis*, 2012, **224**, 136–142.
40. E. Gerhardt, S. Graber, E. M. Szego, N. Moisoi, L. M. Martins, T. F. Outeiro and P. Kermer, *PLoS One*, 2011, **6**, e28855.
41. D. Porquet, G. Casadesus, S. Bayod, A. Vicente, A. M. Canudas, J. Vilaplana, C. Pelegri, C. Sanfeliu, A. Camins, M. Pallas and J. del Valle, *Age (Dordrecht, Neth.)*, 2013, **35**, 1851.
42. L. Song, L. Chen, X. Zhang, J. Li and W. Le, *BioMed Res. Int.*, 2014, **2014**, 483501.
43. R. Mancuso, J. del Valle, L. Modol, A. Martinez, A. B. Granado-Serrano, O. Ramirez-Nunez, M. Pallas, M. Portero-Otin, R. Osta and X. Navarro, *Neurotherapeutics*, 2014, **11**, 419.
44. S. Rimbaud, M. Ruiz, J. Piquereau, P. Mateo, D. Fortin, V. Veksler, A. Garnier and R. Ventura-Clapier, *PLoS One*, 2011, **6**, e26391.
45. J. C. Klink, A. K. Tewari, E. M. Masko, J. Antonelli, P. G. Febbo, P. Cohen, M. W. Dewhirst, S. V. Pizzo and S. J. Freedland, *Prostate*, 2013, **73**, 754.
46. Y. Jimenez-Gomez, J. A. Mattison, K. J. Pearson, A. Martin-Montalvo, H. H. Palacios, A. M. Sossong, T. M. Ward, C. M. Younts, K. Lewis, J. S. Allard, D. L. Longo, J. P. Belman, M. M. Malagon, P. Navas, M. Sanghvi, R. Moaddel, E. M. Tilmont, R. L. Herbert, C. H. Morrell, J. M. Egan, J. A. Baur, L. Ferrucci, J. S. Bogan, M. Bernier and R. de Cabo, *Cell Metab.*, 2013, **18**, 533.
47. J. A. Mattison, M. Wang, M. Bernier, J. Zhang, S. S. Park, S. Maudsley, S. S. An, L. Santhanam, B. Martin, S. Faulkner, C. Morrell, J. A. Baur, L. Peshkin, D. Sosnowska, A. Csiszar, R. L. Herbert, E. M. Tilmont, Z. Ungvari, K. J. Pearson, E. G. Lakatta and R. de Cabo, *Cell Metab.*, 2014, **20**, 183.
48. F. Faghihzadeh, P. Adibi and A. Hekmatdoost, *Br. J. Nutr.*, 2015, **114**, 796.
49. S. Chen, X. Zhao, L. Ran, J. Wan, X. Wang, Y. Qin, F. Shu, Y. Gao, L. Yuan, Q. Zhang and M. Mi, *Dig. Liver Dis.*, 2015, **47**, 226.
50. S. Heeboll, M. Kreuzfeldt, S. Hamilton-Dutoit, M. Kjaer Poulsen, H. Stodkilde-Jorgensen, H. J. Moller, N. Jessen, K. Thorsen, Y. Kristina Hellberg, S. Bonlokke Pedersen and H. Gronbaek, *Scand. J. Gastroenterol.*, 2016, **51**, 456.
51. V. S. Chachay, G. A. Macdonald, J. H. Martin, J. P. Whitehead, T. M. O'Moore-Sullivan, P. Lee, M. Franklin, K. Klein, P. J. Taylor, M. Ferguson, J. S. Coombes, G. P. Thomas, G. J. Cowin, C. M. Kirkpatrick, J. B. Prins and I. J. Hickman, *Clin. Gastroenterol. Hepatol.*, 2014, **12**, 2092–2103 e2091–2096.

52. K. P. Goh, H. Y. Lee, D. P. Lau, W. Supaat, Y. H. Chan and A. F. Koh, *Int. J. Sport Nutr. Exercise Metab.*, 2014, **24**, 2.
53. M. M. Poulsen, P. F. Vestergaard, B. F. Clasen, Y. Radko, L. P. Christensen, H. Stodkilde-Jorgensen, N. Moller, N. Jessen, S. B. Pedersen and J. O. Jorgensen, *Diabetes*, 2012, **62**, 1186.
54. S. Dash, C. Xiao, C. Morgantini, L. Szeto and G. F. Lewis, *Arterioscler., Thromb., Vasc. Biol.*, 2013, **33**, 2895.
55. J. Yoshino, C. Conte, L. Fontana, B. Mittendorfer, S. Imai, K. B. Schechtman, C. Gu, I. Kunz, F. Rossi Fanelli, B. W. Patterson and S. Klein, *Cell Metab.*, 2012, **16**, 658.
56. J. Tome-Carneiro, M. Gonzalvez, M. Larrosa, M. J. Yanez-Gascon, F. J. Garcia-Almagro, J. A. Ruiz-Ros, F. A. Tomas-Barberan, M. T. Garcia-Conesa and J. C. Espin, *Cardiovasc. Drugs Ther.*, 2013, **27**, 37.
57. J. Tome-Carneiro, M. Gonzalvez, M. Larrosa, M. J. Yanez-Gascon, F. J. Garcia-Almagro, J. A. Ruiz-Ros, M. T. Garcia-Conesa, F. A. Tomas-Barberan and J. C. Espin, *Am. J. Cardiol.*, 2012, **110**, 356–363.
58. K. Magyar, R. Halmosi, A. Palfi, G. Feher, L. Czopf, A. Fulop, I. Battyany, B. Sumegi, K. Toth and E. Szabados, *Clin. Hemorheol. Microcirc.*, 2012, **50**, 179.
59. S. Timmers, J. Auwerx and P. Schrauwen, *Aging*, 2012, **4**, 146.
60. J. R. Gledhill, M. G. Montgomery, A. G. Leslie and J. E. Walker, *Proc. Natl. Acad. Sci. U. S. A.*, 2007, **104**, 13632.
61. F. J. Alcain and J. M. Villalba, *Expert Opin. Ther. Pat.*, 2009, **19**, 403.
62. L. Pirola and S. Frojdo, *IUBMB Life*, 2008, **60**, 323.
63. M. Pacholec, J. E. Bleasdale, B. Chrunyk, D. Cunningham, D. Flynn, R. S. Garofalo, D. Griffith, M. Griffor, P. Loulakis, B. Pabst, X. Qiu, B. Stockman, V. Thanabal, A. Varghese, J. Ward, J. Withka and K. Ahn, *J. Biol. Chem.*, 2010, **285**, 8340.
64. S. J. Park, F. Ahmad, A. Philp, K. Baar, T. Williams, H. Luo, H. Ke, H. Rehmann, R. Taussig, A. L. Brown, M. K. Kim, M. A. Beaven, A. B. Burgin, V. Manganiello and J. H. Chung, *Cell*, 2012, **148**, 421.
65. F. Lan, J. M. Cacicedo, N. Ruderman and Y. Ido, *J. Biol. Chem.*, 2008, **283**, 27628.
66. J. L. Barger, T. Kayo, J. M. Vann, E. B. Arias, J. Wang, T. A. Hacker, Y. Wang, D. Raederstorff, J. D. Morrow, C. Leeuwenburgh, D. B. Allison, K. W. Saupe, G. D. Cartee, R. Weindruch and T. A. Prolla, *PLoS One*, 2008, **3**, e2264.
67. W. R. Swindell, *Mech. Ageing Dev.*, 2008, **129**, 138.
68. J. R. Mayers, B. W. Iliff and S. J. Swoap, *FASEB J.*, 2009, **23**, 1032.
69. D. E. Harrison, J. R. Archer and C. M. Astle, *Proc. Natl. Acad. Sci. U. S. A.*, 1984, **81**, 1835.
70. W. J. Jiang, *Biochem. Biophys. Res. Commun.*, 2008, **373**, 341.
71. M. T. Borra, B. C. Smith and J. M. Denu, *J. Biol. Chem.*, 2005, **280**, 17187.
72. M. Kaeberlein, T. McDonagh, B. Heltweg, J. Hixon, E. A. Westman, S. D. Caldwell, A. Napper, R. Curtis, P. S. DiStefano, S. Fields, A. Bedalov and B. K. Kennedy, *J. Biol. Chem.*, 2005, **280**, 17038.

73. C. Canto and J. Auwerx, *Pharmacol. Rev.*, 2011, **64**, 166.
74. B. P. Hubbard, A. P. Gomes, H. Dai, J. Li, A. W. Case, T. Considine, T. V. Riera, J. E. Lee, S. Y. E, D. W. Lamming, B. L. Pentelute, E. R. Schuman, L. A. Stevens, A. J. Ling, S. M. Armour, S. Michan, H. Zhao, Y. Jiang, S. M. Sweitzer, C. A. Blum, J. S. Disch, P. Y. Ng, K. T. Howitz, A. P. Rolo, Y. Hamuro, J. Moss, R. B. Perni, J. L. Ellis, G. P. Vlasuk and D. A. Sinclair, *Science*, 2013, **339**, 1216.
75. M. Lakshminarasimhan, D. Rauh, M. Schutkowski and C. Steegborn, *Aging*, 2013, **5**, 151.
76. K. Jia and B. Levine, *Autophagy*, 2007, **3**, 597.
77. N. Tavernarakis, A. Pasparaki, E. Tasdemir, M. C. Maiuri and G. Kroemer, *Autophagy*, 2008, **4**, 870.
78. T. Eisenberg, H. Knauer, A. Schauer, S. Buttner, C. Ruckenstuhl, D. Carmona-Gutierrez, J. Ring, S. Schroeder, C. Magnes, L. Antonacci, H. Fussi, L. Deszcz, R. Hartl, E. Schraml, A. Criollo, E. Megalou, D. Weiskopf, P. Laun, G. Heeren, M. Breitenbach, B. Grubeck-Loebenstein, E. Herker, B. Fahrenkrog, K. U. Frohlich, F. Sinner, N. Tavernarakis, N. Minois, G. Kroemer and F. Madeo, *Nat. Cell Biol.*, 2009, **11**, 1305.
79. F. Madeo, N. Tavernarakis and G. Kroemer, *Nat. Cell Biol.*, 2010, **12**, 842.
80. A. Simonsen, R. C. Cumming, A. Brech, P. Isakson, D. R. Schubert and K. D. Finley, *Autophagy*, 2008, **4**, 176.
81. A. L. Widlund, J. A. Baur and O. Vang, *Expert Rev. Mol. Med.*, 2013, **15**, e10.
82. S. M. Armour, J. A. Baur, S. N. Hsieh, A. Land-Bracha, S. M. Thomas and D. A. Sinclair, *Aging*, 2009, **1**, 515.
83. D. Park, H. Jeong, M. N. Lee, A. Koh, O. Kwon, Y. R. Yang, J. Noh, P. G. Suh, H. Park and S. H. Ryu, *Sci. Rep.*, 2016, **6**, 21772.
84. I. Kissova, M. Deffieu, V. Samokhvalov, G. Velours, J. J. Bessoule, S. Manon and N. Camougrand, *Free Radical Biol. Med.*, 2006, **41**, 1655.
85. F. Scarlatti, R. Maffei, I. Beau, P. Codogno and R. Ghidoni, *Cell Death Differ.*, 2008, **15**, 1318.
86. B. Wang, Q. Yang, Y. Y. Sun, Y. F. Xing, Y. B. Wang, X. T. Lu, W. W. Bai, X. Q. Liu and Y. X. Zhao, *J. Cell. Mol. Med.*, 2014, **18**, 1599.
87. E. Morselli, M. C. Maiuri, M. Markaki, E. Megalou, A. Pasparaki, K. Palikaras, A. Criollo, L. Galluzzi, S. A. Malik, I. Vitale, M. Michaud, F. Madeo, N. Tavernarakis and G. Kroemer, *Autophagy*, 2010, **6**, 186.
88. D. G. Hardie, F. A. Ross and S. A. Hawley, *Nat. Rev. Mol. Cell Biol.*, 2012, **13**, 251.
89. A. G. Edwards, A. J. Donato, L. A. Lesniewski, R. A. Gioscia, D. R. Seals and R. L. Moore, *Mech. Ageing Dev.*, 2010, **131**, 739.
90. O. M. Palacios, J. J. Carmona, S. Michan, K. Y. Chen, Y. Manabe, J. L. Ward 3rd, L. J. Goodyear and Q. Tong, *Aging*, 2009, **1**, 771.
91. M. Zang, S. Xu, K. A. Maitland-Toolan, A. Zuccollo, X. Hou, B. Jiang, M. Wierzbicki, T. J. Verbeuren and R. A. Cohen, *Diabetes*, 2006, **55**, 2180.
92. S. A. Hawley, F. A. Ross, C. Chevtzoff, K. A. Green, A. Evans, S. Fogarty, M. C. Towler, L. J. Brown, O. A. Ogunbayo, A. M. Evans and D. G. Hardie, *Cell Metab.*, 2010, **11**, 554.

93. B. Dasgupta and J. Milbrandt, *Proc. Natl. Acad. Sci. U. S. A.*, 2007, **104**, 7217.
94. M. Fulco, Y. Cen, P. Zhao, E. P. Hoffman, M. W. McBurney, A. A. Sauve and V. Sartorelli, *Dev. Cell*, 2008, **14**, 661.
95. C. Canto, Z. Gerhart-Hines, J. N. Feige, M. Lagouge, L. Noriega, J. C. Milne, P. J. Elliott, P. Puigserver and J. Auwerx, *Nature*, 2009, **458**, 1056.
96. J. H. Um, S. J. Park, H. Kang, S. Yang, M. Foretz, M. W. McBurney, M. K. Kim, B. Viollet and J. H. Chung, *Diabetes*, 2010, **59**, 554.
97. C. Canto, L. Q. Jiang, A. S. Deshmukh, C. Mataki, A. Coste, M. Lagouge, J. R. Zierath and J. Auwerx, *Cell Metab.*, 2010, **11**, 213.
98. M. Lagouge, C. Argmann, Z. Gerhart-Hines, H. Meziane, C. Lerin, F. Daussin, N. Messadeq, J. Milne, P. Lambert, P. Elliott, B. Geny, M. Laakso, P. Puigserver and J. Auwerx, *Cell*, 2006, **127**, 1109.
99. E. Nisoli, C. Tonello, A. Cardile, V. Cozzi, R. Bracale, L. Tedesco, S. Falcone, A. Valerio, O. Cantoni, E. Clementi, S. Moncada and M. O. Carruba, *Science*, 2005, **310**, 314.
100. A. E. Civitarese, S. Carling, L. K. Heilbronn, M. H. Hulver, B. Ukropcova, W. A. Deutsch, S. R. Smith and E. Ravussin, *PLoS Med.*, 2007, **4**, e76.
101. J. T. Rodgers, C. Lerin, W. Haas, S. P. Gygi, B. M. Spiegelman and P. Puigserver, *Nature*, 2005, **434**, 113.
102. M. Rera, S. Bahadorani, J. Cho, C. L. Koehler, M. Ulgherait, J. H. Hur, W. S. Ansari, T. Lo Jr., D. L. Jones and D. W. Walker, *Cell Metab.*, 2011, **14**, 623.
103. D. J. Boocock, K. R. Patel, G. E. Faust, D. P. Normolle, T. H. Marczylo, J. A. Crowell, D. E. Brenner, T. D. Booth, A. Gescher and W. P. Steward, *J. Chromatogr.*, 2007, **848**, 182.
104. T. Walle, F. Hsieh, M. H. DeLegge, J. E. Oatis Jr. and U. K. Walle, *Drug Metab. Dispos.*, 2004, **32**, 1377.
105. C. Yu, Y. G. Shin, A. Chow, Y. Li, J. W. Kosmeder, Y. S. Lee, W. H. Hirschelman, J. M. Pezzuto, R. G. Mehta and R. B. van Breemen, *Pharm. Res.*, 2002, **19**, 1907.
106. M. C. Menet, J. Marchal, A. Dal-Pan, M. Taghi, V. Nivet-Antoine, D. Dargere, O. Laprevote, J. L. Beaudeux, F. Aujard, J. Epelbaum and C. H. Cottart, *PLoS One*, 2014, **9**, e91932.
107. M. Asensi, I. Medina, A. Ortega, J. Carretero, M. C. Bano, E. Obrador and J. M. Estrela, *Free Radical Biol. Med.*, 2002, **33**, 387.
108. J. F. Marier, P. Vachon, A. Gritsas, J. Zhang, J. P. Moreau and M. P. Ducharme, *J. Pharmacol. Exp. Ther.*, 2002, **302**, 369.
109. A. J. Gescher and W. P. Steward, *Cancer Epidemiol., Biomarkers Prev.*, 2003, **12**, 953.
110. E. Wenzel, T. Soldo, H. Erbersdobler and V. Somoza, *Mol. Nutr. Food Res.*, 2005, **49**, 482.
111. D. J. Boocock, G. E. Faust, K. R. Patel, A. M. Schinas, V. A. Brown, M. P. Ducharme, T. D. Booth, J. A. Crowell, M. Perloff, A. J. Gescher, W. P. Steward and D. E. Brenner, *Cancer Epidemiol., Biomarkers Prev.*, 2007, **16**, 1246.

112. D. M. Goldberg, J. Yan and G. J. Soleas, *Clin. Biochem.*, 2003, **36**, 79.
113. K. R. Patel, C. Andreadi, R. G. Britton, E. Horner-Glister, A. Karmokar, S. Sale, V. A. Brown, D. E. Brenner, R. Singh, W. P. Steward, A. J. Gescher and K. Brown, *Sci. Transl. Med.*, 2013, **5**, 205ra133.
114. L. M. Bode, D. Bunzel, M. Huch, G. S. Cho, D. Ruhland, M. Bunzel, A. Bub, C. M. Franz and S. E. Kulling, *Am. J. Clin. Nutr.*, 2013, **97**, 295.
115. J. J. Johnson, M. Nihal, I. A. Siddiqui, C. O. Scarlett, H. H. Bailey, H. Mukhtar and N. Ahmad, *Mol. Nutr. Food Res.*, 2011, **55**, 1169.
116. M. Sessa, R. Tsao, R. Liu, G. Ferrari and F. Donsi, *J. Agric. Food Chem.*, 2011, **59**, 12352.
117. R. L. Frozza, A. Bernardi, K. Paese, J. B. Hoppe, T. da Silva, A. M. Battastini, A. R. Pohlmann, S. S. Guterres and C. Salbego, *J. Biomed. Nanotechnol.*, 2010, **6**, 694.
118. L. Guo, Y. Peng, J. Yao, L. Sui, A. Gu and J. Wang, *Cancer Biother. Radiopharm.*, 2010, **25**, 471.
119. J. M. Smoliga and O. Blanchard, *Molecules*, 2014, **19**, 17154.
120. B. K. Suckow and M. A. Suckow, *Int. J. Biomed. Sci.*, 2006, **2**, 402.
121. K. S. Lee, B. S. Lee, S. Semnani, A. Avanesian, C. Y. Um, H. J. Jeon, K. M. Seong, K. Yu, K. J. Min and M. Jafari, *Rejuvenation Res.*, 2010, **13**, 561.
122. L. R. Shen, F. Xiao, P. Yuan, Y. Chen, Q. K. Gao, L. D. Parnell, M. Meydani, J. M. Ordovas, D. Li and C. Q. Lai, *Age (Dordrecht, Neth.)*, 2013, **35**, 1133.
123. Y. H. Siddique, F. Naz and S. Jyoti, *BioMed Res. Int.*, 2014, **2014**, 606928.
124. S. Alavez, M. C. Vantipalli, D. J. Zucker, I. M. Klang and G. J. Lithgow, *Nature*, 2011, **472**, 226.
125. V. H. Liao, C. W. Yu, Y. J. Chu, W. H. Li, Y. C. Hsieh and T. T. Wang, *Mech. Ageing Dev.*, 2011, **132**, 480.
126. K. T. Chandrashekara, S. Popli and M. N. Shakarad, *Age (Dordrecht, Neth.)*, 2015, **36**, 9702.
127. K. M. Seong, M. Yu, K. S. Lee, S. Park, Y. W. Jin and K. J. Min, *BioMed Res. Int.*, 2015, **2015**, 425380.
128. J. W. Soh, N. Marowsky, T. J. Nichols, A. M. Rahman, T. Miah, P. Sarao, R. Khasawneh, A. Unnikrishnan, A. R. Heydari, R. B. Silver and R. Arking, *Exp. Gerontol.*, 2013, **48**, 229.
129. K. Kitani, T. Osawa and T. Yokozawa, *Biogerontology*, 2007, **8**, 567.
130. L. Xiang, Y. Nakamura, Y. M. Lim, Y. Yamasaki, Y. Kurokawa-Nose, W. Maruyama, T. Osawa, A. Matsuura, N. Motoyama and L. Tsuda, *Aging*, 2011, **3**, 1098.
131. P. Anand, A. B. Kunnumakkara, R. A. Newman and B. B. Aggarwal, *Mol. Pharmaceutics*, 2007, **4**, 807.
132. V. S. Gota, G. B. Maru, T. G. Soni, T. R. Gandhi, N. Kochar and M. G. Agarwal, *J. Agric. Food Chem.*, 2010, **58**, 2095.
133. N. Dhillon, B. B. Aggarwal, R. A. Newman, R. A. Wolff, A. B. Kunnumakkara, J. L. Abbruzzese, C. S. Ng, V. Badmaev and R. Kurzrock, *Clin. Cancer Res.*, 2008, **14**, 4491.

134. H. Ohori, H. Yamakoshi, M. Tomizawa, M. Shibuya, Y. Kakudo, A. Takahashi, S. Takahashi, S. Kato, T. Suzuki, C. Ishioka, Y. Iwabuchi and H. Shibata, *Mol. Cancer Ther.*, 2006, **5**, 2563.
135. S. F. Chow, K. Y. Wan, K. K. Cheng, K. W. Wong, C. C. Sun, L. Baum and A. H. Chow, *Eur. J. Pharm. Biopharm.*, 2015, **94**, 436.
136. S. Qureshi, A. H. Shah and A. M. Ageel, *Planta Med.*, 1992, **58**, 124.
137. C. D. Lao, M. T. t. Ruffin, D. Normolle, D. D. Heath, S. I. Murray, J. M. Bailey, M. E. Boggs, J. Crowell, C. L. Rock and D. E. Brenner, *BMC Complementary Altern. Med.*, 2006, **6**, 10.
138. A. L. Cheng, C. H. Hsu, J. K. Lin, M. M. Hsu, Y. F. Ho, T. S. Shen, J. Y. Ko, J. T. Lin, B. R. Lin, W. Ming-Shiang, H. S. Yu, S. H. Jee, G. S. Chen, T. M. Chen, C. A. Chen, M. K. Lai, Y. S. Pu, M. H. Pan, Y. J. Wang, C. C. Tsai and C. Y. Hsieh, *Anticancer Res.*, 2001, **21**, 2895.
139. A. Kampkotter, C. Timpel, R. F. Zurawski, S. Ruhl, Y. Chovolou, P. Proksch and W. Watjen, *Comp. Biochem. Physiol.*, 2008, **149**, 314.
140. N. Saul, K. Pietsch, R. Menzel and C. E. Steinberg, *Mech. Ageing Dev.*, 2008, **129**, 611.
141. K. Pietsch, N. Saul, R. Menzel, S. R. Sturzenbaum and C. E. Steinberg, *Biogerontology*, 2009, **10**, 565.
142. M. Duenas, F. Surco-Laos, S. Gonzalez-Manzano, A. M. Gonzalez-Paramas, E. Gomez-Orte, J. Cabello and C. Santos-Buelga, *Pharmacol. Res.*, 2013, **76**, 41.
143. Y. L. Xue, T. Ahiko, T. Miyakawa, H. Amino, F. Hu, K. Furihata, K. Kita, T. Shirasawa, Y. Sawano and M. Tanokura, *J. Agric. Food Chem.*, 2011, **59**, 5927.
144. F. Surco-Laos, J. Cabello, E. Gomez-Orte, S. Gonzalez-Manzano, A. M. Gonzalez-Paramas, C. Santos-Buelga and M. Duenas, *Food Funct.*, 2011, **2**, 445.
145. A. H. Knezevic, D. Dikic, D. Lisicic, N. Kopjar, N. Orsolic, S. Karabeg and V. Benkovic, *Basic Clin. Pharmacol. Toxicol.*, 2011, **109**, 343.
146. K. Pietsch, N. Saul, S. C. Swain, R. Menzel, C. E. Steinberg and S. R. Sturzenbaum, *Front. Genet.*, 2012, **3**, 48.
147. L. Zhang, G. Jie, J. Zhang and B. Zhao, *Free Radical Biol. Med.*, 2009, **46**, 414.
148. H. K. Na and Y. J. Surh, *Food Chem. Toxicol.*, 2008, **46**, 1271.
149. T. Lopez, S. E. Schriner, M. Okoro, D. Lu, B. T. Chiang, J. Huey and M. Jafari, *J. Med. Food*, 2014, **17**, 1314.
150. C. Peng, H. Y. Chan, Y. M. Li, Y. Huang and Z. Y. Chen, *Exp. Gerontol.*, 2009, **44**, 773.
151. K. Kitani, T. Yokozawa and T. Osawa, *Ann. N. Y. Acad. Sci.*, 2004, **1019**, 424.
152. Y. M. Li, H. Y. Chan, Y. Huang and Z. Y. Chen, *Mol. Nutr. Food Res.*, 2007, **51**, 546.
153. N. Saul, K. Pietsch, R. Menzel, S. R. Sturzenbaum and C. E. Steinberg, *Mech. Ageing Dev.*, 2009, **130**, 477.

154. C. R. Archer, A. Kohler, C. W. Pirk, V. Oosthuizen, Z. Apostolides and S. W. Nicolson, *J. Insect Physiol.*, 2014, **71**, 78.
155. Y. Niu, L. Na, R. Feng, L. Gong, Y. Zhao, Q. Li, Y. Li and C. Sun, *Aging Cell*, 2013, **12**, 1041.
156. Y. M. Li, H. Y. Chan, X. Q. Yao, Y. Huang and Z. Y. Chen, *J. Nutr. Biochem.*, 2008, **19**, 376.
157. H. Si, Z. Fu, P. V. Babu, W. Zhen, T. Leroith, M. P. Meaney, K. A. Voelker, Z. Jia, R. W. Grange and D. Liu, *J. Nutr.*, 2011, **141**, 1095.
158. I. M. Ernst, K. Pallauf, J. K. Bendall, L. Paulsen, S. Nikolai, P. Huebbe, T. Roeder and G. Rimbach, *Ageing Res. Rev.*, 2013, **12**, 365.
159. C. Driver and A. Georgeou, *Biogerontology*, 2003, **4**, 91.
160. S. Zou, J. Sinclair, M. A. Wilson, J. R. Carey, P. Liedo, A. Oropeza, A. Kalra, R. de Cabo, D. K. Ingram, D. L. Longo and C. A. Wolkow, *Mech. Ageing Dev.*, 2007, **128**, 222.
161. S. Bahadorani, P. Bahadorani, J. P. Phillips and A. J. Hilliker, *J. Gerontol.*, 2008, **63**, 35.
162. A. D. Blackett and D. A. Hall, *Age Ageing*, 1981, **10**, 191.
163. M. Ledvina and M. Hodanova, *Exp. Gerontol.*, 1980, **15**, 67.
164. R. D. Lipman, R. T. Bronson, D. Wu, D. E. Smith, R. Prior, G. Cao, S. N. Han, K. R. Martin, S. N. Meydani and M. Meydani, *Mech. Ageing Dev.*, 1998, **103**, 269.
165. A. A. Morley and K. J. Trainor, *Biogerontology*, 2001, **2**, 109.
166. E. A. Porta, N. S. Joun and R. T. Nitta, *Mech. Ageing Dev.*, 1980, **13**, 1.
167. C. Selman, J. S. McLaren, A. R. Collins, G. G. Duthie and J. R. Speakman, *Biol. Lett. (London, U. K.)*, 2013, **9**, 20130432.
168. C. C. Hsieh and B. F. Lin, *Nutrition*, 2005, **21**, 940.
169. A. Navarro, C. Gomez, M. J. Sanchez-Pino, H. Gonzalez, M. J. Bandez, A. D. Boveris and A. Boveris, *Am. J. Physiol.*, 2005, **289**, R1392.
170. C. Selman, J. S. McLaren, C. Mayer, J. S. Duncan, A. R. Collins, G. G. Duthie, P. Redman and J. R. Speakman, *Rejuvenation Res.*, 2008, **11**, 83.
171. V. A. Olson and I. P. Owens, *Trends Ecol. Evol.*, 1998, **13**, 510.
172. E. Lashmanova, E. Proshkina, S. Zhikrivetskaya, O. Shevchenko, E. Marusich, S. Leonov, A. Melerzanov, A. Zhavoronkov and A. Moskalev, *Pharmacol. Res.*, 2015, **100**, 228.
173. I. K. Beliaev, A. V. Zaraiskii, V. K. Limberg and L. A. Vakulova, *Vopr. Med. Khim.*, 1992, **38**, 39.
174. Z. Zhang, S. Han, H. Wang and T. Wang, *Arch. Gerontol. Geriatr.*, 2014, **58**, 153.
175. L. J. Martin, *Mar. Drugs*, 2015, **13**, 4784.
176. K. Muradian, A. Vaiserman, K. J. Min and V. E. Fraifeld, *Nutr., Metab. Cardiovasc. Dis.*, 2015, **25**, 891.
177. M. K. Brown, J. L. Evans and Y. Luo, *Pharmacol., Biochem. Behav.*, 2006, **85**, 620.
178. C. K. Lee, T. D. Pugh, R. G. Klopp, J. Edwards, D. B. Allison, R. Weindruch and T. A. Prolla, *Free Radical Biol. Med.*, 2004, **36**, 1043.

179. S. A. Farr, T. O. Price, W. A. Banks, N. Ercal and J. E. Morley, *J. Alzheimer's Dis.*, 2012, **32**, 447.
180. T. A. Dubiley, Y. E. Rushkevich, N. M. Koshel, V. P. Voitenko and A. M. Vaiserman, *Biogerontology*, 2011, **12**, 179.
181. K. Sun, S. Cao, L. Pei, A. Matsuura, L. Xiang and J. Qi, *Int. J. Mol. Sci.*, 2013, **14**, 4461.
182. K. Pietsch, N. Saul, S. Chakrabarti, S. R. Sturzenbaum, R. Menzel and C. E. Steinberg, *Biogerontology*, 2011, **12**, 329.

CHAPTER 14

Extending Lifespan by Inhibiting the Mechanistic Target of Rapamycin (mTOR)

DUDLEY W. LAMMING[a,b]

[a]Department of Medicine, University of Wisconsin–Madison, Madison, Wisconsin, USA; [b]William S. Middleton Memorial Veterans Hospital, Madison, Wisconsin, USA
*E-mail: dlamming@medicine.wisc.edu

14.1 The Discovery of Rapamycin and mTOR

Rapamycin is a polyketide produced by *Streptomyces hygroscopicus* and was originally identified as the active antifungal agent in a soil sample from Easter Island.[1] The ability of rapamycin to also inhibit the proliferation of mammalian cells led to its development in the 1990s as an immunosuppressant, and more recently to the FDA approval of rapamycin (sirolimus) and the rapamycin derivatives everolimus, temsirolimus, and zotarolimus for the treatment of specific cancers and in drug-eluting stents. The robust biological effects of rapamycin have spurred interest in understanding the molecular basis underlying its incredible potency. Rapid progress in the area was made in the 1990s following the identification of the immunophilin FK506-binding protein 12 (FKBP12) as a protein with extremely high, sub-nanomolar affinity for rapamycin.[2] Genetic screens in yeast concluded that the FKBP-rapamycin complex was the active agent as yeast mutants lacking the gene encoding

RSC Drug Discovery Series No. 57
Anti-aging Drugs: From Basic Research to Clinical Practice
Edited by Alexander M. Vaiserman

Published by the Royal Society of Chemistry, www.rsc.org

FKBP were fully resistant to rapamycin, and also identified two yeast TOR (Target of Rapamycin) genes.[3] Subsequent work in the early 1990s identified the mammalian homologue of these genes, now known as the mechanistic Target of Rapamycin (mTOR), a large 289 kD protein with approximately 40% homology to the two yeast TOR proteins.[4–6]

mTOR is a serine/threonine phosphatidylinositol 3-kinase (PI3K)-like kinase (PIKK).[7] Since the initial discovery of mTOR, it has become clear that mTOR participates in two distinct protein complexes, mTOR complex 1 (mTORC1) and mTORC2. mTORC1, which is acutely sensitive to rapamycin, is defined by the association of mTOR with the adaptor protein Raptor,[8] while mTORC2, which is relatively insensitive to rapamycin, is defined by the association of mTOR with the adaptor protein Rictor.[9] Several other proteins that associate with both mTORC1 and mTORC2 have been identified, including mLST8/GβL, which is required for complex assembly and stability of both mTORC1 and mTORC2; Tti1 and Tel2, which are likewise important for the assembly of both complexes; and the regulatory protein DEPTOR, which is found exclusively in vertebrates.[10–12] Additional complex-specific subunits include PRAS40, which interacts specifically with mTORC1, and mSin1 and Protor, which are specific to mTORC2.[13–15] mTORC1 is an obligate dimer,[16] with Raptor and mLST8 limiting access to the active site.[17] The structure of mTORC2 has not yet been determined, but as mTORC2 does not interact with FKBP12–rapamycin *in vitro*, it is believed that the rapamycin-interacting FRB domain of mTOR is not accessible.[9] This hypothesis is supported by a recent structure of the yeast mTORC2 homologue TORC2.[18]

The two mTOR complexes have distinct substrates and regulate different biological processes (Figure 14.1). The classical substrates of mTORC1, which are better understood due to the availability of rapamycin as a tool compound, include S6 kinase 1 (S6K1) T389, multiple serine and threonine residues on the eIF-4E binding proteins (4E-BPs), including 4E-BP1, and ULK1 S757.[19] Other more recently identified mTORC1 substrates include specific residues on S6K2, Lipin1, the transcription factor EB (TFEB), and La-related protein 1 (LARP1).[20–24] The regulation of protein translation, one of the most significant biological effects of rapamycin, is specifically mediated by mTORC1 through regulation of the 4E-BPs and LARP1.[24,25]

As rapamycin is a relatively poor inhibitor of mTORC2, much less is understood about the substrates of this complex. mTORC2 primarily functions as a downstream effector of insulin/IGF-1/PI3K signaling, and its substrates include several residues of AKT that are important to its activity, including AKT T450, S473, and S477/479.[26–28] Many other kinases are also substrates of mTORC2, including the PKC alpha and other PKC family members, the serum- and glucocorticoid-induced protein kinase 1 (SGK1), the oxidative stress-responsive 1 (OSR1) and the Hippo pathway signaling kinase MST1, through which mTORC2 is linked to many diverse biological processes (Figure 14.1).[27–33] More substrates of both mTOR complexes continue to be identified through phosphoproteomic analysis of cells lacking mTOR complex subunits or treated with either rapamycin or mTOR kinase inhibitors.[34–36]

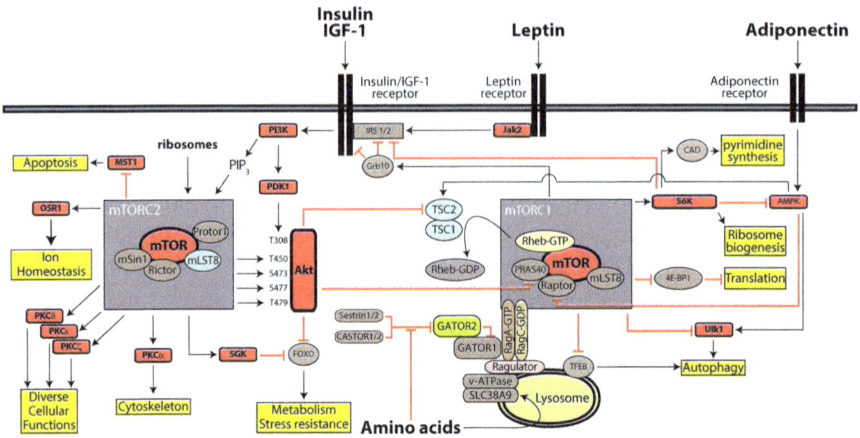

Figure 14.1 The mTOR signaling pathway. The mTOR pathway signaling is mediated by numerous environmental and hormonal cues, including amino acids and insulin, IGF-1, adiponectin, and leptin. mTOR, particularly mTOR complex 1 (mTORC1) integrates these signaling to determine if conditions are permissive for growth, and then promotes anabolic processes including protein synthesis, ribosome biogenesis and nucleotide biosynthesis, while suppressing autophagy. mTORC2, which functions as an effector of PI3K signaling, has been shown to regulate AKT, SGK, multiple PKC family members, and the Hippo pathway kinase MST1, placing it upstream of many diverse processes, including metabolism.

14.2 mTOR Regulates Longevity in Model Organisms

The first connection of mTOR signaling to aging was found in 2003, with the discovery that RNAi against *mTOR* in *Caenorhabditis elegans* significantly extended lifespan.[37] Over the subsequent two years, this was followed by the discovery that inhibition of mTOR signaling could also extend the lifespan of *Drosophila melanogaster* and the budding yeast *Saccharomyces cerevisiae*.[38,39] From the first, it was clear that lifespan extension resulting from reduced mTOR signaling was distinct from previously identified aging pathways. In *C. elegans*, lifespan extension by RNAi against *mTOR* was found to be independent from the FOXO homologue *daf-16*, which is critical for the extended lifespan of *daf-2* mutants,[37] while in *D. melanogaster*, inhibition of signaling downstream of mTORC1 through expression of either dominant-negative mTOR or S6K had similar effects on lifespan.[38]

The study of mTOR in *D. melanogaster* by Kapahi and colleagues also provided the first hints that the mTOR pathway was involved in the response to a calorie restricted (CR) diet as inhibition of mTOR failed to extend the lifespan of flies fed a low-calorie diet.[38] Extensive genetic work in *S. cerevisiae* elaborated on this possibility, demonstrating the induction of a starvation-like phenotype in yeast with extended chronological lifespan, and that removal of specific amino acids from the media could also promote survival.[39] The interaction of yeast TOR with a CR diet was later examined more extensively in the context of yeast replicative lifespan. Deletion of either *TOR1* or the S6K1 homologue *SCH9* significantly extends yeast lifespan,[40] but CR is unable to further extend the lifespan of these mutants.[41]

A major advantage of studying aging in yeast, worms, and flies is the relative ease of conducting genetic screens for organisms with altered longevity. A number of experiments conducted in these organisms suggest that decreased or altered translation underlies the beneficial effects of both mTOR inhibition and CR on lifespan. One such study in yeast demonstrated that a reduced level of 60S ribosomal subunits significantly extends lifespan and is epistatic with CR.[42] Reduction of either ribosomal subunits or translation initiation factors significantly extends the lifespan of *C. elegans*, and the lifespan of CR worms is not further extended by inhibition of mTOR.[43] However, rapamycin treatment can extend the lifespan of flies on a CR diet, suggesting that although there may be significant overlap between CR and mTOR inhibition, there may also be additional mechanisms unique to each regimen.[44] Subsequent studies that compared mice placed on either CR or rapamycin have identified both overlapping and unique changes in the liver and white adipose transcriptome and the liver metabolome, supporting a model in which the effects of CR and mTOR inhibition do not completely overlap.[45–47]

14.3 Rapamycin Extends the Lifespan and Healthspan of Mice

In mammals, a seminal study conducted by the National Institute on Aging Interventions Testing Program in 2009 demonstrated that treatment with rapamycin could significantly extend the lifespan of genetically heterogeneous mice of both sexes.[48] Since that time, there have been at least 9 additional published studies demonstrating that rapamycin can extend the lifespan of wild-type inbred and outbred strains, conducted by many laboratories and using a variety of dosing regimens (Table 14.1).[49] Rapamycin shows efficacy at extending median as well as maximum lifespan in both sexes, even when begun late in life. Numerous studies in disease models have also been performed, also typically (but not always) resulting in a significant increase in lifespan (Table 14.2).

One concern about rapamycin has been the possibility that its effects on longevity may result from an anticancer effect rather than an effect on aging itself.[50] While rapamycin robustly extends the lifespan of genetically heterogeneous mice, which should avoid any confounding effects of rapamycin on lifespan resulting from an effect against strain-specific pathologies, a susceptibility to cancer is broadly shared by inbred mice. In comparison to calorie restriction (CR), which dramatically improves many different age-related phenotypes, rapamycin has modest results.[50,51]

Table 14.1 The impact of rapamycin on the lifespan of wild-type mice in mouse studies since 2009 where longevity or mortality rate was determined. Sex is listed separately for males and females where sex-specific data exist. The rapamycin dose listed for dietary administration indicates the drug concentration in the *ad libitum* fed diet; the dose listed for administration in water, or administered intraperitoneally (IP) or subcutaneously (SC), indicates the dose in mg per kg of body weight. Δ lifespan is the percentage change in median lifespan (* indicates that mean is reported instead). *MF* indicates that the lifespan results were not broken down by sex or that sex was not reported; *a*: Lifespan study % increase was not determined; *NS*: Not Statistically Significant. Percentage change is estimated where precise information is not listed in the referenced study. Adapted from Arriola Apelo and Lamming, 2016, *J. Gerontol. A Biol. Sci. Med. Sci.*, used with permission.

Strain	Sex	Starting age	Rapa dose	Route	Δ lifespan (%)	Reference
UM-HET3	Male	20 months	14 ppm	Diet	9	48
UM-HET3	Female	20 months	14 ppm	Diet	14	48
C57BL/6J.Nia	*MF*	22–24 months	4 mg kg^{-1}	IP 1×/2 days	>14[a]	63
UM-HET3	Female	9 months	14 ppm	Diet	18	169
UM-HET3	Male	9 months	14 ppm	Diet	10	169
129/Sv	Female	2 months	1.5 mg kg^{-1}	SC 3×/week 2 weeks per 4	10	170
C57BL/6J.Rj	Male	4, 13, 20 months	14 ppm	Diet	~10[a]	50
UM-HET3	Male	9 months	4.7 ppm	Diet	3[NS]	171
UM-HET3	Male	9 months	14 ppm	Diet	13	171
UM-HET3	Male	9 months	42 ppm	Diet	23	171
UM-HET3	Female	9 months	4.7 ppm	Diet	16	171
UM-HET3	Female	9 months	14 ppm	Diet	21	171
UM-HET3	Female	9 months	14 ppm	Diet	26	140
C57BL/6J.Nia	Male	4 months	14 ppm	Diet	11*	172
C57BL/6J.Nia	Female	4 months	14 ppm	Diet	16*	172
C57BL/6J.Nia	Female	20 months	2 mg kg^{-1}	IP 1×/5 days	7	104

In addition to extending lifespan, rapamycin also promotes healthspan—which, from the perspective of improving the health of an elderly and graying population, may be even more important than simply prolonging life.[52] The potential benefits of rapamycin have been most clearly demonstrated with respect to two of the most deadly age-related diseases, Alzheimer's disease and cardiovascular disease. Rapamycin prevents or delays the onset of Alzheimer's disease as well as ameliorating age-related cognitive decline.[53–55] Early intervention with rapamycin also preserves cerebral vascular blood flow and metabolism, and rescues learning defects, in mice expressing a transgenic Apolipoprotein E ε4 allele.[56]

With respect to cardiovascular disease, rapamycin and its analogs everolimus and zotarolimus have been used extensively for more than a decade

Table 14.2 The impact of rapamycin on the survival of mouse disease models. Sex is listed separately for males and females where sex-specific data exists. The rapamycin dose listed for dietary administration indicates the drug concentration in the *ad libitum* fed diet; the dose listed for administration in water, or administered intraperitoneally (IP) or subcutaneous (SC) indicates the dose in mg per kg of body weight. Δ lifespan is the percentage change in median lifespan (* indicates that mean is reported instead). *MF* indicates that the lifespan results were not broken down by sex or that sex was not reported; *a*: Lifespan study % increase was not determined; *b*: 100% of rapamycin-treated mice survived to 2 years of age *vs.* 40% of control mice; *NS*: Not Statistically Significant. Percentage change is estimated where precise information is not listed in the referenced study. Adapted from Arriola Apelo and Lamming, 2016, *J. Gerontol. A Biol. Sci. Med. Sci.*, used with permission.

Strain	Sex	Starting age	Rapa dose	Route	Δ lifespan (%)	Reference
$Pten^{-/-}$	*MF*	1 month	10 mg kg^{-1} (everolimus)	Oral	>292*,[a]	173
FVB/N HER-2/neu	Female	2 months	1.5 mg kg^{-1}	SC 3×/week 2 weeks per 4	13.6	174
$SOD1^{H46R/HR8Q}$	*MF*	1.5 months	14 ppm	Diet	*NS*	175
$p53^{+/-}$	Male	<5 months	1.5 mg kg^{-1}	Water	28*	176*
$p53^{+/-}$	Male	>5 months	1.5 mg kg^{-1}	Water	10*	176
$p53^{-/-}$	Male	2 months	0.5 mg kg^{-1}	Oral 1×/day 5 d on/9 d off	35	177
$Lmna^{-/-}$	*MF*	1 month	14 ppm	Diet	35	62
$Lmna^{-/-}$	*MF*	1 month	8 mg kg^{-1}	IP 1×/2 days	56	62
$Rb1^{+/-}$	Male	2 months	14 ppm	Diet	13.8	178
$Rb1^{+/-}$	Female	2 months	14 ppm	Diet	8.9	178
$Bmal1^{-/-}$	*MF*	16 weeks	0.5 mg kg^{-1}	Water	47	179
HER-2/neu	Female	2, 4, or 5 months	0.45 mg kg^{-1}	SC 3×/week 2 weeks per 4	5.7[NS], 6.1, 5.5	180
C57BL/6NCr HFD	Male	12 months	1.5 mg kg^{-1}	IP 1×/week	[b]	106
$Ndufs4^{-/-}$	*MF*	<1 month	42 ppm	Diet	29[NS]	181
$Ndufs4^{-/-}$	*MF*	<1 month	378 ppm	Diet	92	181
$Ndufs4^{-/-}$	*MF*	<1 month	8 mg kg^{-1}	IP 1×/day	119	181
$Rag2^{-/-}$	*MF*	3 months	14 ppm	Diet	121	182
IFN-$\gamma^{-/-}$	*MF*	5 months	14 ppm	Diet	34	182
C57BLKS/J $lepr^{db/db}$	Male	4 months	14 ppm	Diet	−16	183
C57BLKS/J $lepr^{db/db}$	Female	4 months	14 ppm	Diet	−18	183

in the treatment of atherosclerosis. These compounds are used as antiproliferative agents in drug-eluting stents, and are placed in arteries in order to prevent restenosis following angioplasty.[57] When used in a drug-eluting stent, the drug concentration is high at the stent site, yet the levels of these compounds in the general circulation are negligible and side effects rare. However, systemic treatment with rapamycin has also attracted attention as a potential therapy to prevent or delay atherosclerosis. Systemic treatment of genetic mouse models of atherosclerosis or of rabbits significantly reduces plaque formation, despite not improving cholesterol or other serum lipids (reviewed in ref. 58). A pair of recent studies has also shown direct beneficial effects of rapamycin on the heart, with approximately three weeks of rapamycin treatment reversing age-dependent cardiac hypertrophy and diastolic dysfunction, with beneficial shifts in the transcriptome and proteome.[59,60] Rapamycin also significantly extends survival the survival of short-lived *Lmna*$^{-/-}$ mice, which is limited by cardiac pathology.[61,62]

Rapamycin has been demonstrated to have potentially rejuvenating effects on other cells and organ systems. In 2009, the transient treatment of aged mice with rapamycin for only 6 weeks was shown to boost the self-renewal and hematopoiesis of hematopoietic stem cells, improving the immune response to vaccination as well as increasing lifespan.[63] Hematopoietic stem cell function may also be preserved in *S6K1*$^{-/-}$ mice.[64] A similar improved response to influenza vaccination following a short course of the rapamycin analog everolimus was recently observed in humans.[65] Treatment with rapamycin also increases the clonogenicity of mouse intestinal crypts by enhancing the renewal of intestinal stem cells.[66]

14.4 How Does Rapamycin Increase Longevity?

Despite this potent effect of rapamycin on mouse lifespan, there is extremely little known about the mechanism by which rapamycin promotes longevity in mammals. Genetic inhibition of mTOR *via* expression of a hypomorphic allele extends the lifespan of both male and female mice,[67] but these mice have extremely reduced mTORC1 and mTORC2 signaling and are born at sub-Mendelian ratios. More modest genetic interventions, such as the deletion of *S6K1* or deletion of a single copy of both *mTOR* and *mLST8*, specifically reduce mTORC1 signaling through S6K but have only been demonstrated to extend female lifespan.[68,69] Genetic reduction in 4E-BP1 activity in skeletal muscle improves metabolic health, but it is not known if lifespan will also be extended.[70]

While studies in yeast, worms and flies link the beneficial effects of mTOR inhibition on lifespan to decreased translation, it has recently become apparent that neither chronic treatment with rapamycin or deletion of *S6K1* reduce ribosome activity *in vivo* in either liver or skeletal muscle,[71] casting doubt on this simplistic model. A more refined variant of this idea is based on the observation that mTOR inhibition shifts the type of mRNAs that are translated, rather than simply the amount. In yeast, deletion of 14 distinct ribosomal 60S subunits promotes longevity while deletions of 40S ribosomal subunits that have similar effects on translation do not,[42] suggesting that the specific composition of the ribosome may impact which mRNAs are translated. This idea of a ribosome code is supported by yeast studies, which identify distinct roles for ribosomal paralogs in the translation of localized mRNAs.[72] While it is not yet clear if mammals have a similar ribosome code, it is now clear that mTOR specifically promotes the translation of mRNAs with 5' terminal oligopyrimidine (TOP) motifs,[73] and thus mTOR inhibition shifts the set of mRNAs being translated.

A different theory is that mTOR inhibition may promote longevity by decreasing insulin/IGF-1/PI3K signaling.[74,75] This is supported by correlative evidence from numerous other mouse models with both decreased insulin/IGF-1/PI3K/mTOR signaling and extended lifespan, including mice heterozygous for either *Igf1r* or *Akt1* and mice lacking *IRS1*.[76–78] Deletion of the insulin receptor specifically in adipose tissue or deletion of *IRS2* specifically in the brain also promotes longevity.[79,80] Notably, just as with rapamycin treated animals, many of these mutants display either systemic or tissue-specific insulin resistance, leading to the suggestion that insulin resistance that decreases insulin/IGF-1/PI3K/mTOR signaling is beneficial.[74] In contrast to this view, depletion of *Rictor*, an essential component of mTORC2 signaling, either specifically in the liver or in the whole body of mice, greatly impairs male longevity.[81] This divergent phenotype suggests the possibility that male survival may require signaling downstream of mTORC2 that is mediated by other substrates, such as SGK1.

14.5 Side Effects of Rapamycin Treatment—The Role of mTORC2

While rapamycin and its analogs (rapalogs) are FDA approved for specific indications, they are far from ideal drugs when considered from the perspective of using rapalogs as anti-aging compounds. The side effects of rapamycin are as diverse as they are numerous, and include: metabolic disturbances (*e.g.* hyperglycemia, hyperlipidemia, insulin resistance); dermatological events, including painful ulcers; in males, testicular dysfunction; and most acutely serious, an increase in infections, stemming from the potent immunosuppressive effects of rapalogs (reviewed in ref. 49). These side effects do not outweigh the potential benefits of rapalogs during cancer treatment or in the context of organ transplantation, but must be weighed heavily in the context of an anti-aging medication that may need to be taken prophylactically for a long period of time by otherwise healthy individuals.

Over the last several years, our laboratory and others have demonstrated that many of these side effects result not from inhibition of mTORC1, the canonical target of rapamycin, but from inhibition of the 'rapamycin-resistant' mTORC2.[69] In brief, mTORC2 is resistant to acute treatment with rapamycin in *vitro* and i*n vivo*, yet chronic, prolonged treatment with rapamycin *in vivo* disrupts mTORC2 in the majority of mouse tissues.[69,82,83] Using genetic mouse models, important metabolic roles for mTORC2 in many tissues have been identified (reviewed in ref. 19). mTORC2 promotes insulin sensitivity in the liver,[35,84,85] white and brown adipose tissue,[86–89] and skeletal muscle.[90,91] mTORC2 plays a critical role in the regulation of lipid homeostasis, regulating lipolysis, lipogenesis and adipogenesis (reviewed in ref. 92). A growing body of work suggests that disruption of mTORC2 by rapamycin is responsible for some of the immunosuppressive effects of prolonged rapamycin treatment.[93–98] Finally, although the mechanism for this effect remains unknown, depletion of *Rictor*, a key subunit of mTORC2, significantly decreases the lifespan of male but not female mice.[81]

Both mice and humans show significant metabolic and immunological side effects when exposed to high doses of rapamycin or its analogs,[99] but these effects appear to be largely dose-dependent. A recent human study of rapamycin, which demonstrated positive effects on rejuvenation of the immune response to vaccination, used very low doses for a short period of time with few serious side effects.[65] Notably, marmosets treated with rapamycin did not experience significant negative side effects, although the low number of animals used in this initial study precludes making definitive conclusions.[100,101] New larger-scale studies now underway in companion animals, including dogs,[102] should answer many questions about both the efficacy and safety of rapamycin in healthy mammals outside the laboratory environment.

The overwhelming negative consequences of mTORC2 inhibition on metabolism and immunity suggest that specifically inhibiting mTORC1 will promote health and longevity with fewer negative side effects.[103] We recently hypothesized that the differential kinetics of mTORC1 and mTORC2 inhibition by rapamycin might create a therapeutic window through which acute, intermittent dosage of rapamycin could specifically inhibit mTORC1, and we demonstrated that not only did this regimen have reduced effects on glucose homeostasis and the immune system, it remained able to extend the lifespan of mice.[104,105] Weekly treatment with rapamycin has similarly promoted survival of mice fed a high-fat diet,[106] and intermittent administration of rapamycin has also been shown to promote weight loss with reduced side effects on glucose metabolism in rats.[107] While intermittent administration of rapamycin may be a clinically useful technique to reduce rapamycin-associated side effects for the treatment of severe age-related diseases, in the long term a true mTORC1-specific inhibitor has the potential to be used much more widely.

14.6 mTORC1 Is a Key Integrator of Nutrient and Hormonal Signaling

In order to discuss potential mechanisms by which mTORC1 can be specifically inhibited, it is necessary for us to discuss how mTORC1 signaling is regulated. Initial studies of mTORC1 determined that the phosphorylation of S6K1 and 4E-BP1 was sensitive to nutrients including amino acids and glucose, promoting translation.[8] Subsequent work has shown that mTORC1 is sensitive to many other environmental cues, including cellular energy, oxygen availability, reactive oxygen levels, and phosphatidic acid,[108–111] and (in part *via* mTORC2 and Akt) hormonal signals including insulin, IGF-1, and leptin (Figure 14.1). mTORC1 integrates all of these signals in order to determine if environmental and hormonal cues are permissive for growth and proliferation. In order to successfully respond to all of these diverse cues, the regulation of mTORC1 signaling is extremely complex.

The activity of mTORC1 is strictly dependent upon interaction with a small GTPase, Rheb, and the requirement for mTORC1 to localize with GTP-bound Rheb lies at the heart of the molecular mechanism by which mTORC1 responds to both nutrient and environmental cues.[112] While the recent development of fluorescence resonance energy transfer (FRET)-based mTORC1 sensors suggests that mTORC1 activity may occur in multiple subcellular compartments, including the nucleus and plasma membrane,[113] the activation of mTORC1 has only been fully characterized at the surface of the lysosome.

In response to stimulation with amino acids, mTORC1 is recruited to the surface of the lysosome *via* a series of steps that have been identified over the last several years (reviewed in ref. 114). Briefly, recruitment of mTORC1 to the lysosome requires the Rag family of small GTPases, which interact with mTORC1 and recruit it to the Ragulator, a protein complex on the surface of the lysosome.[115,116] The GTP-bound state of the Rags mediates their ability to recruit mTORC1 to the lysosome, and recent work has highlighted the essential regulator role of the Ragulator as well as a Rag-interactor complex, GATOR, in signaling amino acid sufficiency. GATOR has two distinct subunits: GATOR1 has GTPase-activating protein (GAP) activity for RagA and RagB, while GATOR2 inhibits GATOR1 activity.[117,118]

Several distinct mechanisms have been uncovered by which amino acids regulate mTORC1 activity. First, mTORC1 senses amino acid levels in the lysosome through a mechanism that requires the interaction of Ragulator with a functional vacuolar H(+)-adenosine triphosphatase ATPase (v-ATPase) as well as SLC38A9, a low-affinity amino acid transporter for amino acids including arginine.[119–122] Secondly, Sestrin2 binds to and inhibits the activity of GATOR2, an interaction that is relieved by leucine binding directly to a pocket on Sestrin2; the homologous protein Sestrin1 interacts with GATOR2 and leucine in a similar fashion.[123,124] CASTOR1 and its homolog CASTOR2 also act to inhibit GATOR2 activity, but in response to arginine.[125] Other regulators of mTORC1 activation in response to amino acids continue to be identified, including the E3 ubiquitin ligases Skp2 and RNF152, which were recently identified as negative regulators of mTORC1 that ubiquitinate RagA K63.[126,127] Finally, at the lysosomal surface, activation of mTORC1 requires the interaction of mTORC1 with Rheb-GTP, and the localization and GTP-bound status of Rheb is at least partially dependent upon amino acids as well. The microspherule protein 1 (MCRS1) binds to Rheb and, in the presence of amino acids, localizes Rheb to the lysosome, where it can interact with mTORC1, Ragulator and the v-ATPase.[128]

In contrast to amino acid signaling, which regulates the localization of mTORC1, growth factor signaling is primarily mediated by the tuberous sclerosis complex (TSC). TSC is a GAP for Rheb, which inhibits the ability of Rheb to activate mTORC1; this inhibition is relieved by the action of insulin, which inhibits TSC activity.[129,130] TSC is phosphorylated by a number of kinases, including AKT, which regulates its activity. In the absence of growth factor signaling, TSC is localized to the lysosome; insulin stimulates the disassociation of TSC from the lysosomal surface.[131] While it was originally reported that amino acids do not alter the subcellular localization of TSC,[128] subsequent work suggests that numerous other stresses, including a lack of amino acids, promote the lysosomal accumulation of TSC.[132,133]

At least some of the many stimuli that regulate mTORC1 activity function by co-opting parts of these two regulatory mechanisms. For example, glucose activates mTORC1, an effect mediated by activation of the Rag proteins and that requires both Ragulator and the v-ATPase.[134] At least some other stimuli that regulate TSC, including oxygen and cellular energy levels, are likely to regulate TSC localization. Some stimuli regulate mTORC1 at multiple levels; for example, AMPK phosphorylates both Raptor and TSC2, which physically destabilizes mTORC1 while activating TSC to inhibit mTORC1 activity.[135,136] Adiponectin is an example of a hormonal stimulus that inhibits mTORC1 *via* activation of AMPK and the subsequent post-translational modification of TSC2 and Raptor (Figure 14.1).[137]

AKT is a major effector of PI3K signaling and regulates mTORC1 signaling *via* an inactivating phosphorylation of TSC2 as well as phosphorylation of PRAS40, which frees mTORC1 from inhibition by this protein.[13] As such, mTORC2, which phosphorylates AKT at three separate sites, is putatively upstream of mTORC1 in the insulin/IGF-1/PI3K signaling pathway.[75] mTORC2 is an effector of PI3K signaling—it was recently shown that mTORC2 is directly stimulated by phosphatidylinositol (3,4,5)-trisphosphate (PIP_3)[138]—and as such mTORC2 is activated by stimuli including insulin, IGF-1, and leptin (Figure 14.1).[28,139] mTORC2 is also stimulated by palmitoleic acid, a substrate for fatty acid elongase-5 (Elovl5),[140] by substrates of glycerol-3-phosphate acyltransferase-1 (Gpat1),[141] and by association with ribosomal protein subunits.[142] While mTORC2 has been shown to localize to the ribosome-rich mitochondria-associated endoplasmic reticulum,[143] mTORC2 was also recently identified at the lysosome,[144] suggesting the possibility that both mTOR complexes might be regulated by lysosomal-mediated nutrient sensing.

14.7 How Can mTORC1 Be Specifically Targeted?

One possible way to approach the development of mTORC1-specific inhibitors is to focus on the regulatory steps required to activate mTORC1. Notably, mTORC1 activation seems to require the lysosomal localization of mTORC1, as well as the departure of TSC, and in the last few years the molecular mechanisms that mediate the lysosomal activation of mTORC1 by amino acids have been defined in great detail. It was recently shown that leucine as well as isoleucine, valine and methionine disrupt an inhibitory interaction between Sestrin2 and GATOR2.[124] Subsequent structural work shows that leucine binds to a specific pocket on Sestrin2; binding of leucine likely then transmits a conformational change to an adjacent, GATOR2 binding domain of Sestrin2.[123] Conceivably, understanding the molecular mechanism of leucine sensing by Sestrin2/GATOR2 will permit the design of pharmaceuticals that either agonize or antagonize the Sestrin2/GATOR2 interaction and thus specifically regulate mTORC1 signaling. Other potential pharmaceutical targets that specifically mediate mTORC1 activity may include the sensing of amino acids by the CASTOR proteins or by the lysosomal transporter SLC38A9.[121,122,125] Finally, while specific amino acids may be sensed by mTORC1 *via* CASTOR, GATOR and SLC38A9, the GCN2 (general control nonderepressible 2) kinase is also a major sensor of amino acids. GCN2 is normally activated by uncharged tRNAs, and acts to inhibit translation through phosphorylation of eIF2α. It was recently demonstrated that GNC2 acts to inhibit mTORC1 by promoting expression of Sestrin2;[145] activating GCN2 activity, perhaps with small molecules which mimic uncharged tRNAs, could therefore be a potential means of specifically inhibiting mTORC1.

Notably, these potential pharmaceutical targets all rely on a common theme—the sensitivity of mTORC1, but not mTORC2, to amino acids. As discussed above, it is widely believed that inhibition of mTOR signaling may play a role in the response to a calorie restricted (CR) diet, and a CR diet necessarily entails a sharp restriction in all macronutrients, including amino acids. Intriguingly, protein restricted (PR) diets have recently been shown to significantly improve the metabolic health and longevity of rodents.[146–148] Conversely, recent epidemiological studies in humans have found that protein intake correlates with increased mortality and an increased risk of metabolic disease.[149–153] We recently considered the possibility that protein restriction (PR) would specifically reduce mTORC1 signaling. Indeed, mTORC1 signaling is reduced in multiple tissues of mice fed a PR diet, including the liver, heart, skeletal muscle and white adipose tissue, while mTORC2 signaling is unaffected by a PR diet.[154]

While it is not clear which specific components altered in a PR diet regulate mTOR activity, a reduction in certain specific amino acids that are potent mTORC1 agonists could mediate this effect. In rodents, the three branched-chain amino acids (BCAAs)—leucine, isoleucine, and valine—have been demonstrated to promote mTORC1 activity in the liver, skeletal muscle, adipose tissue and the pancreas.[155–158] The same BCAAs have been linked to insulin-resistance in humans and rodents.[159,160] A PR diet lowers circulating levels of BCAAs in humans, and consuming a diet in which the BCAAs are specifically reduced improves metabolic health in mice.[161] It remains to be determined if a PR diet decreases mTORC1 signaling in humans, and if so, if this can be attributed specifically to reduced consumption of BCAAs. If a partial reduction of specific dietary amino acids can mediate a systemic reduction in mTORC1 signaling, chemicals that partially block the uptake of these specific amino acids from the intestine could, at least in theory, reduce mTORC1 signaling and promote health and longevity. While these types of interventions are being developed, an alternative and sustainable way to lower mTORC1 activity may be to eat a plant-based or vegan diet, which are low in methionine, an essential amino acid that when restricted significantly extends lifespan.[162,163]

An alternative approach to specifically inhibiting mTORC1 could rely on understanding why the sensitivity of mTORC2 to rapamycin varies by cell type.[82] The canonical mechanism for mTORC1 inhibition by rapamycin is that a rapamycin first binds to FK506-binding protein 12 (FKBP12), forming a complex that then interacts with and inhibits mTORC1. However, FKBP12–rapamycin specifically disrupts mTORC1 but not mTORC2 *in vitro*, suggesting that the FKBP–rapamycin-binding (FRB) domain of mTOR is not accessible in mTORC2, a hypothesis supported by the structure of the yeast mTORC2 homologue TORC2.[9,18] Thus, rapamycin–FKBP12 likely inhibits mTORC2 indirectly by sequestering newly synthesized mTOR, with an accessible FRB domain, before it can assemble into mTORC2 (Figure 14.2).[82] As the sensitivity of mTORC2 to rapamycin varies by cell and tissue type, but does not directly correlate with the expression of mTOR complex subunits, it has long been suspected that some additional factor contributes to the ability of

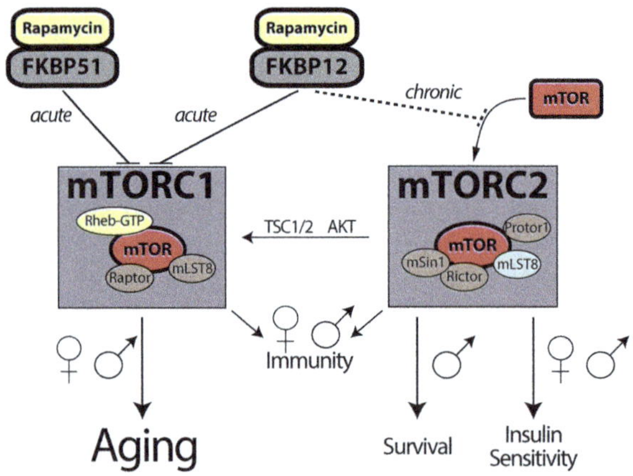

Figure 14.2 The action of rapamycin against mTORC1 or mTORC2 is dependent upon FK506-binding proteins. mTORC1 activity promotes aging, while mTORC2 activity promotes insulin sensitivity and male survival. Rapamycin binding to either FKBP12 or FKBP51 forms a complex that can then act to inhibit mTORC1. However, inhibition of mTORC2 activity by rapamycin is dependent upon a rapamycin-FKBP12 complex that can prevent the formation of mTORC2 by binding to free mTOR, preventing the incorporation of mTOR into mTORC2. Rapamycin–FKBP51 does not inhibit mTORC2. The relative expression level of FKBP12 and FKBP51 determines the rapamycin sensitivity of mTORC2 in each cell line or tissue. Adapted from Cell Metabolism, vol. 23, B. K. Kennedy and D. W. Lamming, The Mechanistic Target of Rapamycin: The Grand ConducTOR of Metabolism and Aging, 990–1003., Copyright (2016), with permission from Elsevier.[19]

rapamycin to disrupt mTORC2. Schreiber and colleagues recently solved this mystery, by determining that other FKBPs (*e.g.*, FKBP51) that interact with rapamycin and can inhibit mTORC1 do not necessarily inhibit mTORC2 (Figure 14.2).[83] While it is not yet clear why FKBP51–rapamycin does not inhibit mTORC2 formation, the authors note that this "work gives us insights into how we might alter rapamycin in order to get the mTORC1 specificity needed to get the longevity effects with reduced side effects."[164]

It is worth noting that there are numerous interventions in the scientific literature that purportedly inhibit mTORC1 signaling *in vitro* or even *in vivo*. These include a broad range of chemical compounds, some of which are medications in common use. Most notably, this includes the anti-inflammatory drug aspirin[165] as well as the most widely prescribed drug for type 2 diabetics, metformin.[166] Notably, while both of these compounds activate AMPK and thus would be expected to decreased mTORC1 activity through several distinct mechanisms (Figure 14.1), both of these medications are believed to also have AMPK-independent effects on mTORC1 activity, and both compounds may promote longevity in mice.[167,168] As both of these molecules are widely used by humans and are (relatively) safe, understanding how compounds like metformin inhibit mTORC1 signaling may provide much-needed mechanistic insight into how mTORC1 can be safely and beneficially inhibited to promote health and longevity.

14.8 Conclusions

The past decade has seen significant advances in our understanding of how genetic and pharmaceutical inhibition of the mTOR can extend lifespan, with initial studies in yeast, worms and flies being extended to mice and now other mammals. Rapamycin is potentially a very powerful anti-aging drug, with rapamycin-treated animals showing increased lifespan and healthspan as well as "rejuvenated" tissues. While the side effect profile of rapamycin is reason to be cautious, the use of intermittent or acute rejuvenative treatment regimens may permit the cautious clinical use of rapamycin for age-related diseases in the near future. In the long term, recent discoveries regarding how mTORC1 is regulated by amino acids at the molecular level and the discovery of the mechanistic basis for mTORC2 inhibition by rapamycin may permit the rational design of molecules that more selectively inhibit mTORC1, promoting healthy aging with reduced side effects. While the dream for a "fountain of youth" that restores health and vigor to the aged remains out of reach, the powerful results achieved with rapamycin suggest that we will soon at least be able to promote healthy aging.

Acknowledgements

D.W.L. is supported in part by an NIH K99/R00 Pathway to Independence Award (R00 AG041765), a Glenn Foundation Award for Research in the Biological Mechanisms of Aging, and startup funds from the UW–Madison

School of Medicine and Public Health and the UW–Madison Department of Medicine. This review was prepared while D.W.L. was an AFAR Research Grant recipient from the American Federation for Aging Research. This work was supported using facilities and resources from the William S. Middleton Memorial Veterans Hospital. This work does not represent the views of the Department of Veterans Affairs or the United States Government.

References

1. C. Vezina, A. Kudelski and S. N. Sehgal, *J Antibiot. (Tokyo)*, 1975, **28**, 721.
2. W. S. Lane, A. Galat, M. W. Harding and S. L. Schreiber, *J. Protein Chem.*, 1991, **10**, 151.
3. J. Heitman, N. R. Movva and M. N. Hall, *Science*, 1991, **253**, 905.
4. E. J. Brown, M. W. Albers, T. B. Shin, K. Ichikawa, C. T. Keith, W. S. Lane and S. L. Schreiber, *Nature*, 1994, **369**, 756.
5. D. M. Sabatini, H. Erdjument-Bromage, M. Lui, P. Tempst and S. H. Snyder, *Cell*, 1994, **78**, 35.
6. C. J. Sabers, M. M. Martin, G. J. Brunn, J. M. Williams, F. J. Dumont, G. Wiederrecht and R. T. Abraham, *J. Biol. Chem.*, 1995, **270**, 815.
7. J. A. Engelman, *Nat. Rev. Cancer*, 2009, **9**, 550.
8. D. H. Kim, D. D. Sarbassov, S. M. Ali, J. E. King, R. R. Latek, H. Erdjument-Bromage, P. Tempst and D. M. Sabatini, *Cell*, 2002, **110**, 163.
9. D. D. Sarbassov, S. M. Ali, D. H. Kim, D. A. Guertin, R. R. Latek, H. Erdjument-Bromage, P. Tempst and D. M. Sabatini, *Curr. Biol.*, 2004, **14**, 1296.
10. D. H. Kim, D. D. Sarbassov, S. M. Ali, R. R. Latek, K. V. Guntur, H. Erdjument-Bromage, P. Tempst and D. M. Sabatini, *Mol. Cell*, 2003, **11**, 895.
11. T. R. Peterson, M. Laplante, C. C. Thoreen, Y. Sancak, S. A. Kang, W. M. Kuehl, N. S. Gray and D. M. Sabatini, *Cell*, 2009, **137**, 873.
12. T. Kaizuka, T. Hara, N. Oshiro, U. Kikkawa, K. Yonezawa, K. Takehana, S. Iemura, T. Natsume and N. Mizushima, *J. Biol. Chem.*, 2010, **285**, 20109.
13. Y. Sancak, C. C. Thoreen, T. R. Peterson, R. A. Lindquist, S. A. Kang, E. Spooner, S. A. Carr and D. M. Sabatini, *Mol. Cell*, 2007, **25**, 903.
14. M. A. Frias, C. C. Thoreen, J. D. Jaffe, W. Schroder, T. Sculley, S. A. Carr and D. M. Sabatini, *Curr. Biol.*, 2006, **16**, 1865.
15. L. R. Pearce, X. Huang, J. Boudeau, R. Pawlowski, S. Wullschleger, M. Deak, A. F. Ibrahim, R. Gourlay, M. A. Magnuson and D. R. Alessi, *Biochem. J.*, 2007, **405**, 513.
16. C. K. Yip, K. Murata, T. Walz, D. M. Sabatini and S. A. Kang, *Mol. Cell*, 2010, **38**, 768.
17. C. H. Aylett, E. Sauer, S. Imseng, D. Boehringer, M. N. Hall, N. Ban and T. Maier, *Science*, 2016, **351**, 48.
18. C. Gaubitz, T. M. Oliveira, M. Prouteau, A. Leitner, M. Karuppasamy, G. Konstantinidou, D. Rispal, S. Eltschinger, G. C. Robinson, S. Thore, R. Aebersold, C. Schaffitzel and R. Loewith, *Mol. Cell*, 2015, **58**, 977.
19. B. K. Kennedy and D. W. Lamming, *Cell Metab.*, 2016, **23**, 990–1003.

20. I. H. Park, R. Bachmann, H. Shirazi and J. Chen, *J. Biol. Chem.*, 2002, **277**, 31423.
21. S. Phin, D. Kupferwasser, J. Lam and K. K. Lee-Fruman, *Biochem. J.*, 2003, **373**, 583.
22. T. R. Peterson, S. S. Sengupta, T. E. Harris, A. E. Carmack, S. A. Kang, E. Balderas, D. A. Guertin, K. L. Madden, A. E. Carpenter, B. N. Finck and D. M. Sabatini, *Cell*, 2011, **146**, 408.
23. S. Pena-Llopis, S. Vega-Rubin-de-Celis, J. C. Schwartz, N. C. Wolff, T. A. Tran, L. Zou, X. J. Xie, D. R. Corey and J. Brugarolas, *EMBO J.*, 2011, **30**, 3242.
24. B. D. Fonseca, C. Zakaria, J. J. Jia, T. E. Graber, Y. Svitkin, S. Tahmasebi, D. Healy, H. D. Hoang, J. M. Jensen, I. T. Diao, A. Lussier, C. Dajadian, N. Padmanabhan, W. Wang, E. Matta-Camacho, J. Hearnden, E. M. Smith, Y. Tsukumo, A. Yanagiya, M. Morita, E. Petroulakis, J. L. Gonzalez, G. Hernandez, T. Alain and C. K. Damgaard, *J. Biol. Chem.*, 2015, **290**, 15996–16020.
25. P. E. Burnett, R. K. Barrow, N. A. Cohen, S. H. Snyder and D. M. Sabatini, *Proc. Natl. Acad. Sci. U. S. A.*, 1998, **95**, 1432.
26. P. Liu, Z. Wang and W. Wei, *Cell Cycle*, 2014, **13**, 2162.
27. T. Ikenoue, K. Inoki, Q. Yang, X. Zhou and K. L. Guan, *EMBO J.*, 2008, **27**, 1919.
28. D. D. Sarbassov, D. A. Guertin, S. M. Ali and D. M. Sabatini, *Science*, 2005, **307**, 1098.
29. X. Gan, J. Wang, C. Wang, E. Sommer, T. Kozasa, S. Srinivasula, D. Alessi, S. Offermanns, M. I. Simon and D. Wu, *Nat. Cell Biol.*, 2012, **14**, 686.
30. X. Li and T. Gao, *EMBO Rep.*, 2014, **15**, 191.
31. S. Sciarretta, P. Zhai, Y. Maejima, D. P. Del Re, N. Nagarajan, D. Yee, T. Liu, M. A. Magnuson, M. Volpe, G. Frati, H. Li and J. Sadoshima, *Cell Rep.*, 2015, **11**, 125.
32. J. M. Garcia-Martinez and D. R. Alessi, *Biochem. J.*, 2008, **416**, 375.
33. S. Sengupta, A. Lorente-Rodriguez, S. Earnest, S. Stippec, X. Guo, D. C. Trudgian, H. Mirzaei and M. H. Cobb, *Proc. Natl. Acad. Sci. U. S. A.*, 2013, **110**, 18826.
34. P. P. Hsu, S. A. Kang, J. Rameseder, Y. Zhang, K. A. Ottina, D. Lim, T. R. Peterson, Y. Choi, N. S. Gray, M. B. Yaffe, J. A. Marto and D. M. Sabatini, *Science*, 2011, **332**, 1317.
35. D. W. Lamming, G. Demirkan, J. M. Boylan, M. M. Mihaylova, T. Peng, J. Ferreira, N. Neretti, A. Salomon, D. M. Sabatini and P. A. Gruppuso, *FASEB J.*, 2014, **28**, 300.
36. A. M. Robitaille, S. Christen, M. Shimobayashi, M. Cornu, L. L. Fava, S. Moes, C. Prescianotto-Baschong, U. Sauer, P. Jenoe and M. N. Hall, *Science*, 2013, **339**, 1320.
37. T. Vellai, K. Takacs-Vellai, Y. Zhang, A. L. Kovacs, L. Orosz and F. Muller, *Nature*, 2003, **426**, 620.
38. P. Kapahi, B. M. Zid, T. Harper, D. Koslover, V. Sapin and S. Benzer, *Curr. Biol.*, 2004, **14**, 885.

39. R. W. Powers 3rd, M. Kaeberlein, S. D. Caldwell, B. K. Kennedy and S. Fields, *Genes Dev.*, 2006, **20**, 174.
40. J. Urban, A. Soulard, A. Huber, S. Lippman, D. Mukhopadhyay, O. Deloche, V. Wanke, D. Anrather, G. Ammerer, H. Riezman, J. R. Broach, C. De Virgilio, M. N. Hall and R. Loewith, *Mol. Cell*, 2007, **26**, 663.
41. M. Kaeberlein, R. W. Powers 3rd, K. K. Steffen, E. A. Westman, D. Hu, N. Dang, E. O. Kerr, K. T. Kirkland, S. Fields and B. K. Kennedy, *Science*, 2005, **310**, 1193.
42. K. K. Steffen, V. L. MacKay, E. O. Kerr, M. Tsuchiya, D. Hu, L. A. Fox, N. Dang, E. D. Johnston, J. A. Oakes, B. N. Tchao, D. N. Pak, S. Fields, B. K. Kennedy and M. Kaeberlein, *Cell*, 2008, **133**, 292.
43. M. Hansen, S. Taubert, D. Crawford, N. Libina, S. J. Lee and C. Kenyon, *Aging Cell*, 2007, **6**, 95.
44. I. Bjedov, J. M. Toivonen, F. Kerr, C. Slack, J. Jacobson, A. Foley and L. Partridge, *Cell Metab.*, 2010, **11**, 35.
45. W. C. Fok, A. Bokov, J. Gelfond, Z. Yu, Y. Zhang, M. Doderer, Y. Chen, M. Javors, W. H. Wood 3rd, Y. Zhang, K. G. Becker, A. Richardson and V. I. Perez, *Aging Cell*, 2014, **13**, 311.
46. W. C. Fok, C. Livi, A. Bokov, Z. Yu, Y. Chen, A. Richardson and V. I. Perez, *Mech. Ageing Dev.*, 2014, **140**, 23.
47. W. C. Fok, Y. Zhang, A. B. Salmon, A. Bhattacharya, R. Gunda, D. Jones, W. Ward, K. Fisher, A. Richardson and V. I. Perez, *J. Gerontol., Ser. A*, 2013, **68**, 108.
48. D. E. Harrison, R. Strong, Z. D. Sharp, J. F. Nelson, C. M. Astle, K. Flurkey, N. L. Nadon, J. E. Wilkinson, K. Frenkel, C. S. Carter, M. Pahor, M. A. Javors, E. Fernandez and R. A. Miller, *Nature*, 2009, **460**, 392.
49. S. I. Arriola Apelo and D. W. Lamming, *J. Gerontol., Ser. A*, 2016, **71**, 841–849.
50. F. Neff, D. Flores-Dominguez, D. P. Ryan, M. Horsch, S. Schroder, T. Adler, L. C. Afonso, J. A. Aguilar-Pimentel, L. Becker, L. Garrett, W. Hans, M. M. Hettich, R. Holtmeier, S. M. Holter, K. Moreth, C. Prehn, O. Puk, I. Racz, B. Rathkolb, J. Rozman, B. Naton, R. Ordemann, J. Adamski, J. Beckers, R. Bekeredjian, D. H. Busch, G. Ehninger, J. Graw, H. Hofler, M. Klingenspor, T. Klopstock, M. Ollert, J. Stypmann, E. Wolf, W. Wurst, A. Zimmer, H. Fuchs, V. Gailus-Durner, M. Hrabe de Angelis and D. Ehninger, *J. Clin. Invest.*, 2013, **123**, 3272.
51. J. E. Wilkinson, L. Burmeister, S. V. Brooks, C. C. Chan, S. Friedline, D. E. Harrison, J. F. Hejtmancik, N. Nadon, R. Strong, L. K. Wood, M. A. Woodward and R. A. Miller, *Aging Cell*, 2012, **11**, 675.
52. F. Sierra, *Cold Spring Harbor Perspect. Med.*, 2016, **6**, a025163.
53. S. Majumder, A. Caccamo, D. X. Medina, A. D. Benavides, M. A. Javors, E. Kraig, R. Strong, A. Richardson and S. Oddo, *Aging Cell*, 2012, **11**, 326.
54. P. Spilman, N. Podlutskaya, M. J. Hart, J. Debnath, O. Gorostiza, D. Bredesen, A. Richardson, R. Strong and V. Galvan, *PLoS One*, 2010, **5**, e9979.
55. S. Ozcelik, G. Fraser, P. Castets, V. Schaeffer, Z. Skachokova, K. Breu, F. Clavaguera, M. Sinnreich, L. Kappos, M. Goedert, M. Tolnay and D. T. Winkler, *PLoS One*, 2013, **8**, e62459.

56. A. L. Lin, J. B. Jahrling, W. Zhang, N. DeRosa, V. Bakshi, P. Romero, V. Galvan and A. Richardson, *J. Cereb. Blood Flow Metab.*, 2015, DOI: 10.1177/0271678X15621575.
57. M. Bikkina and J. Koneru, *Med. Devices (Auckl.)*, 2011, **4**, 117.
58. W. Martinet, H. De Loof and G. R. De Meyer, *Atherosclerosis*, 2014, **233**, 601.
59. D. F. Dai, P. P. Karunadharma, Y. A. Chiao, N. Basisty, D. Crispin, E. J. Hsieh, T. Chen, H. Gu, D. Djukovic, D. Raftery, R. P. Beyer, M. J. MacCoss and P. S. Rabinovitch, *Aging Cell*, 2014, **13**, 529.
60. J. M. Flynn, M. N. O'Leary, C. A. Zambataro, E. C. Academia, M. P. Presley, B. J. Garrett, A. Zykovich, S. D. Mooney, R. Strong, C. J. Rosen, P. Kapahi, M. D. Nelson, B. K. Kennedy and S. Melov, *Aging Cell*, 2013, **12**, 851.
61. R. L. Frock, S. C. Chen, D. F. Da, E. Frett, C. Lau, C. Brown, D. N. Pak, Y. Wang, A. Muchir, H. J. Worman, L. F. Santana, W. C. Ladiges, P. S. Rabinovitch and B. K. Kennedy, *PLoS One*, 2012, **7**, e42918.
62. F. J. Ramos, S. C. Chen, M. G. Garelick, D. F. Dai, C. Y. Liao, K. H. Schreiber, V. L. MacKay, E. H. An, R. Strong, W. C. Ladiges, P. S. Rabinovitch, M. Kaeberlein and B. K. Kennedy, *Sci. Transl. Med.*, 2012, **4**, 144ra103.
63. C. Chen, Y. Liu, Y. Liu and P. Zheng, *Sci. Signaling*, 2009, **2**, ra75.
64. C. Selman, A. Sinclair, S. M. Pedroni, E. E. Irvine, A. M. Michie and D. J. Withers, *Oncotarget*, 2016, **7**, 29937–29943.
65. J. B. Mannick, G. Del Giudice, M. Lattanzi, N. M. Valiante, J. Praestgaard, B. Huang, M. A. Lonetto, H. T. Maecker, J. Kovarik, S. Carson, D. J. Glass and L. B. Klickstein, *Sci. Transl. Med.*, 2014, **6**, 268ra179.
66. O. H. Yilmaz, P. Katajisto, D. W. Lamming, Y. Gultekin, K. E. Bauer-Rowe, S. Sengupta, K. Birsoy, A. Dursun, V. O. Yilmaz, M. Selig, G. P. Nielsen, M. Mino-Kenudson, L. R. Zukerberg, A. K. Bhan, V. Deshpande and D. M. Sabatini, *Nature*, 2012, **486**, 490.
67. J. J. Wu, J. Liu, E. B. Chen, J. J. Wang, L. Cao, N. Narayan, M. M. Fergusson, I. I. Rovira, M. Allen, D. A. Springer, C. U. Lago, S. Zhang, W. DuBois, T. Ward, R. deCabo, O. Gavrilova, B. Mock and T. Finkel, *Cell Rep.*, 2013, **4**, 913.
68. C. Selman, J. M. Tullet, D. Wieser, E. Irvine, S. J. Lingard, A. I. Choudhury, M. Claret, H. Al-Qassab, D. Carmignac, F. Ramadani, A. Woods, I. C. Robinson, E. Schuster, R. L. Batterham, S. C. Kozma, G. Thomas, D. Carling, K. Okkenhaug, J. M. Thornton, L. Partridge, D. Gems and D. J. Withers, *Science*, 2009, **326**, 140.
69. D. W. Lamming, L. Ye, P. Katajisto, M. D. Goncalves, M. Saitoh, D. M. Stevens, J. G. Davis, A. B. Salmon, A. Richardson, R. S. Ahima, D. A. Guertin, D. M. Sabatini and J. A. Baur, *Science*, 2012, **335**, 1638.
70. S. Tsai, J. M. Sitzmann, S. G. Dastidar, A. A. Rodriguez, S. L. Vu, C. E. McDonald, E. C. Academia, M. N. O'Leary, T. D. Ashe, A. R. La Spada and B. K. Kennedy, *J. Clin. Invest.*, 2015, **125**, 2952.
71. M. G. Garelick, V. L. Mackay, A. Yanagida, E. C. Academia, K. H. Schreiber, W. C. Ladiges and B. K. Kennedy, *Cell Cycle*, 2013, **12**, 2493.
72. S. Komili, N. G. Farny, F. P. Roth and P. A. Silver, *Cell*, 2007, **131**, 557.

73. C. C. Thoreen, L. Chantranupong, H. R. Keys, T. Wang, N. S. Gray and D. M. Sabatini, *Nature*, 2012, **485**, 109.
74. M. V. Blagosklonny, *Aging (Albany NY)*, 2012, **4**, 350.
75. D. W. Lamming, *SpringerPlus*, 2014, **3**, 735.
76. C. Selman, L. Partridge and D. J. Withers, *PLoS One*, 2011, **6**, e16144.
77. A. F. Bokov, N. Garg, Y. Ikeno, S. Thakur, N. Musi, R. A. DeFronzo, N. Zhang, R. C. Erickson, J. Gelfond, G. B. Hubbard, M. L. Adamo and A. Richardson, *PLoS One*, 2011, **6**, e26891.
78. A. Nojima, M. Yamashita, Y. Yoshida, I. Shimizu, H. Ichimiya, N. Kamimura, Y. Kobayashi, S. Ohta, N. Ishii and T. Minamino, *PLoS One*, 2013, **8**, e69178.
79. M. Bluher, B. B. Kahn and C. R. Kahn, *Science*, 2003, **299**, 572.
80. A. Taguchi, L. M. Wartschow and M. F. White, *Science*, 2007, **317**, 369.
81. D. W. Lamming, M. M. Mihaylova, P. Katajisto, E. L. Baar, O. H. Yilmaz, A. Hutchins, Y. Gultekin, R. Gaither and D. M. Sabatini, *Aging Cell*, 2014, **13**, 911.
82. D. D. Sarbassov, S. M. Ali, S. Sengupta, J. H. Sheen, P. P. Hsu, A. F. Bagley, A. L. Markhard and D. M. Sabatini, *Mol. Cell*, 2006, **22**, 159.
83. K. H. Schreiber, D. Ortiz, E. C. Academia, A. C. Anies, C. Y. Liao and B. K. Kennedy, *Aging Cell*, 2015, **14**, 265.
84. A. Hagiwara, M. Cornu, N. Cybulski, P. Polak, C. Betz, F. Trapani, L. Terracciano, M. H. Heim, M. A. Ruegg and M. N. Hall, *Cell Metab.*, 2012, **15**, 725.
85. M. Yuan, E. Pino, L. Wu, M. Kacergis and A. A. Soukas, *J. Biol. Chem.*, 2012, **287**, 29579.
86. Y. Tang, M. Wallace, J. Sanchez-Gurmaches, W. Y. Hsiao, H. Li, P. L. Lee, S. Vernia, C. M. Metallo and D. A. Guertin, *Nat. Commun.*, 2016, **7**, 11365.
87. A. Kumar, J. C. Lawrence Jr, D. Y. Jung, H. J. Ko, S. R. Keller, J. K. Kim, M. A. Magnuson and T. E. Harris, *Diabetes*, 2010, **59**, 1397.
88. V. Albert, K. Svensson, M. Shimobayashi, M. Colombi, S. Munoz, V. Jimenez, C. Handschin, F. Bosch and M. N. Hall, *EMBO Mol. Med.*, 2016, **8**, 232.
89. C. M. Hung, C. M. Calejman, J. Sanchez-Gurmaches, H. Li, C. B. Clish, S. Hettmer, A. J. Wagers and D. A. Guertin, *Cell Rep.*, 2014, **8**, 256.
90. A. Kumar, T. E. Harris, S. R. Keller, K. M. Choi, M. A. Magnuson and J. C. Lawrence Jr, *Mol. Cell. Biol.*, 2008, **28**, 61.
91. M. Kleinert, L. Sylow, D. J. Fazakerley, J. R. Krycer, K. C. Thomas, A. J. Oxboll, A. B. Jordy, T. E. Jensen, G. Yang, P. Schjerling, B. Kiens, D. E. James, M. A. Ruegg and E. A. Richter, *Mol. Metab.*, 2014, **3**, 630.
92. D. W. Lamming and D. M. Sabatini, *Cell Metab.*, 2013, **18**, 465.
93. E. Maier, A. Duschl and J. Horejs-Hoeck, *Eur. J. Immunol.*, 2012, **42**, 2827.
94. A. S. Lazorchak and B. Su, *Protein Cell*, 2011, **2**, 523.
95. V. Byles, A. J. Covarrubias, I. Ben-Sahra, D. W. Lamming, D. M. Sabatini, B. D. Manning and T. Horng, *Nat. Commun.*, 2013, **4**, 2834.
96. J. D. Powell, K. N. Pollizzi, E. B. Heikamp and M. R. Horton, *Annu. Rev. Immunol.*, 2012, **30**, 39.

97. A. S. Lazorchak, D. Liu, V. Facchinetti, A. Di Lorenzo, W. C. Sessa, D. G. Schatz and B. Su, *Mol. Cell*, 2010, **39**, 433.
98. W. T. Festuccia, P. Pouliot, I. Bakan, D. M. Sabatini and M. Laplante, *PLoS One*, 2014, **9**, e95432.
99. J. Trelinska, I. Dachowska, K. Kotulska, W. Fendler, S. Jozwiak and W. Mlynarski, *Anti-Cancer Drugs*, 2015, **26**, 437.
100. C. Ross, A. Salmon, R. Strong, E. Fernandez, M. Javors, A. Richardson and S. Tardif, *Aging (Albany NY)*, 2015, **7**, 964.
101. S. Tardif, C. Ross, P. Bergman, E. Fernandez, M. Javors, A. Salmon, J. Spross, R. Strong and A. Richardson, *J. Gerontol., Ser. A*, 2015, **70**, 577.
102. M. Kaeberlein, *Vet. Pathol.*, 2016, **53**, 291.
103. D. W. Lamming, L. Ye, D. M. Sabatini and J. A. Baur, *J. Clin. Invest.*, 2013, **123**, 980.
104. S. I. Arriola Apelo, J. C. Neuman, E. L. Baar, F. A. Syed, N. E. Cummings, H. K. Brar, C. P. Pumper, M. E. Kimple and D. W. Lamming, *Aging Cell*, 2016, **15**, 28.
105. S. I. Arriola Apelo, C. P. Pumper, E. L. Baar, N. E. Cummings and D. W. Lamming, *J. Gerontol., Ser. A*, 2016, **71**, 876–881.
106. O. V. Leontieva, G. M. Paszkiewicz and M. V. Blagosklonny, *Aging Cell*, 2014, **13**, 616.
107. P. J. Scarpace, M. Matheny, K. Y. Strehler, H. Z. Toklu, N. Kirichenko, C. S. Carter, D. Morgan and N. Tumer, *J. Gerontol., Ser. A*, 2015, **71**, 891–899.
108. Y. Fang, I. H. Park, A. L. Wu, G. Du, P. Huang, M. A. Frohman, S. J. Walker, H. A. Brown and J. Chen, *Curr. Biol.*, 2003, **13**, 2037.
109. J. Brugarolas, K. Lei, R. L. Hurley, B. D. Manning, J. H. Reiling, E. Hafen, L. A. Witters, L. W. Ellisen and W. G. Kaelin Jr, *Genes Dev.*, 2004, **18**, 2893.
110. A. Alexander, S. L. Cai, J. Kim, A. Nanez, M. Sahin, K. H. MacLean, K. Inoki, K. L. Guan, J. Shen, M. D. Person, D. Kusewitt, G. B. Mills, M. B. Kastan and C. L. Walker, *Proc. Natl. Acad. Sci. U. S. A.*, 2010, **107**, 4153.
111. J. Kim, M. Kundu, B. Viollet and K. L. Guan, *Nat. Cell Biol.*, 2011, **13**, 132.
112. A. R. Tee, B. D. Manning, P. P. Roux, L. C. Cantley and J. Blenis, *Curr. Biol.*, 2003, **13**, 1259.
113. X. Zhou, T. L. Clister, P. R. Lowry, M. M. Seldin, G. W. Wong and J. Zhang, *Cell Rep.*, 2015, DOI: 10.1016/j.celrep.2015.02.031.
114. D. C. Goberdhan, C. Wilson and A. L. Harris, *Cell Metab.*, 2016, **23**, 580.
115. Y. Sancak, L. Bar-Peled, R. Zoncu, A. L. Markhard, S. Nada and D. M. Sabatini, *Cell*, 2010, **141**, 290.
116. Y. Sancak, T. R. Peterson, Y. D. Shaul, R. A. Lindquist, C. C. Thoreen, L. Bar-Peled and D. M. Sabatini, *Science*, 2008, **320**, 1496.
117. L. Bar-Peled, L. Chantranupong, A. D. Cherniack, W. W. Chen, K. A. Ottina, B. C. Grabiner, E. D. Spear, S. L. Carter, M. Meyerson and D. M. Sabatini, *Science*, 2013, **340**, 1100.
118. L. Bar-Peled, L. D. Schweitzer, R. Zoncu and D. M. Sabatini, *Cell*, 2012, **150**, 1196.

119. R. Zoncu, L. Bar-Peled, A. Efeyan, S. Wang, Y. Sancak and D. M. Sabatini, *Science*, 2011, **334**, 678.
120. J. Jung, H. M. Genau and C. Behrends, *Mol. Cell. Biol.*, 2015, **35**, 2479.
121. M. Rebsamen, L. Pochini, T. Stasyk, M. E. de Araujo, M. Galluccio, R. K. Kandasamy, B. Snijder, A. Fauster, E. L. Rudashevskaya, M. Bruckner, S. Scorzoni, P. A. Filipek, K. V. Huber, J. W. Bigenzahn, L. X. Heinz, C. Kraft, K. L. Bennett, C. Indiveri, L. A. Huber and G. Superti-Furga, *Nature*, 2015, **519**, 477.
122. S. Wang, Z. Y. Tsun, R. L. Wolfson, K. Shen, G. A. Wyant, M. E. Plovanich, E. D. Yuan, T. D. Jones, L. Chantranupong, W. Comb, T. Wang, L. Bar-Peled, R. Zoncu, C. Straub, C. Kim, J. Park, B. L. Sabatini and D. M. Sabatini, *Science*, 2015, **347**, 188.
123. R. A. Saxton, K. E. Knockenhauer, R. L. Wolfson, L. Chantranupong, M. E. Pacold, T. Wang, T. U. Schwartz and D. M. Sabatini, *Science*, 2016, **351**, 53.
124. R. L. Wolfson, L. Chantranupong, R. A. Saxton, K. Shen, S. M. Scaria, J. R. Cantor and D. M. Sabatini, *Science*, 2016, **351**, 43.
125. L. Chantranupong, S. M. Scaria, R. A. Saxton, M. P. Gygi, K. Shen, G. A. Wyant, T. Wang, J. W. Harper, S. P. Gygi and D. M. Sabatini, *Cell*, 2016, **165**, 153.
126. G. Jin, S. W. Lee, X. Zhang, Z. Cai, Y. Gao, P. C. Chou, A. H. Rezaeian, F. Han, C. Y. Wang, J. C. Yao, Z. Gong, C. H. Chan, C. Y. Huang, F. J. Tsai, C. H. Tsai, S. H. Tu, C. H. Wu, D. Sarbassov dos, Y. S. Ho and H. K. Lin, *Mol. Cell*, 2015, **58**, 989.
127. L. Deng, C. Jiang, L. Chen, J. Jin, J. Wei, L. Zhao, M. Chen, W. Pan, Y. Xu, H. Chu, X. Wang, X. Ge, D. Li, L. Liao, M. Liu, L. Li and P. Wang, *Mol. Cell*, 2015, **58**, 804.
128. M. A. Fawal, M. Brandt and N. Djouder, *Dev. Cell*, 2015, **33**, 67.
129. A. Garami, F. J. Zwartkruis, T. Nobukuni, M. Joaquin, M. Roccio, H. Stocker, S. C. Kozma, E. Hafen, J. L. Bos and G. Thomas, *Mol. Cell*, 2003, **11**, 1457.
130. A. R. Tee, R. Anjum and J. Blenis, *J. Biol. Chem.*, 2003, **278**, 37288.
131. S. Menon, C. C. Dibble, G. Talbott, G. Hoxhaj, A. J. Valvezan, H. Takahashi, L. C. Cantley and B. D. Manning, *Cell*, 2014, **156**, 771.
132. C. Demetriades, N. Doumpas and A. A. Teleman, *Cell*, 2014, **156**, 786.
133. C. Demetriades, M. Plescher and A. A. Teleman, *Nat. Commun.*, 2016, 7, 10662.
134. A. Efeyan, R. Zoncu, S. Chang, I. Gumper, H. Snitkin, R. L. Wolfson, O. Kirak, D. D. Sabatini and D. M. Sabatini, *Nature*, 2013, **493**, 679.
135. D. M. Gwinn, D. B. Shackelford, D. F. Egan, M. M. Mihaylova, A. Mery, D. S. Vasquez, B. E. Turk and R. J. Shaw, *Mol. Cell*, 2008, **30**, 214.
136. K. Inoki, Y. Li, T. Zhu, J. Wu and K. L. Guan, *Nat. Cell Biol.*, 2002, **4**, 648.
137. C. Wang, X. Mao, L. Wang, M. Liu, M. D. Wetzel, K. L. Guan, L. Q. Dong and F. Liu, *J. Biol. Chem.*, 2007, **282**, 7991.
138. P. Liu, W. Gan, Y. R. Chin, K. Ogura, J. Guo, J. Zhang, B. Wang, J. Blenis, L. C. Cantley, A. Toker, B. Su and W. Wei, *Cancer Discovery*, 2015, **5**, 1194.

139. H. K. Park and R. S. Ahima, *F1000Prime Reports*, 2014, **6**, 73.
140. S. Tripathy and D. B. Jump, *J. Lipid Res.*, 2013, **54**, 71.
141. C. Zhang, A. A. Wendel, M. R. Keogh, T. E. Harris, J. Chen and R. A. Coleman, *Proc. Natl. Acad. Sci. U. S. A.*, 2012, **109**, 1667.
142. V. Zinzalla, D. Stracka, W. Oppliger and M. N. Hall, *Cell*, 2011, **144**, 757.
143. C. Betz, D. Stracka, C. Prescianotto-Baschong, M. Frieden, N. Demaurex and M. N. Hall, *Proc. Natl. Acad. Sci. U. S. A.*, 2013, **110**, 12526.
144. E. Arias, H. Koga, A. Diaz, E. Mocholi, B. Patel and A. M. Cuervo, *Mol. Cell*, 2015, **59**, 270.
145. J. Ye, W. Palm, M. Peng, B. King, T. Lindsten, M. O. Li, C. Koumenis and C. B. Thompson, *Genes Dev.*, 2015, **29**, 2331.
146. S. M. Solon-Biet, A. C. McMahon, J. W. O. Ballard, K. Ruohonen, L. E. Wu, V. C. Cogger, A. Warren, X. Huang, N. Pichaud, R. G. Melvin, R. Gokarn, M. Khalil, N. Turner, G. J. Cooney, D. A. Sinclair, D. Raubenheimer, D. G. Le Couteur and S. J. Simpson, *Cell Metab.*, 2014, **19**, 418.
147. S. M. Solon-Biet, S. J. Mitchell, S. C. Coogan, V. C. Cogger, R. Gokarn, A. C. McMahon, D. Raubenheimer, R. de Cabo, S. J. Simpson and D. G. Le Couteur, *Cell Rep.*, 2015, **11**, 1529.
148. L. Fontana and L. Partridge, *Cell*, 2015, **161**, 106.
149. P. Lagiou, S. Sandin, E. Weiderpass, A. Lagiou, L. Mucci, D. Trichopoulos and H. O. Adami, *J. Intern. Med.*, 2007, **261**, 366.
150. M. E. Levine, J. A. Suarez, S. Brandhorst, P. Balasubramanian, C. W. Cheng, F. Madia, L. Fontana, M. G. Mirisola, J. Guevara-Aguirre, J. Wan, G. Passarino, B. K. Kennedy, M. Wei, P. Cohen, E. M. Crimmins and V. D. Longo, *Cell Metab.*, 2014, **19**, 407.
151. J. Halkjaer, A. Olsen, K. Overvad, M. U. Jakobsen, H. Boeing, B. Buijsse, D. Palli, G. Tognon, H. Du, A. D. van der, N. G. Forouhi, N. J. Wareham, E. J. Feskens, T. I. Sorensen and A. Tjonneland, *Int. J. Obes. (London)*, 2011, **35**, 1104.
152. A. C. Vergnaud, T. Norat, T. Mouw, D. Romaguera, A. M. May, H. B. Bueno-de-Mesquita, A. D. van der, A. Agudo, N. Wareham, K. T. Khaw, I. Romieu, H. Freisling, N. Slimani, F. Perquier, M. C. Boutron-Ruault, F. Clavel-Chapelon, D. Palli, F. Berrino, A. Mattiello, R. Tumino, F. Ricceri, L. Rodriguez, E. Molina-Montes, P. Amiano, A. Barricarte, M. D. Chirlaque, F. L. Crowe, P. Orfanos, A. Naska, A. Trichopoulou, B. Teucher, R. Kaaks, H. Boeing, B. Buijsse, I. Johansson, G. Hallmans, I. Drake, E. Sonestedt, M. U. Jakobsen, K. Overvad, A. Tjonneland, J. Halkjaer, G. Skeie, T. Braaten, E. Lund, E. Riboli and P. H. Peeters, *PLoS One*, 2013, **8**, e57300.
153. I. Sluijs, J. W. Beulens, A. D. van der, A. M. Spijkerman, D. E. Grobbee and Y. T. van der Schouw, *Diabetes Care*, 2010, **33**, 43.
154. D. W. Lamming, N. E. Cummings, A. L. Rastelli, F. Gao, E. Cava, B. Bertozzi, F. Spelta, R. Pili and L. Fontana, *Oncotarget*, 2015, **6**, 31233.
155. F. Li, Y. Yin, B. Tan, X. Kong and G. Wu, *Amino Acids*, 2011, **41**, 1185.
156. M. D. Sans, M. Tashiro, N. L. Vogel, S. R. Kimball, L. G. D'Alecy and J. A. Williams, *J. Nutr.*, 2006, **136**, 1792.

157. E. Blomstrand, J. Eliasson, H. K. Karlsson and R. Kohnke, *J. Nutr.*, 2006, **136**, 269S.
158. F. Xiao, Z. Huang, H. Li, J. Yu, C. Wang, S. Chen, Q. Meng, Y. Cheng, X. Gao, J. Li, Y. Liu and F. Guo, *Diabetes*, 2011, **60**, 746.
159. C. J. Lynch and S. H. Adams, *Nat. Rev. Endocrinol.*, 2014, **10**, 723.
160. C. B. Newgard, J. An, J. R. Bain, M. J. Muehlbauer, R. D. Stevens, L. F. Lien, A. M. Haqq, S. H. Shah, M. Arlotto, C. A. Slentz, J. Rochon, D. Gallup, O. Ilkayeva, B. R. Wenner, W. S. Yancy Jr, H. Eisenson, G. Musante, R. S. Surwit, D. S. Millington, M. D. Butler and L. P. Svetkey, *Cell Metab.*, 2009, **9**, 311.
161. L. Fontana, N. E. Cummings, S. I. Arriola Apelo, J. C. Neuman, I. Kasza, B. A. Schmidt, E. Cava, F. Spelta, V. Tosti, F. A. Syed, E. L. Baar, N. Veronese, S. E. Cottrell, R. J. Fenske, B. Bertozzi, H. K. Brar, T. Pietka, A. D. Bullock, R. S. Figenshau, G. L. Andriole, M. J. Merrins, C. M. Alexander, M. E. Kimple and D. W. Lamming, *Cell Rep.*, 2016, **16**, 520–530.
162. M. F. McCarty, J. Barroso-Aranda and F. Contreras, *Med. Hypotheses*, 2009, **72**, 125.
163. R. A. Miller, G. Buehner, Y. Chang, J. M. Harper, R. Sigler and M. Smith-Wheelock, *Aging Cell*, 2005, **4**, 119.
164. Buck Institute for Research on Aging, http://www.buckinstitute.org/buck-news/mechanistic-finding-may-provide-means-deal-side-effects-lifespan-extending-drug-rapamycin, *Journal*, 2015.
165. F. V. Din, A. Valanciute, V. P. Houde, D. Zibrova, K. A. Green, K. Sakamoto, D. R. Alessi and M. G. Dunlop, *Gastroenterology*, 2012, **142**, 1504–1515 e1503.
166. A. Kalender, A. Selvaraj, S. Y. Kim, P. Gulati, S. Brule, B. Viollet, B. E. Kemp, N. Bardeesy, P. Dennis, J. J. Schlager, A. Marette, S. C. Kozma and G. Thomas, *Cell Metab.*, 2010, **11**, 390.
167. A. Martin-Montalvo, E. M. Mercken, S. J. Mitchell, H. H. Palacios, P. L. Mote, M. Scheibye-Knudsen, A. P. Gomes, T. M. Ward, R. K. Minor, M. J. Blouin, M. Schwab, M. Pollak, Y. Zhang, Y. Yu, K. G. Becker, V. A. Bohr, D. K. Ingram, D. A. Sinclair, N. S. Wolf, S. R. Spindler, M. Bernier and R. de Cabo, *Nat. Commun.*, 2013, **4**, 2192.
168. R. Strong, R. A. Miller, C. M. Astle, R. A. Floyd, K. Flurkey, K. L. Hensley, M. A. Javors, C. Leeuwenburgh, J. F. Nelson, E. Ongini, N. L. Nadon, H. R. Warner and D. E. Harrison, *Aging Cell*, 2008, **7**, 641.
169. R. A. Miller, D. E. Harrison, C. M. Astle, J. A. Baur, A. R. Boyd, R. de Cabo, E. Fernandez, K. Flurkey, M. A. Javors, J. F. Nelson, C. J. Orihuela, S. Pletcher, Z. D. Sharp, D. Sinclair, J. W. Starnes, J. E. Wilkinson, N. L. Nadon and R. Strong, *J. Gerontol., Ser. A*, 2011, **66**, 191.
170. V. N. Anisimov, M. A. Zabezhinski, I. G. Popovich, T. S. Piskunova, A. V. Semenchenko, M. L. Tyndyk, M. N. Yurova, S. V. Rosenfeld and M. V. Blagosklonny, *Cell Cycle*, 2011, **10**, 4230.
171. R. A. Miller, D. E. Harrison, C. M. Astle, E. Fernandez, K. Flurkey, M. Han, M. A. Javors, X. Li, N. L. Nadon, J. F. Nelson, S. Pletcher, A. B. Salmon, Z. D. Sharp, S. Van Roekel, L. Winkleman and R. Strong, *Aging Cell*, 2014, **13**, 468.

172. W. C. Fok, Y. Chen, A. Bokov, Y. Zhang, A. B. Salmon, V. Diaz, M. Javors, W. H. Wood 3rd, Y. Zhang, K. G. Becker, V. I. Perez and A. Richardson, *PLoS One*, 2014, **9**, e83988.
173. E. Hernando, E. Charytonowicz, M. E. Dudas, S. Menendez, I. Matushansky, J. Mills, N. D. Socci, N. Behrendt, L. Ma, R. G. Maki, P. P. Pandolfi and C. Cordon-Cardo, *Nat. Med.*, 2007, **13**, 748.
174. V. N. Anisimov, M. A. Zabezhinski, I. G. Popovich, T. S. Piskunova, A. V. Semenchenko, M. L. Tyndyk, M. N. Yurova, M. P. Antoch and M. V. Blagosklonny, *Am. J. Pathol.*, 2010, **176**, 2092.
175. A. Bhattacharya, A. Bokov, F. L. Muller, A. L. Jernigan, K. Maslin, V. Diaz, A. Richardson and H. Van Remmen, *Neurobiol. Aging*, 2012, **33**, 1829.
176. E. A. Komarova, M. P. Antoch, L. R. Novototskaya, O. B. Chernova, G. Paszkiewicz, O. V. Leontieva, M. V. Blagosklonny and A. V. Gudkov, *Aging (Albany NY)*, 2012, **4**, 709.
177. M. Comas, I. Toshkov, K. K. Kuropatwinski, O. B. Chernova, A. Polinsky, M. V. Blagosklonny, A. V. Gudkov and M. P. Antoch, *Aging (Albany NY)*, 2012, **4**, 715.
178. C. B. Livi, R. L. Hardman, B. A. Christy, S. G. Dodds, D. Jones, C. Williams, R. Strong, A. Bokov, M. A. Javors, Y. Ikeno, G. Hubbard, P. Hasty and Z. D. Sharp, *Aging (Albany NY)*, 2013, **5**, 100.
179. R. V. Khapre, A. A. Kondratova, S. Patel, Y. Dubrovsky, M. Wrobel, M. P. Antoch and R. V. Kondratov, *Aging (Albany NY)*, 2014, **6**, 48.
180. I. G. Popovich, V. N. Anisimov, M. A. Zabezhinski, A. V. Semenchenko, M. L. Tyndyk, M. N. Yurova and M. V. Blagosklonny, *Cancer Biol. Ther.*, 2014, **15**, 586.
181. S. C. Johnson, M. E. Yanos, A. Bitto, A. Castanza, A. Gagnidze, B. Gonzalez, K. Gupta, J. Hui, C. Jarvie, B. M. Johnson, N. Letexier, L. McCanta, M. Sangesland, O. Tamis, L. Uhde, A. Van Den Ende, P. S. Rabinovitch, Y. Suh and M. Kaeberlein, *Front. Genet.*, 2015, **6**, 247.
182. V. Hurez, V. Dao, A. Liu, S. Pandeswara, J. Gelfond, L. Sun, M. Bergman, C. J. Orihuela, V. Galvan, A. Padron, J. Drerup, Y. Liu, P. Hasty, Z. D. Sharp and T. J. Curiel, *Aging Cell*, 2015, **14**, 945.
183. K. Sataranatarajan, Y. Ikeno, A. Bokov, D. Feliers, H. Yalamanchili, H. J. Lee, M. M. Mariappan, H. Tabatabai-Mir, V. Diaz, S. Prasad, M. A. Javors, G. Ghosh Choudhury, G. B. Hubbard, J. L. Barnes, A. Richardson and B. S. Kasinath, *J. Gerontol., Ser. A*, 2015, **71**, 850–857.

CHAPTER 15

mTOR, Aging and Cancer: Prospects for Pharmacological Interventions

Z. D. SHARP[a]

[a]University of Texas Health Science Center San Antonio, Department of Molecular Medicine, Institute of Biotechnology, Barshop Institute for Longevity and Aging Studies and Cancer and Therapy Research Center, San Antonio, Texas 78245, USA
*E-mail: sharp@uthscsa.edu

15.1 Rapamycin: A Brief History

This now familiar, but nevertheless remarkable, story begins at Ayerst Labs in Montreal, where scientists in the 1970s identified a lipophilic macrocyclic lactone antibiotic produced by *Streptomyces hygroscopicus* in a soil sample collected from Rapa Nui (Easter Island). Scientists dropped its development as a fungicide[1] upon learning that it inhibited the immune response in various cell culture and animal model settings (reviewed by Sehgal[2]). Assigned the generic name, sirolimus, Wyeth Ayerst marketed it as Rapamune to prevent host rejection of transplants, only later to discover rapamycin has anti-tumor activity. Pharmaceutical companies have developed a number of derivative compounds (rapalogs) with indications that include prevention of allograft rejection, anti-cancer and anti-restenosis.[3] Even at this early stage, the mystery was there but not mentioned—how could a drug that suppresses

RSC Drug Discovery Series No. 57
Anti-aging Drugs: From Basic Research to Clinical Practice
Edited by Alexander M. Vaiserman

Published by the Royal Society of Chemistry, www.rsc.org

immunity (*e.g.*, surveillance of cancer[4]) be safe and effective as an anti-cancer drug? The rapamycin mystery deepens but first it will be useful for the reader to have a basic understanding of its mode of action in cells.

15.2 The Target of Rapamycin

Since this drug was first identified as a fungicide, yeast was the first organism that investigators used to uncover its mechanism of action. Heitman *et al.*[5] showed that rapamycin, when bound to the product of the *FPR1* gene (encoding an FK506-binding protein [FKBP], a proline rotamase), inhibited cell cycle progression. These authors also showed that rapamycin sensitivity depended on two other genes (aptly named the target of rapamycin (*TOR*) *1* and *TOR2*) both of which encoded phosphatidylinositol kinase homologues.[6] Subsequent years of fruitful research by many labs studying budding[7] and fission[8] yeast revealed many of the fundamentals of its biological function and significance. Of significance for this chapter, inhibition of TOR in budding yeast increased chronological life span.[9]

Subsequent to the identification of TOR in budding yeast, multiple labs reported identification of one gene encoding the mammalian target of rapamycin.[10–13] Each dubbed with different names, these coalesced first into mammalian TOR (mTOR), then mechanistic TOR, with some reservation.[14] mTOR is conserved in structure and function in eukaryotes, including plants.[15] It belongs to a family of complexes that Smerdon[16] referred to as "'giant' phosphatidylinositol 3-kinase (PI3K)-like protein kinases (PIKKs)." The conserved structure function domains of this family and the one specific for mTOR are shown linearly in Figure 15.1. Aylett *et al.*[17] identified the horn

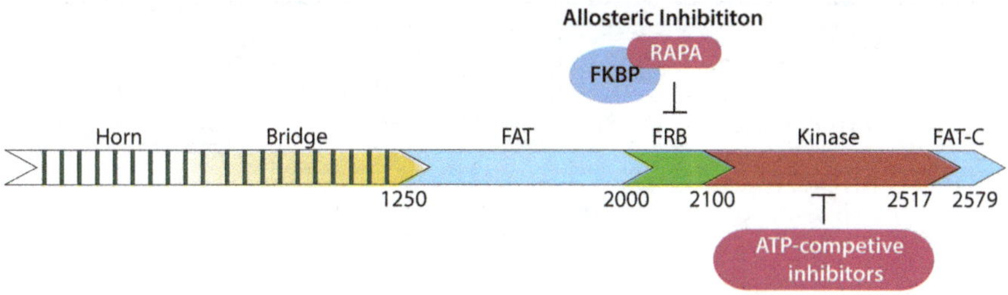

Figure 15.1 Schematic of mTOR structure. Indicated are identified domains. The kinase domain is located near the C-terminus following an N-terminal expanse of helical repeat motifs (HEAT repeats), which are structurally subdivided into a curved solenoid (called the horn) and a straight solenoid (referred to as the bridge).[17] Another tetratricopeptide (TPR)[19] repeat-containing domain named FAT (Frap, ATM and TRRAP). A C-terminal FAT domain (FAT-C) is structurally different from the FAT domain, but was also named after Frap, ATM, TRRAP. The N-terminus of the kinase domain forms the FK506 binding protein (FKBP)–rapamycin binding (FRB) domain, which is necessary for rapamycin allosteric inhibition of mTOR. ATP-competitive inhibitors have been developed that act independently of FRB.

and bridge areas of the heat-repeat-containing N-terminal domain, which is common to other members of the PIKK family as are the FAT, FAT-C and kinase domains.[16] The unusual (and defining) feature of mTOR is the FRB region located on the N-terminus of the kinase domain, with which the FKBP12-rapamycin complex interacts. Evidently, the FRB domain evolved, at least in part, to interact with phosphatidic acid (PA) thereby stabilizing and activating one of the complexes containing mTOR (mTORC1), reviewed by Foster[18]. Rapamycin–FKBP12 competes with PA for mTOR binding. PA also stabilizes the other mTOR complex (mTORC2), which is less sensitive to acute rapamycin-FKBP12–competition.

The two complexes containing mTOR (mTORC1 and mTORC2) each promote diverse cell autonomous and non-cell autonomous functions. Initially mTORC1 represented the major focus with respect to aging with many studies indicating that it is a key modulator of aging and its associated diseases (reviewed by[20]). In replete conditions (including nutrients and growth factor signaling) mTORC1 regulates anabolic pathways for cell mass accumulation. In opposite settings, mTORC1 promotes catabolic processes for survival of cells. In addition to anabolic inputs, shown in Figure 15.2, a wide variety of stresses that cells encounter lead to repression of mTORC1 and to its downstream effectors that function in aging and cancer. Figure 15.2 is highly simplified, and there are excellent reviews that provide detailed discussions of mTORC1 and mTORC2 structure, function and signaling networks.[21–29]

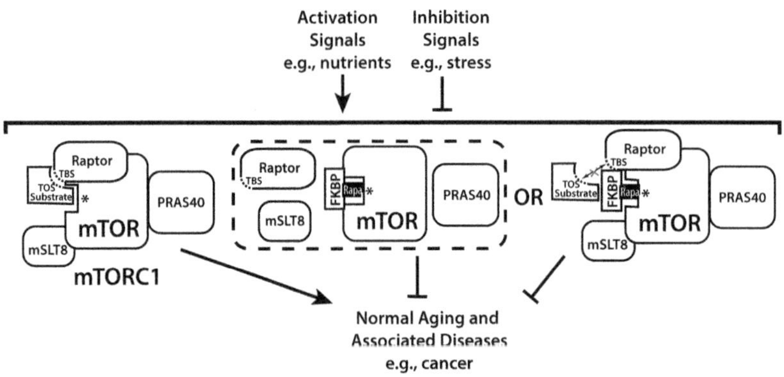

Figure 15.2 Generalized model of rapamycin effects on aging through inhibition of mTORC1. mTORC2 is not considered in this model. mTORC1 responds to various activation signals (nutrients, growth factors, *etc.*) and to inhibitory signals (stresses such as genotoxic, oxidative, *etc.*). Inhibition of mTORC1 by rapamycin inhibits normal aging and associated diseases by currently unknown mechanisms. Two mechanisms currently explain rapamycin inhibitory effects on mTORC1. Yip *et al.*[31] proposed that rapamycin destabilized the complex (middle schematic surrounded by dashed lines) thereby decreasing its activity. Alternately, Aylett *et al.*[17] found no evidence of complex instability, but rather a structural change whereby the TOR signaling motif (TOS) in raptor was displaced thereby limiting access of the kinase active site (asterisk) to TOS-containing substrates such as S6 kinase 1 (right schematic).

Based on depletion of rictor (an mTORC2-specific component), Lamming *et al.*[30] recently proposed that active mTORC1 represses longevity, while a functioning mTORC2 promotes longevity in males.

15.3 Rapamycin's Mysterious Effects on Aging

Over the years leading up to 2004, there were many approaches advertised to promote rejuvenation and longevity. Mainstream researchers were quite skeptical that a pill could be developed that would allay the deleterious effects of aging. A thought would go like this: aging is not a disease that can be treated. Given the safety profile of rapamycin at that time, it is remarkable that the Interventions Testing Program[32] (ITP) accepted the author's proposal in 2005 to test chronic treatment with rapamycin for longevity effects in mice. A key to these experiments proceeding was Randy Strong's development of encapsulation of rapamycin in Eudragit S100 (called eRapa), which accomplished two critical things; it stabilized the drug in food (around 80% is lost to degradation) and released rapamycin when the pH of the gut approached 7 (*i.e.*, the lower part of small intestine and colon), resulting in stable blood levels.[32] The ITP performed the first test of eRapa in an awaiting (and now older) cohort or UM-HET3 mice at each of test centers. Would it kill these old (60 in human years) mice? The results in 2009[33] were a major surprise: eRapa treatment beginning at 20 months of age resulted in extension of maximum life span in both males and females. The ITP has now repeated tests of eRapa in younger mice with the same results[34] and in the latest trial that showed dose- and sex-dependent effects on longevity.[35]

This is now very mysterious: the more that mice consume of this drug, previously thought to suppress immunity and increase risk of cancer, the longer they lived, especially females. It should be noted that eRapa chronically resulted in an extension of median and maximum life span, the latter result indicating that all causes of mortality were prevented, delayed or reduced in severity. Johnson *et al.* explored the limits of dosing by treating mice with nine times the original life extending dose of eRapa, and observed only a decrease in body weight (BW).[36]

15.4 Effects of Chronic Rapamycin on Age-Associated Diseases

These results prompted investigators around the world to test eRapa and other formulations on numerous age-associated diseases. The ITP and other groups investigated the question: does eRapa delay aging-associated traits? Wilkinson *et al.*[37] showed that eRapa beginning at 9 months of age "slows aging in (UM-HET3) mice," although noting a higher incidence of testicular degeneration and cataracts. Zhang *et al.*[38] investigated this question in an inbred strain, C57BL/6, starting the diet at 19 months of age and concluded that eRapa extended life and health span, with no differences noted in testicular degeneration or cataracts in older mice. This group followed up with a study of longer term effects of chronic eRapa treatment (starting at four months of age and continued throughout life), and reported sex-dependent differences in effects that included absence in one sex or a change in the opposite direction.[39] Neff *et al.*[40] also tested eRapa in C57BL/6J mice and observed an extension of life span, but ameliorated only a subset of a wide range of age-associated phenotypes and exhibited two toxicities; testicular degeneration and nephrotoxicity. These authors concluded that chronic rapamycin does not slow aging but rather suppresses cancer.[41] Johnson *et al.*[42] responded to this paper and concluded that the result "supports the model that rapamycin promotes longevity by targeting some, but perhaps not all, core molecular processes that drive cellular and systemic aging." Regarding age-associated diseases, Table 15.1 lists selected papers showing

Table 15.1 Selected papers showing results for various formulations of rapamycin.

Disease or age associated phenotype	Formulation	Ref.
Neurodegenerative		
Attenuation of synaptic injury in a mouse model of synucleinopathy	14 ppm eRapa diet	44
Protects against Parkinson's in a mouse model	0.75 mg ml^{-1}	45
Prevents Parkinsonian dopaminergic neuron loss in *Drosophila*	0.2 or 200 μM	46
Abolishes cognitive deficits in a mouse model of Alzheimer's	14 ppm eRapa diet	47
Improves cognitive function in mice	14 ppm eRapa diet	48
Amelioration of age-dependent cognitive deficits in Alzheimer's mouse model	14 ppm eRapa diet	49
Amelioration of age-dependent cognitive deficits in mice	14 ppm eRapa diet	50
Restores brain vascular integrity and function and improves memory in symptomatic mice modeling Alzheimer's disease	14 ppm eRapa diet	51
Suppression of brain aging in OXYS rats	0.1 or 0.5 mg kg^{-1} BW rapamycin	52
Protection against frontotemploral lobar dementia (TDP-43 proteinopathies)	10 mg kg^{-1} BW	53
Cancer		
Prevention dermal cancer in mice	14 ppm eRapa diet	54
Extend life span in *p53*$^{+/-}$ and *p53*$^{+/+}$ mice	14 or 42 ppm eRapa	55
Prolongs life span of *Rb1*$^{+/-}$ mice	14 ppm eRapa diet	56
Normal life span of *Apc*$^{Min/+}$ mice	14 or 42 ppm eRapa	57
Prolongation of life span in *p53*$^{-/-}$ mice	0.5–4 mg kg^{-1} Rapatar	58
Inhibits growth and progression of prostate cancer	14 ppm eRapa diet	59
Used to treat several types of cancer—an evolving art	Various	60
Metabolic including mitochondrial		
Mitochondrial disease in mice	14–378 ppm eRapa 8 mg kg^{-1} rapamycin	36
Ameliorates age-dependent obesity in aged mice	0.11 μl h^{-1}; 10 mg ml^{-1} rapamycin	61
Prolonged rapamycin treatment led to beneficial metabolic alterations in mice	4 mg kg^{-1} BW rapamycin	62
Insulin resistance by rapamycin uncoupled from longevity	2 mg kg^{-1} BW rapamycin	63

Glucose intolerant but insulin sensitive rapamycin-treated mice	14 ppm eRapa diet	64
Increases mortality in mouse model of type II diabetes	14 ppm eRapa diet	43
Metabolically distinct phenotypes by rapamycin	14 ppm eRapa diet	65
Modulation of mitochondrial biogenesis and fatty acid oxidation in adipose of *db/db* mice	14 ppm eRapa diet	66
Improves survival and biomarkers in obese male mice on high-fat diet	1.5 mg kg^{-1} BW rapamycin	67
Blocks induction of the thermogenic program in white adipose tissue	2 mg kg^{-1} rapamycin 42 ppm eRapa diet	68
Heart and lung		
Attenuates load-induced cardiac hypertrophy in mice	2 mg kg^{-1} day^{-1} rapamycin	69
Regresses established cardiac hypertrophy in mice	2 mg kg^{-1} day^{-1} rapamycin	70
Attenuates cardiac enlargement in zebrafish	0.2–0.4 μmol L^{-1} rapamycin	71
Rescued cardiac and skeletal function in Lamin A/C deficient mice	8 mg kg^{-1} BW	72
Reversed age-related heart dysfunction in late life of mice.	14 ppm eRapa diet	73
Rejuvenation of aging heart in mice	14 ppm eRapa diet	74
Mitochondrial remodeling in old heart	14 or 42 ppm eRapa	75
Remodeling of aged lungs	14 or 42 ppm eRapa	76
Improves cardiac function in type 2 diabetic mice	0.25 mg kg^{-1} rapamycin	77
S6K1 inhibition and rapamycin protects against myocardial infarction	0.17 mg kg^{-1} rapamycin	78
Other effects (including anti-aging)		
Longevity in genetically heterogeneous mice started in late life	14 ppm eRapa diet	33
Longevity in genetically heterogeneous mice started in mid life	14 ppm eRapa diet	34
Longevity in genetically heterogeneous mice for dose response	4.2–42 ppm eRapa diet	35
Longevity of inbred C23682161 57BL6/J male and female mice	14 ppm eRapa diet	38
Longevity in inbred C23682161 57BL6/J male mice	14 ppm eRapa diet	40
Prevention of age-related macular degeneration in mice	0.1 or 0.5 mg kg^{-1} rapamycin	79
Contraindicated in ADPKD[a] and CKD[b] stages 3b–4	3 mg day^{-1} sirolimus	80
Immunoproteasome and age effects	14 ppm eRapa diet	81
Prolongs life in immune-deficient mice	14 ppm eRapa diet	82
Cancer protective effects of mTOR inhibition in kidney transplants	Not identified	83
Attenuation of age-associated changes in tibialis anterior tendon	4.2–44 ppm eRapa	84
Resistance to pneumococcal pneumonia in mice	14 ppm eRapa diet	85
Effects on adult neural progenitor cells in mice	75 μg kg^{-1} or 2.5 mg kg^{-1} BW rapamycin	86
Suppression of age-associated changes in ovarian surface epithelium	14 ppm eRapa diet	87

[a]Autosomal dominant polycystic kidney disease.
[b]Chronic kidney disease.

results for various formulations of rapamycin. While most are anti-aging, results in one are noteworthy for its opposite effect and relevance to a large patient population in the elderly, type-2 diabetes. Chronic eRapa resulted in an increase in mortality in a mouse model (*db/db*) of this disease due to suppurative inflammation.[43] The use of mTOR inhibitors in diseases not associated with aging is increasingly widespread (summarized in ref. 20).

15.5 TOR Reductions and Rapamycin Increase Longevity in Other Organisms

Although mTOR has multiple vital roles in development, accruing evidence suggests its early stage level of activity continued into life's later years may be detrimental to adult somatic tissues/organs. Although the connection between mTOR and nutrient sensing was not known at the time, a hint of this antagonistic pleotropic relationship was evident as early as the 1930s in McCay *et al.*'s well executed survival studies showing that food restriction extended life span in rats.[88] This relationship was also hinted at by extension of life span under conditions of reduced growth hormone/IGF-1,[89] evidence for which was shown by Sharp and Bartke.[90]

In smaller organisms, evidence for this relationship began with reports showing reductions in Sch9 (yeast orthologue of S6K1) associated with chronological life span extension.[91] Similarly, reductions of Sch9 or TOR were associated with extended replicative life span in *Saccharomyces cerevisiae*.[92] A recent resource publication of a comprehensive analysis of 4698 gene deletions in *S. cerevisiae* by McCormick *et al.*[93] revealed 238 with increased life span, including 60S ribosome components, TOR and a tRNA transporter (*LOS1*). Dietary restriction and rapamycin exclude Los1 from the nucleus in a Rad53-dependent manner, suggesting that DNA damage response and mTOR converge on Los1 to regulate aging through Gcn4 activity.[94]

In the nematode, *Caenorhabditis elegans*, knockdown of TOR (*let-363*) or the mTORC1 component raptor (*daf-15*) led to extended longevity.[95,96] Syntichaki *et al.*[97] showed that a somatic tissue-specific loss of eIF4E (a eukaryotic translation initiation factor regulated by mTORC1[98]) reduced global protein synthesis, protected against oxidative stress and extended life span of *C. elegans*. In a similar vein, inhibition of mRNA translation (by inhibition of *ifg-1*, worm homologue of eIF4G in mRNA cap binding complex[98]) extended life span in *C. elegans*.[99] Additionally these authors found the inhibition of *rsks-1* (worm homologue of S6K1 regulated by mTORC1[100]) increased life span. An RNAi screen for longevity genes in *C. elegans* identified 89 genes,[101] among them the eukaryotic translation initiation factor 5 (eIF5[102]). In keeping with reduced protein synthesis associating with extended life span, Essers, *et al.*[103] found that a long non-coding RNA (*tts-1*) associates with reduced levels of ribosomes, which was required for longevity extension by *daf-2* (worm insulin receptor) and *clk-1* (mitochondrial gene) mutations. Lysosomal signaling molecules appear to regulate life span in *C. elegans*,[104] a process that might be regulated by TOR.[105]

In the fruit fly, *Drosophila melanogaster*, modulation of genes in the TOR signaling pathway extended longevity.[106] Zid *et al.*[107] found that 4E-BP, a translation repressor regulated by mTORC1,[98] extended life span by enhancing mitochondrial activity in *Drosophila*. A recent study linked peripheral circadian clocks with dietary restriction-mediated life span in *Drosophila*, potentially through decreased TOR.[108] Rapamycin fed to flies extended life span analogous to TOR mutants, hypothetically through TOR complex 1.[109]

15.6 Genetic mTOR Inhibition in Mice that Extends Life Span

Lamming *et al.*[63] reported that $Mtor^{+/-}$; $Mlst8^{+/-}$ females, but not males, had extended longevity, which is interesting in light of the greater effects of chronic rapamycin in females.[35] In mice heterozygous for *Rictor*, a defining component of mTORC2 ($Rictor^{+/-}$), or those with liver-specific knockout of *Rictor* (L-RKO), males, but not females, had a shortened life span.[30] This led these authors to propose that mTORC2 promotes longevity (at least in males), while mTORC1 is anti-longevity. Since chronic rapamycin has been reported to destabilize mTORC2 in liver, muscle and adipose, concomitant with reduced S437 phosphorylation of Akt, (after refeeding),[63] Lamming *et al.* suggested that this may also explain the greater beneficial effects of rapamycin in female life span.[30] Hasty *et al.* did not observe this effect on Akt phosphorylation under chronically high rapamycin concentrations in the colon.[57] Mice with two hypomorphic ($Mtor^{\Delta/\Delta}$)[110] alleles have an increased life span and tissue-specific slowing of aging.[111] Selman *et al.*[112] showed that the knockout of S6 kinase 1 ($S6k1^{-/-}$, a downstream substrate of mTORC1) extends life span in mice. Thus, the preponderance of evidence, both interventional (food restriction and pharmacological) and genetic studies strongly indicates that the mTOR network plays a key role in health and aging from yeast to mammals. Evidence also clearly indicates that a down regulation of mTORC1 leads to a longer life span. Does this result in a greater length of healthy living? The gold standard intervention for achieving a longer health span is food (or diet) restriction.[113] How does rapamycin compare? Johnson *et al.*[42] argue that chronic rapamycin represents a starting point in the quest for pharmacological interventions that preserve youth similarly to diet restriction. How does it work?

15.7 Composite Picture of mTOR Signaling Pathways in Aging

It is very impressive that we have seen the mTOR signaling go from the simple nutrient sensing pathway first proposed by Barbet *et al.*[6] to the daunting picture of this system today. Figure 15.3 attempts to place our current

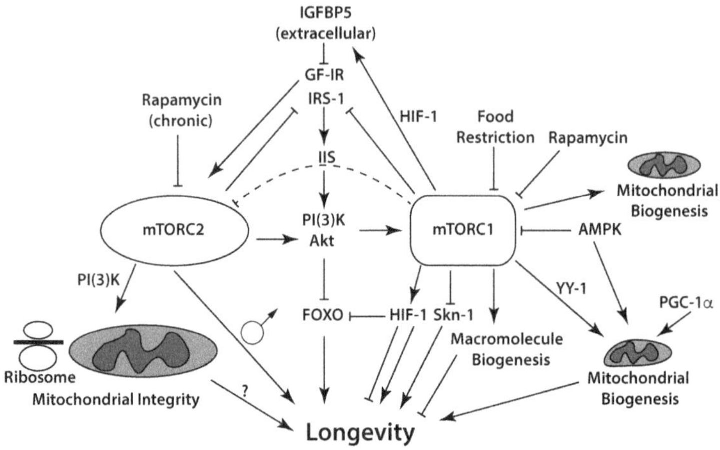

Figure 15.3 mTOR and longevity pathways. The FOXO family of transcription factors mediates the life span extension by reduced insulin/IGF-I (IIS) signaling, which connects to mTORC1. As a negative feedback, IIS-activated mTORC1 decreases IIS signaling through S6K1 and IRS-1. Both an upstream (tuberous sclerosis complex protein 1, TSC1; part of an mTORC1 negative regulatory complex) and a downstream (4E-BP1, a translation repressor) member of the mTORC1 pathway are regulated by FOXO factors. Genetic studies in *C. elegans* partially uncouple these pathways leading to suggestions of overlapping mechanisms involved in IIS- and mTORC1-mediated longevity extension by food restriction and chronic rapamycin.[20] mTORC2 also connects to mTORC1 *via* Akt. mTORC1 can also be inhibitory to mTORC2 (dashed block) *via* S6K1 phosphorylation of two complex components, mSIN1 and Rictor.[114] HIF-1 has a complicated relationship with mTOR. Stimulated by low oxygen levels, mTORC1 activates the hypoxic response by enhancing translation of HIF-1 that inhibits FOXO to increase longevity. To further complicate the issue, HIF-1 is thought to extend longevity at higher temperatures and inhibit it at lower temperatures.[115] As a negative feedback through HIF-1, mTORC1 increases secreted IGF-1 binding protein 5 (IGFBP5), which has the effect of decreasing IGF-1 signaling.[116] Rapamycin inhibition of mTORC1 de-represses Skn-1, which, together with DAF16/FOXO, activates protective genes for longevity extension.[117] mTORC1 also has a complicated relationship with mitochondria biogenesis and respiration. Mitochondria proteins translation is promoted by mTORC1 *via* eIF4E sensitive translation.[118] Food restriction is likely to activate AMPK, which negatively regulates mTORC1 to promote longevity, which promotes mitochondria biogenesis through YY-1 and pGC-1α.[118] Growth factors also promote Akt-dependent mTORC2 activation resulting in ribosome association for mitochondrial integrity, and for co-translational substrate phosphorylation.[118] Finally, mTORC1 regulates the biosynthesis of macromolecules needed for cell growth through the eIF4E sensitive pathway, the S6K1/rpS6 pathway for RNA polymerase II transcription of ribosome subunit mRNAs,[119] the regulation of RNA Polymerases I and III for transcription of ribosomal RNAs (ribosome subunit RNAs) for protein synthesis,[119] lipid biosynthesis, storage and adipogenesis,[120] and pyrimidine[121] and purine[122] synthesis for nucleic acids.

knowledge of this system into an aging context. Since mTORC1 and mTORC2 are central to a myriad of vital cellular and organism function, elucidating the mechanism(s) of action of chronic rapamycin in longevity extension is going to be a difficult task. We still do not know how food restriction affects longevity in spite of decades of intense investigation. Presumably, a reduction in mTORC1 activity (analogous to rapamycin) has a role, although this is far from certain. Miller *et al.*[35] reported distinct changes in food-restricted mice compared to those treated chronically with eRapa. Reduced protein synthesis, autophagy activation, mitochondrial regulation, inflammation reduction, and preservation of stem cells are potential mechanisms (reviewed by Johnson *et al.*[20]) that have been proposed to play a role in longevity extension by mTORC1 inhibition.

Although rapamycin is a curious drug, it is also a wonderful one because of all that we have learned and will continue to learn about it and its cellular target and associated pathways that regulate aging, which, in the opinion of this author, is the hardest problem in biology.

15.8 Why This Is Important

According to the United Nations, the number of people 60 years or older in 2012 was 809 743 000 (one out of nine) (http://www.un.org/en/development/desa/population/publications/ageing/population-ageing-development-2012.shtml). In 2050, that number expands to 2 031 337 000 (one in five). If nothing is done, there is little doubt that this situation will have a huge impact on health care for the elderly and a significant financial burden on societies worldwide. Adult cancer, a disease of aging, exemplifies the fundamental flaw in our current approach to health care. In 2011, Siegel *et al.*[123] projected that there would be the diagnosis of 1 596 670 new cancer cases resulting in 571 950 deaths. Edwards *et al.*[124] also studied how these demographics would affect cancer and ominously reported: (a) the number of cancer patients will double between 2000 and 2050; (b) a dramatic increase in the percentage of elderly from 30% (389 000 in 2000) to 42% (1 102 000 in 2050) in 2050; (c) a four-fold increase in cancer patients 85 years of age; and (d) a doubling of the absolute number of cancers in people 65 or older. Since people over 65 have an age-adjusted cancer mortality rate 15 times greater than young people, the risk of developing and dying of cancer becomes highly significant as the population ages. But here is the big question: if we could by some new miracle intervention prevent and/or cure all cancer, would this have a significant impact on the aging problem? There is a compelling argument that without mitigating other effects of aging, the economic effects would be significantly worsened. Here's why. Bonneux[125] estimated that eliminating all adult cancer would add four years to life, but would raise health care costs 8.3% due to the costs of treating other age-caused diseases (*e.g.*, dementias, sarcopenia, frailty and diseases associated with immune senescence). Another example is cardiovascular diseases, the elimination of which would increase life span by 5.3 year and raise health care costs by 5.2%.[125] In the

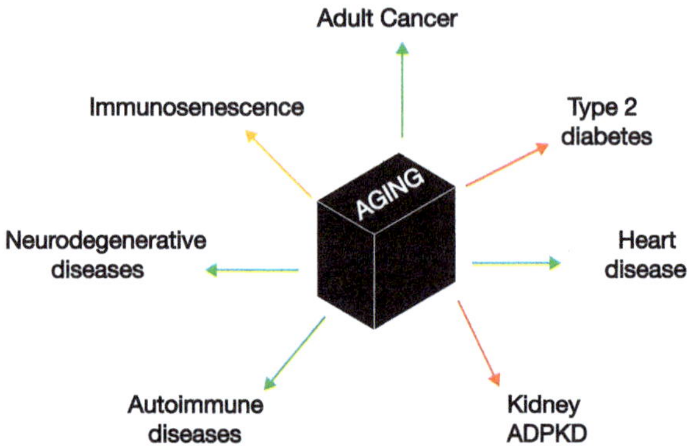

Figure 15.4 Aging, represented by the black box, is one of, if not the, hardest problems in biology. We do know it causes or at least contributes to a wide variety of late adult stage diseases. Rapamycin has variable effects on these diseases. Some it helps (green arrows) and others it hurts (red arrows). It appears to have both good and (not so) bad effects on the immune system (gold arrow) and might be better termed an immune modulator.[82] This indicates that, while rapamycin might be an effective approach for translational gerontology, each patient will need to be evaluated in light of these differential effects.

context of aging, curing single diseases, although desirable and necessary, is a catch-22 approach. As illustrated in Figure 15.4, aging is by far the most significant risk factor for a large number of diseases,[20] including cancer.[126]. An ideal strategy would be to develop interventions that target "aging", which would ideally also reduce the incidence or ameliorate the impact of multiple diseases simultaneously.

Biogerontological research has provided proof-of-principal approaches based on nutrients, genetic and, now, pharmacological studies in animal models strongly indicating that aging can be "treated" with clinical interventions. These studies show that the interventions achieve both age extension and disease delay or alleviation, a response referred to as the "longevity dividend".[127] For example, consider the huge (and still growing) body of studies showing that diet restriction increases *maximum* life span[128] and improves most measures of health.[113,129–131] Maximum extension of life span can only be achieved by reducing all competing causes of mortality.[132] Genetic mutations, such as those that reduce growth factors (*e.g.*, pituitary dwarfism), also result in maximum life span extension in the laboratory (reviewed by Richardson[133]), reduce cancer[134,135] and delay other age-sensitive traits.[136]

15.9 Summary

As elegantly argued by Kaeberlein, Rabinovitch, and Martin,[137] I also contend that the time is ripe for serious consideration of the use of mTOR inhibitors (or other equally effective approaches) as a preventive intervention for the debilitating and costly effects of aging. These approaches are not theoretical any more, but are based on a substantial amount of high-quality science by numerous laboratories around the world. Even small effects on increased age-related disabilities would have an enormous positive effect economically and, importantly, on the overall improvement in the quality of our lives. Will these approaches achieve "morbidity compression" in patients is a huge question that deserves close scrutiny as we enter this new age of "translational geroscience".[137]

Potential Financial Conflict of Interest

Under a licensing agreement between Rapamycin Holdings, Inc. and the University of Texas Health Science Center San Antonio, Z. D. Sharp and the University are entitled to milestone payments and royalty on sales of the rapamycin formulation mentioned in this chapter.

Acknowledgements

Work in the author's laboratories is supported by a grant from the NIH, CA193835-01A1, and pilot funding from the Nathan Shock Center of Excellence in the Biology of Aging.

References

1. C. Vezina, A. Kudelski and S. N. Sehgal, *J. Antibiot.*, 1975, **28**, 721.
2. S. N. Sehgal, *Transplant. Proc.*, 2003, **35**(suppl. 3), 7S.
3. D. Benjamin, M. Colombi, C. Moroni and M. N. Hall, *Nat. Rev. Drug Discovery*, 2011, **10**, 868.
4. C. M. Koebel, W. Vermi, J. B. Swann, N. Zerafa, S. J. Rodig and L. J. Old, *et al.*, *Nature*, 2007, **450**, 903.
5. J. Heitman, N. R. Movva and M. N. Hall, *Science*, 1991, **253**, 905.
6. N. C. Barbet, U. Schneider, S. B. Helliwell, I. Stansfield, M. F. Tuite and M. N. Hall, *Mol. Biol. Cell*, 1996, **7**, 25.
7. M. Kaeberlein, *Nature*, 2010, **464**, 513.
8. Y. Otsubo and M. Yamamato, *Crit. Rev. Biochem. Mol. Biol.*, 2008, **43**, 277.
9. R. W. Powers, M. Kaeberlein, S. D. Caldwell, B. K. Kennedy and S. Fields, *Genes Dev.*, 2006, **20**, 174.
10. E. J. Brown, M. W. Albers, T. B. Shin, K. Ichikawa, C. T. Keith and W. S. Lane, *et al.*, *Nature*, 1994, **369**, 756.
11. M. I. Chiu, H. Katz and V. Berlin, *Proc. Natl. Acad. Sci. U. S. A.*, 1994, **91**, 12574.
12. D. M. Sabatini, H. Erdjument-Bromage, M. Lui, P. Tempst and S. H. Snyder, *Cell*, 1994, **78**, 35.
13. C. J. Sabers, M. M. Martin, G. J. Brunn, J. M. Williams, F. J. Dumont and G. Wiederrecht, *et al.*, *J. Biol. Chem.*, 1995, **270**, 815.

14. M. N. Hall, *Biochem. Soc. Trans.*, 2013, **41**, 887.
15. F. John, S. Roffler, T. Wicker and C. Ringli, *Plant Signaling Behav.*, 2011, **6**, 1700.
16. S. J. Smerdon, *Sci. Signaling*, 2014, **7**, pe6.
17. C. H. S. Aylett, E. Sauer, S. Imseng, D. Boehringer, M. N. Hall and N. Ban, *et al.*, *Science*, 2016, **351**, 48.
18. D. A. Foster, *Trends Endocrinol. Metab.*, 2013, **24**, 272.
19. B. A. Knutson, *J. Struct. Biol.*, 2010, **170**, 354.
20. S. C. Johnson, P. S. Rabinovitch and M. Kaeberlein, *Nature*, 2013, **493**, 338.
21. M. Laplante and D. M. Sabatini, *Cell*, 2012, **149**, 274.
22. J. L. Jewell and K. L. Guan, *Trends Biochem. Sci.*, 2013, **38**, 233.
23. C. C. Dibble and B. D. Manning, *Nat. Cell Biol.*, 2013, **15**, 555.
24. M. Cornu, V. Albert and M. N. Hall, *Curr. Opin. Genet. Dev.*, 2013, **23**, 53.
25. L. Bar-Peled and D. M. Sabatini, *Trends Cell Biol.*, 2014, **24**, 400.
26. C. Betz and M. N. Hall, *J. Cell Biol.*, 2013, **203**, 563.
27. D. Rexin, C. Meyer, C. Robaglia and B. Veit, *Biochem. J.*, 2015, **470**, 1.
28. J. D. Ochocki and M. C. Simon, *J. Cell Biol.*, 2013, **203**, 23.
29. K. Huang and D. C. Fingar, *Semin. Cell Dev. Biol.*, 2014, **36**, 79.
30. D. W. Lamming, M. M. Mihaylova, P. Katajisto, E. L. Baar, O. H. Yilmaz and A. Hutchins, *et al.*, *Aging Cell*, 2014, **13**, 911.
31. C. K. Yip, K. Murata, T. Walz, D. M. Sabatini and S. A. Kang, *Mol. Cell*, 2010, **38**, 768.
32. N. L. Nadon, R. Strong, R. Miller, J. Nelson, M. Javors and Z. D. Sharp, *et al.*, *Age*, 2008, **30**, 187.
33. D. E. Harrison, R. Strong, Z. D. Sharp, J. F. Nelson, C. M. Astle and K. Flurkey, *et al.*, *Nature*, 2009, **460**, 392.
34. R. A. Miller, D. E. Harrison, C. M. Astle, J. A. Baur, A. R. Boyd and R. De Cabo, *et al.*, *J. Gerontol., Ser. A*, 2011, **66**, 191.
35. R. A. Miller, D. E. Harrison, C. M. Astle, E. Fernandez, K. Flurkey and M. Han, *et al.*, *Aging Cell*, 2014, **13**, 468.
36. S. C. Johnson, M. E. Yanos, A. Bitto, A. Castanza, A. Gagnidze and B. Gonzalez, *et al.*, *Front. Genet.*, 2015, **6**, 247.
37. J. E. Wilkinson, L. Burmeister, S. V. Brooks, C. C. Chan, S. Friedline and D. E. Harrison, *et al.*, *Aging Cell*, 2012, **11**, 675.
38. Y. Zhang, A. Bokov, J. Gelfond, V. Soto, Y. Ikeno and G. Hubbard, *et al.*, *J. Gerontol., Ser. A*, 2014, **69**, 119.
39. K. E. Fischer, J. A. L. Gelfond, V. Y. Soto, C. Han, S. Someya and A. Richardson, *et al.*, *PLoS One*, 2015, **10**, e0126644.
40. F. Neff, D. Flores-dominguez, D. P. Ryan, M. Horsch, S. Schroder and T. Adler, *et al.*, *J. Clin. Invest.*, 2013, **123**, 1.
41. D. Ehninger, F. Neff, K. Xie, *Cell. Mol. Life Sci.*, 2014, **71**, 4325.
42. S. C. Johnson, G. M. Martin, P. S. Rabinovitch and M. Kaeberlein, *Sci. Transl. Med.*, 2013, **5**, 211fs40.
43. K. Sataranatarajan, Y. Ikeno, A. Bokov, D. Feliers, H. Yalamanchili and H. J. Lee, *et al.*, *J. Gerontol., Ser. A*, DOI: 10.1093/gerona/glv170, PMID: 26442901.

44. X. Bai, M. C.-Y. Wey, E. Fernandez, M. J. Hart, J. Gelfond and A. F. Bokov, *et al.*, *Pathobiol. Aging Age-Relat. Dis.*, 2015, **5**, 28743.
45. C. Malagelada, Z. H. Jin, V. Jackson-Lewis, S. Przedborski and L. A. Greene, *J. Neurosci.*, 2010, **30**, 1166.
46. L. S. Tain, H. Mortiboys, R. N. Tao, E. Ziviani, O. Bandmann and A. J. Whitworth, *Nat. Neurosci.*, 2009, **12**, 1129.
47. P. Spilman, N. Podlutskaya, M. J. Hart, J. Debnath, O. Gorostiza and D. Bredesen, *et al.*, *PLoS One*, 2010, **5**, 1.
48. J. Halloran, S. A. Hussong, R. Burbank, N. Podlutskaya, K. E. Fischer and L. B. Sloane, *et al.*, *Neuroscience*, 2012, **223**, 102.
49. A. Caccamo, S. Majumder, A. Richardson, R. Strong and S. Oddo, *J. Biol. Chem.*, 2010, **285**, 13107.
50. S. Majumder, A. Caccamo, D. X. Medina, A. D. Benavides, M. A. Javors and E. Kraig, *et al.*, *Aging Cell*, 2012, **11**, 326.
51. A.-L. Lin, W. Zheng, J. J. Halloran, R. R. Burbank, S. A. Hussong and M. J. Hart, *et al.*, *J. Cereb. Blood Flow Metab.*, 2013, **33**, 1412.
52. N. G. Kolosova, A. O. Vitovtov, N. A. Muraleva, A. E. Akulov, N. A. Stefanova and M. V. Blagosklonny, *Aging*, 2013, **5**, 474.
53. I.-F. Wang, B.-S. Guo, Y.-C. Liu, C.-C. Wu, C.-H. Yang and K.-J. Tsai, *et al.*, *Proc. Natl. Acad. Sci. U. S. A.*, 2012, **109**, 15024.
54. V. Dao, S. Pandeswara, Y. Liu, V. Hurez, S. Dodds and D. Callaway, *et al.*, *Cancer Prev. Res. (Phila)*, 2015, **8**, 400.
55. B. Christy, M. Demaria, J. Campisi, J. Huang, D. Jones and S. G. Dodds, *et al.*, *Oncotarget*, 2015, **6**, 15802.
56. C. B. Livi, R. L. Hardman, B. Christy, S. G. Dodds, D. Jones and C. Williams, *et al.*, *Aging (Albany NY)*, 2013, **5**, 100.
57. P. Hasty, C. B. Livi, S. G. Dodds, D. Jones, R. Strong and M. Javors, *et al.*, *Cancer Prev. Res.*, 2014, **7**, 169.
58. M. Comas, I. Toshkov, K. K. Kuropatwinski, O. B. Chernova, A. Polinsky and M. V. Blagosklonny, *et al.*, *Aging (Albany NY)*, 2012, **4**, 715.
59. A. Saha, J. Blando, L. Tremmel and J. DiGiovanni, *Cancer Prev. Res. (Phila)*, 2015, **8**, 597.
60. F. Chiarini, C. Evangelisti, J. McCubrey and A. M. Martelli, *Trends Pharmacol. Sci.*, 2015, **36**, 124.
61. S. B. Yang, A. C. Tien, G. Boddupalli, A. W. Xu, Y. N. Jan and L. Y. Jan, *Neuron*, 2012, **75**, 425.
62. Y. Fang, R. Westbrook, C. Hill, R. K. Boparai, O. Arum and A. Spong, *et al.*, *Cell Metab.*, 2013, **17**, 456.
63. D. W. Lamming, L. Ye, P. Katajisto, M. D. Goncalves, M. Saitoh and D. M. Stevens, *et al.*, *Science*, 2012, **335**, 1638.
64. D. W. Lamming, L. Ye, C. M. Astle, J. A. Baur, D. M. Sabatini and D. E. Harrison, *Aging Cell*, 2013, **12**, 712.
65. Z. Yu, R. Wang, W. C. Fok, A. Coles, A. B. Salmon and V. I. Perez, *J. Gerontol., Ser. A*, 2015, **70**, 410.
66. S. S. Deepa, M. E. Walsh, R. T. Hamilton, D. Pulliam, Y. Shi and S. Hill, *et al.*, *J. Biochem. Pharmacol. Res.*, 2013, **1**, 114.

67. O. V. Leontieva, G. M. Paszkiewicz and M. V. Blagosklonny, *Aging Cell*, 2014, **13**, 616.
68. C. M. Tran, S. Mukherjee, L. Ye, D. W. Frederick, M. Kissig and J. G. Davis, *et al.*, *Diabetes*, 2016, **65**, 927.
69. T. Shioi, J. R. McMullen, O. Tarnavski, K. Converso, M. C. Sherwood and W. J. Manning, *et al.*, *Circulation*, 2003, **107**, 1664.
70. J. R. McMullen, M. C. Sherwood, O. Tarnavski, L. Zhang, A. L. Dorfman and T. Shioi, *et al.*, *Circulation*, 2004, **109**, 3050.
71. Y. Ding, X. Sun, W. Huang, T. Hoage, M. Redfield and S. Kushwaha, *et al.*, *Circ. Res.*, 2011, **109**, 658.
72. F. J. Ramos, S. C. Chen, M. G. Garelick, D.-F. Dai, C.-Y. Liao and K. H. Schreiber, *et al.*, *Sci. Transl. Med.*, 2012, **4**, 144ra103.
73. J. M. Flynn, M. N. O'Leary, C. Zambataro, E. C. Academia, M. P. Presley and B. J. Garrett, *et al.*, *Aging Cell*, 2013, **12**, 851.
74. D. F. Dai, P. P. Karunadharma, Y. A. Chiao, N. Basisty, D. Crispin and E. J. Hsieh, *et al.*, *Aging Cell*, 2014, **13**, 529.
75. Y. A. Chiao, S. C. Kolwicz, N. Basisty, A. Gagnidze, J. Zhang and H. Gu, *et al.*, *Aging (Albany NY)*, 2016, **8**, 314.
76. C. Calhoun, P. Shivshankar, M. Saker, L. B. Sloane, C. B. Livi and Z. D. Sharp, *et al.*, *J. Gerontol., Ser. A*, 2016, **71**, 153.
77. A. Das, D. Durrant, S. Koka, F. N. Salloum, L. Xi and R. C. Kukreja, *J. Biol. Chem.*, 2014, **289**, 4145.
78. R. Di, X. Wu, Z. Chang, X. Zhao, Q. Feng and S. Lu, *et al.*, *Biochem. J.*, 2011, **441**, 199.
79. N. G. Kolosova, N. A. Muraleva, A. A. Zhdankina, N. A. Stefanova, A. Z. Fursova and M. V. Blagosklonny, *Am. J. Pathol.*, 2012, **181**, 472.
80. P. Ruggenenti, G. Gentile, N. Perico, A. Perna, L. Barcella and M. Trillini, *et al.*, *Clin. J. Am. Soc. Nephrol.*, 2016, **11**, 785.
81. A. M. Pickering, M. Lehr and R. A. Miller, *J. Clin. Invest.*, 2015, **125**, 2059.
82. V. Hurez, V. Dao, A. Liu, S. Pandeswara, J. Gelfond and L. Sun, *et al.*, *Aging Cell*, 2015, **14**, 945.
83. P. Piselli, D. Serraino, G. P. Segoloni, S. Sandrini, G. B. Piredda and M. P. Scolari, *et al.*, *Eur. J. Cancer*, 2013, **49**, 336.
84. L. W. Zaseck, R. A. Miller and S. V. Brooks, *J. Gerontol., Ser. A*, 2016, **71**, 858–865.
85. C. A. Hinojosa, V. Mgbemena, S. Van Roekel, S. N. Austad, R. A. Miller and S. Bose, *et al.*, *Exp. Gerontol.*, 2012, **47**, 958.
86. Y. Kusne, E. L. Goldberg, S. S. Parker, S. M. Hapak, I. Y. Maskaykina and W. M. Chew, *et al.*, *Age*, 2014, **36**, 199.
87. P. Bajwa, P. B. Nagendra, S. Nielsen, S. S. Sahoo, A. Bielanowicz and J. M. Lombard, *et al.*, *Oncotarget*, 2016, 7, 19214–19227.
88. C. M. McCay, M. F. Crowell and L. A. Maynard, *J. Nutr.*, 1935, **10**, 63.
89. H. M. Brown-Borg, K. E. Borg, C. J. Meliska and A. Bartke, *Nature*, 1996, **384**, 33.
90. Z. D. Sharp and A. Bartke, *J. Gerontol., Ser. A*, 2005, **60**, 293.

91. P. Fabrizio, F. Pozza, S. D. Pletcher, C. M. Gendron and V. D. Longo, *Science*, 2001, **292**, 288.
92. M. Kaeberlein, R. W. Powers, K. K. Steffen, E. A. Westman, D. Hu and N. Dang, *et al.*, *Science*, 2005, **310**, 1193.
93. M. A. McCormick, J. R. Delaney, M. Tsuchiya, S. Tsuchiyama, A. Shemorry and S. Sim, *et al.*, *Cell Metab.*, 2015, **22**, 895.
94. A. G. Hinnebusch, *Annu. Rev. Microbiol.*, 2005, **59**, 407.
95. K. Jia, D. Chen and D. L. Riddle, *Development*, 2004, **131**, 3897.
96. T. Vellai, K. Takacs-Vellai, Y. Zhang, A. L. Kovacs, L. Orosz and F. Muller, *Nature*, 2003, **426**, 620.
97. P. Syntichaki, K. Troulinaki and N. Tavernarakis, *Nature*, 2007, **445**, 922.
98. J. Pelletier, J. Graff, D. Ruggero and N. Sonenberg, *Cancer Res.*, 2015, **75**, 250.
99. K. Z. Pan, J. E. Palter, A. N. Rogers, A. Olsen, D. Chen and G. J. Lithgow, *et al.*, *Aging Cell*, 2007, **6**, 111.
100. M. R. Tavares, I. C. B. Pavan, C. L. Amaral, L. Meneguello, A. D. Luchessi and F. M. Simabuco, *Life sci*, 2015, **131**, 1.
101. B. Hamilton, Y. Dong, M. Shindo, W. Liu, I. Odell and G. Ruvkun, *et al.*, *Genes Dev.*, 2005, **19**, 1544.
102. N. Sonenberg and A. G. Hinnebusch, *Cell*, 2009, **136**, 731.
103. P. B. Essers, J. Nonnekens, A. B. Brenkman, A. W. Macinnes, P. B. Essers and J. Nonnekens, *et al.*, *Cell Rep.*, 2015, **10**, 339.
104. A. Folick, H. D. Oakley, Y. Yu, E. H. Armstrong, M. Kumari and L. Sanor, *et al.*, *Science*, 2015, **347**, 83.
105. B. S. Han and A. Brunet, *Science*, 2015, **347**, 32.
106. P. Kapahi, B. M. Zid, T. Harper, D. Koslover, V. Sapin and S. Benzer, *Curr. Biol.*, 2004, **14**, 885.
107. B. M. Zid, A. N. Rogers, S. D. Katewa, M. A. Vargas, M. C. Kolipinski and T. A. Lu, *et al.*, *Cell*, 2009, **139**, 149.
108. S. D. Katewa, K. Akagi, N. Bose, K. Rakshit, T. Camarella and X. Zheng, *et al.*, *Cell Metab.*, 2015, **23**, 143.
109. I. Bjedov, J. M. Toivonen, F. Kerr, C. Slack, J. Jacobson and A. Foley, *et al.*, *Cell Metab.*, 2010, **11**, 35.
110. S. Zhang, J. A. Readinger, W. DuBois, M. Janka-Junttila, R. Robinson and M. Pruitt, *et al.*, *Blood*, 2011, **117**, 1228.
111. J. J. Wu, J. Liu, E. Chen, J. Wang, L. Cao and N. Narayan, *et al.*, *Cell Rep.*, 2013, **4**, 913.
112. C. Selman, J. M. A. Tullet, D. Wieser, E. Irvine, S. J. Lingard and A. I. Choudhury, *et al.*, *Science*, 2009, **326**, 140.
113. L. Fontana, L. Partridge and V. D. Longo, *Science*, 2010, **328**, 321.
114. K. Masui, W. K. Cavenee and P. S. Mischel, *Trends Endocrinol. Metab.*, 2014, **25**, 364.
115. S. F. Leiser, A. Begun and M. Kaeberlein, *Aging Cell*, 2011, **10**, 318.
116. M. Ding, R. K. Bruick and Y. Yu, *Nat. Cell Biol.*, 2016, **18**, 319.
117. S. Robida-Stubbs, K. Glover-Cutter, D. W. Lamming, M. Mizunuma, S. D. Narasimhan and E. Neumann-Haefelin, *et al.*, *Cell Metab.*, 2012, **15**, 713.

118. V. Albert and M. N. Hall, *Curr. Opin. Cell Biol.*, 2015, **33**, 55.
119. C. Chauvin, V. Koka, A. Nouschi, V. Mieulet, C. Hoareau-Aveilla and A. Dreazen, *et al.*, *Oncogene*, 2014, **33**, 474.
120. S. J. H. Ricoult and B. D. Manning, *EMBO Rep.*, 2013, **14**(3), 242–251.
121. I. Ben-Sahra, J. J. Howell, J. M. Asara and B. D. Manning, *Science*, 2013, **339**, 1323.
122. I. Ben-Sahra, G. Hoxhaj, S. J. H. Ricoult, J. M. Asara and B. D. Manning, *Science*, 2016, **351**, 728.
123. R. Siegel, E. Ward, O. Brawley and A. Jemal, *Cancer J. Clin.*, 2011, **61**, 212.
124. B. K. Edwards, H. L. Howe, L. A. Ries, M. J. Thun, H. M. Rosenberg and R. Yancik, *et al.*, *Cancer*, 2002, **94**, 2766.
125. L. Bonneux, J. J. Barendregt, W. J. Nusselder and P. J. der Maas, *BMJ*, 1998, **316**, 26.
126. D. A. Juckett, *Age*, 2010, **32**, 155.
127. R. N. Butler, R. A. Miller, D. Perry, B. A. Carnes, T. F. Williams and C. Cassel, *et al.*, *BMJ*, 2008, **337**, a399.
128. R. Weindruch and R. S. Sohal, *N. Engl. J. Med.*, 1997, **337**, 986.
129. J. A. Mattison, G. S. Roth, T. M. Beasley, E. M. Tilmont, A. M. Handy and R. L. Herbert, *et al.*, *Nature*, 2012, **489**, 318.
130. C. López-Otín, M. A. Blasco, L. Partridge, M. Serrano and G. Kroemer, *Cell*, 2013, **153**, 1194.
131. R. J. Colman, R. M. Anderson, S. C. Johnson, E. K. Kastman, K. J. Kosmatka and T. M. Beasley, *et al.*, *Science*, 2009, **325**, 201.
132. K. Flurkey, J. M. Currer and D. E. Harrison, in *The Mouse In Biomedical Research*, ed. J. Fox, 2007, pp. 639–672.
133. A. Richardson, A. R. Heydari, W. W. Morgan, J. F. Nelson, Z. D. Sharp and C. A. Walter, *ILAR J.*, 1997, **38**, 125.
134. Y. Ikeno, R. T. Bronson, G. B. Hubbard, S. Lee and A. Bartke, *J. Gerontol., Ser. A*, 2003, **58**, 291.
135. Y. Ikeno, G. B. Hubbard, S. Lee, L. A. Cortez, C. M. Lew and C. R. Webb, *et al.*, *J. Gerontol., Ser. A*, 2009, **64**, 522.
136. A. Bartke, *Endocrinology*, 2005, **146**, 3718.
137. M. Kaeberlein, P. S. Rabinovitch and G. M. Martin, *Science*, 2015, **350**, 1191.

CHAPTER 16

Anti-Aging Action of PPARs: Potential Therapeutic Targets

KI WUNG CHUNG[a], BYUNG PAL YU[b] AND HAE YOUNG CHUNG*[a]

[a]Molecular Inflammation Research Center for Aging Intervention (MRCA), College of Pharmacy, Pusan National University, Busan, Republic of Korea; [b]Department of Physiology, The University of Texas Health Science Center at San Antonio, San Antonio, TX 78229-3900, USA
*E-mail: hyjung@pusan.ac.kr

16.1 Introduction

Aging is a biological process characterized by a time-dependent loss of physiological integrity, accompanied with increased risk for various diseases, including cancer, dementia, cardiovascular diseases, and other disorders.[1] Although extensive aging research brought great advances over recent years in healthy longevity, the underlying major causes of aging are not well defined. Because there are multiple causes and effects of aging, many theories have been forwarded.[2] According to a recent review article, nine candidate hallmarks are considered to contribute to the aging process.[3] These nine hallmarks of aging includes genomic instability, telomere attrition, epigenetic modification, loss of proteostasis, changes in nutrient sensing signaling, mitochondrial dysfunctions, cellular senescence, stem cell exhaustion, and altered intercellular communications.

RSC Drug Discovery Series No. 57
Anti-aging Drugs: From Basic Research to Clinical Practice
Edited by Alexander M. Vaiserman

Published by the Royal Society of Chemistry, www.rsc.org

Among many potential culprits of aging, changes in metabolic pathways and age-related chronic diseases due to low-grade inflammation seem to be a common occurrence.[4–7] A good example is insulin resistance (IR), which represents a major metabolic problem that is commonly observed in aged people.[8] Along with IR, major metabolic impairments including obesity, type 2 diabetes, fatty liver diseases, and atherosclerosis also increase with aging.[6,9,10] Although metabolic syndromes are associated with many biologic processes, including genetics and epigenetics, it has become clear that inflammation is a key feature.[11] The term "meta-inflammation" describes the importance of inflammation on the onset and progression of metabolic syndrome.[12,13] Collectively, these age-associated alterations in metabolism and inflammation are intricately involved and connected. The identification of signal pathways that control age-related metabolism dysfunction and dysregulate inflammation is therefore crucial for a better understanding of the factors involved in regulation of the aging process.

In this chapter, peroxisome proliferator-activated receptors (PPARs) will be highlighted as important transcriptional factors with substantial potential in the regulation of aging process. First, changes in inflammation and metabolism during aging will be described, especially focusing on their mutual relationship. Then functions of PPARs will be briefly reviewed in the context of metabolism and inflammation. Thus, the involvement of PPARs in aging and age-related diseases will be discussed based on the roles they play in metabolism and inflammation according to recent evidence. Furthermore, newly synthesized PPAR agonists will be suggested as anti-aging drugs with therapeutic potential.

16.2 Age-Related Changes in Inflammation and Their Role in Metabolic Diseases

16.2.1 Chronic Inflammation and Aging

Current aging research focuses on chronic inflammation as a potent casual mediator underlying the process of aging. This age-related, low grade inflammation is different from classical views on inflammation as traditional inflammation was defined as a part of the body's complex biological response to harmful stimuli, such as pathogens, damaged cells, or other irritants that may induce the acute phase response. The initial recognition of harmful stimuli in the body is mediated by receptors of the innate immune system, such as Toll-like receptor (TLRs) and NOD-like receptors. The most powerful players in this process are tissue resident macrophages, mast cells, and neutrophils.[14] They effectively cope with the initial injury or infection by production of various inflammatory mediators, including cytokines, chemokines, eicosanoids, and other physiologically active molecules.[14] A successful acute inflammatory response leads to elimination of the cause of inflammation by a resolution process. In this context, inflammation is a protective process that maintains the homeostasis of individuals.

In contrast, unresolved low-grade inflammation is different from acute inflammation in several aspects. When acute inflammation is not resolved, the inflammatory process persists and acquires new characteristics. Infiltration of macrophages and T cells replaces the neutrophils in the acute phase inflammation.[15] If these secondary immune cells fail to eliminate the cause of inflammation, a chronic inflammatory state continues with the formation of massive lymphoid infiltrate-like structures, such as granulomas. Although the general processes and mechanisms of chronic inflammation are not fully understood, its consequences are generally associated with many pathological conditions including autoimmunity problems, inflammatory tissue damage, fibrosis, metaplasia, and other tissue degenerative diseases.[16]

Recent studies have revealed the importance of inflammation as a major risk factor of aging. Our lab has proposed the molecular inflammation hypothesis and presented evidence supporting that the inflammatory process may play a major role in the aging process.[17] Accumulated data strongly suggest that continuous up-regulation of pro-inflammatory mediators is induced during the aging process.[18,19] These increases of inflammatory mediators and accumulation of pro-inflammatory tissue damages may result from multiple reasons, such as enhanced activation of the NF-κB transcriptional factor, the failure of the immune system to effectively clear pathogens or dysfunctional host cells (immunosenescence), the propensity of senescent cells to secrete pro-inflammatory cytokines, and other homeostatic unbalances.[7,20–23] Furthermore, these chronic inflammatory conditions are significantly associated with increased mortality and morbidity in elderly people.[24] Thus, importantly, the molecular inflammation hypothesis of aging may serve not only as a molecular basis for a link between normal aging and age-related pathological processes, but also aid in the identification of pathways that control age-related inflammation.

16.2.2 Roles of Inflammation in Metabolic Diseases During Aging

Aging is undoubtedly the most potent contributor to the etiologies of metabolic diseases.[9] Especially in industrialized and westernized society, it is easy to acquire various metabolic diseases from the current life style, like overnutrition and lack of exercise. These metabolic diseases include type 2 diabetes, cardiovascular diseases, and stroke. Along with these metabolic syndromes (MS), IR represents a major component of aging.[11] Persistence of these metabolic alterations leads to impairments of metabolic organs, including de-regulated hepatic gluconeogenesis, adipose lipogenesis, and defective glucose uptake and glycogen synthesis in skeletal muscle. These age-related alterations were once thought to be passive players in the aging process, now they are considered active participants in a vicious cycle that can accelerate the aging process.[10]

Changes in body composition account for the vast majority of declines in metabolic function. Increased visceral fat, which is commonly named as abdominal obesity, is a major contributor to IR and MS.[25] As it turns out, adipose tissue participates in many biological processes.[26] For example, adipose tissue is now recognized as an immune organ that secretes various adipokines and pro-inflammatory cytokines that contribute to the pathogenesis of IR and age-associated chronic diseases.[27] In addition, a decline in endocrine function also contributes to age-related metabolic dysregulation.

Age-related changes in growth hormone (GH), insulin-like growth factor-1 (IGF-1), and sex steroids are known to be linked to the aging process. GH/IGF-1 signaling pathways are particularly of interest because they elicit the same effect as insulin. GH/IGF-1 secretion markedly decreases with aging, and decreased levels of IGF-1 are associated with an increased risk of metabolic disorders.[28] Since nutrient sensing signaling is deregulated with aging, age-related decline of GH/IGF-1 signaling should also be considered important. Other changes including changes in energy sensing systems (mTOR, AMPK, and sirtuins), mitochondria dysfunction, and epigenetic modification also cause metabolic problems during aging.[9,10]

One recent study showed the importance of hyperthalamic inflammation (especially involved in NF-κB signaling) in aging.[23] Because study results show that the hypothalamus plays a pivotal role in the regulation of whole-body metabolism including appetite, glucose metabolism and lipid metabolism *via* endocrine system regulation, these findings suggest that uncontrolled inflammation in the hypothalamus may modulate systemic aging and metabolic disorders. Moreover, abnormal increases in several metabolites accompanying metabolic syndrome can directly influence inflammation. Among the numerous metabolites increased in metabolic diseases, saturated fatty acids (SFAs) and ceramides are potent inflammation inducers.[29] Collectively, these age-related alterations in inflammation and metabolism are active participants in a vicious cycle that can accelerate the aging process and onset of metabolic diseases.

16.3 Functions of PPARs in the Regulation of Metabolism and Inflammation

16.3.1 PPAR Signaling and Metabolism

PPARs are ligand-activated transcriptional factors belonging to the nuclear receptor superfamily.[30] PPARs are classified into three types: PPARα, PPARβ/δ, and PPARγ. By binding to PPAR-responsive regulatory elements (PPRE) with heterodimeric formation with retinoid X receptor (RXR), PPARs control the expression of networks of genes involved in a broad spectrum of biological processes including metabolism, inflammation, cellular proliferation, and tissue remodeling.[31] Although PPARs regulate various cellular processes, the most important role of PPARs is regulation of energy metabolism. Thus, PPARs agonists/antagonists are proposed to be promising therapeutic targets for the treatment of various metabolic diseases (Figure 16.1).[30]

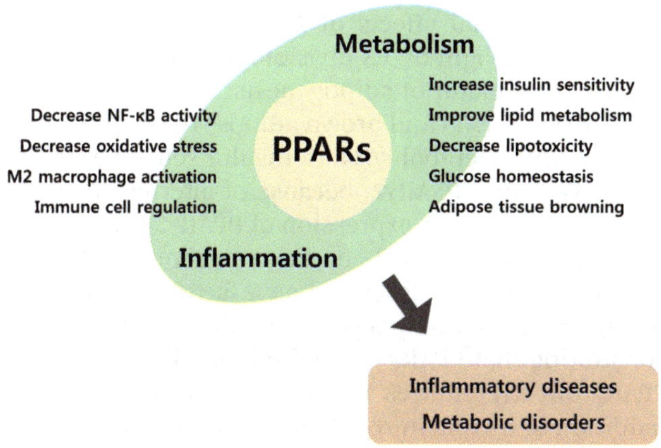

Figure 16.1 Effects of PPARs activation on metabolism and inflammation. Activation of PPARs results in beneficial effects on metabolism and inflammation. PPAR activation improves metabolic parameters leading to decreased metabolic syndrome severity. In addition, activation of PPARs also regulates inflammation and immune systems, which can suppress inflammatory diseases.

PPARα is highly expressed in liver, kidney, brown adipose tissue, and to a lesser extent, in muscle, heart, intestines, and cells from the vascular walls.[32] PPARα can be activated by certain endogenous and synthetic ligands, such as PUFAs, eicosanoids, and fibrates.[31,33] When ligands bind to PPARα, a conformational change of PPARα induces its interaction with RXR, which facilitates nuclear translocation. The PPARα-RXR heterodimer binds with coactivator proteins and recruits RNA polymerase for transcription of specific genes.[34] The major function of PPARα is to promote fatty acid utilization in the liver.[32] Major genes regulated by PPARα include genes involved in mitochondrial and peroxisomal β-oxidation, fatty acid uptake and binding, lipoprotein assembly and transport. PPARα null mice display a fatty liver-like phenotype resulting from shortage of fatty acid utilization.[35,36] In conditions such as starvation or fasting that require fatty acids as energy source, PPARα plays a pivotal role for the up-regulation of gene expression necessary to fulfill the energy needs in these situations.[36] PPARα also regulates ketogenesis in the liver, which is critical to the adaptive response to fasting.[37] Furthermore, the induction of the pro-longevity hormone, fibroblast growth factor 21 (FGF21), by PPARα during fasting appears to play an important role in regulating whole body lipid metabolism.[38,39]

PPARβ/δ shares similar functions with PPARα.[40] PPARβ/δ plays a key role in lipid and cholesterol metabolism. It is ubiquitously expressed and has a crucial role in fatty acid oxidation in key metabolic tissues such as muscle, liver, adipose tissue, and heart. Recent evidence strongly suggests that PPARβ/δ is an integral component in a transcriptional network that regulates brown fat metabolism.[41] Through regulation of PGC1α and UCP-1, PPARβ/δ showed its powerful fat burning activity in adipose tissues.[41] Although the regulatory mechanisms and beneficial effects of PPARβ/δ are not fully understood, recent studies have shed light on its physiological roles.[31]

PPARγ is a master regulator of adipogenesis.[42] It is most highly expressed in white adipose tissue (WAT) and brown adipose tissue (BAT), where it modulates whole-body lipid metabolism and insulin sensitivity. PPARγ exists as two isoforms, PPARγ1 and PPARγ2, because of alternative splicing and differential promoter usage.[43] The expression of PPARγ2 is restricted to adipose tissue under normal physiological condition, whereas PPARγ1 is expressed widely. Originally, PPARγ was described as a factor that permits differentiation of preadipocytes into adipocytes.[44] PPARγ null mice lacked adipose tissue, demonstrating that PPARγ is required for adipocyte differentiation.[45] Likewise, PPARγ directly induces many genes involved in adipocyte lipid storage. In addition to its role in regulation of adipose tissue differentiation and lipid metabolism, PPARγ also regulates glucose homeostasis.[46] PPARγ directly regulates the expression of glucose transporter type 4 (Glut4) and c-Cbl-associated protein (CAP), which is important for glucose homeostasis. PPARγ also controls the expression of numerous factors secreted from adipose tissue that influence insulin sensitivity. Therefore, PPARγ can play an important role in regulating metabolism and may be involved in achieving the insulin sensitizing effect.

16.3.2 PPARs and Inflammation

Although the main role of PPARs is regulating metabolic homeostasis in various pathophysiologic conditions, recent evidence has revealed a new role of PPARs as regulators of inflammation (Table 16.1).[47–49] This newly discovered anti-inflammatory feature of PPARs potentiated its role as an important metabolic regulator because many metabolic disorders are accompanied by pro-inflammatory states, as discussed in the previous section.

The well-characterized anti-inflammatory effects of all three PPAR isotypes are shown through the trans-repression of important transcription factors in inflammation. The most important regulators of inflammation, including NF-κB, activator protein-1 (AP-1), ATF family, and signal transducer and activator of transcription (STAT) family, are known to be regulated by PPARs through trans-repression mechanisms.[50,51] Activation of PPARs leads to repression of inflammatory cytokines and molecules through repression of these transcriptional factors. There are several proposed mechanisms for the possible interaction between PPARs and several transcription factors. First, the co-activator competition model proposes that NF-κB and PPARs use an overlapping set of co-activator proteins. In this scenario, PPARs compete with NF-κB for binding to the co-activators, thereby regulating its transcriptional activities.[50] The second model proposes direct interactions between PPARs and other transcriptional factors, resulting in the inhibition of transcriptional activity of one or both factors.[50,52,53] Lastly, the co-repressor-dependent model explains that PPAR ligands mediate the trans-repression of inflammatory genes by preventing the clearance of co-repressor complexes.[51]

Table 16.1 Synthetic anti-inflammatory/anti-aging PPAR ligands.

Compounds	Target	Model	Effects	Ref.
Wy-14,643	PPARα	Obese diabetic KKAy mice	Increase in adiponectin receptor, reduce inflammation	144
CP900691	PPARα	Diabetic cynomolgus monkey	Improvement of lipid, glucose metabolism	145
GW501516	PPARβ/δ	Microglial and astrocytes brain inflammation model	Anti-inflammation effects	146
		C2C12 and human skeletal muscle cells	Attenuation of inflammation and improvement of insulin resistance	147
		$ApoE^{-/-}$ atherosclerosis mice	Anti-inflammation effects	80
MBX8025	PPARβ/δ	Combined dyslipidemic patients	Improvement of various metabolic parameters	148
Pioglitazone	PPARγ	Parkinsonian monkey	Anti-inflammatory and neuroprotective effects	149
		Diabetic patients	Anti-inflammation	150
		Aged rat	Retards age-related renal function	128
Rosiglitazone	PPARγ	Aged rat	Attenuates age-related neuroinflammation	151
MHY-908	PPARα/γ dual agonist	Aged rat	Reduces inflammation and insulin resistance	129
		db/db obese mice	Anti-diabetic effects	152
Tesaglitazar	PPARα/γ dual agonist	ApoE*3Leiden mice	Anti-atherosclerosis, anti-hypercholesterolemic and anti-inflammation	153
CG301269	PPARα/γ dual agonist	db/db obese mice	Improvement of glucose and lipid metabolism	154
Cevoglitazar	PPARα/γ dual agonist	Obese mice; diabetic cynomolgus monkey	Reduction of body weight; improvement of overall metabolic parameters	155

These trans-repression mechanisms generally explain the anti-inflammatory effects of all three PPAR isotypes, but differ in the specific subtype and its individual interaction transcriptional factors.

Activated PPARα interferes with the recruitment of glucocorticoid receptor alpha (GRα) to glucocorticoid response element (GRE) and blocks the expression of GRE-associated genes.[54,55] Although GRE-associated genes are repressed, PPARα cooperates with activated GRα for trans-repression on NF-κB.[54] Estrogen receptor-mediated anti-inflammatory effects are also associated with PPARα, although further analyses are needed. PPARα also inhibits pro-inflammatory response by direct interaction with p65 subunit of NF-κB.[52,53] There are also other possible mechanisms that could explain the anti-inflammatory effects of PPARα. PPARα, by directly increasing IκBα, can inhibit translocation of NF-κB to the nucleus and subsequently suppress transcriptional activity.[56] Furthermore, PPARα exerts its anti-inflammatory effect through regulation of leukotriene B_4 (LTB_4). LTB_4 is a potent pro-inflammatory lipid mediator that is increased in various forms of inflammation through lipoxygenase (LOX) activity. Since LTB_4 is a direct ligand for PPARα and since PPARα increases the expression of cytochrome P450 and β-oxidation enzymes responsible for the breakdown of LTB_4, PPARα can likely contribute to the resolution of inflammation through this mechanism.[57,58]

PPARγ also has several anti-inflammatory effects through trans-repression or other mechanisms. PPARγ agonists were shown to decrease inflammatory responses in several innate immune cells.[59,60] Pascual *et al.*[51] reported that PPARγ interacts with a protein inhibitor of the activated transcription factor (PIAS1), the physiological role of which is to facilitate the localization of PPARγ to the NCoR complexes on the promoters of inflammatory genes. Consequently, PPARγ inhibits NF-κB-mediated inflammatory gene expressions in a trans-repression manner. Furthermore, PPARγ is an important transcriptional factor for the alternative macrophage activation that shows anti-inflammatory properties needed for resolution of inflammation.[61] Although the Th2 cytokines (IL-4, IL-13) are needed for the alternative macrophage activation, the acquisition and maintenance of this phenotype require PPARγ activation.[61]

To understand the action of PPARβ/δ on inflammation, a description on Bcl6 is needed. The release of Bcl6 is known to contribute to several of the anti-inflammatory actions.[62] The dissociation of Bcl6 from activated PPARβ/δ renders this cofactor available for gene repression, such as MCP1 and IL-1β.[62] The inhibition of NF-κB signaling is another common mechanism for the anti-inflammatory actions of PPARβ/δ, although the clear mechanisms are not fully understood yet.[63,64] Induction of anti-inflammatory genes may also be another mechanism for the PPARβ/δ's anti-inflammatory actions. Some anti-oxidant genes and anti-apoptosis genes are induced by PPARβ/δ, which can indirectly suppress inflammation.[65,66] More directly, PPARβ/δ induces the well-known anti-inflammatory mediator TGFβ.[67]

16.4 Evidence for Involvement of PPARs in Age-Related Inflammatory Diseases and Aging

As mentioned above, aging is characterized by time-dependent changes in physiological functions accompanied by pathological diseases. Evidence from many recent studies has linked chronic inflammation to the progression of age-related diseases including arthritis, cardiovascular diseases, dementia, inflammatory bowel diseases, and metabolic syndrome. In this regard, it is evident that the ability of PPARs as regulators of inflammation and metabolism will retard age-associated diseases. Several reviews already cover a great deal of information regarding the role of PPARs in regulation of metabolism.[42,68–70]

16.4.1 The Role of PPARs in Age-Related Inflammatory Diseases

16.4.1.1 Atherosclerosis and Cardiovascular Diseases

It has been suggested that chronic vascular inflammation underlies the pathogenesis of atherosclerosis.[71] Although atherosclerosis is traditionally viewed as a sole metabolic disease with abnormal lipids in systemic circulation, the inflammatory component of atherogenesis is increasingly being recognized.[72] It is now known that inflammation participates in all stages of atherosclerosis including initiation and progression of atherogenic dyslipidemia. Chronic low-grade inflammation also participates in cardiac hypertrophy and heart failure.[73] In this respect, PPARs are involved in the progression of atherosclerosis and other cardiovascular diseases and therefore they are promising targets for treatment of these diseases.

Roles for PPARα in the progression of atherosclerosis are well documented. PPARα agonists suppress the progression of atherosclerosis by inhibiting foam cell formation in a low-density lipoprotein receptor (LDLR)-deficient mice model.[74] Furthermore, PPARα expression in macrophages showed strong anti-atherogenic effects in an LDLR-deficient mice model *via* modulation of cell cholesterol trafficking and inflammation.[75] Interestingly, miRNA-21 induced by shear stress decreased PPARα expression and activated pro-inflammatory AP-1, demonstrating the anti-atherogenic role of PPARα.[76] PPARα also controls cardiac hypertrophy by reducing inflammation and regulating metabolism. A PPARα agonist inhibited hypertrophy of neonatal rat cardiac myocytes.[77] The role of PPARα in hypertrophy is also demonstrated using PPARα deficient mice in response to chronic pressure overload.[78] In addition, PPARα in association with NAD+-dependent deacetylase, SIRT1, reduces inflammation and cardiac hypertrophy.[79] PPARβ/δ also inhibits atherosclerosis and cardiac hypertrophy.[80] The synthetic PPARβ/δ agonists reduced atherosclerosis in LDLR-deficient mice by decreasing pro-inflammatory mediators.[81,82] Activation of anti-inflammatory mediator Bcl-6 seems to contribute to the anti-atherogenic role of PPARβ/δ.[82,83] Similarly, PPARβ/δ agonists normalize cardiac substrate metabolism and reduce cardiac hypertrophy.[84] Activated PPARβ/δ also dampens LPS-induced inflammatory signaling in cardiomyocytes, and it blocks lipid-induced inflammatory pathways in mouse heart and human cardiac cells.[85,86] PPARγ shows its anti-atherogenic activity in various cells, including endothelial cells, macrophages, and smooth muscle cells. The disruption of PPARγ in macrophages and smooth muscle cells increases atherosclerosis.[87,88] Endothelial PPARγ prevents the initiation of atherosclerosis by enhancing endothelial cell function.[89] There are conflicting outcomes on the role of PPARγ in cardiac physiology. Expression of PPARγ in macrophages attenuates progressive cardiac fibrosis occurring in diabetic cardiomyopathy.[90] However, cardiomyocyte expression of PPARγ can lead to cardiac dysfunction implying cell-specific functions for PPARγ.[90,91]

16.4.1.2 Alzheimer's Disease

The number of individuals with Alzheimer's disease (AD) is dramatically increasing with the aging of the population. Analyses of genetic background and animal models suggest a pivotal role of amyloid β peptide (Aβ) and neurofibrillary tangles in AD, although the biological basis of AD is still poorly understood. One key hallmark of the AD brain is the inflammatory processes that exert neurotoxicity during aging.[92,93] Excessively activated microglia and astrocytes in the brain accelerate amyloid plaque formation. Microglial activation along with increased inflammatory cytokines and chemokines may deteriorate neuronal loss and accelerate progression of AD. This extensive innate immune gene activation accompanies brain aging, increasing vulnerability to cognitive decline and neurodegeneration.[93]

There are several recent investigations supporting the anti-inflammatory role of PPARs in AD and brain aging. Among the three PPAR isotypes, PPARγ showed prominent effects on the delay of AD onset. The initial studies exploring the roles of PPARγ in AD were based on people with long-term intake of non-steroidal anti-inflammatory drugs (NSAIDs) and PPARγ ligands.[94,95] Long-term NSAID treatment reduces AD risk, and it was suggested that PPARγ stimulation may be involved in this beneficial effect.[96] Various *in vivo* and *in vitro* investigations further demonstrated the anti-inflammatory roles of PPARγ activation.[97] In addition, PPARγ agonists have been demonstrated to directly suppress the Aβ-mediated activation of microglia and prevent neuronal cell death.[98] Further, animal and clinical studies using pioglitazone and rosiglitazone demonstrated the role of PPARγ activation in the prevention of AD.[97,99] Although the beneficial effects of PPARγ agonists on AD are clear and evident, the underlying molecular mechanisms must be elucidated in future studies. Interestingly, recent studies revealed the beneficial role of PPARβ/δ on AD. Activation of PPARβ/δ agonists reduced the amyloid burden and exerted neuroprotective effects in a mouse model of AD.[100,101]

16.4.1.3 Inflammatory Bowel Diseases

Inflammatory bowel diseases (IBD) are generally thought to be diseases of young individuals. However, IBD among the elderly are becoming common with growing incidence and prevalence rates.[102] Compared with younger IBD patients, dysregulation of the immune system and chronic inflammation play more important roles in older-onset IBD.[103] In this respect, the age-associated increase in inflammation should be considered as an important factor for older-onset IBD. PPARs are broadly associated with onset and progression of IBD.

PPARγ exerts its most powerful effects on suppression of IBD. Natural and synthetic ligands of PPARγ reduce colitis in experimental animal models.[104,105] Furthermore, cell-specific deletion of PPARγ in epithelial cells, macrophages, and T cells has demonstrated the role of PPARγ in a colitis model.[106–108] In addition, PPARγ also maintains antibacterial effects of epithelial cells by expression of a subset of β-defensins.[104] Accordingly, it is important to maintain proper activity of PPARγ by therapeutic or nutritional means to prevent colonic inflammation, which contributes to the progression of IBD. Although experimental evidence and precise mechanism studies are insufficient, PPARα agonist also showed anti-inflammatory effects in a colitis model. The deletion of PPARα or supply of exogenous PPARα ligands reduce the degree of colitis induced by dinitrobenzene sulfonic acid (DNBS) and dextran sulfate sodium (DSS).[109,110] The role of PPARβ/δ in colitis was controversial in mouse models of IBD, so additional studies are needed to fully address its exact roles.[111,112]

16.4.2 PPARs in Aging and Longevity

Increasing evidence demonstrates the PPARs' intimate involvement in the aging process. Whether or not associated with age-related inflammation, the decreased expression or activity of PPARs during aging is detected commonly in various tissues. In aged rat kidney, decreased PPARα and PPARγ expression were detected both in mRNA and protein levels.[113] Furthermore, binding activity of PPARs was also decreased during the aging process in the kidney.[113] Aged rat heart also showed decreased PPARα expression followed by target-gene reduction.[114] PPARγ expression during aging in metabolically important tissues (adipose tissue, muscle) also decreased in aged rodents, indicating possible association of PPARγ with insulin resistance during aging.[115,116] Interestingly, spleen PPARα levels were also decreased during aging, implying the possible relevance of the immune system and PPARs during aging.[117]

More direct evidence on the important role of PPARs in the aging process comes from observation of PPARα knockout mice. As explained in a previous chapter about the role of PPARα in lipid metabolism, PPARα null mice exhibited a number of defects in lipid metabolism and lipid accumulation in the liver.[35,36,118] In addition to altered lipid metabolism, Poynter and Daynes first reported a premature and enhanced age-dependent increase of oxidative stress and NF-κB activation in PPARα knockout mice.[53] The administration of PPARα ligands to aged mice restored cellular redox balance as well as highly activated inflammatory response, which was not detected in PPARα knockout mice.[53] These results suggested that PPARα plays an important role in maintaining proper levels of oxidative stress and inflammation during aging. Howroyd *et al.* further expanded the role of PPARα during aging. The authors showed that PPARα knockout mice had decreased longevity compared with wild type mice. Reduced longevity in PPARα-null mice was associated with increased levels of various non-neoplastic spontaneous aging lesions.[119] Although direct evidence for the link between PPARγ and longevity are lacking because of the lethality of PPARγ knockout mice during embryogenesis, various studies suggest the important role of PPARγ in aging. PPARγ variants were reported to have an essential role in aging in humans with low insulin resistance.[120,121] In addition, Klotho, which is a transmembrane protein that suppresses aging, is directly regulated by PPARγ, suggesting possible links between PPARγ and aging.[122] Interestingly, one recent study indicated PPARγ as an important longevity gene especially in white adipose tissue (WAT).[123] *In silico*, these authors found that both transcriptional signatures of the PPARγ signaling pathway and of the PPARγ agonist rosiglitazone overlapped in the subnetwork of their longevity-associated genes. Further *in vivo* experiments demonstrated the role of PPARγ in life span using WAT-specific PPARγ knockout mice.[123] These genetic approaches and other available biological evidence strongly suggest the very complicated involvement of PPARs in age-related diseases and aging processes.

16.5 Anti-Aging and Therapeutic Potentials of New PPAR Agonists

To date, many endogenous ligands and natural compounds have been discovered as PPARs activators, and new synthetic chemical compounds are being developed to activate PPARs.[124] Endogenous ligands, including polyunsaturated fatty acids and some eicosanoids like prostaglandins and leukotriene, are produced in the metabolic pathways of fatty acids and regulate an individual's metabolism.[125] In addition to endogenous PPARs ligands, many PPAR agonists are developed based on a ligand binding site. The hypolipidemic fibrate and antidiabetic thiazolidinedione (TZD) classes of drugs are two representative PPAR agonists that activate PPARα and PPARγ, respectively.[126]

As activation of PPARα is known to increase β-oxidation-associated gene expression, fibrates decrease high triglyceride-containing lipoproteins and improve overall lipid profiles. Fibrates also increase insulin sensitivity and reduce plasma glucose levels. TZD drugs are used in the treatment of type 2 diabetes mellitus as TZD increases insulin sensitivity. In particular, PPARγ activation by TZD prevents lipotoxicity by regulating adipose tissue lipid accumulation and protects non-adipose tissues (liver, skeletal muscle) against excessive lipid overload. Furthermore, activated PPARγ by TZD also permits adequate secretion of leptin and adiponectin, which are mediators of insulin action in peripheral tissues. Collectively, considering the wide range of actions of PPAR agonists, PPAR modulators are suggested to be promising agents for the treatment of metabolic disorders, including hyperlipidemia, hyperglycemia, and type 2 diabetes.

As mentioned earlier, the PPAR family has dynamic roles, including regulation of inflammation, immunity, cell proliferation, and tissue remodeling. As a consequence, some PPAR agonists have shown beneficial effects in various diseases, including atherosclerosis, cardiovascular diseases, Alzheimer's disease, inflammatory bowel diseases, and renal diseases.[126,127] Although the roles of PPAR activation in various disease models are fairly well demonstrated, their beneficial roles in the aging process are not fully verified. Recently, several studies have investigated the roles of PPAR agonists in pathophysiological changes during the aging process. Yang *et al.*

investigated the role of a PPARγ agonist, pioglitazone, in renal injury during aging.[128] Pioglitazone effectively reduced proteinuria, sclerosis, and cellular senescence and improved GFR by activating PPARγ. The authors found that increased oxidative stress associated with mitochondrial dysfunction can contribute to renal injury during aging, and pioglitazone can protect the aged kidney by increasing klotho and reducing protein kinase C-β and p66Shc phosphorylation.[128] More recently, the effects of a newly synthesized PPARα/γ dual agonist, MHY-908, were evaluated in aged rat liver and kidney.[129] In this study, MHY-908 effectively improved serum metabolic profiles and insulin resistance and reduced lipid accumulation and endoplasmic reticulum (ER) stresses in aged rats. It also suppressed age-related renal inflammation by inhibiting NF-κB signaling pathway. Although the effects of PPARs agonists still need to be further demonstrated in aging, current experimental data strongly suggest a beneficial role of PPARs agonists during aging.[130]

The clinical use of the PPARα agonists fibrates and PPARγ agonist TZDs, however, was associated with a number of adverse effects. Although fibrates are generally well tolerated, continuous PPARα activation by agonists is known to cause hyperproliferations of hepatocytes leading in several cases to liver cancers in rodent studies.[131] However, in most of the studies, the doses shown to increase the risk of cancer development were higher than the recommended doses for humans.[132] TZDs commonly show more severe adverse effects, including weight gain, fluid retention, and bone fracture.[125] Furthermore, the two most widely used TZDs, rosiglitazone and pioglitazone, also show adverse side effects owing to their off-target activity.[125] These undesirable effects of PPAR agonists are an unresolved problem that limits the use of these drugs clinically. In addition, numerous PPAR agonists have been dropped from the market due to their severe adverse effects.[126]

Presently, some clinical trials with more specific PPAR subtypes are in progress, and investigations are ongoing to develop new types of PPAR agonist. New selective PPARγ agonists are currently being developed to minimize the side effects of existing PPARγ agonists.[133,134] Dual PPAR agonists, *i.e.*, glitazars, are currently under active investigation for co-treatment of hyperlipidemia and type 2 diabetes. These dual agonists are designed to provide a better balance between efficacy and side effects. Although some dual agonists showed similar adverse effects with PPARγ agonists, the further investigation on more balanced dual agonists is very promising.[125] In addition, there is continuing research and development of new dual α/γ and α/δ agonists and α/γ/δ pan agonists for additional therapeutic indications.[135] These newly developed agonists may have not only anti-diabetic effects but also exhibit anti-inflammatory, anti-coagulation action, and beneficial effects on the cardiovascular circulation system.[126]

Although some PPAR agonists show beneficial effects on age-related metabolic/inflammatory changes, their side effects have not been fully considered and the extent of their efficiency is still questionable. Overall, it can be suggested that newly developed PPAR agonists with minimum adverse side effects will have beneficial effects on aging and age-related diseases, and it is important to evaluate their anti-aging effects experimentally and clinically.

16.6 Effects of Anti-Aging Calorie Restriction on PPAR Modulation

One interesting aspect of PPARs is their involvement in calorie restriction (CR), which is well known as an effective anti-aging treatment. It was firmly established that CR markedly increases median and maximum life span in several species including mammals.[136] Although the precise mechanisms of the action of CR on aging and longevity are not fully established, several plausible mechanisms have been proposed. Generally, CR is known to alter various physiological functions, including lipid and carbohydrate metabolism, the immune system, and inflammation.[137] Among the many changes occurring during CR, various metabolic processes are known to be changed most substantially. CR reduces overall energy expenditure and also alters insulin sensitivity and insulin signaling, neuroendocrine functions, and stress response.[138] In addition, CR also exerts its benefits through inducing the antioxidative defense mechanisms to suppress age-related oxidative stress. Interestingly, some changes occurring during CR seem to be similar to those induced by PPAR activation. Several pieces of evidence suggest the possibility that PPARs mediate the effects of CR by modulating similar signaling pathways.

Although PPARs are regulated by CR and mediate some beneficial effects of CR, the effects are organ-specific.[139] Furthermore, depending on the experimental designs and animals used, a decrease, an increase, or no changes are observed in the expression of PPARs in response to CR.[139] One most reasonable explanation of the relationship between CR and PPAR comes from changes of hepatic PPARα levels by CR in comparison to *ad libitum* (AL, free access to food)-fed mice.[140,141] Increased hepatic PPARα expression may play an adaptive role in regulating glucose homeostasis to prevent hypoglycemia during CR.[36] In addition, Corton and Brown-Borg demonstrated that 19% of genes (mostly involved in metabolism and inflammation) changed by CR were dependent on PPARα[142] as the protective effects of CR were lost in PPARα null mice. These findings indicate that PPARα plays an important role in mediating the action of CR and suggest that PPARα agonists may act as plausible CR mimetics.

CR experiments in aged rats also suggest the relationship with PPAR. Sung *et al.* first reported that the levels of PPARα and PPARγ decreased in aged rat kidneys.[113] They found that in young healthy rats, CR did not have any effects on the levels of PPARα and PPARγ expression. However, CR in aged rat kidneys increased PPARα and PPARγ expression compare to *ad libitum*-fed aged rat kidneys. An age-related decrease in PPAR binding activity was also slowed by CR in aged rat kidneys. To further examine the possible role of PPARα and PPARγ in age-related inflammation, lipopolysaccharide (LPS) treatments were administered to young and aged rats. Treatment with LPS deceased the levels of both PPARs in young and aged rat kidneys, but the extent of this decrease was greater in aged rat kidneys.[113] The authors further compared the effect of the PPARγ agonist with CR on the aging process of rat kidneys.[143] PPARγ activation by its agonist suppressed age-related oxidative stress and inflammation through inhibiting the NF-κB signaling, just like the CR effects. Collectively, the authors concluded that down-regulation of PPARs in the aged rat kidneys might be related to age-related oxidative stress and inflammation, and those conditions could be reversed by CR or PPAR activation.

16.7 Conclusion

PPARs have been extensively investigated since their discovery as ligand-dependent nuclear transcriptional receptors. The roles of PPARs are also well characterized. Because PPARs control patterns of gene expression involved in a broad spectrum of biological processes, including metabolism and inflammation, the PPAR family has been proposed to be an attractive target for various pharmacological interventions. Through continued efforts, several PPAR agonists (fibrates for PPARα, TZDs for PPARγ) have been developed and used for the treatment of metabolic diseases.

More recently, PPARs have been shown to be associated with aging in many aspects. In the aging process, increased low-grade chronic inflammation is commonly observed and its association with various age-related diseases is well documented. In particular, an age-related increase in inflammation is strongly associated with progressive deterioration of metabolic function. Recent evidence also strongly suggests PPARs as key modulatory transcription factors responsible for the suppression of inflammation through regulation of NF-κB. The anti-inflammatory actions of PPARs were further verified by *in vitro* and *in vivo* studies that indicate the importance of PPARs as major players in the pathogenesis of many inflammatory diseases. Furthermore, more recent studies suggest that PPARs agonists can directly reduce age-related inflammation and thereby modulate the pathophysiology of aging. Because currently available PPAR agonists have unwanted adverse effects, great efforts are being made to develop more selective PPAR agonists without adverse effects. In addition, balanced activation of PPARs through dual- or pan-agonists provides a better strategy in controlling age-related diseases. Concluding, it can be assumed that better understanding the association between aging and PPARs can further lead to the development of new therapeutic agents (including PPAR agonists) that modulate aging and age-associated diseases (Figure 16.2).

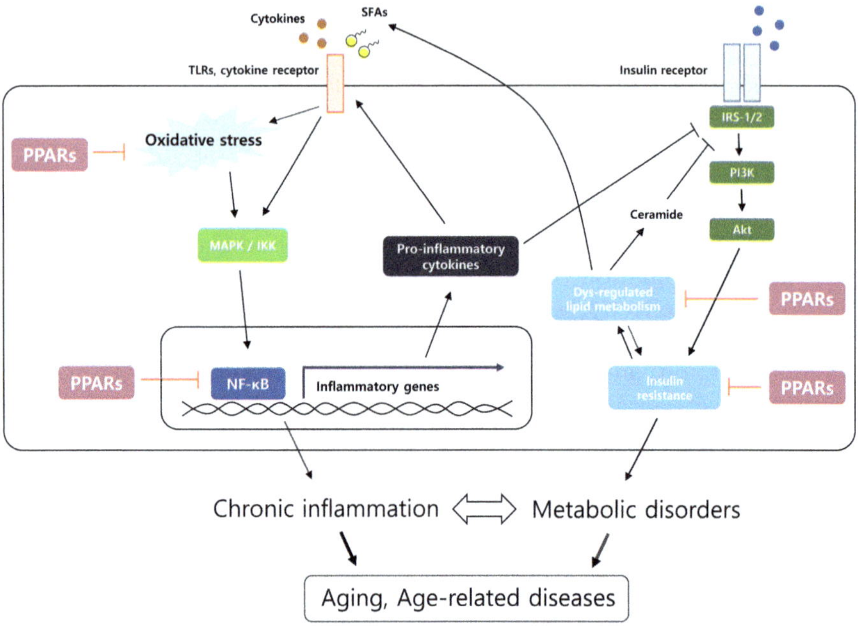

Figure 16.2 Molecular mechanisms of PPARs in aging processes. Aging is accompanied by increased chronic inflammatory states with metabolic disorders. Chronic inflammation and metabolic abnormalities can affect each other and form a vicious cycle that worsens physiological integrity during the aging process. Activation of PPARs can influence these age-related vicious cycles between inflammation and metabolic diseases.

References

1. J. Campisi, *Annu. Rev. Physiol.*, 2013, **75**, 685.
2. B. T. Weinert and P. S. Timiras, *J. Appl. Physiol. (1985)*, 2003, **95**, 1706.
3. C. Lopez-Otin, M. A. Blasco, L. Partridge, M. Serrano and G. Kroemer, *Cell*, 2013, **153**, 1194.
4. V. Guarner and M. E. Rubio-Ruiz, *Interdiscip. Top. Gerontol.*, 2015, **40**, 99.
5. N. S. Jenny, *Discovery Med.*, 2012, **13**, 451.
6. T. Finkel, *Nat. Med.*, 2015, **21**, 1416.
7. A. C. Shaw, D. R. Goldstein and R. R. Montgomery, *Nat. Rev. Immunol.*, 2013, **13**, 875.
8. A. S. Ryan, *Sports Med.*, 2000, **30**, 327.
9. N. Barzilai, D. M. Huffman, R. H. Muzumdar and A. Bartke, *Diabetes*, 2012, **61**, 1315.
10. R. Curtis, B. J. Geesaman and P. S. DiStefano, *Nat. Rev. Drug Discovery*, 2005, **4**, 569.
11. N. Dali-Youcef, M. Mecili, R. Ricci and E. Andres, *Ann. Med.*, 2013, **45**, 242.
12. G. S. Hotamisligil, *Nature*, 2006, **444**, 860.

13. C. N. Lumeng and A. R. Saltiel, *J. Clin. Invest.*, 2011, **121**, 2111.
14. R. Medzhitov, *Nature*, 2008, **454**, 428.
15. M. Murakami and T. Hirano, *Front. Immunol.*, 2012, **3**, 323.
16. D. Laveti, M. Kumar, R. Hemalatha, R. Sistla, V. G. Naidu, V. Talla, V. Verma, N. Kaur and R. Nagpal, *Inflammation Allergy: Drug Targets*, 2013, **12**, 349.
17. H. Y. Chung, M. Cesari, S. Anton, E. Marzetti, S. Giovannini, A. Y. Seo, C. Carter, B. P. Yu and C. Leeuwenburgh, *Ageing Res. Rev.*, 2009, **8**, 18.
18. H. Bruunsgaard, M. Pedersen and B. K. Pedersen, *Curr. Opin. Hematol.*, 2001, **8**, 131.
19. J. E. Morley and R. N. Baumgartner, *J. Gerontol., Ser. A*, 2004, **59**, M924.
20. A. L. Gruver, L. L. Hudson and G. D. Sempowski, *J. Pathol.*, 2007, **211**, 144.
21. A. Freund, C. K. Patil and J. Campisi, *EMBO J.*, 2011, **30**, 1536.
22. A. S. Adler, S. Sinha, T. L. Kawahara, J. Y. Zhang, E. Segal and H. Y. Chang, *Genes Dev.*, 2007, **21**, 3244.
23. G. Zhang, J. Li, S. Purkayastha, Y. Tang, H. Zhang, Y. Yin, B. Li, G. Liu and D. Cai, *Nature*, 2013, **497**, 211.
24. C. Franceschi, M. Bonafe, S. Valensin, F. Olivieri, M. De Luca, E. Ottaviani and G. De Benedictis, *Ann. N. Y. Acad. Sci.*, 2000, **908**, 244.
25. M. Qatanani and M. A. Lazar, *Genes Dev.*, 2007, **21**, 1443.
26. A. Maiorana, C. Del Bianco and S. Cianfarani, *Rev. Diabet. Stud.*, 2007, **4**, 134.
27. H. Kwon and J. E. Pessin, *Front. Endocrinol.*, 2013, **4**, 71.
28. M. Rincon, R. Muzumdar, G. Atzmon and N. Barzilai, *Mech. Ageing Dev.*, 2004, **125**, 397.
29. C. K. Glass and J. M. Olefsky, *Cell Metab.*, 2012, **15**, 635.
30. J. Berger and D. E. Moller, *Annu. Rev. Med.*, 2002, **53**, 409.
31. F. A. Monsalve, R. D. Pyarasani, F. Delgado-Lopez and R. Moore-Carrasco, *Mediators Inflammation*, 2013, **2013**, 549627.
32. P. Lefebvre, G. Chinetti, J. C. Fruchart and B. Staels, *J. Clin. Invest.*, 2006, **116**, 571.
33. M. V. Chakravarthy, I. J. Lodhi, L. Yin, R. R. Malapaka, H. E. Xu, J. Turk and C. F. Semenkovich, *Cell*, 2009, **138**, 476.
34. N. Viswakarma, Y. Jia, L. Bai, A. Vluggens, J. Borensztajn, J. Xu and J. K. Reddy, *PPAR Res.*, 2010, **2010**, 250126.
35. T. C. Leone, C. J. Weinheimer and D. P. Kelly, *Proc. Natl. Acad. Sci. U. S. A.*, 1999, **96**, 7473.
36. S. Kersten, J. Seydoux, J. M. Peters, F. J. Gonzalez, B. Desvergne and W. Wahli, *J. Clin. Invest.*, 1999, **103**, 1489.
37. J. C. Rodriguez, G. Gil-Gomez, F. G. Hegardt and D. Haro, *J. Biol. Chem.*, 1994, **269**, 18767.
38. T. Inagaki, P. Dutchak, G. Zhao, X. Ding, L. Gautron, V. Parameswara, Y. Li, R. Goetz, M. Mohammadi, V. Esser, J. K. Elmquist, R. D. Gerard, S. C. Burgess, R. E. Hammer, D. J. Mangelsdorf and S. A. Kliewer, *Cell Metab.*, 2007, **5**, 415.

39. M. K. Badman, P. Pissios, A. R. Kennedy, G. Koukos, J. S. Flier and E. Maratos-Flier, *Cell Metab.*, 2007, **5**, 426.
40. S. M. Reilly and C. H. Lee, *FEBS Lett.*, 2008, **582**, 26.
41. D. Pan, M. Fujimoto, A. Lopes and Y. X. Wang, *Cell*, 2009, **137**, 73.
42. M. Ahmadian, J. M. Suh, N. Hah, C. Liddle, A. R. Atkins, M. Downes and R. M. Evans, *Nat. Med.*, 2013, **19**, 557.
43. A. Zieleniak, M. Wojcik and L. A. Wozniak, *Arch. Immunol. Ther. Exp.*, 2008, **56**, 331.
44. R. P. Brun, P. Tontonoz, B. M. Forman, R. Ellis, J. Chen, R. M. Evans and B. M. Spiegelman, *Genes Dev.*, 1996, **10**, 974.
45. Y. Barak, M. C. Nelson, E. S. Ong, Y. Z. Jones, P. Ruiz-Lozano, K. R. Chien, A. Koder and R. M. Evans, *Mol. Cell*, 1999, **4**, 585.
46. F. Picard and J. Auwerx, *Annu. Rev. Nutr.*, 2002, **22**, 167.
47. R. B. Clark, *J. Leukocyte Biol.*, 2002, **71**, 388.
48. H. Martin, *Mutat. Res.*, 2009, **669**, 1.
49. J. H. Chung, A. Y. Seo, S. W. Chung, M. K. Kim, C. Leeuwenburgh, B. P. Yu and H. Y. Chung, *Ageing Res. Rev.*, 2008, **7**, 126.
50. M. Ricote and C. K. Glass, *Biochim. Biophys. Acta*, 2007, **1771**, 926.
51. G. Pascual, A. L. Fong, S. Ogawa, A. Gamliel, A. C. Li, V. Perissi, D. W. Rose, T. M. Willson, M. G. Rosenfeld and C. K. Glass, *Nature*, 2005, **437**, 759.
52. S. W. Chung, B. Y. Kang, S. H. Kim, Y. K. Pak, D. Cho, G. Trinchieri and T. S. Kim, *J. Biol. Chem.*, 2000, **275**, 32681.
53. M. E. Poynter and R. A. Daynes, *J. Biol. Chem.*, 1998, **273**, 32833.
54. N. Bougarne, R. Paumelle, S. Caron, N. Hennuyer, R. Mansouri, P. Gervois, B. Staels, G. Haegeman and K. De Bosscher, *Proc. Natl. Acad. Sci. U. S. A.*, 2009, **106**, 7397.
55. C. Crisafulli, S. Bruscoli, E. Esposito, E. Mazzon, R. Di Paola, T. Genovese, P. Bramanti, G. Migliorati and S. Cuzzocrea, *J. Pharmacol. Exp. Ther.*, 2009, **331**, 796.
56. P. Delerive, P. Gervois, J. C. Fruchart and B. Staels, *J. Biol. Chem.*, 2000, **275**, 36703.
57. P. R. Devchand, H. Keller, J. M. Peters, M. Vazquez, F. J. Gonzalez and W. Wahli, *Nature*, 1996, **384**, 39.
58. V. R. Narala, R. K. Adapala, M. V. Suresh, T. G. Brock, M. Peters-Golden and R. C. Reddy, *J. Biol. Chem.*, 2010, **285**, 22067.
59. A. Chawla, Y. Barak, L. Nagy, D. Liao, P. Tontonoz and R. M. Evans, *Nat. Med.*, 2001, **7**, 48.
60. C. Jiang, A. T. Ting and B. Seed, *Nature*, 1998, **391**, 82.
61. M. A. Bouhlel, B. Derudas, E. Rigamonti, R. Dievart, J. Brozek, S. Haulon, C. Zawadzki, B. Jude, G. Torpier, N. Marx, B. Staels and G. Chinetti-Gbaguidi, *Cell Metab.*, 2007, **6**, 137.
62. D. Bishop-Bailey and J. Bystrom, *Pharmacol. Ther.*, 2009, **124**, 141.
63. Y. Rival, N. Beneteau, T. Taillandier, M. Pezet, E. Dupont-Passelaigue, J. F. Patoiseau, D. Junquero, F. C. Colpaert and A. Delhon, *Eur. J. Pharmacol.*, 2002, **435**, 143.

64. B. Zingarelli, G. Piraino, P. W. Hake, M. O'Connor, A. Denenberg, H. Fan and J. A. Cook, *Am. J. Pathol.*, 2010, **177**, 1834.
65. N. S. Tan, L. Michalik, N. Noy, R. Yasmin, C. Pacot, M. Heim, B. Fluhmann, B. Desvergne and W. Wahli, *Genes Dev.*, 2001, **15**, 3263.
66. R. Di Paola, E. Esposito, E. Mazzon, I. Paterniti, M. Galuppo and S. Cuzzocrea, *J. Leukocyte Biol.*, 2010, **88**, 291.
67. H. J. Kim, S. A. Ham, S. U. Kim, J. Y. Hwang, J. H. Kim, K. C. Chang, C. Yabe-Nishimura, J. H. Kim and H. G. Seo, *Circ. Res.*, 2008, **102**, 193.
68. Y. X. Wang, *Cell Res.*, 2010, **20**, 124.
69. B. Desvergne and W. Wahli, *Endocr. Rev.*, 1999, **20**, 649.
70. J. A. Youssef and M. Z. Badr, *Peroxisome proliferator-activated receptors: discovery and recent advances*, Humana Press, New York, 2013.
71. P. Libby, *Nature*, 2002, **420**, 868.
72. R. Paoletti, C. Bolego, A. Poli and A. Cignarella, *Vasc. Health Risk Manage.*, 2006, **2**, 145.
73. P. Libby, *Am. J. Clin. Nutr.*, 2006, **83**, 456S.
74. A. C. Li, C. J. Binder, A. Gutierrez, K. K. Brown, C. R. Plotkin, J. W. Pattison, A. F. Valledor, R. A. Davis, T. M. Willson, J. L. Witztum, W. Palinski and C. K. Glass, *J. Clin. Invest.*, 2004, **114**, 1564.
75. V. R. Babaev, H. Ishiguro, L. Ding, P. G. Yancey, D. E. Dove, W. J. Kovacs, C. F. Semenkovich, S. Fazio and M. F. Linton, *Circulation*, 2007, **116**, 1404.
76. J. Zhou, K. C. Wang, W. Wu, S. Subramaniam, J. Y. Shyy, J. J. Chiu, J. Y. Li and S. Chien, *Proc. Natl. Acad. Sci. U. S. A.*, 2011, **108**, 10355.
77. F. Liang, F. Wang, S. Zhang and D. G. Gardner, *Endocrinology*, 2003, **144**, 4187.
78. P. J. Smeets, B. E. Teunissen, P. H. Willemsen, F. A. van Nieuwenhoven, A. E. Brouns, B. J. Janssen, J. P. Cleutjens, B. Staels, G. J. van der Vusse and M. van Bilsen, *Cardiovasc. Res.*, 2008, **78**, 79.
79. A. Planavila, R. Iglesias, M. Giralt and F. Villarroya, *Cardiovasc. Res.*, 2011, **90**, 276.
80. G. D. Barish, A. R. Atkins, M. Downes, P. Olson, L. W. Chong, M. Nelson, Y. Zou, H. Hwang, H. Kang, L. Curtiss, R. M. Evans and C. H. Lee, *Proc. Natl. Acad. Sci. U. S. A.*, 2008, **105**, 4271.
81. C. H. Lee, A. Chawla, N. Urbiztondo, D. Liao, W. A. Boisvert, R. M. Evans and L. K. Curtiss, *Science*, 2003, **302**, 453.
82. T. L. Graham, C. Mookherjee, K. E. Suckling, C. N. Palmer and L. Patel, *Atherosclerosis*, 2005, **181**, 29.
83. Y. Takata, J. Liu, F. Yin, A. R. Collins, C. J. Lyon, C. H. Lee, A. R. Atkins, M. Downes, G. D. Barish, R. M. Evans, W. A. Hsueh and R. K. Tangirala, *Proc. Natl. Acad. Sci. U. S. A.*, 2008, **105**, 4277.
84. B. M. Jucker, C. P. Doe, C. G. Schnackenberg, A. R. Olzinski, K. Maniscalco, C. Williams, T. C. Hu, S. C. Lenhard, M. Costell, R. Bernard, L. Sarov-Blat, K. Steplewski and R. N. Willette, *J. Cardiovasc. Pharmacol.*, 2007, **50**, 25.

85. G. Ding, L. Cheng, Q. Qin, S. Frontin and Q. Yang, *J. Mol. Cell. Cardiol.*, 2006, **40**, 821.
86. D. Alvarez-Guardia, X. Palomer, T. Coll, L. Serrano, R. Rodriguez-Calvo, M. M. Davidson, M. Merlos, I. El Kochairi, L. Michalik, W. Wahli and M. Vazquez-Carrera, *Biochim. Biophys. Acta*, 2011, **1811**, 59.
87. V. R. Babaev, P. G. Yancey, S. V. Ryzhov, V. Kon, M. D. Breyer, M. A. Magnuson, S. Fazio and M. F. Linton, *Arterioscler., Thromb., Vasc. Biol.*, 2005, **25**, 1647.
88. M. Hamblin, L. Chang, H. Zhang, K. Yang, J. Zhang and Y. E. Chen, *Arterioscler., Thromb., Vasc. Biol.*, 2011, **31**, 352.
89. A. Qu, Y. M. Shah, S. K. Manna and F. J. Gonzalez, *Arterioscler., Thromb., Vasc. Biol.*, 2012, **32**, 65.
90. E. Caglayan, B. Stauber, A. R. Collins, C. J. Lyon, F. Yin, J. Liu, S. Rosenkranz, E. Erdmann, L. E. Peterson, R. S. Ross, R. K. Tangirala and W. A. Hsueh, *Diabetes*, 2008, **57**, 2470.
91. N. H. Son, T. S. Park, H. Yamashita, M. Yokoyama, L. A. Huggins, K. Okajima, S. Homma, M. J. Szabolcs, L. S. Huang and I. J. Goldberg, *J. Clin. Invest.*, 2007, **117**, 2791.
92. D. H. Cribbs, N. C. Berchtold, V. Perreau, P. D. Coleman, J. Rogers, A. J. Tenner and C. W. Cotman, *J. Neuroinflammation*, 2012, **9**, 179.
93. F. L. Heppner, R. M. Ransohoff and B. Becher, *Nat. Rev. Neurosci.*, 2015, **16**, 358.
94. B. A. in t' Veld, A. Ruitenberg, A. Hofman, L. J. Launer, C. M. van Duijn, T. Stijnen, M. M. Breteler and B. H. Stricker, *N. Engl. J. Med.*, 2001, **345**, 1515.
95. G. E. Landreth and M. T. Heneka, *Neurobiol. Aging*, 2001, **22**, 937.
96. J. M. Lehmann, J. M. Lenhard, B. B. Oliver, G. M. Ringold and S. A. Kliewer, *J. Biol. Chem.*, 1997, **272**, 3406.
97. M. T. Heneka, E. Reyes-Irisarri, M. Hull and M. P. Kummer, *Curr. Neuropharmacol.*, 2011, **9**, 643.
98. C. K. Combs, D. E. Johnson, J. C. Karlo, S. B. Cannady and G. E. Landreth, *J. Neurosci.*, 2000, **20**, 558.
99. B. W. Miller, K. C. Willett and A. R. Desilets, *Ann. Pharmacother.*, 2011, **45**, 1416.
100. S. Kalinin, J. C. Richardson and D. L. Feinstein, *Curr. Alzheimer Res.*, 2009, **6**, 431.
101. T. Malm, M. Mariani, L. J. Donovan, L. Neilson and G. E. Landreth, *J. Neuroinflammation*, 2015, **12**, 7.
102. S. Taleban, J. F. Colombel, M. J. Mohler and M. J. Fain, *J. Crohns Colitis*, 2015, **9**, 507.
103. S. Katz and D. S. Pardi, *Am. J. Gastroenterol.*, 2011, **106**, 1889.
104. L. Peyrin-Biroulet, J. Beisner, G. Wang, S. Nuding, S. T. Oommen, D. Kelly, E. Parmentier-Decrucq, R. Dessein, E. Merour, P. Chavatte, T. Grandjean, A. Bressenot, P. Desreumaux, J. F. Colombel, B. Desvergne, E. F. Stange, J. Wehkamp and M. Chamaillard, *Proc. Natl. Acad. Sci. U. S. A.*, 2010, **107**, 8772.

105. R. Marion-Letellier, P. Dechelotte, M. Iacucci and S. Ghosh, *Gut*, 2009, **58**, 586.
106. A. J. Guri, N. P. Evans, R. Hontecillas and J. Bassaganya-Riera, *J. Nutr. Biochem.*, 2011, **22**, 812.
107. Y. M. Shah, K. Morimura and F. J. Gonzalez, *Am. J. Physiol.: Gastrointest. Liver Physiol.*, 2007, **292**, G657.
108. M. Adachi, R. Kurotani, K. Morimura, Y. Shah, M. Sanford, B. B. Madison, D. L. Gumucio, H. E. Marin, J. M. Peters, H. A. Young and F. J. Gonzalez, *Gut*, 2006, **55**, 1104.
109. Y. T. Azuma, K. Nishiyama, Y. Matsuo, M. Kuwamura, A. Morioka, H. Nakajima and T. Takeuchi, *Int. Immunopharmacol.*, 2010, **10**, 1261.
110. S. Cuzzocrea, R. Di Paola, E. Mazzon, T. Genovese, C. Muia, T. Centorrino and A. P. Caputi, *Lab. Invest.*, 2004, **84**, 1643.
111. F. Varnat, B. B. Heggeler, P. Grisel, N. Boucard, I. Corthesy-Theulaz, W. Wahli and B. Desvergne, *Gastroenterology*, 2006, **131**, 538.
112. J. W. Lee, P. J. Bajwa, M. J. Carson, D. R. Jeske, Y. Cong, C. O. Elson, C. Lytle and D. S. Straus, *Gastroenterology*, 2007, **133**, 108.
113. B. Sung, S. Park, B. P. Yu and H. Y. Chung, *J. Gerontol., Ser. A*, 2004, **59**, 997.
114. M. Iemitsu, T. Miyauchi, S. Maeda, T. Tanabe, M. Takanashi, Y. Irukayama-Tomobe, S. Sakai, H. Ohmori, M. Matsuda and I. Yamaguchi, *Am. J. Physiol.: Heart Circ. Physiol.*, 2002, **283**, H1750.
115. P. Ye, X. J. Zhang, Z. J. Wang and C. Zhang, *Gerontology*, 2006, **52**, 69.
116. D. Wu, Z. Ren, M. Pae, W. Guo, X. Cui, A. H. Merrill and S. N. Meydani, *J. Immunol.*, 2007, **179**, 4829.
117. D. S. Gelinas and J. McLaurin, *Neurochem. Res.*, 2005, **30**, 1369.
118. T. Aoyama, J. M. Peters, N. Iritani, T. Nakajima, K. Furihata, T. Hashimoto and F. J. Gonzalez, *J. Biol. Chem.*, 1998, **273**, 5678.
119. P. Howroyd, C. Swanson, C. Dunn, R. C. Cattley and J. C. Corton, *Toxicol. Pathol.*, 2004, **32**, 591.
120. M. Barbieri, M. R. Rizzo, M. Papa, R. Acampora, L. De Angelis, F. Olivieri, F. Marchegiani, C. Franceschi and G. Paolisso, *Exp. Gerontol.*, 2005, **40**, 599.
121. M. Capri, S. Salvioli, F. Sevini, S. Valensin, L. Celani, D. Monti, G. Pawelec, G. De Benedictis, E. S. Gonos and C. Franceschi, *Ann. N. Y. Acad. Sci.*, 2006, **1067**, 252–263.
122. H. Zhang, Y. Li, Y. Fan, J. Wu, B. Zhao, Y. Guan, S. Chien and N. Wang, *Kidney Int.*, 2008, **74**, 732.
123. C. Argmann, R. Dobrin, S. Heikkinen, A. Auburtin, L. Pouilly, T. A. Cock, H. Koutnikova, J. Zhu, E. E. Schadt and J. Auwerx, *PLoS Genet.*, 2009, **5**, e1000752.
124. B. Staels and J. C. Fruchart, *Diabetes*, 2005, **54**, 2460.
125. M. B. Wright, M. Bortolini, M. Tadayyon and M. Bopst, *Mol. Endocrinol.*, 2014, **28**, 1756.
126. B. Grygiel-Gorniak, *Nutr. J.*, 2014, **13**, 17.
127. S. Tyagi, P. Gupta, A. S. Saini, C. Kaushal and S. Sharma, *J. Adv. Pharm. Technol. Res.*, 2011, **2**, 236.

128. H. C. Yang, S. Deleuze, Y. Zuo, S. A. Potthoff, L. J. Ma and A. B. Fogo, *J. Am. Soc. Nephrol.*, 2009, **20**, 2380.
129. M. H. Park, D. H. Kim, M. J. Kim, E. K. Lee, H. J. An, J. W. Jeong, H. R. Kim, S. J. Kim, B. P. Yu, H. R. Moon and H. Y. Chung, *J. Gerontol., Ser. A*, 2016, **71**, 300.
130. M. M. Speeckaert, C. Vanfraechem, R. Speeckaert and J. R. Delanghe, *Ageing Res. Rev.*, 2014, **14**, 1.
131. T. B. Newman and S. B. Hulley, *JAMA*, 1996, **275**, 55.
132. J. E. Dalen and W. S. Dalton, *JAMA*, 1996, **275**, 67.
133. M. C. Carmona, K. Louche, B. Lefebvre, A. Pilon, N. Hennuyer, V. Audinot-Bouchez, C. Fievet, G. Torpier, P. Formstecher, P. Renard, P. Lefebvre, C. Dacquet, B. Staels, L. Casteilla, L. Penicaud and R. Consortium of the French Ministry of and Technology, *Diabetes*, 2007, **56**, 2797.
134. F. L. Dunn, L. S. Higgins, J. Fredrickson, A. M. DePaoli and I.N.T.s. group, *J. Diabetes Complications*, 2011, **25**, 151.
135. P. L. Feldman, M. H. Lambert and B. R. Henke, *Curr. Top. Med. Chem.*, 2008, **8**, 728.
136. K. W. Chung, D. H. Kim, M. H. Park, Y. J. Choi, N. D. Kim, J. Lee, B. P. Yu and H. Y. Chung, *Exp. Gerontol.*, 2013, **48**, 1049.
137. L. Bordone and L. Guarente, *Nat. Rev. Mol. Cell Biol.*, 2005, **6**, 298.
138. S. Libert and L. Guarente, *Annu. Rev. Physiol.*, 2013, **75**, 669.
139. M. M. Masternak and A. Bartke, *PPAR Res.*, 2007, **2007**, 28436.
140. M. M. Masternak, K. A. Al-Regaiey, M. M. Del Rosario Lim, V. Jimenez-Ortega, J. A. Panici, M. S. Bonkowski, J. J. Kopchick and A. Bartke, *J. Gerontol., Ser. A*, 2005, **60**, 1394.
141. M. M. Masternak, K. Al-Regaiey, M. S. Bonkowski, J. Panici, L. Sun, J. Wang, G. K. Przybylski and A. Bartke, *J. Gerontol., Ser. A*, 2004, **59**, 784.
142. J. C. Corton and H. M. Brown-Borg, *J. Gerontol., Ser. A*, 2005, **60**, 1494.
143. B. Sung, S. Park, B. P. Yu and H. Y. Chung, *Exp. Gerontol.*, 2006, **41**, 590.
144. A. Tsuchida, T. Yamauchi, S. Takekawa, Y. Hada, Y. Ito, T. Maki and T. Kadowaki, *Diabetes*, 2005, **54**, 3358.
145. J. D. Wagner, M. K. Shadoan, L. Zhang, G. M. Ward, L. J. Royer, K. Kavanagh, O. L. Francone, B. J. Auerbach and H. J. Harwood Jr., *J. Pharmacol. Exp. Ther.*, 2010, **333**, 844.
146. A. Defaux, M. G. Zurich, O. Braissant, P. Honegger and F. Monnet-Tschudi, *J. Neuroinflammation*, 2009, **6**, 15.
147. T. Coll, D. Alvarez-Guardia, E. Barroso, A. M. Gomez-Foix, X. Palomer, J. C. Laguna and M. Vazquez-Carrera, *Endocrinology*, 2010, **151**, 1560.
148. H. E. Bays, S. Schwartz, T. Littlejohn, 3rd, B. Kerzner, R. M. Krauss, D. B. Karpf, Y. J. Choi, X. Wang, S. Naim and B. K. Roberts, *J. Clin. Endocrinol. Metab.*, 2011, **96**, 2889.
149. C. R. Swanson, V. Joers, V. Bondarenko, K. Brunner, H. A. Simmons, T. E. Ziegler, J. W. Kemnitz, J. A. Johnson and M. E. Emborg, *J. Neuroinflammation*, 2011, **8**, 91.
150. R. Agarwal, *Am. J. Physiol.: Renal Physiol.*, 2006, **290**, F600.

151. T. R. Cowley, J. O'Sullivan, C. Blau, B. F. Deighan, R. Jones, C. Kerskens, J. C. Richardson, D. Virley, N. Upton and M. A. Lynch, *Neurobiol. Aging*, 2012, **33**, 162.
152. M. H. Park, J. Y. Park, H. J. Lee, D. H. Kim, D. Park, H. O. Jeong, C. H. Park, P. Chun, H. R. Moon and H. Y. Chung, *PLoS One*, 2013, **8**, e78815.
153. A. S. Zadelaar, L. S. Boesten, J. W. Jukema, B. J. van Vlijmen, T. Kooistra, J. J. Emeis, E. Lundholm, G. Camejo and L. M. Havekes, *Arterioscler., Thromb., Vasc. Biol.*, 2006, **26**, 2560.
154. H. W. Jeong, J. W. Lee, W. S. Kim, S. S. Choe, K. H. Kim, H. S. Park, H. J. Shin, G. Y. Lee, D. Shin, H. Lee, J. H. Lee, E. B. Choi, H. K. Lee, H. Chung, S. B. Park, K. S. Park, H. S. Kim, S. Ro and J. B. Kim, *Diabetes*, 2011, **60**, 496.
155. H. Chen, B. Dardik, L. Qiu, X. Ren, S. L. Caplan, B. Burkey, B. R. Boettcher and J. Gromada, *Endocrinology*, 2010, **151**, 3115.

CHAPTER 17

Antidiabetic Biguanides as Anti-Aging Drugs

VLADIMIR N. ANISIMOV[a,b,c]

[a]Department of Carcinogenesis and Oncogerontology, N.N. Petrov Research Institute of Oncology, Leningradskaya str., 68, Pesochny-2, St. Petersburg, 197758, Russian Ferderation; [b]Institute of Cytology, The Russian Academy of Sciences, Tikhoretsky prosp. 4, St.Petersburg, 194064, Russian Federation; [c]People Friendship University of Russia, Miklukho-Maklay str. 6, Moscow, 117198, Russian Federation
*E-mail: aging@mail.ru

17.1 Introduction

During the last decade, there has been a burst of interest in the geroprotective, anti-aging and cancer prevention potential of the antidiabetic biguanide metformin. More than 120 million prescriptions of metformin are written yearly for treatment of type 2 diabetes mellitus and this may have already saved more people from cancer death than any other drug in history.[1] Recently, an impressive project of clinical trial called TAME (Targeting Aging with Metformin) was proposed by Nir Barzilai and colleagues (Albert Einstein College of Medicine in New York).[2,3] Metformin will be administered for 5–7 years to 3000 people aged 70–80 years who already have one or two of three age-associated diseases (heart disease, cancer, cognitive decline). The authors expect that metformin will delay these pathologies and death. Persons with type 2 diabetes cannot be enrolled because they were already being

RSC Drug Discovery Series No. 57
Anti-aging Drugs: From Basic Research to Clinical Practice
Edited by Alexander M. Vaiserman

Published by the Royal Society of Chemistry, www.rsc.org

treated with metformin. This project is based on data showing increased life span of mice and rats supplemented with metformin and on results of clinical observations that demonstrated decrease of mortality in diabetic patients treated with metformin.[4–7] Moreover, a 15% increase was reported in survival of type 2 diabetic patients primarily treated with metformin as compared with healthy people without diabetes.[8] These findings and the planning of the TAME clinical trial[2] raise the question of the safety of long-term administration of metformin in non-diabetic people. In this chapter, we mainly evaluate the available results of preclinical studies on the geroprotective effects of metformin (*N*,*N*-dimethylbiguanide) and other antidiabetic biguanides, phenformin (1-phenylethylbiguanide), and buformin (1-butylbiguanide hydrochloride), and give perspectives on their wide introduction in clinical practice. We focus mainly on end-point results of studies to get answers to two critical questions: (1) could biguanides promote life span extension in non-diabetic individuals? and (2) are they safe for long-term treatment? The mechanisms of the geroprotective, anti-carcinogenic and antitumor effects of biguanides are being intensively studied at present. The findings of these investigations are reported in a lot of comprehensive reviews.[9–12]

17.2 Milestones in Research on Biguanides as Drugs for Aging Prevention in Rodents

In the early 1900s, guanidine was identified as an active compound of the botanic medicine plant *Galega officinalis* (French Lilac), which was commonly used in medieval Europe for the treatment of polyurea in diabetic patients.[13] However, due to the discovery of insulin in 1921, only 30 years later the first biguanides (phenformin, buformin and metformin) were synthesized. The drugs were approved in the middle of the last century in the USA and Europe for the treatment of type 2 diabetes mellitus.

In 1971, Dilman[14] proposed that antidiabetic biguanides may be promising as potent anti-aging and anti-cancer drugs. In the middle of the 1970s, he initiated a series of experiments in mice and rats in the N.N. Petrov Research Institute of Oncology to prove this suggestion. In 1974, it was shown for the first time that phenformin inhibits 7,12-dimethylbenz(*a*)anthracene (DMBA)-induced mammary carcinogenesis in female rats.[15] Five years later, the first article on the inhibitory effect of phenformin on spontaneous mammary carcinogenesis and life span extension in female C3H/Sn mice was published.[16,17] At the same time, we reported the results of studies of the impact of buformin and phenformin on the aging of the reproductive system, life span and incidence of spontaneous tumors in female rats.[18–20] In the same period, a lot of research studies showed the capacity of biguanides to prevent chemically- and irradiation-induced carcinogenesis in rodents. Data on the cancer preventive and anti-tumor effects of biguanides have been analyzed in some recent papers.[21–23] In 2005, it was shown that metformin prolongs the life span and inhibits the development of mammary adenocarcinomas of cancer-prone female transgenic HER-2/neu mice.[24] In 2008, we observed a 38% increase in life span delay in female SHR mice treated with metformin starting from the age of 3 months.[25] Then, it was for the first time shown that metformin is less effective as a geroprotector in adult and old females of the same strain.[26] We also observed that the effects of metformin depend on the sex of the animals—it extended the life span in female but reduced it in male 129/Sv mice.[27] At the same time, Smith *et al.*[28] failed to increase life span in male F344 rats by treatment with metformin. Notably, the neonatal treatment of 129/Sv mice with metformin inverted its gender-dependent effect: in males an increase of life span was observed whereas the female longevity was reduced.[29]

17.3 Effect of Antidiabetic Biguanides on Aging and Life Span in Rats

F344 rats of both sexes were used in the National Cancer Institute Bioassay of Phenformin for Possible Carcinogenicity.[30] The matched control groups included 15 animals each and groups exposed to phenformin consisted of 35 animals each. Phenformin was given in doses 400 and 800 ppm in diet for 78 weeks starting at the age of 8 weeks. Measurement of food consumption allowed doses of phenformin to be estimated as 300–625 mg kg^{-1} day^{-1}. The treatment was followed by an observation period of 26 weeks, than all survived animals were sacrificed. The mean body weights were consistently lower as compared with the controls during the treatment period, while the body weights of males were unchanged by the drug. 53% of control male rats survived until the age of 105 weeks, among them 67% high-dose and 91% low-dose treated males. There were no significant differences between the mortality in the different groups of female rats. 83% of the high-dose group, 68% in the low-dose group and 67% in the control group survived to the end of the study. The incidences of tumors of any localization in males as well as of the majority of tumors in female rats were similar in the control and in the phenformin-treated groups. However, the incidence of tumors of reproductive system was statistically less (21% and 17%) than that in the matched controls (47%) ($p = 0.027$).

Buformin was given 5 times a week in a single dose of 5 mg rat^{-1} day^{-1} orally to female Leningrad Institute Oncology (LIO) rats starting from the age of 3.5 month until a natural death.[18,20] The treatment slightly increased mean life span of rats (by 7%; $p > 0.05$). The mean life span of the last 10% survivors increased by 12% ($p < 0.05$), and the maximum life span increased by 2 months (+5.5%) as compared with controls. The body weight of rats treated with buformin was slightly (5.2 to 9.4%) but statistically significantly ($p < 0.05$) decreased in comparison with the control. At the age of 16–18 months, 38% of control rats revealed disturbances in the estrus cycle (persistent estrus, repetitive pseudopregnancies or anestrous), whereas in females treated with buformin these disturbances were observed only in 9% of rats ($p < 0.05$). The cumulative incidence of spontaneous tumors was decreased by 1.6 times in buformin-treated rats as compared with the control rats and the multiplicity of spontaneous neoplasms was decreased almost 2-fold.

Phenformin was given to female outbred LIO rats intragastrally 5 times a week starting from the age of 3.5 months in a single dose of 5 mg rat^{-1} day until natural death.[19,20] Administration of phenformin failed to influence the mean life span in rats. At the same time, the mean life span of the last 10% survivors was increased by 10% ($p < 0.05$), and maximum life span was increased by 3 months (+10%) in comparison with the controls. The treatment with phenformin slightly decreased the body weight of rats in comparison with the control ($p > 0.05$). Disturbances in the estrous function were observed in 36% of 15–16 month-old rats of the control group and only in 7% of rats in phenformin-treated group ($p < 0.05$). The incidence of spontaneous tumors was decreased by 1.3 times in penformin-treated rats as compared with the control group.[19,20]

Six month-old male F344 rats were randomly subdivided into four groups and were maintained on one of four diets: control, calorie restricted (CR), metformin (300 mg kg^{-1} day^{-1}) and pair-fed to metformin.[28] The CR group had significantly reduced food intake and body weight throughout the study. Body weight was significantly reduced in the metformin group compared with the control group during the middle of the study, despite the similar weekly food intake. There were no significant differences in the mean life span or the mean life span of the last surviving 10% of each group in the CR, metformin and pair-fed groups compared with the control. However, the aging rate estimate (α – slope, rate of increase of mortality) of the Gompertz model in the control group alone was significantly different from the three other groups, reflecting the early deaths in the CR, metformin and pair-fed groups. CR significantly increased life span in the 25th quantile but not the 50th, 75th, or 90th quantiles. The survival of rats in groups exposed to metformin or to the pair-feeding were not significantly different from the controls at any quantile.[28] The authors stressed the one limitation of this study—the lack of a robust CR response for extension of maximum life span, which has been observed in another CR study using the same rat strain.[31] The reduced efficacy of CR in this study might provide a partial explanation for the lack of a significant increase with metformin treatment. In addition to the dampened CR response, metformin treatment did not significantly affect glucose/insulin levels in this study. The metformin concentration utilized in the diet was approximately 10 times greater than the highest dose used in human treatments, implying that any increase necessary to observe life span benefits is questionable for a human application.[28]

Thus, the available data on the effects of biguanides in rats are very scarce. The early NCI study was terminated before the natural death of the majority of animals. Moreover, the sample sizes (15 rats in the control group) were too small in this study. In our long-term studies, buformin and phenformin were tested only in females and in a single dose, whereas metformin was tested only in male rats and also in a single dose.

17.4 Effect of Antidiabetic Biguanides on Aging and Life Span in Mice

In the National Cancer Institute Bioassay of Phenformin for Possible Carcinogenicity, B6C3F1 mice of both sexes were used.[30] Similar to the rat study, the matched control groups included 15 animals each whereas groups exposed to phenformin comprised 35 animals each. Phenformin was given in low and high doses (400 and 800 ppm) in diet for 78 weeks starting at the age of 8 weeks. The treatment was followed by an observation period of 26 weeks, than all surviving animals were sacrificed. The mean body weights of both male and female B6C3F1 mice were markedly lower than those in controls during the first 60–78 weeks of administration of phenformin at the low and high doses. Until the age of 105 weeks, 13% controls, 19% low-dose and 29% high-dose treated male mice and 33%, 59% and 52% of female mice, respectively, were surviving. The incidence of hematopoietic tumors was 33% of the matched controls of both male and female mice, compared to only 1.5% hematopoietic tumors in the male and 5.4% of the female controls. The conclusion was that there is no evidence that under these conditions phenformin was carcinogenic. It's worthy to note that in female B6C3F1 mice the survival was rather increased in the phenformin-treated groups. The treatment with this drug also decreased the incidence of lymphomas in male and female mice. Unfortunately, since this study was terminated after the age of 112 weeks, animals could not survive until their natural death and it did not allow the evaluation of the effect of phenformin on the mean and maximal life span.

The geroptotector effect of biguanides was first demonstrated in our studies with phenformin orally given to mice.[16,17] Long-term administration of phenformin to female C3H/Sn mice (2 mg day^{-1} $mouse^{-1}$ orally) started at the age of 3.5 months was followed by a 21% increase in mean life span and a 26% increase in the maximum life span (Table 17.1). The incidence of spontaneous mammary adenocarcinomas as well as leukemias was reduced by four times under the treatment with phenformin as compared with control mice given tap water. It is worthy to note that the treatment with phenformin significantly—to more than 6 months (+53.9%)—increased the mean life span of tumor-free C3H/Sn mice.

Available data on the effects of antidiabetic biguanides in rodents are summarized in the Table 17.2.

In two sets of our experiments, administration of metformin with drinking water (100 mg kg^{-1} 5 times a week starting at the age of 2 months) to transgenic HER-2/neu female mice slightly increased the mean life span by 4–8%, and decreased the size and multiplicity of mammary adenocarcinomas.[24,32] The treatment reduced food consumption but did not have influence on the dynamics of body weight. In tumor mice treated with metformin, the expression of mRNA coding for lymphocyte-associated proteins granzyme-b and perforin mRNA was explored. Expression of these cytolytic molecules was not detected in the control, but it was significantly increased in mice treated

Table 17.1 Effect of phenformin on life span and spontaneous carcinogenesis in female C3H/Sn mice.[a]

Parameters		Control	Phenformin	Δ%; p
Number of mice		30	24	
Life span, days:	Mean	450 ± 23.4	545 ± 39.2	+21.1%, $p < 0.05$
	Last 10% of survivors	631 ± 11.4	810 ± 0	+28.4%, $p < 0.05$
	Maximum	643	810	+26.0%
Aging rate $\alpha \times 10^3$, days^{-1}		7.64 (7.59; 8.10)	5.26 (4.94; 5.51)	$p < 0.05$
MRDT, days		90.7 (85.6; 92.5)	131.8 (125.8; 140.3)	$p < 0.05$
Number of tumor-free mice		6 (20%)	19 (79.2%)	$p < 0.05$
Mean life span of tumor-free mice		362 ± 49.0	557 ± 41.6	+53.9%, $p < 0.05$
Number of tumor-bearing mice (TBM)		24 (80%)	5 (20.8%)	−3.8 fold, $p < 0.05$
Mean life span of TBM, days		472 ± 25.1	499 ± 111.6	+5.7%
Total number of tumors		41	5	
Number of tumors per TBM		1.7	1.0	−41.2%
Number of mice with mammary adenocarcinomas (MAC)		19 (63.3%)	4 (16.0%)	−4.0 fold, $p < 0.05$
Number of mice with leukemia		5 (16.7%)	1 (4.2%)	−4.0 fold, $p < 0.05$

[a]Life spans are given as means ± standard errors; 95% confidence limits are given in parentheses; MRDT = mortality rate doubling time.

with metformin. Treatment with metformin has been accompanied by a slow down of the age-related increase in blood glucose levels, as well as by a reduction in the levels of insulin, triglycerides and lipoproteins in the blood serum, compared with controls.[24]

In experiments on female SHR mice, administration of metformin in a dose of 100 mg kg^{-1} started at the age of 2 months shifted survival curve to the right, and increased mean life span by 38%.[25] In another experiment, the same strain of mice were given metformin in drinking water from the age of 3, 9 or 15 months.[18] The mean life span of mice given the drug started at the age of 3 months increased by 14%, and the maximum one increased by 1 month. If treatment was started at the age of 9 months, the mean life span increased by only 6%, while in the older group it was not changed. The average life expectancy of mice without tumors increased by 21% and 7% in young and middle-age groups, respectively, whereas in old group it was decreased by 13%. It is important to note that in all age groups the use of metformin was accompanied by a decrease in the body temperature of mice and postponed age-related switching of the estrous cycle. The accumulation of senescent cells was slowed down in primary cultures of skin fibroblasts derived from mice injected with metformin from the age of 3, 9 or 15 months.[33]

In inbred 129/Sv mice, the mean life span of males treated with metformin started at the age of 3 months was reduced by 13.4% in males and slightly

Table 17.2 Effect of antidiabetic biguanides on life span and spontaneous carcinogenesis in rodents. « means the same dose as indicated in the same column at the previous line.

Strain	Sex[a]	Number of animals: control/ treatment	Age at start of treatment, months	Drug[b]	Dose and route of treatment[c]	Effect on mean life span, Δ%	Effect on carcinogenesis[d]	Refe- rences
Rats								
F344	M	15/35	2	PF	400 ppm, in diet	$+14^e$	↓	30
					800 ppm	$+38^e$	↓	
F344	F	15/35	2	PF	400 ppm, in diet	$+1^e$	↓	
					800 ppm	$+16^e$	↓	
LIO	F	41/44	3.5	PF	5 mg rat^{-1}, p.o.	0	↓	19,20
LIO	F	74/42	3.5	BF	«	+7	↓	18,20
F344	M	31/40	6	PF	300 mg kg^{-1}, in diet	0	ND	28
Mice								
B3C6F1	M	15/35	2	PF	400 ppm, in diet	$+6^e$	↓	30
					800 ppm	$+16^e$	↓	
B3C6F1	F	15/35	2	PF	400 ppm, in diet	$+26^e$	↓	
					800 ppm	$+19^e$	↓	
C3H/Sn	F	20/24	3.5	PF	2 mg mouse^{-1}, p.o.	+21	↓	16,17
HER-2/neu	F	34/32	2	MF	100 mg kg^{-1}, d.w.	+8	↓	24
HER-2/neu	F	31/35	2	MF	«	+4	↓	31
SHR	F	50/50	3	MF	«	+38	=	25
SHR	F	119/51	3	MF	«	+14	=	26
		97/45	9	MF	«	+6	=	
		69/33	15	MF	«	0	=	
129/Sv	M	50/39	3rd, 5th,	MF	100 mg kg^{-1}, s.c.	+20	=	29
129/Sv	F	35/30	7th days	MF	100 mg kg^{-1}, s.c.	−9	=	
129/Sv	M	41/46	3	MF	100 mg kg^{-1}, d.w.	−13	=	27
129/Sv	F	47/41	3	MF	100 mg kg^{-1}, d.w.	+5	↓	
C57BL/6	M	64/83	12	MF	10 mg kg^{-1}, in diet	+6	=	32
C57BL/6	M	90/88	12	MF	100 mg kg^{-1}, in diet	−14	↓	
B3C6F1	M	297/36	12	MF	10 mg kg^{-1}, p.o.	+6	=	

[a]F: female; M: male.
[b]BF: buformin; MF: metfromin; PF: phenformin.
[c]d.w.: drinking water; i.p.: intraperitoneally; p.o.: (orally) gavage; s.c.: subcutaneously.
[d]↑: increase; =: no effect; ND: not detected.
[e]survival at the 105th week.

increased in females. In addition, treatment with metformin resulted in a 3.5 times reduction of the incidence of spontaneous malignant tumors in females, and did not affect this in males.[27]

Martin-Montalvo *et al.*[34] treated male C57BL/6 mice with metformin (0.1% and 1% w/w) starting at the age of 12 months until the natural death of mice. Administration of a small dose of metformin was followed by a 5.83% increase in mean life span, whereas 1% of metformin in diet reduced it by 14.4%. A small dose of the drug leads to a decrease in liver cancer incidence (3.3% *vs.* 26.5%). Authors also reported improved hormone-metabolic parameters in mice exposed to a small dose of metformin. The same effect of a small dose of metformin was also observed in mice of another strain—B6C3F1.[34] Remarkably, the significant similarity was observed between patterns of gene expression in skeletal muscles and liver in mice maintained on a 40% calorie restricted diet and those that received the 0.1% metformin diet.[35,36]

The perinatal (prenatal and early neonatal) period is a critical stage for hypothalamic programming of sexual differentiation as well as for the development of energy and metabolic homeostasis. We hypothesized that neonatal treatment with metformin would positively modify regulation of the growth hormone/IGF-1/insulin signaling pathway, slowing down aging and improving cancer preventive pathways in rodents. To test this hypothesis male and female 129/Sv mice were injected with metformin (100 mg kg^{-1}) at the 3rd, 5th and 7th days after birth.[29] Metformin-treated males consumed less food and water and their body weight was decreased as compared with control mice over their entire life span. There were, however, no significant differences in the age-related dynamics of food and water consumption in females and they were heavier than the controls. The fraction of mice with regular estrous cycles decreased with age and demonstrated a tendency to decrease in the females neonatally treated with metformin. Serum insulin levels were reduced whereas the levels of serum IGF-1, cholesterol and nitric oxide were increased in 3 month-old control females in comparison to control males. No age-related difference in levels of glucose, total cholesterol, triglycerides, insulin and some other metabolic and hormonal parameters between 3 and 9 month-old male control mice was revealed. The malonic dialdehyde, IGF-1 and NO levels in 9 month-old control males were, however, higher in comparison to those in 3 month-old males. In 9 month-old control females, malonic dialdehyde and IGF-1 levels were increased as compared to those in 3 month-old females, whereas the level of cholesterol was decreased. Neonatal exposure to metformin failed to change most hormonal and metabolic parameters in the blood serum of male and female mice. In the male group, neonatal metformin treatment significantly increased the mean life span (+20%, $p < 0.05$) and slightly increased the maximum life span (+3.5%). In females, the mean life span in metformin-treated groups was slightly decreased (−9.1%, $p > 0.05$) in comparison to controls, whereas the median was increased by 13.8%. The mean life span of the last 10% survivors and maximum life span were the same as in the controls. 45% of control male mice and 71.8% of male mice who were neonatally exposed to metformin survived up to 800 days of age, and the same age was achieved by 54.3% of mice in the control female group and 30% of metformin-treated females ($p < 0.03$; Fischer's exact test). According to the log-rank test, the life span distribution of 129/Sv mice treated with metformin differed significantly from the control population. The difference was much more significant in male mice ($p = 0.0006$) than in females ($p = 0.0555$). The Cox's regression model has shown that neonatal metformin treatment increased the relative risk of death in female mice and decreased it in males compared to the respective intact control groups. Thus, neonatal metformin exposure slows down aging and prolongs life span in male but not in female mice.[29]

17.5 Antidiabetic Biguanides in Prevention of Age-Associated Diseases in Mouse Models

Transgenic mice with Hungtington's disease (HD) (the R6/2 line expressing exon 1 of the Huntington protein including ~130 glutamine repeats) were given metformin in drinking water (2 or 5 mg ml^{-1}) starting from the age of 5 weeks.[37] Metformin treatment significantly prolonged (by 20.1%) the survival time of male (but not female) HD mice at the 2 mg ml^{-1} dose (~300 mg kg^{-1} day^{-1}) without affecting the fasting blood glucose level. This dose of the drug also decreased hind limb clasping time in 11 week-old mice. The higher dose of metformin did not prolong life span, and neither dose was effective in female HD mice. Recently, additional evidence of a protective effect of low-dose metformin on neuronal dysfunction has been reported in mouse model of Huntington's disease.[38]

In another study, SOD1^{G92A} mice of both sexes with a transgenic model of amyotrophic lateral sclerosis (ALS) were given metformin with their drinking water in various doses (0.5, 2, and 5 mg ml^{-1}) starting from the 35th day of age.[39] Administration of metformin failed to have any effect on the disease onset, progression or survival in male SOD1^{G92A} mice at any doses. Moreover, in females authors observed a dose-dependent negative effect of metformin on neurological response. All groups exposed to metformin exhibited weight loss and significant life extension. The authors noted, however, a trend toward increased survival with a decreasing dose of metformin.[39]

17.6 Antidiabetic Biguanides as Anti-Carcinogens and Inhibitors of Tumor Growth in Rodents

The available data on the results of *in vivo* studies of effects of biguanides involving more than 20 experimental models of carcinogenesis were recently analyzed.[22,23,40] They included models of spontaneous carcinogenesis (in rats and mice), chemical carcinogenesis induced by 18 different agents, 4 viruses, 2 dietetic modifications, and 2 types of ionizing radiation (X-rays and gamma-rays) (Table 17.3). Antidiabetic biguanides were given with diet, drinking

Table 17.3 List of carcinogenic agents used in studies on preventive effect of antidiabetic biguanides in rodents.

Type of carcinogens	Carcinogenic agent
Chemical carcinogens	Azoxymethane (AOM)
	Benzo(*a*)pyrene (BP)
	7,12-Dimethylbenz(*a*)anthracene (DMBA)
	1,2-Dimethyhydrazine (DMH)
	Dextrane sodium sulfur (DSS)
	20-Methylcholanthrene (MCA)
	4-Methylnitrosamino-1-(3-pyridyl)-1-butanone (NKK)
	4-Nitroquinoline-1-oxide (4-NQO)
	N-Nitrosobis(2-oxopropyl)amine (NBOPA)
	N-Nitrosodiethylamin (NDEA)
	N-Nitrosomethylurea (NMU)
	N-Nitrosoethylurea (NEU)
	Streptozotocin (STZ)
	Tobacco smoking condensate
	Tamoxifen
	12-*O*-Tetradecanoylphorbol-13-acetate (TPA)
	Urethan
	Estradiol-17β
Ionizing radiation	Total-body X-rays
	Gamma-rays
Viruses	Adenovirus LSA^{+}L-Kras; trp53
	MMTV (murine mammary tumor virus)
	MMTV-neu; p53±
	MMTV-PyVT
Genetically modified mice	HER-2/neu
	$Apc^{Min/+}$
	HBx
	LID, liver-IGF-1-deficient mice
	$PTEN^{+/-}$
Diets	High fat diet
	High carbohydrate fat diet

water, orally (gavage), intraperitoneally or subcutaneously. Experiments were performed on more than 20 strains of mice (inbred, outbred, genetically modified mouse strains), 4 rat strains and 1 hamster strains (Table 17.4). The effect of the antidiabetic biguanides has been studied in 17 target organs/tissues. The majority of these studies were focused on mammary gland (19 articles) and colon tumors (11), the liver (7) and the uterus (5), which reflects the importance of these localizations for clinical practice. The effect of biguanides on total tumor incidence was evaluated in 11 articles. Positive (inhibitory) effects induced by treatment with biguanides have been observed in 77.9% of cases (Table 17.5). It is worthy to note that there were no cases of stimulation of any type of carcinogenesis with antidiabetic biguanides. In the majority of studies on the effect of biguanides on induced carcinogenesis, young adult rodents were used, and only in one article the results of the treatment with metformin started at the young, adult and old age have been

Table 17.4 List of rodent strains used in studies on preventive effect of antidiabetic biguanides on carcinogenesis.

Species	Strain
Mice	A/J; *Apc*$^{Min/+}$; Balb/c; B6C3F1; CD1; C3H/S n; C57BL/6; C57BL/6(HBxTg); db/db; FVB/N; FVB/N-Tg; HBx; ICR; KPC; LID; LSA^{+}L-Kras; trp53; p48Cre$^{/+}$. LSL-ras G12D; Swiss; Swiss H; outbred; PTEN$^{+/-}$; SHR; SOD1^{G92A}; Swiss; Swiss-H; 129/Sv; Thrb$^{PV/PV}$Pten$^{+/-}$
Rats	F344; LIO; Sprague-Dawley; Wistar
Hamsters	Not shown

Table 17.5 Effect of antidiabetic biguanides on spontaneous and induced carcinogenesis in various organs and tissues in rodents.

		Number of studies		
No.	Target organ/tissue	Total	Positive effect	No effect
1.	Mammary gland	19	13	6
2.	Pituitary	2	2	0
3.	Thyroid gland	3	3	0
4.	Skin	3	3	0
5.	Soft tissues	2	2	0
6.	Uterus	5	5	0
7.	Cervix utery	1	1	0
8.	Lungs	4	4	0
9.	Oral mucosa	1	1	0
10.	Pancreas	4	4	0
11.	Pancreatic islets	1	1	0
12.	Liver	7	6	1
13.	Small interstines	3	1	2
14.	Colon	11	8	3
15.	Kidney	2	2	0
16.	Lymphoid tissues	5	4	1
17.	Nervous tissues	2	2	0
18.	Total tumor incidence	11	5	6
	Total studies	86	67 (77.9%)	19 (22.1%)

published.[26] Furthermore, only one study was focused on tumorigenesis in mice neonatally exposed to metformin. A detailed analysis of the experimental studies on anti-carcinogenic effect of antidiabetic biguanides has been done recently.[22,23,40]

The biguanides suppressed tumor growth in the majority of *in vitro* studies conducted in 46 different cell lines originated from malignant tumors of 15 localization as well as in athymic mice with xenografts of 31 tumor lines.[40] It was concluded that there is sufficient experimental evidence of the anti-carcinogenic and antitumor effects of antidiabetic biguanides in a number of models of induced and spontaneous carcinogenesis.[23,40] There are several hundred excellent reviews and meta-analyses of epidemiological data on the effect of metformin on cancer risk in type 2 diabetes patients(see, *e.g.*, ref. 10,41–45).

17.7 Conclusion

One of the hot spots in current gerontology is the question about the role of age-related changes in the system of growth hormone (GH)/insulin-like growth factor-1 (IGF-1)/insulin in aging and age-related pathology, including cancer.[21,42–44] The night peak of secreting GH accompanied by the decline in the concentration of serum IGF-1 are known to be reduced with age in both humans and laboratory rodents. It is generally believed that the age-associated reduction in GH secretion is due to a lower response of the pituitary gland on the action of growth hormone releasing hormone (GHRH) which, in turn, also decreases with age. An important role in reducing the significance of GH with aging may also be played by an age-related increase of tonic production of somatostatin and reduced sensitivity of neurons of the hypothalamus, providing homeostatic effects of GH. It should be noted that the aging of the pineal gland and hypothalamus affects important regulatory mechanisms for monitoring food intake, *e.g.*, centers of appetite and satiety.[21,45–47]

Another "hot spot" in the field is the role of the mammalian target of rapamycin (mTOR) signaling pathway in the control of aging and carcinogenesis.[21,48,49] mTORC1 (mammalian target of rapamycin complex 1) is activated by insulin and related growth factors through phosphatidylinositol-3-OH kinase and AKT kinase signaling and it is repressed by AMP-activated protein kinase, a key sensor of cellular energy status.[47] mTORC1 is known to be involved into promotion of messenger RNA translation and protein synthesis through ribosomal protein S6 kinases (S6Ks) and 4E-BP protein. mTORC1 also stimulates lipid biosynthesis, inhibits autophagy, and regulates mitochondrial function and glucose metabolism through hypoxic response transcription factor HIF-1α. The life span of S6K1-deficient female mice increased by 19% without an effect on tumor development.[50] These data suggest that S6K1 plays a substantial role in life span regulation downstream of TORC1. It has been shown that decreased mTORC1 signaling is sufficient for life span prolongation independently from changes in glucose homeostasis.[48–50] Rapamycin suppresses mTORC1 and indirectly mTORC2 that leads to metabolic lesions like glucose intolerance and abnormal lipid profile.[49–51] Treatment with rapamycin or its more soluble form rapatar increased the mean life span in various strain of mice.[52–56] It can be assumed that the regulation of GH and IGF-1, oxidative stress, DNA damage, and metabolic pathways by calorie restriction could simultaneously lead to the anti-aging and anti-tumor activities as well as to reduction of the number of senescent cells in some tissues.[33,50,51]

In Table 17.6, data on patterns of changes during aging, calorie restriction, carcinogenesis and metformin treatment at molecular, cellular, tissue, systemic and organism levels are presented. There are obviously a lot of similarities between aging and carcinogenesis, on the one hand, and between effects of calorie restriction and treatment with metformin, on the other hand. At the same time, both calorie restriction and metformin alleviate effects of aging and carcinogenesis in a similar way.

Table 17.6 Changes developing in organisms during natural aging and carcinogenesis: effects of calorie restriction and metformin.

Parameters	Aging	Carcinogenesis	Calorie restriction	Biguanides (metformin)
Molecular level				
Free radical generation	↑	↑	↓	↓
AGEs formation	↑	↑	↓	↓
DNA adducts formation	↑	↑	↓	↓
DNA repair efficacy	↓	↓	↓	↓
Genomic instability	↑	↑	↓	↓
Telomerase activity	↓	↑	↓	↓
Telomere length	↓	↑	↑	↑
mTOR activity	↑	↑	↓	↓
IKK-β/NF-κB activity	↑	↑	↓	↓
Clock gene expression (*Per1*, *Per2*)	↓	↓	↓	↓
Mutation rate	↑	↑	↓	↓
Oncogene expression	↑	↑	↓	↓
p53 mutations	↓	↑	?	?
Cellular/tissue level				
Oxidative stress	↑	↑	↓	↓
Chromosome aberrations	↑	↑	↓	↓
Induced pluripotent stem cells (iPSC)	↓	↓	↓	↓
Proliferative activity	↓	↑	↓	↓
Focal hyperplasia	↑	↑	↓	↓
Apoptosis	↓	↓	↑	↑
Autophagy	↓	↓	↑	↑
Angiogenesis	↓	↓	↓	↓
Cell-to-cell communication	↓	↓	↑	↑
Senescent cells number	↑	↑	↓	↓
Latent (dormant) tumor cells number	↑	↑	↓	↓
Systemic/organism level				
Melatonin circadian rhythm	↓	↓	↑	↑
Serum melatonin level	↓	↓	↑	↑
Hypothalamic threshold of scnisitivity to homcostatic inhibition by steroids	↑	↑	↓	↓
Tolerance to glucose	↓	↓	↑	↑
Serum insulin level	↑	↑	↓	↓
Susceptibility to insulin	↓	↓	↑	↑
LDL and cholesterol level	↑	↑	↓	↓
Ovulatory function	↓	↓	↓	↑
Fertility	↓	↓	↓	↑
T-cell immunity	↓	↓	↑	↑
Inflammation	↑	↑	↓	↓
Cancer risk	↑	↑	↓	↓
Life span	↓	↓	↑	↑

The available data allow us to examine metformin and other antidiabetic biguanides as promising drugs to prevent age-associated pathologies. In many clinical trials, it was found that the use of metformin and other biguanides can reduce by more than a third the total mortality, myocardial infarctions and mortality from complications of diabetes, improve the survival rate of cancer patients and reduce the risk of breast cancer in patients with type 2 diabetes mellitus.[8–13] Experiments on laboratory rodents presented sufficient evidence of capacity of antidiabetic biguanides to slow down aging, to increase the life span and to prevent development of spontaneous and induced tumors. As was recently noted by Michael Pollack: "The problem with metformin is it's cheap, it's widely available, it has a great safety profile, and anyone can use it".[57] Finally, I believe that the title of the recently published article "Metformin: a hopeful promise in aging research"[58] comprehensively reflects the situation with this rather old drug, which is a novel promising candidate to keep us youthful...

Acknowledgement

The support from The Russian Foundation for Basic Research (14-04-01653) is acknowledged.

References

1. G. Taubes, *Science*, 2012, **335**, 29.
2. S. S. Hall, *Science*, 2015, **349**, 1275.
3. E. C. Hayden, *Nature*, 2015, **522**, 265.
4. S. C. Johnson, P. S. Rabinovitch and M. Kaeberlein, *Nature*, 2013, **493**, 338.
5. E. O. Cosmanova, R. B. Canada, T. A. S. Magbold, W. N. Rawn and B. M. Wall, *Am. J. Med. Sci.*, 2008, **336**, 241.
6. L. Azoulay, V. Schneider-Lindner, S. Dell'aniello, A. Schiffrin and S. Susa, *Pharmacoepidemiol. Drug Saf.*, 2010, **19**, 335.
7. C.-P. Wang, C. Lorenzo and S. E. Espinosa, *J. Endocrinol. Diabetes Obes.*, 2014, **2**, 1031.
8. C. A. Bannister, S. E. Holden, S. Jenkins-Jones, C. L. Morgan, J. P. Halcox, G. Schernthaner, J. Mukherjee and C. J. Currie, *Diabetes Obes. Metab.*, 2014, **16**, 1165.
9. S. Del Barco, A. Vazquez-Martin, S. Cufí, C. Oliveras-Ferraros, J. Bosch-Barrera, J. Joven, B. Martin-Castillo and J. A. Menendez, *Oncotarget*, 2011, **2**, 896.
10. B. Martin-Castillo, A. Vazquez-Martin, C. Oliveras-Ferraros and J. A. Menendez, *Cell Cycle*, 2010, **9**, 1057.
11. M. V. Blagosklonny, *Cell Cycle*, 2010, **9**, 3151.
12. J. A. Menendez, S. Cufí, C. Oliveras-Ferraros, L. Vellon, J. Joven and A. Vazquez-Martin, *Aging (Albany NY)*, 2011, **3**, 348.

13. M. Pollak, *J. Clin. Invest.*, 2013, **123**, 3693.
14. V. M. Dilman, *Lancet*, 1971, **1**, 1211.
15. V. M. Dilman, L. M. Berstein, M. A. Zabezhinski and V. A. Alexandrov, *Vopr. Onkol.*, 1974, **20**, 94.
16. V. M. Dilman and V. N. Anisimov, *Dokl. Akad. Nauk SSSR*, 1979, **245**, 753.
17. V. M. Dilman and V. N. Anisimov, *Gerontology*, 1980, **26**, 241.
18. V. N. Anisimov, *Vopr. Onkol.*, 1980, **26**, 42.
19. V. N. Anisimov, *Farmakol. Toksikol.*, 1982, **45**, 127.
20. V. N. Anisimov, A. V. Semenchenko and A. I. Yashin, *Biogerontology*, 2003, **4**, 297.
21. V. N. Anisimov and A. Bartke, *Crit. Rev. Oncol. Hematol.*, 2013, **87**, 201.
22. V. N. Anisimov, *Ann. Transl. Med.*, 2014, **2**, 60.
23. V. N. Anisimov, *Oncotarget*, 2015, **6**, 39398.
24. V. N. Anisimov, L. M. Berstein, P. A. Egormin, T. S. Piskunova, I. G. Popovich, M. A. Zabezhinski, I. G. Kovalenko, T. E. Poroshina, A. V. Semenchenko, M. Provinciali, F. Re and C. Franceschi, *Exp. Gerontol.*, 2005, **40**, 685.
25. V. N. Anisimov, L. M. Berstein, P. A. Egormin, T. S. Piskunova, I. G. Popovich, M. A. Zabezhinski, M. L. Tyndyk, M. N. Yurova, I. G. Kovalenko, T. E. Poroshina and A. V. Semenchenko, *Cell Cycle*, 2008, **7**, 2769.
26. V. N. Anisimov, L. M. Berstein, I. G. Popovich, M. A. Zabezhinski, M. L. Tyndyk, P. A. Egormin, T. S. Piskunova, M. N. Yurova, A. V. Semenchenko, I. G. Kovalenko and T. E. Poroshina, *Aging (Albany NY)*, 2011, **3**, 148.
27. V. N. Anisimov, T. S. Piskunova, I. G. Popovich, M. A. Zabezhinski, M. L. Tyndyk, P. A. Egormin, M. N. Yurova, A. V. Semenchenko, I. G. Kovalenko, T. E. Poroshina and L. M. Berstein, *Aging (Albany NY)*, 2011, **2**, 945.
28. D. L. Smith, C. F. Elam, J. A. Mattison, M. A. Lane, G. S. Roth, D. K. Ingram and B. Allison, *J. Gerontol., Ser. A*, 2010, **65**, 468.
29. V. N. Anisimov, I. G. Popovich, M. A. Zabezhinski, P. A. Egormin, N. N. Yurova, A. V. Semenchenko, M. L. Tyndyk, A. V. Panchenko, A. P. Trashkov, A. G. Vasilieva and N. V. Khaitsev, *Cell Cycle*, 2015, **14**, 46.
30. *Bioassay of Phenformin for Possible Carcinogenicity. CAS No. 114-86-3; NCI-CG-TR-7. Natl Cancer Institute. Carcinogenesis. Technical Report Series No. 7*, US DHEW. Public Health Service, NIH, January 1977.
31. B. P. Yu, *Biogerontology*, 2006, **7**, 179.
32. V. N. Anisimov, P. A. Egormin, T. S. Piskunova, I. G. Popovich, M. L. Tyndyk, M. N. Yurova, M. A. Zabezhinski, I. V. Anikin, A. S. Karkach and A. A. Romanyukha, *Cell Cycle*, 2010, **9**, 188.
33. A. V. Arkadieva, A. A. MAmonov, I. G. Popovich, V. N. Anisimov and V. M. Mikhelson, *Tsytologia*, 2011, **53**, 166.
34. A. Martin-Montalvo, E. M. Mercken, S. J. Mitchell, H. H. Palacios, P. L. Mote, M. Scheibye-Knudsen, A. P. Gomes, T. M. Ward, R. K. Minor, M. J. Blouin, M. Schwab, M. Pollak, Y. Zhang, Y. Yu, K. G. Becker, V. A. Bohr, D. K. Ingram, D. A. Sinclair, N. S. Wolf, S. R. Spindler, M. Bernier and R. de Cabo, *Nat. Commun.*, 2013, **4**, 2192.
35. R. de Cabo, D. Carmona-Gutierrez, M. Bernier, M. N. Hall and E. Madeo, *Cell*, 2014, **157**, 1515.

36. S. R. Spindler, *Ageing Res. Rev.*, 2010, **9**, 324.
37. T. C. Ma, J. L. Buescher, B. Oatis, J. A. Funk, A. J. Nash, R. L. Carrie and K. R. Hoyt, *Neurosci. Lett.*, 2007, **411**, 98.
38. R. P. Vázquez-Manrique, F. Farina, K. Cambon, M. Dolores Sequedo, A. J. Parker, J. M. Millán, A. Weiss, N. Déglon and C. Neri, *Hum. Mol. Genet.*, 2016, **25**, 1043.
39. H. M. Kaneb, P. S. Sharp, N. Rahmani-Kondori and D. J. Wells, *PLoS One*, 2011, **6**, e24189.
40. V. N. Anisimov, I. G. Popovich, P. A. Egormin, M. L. Tyndyk, M. N. Yurova and M. A. Osipov, *Vopr. Onkol.*, 2016, **62**, 234.
41. L. Wu, L. Zhu, L. J. Prokop and M. H. Murad, *Sci. Rep.*, 2015, **5**, 10147.
42. L. Jacob, K. Kostev, W. Rathmann and M. Kalder, *J. Diabetes Complications*, 2016, **30**, 1056.
43. M. V. Blagosklonny, *Oncotarget*, 2012, **3**, 1711.
44. M. V. Blagosklonny, *Cancer Biol. Ther.*, 2013, **14**, 1092.
45. L. M. Berstein, *Biguanides: An Expansion to Practical Oncology (past and present)*, Aesculap, St.Petersburg, 2010.
46. G. Zhang, J. Li, S. Purkayastha, Y. Tang, H. Zhang, Y. Yin, B. Li, G. Liu and D. Cai, *Nature*, 2013, **497**, 211.
47. V. N. Anisimov, *Molecular and Physioslogical Mechanisms of Aging*, Nauka, St.-Petersburg, 2nd edn, 2008, vol. 1 and 2.
48. S. C. Johnson, M. Sangesland, M. Kaeberlein and P. S. Rabinovitch, *Interdiscip. Top. Gerontol.*, 2015, **40**, 107.
49. M. V. Blagosklonny, *Aging (Albany NY)*, 2013, **5**, 592.
50. D. W. Lamming, *Cold Spring Harbor Perspect. Med.*, 2016, **6**, a025924.
51. H. D. Halicka, H. Zhao, J. Li, Y. S. Lee, T. C. Hsieh, J. M. Wu and Z. Darzynkiewicz, *Aging (Albany NY)*, 2012, **4**, 952.
52. D. E. Harrison, R. Strong, Z. D. Sharp, J. F. Nelson, C. M. Astle, K. Flurkey, N. L. Nadon, J. E. Wilkinson, K. Frenkel, C. S. Carter, N. Pahor, M. A. Javors, E. Fernandez and R. A. Miller, *Nature*, 2009, **460**, 392.
53. V. N. Anisimov, M. A. Zabezhinski, I. G. Popovich, T. S. Piskunova, A. V. Semenchenko, M. L. Tyndyk, M. N. Yurova, M. P. Antoch and M. V. Blagosklonny, *Am. J. Pathol.*, 2010, **176**, 1092.
54. V. N. Anisimov, M. A. Zabezhinski, I. G. Popovich, T. S. Piskunova, M. N. Yurova, A. V. Semenchenko, M. L. Tyndyk, S. V. Rosenfeld and V. M. Blagosklonny, *Cell Cycle*, 2011, **10**, 4230.
55. E. A. Komarova, M. P. Antoch, L. R. Novototskaya, O. B. Chernova, G. Paszkiewicz, O. V. Leontieva, M. V. Blagosklonny and A. V. Gudkov, *Aging (Albany NY)*, 2012, **4**, 709.
56. M. Comas, I. Toshkov, K. K. Kuropatwinski, O. B. Chernova, A. Polinsky, M. V. Blagosklonny, A. V. Gudkov and M. P. Antoch, *Aging (Albany NY)*, 2012, **4**, 715.
57. N. Chandel, *BMC Biol.*, 2014, **12**, 85.
58. M. G. Novelle, A. Ali, m. C. Diéguez, M. Bernier and R. de Cabo, *Cold Spring Harbor Perspect. Med.*, 2016, **6**, a025932.

Section V

Other Pharmacological Approaches

CHAPTER 18

S-Adenosylmethionine Metabolism: A Promising Avenue in Anti-Aging Medicine?

W. A. M. LOENEN[a]

[a]Leiden University Medical Center, Department D.O., Schapenrustweg 7, 2351 NK Leiderdorp, The Netherlands
*E-mail: w.a.m.loenen@lumc.nl

18.1 Introduction

18.1.1 Discovery of *S*-Adenosylmethionine

In 1951, Cantoni[1,2] identified methylation activity of nicotinamide in rat liver extracts, in which reaction *S*-adenosylmethionine (SAM, also called AdoMet) proved to be the methyl donor (reviewed in ref. 3). Methionine adenosyltransferase (MAT or SAM synthetase) generates SAM by linking the sulfur moiety of methionine with adenosine (derived from ATP; Figure 18.1).

The role of SAM is not confined to methylation, but it is a cofactor to various nucleases, which are implicated in bacterial chromosome integrity.[4–16] SAM plays a pivotal role in the methionine cycle, the polyamine pathway, and the transsulfuration route to glutathione, placing SAM at the heart of metabolism.[7] A proper balance is required between its well-known task as a methyl donor and its availability to these and other biosynthetic routes in order to live to a healthy old age.[17] In 2001, a novel SAM-dependent superfamily was

RSC Drug Discovery Series No. 57
Anti-aging Drugs: From Basic Research to Clinical Practice
Edited by Alexander M. Vaiserman

Published by the Royal Society of Chemistry, www.rsc.org

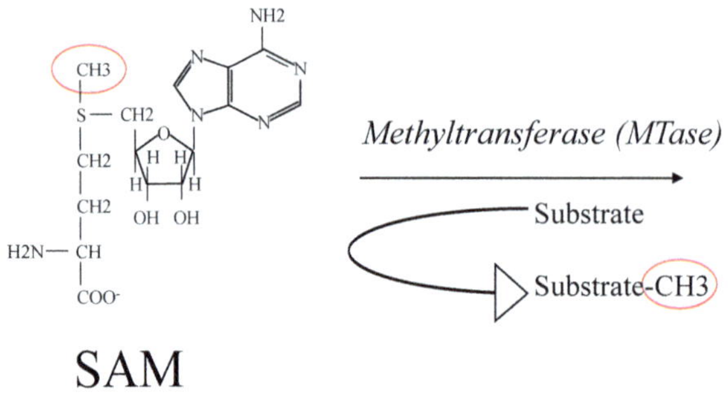

Figure 18.1 *S*-Adenosylmethionine (SAM) is the universal methyl donor. SAM is the major methyl donor used for transfer by methyltransferases (MTases) to DNA, RNA, protein, lipids, small molecules, arsenic, *etc.* SAM is converted to *S*-adenosylhomocysteine (SAH). The methyl group is indicated with a circle.

classified, that of the radical SAM enzymes,[18] which use an FeS cluster for radical chemistry for many purposes.[1,20] As a universal multi-tasker in biochemical transfer reactions, SAM is one of the most frequently used substrates after ATP.

18.1.2 SAM and Aging

What determines aging and life span? Life span may range from days to >100 years, and depends on both genes and environment, and the number of cell divisions.[21–26] In *Escherichia coli* SAM is essential for the assembly of the septal ring during cell division,[27–30] while extracellular communication and longevity are also linked to SAM *via* a process called quorum sensing, which affects virulence and involves *e.g.* AI-2 synthase.[17,31–34] Environmental sensing may also contribute to longevity by triggering pathways involving *e.g.* insulin and Daf-16/FOXO in *Caenorhabditis elegans*.[31] Aging depends on maintenance of genome stability and epigenetic markings, and shows conserved stress-related features. How this is orchestrated is becoming increasingly clear.[35] In eukaryotes, stem cells contribute to the aging process as numbers and the ability to self-renew decline over time in adult tissues.[36–40] Disruption of maintenance methylation causes disease by affecting expression of oncogenes and tumor suppressor genes, and interactions with histones. In addition, endogenous metabolism causes stress by errors during growth and replication, which results in damage to DNA, which indirectly enhances damage to other cellular components, impairs maintenance, and reduces life span.[41–48] Mitochondria play a major role in this redox balance, which shifts to a more oxidized state during aging. Thioredoxin is the major reductant to keep the redox balance, reduce stress and enhance longevity by harnessing the generation of free radicals (ROS) while generating energy.[49] Minor DNA damage provides time for repair, but if repair is too slow or impossible, this results in metabolic malfunction, and either apoptosis ('programmed cell death') or cellular senescence (permanent cell cycle arrest). Senescence often induces prolonged stress signaling *via* cell–cell contact and soluble growth or inflammatory factors, which alter nutrient sensing, energy metabolism and redox status, impair tissues and stem cells, and contribute to aging, neurodegeneration, and cancer.[50–64] There is increasing evidence that levels of food intake and exercise are important factors in health and disease. Diet-induced senescence may be reduced by exercise,[65] and removal of senescent cells could be beneficial, though indiscriminate targeting of senescent cells may be harmful.[66–71] Senescence also occurs in plants as older leaves become senescent in an orderly, SAM-dependent way,[72] which may be nature's way to prepare for programmed renewal.

Anti-aging processes to prevent DNA damage and enhance repair require energy (a human body receives ~10 000 hits in the DNA per day) and nutrient sensors (*e.g.* mammalian target of rapamycin (mTOR)), which is supported by studies on premature aging.[52,55,73–76] SAM is implicated in two mouse models on premature aging (mentioned in ref. 17). Caloric restriction influences longevity and delays age-related diseases, *e.g.* arthritis, cardiovascular disease (CVD), diabetes, obesity, neurodegeneration, muscle atrophy, and osteoporosis.[77–83] It also helps to maintain colon health and lowers the incidence and progression of cancer, which implicates the microbiome, a topic of much current interest (see Sections 18.6.4 and 18.6.5).

18.1.3 This Review

This review starts with the three well-known functions of SAM in the methionine cycle, transsulfuration and polyamine pathways.[7,17] These routes are targets for pharmacological interventions of key enzymes in these pathways *via e.g.* difluoromethylornithine (DFMO) or hydroxylamines (see Section 18.7). The review focuses on recent developments of radical SAM enzymes in central metabolism that affect aging. Exciting new findings suggest novel roles for SAM in RNA metabolism and control of vital functions, more than could have been anticipated a decade ago. These SAM-dependent enzymes are found in ancient processes in all organisms looked at: (eu)bacteria, archaea, yeast, (in)vertebrates, and mammalian cells, while some other functions appear to be newly designed in higher eukaryotes. An important common theme of these routes is the role of sulfur, a pivotal compound before the advent of atmospheric oxygen and more sophisticated macromolecules. These pathways are under very tight control and are carefully monitored by multilayered maintenance and repair routes to prevent disease and aging. Due to lack of space, original papers are sparingly listed and the reader is referred to reviews for further reading.

18.2 SAM-Dependent Enzymes

18.2.1 Parts of SAM Used by SAM-Dependent Enzymes

Different parts of SAM are used in transfer reactions: methyl, methylene, amino, and aminopropyl groups, as well as radicals.[17] Some of these reactions have been identified in bacteria and await unraveling in other organisms, including mammalian cells.[19,20,84–87] SAM superfamilies have many different structural domains and folds without obvious homology that perform all these varied chemical reactions. These families probably date back to a common ancestor of bacteria, archaea, and eukaryotes.[19,20] It has been estimated that ~95% of SAM is used for methylation and 3–5% for the generation of decarboxylated SAM (dcSAM) for polyamine synthesis,[88] leaving only ~1–2% for use by many other crucial enzymes with perhaps only a few copies per cell, but with major impact.[20,89] In humans, 85% of all of methylation reactions and 50% of all methionine metabolism takes place in the liver.[90]

18.2.2 Structures of SAM-Dependent Enzymes

The majority of SAM-dependent methyltransferases (MTases) share a common structure, the Rossmann fold, which is conserved in evolution, though the residues that contact SAM are not (see ref. 14,19,84–86 for reviews). Classification of this large superfamily is based on substrate specificity, and on the atom targeted for methylation (*e.g.* N, O, C, or S). Several proteins contain the Rossmann fold despite being inactive as MTase, *e.g.* spermidine synthase.[84] A triose phosphate isomerase-like domain (called the TIM barrel) is present in radical SAM enzymes, which use SAM to generate methionine and a 5′-deoxyadenosyl radical that can be used to generate further radicals on the same or another protein (see Section 18.5.2).[19,20] A third class of SAM-dependent enzymes contains the SET domain. This domain was discovered as a conserved domain shared by the chromatin remodeling protein suppressor of variegation 3-9 (Su(var)3-9), enhancer of zest and trithorax. These enzymes affect chromatin function and transcription by methylating lysines in *e.g.* histones and P53.[19] The SPOUT fold was originally identified as a domain shared by the SpoU and TrmD MTases in a superfamily of enzymes that methylate tRNA and rRNA.[91,92] Archaeal Tsr3 has a distinct SAM binding mode and modifies a conserved hypermodified nucleotide in eukaryotic 18S rRNA.[93] Some enzymes in the methionine cycle have unusual folds: *e.g.* MAT proteins (involved in *de novo* synthesis of SAM), methionine synthetase (MS, which regenerates methionine using the methyl group from methyltetrahydrofolate (MTHF)), and the repressor of the methionine operon, MetJ.[19]

18.3 Well-Known Pathways of SAM in Central Metabolism

18.3.1 The Methionine Cycle

As mentioned in Section 18.2.1, ~95% of SAM is used for methylation. In the methionine cycle (also called one-carbon or SAM cycle) demethylation of SAM yields *S*-adenosylhomocysteine (SAH). SAM is regenerated *via* a homocysteine (HCY) intermediate, in turn the substrate for MS with vitamin B12 as the cofactor and MTHF as the methyl donor, and the cycle is completed by MAT (Figure 18.2). The name 'folate,' the precursor for MTHF, describes a family of related molecules capable of one-carbon transfer. Folate-derived

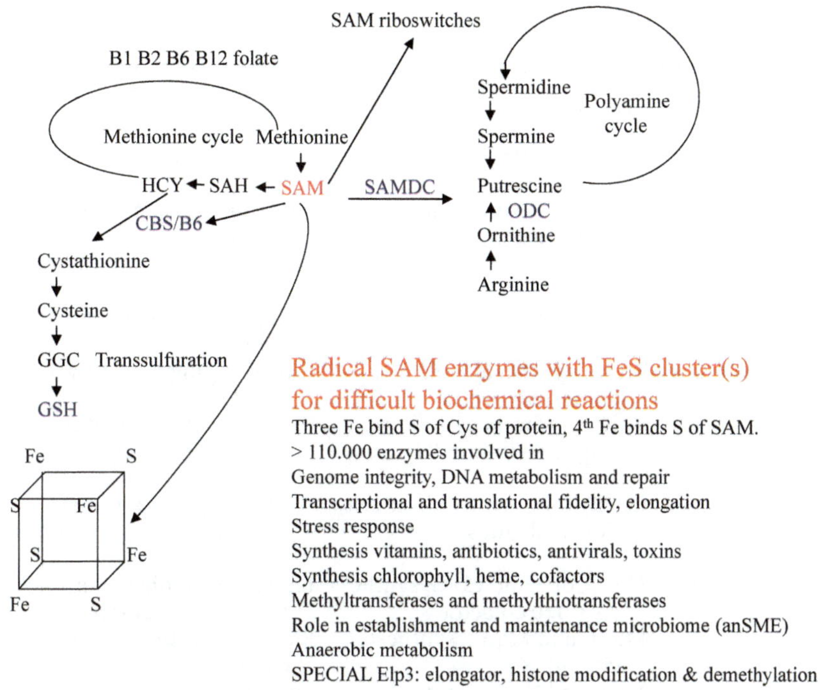

Figure 18.2 Simplified diagram of the manifold uses of SAM. The well-known uses of SAM in the methionine cycle, transsulfuration and polyamine synthesis, and in RNA-based riboswitches are shown, as well as the more recently emerging important role of radical SAM enzymes. SAM donates the methyl group (derived from food and/or from recycling) in the methionine cycle (also called SAM cycle or one-carbon cycle, Section 18.3.1). SAM is generated from methionine by methionine adenosyltransferase (MAT or SAM synthetase). SAM is converted to *S*-adenosylhomocysteine (SAH) by methyltransferases (MTases). SAH is hydrolyzed to homocysteine (HCY). The latter is the substrate for methionine synthase (MS), which uses a derivative of folate, MTHF, as a methyl donor to regenerate methionine, producing tetrahydrofolate (THF). HCY is also the substrate for the route to glutathione *via* cystathionine beta-synthase (CBS or cystathionine synthase), which is converted to cysteine (Section 18.3.2). This amino acid is used to generate the tripeptide glutathione, an important antioxidant and detoxifier in the cell *via* glutamate cysteine ligase (GCL), with gamma-L-glutamyl-L-cysteine (GGC) as an intermediate. Other conversion routes are not shown for clarity. About 3–5% of SAM is diverted to the polyamine pathway *via* decarboxylation (to dcSAM) by SAM decarboxylase (SAMDC, Section 18.3.3). The methionine backbone of dcSAM is used by spermidine synthetase to convert putrescine into spermidine. Spermine synthase converts spermidine into spermine using a second molecule of dcSAM (back-conversion routes to putrescine involve spermidine/spermine-N1-acetyltransferase (SSAT) and other enzymes, not shown). *De novo* putrescine production requires the activity of ornithine decarboxylase (ODC), the rate-limiting step in this pathway. SAM riboswitches are small highly looped RNA molecules, which change configuration upon SAM binding, thus opening up or closing regions involved in transcription or translation (Section 18.4.1). Radical SAM enzymes with FeS clusters: see text and Table 18.1 (Section 18.5).[7]

Table 18.1 Selected examples of radical enzymes using data from ref. 20.

Type	Example	Function
Glycyl radical	PFL-AE, anaerobic glucose metabolism in microbes	Central role in generation of energy
	aRNR, anaerobic reduction CTP to dCTP	Rate limiting step in DNA metabolism
Sulfur insertion	BS, synthesis vitamin B1	Essential cofactor methionine cycle
Mutases	PylB, synthesis pyrrolysine	UAG stop codon of some MTases
Complex changes	MoaA, synthesis molybdopterin cofactor	Global cycling C, N, and S
	ThiC, synthesis pyrimidine ring and vitamin B1	Essential cofactor methionine cycle and central metabolism
	QueE, ToyC, synthesis nucleotide analogues	Modified tRNA for translational fidelity; antibiotic and antineoplastic
	FbiC/CofH and CofG, synthesis F 420 cofactor	Energy metabolism, antibiotic synthesis, DNA repair
	MqnC and MqnE, synthesis vitamin K2	Electron shuttle in respiratory chain
	NosL and NocL, synthesis modified thiopeptides	Antibiotic against drug-resistant pathogens
Methylation	Radical SAM MTases (RSMT) class A. Cfr, chloramphenicol-florfenicol resistance	Methylation 23s rRNA and antibiotic
	Radical SAM MTases (RSMT) class B: SAM + Cobalamine domain. TsrM: synthesis thiostrepton A	Antibiotic against drug-resistant pathogens, fungi, malaria, and (and antineoplastic?)
	Radical SAM MTases (RSMT) class C: SAM + HemN domain. NocN thiopeptide synthesis/bleomycin family	Antibiotic against drug-resistant pathogens
Thiomethylation	Radical SAM methylthiotransferases (MMTases). Five groups	Posttranslational modifications, stability codon-anticodon and fidelity, antibiotic
	(a) RimO: Asp on ribosomomal S12	Bacteria
	(b) MiaB: tRNAphe, anticodon reading codon with U start	Bacteria and eukaryotic organelles
	(c) MtaB	Eubacteria
	(d) e-MtaB, *e.g.* CDKAL1 involved in protein folding	Archaea and eukaryotes. Link diabetes
	(e) MTL1	Epsilon-proteobacteria
Dehydrogenation	Sulfatase maturating enzymes (anSME): Anaerobic oxidation Cys or Ser to generate electrons. BtrN	Microbiome establishment and maintenance. Antibiotic butirosine
New bonds	C–C, C–N, C–S	Energy metabolism, translational fidelity, antibiotics

	PqqE, cofactor	Energy metabolism
	TYW1, modification guanosine to wybutosine	Translational fidelity, stability codon-anticodon
	Sectipeptides, TrnC and TrnD: two-component Thuricin D; AlbA: subtilosin A	Posttranslational modifications, antimicrobial and/or spermicidal
Bond cleavage	C–C, C–N, and C–P	Methionine cycle, DNA repair, antibiotic synthesis
	SPL, spore photoproduct lyase, DNA repair	Repair UV-induced T dimers in absence of light DNA
	DesII, synthesis D-desosamine	Antibiotic
	C–P lyase cluster, PhnJ conversion alkylphosphonates to phosphates	Central metabolism
	BslE, synthesis blasticidin S, inhibitor protein synthesis	Antibiotic inhibiting peptide bond formation in ribosome
	Elp3, component of elongator complex. Multifunctional: histone acetyltransferase; modification of stress response tRNA; regulation mitosis and cytokinesis, demethylation	Linked to mitochondrial dysfunction, neurodegeneration, viability of *Drosophila*, plants, and parasites; cancer
Modified tetrapyrroles	HemN and BchE: Synthesis heme, chlorophyll, cobalamine *etc.*	Central metabolic processes
Complex cluster formation	NifB, insertion large FeS cluster (FeMoco) in nitrogenase precursor for; HydG, synthesis H-cluster in [FeFe] enzymes	Reduction of N_2 in the air; many metabolic pathways as e-acceptor or use in energetic processes
Others	Human viperin/RSAD2, interferon-stimulated gene	Antiviral (dsRNA/DNA), antibacterial
	Dph, modified His in Elp2 in archaea and eukaryotes	Translational fidelity and target of diphtheria toxin; cancer, mouse viability

tetrahydrofolate (THF) can be synthesized by plants in mitochondria, and by microorganisms, but animals and humans must ingest this with food.[94] Impaired mitochondrial activity has been linked to the methionine cycle in *e.g.* diabetes type 2.[95,96]

Methionine can also be regenerated *via* betaine-homocysteine MTase 1, BHMT1. In contrast to most MTases, BHMT1 is insensitive to feedback inhibition by SAH, which may perhaps prevent high HCY levels in plasma (linked to CVD and diabetes[97]). Deregulation of SAM causes chromosome instability, imprinting problems, cancer and disease, due to *e.g.* hypomethylation of promoter regions, activation of transposons and oncogenes, inactivation of tumor suppressor genes, incorporation of uracil in DNA, and/or futile cycles of DNA repair and chromosome breaks.[98–104] Many human diseases are linked to faulty (de)methylation, *e.g.* systemic lupus erythematosus (SLE) is linked to overexpression of the ligand of CD27,[105] as a result of decreased promoter methylation.[106–120] SAM supplements sometimes ameliorate disease,[121] but may contain the biologically active *S,S* form of SAM as well as the inactive and possibly toxic *R,S* molecules.[122] Finally, SAM methylates and detoxifies compounds such as arsenic.[123–125]

18.3.2 The Transsulfuration Pathway to Glutathione

Cystathionine beta-synthase (CBS) can divert HCY from the methionine cycle to generate glutathione (GSH) *via* the transsulfuration pathway with vitamin B6 as the cofactor.[124,126,127] (Figure 18.2). GSH is a ubiquitous tripeptide with antioxidant and detoxification properties. Keeping the balance between methylation and transsulfuration is important to prevent disease, and depends on methionine levels and allosteric activation of CBS by SAM. High methionine levels favor transsulfuration to cysteine and GSH, while low levels favor methylation. In the latter case, decreased binding of SAM to CBS destabilizes the protein, and thus CBS affects viability under conditions of oxidative stress. The conserved SAM-binding domain of CBS probably functions as a metabolic sensor, and mutations in this CBS domain are linked to human disease, which may benefit from manipulation of GSH synthesis.[127–129] The methionine flux to transsulfuration is involved in longevity of certain rodents, *e.g.* the Ames dwarf mouse and the naked mole rat (see Section 18.6.3).[130,131]

18.3.3 The Polyamine Pathway

An estimated 3–5% of SAM in the cell is used by SAM decarboxylase (SAMDC) to generate dcSAM, the substrate for the synthesis of the polyamines spermidine and spermine from putrescine (Figure 18.2). Already observed by Van Leeuwenhoeck in 1678 as crystals in semen, these positively charged molecules bind negatively charged molecules in the cell, including DNA, RNA, proteins, phospholipids, and many other molecules. Ornithine decarboxylase (ODC) and SAMDC are rate-limiting in this carefully regulated pathway, which is important to maintain proper polyamine levels in the cell to remain alive and healthy (see *e.g.* ref. 19,132–146). Regulation involves polyamine-dependent programmed frame shifting, proteasome-independent degradation, control by *e.g.* c-Myc, NQO1, APC, and an internal ribosome entry site (IRES), to name just a few.[147–150] Ornithine, derived from arginine in the urea cycle, is expressed primarily in the liver and intestine, while polyamines are made in all tissues and are present in cheese and red meat. Spermidine synthase and spermine synthase use dcSAM to convert putrescine to spermidine, and spermidine to spermine, respectively, while fusion of the methionine backbone of dcSAM to putrescine generates methylthioadenosine (MTA), which can be recycled to methionine.[88,151] Both methylation and polyamine levels decline during life, instigating early trials to prevent age-related senescence and/or cancer.[138,152,153]

18.4 SAM and RNA-Based Control by Riboswitches

18.4.1 Discovery of SAM Riboswitches

Several decades ago Tina Henkin reported a common mechanism with respect to the regulation of some aminoacyl-tRNA synthetase and amino acid biosynthesis genes in gram-positive bacteria, which involved transcription antitermination at a conserved region in 5′ mRNA leader regions.[154,155] This led to the identification of conserved motifs in genes for synthesis of methionine and cysteine.[156] These motifs were recognized by a small SAM-dependent, highly structured, RNA called riboswitch or aptamer, which prevented binding of the 30S ribosomal subunit to the mRNA in the presence of SAM.[157–159] Currently seven different SAM riboswitches, called S-boxes or SMK, are known and are involved in methionine synthesis, as expected, but also sulfur metabolism and other pathways.[158,160–178] Riboswitches may be sensitive to SAM or its metabolite SAH,[177,179] or work in tandem, *e.g.* SAM-I and a B12 riboswitch.[171,173,180,181] Other metabolite-dependent riboswitches for *e.g.* thiamine, purine, glycine, THF, vitamin B12, uncharged tRNA, and SAM are common in bacteria (reviewed in ref. 170,172,182–186) and some have been described in plants, fungi, and marine eukaryotes.[158,170,187] The thiamine pyrophosphate (TPP/vitamin B1) riboswitch occurs in all three domains of life.[188–193] Interestingly, this riboswitch blocks ribosome binding or terminates transcription in bacteria, but appears to regulate gene expression in eukaryotes *via* alternative RNA splicing.[188,194] Will there be SAM-dependent (alternative) splicing to be discovered in humans?

18.4.2 SAM and Other Riboswitches

Several riboswitches affect SAM metabolism indirectly, *e.g.* ZTP riboswitches are members of a large family of regulatory RNAs that upregulate *de novo* purine synthesis in response to increased intracellular levels of ZMP or ZTP (5-aminoimidazole-4-carboxamide riboside 5′-triphosphate). ZMP is an important intermediate in purine biosynthesis and is linked to folate stress *via* the regulation of the levels of a key component of one-carbon metabolism, N10-formyl-THF.[195,196] By 2006, the first structures of riboswitches had become available, as well as riboswitches that *e.g.* control expression of a reporter gene or splicing in yeast,[174,189,197–212] though the latter use of reporters controlled by antibiotics such as tetracyclin needs careful evaluation.[213–215] There is probably no end to the amazing roles of these small RNAs, as they may sense magnesium,[189] or link to repeat expansion, and neurological disease in mammals (see *e.g.* ref. 216 and 217). Batey[172] compared the riboswitch with the IRES, the highly structured mRNA region in important mammalian genes such as ODC, insulin-like growth factor (IGF)-2, and c-myc.[218–220] The question raised by the story of the TPP riboswitches is inevitable: Will RNA structures that bind SAM and thus control gene expression appear on the human horizon and dictate the activity of non-coding and microRNAs?

18.5 'Radical SAM' Proteins with Iron–Sulfur (FeS) Clusters

18.5.1 Discovery of Radical SAM Enzymes

SAM-dependent 'radical SAM' enzymes with FeS clusters were recognized as a superfamily as recent as 15 years ago.[18,89] These enzymes use a novel common mechanism of catalysis in all kingdoms in many metabolic pathways: a reduced $[4Fe-4S]^+$ cluster transfers an electron to SAM to generate methionine and a 5′-deoxyadenosyl radical,[221,222] (see ref. 20 for details). Radical SAM proteins share a CX_3CX_2C motif, or variations thereof, which forms the FeS cluster (Figure 18.2). Three cysteines bind three of the four irons of [4Fe–4S] at the active site of the enzyme, while the cluster requires the sulfur moiety of SAM at the fourth iron for the generation of a 5′-deoxyadenosyl radical derived from SAM. The lack of a fourth cysteine makes these proteins hard to work with, as the cluster is oxygen-sensitive resulting in inactive [3Fe–4S] (which can be reverted back to [4Fe–4S] by reducing agents in an anaerobic atmosphere).[20]

The mechanism of radical SAM enzymes is similar to that of B12 enzymes.[20,223] The radical produced can be the end product or an intermediate in a complex chain of reactions, while SAM itself can be used up, or regenerated to SAM *via* methionine.[20] The review by Broderick *et al.*[20] gives extensive coverage of common features of these enzymes, as well as biochemical, spectroscopic, structural, and mechanistic details. Some examples are presented in Table 18.1 and in the next section, in view of their relevance to the role of SAM in central metabolism and aging. By the end of May, 2016, ~114 000 enzymes had been found, eight of these in humans.[224] Most of these enzymes have been identified in bacteria, and hence may be relevant in microbiota that colonize our gut (see Section 18.6.5).

18.5.2 The Radical SAM-Binding Domain

In 2008, the first structure of a radical SAM enzyme was reported.[225] By 2014, the crystal structures of 14 radical SAM enzymes were known and supported the notion of a common fold composed of a full or partial TIM barrel (Section 18.2.2). A full barrel consists of eight alternating alpha helices and beta strands with the beta strands on the inside. The size of the barrel varies depending on the size of the substrate, which binds within the TIM barrel.[20,226] SAM associates with the fourth iron of the [4Fe–4S] cluster in the same way in these 14 crystal structures.[20]

Many parallels exist between radical SAM and B12 enzymes, but also striking contrasts: the cofactor for B12 enzymes binds outside the barrel, and the dozen or so known B12 enzymes are mainly bacterial, while radical SAM enzymes occur in all kingdoms with an astounding rise in numbers from 600 in 2001 to ~48 000 in 2014, with the latest figure >110 000.[20,224,227] Broderick *et al.*[20] state that "ultimately, the use of the TIM barrel fold by B12 and radical

SAM systems speaks toward the evolutionary development of these enzymes and the requirement for a protein architecture that was inherently not complex and in regards to radical SAM proteins allowed for the diversification of chemical reactions through the acquisition of additional modular protein domains." Does this support the idea that radical SAM enzymes with their FeS clusters predate the B12 enzymes in protein-based catalysis?[19,20,228,229]

18.5.3 Types of SAM Radical Enzymes

At least a dozen types of radical SAM enzymes perform a wide variety of reactions.[20] These enzymes are involved in central metabolism, energy generation and transfer, in the synthesis of many essential cofactors and antibiotics, but also DNA repair, and RNA modifications that prevent inaccurate translation or generate antibiotic resistance[20,230] (see Table 18.1; data derived from ref. 20). Many enzymes synthesize antibiotics and toxins that may affect functioning of mitochondria and chloroplasts.[213,214,231–233] Well-known cofactors include modified tetrapyrroles, such as (bacterio)chlorophyll, heme, and cobalamins, to name just a few.[18] Anaerobic ribonucleotide reductase (aRNR) reduces CTP to dCTP, a rate-limiting step in DNA metabolism strictly dependent on SAM.[234,235] The sulfur-inserting enzyme biotin synthase (BS/BioB) uses two FeS clusters in a difficult final step in the synthesis of vitamin B1 (essential in the methionine cycle (Figure 18.2)). The reduced SAM-dependent $[4Fe-4S]^+$ cluster donates one electron to SAM producing a 5′-deoxyadenosine radical, which then requires a second half cluster, [2Fe–2S], to insert a sulfur atom into the biotin precursor.[20,236,237] PylB is involved in the synthesis of pyrrolysine, present in the in-frame UAG amber codon in SAM-dependent MTases in certain archaea that use these MTases to generate methane.[20] Complex changes involve formation of the pyrimidine ring and synthesis of vitamin B1. Nucleotide analogues act as antibiotics, neoplastic agents, enhance translational fidelity, and synthesize F420, the cofactor for hydride transfer in energy metabolism, antibiotic biosynthesis, and DNA repair. MqnC and MqnE are involved in the synthesis of vitamin K2, which serves as an electron shuttle between membrane-bound proteins in the respiratory chain; NosL/NocL are involved in antibiotics of clinical interest against drug-resistant bacterial pathogens (see ref. 20 for details).

Of particular interest are anaerobic sulfatase maturating enzymes (anSME) that can oxidize cysteine or serine residues in proteins to generate electrons, and play a key role in the microbiome (see Section 22.6.5). An SME is a group of at least 1400 enzymes with a SPASM domain and an amazing 7-cysteine motif ($CX_{9-15}GX_4C$-gap-$CX_2CX_5CX_3C$-gap-C), which coordinates additional FeS clusters in these enzymes.[20,238]

Synthesis of enzymes that form C–C, C–N, and C–S bonds includes enzymes that regulate (post)translational modifications and fidelity, and synthesis of antibiotics that target feared *Clostridium difficile* and other pathogens.[20,239] PqqE is involved in the synthesis of pyrroloquinoline quinone (PQQ), a cofactor found in alcohol dehydrogenases and other bacterial enzymes.[240–242] Not all prokaryotes synthesize PQQ, but *e.g. E. coli* can use it in an alternative sugar transport system.[243] PQQ is available as a probiotic and food supplement.[244] In contrast to most radical SAM enzymes, PqqE of AM1 is markedly oxygen-tolerant.[245] Recently, a PQQ-dependent enzyme was identified in mushrooms, with possible homologues in bacteria, archaea, fungi and other organisms.[246] From its humble activity in sugar transport in *E. coli* thirty years ago,[243] current reports link PQQ to oxidative stress and diseases that affect life span and aging, *e.g.* inflammation, liver disease, neurological disease, osteoarthritis, diabetes, and mitochondrial functioning, claims that warrant further investigation.[247–262]

Synthesis of enzymes that cleave C–C, C–N, and C–S bonds include proteins involved in central metabolism and antibiotic synthesis, but also SPL (spore photoproduct lyase), which can repair UV-induced thymine dimers in the absence of light.[20] Biosynthesis of chlorophyll photosynthetic pigments requires additional FeS proteins, including methylation by radical RSMT of the tetrapyrrole ring.[20] Another example of a complex formation is the molybdene-containing 'FeMo-co' [$MoFe_7S_9C$] cluster in nitrogenase that catalyzes the reduction of nitrogen in our atmosphere to NH_3.[20] On a different track, human viperin (RSAD2/Cig5) belongs to the group of interferon-stimulated genes, which act as defense against (dsRNA and DNA) viral and bacterial infections, but may also be involved in pregnancy and atherosclerosis.[20] Diphthamide is a modified histidine in elongator protein Elp2 in archaea and eukaryotes, and the target of diphtheria toxin.[20] The *dph* gene products are linked to various cancers, and Dph1 knocked out mice are inviable.[263]

18.5.4 Radical SAM Methyltransferases (RSMT)

Some SAM-dependent MTases do not have the Rossmann fold (Section 18.2.2), but are radical SAM MTases (RSMT). Three classes, A, B, and C, have been identified.[20] Class A Cfr (chloramphenicol-florfenicol resistance) of *Staphylococcus aureus* has a single radical SAM domain. It uses a conserved cysteine to transfer a methyl group of one SAM molecule to 23S rRNA using a radical generated *via* a second SAM molecule. The likely homologue in *E. coli*, RlmN, also methylates 23S rRNA, but does not confer antibiotic resistance.[20] Class B RSMT contain both a radical SAM domain and a cobalamin-binding domain.[20,264] One SAM molecule is probably used to methylate the cobalamin cofactor using a radical generated with a second SAM molecule. The first enzyme identified of this class was TsrM from *Streptomyces laurentii*, a tryptophan MTase involved in the synthesis of thiostrepton A, which targets various pathogenic bacteria, malaria, and possibly even cancer.[20] All may not be what it seems: TsrM is a class B RSMT, but does not catalyze SAM-based radical chemistry, while another enzyme in this class may be involved in plasma membrane rigidity, requiring further investigations.[20] Class C RSMT contain a radical SAM domain and a C-terminal domain resembling the HemN domain, involved in heme synthesis. The bleomycin biosynthetic genes belong to this class, which make thiopeptides and antibiotics, and may also use Trp as a precursor (like TsrM). Both these and Yatakemycin (YtkT), a naturally occurring antitumor agent, are of clinical interest against drug-resistant bacterial pathogens, some fungi and cancer.[20]

18.5.5 Radical SAM Methylthiotransferases (MMTases)

Translational fidelity depends on RNA-post-translational modifications, including conserved methylthio modifications on ribosomal protein S12 (*e.g.* RimO on an aspartic acid residue), and on the anticodon in tRNAs reading codons beginning with U (*e.g.* MiaB on $tRNA^{phe}$).[20] MiaB and RimO use two molecules of SAM (one for the generation of a 5′ adenosyl radical, the other as methyl donor, as above).[20] Sequence analysis reveals five families of MMTases in different kingdoms: MiaB (bacteria and eukaryotic organelles), RimO (bacteria), MtaB (eubacteria), e-MtaB (archaea and eukaryotes) and MTL1 (epsilon-proteobacteria). Studies with a human e-MtaB (CDKAL1) and the corresponding knockout mouse indicate a role for CDKAL1 in the prevention of frame shifts and/or misreading. Improper translation might prevent proper processing and folding of *e.g.* proinsulin to insulin, which will influence the onset of type 2 diabetes (see ref. 20 for details).

18.5.6 The Special Case of Elp3

Elp3 was identified as the histone acetyltransferase (HAT) component of the elongator complex that associates with RNA polymerase II during transcriptional elongation.[265–272] The six-subunit elongator complex is conserved in eukaryotes including plants, while archaeal Elp3 catalyzes the wobble uridine in tRNA on its own in the absence of other Elp proteins.[273] In *Toxoplasma gondii* a single Elp3 protein is found with a C-terminal transmembrane domain, which localizes Elp3 to the mitochondrion, and is essential for parasite viability.[268] In plants, elongator is required for the modification of stress response tRNAs for efficient translation and protection against infection ('plant immunity'), *e.g.* threats caused by the fungus *Fusarium graminearum*, which causes serious loss of cereal crops and is toxic to humans and animals, and the brown planthopper, a major rice pest.[274–278] In addition to the C-terminal HAT domain, Elp3 has an N-terminal radical SAM domain, which is important for the activity and structural integrity of the elongator complex.[279–281] Elp3 acetylates histones in the nucleus but also modifies tRNA.[282–285] The resemblance of Elp3 to other proteins with a similar two-domain structure suggested an additional role for Elp3 as a radical SAM histone demethylase,[279,286] which was supported *in vivo*.[286,287] Considering its manifold roles, it is not surprising that Elp3 mutations are linked to disease. Elp3 has attracted much attention by its link to actin-rich domains and mitochondrial dysfunction in neurodegenerative disease *e.g.* familial dysautonomia, and motor neuron disease (amyotrophic lateral sclerosis, ALS).[283,288–293] Elp3 has also been implicated in colon and breast cancer.[294,295] In *Drosophila melanogaster* Elp3 is essential for viability, normal development and hematopoiesis, and deletion of Elp3 causes morphological defects in neurons, and results in larval lethality at the pupal stage.[296–298] This may link to yet another twist to the elongator story *via* codon-dependent regulation of translation: lysine codon usage bias, coupled to tRNA modifications, influences translation of Cdr2, a central regulator of mitosis and cytokinesis.[299,300]

18.5.7 Lessons from SAM-Independent FeS Proteins?

SAM-independent FeS clusters were discovered, and recognized as such, long before the radical SAM enzymes. Like radical SAM enzymes, FeS proteins are usually oxygen- and redox-sensitive, and are involved in basic processes in all life on earth.[301–308] The FeS cluster of the regulator of fumarate and nitrate reduction (FNR) illustrates the swift reaction time of FeS proteins. Under anaerobic conditions a [4Fe–4S] cluster enables FNR to dimerize and activate anaerobic genes.[309,310] Oxygen results in oxidation of [4Fe–4S] (to a [2Fe–2S] cluster), which inactivates the dimer and results in a swift, and if necessary temporary, switch to aerobic gene expression.[311] Assembly of FeS proteins requires a scaffold complex,[18–20,307,312–320] and mutations in scaffold genes cause severe, often fatal disease, *e.g.* Friedreich's ataxia.[321,322]

Several DNA base excision repair (BER) enzymes are FeS proteins.[18–20,307,314,315] Interestingly, four FeS DNA helicases are associated with severe human disease and aging: XPD, FancJ, RTEL, and DDX11.[316] XPD (Xeroderma pigmentosum (XP) group D) functions in DNA nucleotide exision repair as well as transcription, and is linked to XP, Cockayne syndrome (CS) and trichothiodystrophy (TTD).[323–325] FancJ interacts with BRCA1 and is associated with breast cancer and genomic instability in Fanconi anemia.[326,327] RTEL/Rtel1 has a role in telomere maintenance.[316] DDX11 (also called ChlRI/Chl1/Ctf1) causes Warsaw Breakage Syndrome (WABS), is embryonically lethal in mice, is required for rRNA transcription dependent on histone epigenetic modifications, is involved in chromosome transmission fidelity and sister-chromatid cohesion, is present at the replication fork as a putative replication licensing factor, and is essential for survival of melanoma cells.[328–334] Disruption of the FeS cluster results in clinically relevant mutations, which were confirmed in yeast.[335] Such FeS-dependent diseases raise the question of whether human radical SAM patients exist with similar serious defects and short life span as a result (see Section 22.7). Will PARP inhibitors perhaps be useful for radical SAM studies?[336]

In recent years XPD was shown to be essential for genome integrity and nuclear division in *e.g.* mouse and Drosophila, but this is not the end of the story.[337–344] Like P53 and some DNA repair enzymes,[345–347] XPD can sense redox changes and oxidative stress in DNA, which may enhance detection of lesions or alterations in base stacking over long distances.[348] In the case of P53, oxidative stress leads to DNA-mediated oxidation and disulfide bond formation in P53, which differentially affects binding of P53 to different promoters.[347,349] In the case of DNA lesions, XPD and/or FeS-containing BER enzymes may act alone or together,[307,314,315] and this DNA-mediated signaling of XPD "may reflect a general biological role for DNA charge transport",[350] which may bring new treatments for patients.[348,351] Are radical SAM enzymes perhaps also capable of such lesion detection?

18.6 SAM and Aging

18.6.1 SAM, Mitochondria and Aging

Harmful effects of ROS radicals like superoxide and hydrogen peroxide are particularly bad news for mitochondria, and hence life span.[352,353] Mitochondria play an important role in aging: (a) Mitochondrial DNA is more vulnerable to damage than nuclear DNA, as repair depends on pol-gamma and is lower than in the nucleus. (b) Mitochondria cannot make SAM, which has to be imported by mitochondrial carriers involved in SAM metabolism, including SAM itself, ornithine, folate, and ATP/ADP exchange.[354] SAM depletion affects the production of GSH and increases mitochondrial instability, though unfortunately methionine supplements may enhance ROS.[354–357] (c) ROS crosslinks Cys residues in mitochondrial proteins, resulting in degradation unless rescued by repair enzymes and GSH.[358] In line with this, long-lived mice tend to have higher mitochondrial GSH levels.[359] Increased GSH production and potential decreased availability of SAM for methylation delays aging and affects development, as *e.g.* shown after overexpression of the rate-limiting enzymes in the transsulfuration pathway in *Drosophila* (GCLC and GCLCM).[352] In two of the three MAT enzymes, oxidation of Cys^{150} reduces MAT activity (mentioned in ref. 352), which taken together make a strong case for a link between SAM and longevity due to altered flux through the transsulfuration route.

Fungal PaMTH, which methylates flavonoids in *Podospora anserina*, accumulates in the mitochondrial matrix of senescent cells and may protect against oxidative stress and aging as overexpression increases life span.[360–363] Such post-translational modifications are not confined to this fungus, as *e.g.* mitochondrial ATP synthase was differentially affected in both *P. anserina* and in the brains of young *versus* old rats, as well as in human cells.[364,365]

18.6.2 SAM and Neurodegeneration

Deregulation of the methionine cycle has been reported in Alzheimer's disease (AD) and other neurodegenerative diseases, such as ALS (see ref. 17,366–382 and OMIM at[383] for further details). Lowered expression of the presenilin 1 (PS1) gene in AD is linked to the accumulation of amyloid-beta, characteristic for this disease.[366] Mildly elevated plasma levels of HCY in elderly people not only increased the risk of AD and neurodegeneration, but also cerebrovascular disease.[384–386] Parkinson's disease is a motor disorder due to loss of dopamine-producing cells in the brain, which leads to neurodegeneration due to methylation of dopamine by catechol-*O*-methyltransferase (COMT).[387] Down's syndrome (DS) people suffer from mental retardation and heart problems, as well as premature aging and AD, as a result of an imbalance in the flux through the methionine cycle in favor of transsulfuration.[388–391] Overexpression of superoxide dismutase (SOD1) in DS causes chronic stress and protein instability in red blood cells due to damaged asparagine residues, which cannot be properly repaired by a specific SAM-dependent MTase.[388] Hence in this case it is radical damage to proteins that requires a SAM-dependent MTase for repair. Whether altered SOD1 expression in patients with ALS, or that of SOD2 in tumor cells, also lack such SAM-dependent repair remains to be investigated.[392,393]

18.6.3 SAM and Long-Lived Rodents

The Ames dwarf mouse is a long-lived mouse with an enhanced flux towards the transsulfuration pathway in several organs, which is due to a pituitary gland problem and lack of growth hormone (GH).[130] This affects tissue levels of SAM and SAH *via* glycine-*N*-methyltransferase (GNMT).[130] SAM and folate control GNMT: low food intake of SAM inhibited GNMT and made more SAM available for methylation. When SAM was high, GNMT demethylated SAM, thus reducing SAM levels and fueling the transsulfuration route. This and other data supported the notion that longevity of these mice was linked to a better defense against oxidative stress *via* higher GSH levels. Other recent reports also link GH (and insulin) levels to aging in these mice.[394–400]

The naked mole rat is a small native rat in East Africa with an incredibly long lifespan of >30 years. It has the hallmarks of longevity, being very healthy with a stable genome, and little or no signs of senescence or cancer (see *e.g.* ref. 401–406 for some recent references). As H_2S has been implicated in aging and lifespan in diet-induced longevity models, the blood of this rat was compared with that of five mammalian species with different life spans.[131] This revealed an inverse correlation between blood sulfide levels and longevity, which was linked to SAM *via* CBS. As mentioned before, SAM activates CBS and thus stimulates the transsulfuration route (Section 18.3.2). In the naked mole rat, SAM activated CBS to a higher degree compared to the other species, which warrants further investigation.[131]

18.6.4 SAM, the Microbiome and Aging

Microorganisms (microbiota, collectively called the microbiome) are essential in living beings, from humans to worms, flies, and coral, but also in symbiosis with plants.[407–410] They have a beneficial effect on host metabolism and the immune system by providing nutrients and energy (like mitochondria and chloroplasts, which are symbionts that originated from bacteria), and do jobs the host cannot do: *e.g.* fix nitrogen, degrade cellulose, and harvest light. In humans, microbiota break down dietary fiber, produce essential vitamins and amino acids, and detoxify harmful chemicals. Especially B vitamins, including B1, B6, B12, and folate (B9), are important and are intertwined with SAM-dependent cycles and pathways. Often the microbiome provides some of the enzymes for a particular route, while the host provides others ('metabolic collaboration'), including critical transport proteins.[411] Lactobacillus-based probiotics are implicated in benefiting the body beyond the gastrointestinal (GI) tract, *e.g.* diabetes type 2, CVD, and cancer.[412] The microbiome appears to be sensitive to changes in the immune system and differences in the composition of our microbiome are apparently associated with disorders including colon cancer, inflammatory bowel disease, obesity, the severity of autism spectrum disorders, protection against pathogens and parasites, and differential responses to medical treatments.[407–410,413]

Bacteria can slow down aging in a simplified model for the human–microbiota–diet system, that of the worm *C. elegans* fed with *E. coli.*[49] The beauty of this system is that mutants of both bacteria and worm can be made and tested in the lab as they will be viable on certain media but conditionally lethal on different feeds.[49] On the one hand, *C. elegans* and *E. coli* share biosynthetic routes, and wild type or mutant bacteria affect the lifespan of the worm in different ways. On the other hand, mutant long-lived worms exist that affect the route to chorismate, the precursor for Tyr, Phe, and Trp, folate and other important molecules.[49] The increased life span due to decreased bacterial folate synthesis is one of the few interventions in *E.coli* that slows aging of *C. elegans* (mentioned in ref. 49). The worm is not the only organism affected by the contents of the microbiome. A striking effect of righting 'wrong' microbiota in humans has been observed with fecal transplantation to alleviate infection with *Clostridium difficile*.[414]

18.6.5 SAM and Establishment and Maintenance of the Microbiome

How diet and microbiota influence aging and chronic disease in humans is an important question, but how do microbes actually establish and maintain themselves?[415–417] Microbiota numbers are staggering: the microbiome genomes combined carry ~10^6 genes (300× more than the human genome); the number of cells may be up to 10× more than that of the host, and take up as much as 35% of the total mass in some marine sponges.[410] Transmission of microbes occurs in various ways: *via* cytoplasmic inheritance, eggs, feces, direct contact during and after birth, breastfeeding, insects, environment, while during vegetative or asexual reproduction microbiota automatically transfer to offspring.[410]

Two protein families have been implicated in the establishment ('colonization') and maintenance ('persistence') of bacteria within the GI tract: sulfatases and radical SAM enzymes (Section 18.5.2).[418–423] Microbes may carry up to 100 or more potential sulfatase genes in their genomes,[419,424,425] which require a single anaerobic sulfatase maturing enzyme, called anSME.[418] AnSME proves to be a key radical SAM enzyme for colonization, which acts on Ser and Cys residues *via* a unique oxygen-independent mechanism.[20,226,407,418,420,422,423,426–429] anSME carries three [4Fe–4S] clusters: the SAM-dependent cluster plus two essential helper clusters, and is the founder member of radical SAM enzymes with the 'SPASM' domain,[427,430,431] see ref. 418 for details. These bacterial sulfatases are probably involved in access to carbon sources in the GI tract, cofactor biosynthesis, and post-translational modifications, suggesting a role in disease like that of human sulfatases.[419–421,426,432–434] Taken together, it is "likely that the central role of radical SAM enzymes in the human microbiota is just emerging."[418]

18.7 Conclusions

The involvement of SAM in methylation, transsulfuration and polyamine synthesis already placed SAM at the heart of metabolism and aging a decade ago.[7,17] Epigenetic regulation affects metabolism at the DNA level, but also the flux through the many SAM-dependent pathways, which contributes to (immune) disease, tumorigenesis, and aging.[114,435] The tremendous progress made in recent years now allows us to start to understand how interconnected the SAM routes are with each other and, especially, with ancient biochemistry *via* riboswitches and radical SAM enzymes. With respect to the latter, it will be interesting to see whether some SAM-dependent riboswitches affect eukaryotic splicing like TPP riboswitches (Section 18.4.1), or radical SAM enzymes respond to PARP like DDX11 and FancJ,[336] and/or are capable of sensing DNA damage like XPD (Section 18.5.7). The necessity for growth, maintenance and repair of cellular components requires multilevel control of DNA, RNA, and protein synthesis with a concomitant careful distribution of energy resources. Switching from growth to maintenance to increase health and life span[436–446] appears to be especially linked to a careful balance between the methionine cycle and transsulfuration route, which relies on SAM (see *e.g.* Sections 18.3.2 and 18.6.3). SAM is involved in stress responses (Section 18.5.3) and good stress management, inter- and intra-cellular communication, and mitochondrial functioning will slow down aging and enhance longevity, which may profit from pharmacological intervention.[435,446–452] SAM is implicated in the usefulness of a compound like resveratrol, which affects MAT2B and SIRT1, and mitochondria,[453–455] and may improve insulin sensitivity in obese mice and humans.[456] SIRT1 links mitochondrial respiration with genome stability, immunity, cell death, and energy metabolism (see *e.g.*[457,458] and elsewhere in this volume). SAM is also linked with SIRT1 *via* synthesis of *e.g.* PQQ (Section 18.5.3).[251] Mutants fail to maintain a proper balance between protein synthesis and energy availability.[77,457,459–461]

Many questions remain and certainly new discoveries will be made in the not too distant future. Obesity is a growing problem in modern society with unresolved issues of causality (apart from over-eating). Do humans run a higher risk of becoming obese due to early exposure to antibiotics, as mice appear to do?[462–464] And are artificial sweeteners a risk for glucose intolerance in humans by changing the microbiome, as they appear to do in mice?[465] Medication that supports the observed effects of caloric restriction, changes in lifestyle and food intake, and exercise, will be the best pharmacological intervention strategy to promote, and maintain, the health of the body and brain, and prevent cancer.[466–470] This indicates targeting metabolism of amino acids like Arg and Gln *via* their links with ornithine/ODC, GGC/GSH, and the urea cycle (Figure 18.2, Sections 18.3.3 and 18.3.2), essential for proper brain function.[471,472] Inhibition of ODC, and hence the polyamine pathway, appears to be a common theme in this respect: *e.g.* disruption of Arg metabolism by ODC inhibitor DFMO protects mice from AD symptoms;[471] both Arg and Gln stimulate ODC activity by inhibiting the synthesis of a natural ODC inhibitor (antizyme-1/AZ1), in turn dependent on the mTOR pathway;[473,474] polyamines improve age-associated mitochondrial dysfunction, and exercise upregulates ODC, and reduces CVD.[475] Other ODC inhibitors, such as herbacetin,[476] and hydroxylamine-containing inhibitors of *e.g.* SAMDC or ODC, can lower polyamine levels and slow growth of cancer cells, as well as block parasites that cause diseases such as malaria and sleeping sickness.[477] Interestingly, knockdown of Cantoni's MTase that methylates nicotinamide (vitamin B3, Section 18.2.1) protects against diet-induced obesity by upregulating ODC.[478] Despite the early promise of DFMO in tumor mouse models, and the strong association of polyamines with cancer, progress as a therapeutic agent in clinical trials in humans is slow.[142,479–481] This may perhaps require a combined therapy with *e.g.* polyamine transport inhibitor AMXT-1501[482] but, alternatively, may require strategies that affect the establishment and proper maintenance of the microbiome, in which radical SAM anSME play a key role (Section 18.6.5 and Figure 18.2).

Note Added after Completion of the Manuscript

A review by Landgraf *et al.* just went online in *Annu. Rev. Biochem.* describing human diseases linked to radical SAM enzymes: MoaA/MOCS1A, TYW1, Elp3, RSAD1, LipA/LIAS, MiaB/CDK5RAP, MtaB/CDKAL1, and Viperin/RSAD2; see ref.224 for details.

Abbreviations

aRNR anaerobic ribonucleotide reductase;
[4Fe–4S] iron-sulfur cluster of 4Fe plus 3× Cys + SAM, or 4× Cys;
AD Alzheimer's disease;
ALS amyotrophic lateral sclerosis;
anSME anaerobic sulfatase maturating enzyme;
BHMT1 betaine-homocysteine MTase 1;
BS biotin synthase/BioB;
CBS cystathionine beta-synthase/cystathionine synthase;
CVD cardiovascular disease;
dcSAM decarboxylated SAM;
DS Down's Syndrome;
Elp elongator protein;
FeS iron–sulfur cluster;

Folate	vitamin B9;
GCLC	subunit C of glutamate cysteine ligase;
GGC	gamma-L-glutamyl-L-cysteine;
GH(R)	growth hormone (receptor);
GNMT	glycine-*N*-methyltransferase;
GSH	glutathione;
HCY	homocysteine/HC;
IGF	insulin-like growth factor;
IRES	internal ribosome entry site;
MAT	methionine adenosyltransferase/SAM synthetase;
MMTase	radical SAM methylthiotransferase;
Mqn	menaquinone/vitamin K2;
MS	methionine synthase;
MTase	methyltransferase;
(M)THF(R)	(methyl)tetrahydrofolate (receptor);
Nicotinamide	vitamin B3;
Nos	nosiheptide, thiopeptide;
PFL-AE	pyruvate formate-lyase activating enzyme;
PQQ	pyrroloquinoline quinone;
Que	quesosine, pyrrolopyrimidine;
Riboswitch	ribozyme or aptamer;
ROS	reactive oxygen species;
RSMT	radical SAM methyltransferases;
S-box	SAM-binding box or SMK;
SAH	*S*-adenosylhomocysteine;
SAM	*S*-adenosylmethionine or AdoMet;
SAMDC	SAM decarboxylase or AMD;
SOD1	superoxide dismutase;
SPL	spore photoproduct lyase;
TIM	triose phosphate isomerase;
Toy	toyocamycin, pyrrolopyrimidine;
TPP	thiamin pyrophosphate/vitamin B1;
XP	Xeroderma pigmentosum.

References

1. G. L. Cantoni, *J. Biol. Chem.*, 1951, **189**, 203.
2. G. L. Cantoni, *J. Biol. Chem.*, 1951, **189**, 745.
3. N. Kresge, H. Tabor, R. D. Simoni and R. L. Hill, *J. Biol. Chem.*, 2005, **280**, e35.
4. M. Meselson and R. Yuan, *Nature*, 1968, **217**, 1110.
5. N. E. Murray, *Microbiol. Mol. Biol. Rev.*, 2000, **64**, 412.
6. W. A. Loenen, *Nucleic Acids Res.*, 2003, **31**, 7059.
7. W. A. Loenen, *Biochem. Soc. Trans.*, 2006, **34**, 330.
8. W. A. Loenen, D. T. Dryden, E. A. Raleigh, G. G. Wilson and N. E. Murray, *Nucleic Acids Res.*, 2014, **42**, 3.

9. W. A. Loenen, D. T. Dryden, E. A. Raleigh and G. G. Wilson, *Nucleic Acids Res.*, 2014, **42**, 20.
10. N. E. Murray, *Microbiology*, 2002, **148**, 3.
11. N. K. Raghavendra and D. N. Rao, *Biochem. Biophys. Res. Commun.*, 2005, **334**, 803.
12. S. Sistla and D. N. Rao, *Crit. Rev. Biochem. Mol. Biol.*, 2004, **39**, 1.
13. P. Bist, S. Sistla, V. Krishnamurthy, A. Acharya, B. Chandrakala and D. N. Rao, *J. Mol. Biol.*, 2001, **310**, 93.
14. http://rebase.neb.com.
15. L. I. Fedoreyeva, D. E. Sobolev and B. F. Vanyushin, *Biochemistry (Moscow)*, 2008, **73**, 1000.
16. L. I. Fedoreyeva, D. E. Sobolev and B. F. Vanyushin, *Epigenetics*, 2007, **2**, 50.
17. W. A. Loenen, in *Epigenetics of Aging*, ed. T. O. Tollefsbol, Springer Science+Business Media, LLC, 2010, p. 107.
18. H. J. Sofia, G. Chen, B. G. Hetzler, J. F. Reyes-Spindola and N. E. Miller, *Nucleic Acids Res.*, 2001, **29**, 1097.
19. P. Z. Kozbial and A. R. Mushegian, *BMC Struct. Biol.*, 2005, **5**, 19.
20. J. B. Broderick, B. R. Duffus, K. S. Duschene and E. M. Shepard, *Chem. Rev.*, 2014, **114**, 4229.
21. M. Ackermann, A. Schauerte, S. C. Stearns and U. Jenal, *BMC Evol. Biol.*, 2007, **7**, 126.
22. T. Nystrom, *PLoS Genet.*, 2007, **3**, e224.
23. U. I. Baig, B. J. Bhadbhade and M. G. Watve, *Q. Rev. Biol.*, 2014, **89**, 209.
24. D. H. Kim, *Annu. Rev. Genet.*, 2013, **47**, 233.
25. J. M. Gomez, *Curr. Aging Sci.*, 2010, **3**, 198.
26. J. P. de Magalhaes and R. G. Faragher, *BioEssays*, 2008, **30**, 567.
27. S. Wang, S. J. Arends, D. S. Weiss and E. B. Newman, *Mol. Microbiol.*, 2005, **58**, 791.
28. Z. W. El-Hajj and E. B. Newman, *J. Bacteriol.*, 2015, **197**, 1507.
29. Z. W. El-Hajj and E. B. Newman, *Front. Microbiol.*, 2015, **6**, 309.
30. Z. W. El-Hajj, R. Reyes-Lamothe and E. B. Newman, *Microbiology*, 2013, **159**, 2036.
31. S. Libert and S. D. Pletcher, *Cell*, 2007, **131**, 1231.
32. S. C. De Keersmaecker, K. Sonck and J. Vanderleyden, *Trends Microbiol.*, 2006, **14**, 114.
33. H. Zhu, Y. L. Shen, D. Z. Wei and J. W. Zhu, *Mol. Cell. Biochem.*, 2008, **315**, 25.
34. Q. H. Christensen, T. L. Grove, S. J. Booker and E. P. Greenberg, *Proc. Natl. Acad. Sci. U. S. A.*, 2013, **110**, 13815.
35. V. D. Longo, J. Mitteldorf and V. P. Skulachev, *Nat. Rev. Genet.*, 2005, **6**, 866.
36. R. A. Signer and S. J. Morrison, *Cell Stem Cell*, 2013, **12**, 152.
37. M. Lavasani, A. R. Robinson, A. Lu, M. Song, J. M. Feduska, B. Ahani, J. S. Tilstra, C. H. Feldman, P. D. Robbins, L. J. Niedernhofer and J. Huard, *Nat. Commun.*, 2012, **3**, 608.

38. J. Wang, Q. Sun, Y. Morita, H. Jiang, A. Gross, A. Lechel, K. Hildner, L. M. Guachalla, A. Gompf, D. Hartmann, A. Schambach, T. Wuestefeld, D. Dauch, H. Schrezenmeier, W. K. Hofmann, H. Nakauchi, Z. Ju, H. A. Kestler, L. Zender and K. L. Rudolph, *Cell*, 2012, **148**, 1001.
39. T. Yahata, T. Takanashi, Y. Muguruma, A. A. Ibrahim, H. Matsuzawa, T. Uno, Y. Sheng, M. Onizuka, M. Ito, S. Kato and K. Ando, *Blood*, 2011, **118**, 2941.
40. D. J. Rossi, D. Bryder, J. Seita, A. Nussenzweig, J. Hoeijmakers and I. L. Weissman, *Nature*, 2007, **447**, 725.
41. S. P. Jackson and J. Bartek, *Nature*, 2009, **461**, 1071.
42. D. Gems and L. Partridge, *Annu. Rev. Physiol.*, 2013, **75**, 621.
43. M. E. Dolle, W. K. Snyder, J. A. Gossen, P. H. Lohman and J. Vijg, *Proc. Natl. Acad. Sci. U. S. A.*, 2000, **97**, 8403.
44. F. J. Rang and J. Boonstra, *Biology (Basel)*, 2014, **3**, 403.
45. S. E. Polo, *J. Mol. Biol.*, 2015, **427**, 626.
46. O. Z. Aydin, W. Vermeulen and H. Lans, *Cell Cycle*, 2014, **13**, 3016.
47. P. A. Jeggo and J. A. Downs, *Exp. Cell Res.*, 2014, **329**, 69.
48. N. M. Shanbhag, I. U. Rafalska-Metcalf, C. Balane-Bolivar, S. M. Janicki and R. A. Greenberg, *Cell*, 2010, **141**, 970.
49. D. Weinkove, *Biochemist*, 2015, August Longevity issue, pp 2–15.
50. A. M. Garcia, R. A. Busuttil, R. B. Calder, M. E. Dolle, V. Diaz, C. A. McMahan, A. Bartke, J. Nelson, R. Reddick and J. Vijg, *Mech. Ageing Dev.*, 2008, **129**, 528.
51. F. Rodier and J. Campisi, *J. Cell Biol.*, 2011, **192**, 547.
52. W. P. Vermeij, J. H. Hoeijmakers and J. Pothof, *Annu. Rev. Pharmacol. Toxicol.*, 2016, **56**, 427.
53. D. G. Burton and R. G. Faragher, *Age (Dordrecht)*, 2015, **37**, 27.
54. M. Toulany, T. A. Schickfluss, K. R. Fattah, K. J. Lee, B. P. Chen, B. Fehrenbacher, M. Schaller, D. J. Chen and H. P. Rodemann, *Radiother. Oncol.*, 2011, **101**, 140.
55. W. P. Vermeij, J. H. Hoeijmakers and J. Pothof, *Curr. Opin. Genet. Dev.*, 2014, **26**, 124.
56. F. Rossiello, U. Herbig, M. P. Longhese, M. Fumagalli and F. d'Adda di Fagagna, *Curr. Opin. Genet. Dev.*, 2014, **26**, 89.
57. J. Campisi, *Exp. Gerontol.*, 2001, **36**, 607.
58. F. Rodier, J. P. Coppe, C. K. Patil, W. A. Hoeijmakers, D. P. Munoz, S. R. Raza, A. Freund, E. Campeau, A. R. Davalos and J. Campisi, *Nat. Cell Biol.*, 2009, **11**, 973.
59. G. Aubert and P. M. Lansdorp, *Physiol. Rev.*, 2008, **88**, 557.
60. J. Neves, M. Demaria, J. Campisi and H. Jasper, *Dev. Cell*, 2015, **32**, 9.
61. C. D. Wiley, M. C. Velarde, P. Lecot, S. Liu, E. A. Sarnoski, A. Freund, K. Shirakawa, H. W. Lim, S. S. Davis, A. Ramanathan, A. A. Gerencser, E. Verdin and J. Campisi, *Cell Metab.*, 2016, **23**, 303.
62. S. J. Chinta, G. Woods, A. Rane, M. Demaria, J. Campisi and J. K. Andersen, *Exp. Gerontol.*, 2015, **68**, 3.
63. T. Lindahl, *Nature*, 1993, **362**, 709.

64. J. A. Swenberg, K. Lu, B. C. Moeller, L. Gao, P. B. Upton, J. Nakamura and T. B. Starr, *Toxicol. Sci.*, 2011, **120**(suppl 1), S130.
65. M. J. Schafer, T. A. White, G. Evans, J. M. Tonne, G. C. Verzosa, M. B. Stout, D. L. Mazula, A. K. Palmer, D. J. Baker, M. D. Jensen, M. S. Torbenson, J. D. Miller, Y. Ikeda, T. Tchkonia, J. M. van Deursen, J. L. Kirkland and N. K. LeBrasseur, *Diabetes*, 2016, **65**, 1606–1615.
66. J. Chang, Y. Wang, L. Shao, R. M. Laberge, M. Demaria, J. Campisi, K. Janakiraman, N. E. Sharpless, S. Ding, W. Feng, Y. Luo, X. Wang, N. Aykin-Burns, K. Krager, U. Ponnappan, M. Hauer-Jensen, A. Meng and D. Zhou, *Nat. Med.*, 2016, **22**, 78.
67. B. G. Childs, M. Durik, D. J. Baker and J. M. van Deursen, *Nat. Med.*, 2015, **21**, 1424.
68. R. M. Naylor, D. J. Baker and J. M. van Deursen, *Clin. Pharmacol. Ther.*, 2013, **93**, 105.
69. D. J. Baker, B. G. Childs, M. Durik, M. E. Wijers, C. J. Sieben, J. Zhong, R. A. Saltness, K. B. Jeganathan, G. C. Verzosa, A. Pezeshki, K. Khazaie, J. D. Miller and J. M. van Deursen, *Nature*, 2016, **530**, 184.
70. B. J. North, M. A. Rosenberg, K. B. Jeganathan, A. V. Hafner, S. Michan, J. Dai, D. J. Baker, Y. Cen, L. E. Wu, A. A. Sauve, J. M. van Deursen, A. Rosenzweig and D. A. Sinclair, *EMBO J.*, 2014, **33**, 1438.
71. D. J. Baker, T. Wijshake, T. Tchkonia, N. K. LeBrasseur, B. G. Childs, B. van de Sluis, J. L. Kirkland and J. M. van Deursen, *Nature*, 2011, **479**, 232.
72. J. A. Brusslan, A. M. Rus Alvarez-Canterbury, N. U. Nair, J. C. Rice, M. J. Hitchler and M. Pellegrini, *PLoS One*, 2012, **7**, e33151.
73. J. H. Hoeijmakers, *Nature*, 2001, **411**, 366.
74. H. Li, J. R. Mitchell and P. Hasty, *Mech. Ageing Dev.*, 2008, **129**, 416.
75. L. J. Niedernhofer, *Mech. Ageing Dev.*, 2008, **129**, 408.
76. L. J. Niedernhofer, *DNA Repair*, 2008, **7**, 1180.
77. R. M. Anderson, *Biochemist*, 2015, pp 20–24.
78. J. R. Brestoff and D. Artis, *Cell*, 2015, **161**, 146.
79. O. Osborn and J. M. Olefsky, *Nat. Med.*, 2012, **18**, 363.
80. L. Fontana and L. Partridge, *Cell*, 2015, **161**, 106.
81. E. C. Macaulay, E. L. Donovan, M. P. Leask, F. H. Bloomfield, M. H. Vickers, P. K. Dearden, P. N. Baker and T. Gravida International College Summit, *J. Dev. Origins Health Dis.*, 2014, **5**, 398.
82. A. K. Elshorbagy, F. Jerneren, D. Samocha-Bonet, H. Refsum and L. K. Heilbronn, *Nutr. Diabetes*, 2016, **6**, e192.
83. D. W. Lamming, *F1000Research*, 2014, **3**, 257.
84. J. L. Martin and F. M. McMillan, *Curr. Opin. Struct. Biol.*, 2002, **12**, 783.
85. M. Fontecave, M. Atta and E. Mulliez, *Trends Biochem. Sci.*, 2004, **29**, 243.
86. H. L. Schubert, R. M. Blumenthal and X. Cheng, *Trends Biochem. Sci.*, 2003, **28**, 329.
87. M. A. Grillo and S. Colombatto, *Amino Acids*, 2007, **32**, 197.
88. S. Merali and A. B. Clarkson Jr, *FEMS Microbiol. Lett.*, 2004, **237**, 179.

89. W. Buckel and B. T. Golding, *Annu. Rev. Microbiol.*, 2006, **60**, 27.
90. J. M. Mato, F. J. Corrales, S. C. Lu and M. A. Avila, *FASEB J.*, 2002, **16**, 15.
91. K. L. Tkaczuk, S. Dunin-Horkawicz, E. Purta and J. M. Bujnicki, *BMC Bioinf.*, 2007, **8**, 73.
92. B. D. Young, D. I. Weiss, C. I. Zurita-Lopez, K. J. Webb, S. G. Clarke and A. E. McBride, *Biochemistry*, 2012, **51**, 5091.
93. B. Meyer, J. P. Wurm, S. Sharma, C. Immer, D. Pogoryelov, P. Kotter, D. L. Lafontaine, J. Wohnert and K. D. Entian, *Nucleic Acids Res.*, 2016, **44**, 4304.
94. T. Sahr, S. Ravanel and F. Rebeille, *Biochem. Soc. Trans.*, 2005, **33**, 758.
95. E. Santamaria, M. A. Avila, M. U. Latasa, A. Rubio, A. Martin-Duce, S. C. Lu, J. M. Mato and F. J. Corrales, *Proc. Natl. Acad. Sci. U. S. A.*, 2003, **100**, 3065.
96. Y. Wang, K. S. Lam, J. B. Lam, M. C. Lam, P. T. Leung, M. Zhou and A. Xu, *Mol. Endocrinol.*, 2007, **21**, 972.
97. E. P. Wijekoon, M. E. Brosnan and J. T. Brosnan, *Biochem. Soc. Trans.*, 2007, **35**, 1175.
98. A. Coppen and C. Bolander-Gouaille, *J. Psychopharmacol.*, 2005, **19**, 59.
99. V. Purohit, J. Khalsa and J. Serrano, *Alcohol*, 2005, **35**, 155.
100. I. P. Pogribny, T. V. Prizimirska, G. I. Kulik, M. B. Lutsiuk, O. O. Pentiuk, K. P. Postovitenko, M. A. Artemchuk, L. A. Poirier and V. F. Chekhun, *Ukr. Biokhim. Zh. (1999)*, 2005, (77), 114.
101. C. D. Davis and E. O. Uthus, *Exp. Biol. Med.*, 2004, **229**, 988.
102. S. A. Lamprecht and M. Lipkin, *Nat. Rev. Cancer*, 2003, **3**, 601.
103. D. C. McCabe and M. A. Caudill, *Nutr. Rev.*, 2005, **63**, 183.
104. C. L. Ulrey, L. Liu, L. G. Andrews and T. O. Tollefsbol, *Hum. Mol. Genet.*, 2005, **14**(1), R139.
105. W. A. Loenen, *Semin. Immunol.*, 1998, **10**, 417.
106. Q. Lu, A. Wu and B. C. Richardson, *J. Immunol.*, 2005, **174**, 6212.
107. D. R. Patel and B. C. Richardson, *Curr. Opin. Rheumatol.*, 2010, **22**, 478.
108. H. Wu, M. Zhao, C. Chang and Q. Lu, *Int. J. Mol. Sci.*, 2015, **16**, 11013.
109. A. Perl, R. Hanczko, Z. W. Lai, Z. Oaks, R. Kelly, R. Borsuk, J. M. Asara and P. E. Phillips, *Metabolomics*, 2015, **11**, 1157.
110. C. Miceli-Richard, *Jt., Bone, Spine*, 2015, **82**, 90.
111. H. Zan, C. Tat and P. Casali, *Autoimmunity*, 2014, **47**, 272.
112. H. Zan, *Autoimmunity*, 2014, **47**, 213.
113. E. C. Somers and B. C. Richardson, *Lupus*, 2014, **23**, 568.
114. Z. Oaks and A. Perl, *Autoimmunity*, 2014, **47**, 256.
115. A. Liu and A. La Cava, *Autoimmunity*, 2014, **47**, 215.
116. Y. Li, G. Gorelik, F. M. Strickland and B. C. Richardson, *Arthritis Rheumatol.*, 2014, **66**, 1574.
117. T. Caza, Z. Oaks and A. Perl, *Int. Rev. Immunol.*, 2014, **33**, 330.
118. Y. Zhang, M. Zhao, A. H. Sawalha, B. Richardson and Q. Lu, *J. Autoimmun.*, 2013, **41**, 92.
119. M. L. Yang, A. J. Gee, R. J. Gee, C. I. Zurita-Lopez, S. Khare, S. G. Clarke and M. J. Mamula, *Autoimmunity*, 2013, **46**, 21.

120. D. R. Patel and B. C. Richardson, *Arthritis Res. Ther.*, 2013, **15**, 201.
121. N. Veal, C. L. Hsieh, S. Xiong, J. M. Mato, S. Lu and H. Tsukamoto, *Am. J. Physiol. Gastrointest. Liver Physiol.*, 2004, **287**, G352.
122. C. R. Vinci and S. G. Clarke, *J. Biol. Chem.*, 2007, **282**, 8604.
123. J. Qin, B. P. Rosen, Y. Zhang, G. Wang, S. Franke and C. Rensing, *Proc. Natl. Acad. Sci. U. S. A.*, 2006, **103**, 2075.
124. R. M. Reguera, C. M. Redondo, Y. Perez-Pertejo and R. Balana-Fouce, *Mol. Biochem. Parasitol.*, 2007, **152**, 1.
125. J. F. Coppin, W. Qu and M. P. Waalkes, *J. Biol. Chem.*, 2008, **283**, 19342.
126. A. Prudova, Z. Bauman, A. Braun, V. Vitvitsky, S. C. Lu and R. Banerjee, *Proc. Natl. Acad. Sci. U. S. A.*, 2006, **103**, 6489.
127. S. C. Lu, *Mol. Aspects Med.*, 2009, **30**, 42.
128. S. Ignoul and J. Eggermont, *Am. J. Physiol.: Cell Physiol.*, 2005, **289**, C1369.
129. T. Majtan, A. L. Pey and J. P. Kraus, *Biochimie*, 2016, DOI: 10.1016/j.biochi.2016.01.009.
130. E. O. Uthus and H. M. Brown-Borg, *Mech. Ageing Dev.*, 2006, **127**, 444.
131. M. Dziegelewska, S. Holtze, C. Vole, U. Wachter, U. Menzel, M. Morhart, M. Groth, K. Szafranski, A. Sahm, C. Sponholz, P. Dammann, K. Huse, T. Hildebrandt and M. Platzer, *Redox Biol.*, 2016, **8**, 192.
132. H. M. Wallace, A. V. Fraser and A. Hughes, *Biochem. J.*, 2003, **376**, 1.
133. A. E. Pegg, D. J. Feith, L. Y. Fong, C. S. Coleman, T. G. O'Brien and L. M. Shantz, *Biochem. Soc. Trans.*, 2003, **31**, 356.
134. T. Thomas and T. J. Thomas, *J. Cell Mol. Med.*, 2003, **7**, 113.
135. H. M. Wallace and A. V. Fraser, *Amino Acids*, 2004, **26**, 353.
136. R. A. Casero and A. E. Pegg, *Biochem. J.*, 2009, **421**, 323.
137. O. Perez-Leal and S. Merali, *Amino Acids*, 2012, **42**, 611.
138. I. P. Ivanov and J. F. Atkins, *Nucleic Acids Res.*, 2007, **35**, 1842.
139. A. E. Pegg and D. J. Feith, *Biochem. Soc. Trans.*, 2007, **35**, 295.
140. H. M. Wallace, *Essays Biochem.*, 2009, **46**, 1.
141. A. E. Pegg, *IUBMB Life*, 2009, **61**, 880.
142. L. Miller-Fleming, V. Olin-Sandoval, K. Campbell and M. Ralser, *J. Mol. Biol.*, 2015, **427**, 3389.
143. N. S. Rial, F. L. Meyskens and E. W. Gerner, *Essays Biochem.*, 2009, **46**, 111.
144. C. S. Hayes, M. R. Burns and S. K. Gilmour, *OncoImmunology*, 2014, **3**, e27360.
145. A. E. Pegg, *Chem. Res. Toxicol.*, 2013, **26**, 1782.
146. N. A. Ignatenko, E. W. Gerner and D. G. Besselsen, *J. Carcinog.*, 2011, **10**, 10.
147. E. W. Gerner and F. L. Meyskens Jr, *Nat. Rev. Cancer*, 2004, **4**, 781.
148. A. E. Pegg, *J. Biol. Chem.*, 2006, **281**, 14529.
149. C. Zoumas-Morse, C. L. Rock, E. L. Quintana, M. L. Neuhouser, E. W. Gerner and F. L. Meyskens Jr, *J. Am. Diet. Assoc.*, 2007, **107**, 1024.
150. http://recode.genetics.utah.edu.

151. M. A. Avila, E. R. Garcia-Trevijano, S. C. Lu, F. J. Corrales and J. M. Mato, *Int. J. Biochem. Cell Biol.*, 2004, **36**, 2125.
152. A. Keren-Paz, Z. Bercovich and C. Kahana, *Biochem. Soc. Trans.*, 2007, **35**, 311.
153. A. E. Pegg, *Am. J. Physiol.: Endocrinol. Metab.*, 2008, **294**, E995.
154. T. M. Henkin, *Mol. Microbiol.*, 1994, **13**, 381.
155. T. M. Henkin, *Annu. Rev. Genet.*, 1996, **30**, 35.
156. F. J. Grundy and T. M. Henkin, *Mol. Microbiol.*, 1998, **30**, 737.
157. W. C. Winkler, A. Nahvi, N. Sudarsan, J. E. Barrick and R. R. Breaker, *Nat. Struct. Biol.*, 2003, **10**, 701.
158. B. A. McDaniel, F. J. Grundy and T. M. Henkin, *Mol. Microbiol.*, 2005, **57**, 1008.
159. R. T. Fuchs, F. J. Grundy and T. M. Henkin, *Proc. Natl. Acad. Sci. U. S. A.*, 2007, **104**, 4876.
160. C. Lu, A. M. Smith, R. T. Fuchs, F. Ding, K. Rajashankar, T. M. Henkin and A. Ke, *Nat. Struct. Mol. Biol.*, 2008, **15**, 1076.
161. C. Lu, A. M. Smith, F. Ding, A. Chowdhury, T. M. Henkin and A. Ke, *J. Mol. Biol.*, 2011, **409**, 786.
162. S. D. Gilbert, R. P. Rambo, D. Van Tyne and R. T. Batey, *Nat. Struct. Mol. Biol.*, 2008, **15**, 177.
163. R. T. Batey, *Wiley Interdiscip. Rev.*, 2011, **2**, 299.
164. A. D. Garst, A. L. Edwards and R. T. Batey, *Cold Spring Harbor Perspect. Biol.*, 2011, **3**.
165. J. J. Trausch, Z. Xu, A. L. Edwards, F. E. Reyes, P. E. Ross, R. Knight and R. T. Batey, *Proc. Natl. Acad. Sci. U. S. A.*, 2014, **111**, 6624.
166. C. Wostenberg, P. Ceres, J. T. Polaski and R. T. Batey, *J. Mol. Biol.*, 2015, **427**, 3473.
167. R. T. Batey, *RNA*, 2015, **21**, 560.
168. M. Shivji, S. Burger, C. A. Moncada, A. B. Clarkson Jr and S. Merali, *J. Biol. Chem.*, 2005, **280**, 15219.
169. K. A. Corbino, J. E. Barrick, J. Lim, R. Welz, B. J. Tucker, I. Puskarz, M. Mandal, N. D. Rudnick and R. R. Breaker, *Genome Biol.*, 2005, **6**, R70.
170. W. C. Winkler and R. R. Breaker, *Annu. Rev. Microbiol.*, 2005, **59**, 487.
171. C. D. Stoddard and R. T. Batey, *ACS Chem. Biol.*, 2006, **1**, 751.
172. R. T. Batey, *Curr. Opin. Struct. Biol.*, 2006, **16**, 299.
173. N. Sudarsan, M. C. Hammond, K. F. Block, R. Welz, J. E. Barrick, A. Roth and R. R. Breaker, *Science*, 2006, **314**, 300.
174. R. T. Fuchs, F. J. Grundy and T. M. Henkin, *Nat. Struct. Mol. Biol.*, 2006, **13**, 226.
175. J. C. Cochrane and S. A. Strobel, *RNA*, 2008, **14**, 993.
176. R. K. Montange and R. T. Batey, *Annu. Rev. Biophys.*, 2008, **37**, 117.
177. J. X. Wang and R. R. Breaker, *Biochem. Cell Biol.*, 2008, **86**, 157.
178. E. E. Regulski, R. H. Moy, Z. Weinberg, J. E. Barrick, Z. Yao, W. L. Ruzzo and R. R. Breaker, *Mol. Microbiol.*, 2008, **68**, 918.
179. J. X. Wang, E. R. Lee, D. R. Morales, J. Lim and R. R. Breaker, *Mol. Cell*, 2008, **29**, 691.

180. R. R. Breaker, *Science*, 2008, **319**, 1795.
181. R. Welz and R. R. Breaker, *RNA*, 2007, **13**, 573.
182. M. Mandal, B. Boese, J. E. Barrick, W. C. Winkler and R. R. Breaker, *Cell*, 2003, **113**, 577.
183. M. Mandal and R. R. Breaker, *Nat. Rev. Mol. Cell Biol.*, 2004, **5**, 451.
184. T. M. Henkin, *Curr. Opin. Microbiol.*, 2000, **3**, 149.
185. F. J. Grundy and T. M. Henkin, *Front. Biosci.*, 2003, **8**, d20.
186. B. A. McDaniel, F. J. Grundy, I. Artsimovitch and T. M. Henkin, *Proc. Natl. Acad. Sci. U. S. A.*, 2003, **100**, 3083.
187. D. McRose, J. Guo, A. Monier, S. Sudek, S. Wilken, S. Yan, T. Mock, J. M. Archibald, T. P. Begley, A. Reyes-Prieto and A. Z. Worden, *ISME J.*, 2014, **8**, 2517.
188. S. Li and R. R. Breaker, *Nucleic Acids Res.*, 2013, **41**, 3022.
189. B. Suess, *RNA Biol.*, 2010, **7**, 65.
190. A. Wachter, M. Tunc-Ozdemir, B. C. Grove, P. J. Green, D. K. Shintani and R. R. Breaker, *Plant Cell*, 2007, **19**, 3437.
191. M. T. Cheah, A. Wachter, N. Sudarsan and R. R. Breaker, *Nature*, 2007, **447**, 497.
192. A. Serganov, A. Polonskaia, A. T. Phan, R. R. Breaker and D. J. Patel, *Nature*, 2006, **441**, 1167.
193. A. Wachter, *RNA Biol.*, 2010, **7**, 67.
194. J. G. Marcano-Velazquez and R. T. Batey, *J. Biol. Chem.*, 2015, **290**, 4464.
195. J. J. Trausch, J. G. Marcano-Velazquez, M. M. Matyjasik and R. T. Batey, *Chem. Biol.*, 2015, **22**, 829.
196. P. B. Kim, J. W. Nelson and R. R. Breaker, *Mol. Cell*, 2015, **57**, 317.
197. F. J. Grundy and T. M. Henkin, *Crit. Rev. Biochem. Mol. Biol.*, 2006, **41**, 329.
198. T. M. Henkin and F. J. Grundy, *Cold Spring Harbor Perspect. Biol.*, 2006, **71**, 231.
199. B. J. Tucker and R. R. Breaker, *Curr. Opin. Struct. Biol.*, 2005, **15**, 342.
200. S. Hanson, K. Berthelot, B. Fink, J. E. McCarthy and B. Suess, *Mol. Microbiol.*, 2003, **49**, 1627.
201. B. Suess, S. Hanson, C. Berens, B. Fink, R. Schroeder and W. Hillen, *Nucleic Acids Res.*, 2003, **31**, 1853.
202. B. Suess, *Biochem. Soc. Trans.*, 2005, **33**, 474.
203. G. Bauer and B. Suess, *J. Biotechnol.*, 2006, **124**, 4.
204. R. R. Breaker, *Nature*, 2004, **432**, 838.
205. K. F. Blount and R. R. Breaker, *Nat. Biotechnol.*, 2006, **24**, 1558.
206. J. E. Weigand and B. Suess, *Nucleic Acids Res.*, 2007, **35**, 4179.
207. J. E. Weigand and B. Suess, *Chem. Biol.*, 2007, **14**, 9.
208. B. Suess and J. E. Weigand, *RNA Biol.*, 2008, **5**, 24.
209. J. E. Weigand and B. Suess, *Appl. Microbiol. Biotechnol.*, 2009, **85**, 229.
210. F. Groher and B. Suess, *Biochim. Biophys. Acta*, 2014, **1839**, 964.
211. C. Berens and B. Suess, *Curr. Opin. Biotechnol.*, 2015, **31**, 10.
212. http://aptamer.icmb.utexas.edu.
213. X. Wang, D. Ryu, R. H. Houtkooper and J. Auwerx, *BioEssays*, 2015, **37**, 1045.

214. N. Moullan, L. Mouchiroud, X. Wang, D. Ryu, E. G. Williams, A. Mottis, V. Jovaisaite, M. V. Frochaux, P. M. Quiros, B. Deplancke, R. H. Houtkooper and J. Auwerx, *Cell Rep.*, 2015, DOI: 10.1016/j.celrep.2015.02.034.
215. I. A. Chatzispyrou, N. M. Held, L. Mouchiroud, J. Auwerx and R. H. Houtkooper, *Cancer Res.*, 2015, **75**, 4446.
216. P. K. Todd and H. L. Paulson, *Ann. Neurol.*, 2010, **67**, 291.
217. V. Bernat and M. D. Disney, *Neuron*, 2015, **87**, 28.
218. M. Holcik, N. Sonenberg and R. G. Korneluk, *Trends Genet.*, 2000, **16**, 469.
219. S. Vagner, B. Galy and S. Pyronnet, *EMBO Rep.*, 2001, **2**, 893.
220. http://www.iresite.org.
221. J. Knappe and T. Schmitt, *Biochem. Biophys. Res. Commun.*, 1976, **71**, 1110.
222. J. Cheek and J. B. Broderick, *J. Biol. Inorg. Chem.*, 2001, **6**, 209.
223. M. Horitani, K. Shisler, W. E. Broderick, R. U. Hutcheson, K. S. Duschene, A. R. Marts, B. M. Hoffman and J. B. Broderick, *Science*, 2016, **352**, 822.
224. B. J. Landgraf, E. L. McCarthy and S. J. Booker, *Annu. Rev. Biochem.*, 2016, DOI: 10.1146/annurev-biochem-060713-035504.
225. J. L. Vey, J. Yang, M. Li, W. E. Broderick, J. B. Broderick and C. L. Drennan, *Proc. Natl. Acad. Sci. U. S. A.*, 2008, **105**, 16137.
226. J. L. Vey and C. L. Drennan, *Chem. Rev.*, 2011, **111**, 2487.
227. E. N. Marsh and G. D. Melendez, *Biochim. Biophys. Acta*, 2012, **1824**, 1154.
228. J. Stubbe, *Curr. Opin. Struct. Biol.*, 2000, **10**, 731.
229. B. R. Duffus, T. L. Hamilton, E. M. Shepard, E. S. Boyd, J. W. Peters and J. B. Broderick, *Biochim. Biophys. Acta*, 2012, **1824**, 1254.
230. A. P. Mehta, S. H. Abdelwahed, N. Mahanta, D. Fedoseyenko, B. Philmus, L. E. Cooper, Y. Liu, I. Jhulki, S. E. Ealick and T. P. Begley, *J. Biol. Chem.*, 2015, **290**, 3980.
231. A. Benjdia, S. Pierre, C. Gherasim, A. Guillot, M. Carmona, P. Amara, R. Banerjee and O. Berteau, *Nat. Commun.*, 2015, **6**, 8377.
232. M. F. Freeman, C. Gurgui, M. J. Helf, B. I. Morinaka, A. R. Uria, N. J. Oldham, H. G. Sahl, S. Matsunaga and J. Piel, *Science*, 2012, **338**, 387.
233. L. Huo, S. Rachid, M. Stadler, S. C. Wenzel and R. Muller, *Chem. Biol.*, 2012, **19**, 1278.
234. R. Eliasson, M. Fontecave, H. Jornvall, M. Krook, E. Pontis and P. Reichard, *Proc. Natl. Acad. Sci. U. S. A.*, 1990, **87**, 3314.
235. T. Barlow, *Biochem. Biophys. Res. Commun.*, 1988, **155**, 747.
236. M. Lotierzo, E. Raux, B. Tse Sum Bui, N. Goasdoue, F. Libot, D. Florentin, M. J. Warren and A. Marquet, *Biochemistry*, 2006, **45**, 12274.
237. J. T. Brosnan and M. E. Brosnan, *J. Nutr.*, 2006, **136**, 1636S.
238. T. A. Grell, P. J. Goldman and C. L. Drennan, *J. Biol. Chem.*, 2015, **290**, 3964.
239. H. Mathur, M. C. Rea, P. D. Cotter, C. Hill and R. P. Ross, *Curr. Protein Pept. Sci.*, 2015, **16**, 549.

240. I. Barr, J. A. Latham, A. T. Iavarone, T. Chantarojsiri, J. D. Hwang and J. P. Klinman, *J. Biol. Chem.*, 2016, **291**, 8877.
241. J. A. Latham, A. T. Iavarone, I. Barr, P. V. Juthani and J. P. Klinman, *J. Biol. Chem.*, 2015, **290**, 12908.
242. J. P. Klinman and F. Bonnot, *Chem. Rev.*, 2014, **114**, 4343.
243. J. J. Meulenberg, E. Sellink, W. A. Loenen, N. H. Riegman, M. van Kleef and P. W. Postma, *FEMS Microbiol. Lett.*, 1990, **59**, 337.
244. S. Pandey, A. Singh, N. Chaudhari, L. P. Nampoothiri and G. N. Kumar, *Curr. Microbiol.*, 2015, **70**, 690.
245. N. Saichana, K. Tanizawa, J. Pechousek, P. Novak, T. Yakushi, H. Toyama and J. Frebortova, *J. Biochem.*, 2016, **159**, 87.
246. H. Matsumura, K. Umezawa, K. Takeda, N. Sugimoto, T. Ishida, M. Samejima, H. Ohno, M. Yoshida, K. Igarashi and N. Nakamura, *PLoS One*, 2014, **9**, e104851.
247. J. Z. Wu, J. H. Huang, R. Khanabdali, B. Kalionis, S. J. Xia and W. J. Cai, *Exp. Gerontol.*, 2016, **80**, 43.
248. A. K. Singh, S. K. Pandey, G. Saha and N. K. Gattupalli, *Exp. Gerontol.*, 2015, **66**, 1.
249. Z. Liu, C. Sun, R. Tao, X. Xu, L. Xu, H. Cheng, Y. Wang and D. Zhang, *Inflammation*, 2016, **39**, 248.
250. J. Qin, M. Wu, S. Yu, X. Gao, J. Zhang, X. Dong, J. Ji, Y. Zhang, L. Zhou, Q. Zhang and F. Ding, *Toxicol. Lett.*, 2015, **238**, 70.
251. J. Zhang, S. Meruvu, Y. S. Bedi, J. Chau, A. Arguelles, R. Rucker and M. Choudhury, *Nutr. Res.*, 2015, **35**, 844.
252. S. Kumar, S. K. Guru, V. Venkateswarlu, F. Malik, R. A. Vishwakarma, S. D. Sawant and S. Bhushan, *Anticancer Agents Med. Chem.*, 2015, **15**, 1297.
253. N. Kumar and A. Kar, *Chem.-Biol. Interact.*, 2015, **240**, 278.
254. D. Jia, F. Duan, P. Peng, L. Sun, Y. Ruan and J. Gu, *PLoS One*, 2015, **10**, e0121939.
255. R. Tao, S. Wang, X. Xia, Y. Wang, Y. Cao, Y. Huang, X. Xu, Z. Liu, P. Liu, X. Tang, C. Liu, G. Shen and D. Zhang, *Inflammation*, 2015, **38**, 1546.
256. X. Zhou, D. Liu, R. Zhang, Y. Peng, X. Qin and S. Mao, *Behav. Brain Res.*, 2016, **308**, 217–221.
257. Q. Zhang, S. Chen, S. Yu, J. Qin, J. Zhang, Q. Cheng, K. Ke and F. Ding, *Neuropharmacology*, 2016, DOI: 10.1016/j.neuropharm.2016.04.025.
258. Y. T. Kuo, P. H. Shih, S. H. Kao, G. C. Yeh and H. M. Lee, *PLoS One*, 2015, **10**, e0143600.
259. Y. Huang, N. Chen and D. Miao, *Exp. Ther. Med.*, 2015, **10**, 451.
260. L. Li, Z. Jiao, L. Hale, W. Wu and Y. Guo, *PLoS One*, 2014, **9**, e115010.
261. X. Zhou, Q. Chen, X. Hu, S. Mao and Y. Kong, *Behav. Brain Res.*, 2014, **258**, 153.
262. F. Xu, H. Yu, J. Liu and L. Cheng, *Mol. Cell. Biochem.*, 2014, **386**, 107.
263. https://en.wikipedia.org/wiki/Diphthamide.
264. W. Ding, Q. Li, Y. Jia, X. Ji, H. Qianzhu and Q. Zhang, *ChemBioChem*, 2016.
265. G. Otero, J. Fellows, Y. Li, T. de Bizemont, A. M. Dirac, C. M. Gustafsson, H. Erdjument-Bromage, P. Tempst and J. Q. Svejstrup, *Mol. Cell*, 1999, **3**, 109.

266. J. H. Kim, W. S. Lane and D. Reinberg, *Proc. Natl. Acad. Sci. U. S. A.*, 2002, **99**, 1241.
267. G. S. Winkler, A. Kristjuhan, H. Erdjument-Bromage, P. Tempst and J. Q. Svejstrup, *Proc. Natl. Acad. Sci. U. S. A.*, 2002, **99**, 3517.
268. K. L. Stilger and W. J. Sullivan Jr, *J. Biol. Chem.*, 2013, **288**, 25318.
269. H. T. Tran, M. Nimick, R. G. Uhrig, G. Templeton, N. Morrice, R. Gourlay, A. DeLong and G. B. Moorhead, *Plant J.*, 2012, **71**, 263.
270. T. Matsumoto, C. S. Yun, H. Yoshikawa and H. Nishida, *PLoS One*, 2011, **6**, e16372.
271. F. Li, J. Lu, Q. Han, G. Zhang and B. Huang, *Mol. Genet. Genomics*, 2005, **273**, 264.
272. T. G. Petrakis, B. O. Wittschieben and J. Q. Svejstrup, *J. Biol. Chem.*, 2004, **279**, 32087.
273. K. Selvadurai, P. Wang, J. Seimetz and R. H. Huang, *Nat. Chem. Biol.*, 2014, **10**, 810.
274. M. Tigano, R. Ruotolo, C. Dallabona, F. Fontanesi, A. Barrientos, C. Donnini and S. Ottonello, *Nucleic Acids Res.*, 2015, **43**, 8368.
275. J. Fernandez-Vazquez, I. Vargas-Perez, M. Sanso, K. Buhne, M. Carmona, E. Paulo, D. Hermand, M. Rodriguez-Gabriel, J. Ayte, S. Leidel and E. Hidalgo, *PLoS Genet.*, 2013, **9**, e1003647.
276. C. T. Defraia, Y. Wang, J. Yao and Z. Mou, *BMC Plant Biol.*, 2013, **13**, 102.
277. Y. Lee, K. Min, H. Son, A. R. Park, J. C. Kim, G. J. Choi and Y. W. Lee, *Mol. Plant-Microbe Interact.*, 2014, **27**, 1344.
278. Y. Zhu, Z. Xie, J. Wang, Y. Liu and J. Wang, *Gene*, 2013, **513**, 63.
279. Y. Chinenov, *Trends Biochem. Sci.*, 2002, **27**, 115.
280. C. Paraskevopoulou, S. A. Fairhurst, D. J. Lowe, P. Brick and S. Onesti, *Mol. Microbiol.*, 2006, **59**, 795.
281. C. Greenwood, L. A. Selth, A. B. Dirac-Svejstrup and J. Q. Svejstrup, *J. Biol. Chem.*, 2009, **284**, 141.
282. W. Versees, S. De Groeve and M. Van Lijsebettens, *Mol. Microbiol.*, 2010, **76**, 1065.
283. J. Q. Svejstrup, *Curr. Opin. Cell Biol.*, 2007, **19**, 331.
284. C. Mehlgarten, D. Jablonowski, U. Wrackmeyer, S. Tschitschmann, D. Sondermann, G. Jager, Z. Gong, A. S. Bystrom, R. Schaffrath and K. D. Breunig, *Mol. Microbiol.*, 2010, **76**, 1082.
285. S. Glatt and C. W. Muller, *Curr. Opin. Struct. Biol.*, 2013, **23**, 235.
286. S. C. Wu and Y. Zhang, *Nat. Rev. Mol. Cell Biol.*, 2010, **11**, 607.
287. Y. Okada, K. Yamagata, K. Hong, T. Wakayama and Y. Zhang, *Nature*, 2010, **463**, 554.
288. D. Barton, F. Braet, J. Marc, R. Overall and J. Gardiner, *Neurosci. Lett.*, 2009, **455**, 60.
289. C. L. Simpson, R. Lemmens, K. Miskiewicz, W. J. Broom, V. K. Hansen, P. W. van Vught, J. E. Landers, P. Sapp, L. Van Den Bosch, J. Knight, B. M. Neale, M. R. Turner, J. H. Veldink, R. A. Ophoff, V. B. Tripathi, A. Beleza, M. N. Shah, P. Proitsi, A. Van Hoecke, P. Carmeliet, H. R. Horvitz, P. N. Leigh,

C. E. Shaw, L. H. van den Berg, P. C. Sham, J. F. Powell, P. Verstreken, R. H. Brown Jr, W. Robberecht and A. Al-Chalabi, *Hum. Mol. Genet.*, 2009, **18**, 472.
290. C. Creppe, L. Malinouskaya, M. L. Volvert, M. Gillard, P. Close, O. Malaise, S. Laguesse, I. Cornez, S. Rahmouni, S. Ormenese, S. Belachew, B. Malgrange, J. P. Chapelle, U. Siebenlist, G. Moonen, A. Chariot and L. Nguyen, *Cell*, 2009, **136**, 551.
291. L. Nguyen, S. Humbert, F. Saudou and A. Chariot, *Trends Mol. Med.*, 2010, **16**, 1.
292. C. Chen, S. Tuck and A. S. Bystrom, *PLoS Genet.*, 2009, **5**, e1000561.
293. A. Wynshaw-Boris, *Cell*, 2009, **136**, 393.
294. A. Ladang, F. Rapino, L. C. Heukamp, L. Tharun, K. Shostak, D. Hermand, S. Delaunay, I. Klevernic, Z. Jiang, N. Jacques, D. Jamart, V. Migeot, A. Florin, S. Goktuna, B. Malgrange, O. J. Sansom, L. Nguyen, R. Buttner, P. Close and A. Chariot, *J. Exp. Med.*, 2015, **212**, 2057.
295. D. Wang, P. N. Yang, J. Chen, X. Y. Zhou, Q. J. Liu, H. J. Li and C. L. Li, *Mol. Cell. Biochem.*, 2014, **396**, 67.
296. J. Walker, S. Y. Kwon, P. Badenhorst, P. East, H. McNeill and J. Q. Svejstrup, *Genetics*, 2011, **187**, 1067.
297. K. Miskiewicz, L. E. Jose, A. Bento-Abreu, M. Fislage, I. Taes, J. Kasprowicz, J. Swerts, S. Sigrist, W. Versees, W. Robberecht and P. Verstreken, *Neuron*, 2011, **72**, 776.
298. N. Singh, M. T. Lorbeck, A. Zervos, J. Zimmerman and F. Elefant, *J. Neurochem.*, 2010, **115**, 493.
299. F. Bauer, A. Matsuyama, J. Candiracci, M. Dieu, J. Scheliga, D. A. Wolf, M. Yoshida and D. Hermand, *Cell Rep.*, 2012, **1**, 424.
300. F. Bauer and D. Hermand, *Cell Cycle*, 2012, **11**, 4524.
301. R. Cammack, *Biochemist*, 2012, 14 October issue.
302. D. C. Johnson, D. R. Dean, A. D. Smith and M. K. Johnson, *Annu. Rev. Biochem.*, 2005, **74**, 247.
303. O. Hoffmann-Ostenhof, W. E. Cohn, A. E. Braunstein, B. L. Horecker, P. Karlson, B. Keil, W. Klyne, C. Liebecq, E. C. Webb and W. J. Whelan, IUPAC-IUB commission on biochemical nomenclature (CBN). Nomenclature of iron-sulfur proteins. 1973 recommendations, *Eur. J. Biochem.*, 1973, **35**, 1.
304. B. Py, P. L. Moreau and F. Barras, *Curr. Opin. Microbiol.*, 2011, **14**, 218.
305. B. Roche, L. Aussel, B. Ezraty, P. Mandin, B. Py and F. Barras, *Biochim. Biophys. Acta*, 2013, **1827**, 455.
306. R. Lill and U. Muhlenhoff, *Trends Biochem. Sci.*, 2005, **30**, 133.
307. O. A. Lukianova and S. S. David, *Curr. Opin. Chem. Biol.*, 2005, **9**, 145.
308. E. S. Boyd, K. M. Thomas, Y. Dai, J. M. Boyd and F. W. Outten, *Biochemistry*, 2014, **53**, 5834.
309. P. J. Kiley and H. Beinert, *FEMS Microbiol. Rev.*, 1998, **22**, 341.
310. P. J. Kiley and H. Beinert, *Curr. Opin. Microbiol.*, 2003, **6**, 181.
311. F. W. Outten, *Nat. Chem. Biol.*, 2007, **3**, 206.
312. R. Lill, B. Hoffmann, S. Molik, A. J. Pierik, N. Rietzschel, O. Stehling, M. A. Uzarska, H. Webert, C. Wilbrecht and U. Muhlenhoff, *Biochim. Biophys. Acta*, 2012, **1823**, 1491.

313. R. Lill, V. Srinivasan and U. Muhlenhoff, *Curr. Opin. Microbiol.*, 2014, **22**, 111.
314. E. J. Merino, A. K. Boal and J. K. Barton, *Curr. Opin. Chem. Biol.*, 2008, **12**, 229.
315. P. A. Sontz, T. P. Mui, J. O. Fuss, J. A. Tainer and J. K. Barton, *Proc. Natl. Acad. Sci. U. S. A.*, 2012, **109**, 1856.
316. M. F. White, *Biochem. Soc. Trans.*, 2009, **37**, 547.
317. B. Blanc, C. Gerez and S. Ollagnier de Choudens, *Biochim. Biophys. Acta*, 2015, **1853**, 1436.
318. B. Py and F. Barras, *Biochim. Biophys. Acta*, 2015, **1853**, 1429.
319. D. J. Netz, J. Mascarenhas, O. Stehling, A. J. Pierik and R. Lill, *Trends Cell Biol.*, 2014, **24**, 303.
320. J. H. Kim, J. R. Bothe, T. R. Alderson and J. L. Markley, *Biochim. Biophys. Acta*, 2015, **1853**, 1416.
321. O. Stehling, C. Wilbrecht and R. Lill, *Biochimie*, 2014, **100**, 61.
322. B. Roche, R. Agrebi, A. Huguenot, S. Ollagnier de Choudens, F. Barras and B. Py, *PLoS Genet.*, 2015, **11**, e1005134.
323. M. Berneburg and A. R. Lehmann, *Adv. Genet.*, 2001, **43**, 71.
324. J. M. Egly, *FEBS Lett.*, 2001, **498**, 124.
325. J. M. Egly and F. Coin, *DNA Repair*, 2011, **10**, 714.
326. Y. Wu, A. N. Suhasini and R. M. Brosh Jr, *Cell. Mol. Life Sci.*, 2009, **66**, 1209.
327. S. K. Bharti, J. A. Sommers, F. George, J. Kuper, F. Hamon, K. Shin-ya, M. P. Teulade-Fichou, C. Kisker and R. M. Brosh Jr, *J. Biol. Chem.*, 2013, **288**, 28217.
328. P. van der Lelij, K. H. Chrzanowska, B. C. Godthelp, M. A. Rooimans, A. B. Oostra, M. Stumm, M. Z. Zdzienicka, H. Joenje and J. P. de Winter, *Am. J. Hum. Genet.*, 2010, **86**, 262.
329. X. Sun, H. Chen, Z. Deng, B. Hu, H. Luo, X. Zeng, L. Han, G. Cai and L. Ma, *Hum. Mol. Genet.*, 2015, **24**, 4901.
330. C. D. Cota and M. J. Garcia-Garcia, *Dev. Dyn.*, 2012, **241**, 1249.
331. F. Cali, S. K. Bharti, R. D. Perna, R. M. Brosh Jr and F. M. Pisani, *Nucleic Acids Res.*, 2016, **44**, 705.
332. C. Bhattacharya, X. Wang and D. Becker, *Mol. Cancer*, 2012, **11**, 82.
333. S. K. Bharti, I. Khan, T. Banerjee, J. A. Sommers, Y. Wu and R. M. Brosh Jr, *Cell. Mol. Life Sci.*, 2014, **71**, 2625.
334. A. S. Boyer, D. Walter and C. S. Sorensen, *Semin. Cancer Biol.*, 2016, DOI: 10.1016/j.semcancer.2016.01.001.
335. J. Rudolf, V. Makrantoni, W. J. Ingledew, M. J. Stark and M. F. White, *Mol. Cell*, 2006, **23**, 801.
336. C. Stoepker, A. Faramarz, M. A. Rooimans, S. E. van Mil, J. A. Balk, E. Velleuer, N. Ameziane, H. Te Riele and J. P. de Winter, *DNA Repair*, 2015, **26**, 54.
337. J. O. Fuss and J. A. Tainer, *DNA Repair*, 2011, **10**, 697.
338. S. Ito, L. J. Tan, D. Andoh, T. Narita, M. Seki, Y. Hirano, K. Narita, I. Kuraoka, Y. Hiraoka and K. Tanaka, *Mol. Cell*, 2010, **39**, 632.

339. A. A. Vashisht, C. C. Yu, T. Sharma, K. Ro and J. A. Wohlschlegel, *J. Biol. Chem.*, 2015, **290**, 14218.
340. O. Stehling, A. A. Vashisht, J. Mascarenhas, Z. O. Jonsson, T. Sharma, D. J. Netz, A. J. Pierik, J. A. Wohlschlegel and R. Lill, *Science*, 2012, **337**, 195.
341. K. Gari, A. M. Leon Ortiz, V. Borel, H. Flynn, J. M. Skehel and S. J. Boulton, *Science*, 2012, **337**, 243.
342. N. van Wietmarschen, A. Moradian, G. B. Morin, P. M. Lansdorp and E. J. Uringa, *J. Biol. Chem.*, 2012, **287**, 43351.
343. E. Yeom, S. T. Hong and K. W. Choi, *Oncogene*, 2015, **34**, 2777.
344. E. Yeom, S. T. Hong and K. W. Choi, *Oncoscience*, 2015, **2**, 821.
345. M. A. Grodick, N. B. Muren and J. K. Barton, *Biochemistry*, 2015, **54**, 962.
346. K. N. Schaefer and J. K. Barton, *Biochemistry*, 2014, **53**, 3467.
347. K. E. Augustyn, E. J. Merino and J. K. Barton, *Proc. Natl. Acad. Sci. U. S. A.*, 2007, **104**, 18907.
348. A. R. Arnold, M. A. Grodick and J. K. Barton, *Cell Chem. Biol.*, 2016, **23**, 183.
349. K. N. Schaefer, W. M. Geil, M. J. Sweredoski, A. Moradian, S. Hess and J. K. Barton, *Biochemistry*, 2015, **54**, 932.
350. T. P. Mui, J. O. Fuss, J. P. Ishida, J. A. Tainer and J. K. Barton, *J. Am. Chem. Soc.*, 2011, **133**, 16378.
351. E. Palecek, J. Tkac, M. Bartosik, T. Bertok, V. Ostatna and J. Palecek, *Chem. Rev.*, 2015, **115**, 2045.
352. M. J. Hitchler and F. E. Domann, *Free Radical Biol. Med.*, 2007, **43**, 1023.
353. M. J. Hitchler and F. E. Domann, *Free Radical Biol. Med.*, 2009, **47**, 115.
354. A. D. Arco and J. Satrustegui, *Cell. Mol. Life Sci.*, 2005, **62**, 2204.
355. N. C. de Souza-Pinto, D. M. Wilson 3rd, T. V. Stevnsner and V. A. Bohr, *DNA Repair*, 2008, **7**, 1098.
356. M. Vermulst, J. Wanagat, G. C. Kujoth, J. H. Bielas, P. S. Rabinovitch, T. A. Prolla and L. A. Loeb, *Nat. Genet.*, 2008, **40**, 392.
357. J. Gomez, P. Caro, I. Sanchez, A. Naudi, M. Jove, M. Portero-Otin, M. Lopez-Torres, R. Pamplona and G. Barja, *J. Bioenerg. Biomembr.*, 2009, **41**, 309.
358. A. Naudi, M. Jove, D. Cacabelos, V. Ayala, R. Cabre, P. Caro, J. Gomez, M. Portero-Otin, G. Barja and R. Pamplona, *Amino Acids*, 2013, **44**, 361.
359. B. Moosmann and C. Behl, *Aging Cell*, 2008, **7**, 32.
360. N. B. Averbeck, O. N. Jensen, M. Mann, H. Schagger and H. D. Osiewacz, *Curr. Genet.*, 2000, **37**, 200.
361. B. Kunstmann and H. D. Osiewacz, *Aging Cell*, 2008, **7**, 651.
362. B. Kunstmann and H. D. Osiewacz, *Aging (Albany NY)*, 2009, **1**, 328.
363. D. Chatterjee, D. Kudlinzki, V. Linhard, K. Saxena, U. Schieborr, S. L. Gande, J. P. Wurm, J. Wohnert, R. Abele, V. V. Rogov, V. Dotsch, H. D. Osiewacz, S. Sreeramulu and H. Schwalbe, *J. Biol. Chem.*, 2015, **290**, 16415.
364. K. Groebe, F. Krause, B. Kunstmann, H. Unterluggauer, N. H. Reifschneider, C. Q. Scheckhuber, C. Sastri, W. Stegmann, W. Wozny, G. P. Schwall, S. Poznanovic, N. A. Dencher, P. Jansen-Durr, H. D. Osiewacz and A. Schrattenholz, *Exp. Gerontol.*, 2007, **42**, 887.

365. V. Soskic, K. Groebe and A. Schrattenholz, *Exp. Gerontol.*, 2008, **43**, 247.
366. S. Scarpa, A. Fuso, F. D'Anselmi and R. A. Cavallaro, *FEBS Lett.*, 2003, **541**, 145.
367. A. Fuso, V. Nicolia, A. Pasqualato, M. T. Fiorenza, R. A. Cavallaro and S. Scarpa, *Neurobiol. Aging*, 2011, **32**, 187.
368. A. Fuso, L. Seminara, R. A. Cavallaro, F. D'Anselmi and S. Scarpa, *Mol. Cell. Neurosci.*, 2005, **28**, 195.
369. A. Fuso, V. Nicolia, R. A. Cavallaro, L. Ricceri, F. D'Anselmi, P. Coluccia, G. Calamandrei and S. Scarpa, *Mol. Cell. Neurosci.*, 2008, **37**, 731.
370. A. Chan and T. B. Shea, *J. Neurochem.*, 2007, **102**, 753.
371. A. Chan, F. Tchantchou, V. Graves, R. Rozen and T. B. Shea, *J. Nutr., Health Aging*, 2008, **12**, 252.
372. F. Tchantchou, M. Graves, D. Ortiz, A. Chan, E. Rogers and T. B. Shea, *J. Nutr., Health Aging*, 2006, **10**, 541.
373. F. Tchantchou, M. Graves and T. B. Shea, *Nutr. Neurosci.*, 2006, **9**, 17.
374. F. Tchantchou, M. Graves, D. Falcone and T. B. Shea, *J. Alzheimer's Dis.*, 2008, **14**, 323.
375. T. B. Shea and R. Remington, *Curr. Opin. Psychiatry*, 2015, **28**, 141.
376. T. B. Shea and E. Rogers, *Curr. Opin. Psychiatry*, 2014, **27**, 138.
377. T. B. Shea, E. Rogers and R. Remington, *J. Alzheimer's Dis.*, 2012, **30**, 27.
378. S. Lee, C. A. Lemere, J. L. Frost and T. B. Shea, *J. Alzheimer's Dis.*, 2012, **28**, 423.
379. J. Suchy, S. Lee, A. Ahmed and T. B. Shea, *NeuroMol. Med.*, 2010, **12**, 86.
380. L. Kifle, D. Ortiz and T. B. Shea, *J. Alzheimer's Dis.*, 2009, **16**, 533.
381. A. Chan, F. Tchantchou, E. J. Rogers and T. B. Shea, *J. Neurochem.*, 2009, **110**, 831.
382. A. Chan, E. Rogers and T. B. Shea, *J. Alzheimer's Dis.*, 2009, **17**, 483.
383. http://www.ncbi.nlm.nih.gov/sites/entrez.
384. A. M. Troen, M. Shea-Budgell, B. Shukitt-Hale, D. E. Smith, J. Selhub and I. H. Rosenberg, *Proc. Natl. Acad. Sci. U. S. A.*, 2008, **105**, 12474.
385. A. M. Troen, B. Shukitt-Hale, W. H. Chao, B. Albuquerque, D. E. Smith, J. Selhub and J. Rosenberg, *J. Alzheimer's Dis.*, 2006, **9**, 381.
386. A. Bernardo, M. McCord, A. M. Troen, J. D. Allison and M. P. McDonald, *Neurobiol. Aging*, 2007, **28**, 1195.
387. E. Martignoni, C. Tassorelli, G. Nappi, R. Zangaglia, C. Pacchetti and F. Blandini, *J. Neurol. Sci.*, 2007, **257**, 31.
388. P. Galletti, M. L. De Bonis, A. Sorrentino, M. Raimo, S. D'Angelo, I. Scala, G. Andria, A. D'Aniello, D. Ingrosso and V. Zappia, *FEBS J.*, 2007, **274**, 5263.
389. M. G. Bacalini, D. Gentilini, A. Boattini, E. Giampieri, C. Pirazzini, C. Giuliani, E. Fontanesi, M. Scurti, D. Remondini, M. Capri, G. Cocchi, A. Ghezzo, A. Del Rio, D. Luiselli, G. Vitale, D. Mari, G. Castellani, M. Fraga, A. M. Di Blasio, S. Salvioli, C. Franceschi and P. Garagnani, *Aging (Albany NY)*, 2015, **7**, 82.
390. S. Horvath, P. Garagnani, M. G. Bacalini, C. Pirazzini, S. Salvioli, D. Gentilini, A. M. Di Blasio, C. Giuliani, S. Tung, H. V. Vinters and C. Franceschi, *Aging Cell*, 2015, **14**, 491.

391. L. I. Al-Gazali, R. Padmanabhan, S. Melnyk, P. Yi, I. P. Pogribny, M. Pogribna, M. Bakir, Z. A. Hamid, Y. Abdulrazzaq, A. Dawodu and S. J. James, *Am. J. Med. Genet.*, 2001, **103**, 128.
392. M. J. Hitchler and F. E. Domann, *Antioxid. Redox Signaling*, 2014, **20**, 1590.
393. J. C. Jimenez-Chillaron, M. J. Nijland, A. A. Ascensao, V. A. Sardao, J. Magalhaes, M. J. Hitchler, F. E. Domann and P. J. Oliveira, *Epigenetics*, 2015, **10**, 259.
394. V. L. Armstrong, S. Rakoczy, L. Rojanathammanee and H. M. Brown-Borg, *J. Gerontol., Ser. A*, 2014, **69**, 923.
395. H. M. Brown-Borg, S. Rakoczy, J. A. Wonderlich, V. Armstrong and L. Rojanathammanee, *Longevity Healthspan*, 2014, **3**, 10.
396. L. Rojanathammanee, S. Rakoczy and H. M. Brown-Borg, *J. Gerontol., Ser. A*, 2014, **69**, 1199.
397. D. S. Wiesenborn, J. E. Ayala, E. King and M. M. Masternak, *Age (Dordrecht)*, 2014, **36**, 9709.
398. D. S. Wiesenborn, V. Menon, X. Zhi, A. Do, A. Gesing, Z. Wang, A. Bartke, D. A. Altomare and M. M. Masternak, *Aging (Albany NY)*, 2014, **6**, 900.
399. A. Do, V. Menon, X. Zhi, A. Gesing, D. S. Wiesenborn, A. Spong, L. Sun, A. Bartke and M. M. Masternak, *Aging (Albany NY)*, 2015, **7**, 241.
400. J. Papaconstantinou and C. C. Hsieh, *Oncotarget*, 2015, **6**, 35315.
401. J. C. Triplett, A. Tramutola, A. Swomley, J. Kirk, K. Grimes, K. Lewis, M. Orr, K. Rodriguez, J. Cai, J. B. Klein, M. Perluigi, R. Buffenstein and D. A. Butterfield, *Biochim. Biophys. Acta*, 2015, **1852**, 2213.
402. J. C. Triplett, A. Swomley, J. Kirk, K. Lewis, M. Orr, K. Rodriguez, J. Cai, J. B. Klein, R. Buffenstein and D. A. Butterfield, *J. Neurochem.*, 2015, **134**, 538.
403. S. L. MacRae, Q. Zhang, C. Lemetre, I. Seim, R. B. Calder, J. Hoeijmakers, Y. Suh, V. N. Gladyshev, A. Seluanov, V. Gorbunova, J. Vijg and Z. D. Zhang, *Aging Cell*, 2015, **14**, 288.
404. M. Brohus, V. Gorbunova, C. G. Faulkes, M. T. Overgaard and C. A. Conover, *PLoS One*, 2015, **10**, e0145587.
405. M. Keane, T. Craig, J. Alfoldi, A. M. Berlin, J. Johnson, A. Seluanov, V. Gorbunova, F. Di Palma, K. Lindblad-Toh, G. M. Church and J. P. de Magalhaes, *Bioinformatics*, 2014, **30**, 3558.
406. Z. Yang, Y. Zhang and L. Chen, *BMC Syst. Biol.*, 2013, 7(suppl 2), S5.
407. F. Sommer and F. Backhed, *Nat. Rev. Microbiol.*, 2013, **11**, 227.
408. J. Qin, R. Li, J. Raes, M. Arumugam, K. S. Burgdorf, C. Manichanh, T. Nielsen, N. Pons, F. Levenez, T. Yamada, D. R. Mende, J. Li, J. Xu, S. Li, D. Li, J. Cao, B. Wang, H. Liang, H. Zheng, Y. Xie, J. Tap, P. Lepage, M. Bertalan, J. M. Batto, T. Hansen, D. Le Paslier, A. Linneberg, H. B. Nielsen, E. Pelletier, P. Renault, T. Sicheritz-Ponten, K. Turner, H. Zhu, C. Yu, S. Li, M. Jian, Y. Zhou, Y. Li, X. Zhang, S. Li, N. Qin, H. Yang, J. Wang, S. Brunak, J. Dore, F. Guarner, K. Kristiansen, O. Pedersen, J. Parkhill, J. Weissenbach, H. I. T. C. Meta, P. Bork, S. D. Ehrlich and J. Wang, *Nature*, 2010, **464**, 59.
409. K. Califf, A. Gonzalez, R. Knight and J. G. Caporaso, *Microbe*, 2014, **9**, 410.

410. E. Rosenvberg and I. Zilber-Rosenberg, *Microbe*, 2016, **11**, 27.
411. S. Weiman, *Microbe*, 2016, **11**, 21.
412. S. Weiman, *Microbe*, 2014, **9**, 437.
413. C. Gunther, C. Josenhans and J. Wehkamp, *Int. J. Med. Microbiol.*, 2016, DOI: 10.1016/j.ijmm.2016.03.003.
414. J. S. Bakken, T. Borody, L. J. Brandt, J. V. Brill, D. C. Demarco, M. A. Franzos, C. Kelly, A. Khoruts, T. Louie, L. P. Martinelli, T. A. Moore, G. Russell, C. Surawicz and W. Fecal Microbiota Transplantation, *Clin. Gastroenterol. Hepatol.*, 2011, **9**, 1044.
415. F. Backhed, H. Ding, T. Wang, L. V. Hooper, G. Y. Koh, A. Nagy, C. F. Semenkovich and J. I. Gordon, *Proc. Natl. Acad. Sci. U. S. A.*, 2004, **101**, 15718.
416. H. J. Flint, K. P. Scott, P. Louis and S. H. Duncan, *Nat. Rev. Gastroenterol. Hepatol.*, 2012, **9**, 577.
417. R. E. Ley, P. J. Turnbaugh, S. Klein and J. I. Gordon, *Nature*, 2006, **444**, 1022.
418. A. Benjdia and O. Berteau, *Biochem. Soc. Trans.*, 2016, **44**, 109.
419. A. Benjdia, E. C. Martens, J. I. Gordon and O. Berteau, *J. Biol. Chem.*, 2011, **286**, 25973.
420. O. Berteau, A. Guillot, A. Benjdia and S. Rabot, *J. Biol. Chem.*, 2006, **281**, 22464.
421. J. E. Ulmer, E. M. Vilen, R. B. Namburi, A. Benjdia, J. Beneteau, A. Malleron, D. Bonnaffe, P. A. Driguez, K. Descroix, G. Lassalle, C. Le Narvor, C. Sandstrom, D. Spillmann and O. Berteau, *J. Biol. Chem.*, 2014, **289**, 24289.
422. A. Benjdia, G. Deho, S. Rabot and O. Berteau, *FEBS Lett.*, 2007, **581**, 1009.
423. A. Benjdia, S. Subramanian, J. Leprince, H. Vaudry, M. K. Johnson and O. Berteau, *J. Biol. Chem.*, 2008, **283**, 17815.
424. J. Xu, M. K. Bjursell, J. Himrod, S. Deng, L. K. Carmichael, H. C. Chiang, L. V. Hooper and J. I. Gordon, *Science*, 2003, **299**, 2074.
425. F. O. Glockner, M. Kube, M. Bauer, H. Teeling, T. Lombardot, W. Ludwig, D. Gade, A. Beck, K. Borzym, K. Heitmann, R. Rabus, H. Schlesner, R. Amann and R. Reinhardt, *Proc. Natl. Acad. Sci. U. S. A.*, 2003, **100**, 8298.
426. A. Benjdia, J. Leprince, A. Guillot, H. Vaudry, S. Rabot and O. Berteau, *J. Am. Chem. Soc.*, 2007, **129**, 3462.
427. A. Benjdia, S. Subramanian, J. Leprince, H. Vaudry, M. K. Johnson and O. Berteau, *FEBS J.*, 2010, **277**, 1906.
428. A. Benjdia, *Curr. Opin. Struct. Biol.*, 2012, **22**, 711.
429. A. Benjdia, K. Heil, T. R. Barends, T. Carell and I. Schlichting, *Nucleic Acids Res.*, 2012, **40**, 9308.
430. D. H. Haft and M. K. Basu, *J. Bacteriol.*, 2011, **193**, 2745.
431. P. J. Goldman, T. L. Grove, L. A. Sites, M. I. McLaughlin, S. J. Booker and C. L. Drennan, *Proc. Natl. Acad. Sci. U. S. A.*, 2013, **110**, 8519.
432. G. Diez-Roux and A. Ballabio, *Annu. Rev. Genomics Hum. Genet.*, 2005, **6**, 355.

433. C. A. Hickey, K. A. Kuhn, D. L. Donermeyer, N. T. Porter, C. Jin, E. A. Cameron, H. Jung, G. E. Kaiko, M. Wegorzewska, N. P. Malvin, R. W. Glowacki, G. C. Hansson, P. M. Allen, E. C. Martens and T. S. Stappenbeck, *Cell Host Microbe*, 2015, **17**, 672.
434. L. Decamps, B. Philmus, A. Benjdia, R. White, T. P. Begley and O. Berteau, *J. Am. Chem. Soc.*, 2012, **134**, 18173.
435. M. D. Hirschey, R. J. DeBerardinis, A. M. Diehl, J. E. Drew, C. Frezza, M. F. Green, L. W. Jones, Y. H. Ko, A. Le, M. A. Lea, J. W. Locasale, V. D. Longo, C. A. Lyssiotis, E. McDonnell, M. Mehrmohamadi, G. Michelotti, V. Muralidhar, M. P. Murphy, P. L. Pedersen, B. Poore, L. Raffaghello, J. C. Rathmell, S. Sivanand, M. G. Vander Heiden, K. E. Wellen and T. Target Validation, *Semin. Cancer Biol.*, 2015, **35**, S129.
436. V. D. Longo, A. Antebi, A. Bartke, N. Barzilai, H. M. Brown-Borg, C. Caruso, T. J. Curiel, R. de Cabo, C. Franceschi, D. Gems, D. K. Ingram, T. E. Johnson, B. K. Kennedy, C. Kenyon, S. Klein, J. J. Kopchick, G. Lepperdinger, F. Madeo, M. G. Mirisola, J. R. Mitchell, G. Passarino, K. L. Rudolph, J. M. Sedivy, G. S. Shadel, D. A. Sinclair, S. R. Spindler, Y. Suh, J. Vijg, M. Vinciguerra and L. Fontana, *Aging Cell*, 2015, **14**, 497.
437. A. Bartke, L. Y. Sun and V. Longo, *Physiol. Rev.*, 2013, **93**, 571.
438. M. E. Levine, J. A. Suarez, S. Brandhorst, P. Balasubramanian, C. W. Cheng, F. Madia, L. Fontana, M. G. Mirisola, J. Guevara-Aguirre, J. Wan, G. Passarino, B. K. Kennedy, M. Wei, P. Cohen, E. M. Crimmins and V. D. Longo, *Cell Metab.*, 2014, **19**, 407.
439. E. Parrella, T. Maxim, F. Maialetti, L. Zhang, J. Wan, M. Wei, P. Cohen, L. Fontana and V. D. Longo, *Aging Cell*, 2013, **12**, 257.
440. L. Fontana, L. Partridge and V. D. Longo, *Science*, 2010, **328**, 321.
441. M. V. Blagosklonny, J. Campisi, D. A. Sinclair, A. Bartke, M. A. Blasco, W. M. Bonner, V. A. Bohr, R. M. Brosh Jr, A. Brunet, R. A. Depinho, L. A. Donehower, C. E. Finch, T. Finkel, M. Gorospe, A. V. Gudkov, M. N. Hall, S. Hekimi, S. L. Helfand, J. Karlseder, C. Kenyon, G. Kroemer, V. Longo, A. Nussenzweig, H. D. Osiewacz, D. S. Peeper, T. A. Rando, K. L. Rudolph, P. Sassone-Corsi, M. Serrano, N. E. Sharpless, V. P. Skulachev, J. L. Tilly, J. Tower, E. Verdin and J. Vijg, *Aging (Albany NY)*, 2010, **2**, 111.
442. C. J. Kenyon, *Nature*, 2010, **464**, 504.
443. J. R. Speakman and S. E. Mitchell, *Mol. Aspects Med.*, 2011, **32**, 159.
444. B. Schumacher, I. van der Pluijm, M. J. Moorhouse, T. Kosteas, A. R. Robinson, Y. Suh, T. M. Breit, H. van Steeg, L. J. Niedernhofer, W. van Ijcken, A. Bartke, S. R. Spindler, J. H. Hoeijmakers, G. T. van der Horst and G. A. Garinis, *PLoS Genet.*, 2008, **4**, e1000161.
445. A. Gesing, K. A. Al-Regaiey, A. Bartke and M. M. Masternak, *Exp. Gerontol.*, 2014, **58**, 219.
446. D. W. Lamming, L. Ye, D. M. Sabatini and J. A. Baur, *J. Clin. Invest.*, 2013, **123**, 980.
447. B. K. Kennedy, S. L. Berger, A. Brunet, J. Campisi, A. M. Cuervo, E. S. Epel, C. Franceschi, G. J. Lithgow, R. I. Morimoto, J. E. Pessin, T. A.

Rando, A. Richardson, E. E. Schadt, T. Wyss-Coray and F. Sierra, *Cell*, 2014, **159**, 709.
448. C. Lopez-Otin, M. A. Blasco, L. Partridge, M. Serrano and G. Kroemer, *Cell*, 2013, **153**, 1194.
449. D. C. Chan, *Cell*, 2006, **125**, 1241.
450. J. Nunnari and A. Suomalainen, *Cell*, 2012, **148**, 1145.
451. J. L. Barger, R. M. Anderson, M. A. Newton, C. da Silva, J. A. Vann, T. D. Pugh, S. Someya, T. A. Prolla and R. Weindruch, *PLoS One*, 2015, **10**, e0120738.
452. P. A. Andreux, R. H. Houtkooper and J. Auwerx, *Nat. Rev. Drug Discovery*, 2013, **12**, 465.
453. H. Yang, Y. Zheng, T. W. Li, H. Peng, D. Fernandez-Ramos, M. L. Martinez-Chantar, A. L. Rojas, J. M. Mato and S. C. Lu, *J. Biol. Chem.*, 2013, **288**, 23161.
454. J. M. Villalba and F. J. Alcain, *Biofactors*, 2012, **38**, 349.
455. R. H. Houtkooper, L. Mouchiroud, D. Ryu, N. Moullan, E. Katsyuba, G. Knott, R. W. Williams and J. Auwerx, *Nature*, 2013, **497**, 451.
456. Y. Jimenez-Gomez, J. A. Mattison, K. J. Pearson, A. Martin-Montalvo, H. H. Palacios, A. M. Sossong, T. M. Ward, C. M. Younts, K. Lewis, J. S. Allard, D. L. Longo, J. P. Belman, M. M. Malagon, P. Navas, M. Sanghvi, R. Moaddel, E. M. Tilmont, R. L. Herbert, C. H. Morrell, J. M. Egan, J. A. Baur, L. Ferrucci, J. S. Bogan, M. Bernier and R. de Cabo, *Cell Metab.*, 2013, **18**, 533.
457. R. H. Houtkooper, E. Pirinen and J. Auwerx, *Nat. Rev. Mol. Cell Biol.*, 2012, **13**, 225.
458. K. H. Vousden and K. M. Ryan, *Nat. Rev. Cancer*, 2009, **9**, 691.
459. M. Coelho, *Biochemist*, 2015, August Longevity issue, pp 4–7.
460. P. E. Slagboom, M. Beekman, W. M. Passtoors, J. Deelen, A. A. Vaarhorst, J. M. Boer, E. B. van den Akker, D. van Heemst, A. J. de Craen, A. B. Maier, M. Rozing, S. P. Mooijaart, B. T. Heijmans and R. G. Westendorp, *Philos. Trans. R. Soc. London, Ser. B*, 2011, **366**, 35.
461. P. Kapahi, M. E. Boulton and T. B. Kirkwood, *Free Radical Biol. Med.*, 1999, **26**, 495.
462. M. J. Blaser, *Science*, 2016, **352**, 544.
463. L. M. Cox and M. J. Blaser, *Nat. Rev. Endocrinol.*, 2015, **11**, 182.
464. L. M. Cox, S. Yamanishi, J. Sohn, A. V. Alekseyenko, J. M. Leung, I. Cho, S. G. Kim, H. Li, Z. Gao, D. Mahana, J. G. Zarate Rodriguez, A. B. Rogers, N. Robine, P. Loke and M. J. Blaser, *Cell*, 2014, **158**, 705.
465. J. Suez, T. Korem, D. Zeevi, G. Zilberman-Schapira, C. A. Thaiss, O. Maza, D. Israeli, N. Zmora, S. Gilad, A. Weinberger, Y. Kuperman, A. Harmelin, I. Kolodkin-Gal, H. Shapiro, Z. Halpern, E. Segal and E. Elinav, *Nature*, 2014, **514**, 181.
466. S. Brandhorst, I. Y. Choi, M. Wei, C. W. Cheng, S. Sedrakyan, G. Navarrete, L. Dubeau, L. P. Yap, R. Park, M. Vinciguerra, S. Di Biase, H. Mirzaei, M. G. Mirisola, P. Childress, L. Ji, S. Groshen, F. Penna, P. Odetti, L. Perin, P. S. Conti, Y. Ikeno, B. K. Kennedy, P. Cohen, T. E. Morgan, T. B. Dorff and V. D. Longo, *Cell Metab.*, 2015, **22**, 86.

467. J. A. Mattison, G. S. Roth, T. M. Beasley, E. M. Tilmont, A. M. Handy, R. L. Herbert, D. L. Longo, D. B. Allison, J. E. Young, M. Bryant, D. Barnard, W. F. Ward, W. Qi, D. K. Ingram and R. de Cabo, *Nature*, 2012, **489**, 318.
468. M. P. Mattson, D. B. Allison, L. Fontana, M. Harvie, V. D. Longo, W. J. Malaisse, M. Mosley, L. Notterpek, E. Ravussin, F. A. Scheer, T. N. Seyfried, K. A. Varady and S. Panda, *Proc. Natl. Acad. Sci. U. S. A.*, 2014, **111**, 16647.
469. V. D. Longo and L. Fontana, *Trends Pharmacol. Sci.*, 2010, **31**, 89.
470. M. Linsalata, A. Orlando and F. Russo, *Int. J. Oncol.*, 2014, **45**, 1802.
471. M. J. Kan, J. E. Lee, J. G. Wilson, A. L. Everhart, C. M. Brown, A. N. Hoofnagle, M. Jansen, M. P. Vitek, M. D. Gunn and C. A. Colton, *J. Neurosci.*, 2015, **35**, 5969.
472. M. L. Batshaw, M. Tuchman, M. Summar, J. Seminara and C. Members of the Urea Cycle Disorders, *Mol. Genet. Metab.*, 2014, **113**, 127.
473. R. M. Ray and L. R. Johnson, *Amino Acids*, 2014, **46**, 565.
474. R. M. Ray, M. Bavaria and L. R. Johnson, *Cell. Signalling*, 2015, **27**, 1850.
475. W. Wang, H. Zhang, G. Xue, L. Zhang, W. Zhang, L. Wang, F. Lu, H. Li, S. Bai, Y. Lin, Y. Lou, C. Xu and Y. Zhao, *Oxid. Med. Cell. Longevity*, 2014, **2014**, 457429.
476. D. J. Kim, E. Roh, M. H. Lee, N. Oi, Y. Lim do, M. O. Kim, Y. Y. Cho, A. Pugliese, J. H. Shim, H. Chen, E. J. Cho, J. E. Kim, S. C. Kang, S. Paul, H. E. Kang, J. W. Jung, S. Y. Lee, S. H. Kim, K. Reddy, Y. I. Yeom, A. M. Bode and Z. Dong, *Cancer Res.*, 2016, **76**, 1146.
477. M. A. Khomutov, J. Weisell, M. Hyvonen, T. A. Keinanen, J. Vepsalainen, L. Alhonen, A. R. Khomutov and S. N. Kochetkov, *Biochemistry (Moscow)*, 2013, **78**, 1431.
478. D. Kraus, Q. Yang, D. Kong, A. S. Banks, L. Zhang, J. T. Rodgers, E. Pirinen, T. C. Pulinilkunnil, F. Gong, Y. C. Wang, Y. Cen, A. A. Sauve, J. M. Asara, O. D. Peroni, B. P. Monia, S. Bhanot, L. Alhonen, P. Puigserver and B. B. Kahn, *Nature*, 2014, **508**, 258.
479. K. P. Raj, J. A. Zell, C. L. Rock, C. E. McLaren, C. Zoumas-Morse, E. W. Gerner and F. L. Meyskens, *Br. J. Cancer*, 2013, **108**, 512.
480. H. Bassiri, A. Benavides, M. Haber, S. K. Gilmour, M. D. Norris and M. D. Hogarty, *Transl. Pediatr.*, 2015, **4**, 226.
481. G. L. Saulnier Sholler, E. W. Gerner, G. Bergendahl, R. B. MacArthur, A. VanderWerff, T. Ashikaga, J. P. Bond, W. Ferguson, W. Roberts, R. K. Wada, D. Eslin, J. M. Kraveka, J. Kaplan, D. Mitchell, N. S. Parikh, K. Neville, L. Sender, T. Higgins, M. Kawakita, K. Hiramatsu, S. S. Moriya and A. S. Bachmann, *PLoS One*, 2015, **10**, e0127246.
482. K. Samal, P. Zhao, A. Kendzicky, L. P. Yco, H. McClung, E. Gerner, M. Burns, A. S. Bachmann and G. Sholler, *Int. J. Cancer*, 2013, **133**, 1323.

CHAPTER 19

Melatonin as a Geroprotector: Healthy Aging vs. Extension of Lifespan

R. HARDELAND[a]

[a]University of Göttingen, Johann Friedrich Blumenbach Institute of Zoology and Anthropology, Berliner Strasse 28, Göttingen, D-37073, Germany
*E-mail: rhardel@gwdg.de

19.1 Introduction

Melatonin (*N*-acetyl-5-methoxytryptamine; Figure 19.1) originally became known as a hormone of the pineal gland. Its name was coined with regard to its first-discovered biological function, skin-lightening in frogs and fish,[1] which possess melanocytes that contain melanosomes moved by motor proteins. Since then, a considerably broader spectrum of actions has successively been identified. The next discovery concerned the chronobiological role of this compound, which was shown to oscillate in vertebrates in a circadian fashion, with a prominent nocturnal peak, and to entrain, in many of them including humans, the rhythmicity in the circadian master clock, the suprachiasmatic nucleus (SCN).[2] In seasonally breeding mammals, but not in humans, melatonin acts also as a key regulator of annual rhythms, in which it transmits photoperiodic information to the respective organs that undergo seasonal changes.[2,3] Meanwhile, countless additional functions

RSC Drug Discovery Series No. 57
Anti-aging Drugs: From Basic Research to Clinical Practice
Edited by Alexander M. Vaiserman

Published by the Royal Society of Chemistry, www.rsc.org

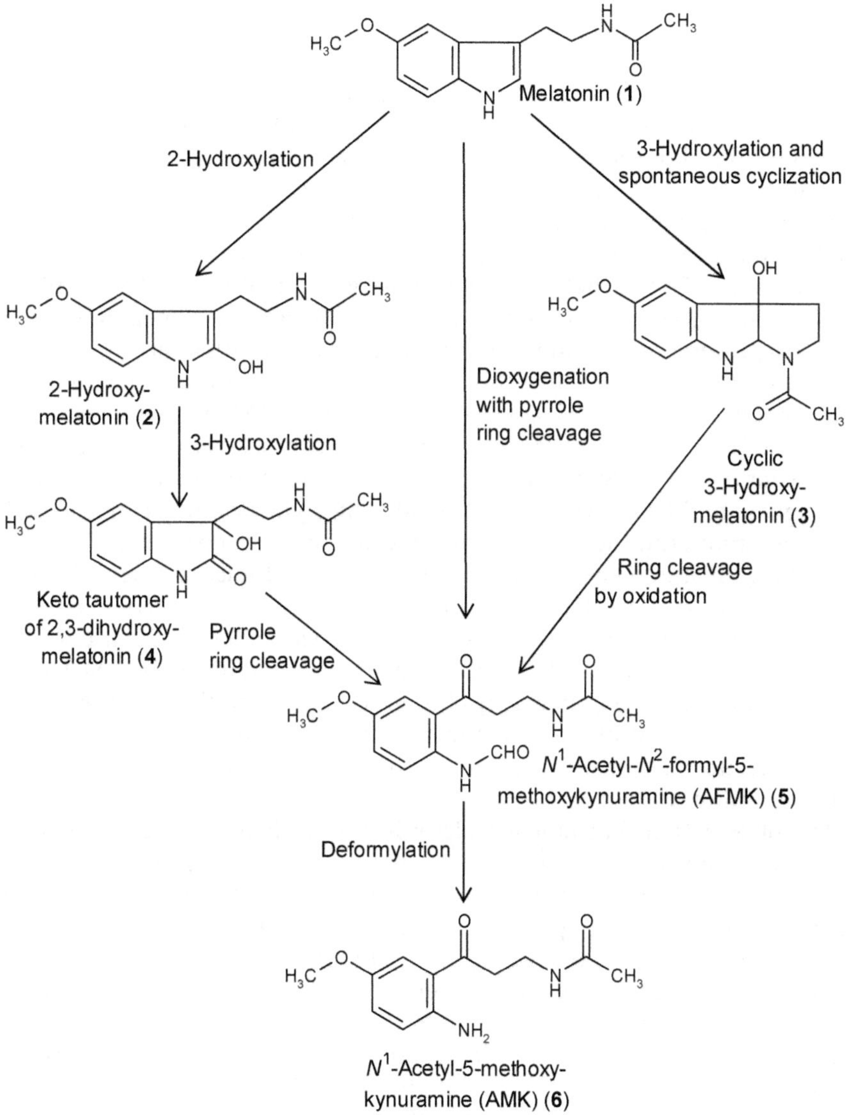

Figure 19.1 Melatonin (**1**) and formation of its protective kynuramine metabolites, AFMK (**5**) and AMK (**6**). Hydroxylations (compounds **1** → **2**, **1** → **3**, **2** → **4**) and oxidative cleavage of **3** → **5** occur under the influence of electron/hydrogen-abstracting free radicals, such as •OH. Pyrrole ring cleavage of **4** → **5** represents a spontaneous rearrangement. Dioxygenation of **1** → **5** is possible by various mechanisms, including enzymatic, pseudo-enzymatic, free-radical, singlet-oxygen and photochemical reactions. Deformylation of **5** → **6** occurs in enzymatic or photochemical reactions. For details see ref. 106 and 107.

have been identified, and melatonin actually appears as an orchestrating, pleiotropic regulator molecule that can act in virtually all cells of a mammalian body.[4,5] This comprises chronobiotic effects in peripheral, sometimes semi-autonomous, oscillators and also direct up- or down-regulation of gene expression.[6,7] Moreover, melatonin is meanwhile known to be formed in numerous extra-pineal organs and cells, in quantities that exceed in total by far those found in the pineal gland, but usually do not contribute much to the circulating hormone.[5]

With regard to the multitude of melatonin's actions, it is not surprising that various forms of melatonergic dysfunction are associated with disorders and diseases.[6–8] Perturbations of melatonergic signalling can have different reasons, such as (i) age- or disease-dependent decreases of melatonin secretion, (ii) presence of melatonin receptor variants (a) with altered or virtually absent binding affinity, (b) with poor surface expression or (c) with imbalances between the parallel signalling pathways of cAMP reduction and MAP kinase activation, and (iii) chronodisruption, which leads to a disturbed, partially suppressed or dysphased melatonin rhythm. Circadian disruption by, *e.g.*, light at night, shift work or unfavorable lifestyle including eating at night can cause numerous health problems, favours the development of metabolic syndrome, insulin resistance as well as sleep deficit-related mood disorders and may represent a risk factor for several types of cancer.[9,10] These observations are also supported by the fact that gene variants of cellular circadian core oscillators are associated with these diseases and disorders.[6] Melatonin levels and patterns are altered by chronodisruption in different ways, (i) by causing shifted and blunted rhythms of secretion, (ii) by photic turnoff of pineal melatonin synthesis, and (iii) by disease-induced reductions of melatonin formation.[9,10]

The observation that pineal melatonin synthesis and secretion typically decrease in the course of aging has been one reason to assume a role of the pineal hormone in the aging process and to speculate whether correction of melatonin deficits might prolong the lifespan. However, two complications concerning the reduction of melatonin secretion should not be overlooked. First, these decreases are inter-individually highly variable and may be caused, at least to a certain extent, by disease-related suppression of melatonin formation.[8] Second, reduced melatonin synthesis and rhythmicity may be preceded and caused by deteriorations of the circadian master clock, SCN, and/or its respective neuronal connections.[6] This is particularly evident in dementias of the Alzheimer's type, but also occurs more mildly in normal senescence. Nevertheless, a good reason for studying the role of melatonin in aging may be seen in the numerous effects of the pineal hormone on key functions of health maintenance and senescence-sensitive physiological processes, such as energy sensing, metabolic regulation, support of mitochondrial electron flux, avoidance of excessive free radical formation and elimination of these reactive compounds, prevention of neuronal over-excitation and microglia activation, and various immunological actions.[11] However, the crucial question raised by the multiplicity of mostly beneficial effects is that of whether melatonin can really extend lifespan, particularly in humans, or whether its value may predominantly be seen in the maintenance of health during aging.

19.2 Overview of Melatonin's Actions in Relation to Aging

Aging is associated with numerous changes. Some of them gave rise to assumptions concerning the primary reasons for aging and led to several partially competing theories of aging.[11] The main lines of argumentation include limitations of age by (i) energy expenditure, (ii) mitochondrial dysfunction that causes increasing damage by free radicals and, often, cell death, (iii) immune remodelling during senescence and mechanisms of inflammaging, (iv) reduction of cell division capacity by progressive telomere attrition and losses of stem cells. A closer look shows that all these processes are not suitable for monocausal explanations, but are, in fact, multiply interconnected. This insight has given rise to network hypotheses that underline the interconnections.[11–14] Moreover, a necessary distinction has to be made between (i) the basic, rather slowly progressing mechanisms of aging that lead to a steady, continuous decline in physical capacity and (ii) the step-wise, discontinuous deteriorations caused by diseases, which bear the potential of considerably accelerating aging.[15] This duality in the dynamics of aging is also relevant to the actions of melatonin because numerous effects have been described that concern either the slow, lingering basic processes or the disease-related, aging-promoting damage.[11]

19.2.1 Energy Balance and Metabolic Sensing

Although a comparison of equally sized endothermic vertebrates clearly demonstrates that energy consumption alone cannot explain the limits of lifespan,[11,15] energy metabolism, balance, sensing and disturbances thereof can certainly contribute to the velocity and course of aging. This may be particularly important for the avoidance of metabolic diseases, especially all aspects of insulin resistance, including obesity, metabolic syndrome and diabetes type 2. Interestingly, calorie restriction not only prolonged lifespan in rodents, but also preserved the functioning and circadian rhythmicity of the pineal gland.[16,17] However, it is still uncertain whether these findings are also applicable to humans because of profound differences to rodents, which continue to grow until senescence and are less facing the problem of malnutrition under food restriction as seen in primates. With regard to melatonin, its administration to rodents in the chow was shown to reduce food intake,[18] whereas intraperitoneal injections favoured carbohydrate consumption after a circadian phase shift.[19] This result indicates a role of the circadian system in the metabolic effects of melatonin. Although strengthening of the circadian system may be beneficial in both rodents and humans,[6] a frequently overlooked difference should be taken as a strong caveat for translating findings obtained in these laboratory animals. In the nocturnally active rats and mice, melatonin is associated with physical exercise, food intake and higher neuronal activity, whereas it is related, in the diurnally active human, to rest, sleep and, thus, a pause in food consumption. Because of this fundamental difference, melatonin can have opposite effects in nocturnal and diurnal species, especially in the field of energy metabolism, something that should be kept in mind when judging the relevance of respective preclinical results, including those discussed in this article.

Several pathways of nutrient sensing and regulation of energy metabolism are affected by melatonin, however, in different ways. One of these concerns the aging-promoting effects of growth hormone (GH), its mediator, insulin-like growth factor 1 (IGF-1), and downstream signalling factors.[11] In rats, melatonin caused an increase in the fronto-parietal density of the somatostatin receptor, which had been interpreted in terms of an inhibition of GH secretion.[20] However, in humans, melatonin decreased somatostatin levels, enhanced GH and also the response to $GRH_{1\text{-}44}$.[21,22] Surprisingly, this does not seem to represent an aging-promoting effect, since the changes were not accompanied by a corresponding increase in IGF-1.

Insulin, an even more important regulator of energy metabolism, is also influenced by melatonin, as indicated by anti-diabetic effects,[23,24] suppression of insulin resistance,[25–29] and variants of the melatonin receptor gene *MTNR1B* that are associated with diabetes type 2.[6,8] A complication may be seen, at first glance, in the existence of common signalling pathways initiated by IFG-1 and insulin. However, as melatonin did not elevate IGF-1, this may be less relevant and effects on insulin actions seem to be of higher importance. In cultured pancreatic islands from rats, melatonin induced tyrosine phosphorylation of IGF and insulin receptors, with subsequent activation of MAP kinase and PI3K/Akt pathways.[30] Although disruption of PI3K/Akt signalling is believed to represent an anti-aging effect,[11,31] its activation may rather represent a health-promoting action in the islets, in terms of diabetes avoidance. Several reports have documented beneficial effects of melatonin in reducing features associated with metabolic syndrome and insulin resistance, such as weight gain, hyperglycemia, dyslipidemia, hyperinsulinemia and hypertension, in different rat models.[32–35] Such findings were not only obtained in nutrient-induced disorders or diabetes-prone animals, but also in aging rats.[36] Moreover, synthetic melatonergic agonists, such as ramelteon and piromelatine, were shown to be effective in improving metabolic parameters, as summarized elsewhere.[11] However, as all these encouraging data had been obtained in nocturnally active rodents, it would be of utmost importance to be certain of the equal suitability of melatonin in humans. To date, respective findings are still controversial. While melatonin secretion was reported to be inversely correlated with insulin resistance in young non-diabetic women,[37] another study reported impairments of glucose tolerance in response to melatonin, regardless of whether it was administered in the morning or in the evening.[38] It may still be possible that results of both studies are correct and that the differences are related to the levels and time course of melatonin after administration, including the possibility of transient receptor desensitization. At least, the unfavourable results should be taken as a caveat concerning the belief that pharmacological levels of melatonin may easily readjust metabolic deviations. However, data on melatonergic treatment of diabetic patients will finally be decisive for judging the suitability of melatonin in this complex of pathologies.

A connection between insulin secretion and another metabolic sensor, AMPK (adenosine 5′-monophosphate kinase), seems to also be influenced by melatonin. This enzyme is activated *via* phosphorylation by AMPK kinase (AMPKK) and pAMPK levels were found to be reduced by melatonin in INS-1E insulinoma cells.[39] Similar reductions were observed in the immortalized hippocampal cell line HT22 exposed to $A\beta_{1\text{-}42}$,[40] but increased expression was reported in steatotic liver[41] as well as in the muscles and liver of aging rats,[42] whereas no changes were observed in pre-myoblastic skeletal muscle cells[43] and in HepG2 hepatoma cells.[44] Therefore, it seems important to discriminate between developmental stages, tissues and transformed *vs.* non-transformed cells. More information is required on melatonin effects on AMPK in the gerontological context. This demand is presumably important under a further aspect of aging concerning circadian rhythmicity.[11,45] AMPK was shown to act as an accessory component of cellular circadian oscillators and, thereby, to phase-shift circadian rhythms.[46] Moreover, the AMPK activator, metformin, otherwise used as an anti-diabetic drug, reduced the amplitude of the melatonin rhythm in ewes.[47] This role is of particular interest with regard to declining rhythm amplitudes in the course of senescence as well as rhythm-supporting effects of melatonin on central and peripheral oscillators.[6] However, the findings on AMPK activation are rather in favour of a negative relationship between melatonin and this metabolic sensor.

Sirtuin-1 (SIRT1) is another metabolic sensor with relevance to circadian oscillators and regulation by melatonin. Primarily known as an aging suppressor with protein deacetylase activity,[48] it is also intertwined with AMPK signalling,[11] calorie restriction,[49] and mitochondrial proliferation.[11,49] Again, SIRT1 acts as an accessory component of circadian oscillators. In this role, it displays a profound regulatory function on the so-called core oscillator and is required for high rhythm amplitudes of *Per2*, *Cry1*, *Bmal1* and *RORγ* transcription.[50–54] In brief, it promotes the degradation of the core oscillator protein PER2 by deacetylation and exerts a crucial effect *via* binding to the BMAL1/CLOCK complex at E-box containing promoters, where it acts as an activator in the presence of its substrate, NAD^+. The ternary protein complex activates E-box-dependent genes, including *Per1*, *Per2*, *Cry1*, *Cry2*, *Rev-erbα* and that of the key enzyme of the NAD^+ salvage pathway, NAMPT (nicotinamide phosphoribosyltransferase). The resulting cycle of NAD^+, thus, drives the expression levels of the mentioned core oscillator components and, indirectly *via* additional feedback loops, those of other components.

With regard to the aging-dependent decreases in circadian rhythm amplitudes, including that of melatonin secretion, an utmost important question was that of whether melatonin would influence the expression of SIRT1. Reports concerning melatonin effects on SIRT1 levels demonstrated strongly contrasting changes of either down- or up-regulations. However, the conditionality of this divergence became readily obvious. When studied in melatonin-responsive cancer cells or tissues, in which apoptosis is induced by this agent, melatonin consistently caused a suppression of SIRT1.[55–57] However, in other systems, the opposite was observed. Melatonin up-regulated SIRT1 in various models of brain injury,[58–61] in myocardial ischemia-reperfusion,[62] in mesenchymal stem cells,[63,64] and, notably, in senescence-accelerated, normally aged, old ovariectomized rodents or in cultured neurons from old animals.[27,65–72] The discrepancy between tumour and non-tumour cells is explained by differences in the circadian oscillator system.[73] In tumour cells, cellular circadian oscillators are strongly dysregulated by epigenetic silencing of several core oscillator genes, especially *Per2*, which otherwise have tumour suppressor properties. Apart from this necessity for being able to exist as a tumour cell, the silencing seems to fix the oscillators in positions favouring cell proliferation, which are characterized by high expression levels of SIRT1 and CLOCK. Melatonin strongly reduces the expression of these two proteins and, thus, proliferative capacity.[73] *Via* further signalling connections, it also allows and promotes apoptosis, which is otherwise inhibited by melatonin in non-tumour cells.[5] In well-operating oscillators, melatonin can only act phase-dependently. Aging-related reductions in the expression levels of SIRT1 and also the core oscillator proteins PER2 and BMAL1 can be reversed by the pineal hormone.[73] In the gerontological context, up-regulation of SIRT1 was accompanied by corresponding changes in acetylated substrates and components of the downstream signalling pathways.[11] These data as well as the changes in interrelated and converging pathways have been summarized in a recent review.[11] These include signalling by NO, peroxisome proliferator-activated receptor-γ coactivator-1α (PGC-1α), peroxisome proliferator-activated receptor-γ (PPARγ), LKB1 (liver kinase B1), PI3K and Akt/PKB. Moreover, it should be briefly mentioned that research has now started to also consider effects of melatonin on other members of the seven mammalian sirtuins. SIRT2 was reported to be increased by melatonin in the aging colonic mucosa.[74] In the dentate gyrus of aging rats, the senescence-dependent decrease of SIRT2 was reversed by growth hormone, but not by melatonin, contrary to stimulatory effects on SIRT1.[70] In hepatocytes, melatonin was shown to up-regulate SIRT3, one of the sirtuins located in mitochondria, in conjunction with elevations of MnSOD, the superoxide dismutase subform of this organelle.[75] The inductions of SIRT3 and MnSOD were also reported to be prevented by knockdown of AMPK,[75] a finding that raises new questions concerning the divergent results on the relationship between melatonin and this kinase. In the context of hepatic cadmium toxicity, another study described melatonin-induced increases in SIRT3 activity, but not SIRT3 expression, along with reduced MnSOD acetylation, *i.e.*, activation of this enzyme, and suppression of autophagic cell death.[76] Further systematic investigation of all sirtuin subforms may be a promising field of future melatonin research.

19.2.2 Counter-Action of Mitochondrial Dysfunction and Anti-Oxidant Actions

Numerous studies have investigated protective effects of melatonin at the level of mitochondria. Various models have followed the concept of inducing mitochondrial dysfunction by applying mitochondrial toxins, high-grade inflammation due to sepsis or endotoxemia, ischemia-reperfusion, excitotoxicity and other means of causing calcium overload, frequently with the aim of inducing apoptosis, autophagic cell death or mitophagy.[5,77] Despite the medical relevance of these approaches, they may not sufficiently cover the changes of mitochondrial function in aging, although there is certainly some overlap.

Mitochondria are protected by melatonin in multiple ways, which comprise (1) reduction of electron dissipation and, thus, free radical formation by modulating electron flux, (2) enhanced *de novo* synthesis of respirasomal proteins, (3) prevention of blockades of the electron transport chain (ETC) by reducing damage caused by oxidation, nitration and nitrosation of ETC components, (4) prevention of long-lasting opening of the mitochondrial permeability transition pore (mtPTP), (5) up-regulation of anti-oxidant enzymes, (6) improvement of the redox balance of glutathione (GSH/GSSG ratio), (7) inhibition of cardiolipin peroxidation, a crucial enzymatic step leading to dysfunction and apoptosis, (8) prevention of Ca^{2+} overload by anti-excitatory and anti-inflammatory actions, (9) increasing the number of mitochondrial DNA (mtDNA) copies, and (10) favouring the maintenance of mitochondrial mass including the inhibition of mitophagy. These numerous actions have been multiply reviewed under various aspects, including their relevance to aging.[5,11,77–85]

Instead of repeating all these details in a general context, several specific aspects of particular importance to aging shall be discussed. The free radical theory of aging assumes progressive damage to mitochondria by radicals that are largely formed in these organelles, *e.g.*, by electron dissipation from the ETC, with the consequence of increasing rates of radical production.[86–91] The primarily formed superoxide anions ($O_2{\bullet}^-$) can lead to free radicals of higher reactivity, either *via* H_2O_2 and the Fenton reaction or by combination with •NO to peroxynitrite ($ONOO^-$), from which hydroxyl radicals (•OH), carbonate radicals ($CO_3{\bullet}^-$) and $\bullet NO_2$ are formed.[11,77,79,83] The mitochondrial formation of CO_2 in the citric acid cycle may indicate an enhanced importance of $CO_3{\bullet}^-$ and $\bullet NO_2$ deriving from the peroxynitrite–CO_2 adduct ($ONOOCO_2^-$).[83] In comparison to •OH, the role of $CO_3{\bullet}^-$ may have been underrated, since it is sufficiently reactive to oxidize many biomolecules but is, by virtue of its resonance stabilization, considerably longer-lived than •OH. Another assumption of the free radical theory of aging, namely the progressive damage of mtDNA, may, however, turn out to be overrated. First, the absence of histones at mtDNA has been misinterpreted in terms of naked, unprotected DNA, although it is, in fact, densely covered by proteins such as the mitochondrial transcription factor A (mtTFA) and other integral components including anti-oxidant enzymes.[92] Second, a study on aging mtDNA mutator mice showed that the production of free radicals was not substantially higher than in age-matched controls, despite accumulated mitochondrial mutations.[93] Although the damage to mtDNA does not appear to be decisive, senescence-associated rises in free radical formation are observed, which should be rather attributed to a higher frequency of partial ETC blockades.

Several actions of melatonin seem to reduce the aging-related increases in mitochondrial free radical generation. Melatonin is capable of efficiently scavenging free radicals of higher reactivity, in particular, •OH[94–96] and $CO_3{\bullet}^-$,[97] but poorly interacts directly, in the absence of catalysts, with $O_2{\bullet}^-$.[98] The real contribution of direct scavenging remains, however, to be convincingly determined with regard to physiologically available melatonin levels, even though melatonin attains higher levels in mitochondria than in the circulation.[5] The reduction of electron leakage by melatonin and, thus, of free radical formation does not require high concentrations and may be more important than direct scavenging.[77] A high-affinity binding site for melatonin located in the amphipathic ramp of Complex I has been assumed to modulate electron flux and to reduce electron backflow.[77] The avoidance of secondary bottlenecks at other respirasomes such as Complexes III and IV, which can be caused by protein nitrosation, nitration and oxidation, is thought to be another means of protecting mitochondria. Notably, the kynuric melatonin metabolites N^1-acetyl-N^2-formyl-5-methoxykynuramine (AFMK) and N^1-acetyl-5-methoxykynuramine (AMK) (Figure 19.1) may contribute to the anti-oxidative and anti-nitrosative protection. AFMK and, even more potently, AMK act as scavengers of oxygen free-radicals.[99,100] AMK is also an efficient scavenger of •NO and its congeners as well as of $\bullet NO_2$, and forms a stable nitrosation product, 3-acetamidomethyl-6-methoxycinnolinone.[101–103] Moreover, it is a potent inhibitor of neuronal nitric oxide synthase (nNOS) and down-regulator of inducible NOS (iNOS), including its mitochondrially located subform.[104–106] The formation of these metabolites, which are poorly apparent in the circulation, is nevertheless relevant to mitochondria, in which melatonin is metabolized to these kynuramines.[107] However, most of the published evidence on respirasomal protection has been obtained under conditions of high-grade inflammation and oxidative stress, but only rarely in the context of aging.[77] On the other hand, the fact that only nanomolar concentrations of melatonin or AMK were needed to enhance activities of Complexes I and IV as well as ATP synthesis[108,109] speaks for a physiological role of melatonin that can be expected to decline during its senescence-associated decrease.

Other actions of melatonin connect mitochondrial protection to prevention of apoptosis. The frequently observed up-regulation of anti-apoptotic factors such as Bcl-2 or Bcl-xL and down-regulation of pro-apoptotic factors such as Bax and Bad[110] represent changes that may halt the progression towards apoptosome formation and activation of caspases. Moreover, melatonin is capable of counter-acting early causes of apoptosis initiation as well. A direct inhibition of the mtPTP opening was described, with an IC_{50} of 0.8 μM.[111] Meanwhile, another potentially important finding has been obtained concerning the duration of the mtPTP opening. So-called "superoxide flashes",[112] which have been observed at elevated rates under oxidative stress, had been suspected to induce apoptosis *via* breakdown of the mitochondrial membrane potential ($\Delta\Psi_{mt}$). However, it turned out that the duration of the mtPTP opening is decisive and that short permeability transitions are more than phenomena of superoxide release and serve the elimination of unfavourable quantities of Ca^{2+} from the mitochondrial matrix.[11] In astrocytes, melatonin was shown to inhibit a prolonged permeability transition, but still allowed short-term openings of the mtPTP.[113] Autophagy and, in particular, mitophagy have been considered as an alternative to apoptosis, although autophagy can also lead to cell death, whereas it also offers the possibility of survival. Again, beneficial effects of melatonin have been described in allowing cell survival.[114] A critical point of progressing mitophagy concerns peripheral mitochondrial depletion, especially in neurons, where it is associated with changes in the fission/fusion balance and causes impairments in transmitter release. Increases in the number of mtDNA copies, numerical density of mitochondria and total mitochondrial mass by melatonin may be indicative of a counter-action of peripheral depletion, as summarized elsewhere.[11] A further effect by melatonin in favour of mitochondrial protection and cell survival concerns the inhibition of cardiolipin peroxidation.[115–117] This process that is crucial to respirasomal dysfunction, cytochrome C release, and apoptosis induction differs from other lipid peroxidation mechanisms because it is catalysed enzymatically rather than by free radicals, but is also prevented by melatonin.

19.2.3 Immunological Actions and Prevention of Inflammaging

Melatonin is an immune modulator with both pro- and anti-inflammatory properties.[5,11,118] Although the conditions under which anti-inflammatory actions prevail over pro-inflammatory responses are not entirely understood, suppression or prevention of inflammation have been almost unanimously reported in the gerontological context.[11,85] For this reason, the aging-associated decline of melatonin levels is of particular relevance. Although anti-inflammatory actions have been also reported in high-grade inflammation caused by sepsis or endotoxemia,[5,85] mechanistic differences to aging have to be considered, because the age-related changes are rather characterized by a low-grade, often lingering inflammation progression. Therefore, it seems important to analyse the role of melatonin in reducing causes of inflammation initiation as far as they contribute to aging.

Leaving apart infectious causes, inflammation-promoting processes can be elicited by the recruitment of macrophages and, in the central nervous system, microglia. This may be favoured by changes towards an immune risk profile (IRP) in the course of the senescence-associated immune remodelling[85,119,120] and by the senescence-associated secretory phenotype (SASP).[121–123] SASP has become known as a response by DNA-damaged, division-arrested non-immune cells, which release various factors including pro-inflammatory cytokines. Meanwhile, SASP has been also found to exist independently of DNA damage in senescent cells by a mechanism driven by p38MAPK.[124,125] Notably, a SASP-like behaviour has also been described in astrocytes[125,126] and should, thus, lead to microglia activation.[85] As all inflammatory actions are accompanied by enhanced formation and release of reactive oxygen and nitrogen species, the anti-oxidant and anti-nitrosant actions of melatonin including their mitochondria-protecting consequences can be assumed to be beneficial. An additional effect concerns signalling *via* NF-κB, which has been shown to drive most SASP-related genes and to be involved in various processes of aging, including the CNS.[124,127,128] These effects can be assumed to be mitigated by melatonin, which is a potent suppressor of NF-κB expression.[129,130]

Especially in the central nervous system (CNS), several actions of melatonin are known that prevent or antagonize the initiation of low-grade inflammation.[11,85] Anti-excitatory actions that prevent excessive •NO formation in neurons and microglia, Ca^{2+} overload in neurons and astrocytes, over-excitation related changes in astrocytes, and microglia activation have recently been summarized.[85] These effects comprise inhibition of nNOS, downregulation of iNOS, decreases in cytosolic Ca^{2+} *via* $GABA_c$ and metabotropic glutamate $mGlu_3$ receptors, GABAergic facilitation, inhibition of high voltage-activated calcium channels, changes in K^+ currents, modulation of the opioid system, and, site-specifically, potentiation of glycine-dependent inhibitory post-synaptic currents.

A particular problem concerning the induction of low-grade inflammation results from the multitude of signals involved and the possible interplay between the different actors. In the CNS, different types of inflammasomes are present in neurons (NLRP1 and AIM2), astrocytes (NLRP2) and microglia (NLRP3),[131] which may cause release of the pro-inflammatory cytokines IL-1β and IL-18 and induce apoptotic or pyroptotic cell death. Dying cells can become another source of local inflammation, *e.g.*, by releasing histone H1, which acts, in addition to cytokines, as a pro-inflammatory signal and chemo-attractant.[132] Another important source of low-grade but serious inflammation are amyloid-β (Aβ) oligomers, with contributions of monomers and amyloid plaques, as summarized elsewhere including counter-actions by melatonin.[85] Aβ monomers and oligomers were shown to induce oxidative stress by microglia activation[133,134] and by stimulating NADPH oxidase in astrocytes and neurons.[135] Microglia was reported to release pro-inflammatory cytokines in response to amyloid plaques.[136] Apart from microglia and astroglia, neurons exposed to Aβ oligomers or polymers are also stimulated to release pro-inflammatory factors such as TNFα, IL-1β, the monocyte attractant chemokine CX3CL1 and to up-regulate cyclooxygenase-2.[137]

Melatonin interferes, in pre-clinical settings, with various processes related to low-grade inflammation, also beyond its classic anti-oxidative and anti-apoptotic properties. It also exerts effects as an anti-fibrillogenic agent and as an inhibitor of tau hyperphosphorylation.[138–140] In microglia exposed to $A\beta_{1\text{-}42}$, it inhibits the activation of NADPH oxidase by preventing the PI3K/Akt-dependent phosphorylation of its p47^{phox} subunit, which blocks the translocation to the plasma membrane and the association with the gp91^{phox} and p67^{phox} subunits, an effect that substantially reduces the formation of superoxide anions.[141] In organotypic brain or hippocampal slices treated with Aβ peptides, melatonin reduced the release of IL-1β and IL-6[142] or TNFα and IL-6,[143] respectively, in conjunction with decreases of tau hyperphosphorylation.[143] The suppression of TNFα by melatonin may also be relevant with regard to shifts from the formation of soluble β-amyloid precursor protein (APP) to amyloidogenic Aβ peptides *via* β- and γ-secretases, as induced by this cytokine.[144] A recent finding that may be more important than previously discussed changes in APP expression concerns a direct stimulation of α-secretase by melatonin, which is responsible for the formation of the non-amyloidogenic and neuroprotective fragment sAPPα.[145] The effect was transmitted *via* ERK1/2 activation and up-regulation of the sheddases ADAM10 and ADAM17. Melatonin was also reported to down-regulate the expression of β- and γ-secretases.[146] However, the results on regulation of the three secretases were obtained either in cells over-expressing the human APP or in human SH-SY5Y neuroblastoma cells. Their applicability to the situation in patients remains to be demonstrated.

Another relationship between inflammation and amyloidogenic Aβ peptides that has recently emerged concerns brain insulin resistance, which has been found to represent an early change in neuro-inflammation and the development of Alzheimer's disease (AD).[147,148] However, with regard to the above-mentioned uncertainties concerning beneficial or detrimental effects of melatonin on insulin sensitivity in humans,[37,38] this facet awaits further clarification. Anyway, the numerous encouraging pre-clinical findings on melatonin in AD have to be seen with caution, not only because of general problems in translation to humans, including the divergent effects in nocturnal and diurnal organisms, but even more with regard to the time spans of disease progression. Even in transgenic AD mice, melatonin was only effective concerning Aβ accumulation and survival when starting with treatment early in life,[149] but not at a later stage.[150] In humans, one may be sceptical as to whether treatments will really be started before the appearance of AD symptoms, a time point at which vicious cycles of neuro-inflammation have led to an aggravation that may no longer be halted.[85] In advanced stages of the disease, melatonin may be not be entirely useless, but its value seems to be restricted to palliative improvements.[139] However, this relatively pessimistic judgment for AD with clinical symptoms should not be generalized. Melatonin treatment may turn out to be useful in individuals with identified AD risk factors prior to disease onset. Whether it may be effective in individuals with mild cognitive impairment (MCI-AD) remains to be studied. It would also be of interest to investigate cognitively asymptomatic elderly persons. Amyloid deposits are rarely found in this group, and the previous assumption that this may be a rather harmless trait of normal aging seems to turn into the conclusion of an early, preclinical AD stage.[151,152] Regardless of whether low-grade neuro-inflammation in normal aging is associated with a certain amount of amyloid deposits, a drug like melatonin that counter-acts in multiple ways the initiation of inflammatory responses and seems to possess additional anti-inflammatory properties, especially in a gerontological context, could be of considerable value for health maintenance in aging.

19.2.4 Telomere Attrition

The usual irreversibility of telomere attrition in differentiated non-tumour cells has been a reason for concluding on limits of lifespan by this process. However, the decisive question is that of whether telomere attrition is exclusively a matter of replication rounds, as originally assumed. Although it may be precocious to seek firm conclusions on a relationship between melatonin and telomere length, a few indications for this shall be mentioned. For instance, increased formation of reactive oxygen species has been observed in mutants of clock genes and these changes were associated with advanced telomere attrition.[153] With regard to melatonin's properties as an anti-oxidant agent and a regulator of circadian oscillators, one might assume an influence of the pineal hormone on telomere length. Moreover, SIRT1 was reported to attenuate telomere shortening.[153] The repeatedly described upregulation of SIRT1 by melatonin in non-tumour cells, as discussed above, might be interpreted in a corresponding way.[11] Telomere attrition is particularly relevant to immuno-senescence. However, lymphocytes have been reported to be capable of up-regulating telomerase expression, which might contribute to a delay in the aging-associated deterioration of the immune system.[154] It remains to be investigated to what extent melatonin's immune-stimulatory properties, which include modulation of number and function of lymphocyte subtypes, may influence telomere length in these cells.[5,11]

19.3 Lifespan, Health, Deceleration and Deacceleration of Aging

An anti-aging drug may be associated with the expectancy of life extension. It seems important to properly distinguish between different processes that may limit lifespan. One of these is the lingering, slowly progressing aging in terms of a basal but poorly reversible change that starts relatively early in life, whereas others represent pathophysiologically relevant alterations that predominantly occur at advanced age and can strongly or even dramatically accelerate aging. Despite a lot of gerontological research, the mechanisms of basal aging are difficult to judge in respect to their relative contribution to the termination of life, perhaps, except for the statement that a well-functioning immune system may be the best predictor of longevity.[120,155] Moreover, drugs have never been convincingly shown to decelerate the basal process of aging in mammals. However, life extension relative to the average population can also be the result of avoiding or counter-acting the pathophysiological alterations that accelerate aging. The maximal possible prevention of age-associated pathologies may find its limits in genetic predispositions, but can be accomplished, at an individually variable basis, by lifestyle and epigenetic modulation resulting thereof. Moreover, this complex of prevention seems to be that field in which the application of anti-aging drugs is most promising.

A compound like melatonin may, therefore, be suitable for reducing age-associated pathologies because of its manifold systemic actions, from anti-nitroxidative protection, over-excitation and brain inflammation preventing properties, support of mitochondrial function and metabolic modulation to its effects on the circadian multi-oscillator system. In the white-toothed shrew, a short-lived mammal, melatonin supported a youthful locomotor activity pattern,[156] a finding of possible interest concerning circadian disturbances in elderly humans. All these findings are well in accordance with the so-called "melatonin Methuselah syndrome" described for melatonin-treated senescent laboratory rodents, which are devoid of osteoporosis and skin inflammation, retain glossy fur and rarely develop cancer.[18] Apart from melatonin's chemo-preventive action against carcinogenesis, its oncostatic, pro-apoptotic and oncocidal effects in tumor cells[11,55,73] may represent another field in which melatonin administration might contribute to healthy aging.

The support of healthy aging, which is, at least, evident in the laboratory rodents, may also exist in humans, although direct evidence is still missing and difficult to obtain for reasons of heterogeneity within populations, deviations in compliance to application rules, and differences in the onset of melatonin intake. Nevertheless, healthy aging seems to be an affordable aim of melatonin treatment, although schedules of administration, release formulations of tablets and optimal doses remain to be developed for this purpose. However, it is important to not confuse health maintenance with deceleration of aging, although the prevention of severe diseases will contribute to lifespan. In pre-clinical experiments, profound extensions of lifespan were only documented in some invertebrate animals, such as the rotifer *Philodina*.[18] Some studies on life extension in laboratory mammals were not convincing for methodological reasons or because they disregard the chemo-preventive action, as in mouse strains that frequently develop cancer.[18] A clear-cut prolongation of lifespan by melatonin was documented in the senescence-accelerated mouse strain SAMP8, with an extension of mean and maximal life-time from 16 to 22 and 23 to 27 months, respectively.[157] However, the effects in the normally aging, widely isogenic strain SAMR1 remained much smaller, with mean and maximal life-time extensions from 20 to 23 and 25 to 26 months, respectively. Therefore, melatonin was considerably more effective in counter-acting the genetically caused and, thus, pathological acceleration of aging than in decelerating normal aging. Again, with regard to humans, such findings may indicate that the gerontological value of melatonin should be sought in the support of healthy aging rather than in the extension of lifespan.

19.4 Conclusion

The pleiotropic, systemic actions of melatonin[4,5] seem to offer an opportunity for correcting malcoordination of physiological functions in elderly persons, in whom the age-associated decrease of its formation and secretion causes deficits in temporal and functional coupling. This concerns both intra-organismal relationships and coordination with the environment. Direct improvements by melatonin supplementation can be expected in the fields of anti-oxidative protection, support of mitochondrial function, with consequences to reduction or avoidance of numerous mitochondrial diseases, and in the prevention of neuronal over-excitation, which should be of value for long-term maintenance of cognitive and motor functions. Additional effects on the expression of neurotrophic factors are assumed to reduce functional decay, as summarized elsewhere.[158] Up-regulation of SIRT1 by melatonin in the CNS[58–61,65–72] may contribute to a gradual reversal of age-related deficits. However, available information on beneficial effects has been mainly obtained in pre-clinical studies and it would be of utmost importance to thoroughly investigate the usefulness of melatonin in humans.

This necessity becomes particularly obvious in the field of metabolic syndrome and insulin resistance, in which controversial findings were obtained in humans[37,38] and in which the different relationships between melatonin and metabolism in nocturnal and diurnal mammals may give rise to the suspicion that the respective effects of melatonin may be opposite in laboratory rodents and humans. Clarification of this point should have highest priority.

Generally, future attention should be focused on the precise role of melatonin in the human circadian system. Its health- and aging-related importance results from the fact that countless physiological functions are controlled by circadian oscillators. Rhythm amplitudes and temporal relationships within the multi-oscillator system are important for the good functioning of the organism.[6] Therefore, the age-related decline of rhythms, which also becomes obvious by sleep disturbances and nocturia, can be expected to contribute to deficits in physical fitness. One of the most impressive rejuvenation effects was obtained by transplanting a juvenile SCN to a senescent hamster, an intervention that not only restored the previously decomposed rhythm patterns, but also improved the physical appearance and extended the lifespan of the recipient.[159] While the aging-associated decline of the melatonin rhythm may largely reflect a progressive malfunctioning of the SCN, melatonin itself acts on the SCN as well as on several peripheral oscillators, not only in terms of phase shifting, but also in supporting high rhythm amplitudes.[6] It will be of importance to investigate in the future more systematically the effects of melatonin on circadian amplitudes generated by both central and peripheral oscillators.

While many functions that decline by age in conjunction with decreasing melatonin levels may be partially reversed or corrected by a melatonin substitution therapy, this is not necessarily the case in the entire field of immuno-senescence because thymic involution and immune remodelling represent irreversible processes. To what extent the immune-stimulatory actions of melatonin may be beneficial in elderly persons remains uncertain with regard to potentially pro-inflammatory effects. However, the other, anti-inflammatory side of melatonin's action spectrum, which has been particularly observed in reducing microglia activation and age-related low-grade inflammation, may represent a most valuable field for protecting the CNS as well as other organs.

In conclusion, melatonin displays many properties of a geroprotector, but mainly in terms of favouring healthy aging. Although a deceleration of the basal aging processes has not been demonstrated in humans and may turn out be marginal, counter-actions against aging-accelerating pathologies, which can be deduced from many pre-clinical data, should be of substantial value. Therefore, a substitution therapy with melatonin may contribute to reaching the individually maximally possible lifespan.

References

1. A. B. Lerner, J. D. Case and Y. Takahashi, *J. Am. Chem. Soc.*, 1958, **80**, 2057.
2. R. J. Reiter, *Experientia*, 1993, **49**, 654.
3. J. Arendt, *Oxford Rev. Reprod. Biol.*, 1986, **8**, 266.
4. S. R. Pandi-Perumal, V. Srinivasan, G. J. M. Maestroni, D. P. Cardinali, B. Poeggeler and R. Hardeland, *FEBS J.*, 2006, **273**, 2813.
5. R. Hardeland, D. P. Cardinali, V. Srinivasan, D. W. Spence, G. M. Brown and S. R. Pandi-Perumal, *Prog. Neurobiol.*, 2011, **93**, 350.
6. R. Hardeland, J. A. Madrid, D.-X. Tan and R. J. Reiter, *J. Pineal Res.*, 2012, **52**, 139.
7. R. Hardeland, *Int. J. Mol. Sci.*, 2013, **14**, 5817.
8. R. Hardeland, *Aging Dis.*, 2012, **3**, 194.
9. T. C. Erren and R. J. Reiter, *J. Pineal Res.*, 2009, **46**, 245.
10. M. A. Bonmati-Carrion, R. Arguelles-Prieto, M. J. Martinez-Madrid, R. Reiter, R. Hardeland, M. A. Rol and J. A. Madrid, *Int. J. Mol. Sci.*, 2014, **15**, 23448.
11. R. Hardeland, *J. Pineal Res.*, 2013, **55**, 325.
12. A. Kowald and T. B. Kirkwood, *Mutat. Res.*, 1996, **316**, 209.
13. W. Ying, *Med. Hypotheses*, 1997, **48**, 143.
14. T. B. Kirkwood and A. Kowald, *Exp. Gerontol.*, 1997, **32**, 395.
15. R. Hardeland and A. Coto-Montes, *Open Biol. J.*, 2010, **3**, 39.
16. K. A. Stokkan, R. J. Reiter, K. O. Nonaka, A. Lerchl, B. P. Yu and M. K. Vaughan, *Brain Res.*, 1991, **545**, 66.
17. G. Huether, *Ann. N. Y. Acad. Sci.*, 1994, **719**, 146.
18. B. Poeggeler, *Endocrine*, 2005, **27**, 201.
19. K. Angers, N. Haddad, B. Selmaoui and L. Thibault, *Physiol. Behav.*, 2003, **80**, 9.
20. R. M. Izquierdo-Claros, M. C. Boyano-Adánez, G. Torrecillas, M. Rodríguez-Puyol and E. Arilla-Ferreiro, *J. Pineal Res.*, 2001, **31**, 46.

21. R. Valcavi, C. Dieguez, C. Azzarito, C. A. Edwards, C. Dotti, M. D. Page, I. Portioli and M. F. Scanlon, *Clin. Endocrinol. (Oxford, U. K.)*, 1987, **26**, 453.
22. E. Nassar, C. Mulligan, L. Taylor, C. Kerksick, M. Galbreath, M. Greenwood, R. Kreider and D. S. Willoughby, *J. Int. Soc. Sports Nutr.*, 2007, **4**, 14.
23. E. Peschke, I. Stumpf, I. Bazwinsky, L. Litvak, H. Dralle and E. Mühlbauer, *J. Pineal Res.*, 2007, **42**, 350.
24. E. Peschke, *J. Pineal Res.*, 2008, **44**, 26.
25. S. Contreras-Alcantara, K. Baba and G. Tosini, *Obesity (Silver Spring)*, 2010, **18**, 1861.
26. T. C. Nogueira, C. Lellis-Santos, D. S. Jesus, M. Taneda, S. C. Rodrigues, F. G. Amaral, A. M. Lopes, J. Cipolla-Neto, S. Bordin and G. F. Anhe, *Endocrinology*, 2011, **152**, 1253.
27. S. Cuesta, R. Kireev, C. Garcia, L. Rancan, E. Vara and J. A. Tresguerres, *Age (Dordr)*, 2013, **35**, 659.
28. M. She, H. Hou, Z. Wang, C. Zhang, M. Laudon and W. Yin, *Biochimie*, 2014, **103**, 126.
29. X. Quan, J. Wang, C. Liang, H. Zheng and L. Zhang, *Biochem. Biophys. Res. Commun.*, 2015, **463**, 1102.
30. M. C. Picinato, A. E. Hirata, J. Cipolla-Neto, R. Curl, C. R. O. Carvalho, G. F. Anhê and A. R. Carpinelli, *J. Pineal Res.*, 2008, **44**, 88.
31. A. Salminen and K. Kaamiranta, *Cell. Signalling*, 2010, **22**, 573.
32. A. Leibowitz, E. Peleg, Y. Sharabi, Z. Shabtai, A. Shamiss and E. Grossman, *Am. J. Hypertens.*, 2008, **21**, 348.
33. A. Kitagawa, Y. Ohta and K. Ohashi, *J. Pineal Res.*, 2012, **52**, 403.
34. D. P. Cardinali, P. A. Bernasconi, R. Reynoso, C. F. Toso and P. Scacchi, *Int. J. Mol. Sci.*, 2013, **14**, 2502.
35. A. Agil, I. Rosado, R. Ruiz, A. Figueroa, N. Zen and G. Fernández-Vázquez, *J. Pineal Res.*, 2012, **52**, 203.
36. I. Vinogradova and V. Anisimov, *Biogerontology*, 2013, **14**, 401.
37. C. J. McMullan, G. C. Curhan, E. S. Schernhammer and J. P. Forman, *Am. J. Epidemiol.*, 2013, **178**, 231.
38. P. Rubio-Sastre, F. A. Scheer, P. Gomez-Abellan, J. A. Madrid and M. Garaulet, *Sleep*, 2014, **37**, 1715.
39. Y. M. Yoo, *J. Pineal Res.*, 2013, **55**, 207.
40. K. J. Kwon, H. J. Kim, C. Y. Shin and S. H. Han, *J. Clin. Neurol.*, 2010, **6**, 127.
41. M. A. Zaouali, E. Boncompagni, R. J. Reiter, M. Bejaoui, I. Freitas, E. Pantazi, E. Folch-Puy, H. B. Abdennebi, F. A. Garcia-Gil and J. Roselló-Catafau, *J. Pineal Res.*, 2013, **55**, 65.
42. C. Mendes, A. M. Lopes, F. G. do Amaral, R. A. Peliciari-Garcia, O. Turati Ade, S. M. Hirabara, J. H. Scialfa Falcão and J. Cipolla-Neto, *J. Pineal Res.*, 2013, **55**, 229.
43. E. Ha, S. V. Yim, J. H. Chung, K. S. Yoon, I. Kang, Y. H. Cho and H. H. Baik, *J. Pineal Res.*, 2006, **41**, 67.

44. J. M. Shieh, H. A. T. Wu, K. C. Cheng and J. T. Cheng, *J. Pineal Res.*, 2009, **47**, 339.
45. R. Hardeland, *Interdiscip. Top. Gerontol.*, 2014, **40**, 128.
46. J. H. Um, S. Yang, S. Yamazaki, H. Kang, B. Viollet, M. Foretz and J. H. Chung, *J. Biol. Chem.*, 2007, **282**, 20794.
47. J. B. Menassol, C. Tautou, A. Collet, D. Chesneau, D. Lomet, J. Dupont, B. Malpaux and R. J. Scaramuzzi, *BMC Neurosci.*, 2011, **12**, 76.
48. L. Guarente, *Cold Spring Harbor Symp. Quant. Biol.*, 2007, **72**, 483.
49. L. Guarente, *Cell*, 2008, **132**, 171.
50. Y. Nakahata, M. Kaluzova, B. Grimaldi, S. Sahar, J. Hirayama, D. Chen, L. P. Guarente and P. Sassone-Corsi, *Cell*, 2008, **134**, 329.
51. B. Grimaldi, Y. Nakahata, M. Kaluzova, S. Masubuchi and P. Sassone-Corsi, *Int. J. Biochem. Cell Biol.*, 2009, **41**, 81.
52. Y. Nakahata, S. Sahar, G. Astarita, M. Kaluzova and P. Sassone-Corsi, *Science*, 2009, **324**, 654.
53. M. M. Bellet, R. Orozco-Solis, S. Sahar, K. Eckel-Mahan and P. Sassone-Corsi, *Cold Spring Harbor Symp. Quant. Biol.*, 2011, **76**, 31.
54. S. Sahar and P. Sassone-Corsi, *Handb. Exp. Pharmacol.*, 2013, **217**, 29.
55. B. Jung-Hynes, T. L. Schmit, S. R. Reagan-Shaw, I. A. Siddiqui, H. Mukhtar and N. Ahmad, *J. Pineal Res.*, 2011, **50**, 140.
56. Y. Cheng, L. Cai, P. Jiang, J. Wang, C. Gao, H. Feng, C. Wang, H. Pan and Y. Yang, *Eur. J. Pharmacol.*, 2013, **715**, 219.
57. M. Bizzarri, S. Proietti, A. Cucina and R. J. Reiter, *Expert Opin. Ther. Targets*, 2013, **17**, 1483.
58. Y. Yang, S. Jiang, Y. Dong, C. Fan, L. Zhao, X. Yang, J. Li, S. Di, L. Yue, G. Liang, R. J. Reiter and Y. Qu, *J. Pineal Res.*, 2015, **58**, 61.
59. L. Zhao, R. An, Y. Yang, X. Yang, H. Liu, L. Yue, X. Li, Y. Lin, R. J. Reiter and Y. Qu, *J. Pineal Res.*, 2015, **59**, 230.
60. S. Emamgholipour, A. Hossein-Nezhad, M. A. Sahraian, F. Askarisadr and M. Ansari, *Life Sci.*, 2016, **145**, 34.
61. L. Zhao, H. Liu, L. Yue, J. Zhang, X. Li, B. Wang, Y. Lin and Y. Qu, *Mol. Neurobiol.*, 2016, DOI: 10.1007/s12035-016-9776-7Epub ahead of print, Feb. 11.
62. L. Yu, H. Liang, X. Dong, G. Zhao, Z. Jin, M. Zhai, Y. Yang, W. Chen, J. Liu, W. Yi, J. Yang, D. Yi, W. Duan and S. Yu, *J. Pineal Res.*, 2015, **59**, 376.
63. L. Zhou, X. Chen, T. Liu, Y. Gong, S. Chen, G. Pan, W. Cui, Z. P. Luo, M. Pei, H. Yang and F. He, *J. Pineal Res.*, 2015, **59**, 190.
64. D. Han, W. Huang, X. Li, L. Gao, T. Su, X. Li, S. Ma, T. Liu, C. Li, J. Chen, E. Gao and F. Cao, *J. Pineal Res.*, 2016, **60**, 178.
65. J. Gutierrez-Cuesta, M. Tajes, A. Jiménez, A. Coto-Montes, A. Camins and M. Pallàs, *J. Pineal Res.*, 2008, **45**, 497.
66. H.-M. Chang, U.-I. Wu and C.-T. Lan, *J. Pineal Res.*, 2009, **47**, 211.
67. M. Tajes, J. Gutierrez-Cuesta, D. Ortuño-Sahagun, A. Camins and M. Pallàs, *J. Pineal Res.*, 2009, **47**, 228.
68. R. Cristòfol, D. Porquet, R. Corpas, A. Coto-Montes, J. Serret, A. Camins, M. Pallàs and C. Sanfeliu, *J. Pineal Res.*, 2012, **52**, 271.

69. L. F. Rodella, G. Favero, C. Rossini, E. Foglio, F. Bonomini, R. J. Reiter and R. Rezzani, *Age (Dordr)*, 2013, **35**, 103.
70. R. A. Kireev, E. Vara and J. A. F. Tresguerres, *Biogerontology*, 2013, **14**, 431.
71. R. A. Kireev, E. Vara, J. Viña and J. A. F. Tresguerres, *Age (Dordr)*, 2014, **36**, 9707.
72. M. R. Ramis, S. Esteban, A. Miralles, D.-X. Tan and R. J. Reiter, *Mech. Ageing Dev.*, 2015, **146–148**, 28–41.
73. R. Hardeland, *Int. J. Mol. Sci.*, 2014, **15**, 18221.
74. K. G. Akbulut, S. H. Aktas and H. Akbulut, *Biogerontology*, 2015, **16**, 99.
75. Y. Chen, W. Qing, M. Sun, L. Lv, D. Guo and Y. Jiang, *Free Radical Res.*, 2015, **49**, 1275.
76. H. Pi, S. Xu, R. J. Reiter, P. Guo, L. Zhang, Y. Li, M. Li, Z. Cao, L. Tian, J. Xie, R. Zhang, M. He, Y. Lu, C. Liu, W. Duan, Z. Yu and Z. Zhou, *Autophagy*, 2015, **11**, 1037.
77. R. Hardeland, *Adv. Stud. Biol.*, 2009, **1**, 207.
78. D. Acuña-Castroviejo, G. Escames, M. I. Rodríguez and L. C. López, *Front. Biosci.*, 2007, **12**, 947.
79. R. Hardeland, *Molecules*, 2009, **14**, 5054.
80. G. Escames, A. López, J. A. García, L. García, D. Acuña-Castroviejo, J. J. García and L. C. López, *Curr. Neuropharmacol.*, 2010, **8**, 182.
81. V. Srinivasan, D. W. Spence, S. R. Pandi-Perumal, G. M. Brown and D. P. Cardinali, *Int. J. Alzheimers Dis.*, 2011, **2011**, 326320.
82. D. Acuña-Castroviejo, L. C. López, G. Escames, A. López, J. A. García and R. J. Reiter, *Curr. Top. Med. Chem.*, 2011, **11**, 221.
83. R. Hardeland, *J. Exp. Integr. Med.*, 2011, **1**, 67.
84. D. P. Cardinali, E. S. Pagano, P. A. Scacchi Bernasconi, R. Reynoso and P. Scacchi, *Horm. Behav.*, 2013, **63**, 322.
85. R. Hardeland, D. P. Cardinali, G. M. Brown and S. R. Pandi-Perumal, *Prog. Neurobiol.*, 2015, **127–128**, 46.
86. D. Harman, *J. Gerontol.*, 1956, **11**, 298.
87. D. Harman, *J. Am. Geriatr. Soc.*, 1972, **20**, 145.
88. J. Miquel, A. C. Economos, J. Fleming and J. E. Johnson Jr, *Exp. Gerontol.*, 1980, **15**, 575.
89. J. Miquel, E. de Juan and I. Sevila, *EXS*, 1992, **62**, 47.
90. D. Harman, *Ann. N. Y. Acad. Sci.*, 2006, **1067**, 10.
91. I. Afanas'ev, *Aging Dis.*, 2010, **1**, 75.
92. J. Kienhöfer, D. J. Häussler, F. Ruckelshausen, E. Muessig, K. Weber, D. Pimentel, V. Ullrich, A. Bürkle and M. M. Bachschmid, *FASEB J.*, 2009, **23**, 2034.
93. L. V. Thompson, *Exp. Gerontol.*, 2006, **41**, 1220.
94. D.-X. Tan, L.-D. Chen, B. Poeggeler, L. C. Manchester and R. J. Reiter, *Endocr. J.*, 1993, **1**, 57.
95. R. J. Reiter, D. Acuña-Castroviejo, D.-X. Tan and S. Burkhardt, *Ann. N. Y. Acad. Sci.*, 2001, **939**, 200.

96. D.-X. Tan, R. J. Reiter, L. C. Manchester, M. T. Yan, M. El-Sawi, R. M. Sainz, J. C. Mayo, R. Kohen, M. Allegra and R. Hardeland, *Curr. Top. Med. Chem.*, 2002, **2**, 181.
97. R. Hardeland, B. Poeggeler, R. Niebergall and V. Zelosko, *J. Pineal Res.*, 2003, **34**, 17.
98. B. Poeggeler, S. Thuermann, A. Dose, M. Schoenke, S. Burkhardt and R. Hardeland, *J. Pineal Res.*, 2002, **33**, 20.
99. D.-X. Tan, L. C. Manchester, S. Burkhardt, R. M. Sainz, J. C. Mayo, R. Kohen, E. Shohami, Y.-S. Huo, R. Hardeland and R. J. Reiter, *FASEB J.*, 2001, **15**, 2294.
100. A.-R. Ressmeyer, J. C. Mayo, V. Zelosko, R. M. Sáinz, D.-X. Tan, B. Poeggeler, I. Antolín, B. K. Zsizsik, R. J. Reiter and R. Hardeland, *Redox Rep.*, 2003, **8**, 205.
101. A. L. Guenther, S. I. Schmidt, H. Laatsch, S. Fotso, H. Ness, A.-R. Ressmeyer, B. Poeggeler and R. Hardeland, *J. Pineal Res.*, 2005, **39**, 251.
102. R. Hardeland, C. Backhaus, A. Fadavi and M. Hess, *J. Pineal Res.*, 2007, **43**, 104.
103. R. Hardeland, C. Backhaus and A. Fadavi, *J. Pineal Res.*, 2007, **43**, 382.
104. J. León, G. Escames, M. I. Rodríguez, L. C. López, V. Tapias, A. Entrena, E. Camacho, M. D. Carrión, M. A. Gallo, A. Espinosa, D.-X. Tan, R. J. Reiter and D. Acuña-Castroviejo, *J. Neurochem.*, 2006, **98**, 2023.
105. V. Tapias, G. Escames, L. C. López, A. López, E. Camacho, M. D. Carrión, A. Entrena, M. A. Gallo, A. Espinosa and D. Acuña-Castroviejo, *J. Neurosci. Res.*, 2009, **87**, 3002.
106. R. Hardeland, D.-X. Tan and R. J. Reiter, *J. Pineal Res.*, 2009, **47**, 109.
107. I. Semak, M. Naumova, E. Korik, V. Terekhovich, J. Wortsman and A. Slominski, *Biochemistry*, 2005, **44**, 9300.
108. M. Martín, M. Macías, J. León, G. Escames, H. Khaldy and D. Acuña-Castroviejo, *Int. J. Biochem. Cell Biol.*, 2002, **34**, 348.
109. D. Acuña-Castroviejo, G. Escames, J. Leon, A. Carazo and H. Khaldy, *Adv. Exp. Med. Biol.*, 2003, **527**, 549.
110. X. Wang, *CNS Neurosci. Ther.*, 2009, **15**, 345.
111. S. A. Andrabi, I. Sayeed, D. Siemen, G. Wolf and T. F. Horn, *FASEB J.*, 2004, **18**, 869.
112. W. Wang, H. Fang, L. Groom, A. Cheng, W. Zhang, J. Liu, X. Wang, K. Li, P. Han, M. Zheng, J. Yin, W. Wang, M. P. Mattson, J. P. Kao, E. G. Lakatta, S. S. Sheu, K. Ouyang, J. Chen, R. T. Dirksen and H. Cheng, *Cell*, 2008, **134**, 279.
113. M. J. Jou, *J. Pineal Res.*, 2011, **50**, 427.
114. A. Coto-Montes, J. A. Boga, S. Rosales-Corral, L. Fuentes-Broto, D.-X. Tan and R. J. Reiter, *Mol. Cell. Endocrinol.*, 2012, **361**, 12.
115. G. Petrosillo, N. Di Venosa, M. Pistolese, G. Casanova, E. Tiravanti, G. Colantuono, A. Federici, G. Paradies and F. M. Ruggiero, *FASEB J.*, 2006, **20**, 269.
116. F. Luchetti, B. Canonico, F. Mannello, C. Masoni, A. D'Emilio, M. Battistelli, S. Papa and E. Falcieri, *Toxicol. In Vitro*, 2007, **21**, 293.

117. G. Petrosillo, P. Fattoretti, M. Matera, F. M. Ruggiero, C. Bertoni-Freddari and G. Paradies, *Rejuvenation Res.*, 2008, **11**, 935.
118. A. Carrillo-Vico, P. J. Lardone, N. Álvarez-Sánchez, A. Rodríguez-Rodríguez and J. M. Guerrero, *Int. J. Mol. Sci.*, 2013, **14**, 8638.
119. G. Candore, C. Caruso and G. Colonna-Romano, *Biogerontology*, 2010, **11**, 565.
120. S. Ponnappan and U. Ponnappan, *Antioxid. Redox Signaling*, 2011, **14**, 1551.
121. J. P. Coppé, C. K. Patil, F. Rodier, Y. Sun, D. P. Muñoz, J. Goldstein, P. S. Nelson, P. Y. Desprez and J. Campisi, *PLoS Biol.*, 2008, **6**, 2853.
122. A. R. J. Young and M. Narita, *EMBO Rep.*, 2009, **10**, 228.
123. J. P. Coppé, P. Y. Desprez, A. Krtolica and J. Campisi, *Annu. Rev. Phytopathol.*, 2010, **5**, 99.
124. A. Freund, C. K. Patil and J. Campisi, *EMBO J.*, 2011, **30**, 1536.
125. J. C. Mombach, B. Vendrusculo and C. A. Bugs, *PLoS One*, 2015, **10**, e0125217.
126. A. Salminen, J. Ojala, K. Kaamiranta, A. Haapasalo, M. Hiltunen and H. Soininen, *Eur. J. Neurosci.*, 2011, **34**, 3.
127. A. S. Adler, S. Sinha, T. L. Kawahara, J. Y. Zhang, E. Segal and H. Y. Chang, *Genes Dev.*, 2007, **21**, 3244.
128. C. Franceschi and J. Campisi, *J. Gerontol., Ser. A*, 2014, **69**, S4.
129. W. G. Deng, S. T. Tang, H. P. Tseng and K. K. Wu, *Blood*, 2006, **108**, 518.
130. A. Korkmaz, S. Rosales-Corral and R. J. Reiter, *Gene*, 2012, **503**, 1.
131. J. P. de Rivero Vaccari, W. D. Dietrich and R. W. Keane, *J. Cereb. Blood Flow Metab.*, 2014, **34**, 369.
132. J. D. Gilthorpe, F. Oozeer, J. Nash, M. Calvo, D. L. Bennett, A. Lumsden and A. Pini, *F1000Research*, 2013, **2**, 148.
133. B. Tan, R. H. Choi, T. J. Chin, C. Kaur and E. A. Ling, *Front. Biosci., Scholar Ed.*, 2012, **4**, 1402.
134. J. G. McLarnon, *BioMed Res. Int.*, 2014, **2014**, 923670.
135. P. Narayan, K. M. Holmström, D. H. Kim, D. J. Whitcomb, M. R. Wilson, P. St George-Hyslop, N. W. Wood, C. M. Dobson, K. Cho, A. Y. Abramov and D. Kleneman, *Biochemistry*, 2014, **53**, 2442.
136. G. A. Rodriguez, L. M. Tai, M. J. LaDu and G. W. Rebeck, *J. Neuroinflammation*, 2014, **11**, 111.
137. C. E. Hanzel, A. Pichet-Binette, L. S. Pomentel, M. F. Iulita, S. Allard, A. Ducatenzeiler, S. Do Carmo and A. C. Cuello, *Neurobiol. Aging*, 2014, **35**, 2249.
138. M. A. Pappolla, Y. J. Chyan, B. Poeggeler, B. Frangione, G. Wilson, J. Ghiso and R. J. Reiter, *J. Neural Transm.*, 2000, **107**, 203.
139. V. Srinivasan, S. R. Pandi-Perumal, D. P. Cardinali, B. Poeggeler and R. Hardeland, *Behav. Brain Funct.*, 2006, **2**, 15.
140. S. A. Rosales-Corral, D. Acuña-Castroviejo, A. Coto-Montes, J. A. Boga, L. C. Manchester, L. Fuentes-Broto, A. Korkmaz, S. Ma, D.-X. Tan and R. J. Reiter, *J. Pineal Res.*, 2012, **52**, 167.
141. J. Zhou, S. Zhang, X. Zhao and T. Wei, *J. Pineal Res.*, 2008, **45**, 157.

142. K. L. Clapp-Lilly, M. A. Smith, G. Perry and L. K. Duffy, *Chem.-Biol. Interact.*, 2001, **134**, 101.
143. J. Bender Hoppe, R. L. Frozza, A. P. Horn, R. Argenta Comiran, A. Bernardi, M. M. Campos, A. M. Oliveira Battastini and C. Salbego, *J. Pineal Res.*, 2010, **48**, 230.
144. I. Blasko, F. Marx, E. Steiner, T. Hartmann and B. Grubeck-Loebenstein, *FASEB J.*, 1999, **13**, 63.
145. M. Shukla, H. H. Htoo, P. Wintachai, J. F. Hernandez, C. Dubois, R. Postina, H. Xu, F. Checler, D. R. Smith, P. Govitrapong and B. Vincent, *J. Pineal Res.*, 2015, **58**, 151.
146. J. Panmanee, C. Nopparat, N. Chavanich, M. Shukla, S. Mukda, W. Song, B. Vincent and P. Govitrapong, *J. Pineal Res.*, 2015, **59**, 308.
147. S. M. de la Monte and M. Tong, *Biochem. Pharmacol.*, 2014, **88**, 548.
148. S. T. Ferreira, J. R. Clarke, T. R. Bomfim and F. G. De Felice, *Alzheimers Dement.*, 2014, **10**(suppl. 1), S76.
149. E. Matsubara, T. Bryant-Thomas, J. Pacheco Quinto, T. L. Henry, B. Poeggeler, D. Herbert, F. Cruz-Sanchez, Y. J. Chyan, M. A. Smith, G. Perry, M. Shoji, K. Abe, A. Leone, I. Grundke-Ikbal, G. L. Wilson, J. Ghiso, C. Williams, L. M. Refolo, M. A. Pappolla, D. G. Chain and E. Neria, *J. Neurochem.*, 2003, **85**, 1101.
150. J. Quinn, D. Kulhanek, J. Nowlin, R. Jones, D. Praticò, J. Rokach and R. Stackman, *Brain Res.*, 2005, **1037**, 209.
151. V. Doré, V. L. Villemagne, P. Bourgeat, J. Fripp, O. Acosta, G. Chetélat, L. Zhou, R. Martins, K. A. Ellis, C. L. Masters, D. Ames, O. Salvado and C. C. Rowe, *JAMA Neurol.*, 2013, **70**, 903.
152. H. Yokoyama, K. Okazaki, D. Imai, Y. Yamashina, R. Takeda, N. Naghavi, A. Ota, Y. Hirasawa and T. Miyagawa, *BMC Geriatr.*, 2015, **15**, 60.
153. Y. Kagawa, *Nutr. Rev.*, 2012, **70**, 459.
154. N. P. Andrews, H. Fujii, J. J. Goronzy and C. M. Weyand, *Gerontology*, 2010, **56**, 390.
155. O. DelaRosa, G. Pawelec, E. Peralbo, A. Wikby, E. Mariani, E. Mocchegiani, R. Tarazona and R. Solana, *Biogerontology*, 2006, **7**, 471.
156. E. Magnanou, J. Attia, R. Fons, G. Boeuf and J. Falcon, *PLoS One*, 2009, **4**, e5904.
157. M. I. Rodríguez, G. Escames, L. C. López, A. López, J. A. García, F. Ortiz, V. Sánchez, M. Romeu and D. Acuña-Castroviejo, *Exp. Gerontol.*, 2008, **43**, 749.
158. R. Hardeland and B. Poeggeler, *Cent. Nerv. Syst. Agents Med. Chem.*, 2012, **12**, 189.
159. M. W. Hurd and M. R. Ralph, *J. Biol. Rhythms*, 1998, **13**, 430.

CHAPTER 20

Short Peptides Regulate Gene Expression, Protein Synthesis and Enhance Life Span

VLADIMIR KHAVINSON[a] AND IRINA POPOVICH[a]

[a]Saint Petersburg Institute of Bioregulation and Gerontology, Dynamo Pr. 3, 197110, Saint Petersburg, Russia
*E-mail: ibg@gerontology.ru

20.1 Introduction

There is accumulated evidence that many of the so-called "diseases of aging", including cancer, are caused by dysregulated immune functions and decreased organism resistance to infections.[1] Peptide extracts of thymus and peptides isolated from thymus were the first preparations proposed for correction of immunodeficiency.[2,3] The origin of short regulatory peptides in the organism became obvious after the discovery of protein degradation in proteasomes.[4,5] The same high-molecular proteins can be differentially hydrolyzed resulting in various short peptides. The peptides produced show different biological functions as compared to the original macromolecules.[6]

Karlin and Altschul demonstrated in their work that protein macromolecules contain several types of recurrent blocks of amino-acid residues with charged side chains. Such blocks are mostly observed in nucleoproteins. Among them, there are transcription factors, centromere proteins and high

RSC Drug Discovery Series No. 57
Anti-aging Drugs: From Basic Research to Clinical Practice
Edited by Alexander M. Vaiserman

Published by the Royal Society of Chemistry, www.rsc.org

mobility group.[7] Proteasome hydrolysis of these nucleoproteins can provide a sufficient amount of peptides with charged side chains. These and some other investigations gave rise to the development of the peptide bioregulation concept.[8,9] This concept suggests that low-molecular peptides are involved in intercellular transfer of information encoded in the amino acid sequence and conformation modifications, thus facilitating regulation of proliferation, differentiation and intercellular interaction.[8,9] Peptide bioregulators were isolated from different tissues. Their major function consists of normalizing the functions of the organs from which they have been isolated. They can also substitute and/or complement biologically active compounds secreted in this morphological structure.[8]

Apart from immunity dysregulation, aging causes other alterations on the cellular level, for example, accumulation of mutations in somatic cells.[10] Although the rate of accumulation of age-specific changes is determined genetically,[11] there are a number of exogenous factors that accelerate this process. Oxidative stress is considered to induce both cell and body aging.[12] Proteins and DNA are known to be damaged by reactive oxygen species (ROS). There is substantial evidence of the role of DNA oxidative damage in organismal senescence.[12–14]

Aging-associated accumulation of somatic mutations is accompanied by the decreasing of DNA repair level, which leads to growing incidence of pathologies including cancer.[15] Higher concentration of damages in heterochromatin regions as compared to active (euchromatic) regions of the DNA can be explained by the fact that reparation can occur only in the DNA regions that are involved in active transcription and are accessible for reparation enzymes.[16] This corresponds to intensive DNA reparative synthesis in the G2 cycle with more active chromosome heterochromatinization, as compared to in the G1 cycle.[17] The frequency of sister chromatid exchanges (SCE) confirms age-related reduction of reparation level. The SCE level in fibroblasts and lymphocytes of the elderly (60–70 y.o.) was found to be lower than that of the younger donors (30–40 y.o.), regardless of their gender.[18] Thus chromosome heterochromatinization and related decrease in the DNA reparation intensity is considered to be a key factor in organismal aging.[19]

Various experimental models have been used for studying preparations with a protective effect against aging and carcinogenesis.[20] In several animal studies, short peptides were demonstrated to be promising in anti-aging medicine. Their geroprotective and anticarcinogenic effects are believed to be mediated by their immunomodulatory and antioxidative properties.[21] The peptides were shown to lead to the reduction of the level of age-related chromosome aberrations (ChA)[22] and to affect the chromosome heterochromatinization,[19] thus retarding the aging process. They can influence the expression of various genes,[23] which is determined by specific short peptides–DNA binding.[24,25] Safety of long-term administration is one of the main advantages of peptide therapy. These properties make them promising candidates for clinical application in old and senile patients.[26] Further investigation of peptide bioregulators appears to be very perspective in modern gerontology, bearing in mind their capacity to inhibit senescence and restore functions of the aging organism.

20.2 Isolated Peptide Complexes

One of the first polypeptide preparations isolated from the calf thymus was Thymalin.[27,28] This polypeptide was able to restore disturbed immunological responsiveness, improve cell metabolism and stimulate cell immunity, regeneration and haemopoiesis (in case of their suppression). It displayed geroprotective properties and increased mean life span in experimental animals.[14,29,30] The important feature of polypeptide preparations is their anti-carcinogenic activity. This property of Thymalin was reported in experiments on induced and spontaneous carcinogenesis. Rats with 7,12-dimethyl-benzantracene (DMBA)-induced carcinogenesis treated with Thymalin revealed decreased tumour incidence by 24% and reduced the number of mammary adenocarcinomas 3.8-fold as compared to the control animals.[21] Administration of Thymalin to irradiated mice and rats for ten days, twice daily, decreased the number of malignant neoplasia. At the same time, the mice exposed to fraction irradiation and treated with Thymalin showed a 3.5-fold decrease in the number of tumors as compared to the irradiated control (Table 20.1).

Prolonged administration of Thymalin to SHR mice starting from 4 months of age resulted in a significant decrease of spontaneous tumor incidence—40% as compared to 55% in the controls. C3H/Sn mice treated with

Table 20.1 Effect of peptides on experimentally induced tumors in rodents.[a,b,c]

Peptides	Animal species/strain	Carcinogenic action	Tumor site/ localization	Tumor incidence%	
				Control	Peptide
Epithalamin complex of pineal peptides	Rats	DMBA	Mammary gland	81	26[d]
		X-ray irradiation	Mammary gland	16	3[d]
Thymalin complex of thymus peptides	Rats	DMBA	Mammary gland	69	18[d]
		X-ray irradiation	Mammary gland	21	3[d]
	C3H mice	X-ray irradiation	Mammary gland	38	14[d]
			Leukemia	46	14[d]
Thymogen/ Glu–Trp (EW)	Rats	Isotopes ^{90}Sr and ^{137}Cs	Any malignant tumors	16	8[d]
Vilon/Lys–Glu (KE)	CBA mice	DMH	Kidney	60	14[d]
Epitalon/ Ala-Glu-Asp-Gly (AEDG)	C3H/He mice	MMTV	Mammary gland	9	5[d]
	Female rats	Constant lighting	Mammary gland	41	27[d]
	Male rats	Constant lighting	Leukemia	12	0

[a]DMBA: 7,12-dimethylbenz(*a*)anthracene;
[b]DMH: 1,2-dimethylhydrazine;
[c]MMTV: mouse mammary tumor virus;
[d]The differences are statistically significant compared to the control by $p < 0.05$.

Thymalin during their life starting from 3.5 months of age exhibited a 2.8-fold decreased spontaneous tumors incidence and a 2.6-fold decreased incidence of mammary adenocarcinomas. At the same time, the experimental mice did not develop leukemia, while in the control group this pathology was registered in 14.3% of female mice (Table 20.1).[31,32]

Long-term administration of Thymalin led to the increase of the life span in mice: mean life span increased by 28% ($p < 0.05$) as compared to the control animals and maximum life span increased by 11% (Table 20.2).

Table 20.2 The effect of peptides on animal life span (LS).

Species of animals or mice strain; effect	Life span indices (days)			
	Mean life span (MLS)	Mean life span of last 10% survivors	Maximum life span	The rate of aging[a] (days^{-1})
Rats				
Control	681 ± 14.5	835	1054	6.8
Epithalamin	852 ± 33.8[d]	1050	1112	3.8
%	+25	+27	+6	−44
SHR mice				
Control	564 ± 22.3	750	843	No differences
Epithalamin	627 ± 20.9[b]	750	827	
%	+11	0	−2	
C3H/Sn mice				
Control	487 ± 29.4	691	776	7.0
Epithalamin	640 ± 33.1[c]	757	885	5.1
%	+31	+20	+14	−27
C3H/Sn mice				
Control	487 ± 29.4	No data	776	No data
Thymalin	623 ± 24.6[b]		863	
%	+28		+11	
Rats				
Control	773 ± 18.4	949	965	7.08
Thymogen	786 ± 26.2	1048	1104	4.12
%	+2	+10	+14	−42
CBA mice				
Control Vilon	685 ± 9.2	737	740	No data
%	694 ± 12.5	761	792	
	+1.3	+3	+7	
SHR mice				
Control	456 ± 29	709	740	4.5
Epitalon	455 ± 31	803	1053	3.2
%	−0.2	+13	+42	−29
CBA mice				
Control	685 ± 9.2	737 ± 1.1	740	6.9
Epitalon	721 ± 11.1[c]	842 ± 58.5[b]	1053	4.1
%	+5.3	+14	+42	−41

[a]The rate of ageing (as per Gompertz equation[45]);
[b]The differences are statistically significant compared to the control by: $p < 0.05$;
[c]$p < 0.01$;
[d]$p < 0.001$.

Anti-carcinogenic and geroprotective properties of Thymalin were supposed to be mediated by its ability to prevent aging-associated decrease of cell immunity in female mice.[31]

Pineal gland preparation was shown to manifest anti-oxidant and geroprotective effects in several studies.[33] Old rats with persisting estrus administered with Epithalamin restored the regular estrus cycles, suggesting prevention of the reproductive function.[34] Old male rats administered with Epithalamin exhibited increased levels of luteinizing hormone and testosterone, which also proves the normalizing effect of Epithalamin on reproductive function in old animals.[35] Female rats treated with Epithalamin starting from the age of 15 months exhibited a 1.6-fold decrease in the incidence of neoplasia and a 2.7-fold decrease in the frequency of malignant tumors. Course administration of Epithalamin to C3H/Sn mice starting from 3.5 months of age led to a 2.1-fold decrease in the incidence of tumors of all kinds, including mammary adenocarcinomas (2.9-fold), as compared to the controls.[31] Epithalamin also contributed to decreased incidence and multiplicity of tumors in the model of DMBA-induced carcinogenesis in rats (Table 20.1).[36,37] Epithalamin application in the models of transplantable tumors resulted in the inhibition of metastatic growth and in tumor size reduction.[9,14]

A significant geroprotective potential of Epithalamin was discovered in various animal models: rats,[38] mice,[31] and *Drosophila melanogaster*.[39] All these animals showed an increase in mean life span under the influence of Epithalamin. Maximum life span of rats increased by 3 months: 23% of animals treated with Epithalamin had a longer life span than the most long-lived control rats.

Collectively, these findings suggest that peptide preparations Thymalin and Epithalamin are able to prevent aging and increase life span, as well as inhibit carcinogenesis in various animal species.

20.3 Short Synthetic Peptides

It was discovered that the extracts isolated from the calf thymus contained peptides with molecular weight less than 1000 Da. One of them is dipeptide Glu–Trp with a molecular weight of 333 Da. It was named Thymogen (Glu–Trp).[28] The effects of Thymogen on spontaneous carcinogenesis and life span in rats have been studied.[40] Thymogen was administered throughout the life span of rats starting from 5 months of age. Like Thymalin, this synthetic peptide inhibited malignant tumors development 2.1-fold. A tendency towards mean and maximum life span increase as well as a decreased aging rate in experimental animals was observed, as compared to the controls.[40] Thus, similarly to Thymalin, the synthetic dipeptide Thymogen has significant geroprotective properties.

Another promising short synthetic peptide is the dipeptide Lys–Glu or Vilon (molecular weight 275 Da). It was shown to be able to stimulate the reparative processes.[41] Prolonged administration of Vilon to CBA mice starting from 6 months of age resulted in the increase of their physical activity and maximum life span, lowering of body temperature and inhibition of spontaneous tumor incidence by 1.5-fold as compared to the control animals. The obtained results prove the safety of chronic administration of Vilon and suggest that its geroprotective properties could be used for prevention of age-related pathologies.[42,43]

The tetrapeptide Ala-Glu-Asp-Gly (molecular weight 390 Da) was synthesized based on the amino acid analysis of Epithalamin. The tetrapeptide obtained was named Epitalon.[44] It showed properties similar to those of Epithalamin: suppression of spontaneous carcinogenesis (Table 20.1) and increase of the life span in experimental animals (Table 20.2).[14] It is important to note that both peptides inhibited aging rate (as per Gompertz equation):[45] Epitalhamin in rats[38] and in C3H/Sn mice,[31] and Epitalon in SHR and CBA mice[14] as compared to the controls.

Epitalon and Epithalamin appeared to be safe alternatives to melatonin in regard to the correction of pineal gland functional insufficiency.[46] Aging leads to decreased production of melatonin, which performs many vital functions.[14,47] Melatonin is involved in the regulation of functions of the central and peripheral nervous systems, endocrine organs and immune system. Decreased melatonin levels caused by the violation of circadian rhythms is considered to be an important factor in reducing life span and causing premature aging and age-related diseases, including cancer.[48,49] Administration of melatonin to experimental animals revealed its geroprotective properties.[14] Melatonin suppressed tumor incidence in chemically or genetically modified animals.[14] Long-term administration of melatonin to CBA mice in spontaneous carcinogenesis models caused the increase of melatonin-mediated malignant tumors (lymphomas) incidence.[50]

Epitalon- and Epithalamin-mediated increases of melatonin levels were recorded in the blood, and also in the pineal gland of old *Macaca mulatta*.[46] Administration of Epitalon to male and female rats stimulated melatonin production during night time, normalized hormonal and metabolic markers and prevented premature aging and tumor development in animals.[51,52]

Thus, administration of short peptides resulted in a number of beneficial effects in different organs and tissues under normal and pathological conditions in experimental studies. However, the mechanisms of their geroprotective and anti-carcinogenic actions are not completely elucidated and require further research.

20.4 Influence of Short Peptides on Immune and Antioxidant Systems

Short peptides of the thymus may produce a specific effect on immunologic responsiveness, homeostasis and metabolism in case of secondary immunodeficiency. Experimental animals administered with Thymogen for 30 days manifested lymphocyte count increase. Remarkably, Thymogen administration resulted in the increase of T-cell count in thymectomy, with its dose 1000 times less than that of Thymalin.[53] Comparative studies of the immunomodulatory effects of Thymalin and Thymogen on the intensity of immune response in rats immunized with sheep red blood cells showed a more significant effect of Thymogen as compared to Thymalin. Thymogen normalized T- and B-immunity in animals under conditions of experimentally induced immunodeficiency. The molecular mechanism of action of this synthesized preparation on T-lymphocytes is suggested to be based on the activity of calcium (Ca^{2+}) transmembrane exchange, as well as redistribution of intracellular cAMP and cGMP concentration.[54] As a result, these processes can induce gene expression followed by proliferation and differentiation of the relevant lymphocyte populations.[54]

Like Thymogen, Vilon was registered to stimulate cell immunity. Animals administrated with Vilon in concentrations from 10 ng l^{-1} to 100 μg l^{-1} showed an increased level of intracellular Ca^{2+} in thymocytes and macrophagocytes, displaying one of the mechanisms of cell activation. In particular, it leads to the stimulation of T-cell RNA and interleukin-2 (IL-2). Vilon was shown to stimulate mRNA IL-2 synthesis in murine spleen lymphocytes after 5 hours of incubation in cell culture.[54] *In vitro* administration of Vilon entailed significant expression of T- and B-lymphocytes in patients with secondary immunodeficiencies. It has also been shown to stimulate IL-1α, IL-1β, IL-8 and TNF-α production. In thymocytes and epithelial cells, Vilon stimulated expression of the argyrophil proteins associated with the nucleolar organizer region responsible for the synthesis, gathering and transportation of ribosomes into the cytoplasm, predetermining the intensity of protein synthesis in these structures.[53]

Vilon administration to irradiated animals promoted regeneration with a revealed differentiation into cortex and medulla in the thymus.[54] Moreover, hyperplasia of mast cells in the thymus was observed under Vilon's influence. Vilon seems to accelerate the proliferative activity of the irradiation-survivor bone marrow stem cells, which are the precursors of T-lymphocytes and mast cells.[54] Administration of another synthesized peptide, Epitalon, to gamma-irradiated rats contributed to ultrastructural manifestation of pinealocytes secretion strengthening, which had been damaged due to irradiation.[55] These results are suggestive of tissue-specificity of peptide bioregulators. Radiation-induced and age-related alterations are known to have many common features. The effects of Vilon and Epitalon were compared in studies on CBA mice injected with these peptides starting from 6 months of age and up to the end of life. Another group of mice was also treated with melatonin in tap water. As a result, Epitalon suppressed reactive oxygen species effectively in the blood serum and brain tissue of the animals. This effect was accompanied by suppression of lipid peroxidation (LP); the Schiff base decrease in the brain and the decrease in the amount of diene conjugates in the liver were also registered.[14] The effect of melatonin was almost the same. Vilon failed to affect any indexes of the free-radical processes studied. Similarly to melatonin, Epitalon was found to be able to stimulate organism antioxidant activity. In a series of experiments, Epitalon has been demonstrated to be more efficient *in vivo* than *in vitro*.[56] *Drosophila melanogaster* larva exposure to Epitalon exhibited the reduction of lipid peroxidation intensity and increase in catalase activity in adult flies.[39] A significant antioxidant effect of Epitalon was found in old rats administered with this compound. It significantly suppressed the formation of LP products in blood serum and brain.[57]

Long-term experimental administration of Epitalon to SHR and SAM mice caused decreased chromosome aberrations of bone marrow cells. The most remarkable effect was seen in SAM mutant mice with accelerated aging.[22] The frequency of chromosome aberrations in SAM mice was higher due to DNA damage with reactive oxygen forms, whose production in SAM mice was enhanced.[58] Administration of Epitalon to these mice resulted in statistically significant reduction (by 20–30%) of ChA frequency, which can be associated with activation of antioxidant defense.

The effect of Epitalon on the number of sister chromatid exchanges (SCE) in lymphocyte culture of humans aged 75–88 was studied by cytogenetic methods. Addition of Epitalon to lymphocyte culture resulted in a 1.4-fold increase in SCE frequency ($p < 0.001$), as compared to the control.[59] Vilon under similar conditions increased SCE frequency to a greater extent than Epitalon and showed a 1.9-fold increase as compared to the control ($p < 0.001$).[60] According to early studies, metabolic processes do not take place in the heterochromatin or heterochromatinized chromosome regions.[61] Thus, SCE frequency increase induced by Vilon indicates decondensation (deheterochromatinization) of the chromosome region condensed with aging followed by the release of functionally inhibited genes located therein.[62] The same research also discovered the ability of both short peptides to activate ribosome genes, as evidenced by the increase of nucleolar organizer regions (NOR) in acrocentric chromosomes, deduced by Ag-staining method,[63] as compared to the control.

Generally, the ability of short peptides to normalize or improve humoral and cellular immunity, reinforce antioxidant defense of the body and affect heterochromatinization—one of the aging factors—is an essential component of the geroprotective mechanism of the short peptides.

20.5 The Influence of Short Peptides on Gene Expression

In research based on DNA microarray technology, the impact of Vilon and Epitalon on gene expression has been observed. In this study, the levels of mRNA of 15 247 genes in mouse heart before and after Vilon and Epitalon administration were studied.[23] Epitalon modulated the expression levels of 98 genes; Vilon changed the expression of 36 genes. Combined treatment with Vilon and Epitalon changed the expression of 114 genes. Among the affected genes, there were genes involved in oncogenesis. Vilon and Epitalon inhibited the expression of genes such as mouse Mybl1 (myeloblastosisoncogene-like1) and proto-oncogene Bcl-3, respectively. Chronic administration of Vilon and Epitalon to female transgene mice led to a 1.9- and 3.7-fold decrease of gene HER-2/neu expression in mammary tumor, as compared to the control group. Moreover, Epitalon reduced the maximum size of mammary tumor and the diameter of lung metastases.[64]

Epitalon-treated culture of human lung fibroblasts manifested the induction of telomerase gene expression, telomerase activity and elongation of telomeres.[65] Activation of telomerase gene expression was accompanied by a 43% increase in the number of cell divisions.[66,67] These results are in accordance with our earlier data demonstrating the impact of particular peptide bioregulators and their complexes on gene expression.[21]

In the rat hypothalamic neurons, Vilon also was shown to stimulate the expression of c-fos gene known to be involved in the organism's stress response. Treatment with Epitalon also led to increased c-fos gene expression in the pineal gland of rats.[9]

One of the essential features of short peptides is their ability to influence cytokines synthesis. The expression of interleukin-2 (IL-2) in lymphocytes is known to decrease with aging.[68] The impact of Vilon on IL-2 gene expression in mouse spleen lymphocytes was studied by *in vitro* hybridization. Lymphocytes were stimulated with Con-A mitogen. Five-hour incubation with Vilon led to increased mRNA synthesis in both lymphoid cells stimulated with Con-A, and in non-stimulated cells. Prolonged Vilon incubation (for 20 hours) promoted IL-2 expression.[69] The effect of Epitalon on subcortex functions has also been found. Administration of Epitalon stimulated IL-2 gene expression in various hypothalamic structures under low stress conditions.[67] In general, our data provide evidence for immunomodulating and stress-protective capabilities of short peptides.[9,67]

The abovementioned experimental data on the mechanisms of action of the short peptides bring us to the conclusion concerning their important role in supporting immune, nervous, endocrine and other systems of the organism throughout the process of aging. These peptide preparations are able to inhibit the development of age-related pathologies, including cancer, thus preventing premature aging. It motivated us to examine their potential for treatment and prevention of age-related diseases in the elderly.

20.6 Application of Peptide Bioregulators in Elderly Patients

The experimental studies of peptide preparations in different animal models proved the safety of those preparations and revealed a wide spectrum of their beneficial effects, making reasonable the application of peptide preparations in humans. Most of the studies were conducted among elderly people and patients with premature aging.

The research was conducted among 106 patients (69 ± 2 years of age) with ischemic heart disease (IHD) and signs of premature aging: blood lipid disorders, low tolerance to carbohydrates, functional decrease of reproductive functions and detoxifying liver function, osteoporosis, mental and physical capacity decrease.[70] All patients were randomly allocated into control and study groups. The patients of the control group received symptomatic therapy while those in the study group were treated with Thymalin in addition to symptomatic treatment, which was administered intramuscularly at a dose of 10 mg every 2–3 days with a total of 5 injections for the whole course and an interval of 5–6 months between the courses. The research lasted for a period of 30 months, during which the patients received 6 courses of Thymalin. An increase in the physical activity threshold by 14% after the first course of injections was found. This was evidenced by the increased ascent along a ramp from 3.4 to 4.8 floors and with decreased fatigue.[70]

Moreover, a significant increase in maximal oxygen intake (MOI) was revealed under the influence of Thymalin during threshold load, which indicates the expansion of the oxygen transport system functionality of the organism.[71] In general, a positive effect of Thymalin was observed in 53% of patients, whereas the same effect was registered only in 7% of the patients in the control group. This slight improvement of MOI registered in patients of the control group could be attributed to the symptomatic therapy they received, while the statistically significant MOI increase in the study group is related to Thymalin treatment.[70]

Thymalin-treated patients exhibited normalized blood lipid markers, *i.e.* significant decrease in cholesterol levels, beta-lipoprotein cholesterol and atherogenicity index. Patients with a high atherogenicity index prior to treatment (over 4) showed a normal level (below 3.5) after the course of Thymalin treatment. At the same time, administration of Thymalin to patients with high levels of circulating immune complexes resulted in their significant decrease, which is believed to be important for reducing the risks of vascular wall damage in IHD patients. Generally, Thymalin-treated patients demonstrated better memory, mood and working capacity. Most patients also exhibited better stress resistance. A lower number of catarrhal diseases and their shorter duration were also observed. No new cases of coronary heart disease, hypertension, heart failure and cardiac arrhythmias were reported during the Thymalin administration period. Thymalin-treated patients demonstrated a decrease in mortality rate during the study period—6.6% as compared to 13.6% in the control group.[70]

A similar research with Epithalamin administration in 46 elderly patients with coronary artery disease and premature aging of the cardiovascular system was conducted. Epithalamin was injected at a dose of 10 mg every 2–3 days (total of 5 injections per one course) for a period of 30 months.[70] The functional age analysis taken prior to the Epithalamin course showed at least a 5 year higher age, as compared to the chronological one, which is clear evidence of premature aging. The first courses of Epithalamin injections resulted in a significant decrease in the functional age of the patients by an average of 7.2 years. Upon completion of a 3 year observation period, the functional age did not differ significantly from initial figures, while the chronological age of the patients non-treated with Epithalamin increased by

3 years.[70] Prior to the treatment undertaken, all elderly patients exhibited lipid profile disorders: low concentrations of high-density lipoproteins, total cholesterol and beta-lipoprotein levels increased, rise in atherogenicity levels (4.0). Epithalamin administration for one year led to a significant decrease of total cholesterol and beta-lipoprotein, as well as the atherogenicity index (3.5). At the same time, the control (conventional therapy) group showed deterioration in the lipid composition. Epithalamin administration also resulted in visual memory and mental capacity improvements, as evidenced by fulfillment of an experimental psychological task in a shorter time period.

The patients in both groups were monitored continuously upon completion of 3 years of peptide therapy. According to the results of a 15 year-long observation of patients in both groups, 66.7% of patients were alive in the Epithalamin-treated group, while only 40% were alive in the control group. A statistically significant decrease of mortality in the Epithalamin-treated patients was demonstrated by the Kaplan–Meier method.[72] The long-term administration of the pineal gland peptide preparation significantly (1.8 times) reduced the number of deaths associated with cardiovascular diseases. Specifically, myocardial infarction- and stroke-caused deaths were observed in 46.2% of the patients in the Epithalamin group, as compared to 83.3% in the control group.

The research also showed that administration of peptide preparations to elderly people can also affect the melatonin-secretion function.[26] Both peptides have a comparable impact on the concentration of melatonin in the blood plasma at night (3:00). The same effect was achieved by administering a significantly lower course dose of Epitalon (0.1 mg) as compared to Epithalamin (50 mg), which indicates the higher biological activity of the synthetic tetrapeptide.

Administration of Thymogen and Vilon was effective in the treatment of different diseases and pathologies accompanied by immune disorders. Administration of Thymogen to elderly patients with secondary immunodeficiencies contributed to normalization of immune markers in 83.6% of cases, as well as to metabolic processes and coagulation improvement.[28,54] Intranasal administration of Thymogen for prevention of influenza and acute respiratory infections (ARI) in patients of all ages, including the elderly, helped to decrease their incidence 3–4-fold, reducing the number of toxic forms by over 30 times.[54] Another synthetic peptide, Vilon, was very effective in the treatment of decreased cellular immunity and phagocytosis-associated diseases. Application of Vilon in addition to conventional therapy to surgical patients accelerated the process of tissue regeneration and restoration of the body functions. Administration of Vilon in elderly and old patients with chronic generalized periodontitis contributed to shortening of the pathological process duration due to the reduction of the periodontal pocket depth, as compared to the patients in the control group.[73]

Thus, administration of peptide preparations is beneficial for the quality of life of elderly patients. Peptides seem to be able to retard age-related functional decline, to improve long-term prognosis and to decrease cardiovascular mortality. Based on the safety, confirmed in experimental studies, peptide preparations can be recommended both for premature aging prevention and for primary or adjunctive therapy in case of various diseases.

20.7 Prospective Cellular and Molecular Mechanism of Action of Short Peptides

It is essential for understanding specific effects of the peptide behavior in the cell and intracellular structures to reveal binding of short peptides with specific DNA sites. In our research, fluorescently labeled short peptides penetrated into cells and intracellular structures.[74] In the HeLa cells, the most intensive fluorescence of the labeled peptides was observed in the nucleus and nucleoli, while the least intensive was observed in the cytoplasm. Investigation of interaction of fluorescence-labeled deoxyribooligonucleotides with short peptides showed that peptides with different primary structures bind with one and the same deoxyribooligonucleotide differently. By using the specific oligonucleotides (FAM–deoxyribooligonucleotides), it has been revealed that Epitalon binds primarily with oligonucleotides that include more cytosine (C) than guanine (G) residues. The constant of binding of Epitalon with FAM–CGC CGC CAG GCG CCG CCG CGC (12 C residues) was almost 2-fold higher than that with FAM–GCG CGG CGG CGC CTG CGC CGC (10 C residues). Introduction of 5-methylcytosine residue into the nucleotide sequence independent of C or G content increased the binding of oligonucleotides with Epitalon. Thus, the binding of peptide Ala-Glu-Asp-Gly is sensitive to the cytosine methylation status of oligonucleotides. Epitalon was shown to preferably bind with single stranded oligonucleotide, containing methylated cytosine.[74] As is commonly known, cytosine DNA methylation is the most extensively studied epigenetic genome modification playing a significant role in stable changes of gene activity upon cell differentiation and aging in mammals.[75–77]

Consequently, there are specific sites for binding of a peptide with a particular amino acid sequence and oligonucleotide with a particular nucleotide sequence. The short peptide may bind to the DNA in various ways depending on its methylation nature; obviously it will cause different effects on the gene functions in various tissues/cells—young and old, normal and cancerous *etc.*[67] Our study shows that unlike the temperature of melting of the DNA double helix (+69.5 °C), in the DNA–tetrapeptide (Ala-Glu-Asp-Gly) system, the melting point occurs at a significantly lower temperature (+28 °C) and is characterized by smaller changes in free energy and an approximately 2-fold decrease in the enthalpy and entropy values.[78] This fact demonstrates that the thermodynamically simplified way of the DNA–peptide complex separation at lower temperature settings is typical of the biochemical processes occurring in living organisms. It also suggests that the mechanism of DNA–Epitalon interaction is based upon the natural mechanism of functioning of a living organism.

A model of complimentary binding of the synthesized peptide Epitalon (Ala-Glu-Asp-Gly) with a DNA double helix was developed.[79] The Ala-Glu-Asp-Gly peptide was found to be located in the major groove of the DNA double helix. A special feature of this model is that the tetrapeptide in the major groove interacts simultaneously with the functional groups of the bases of both DNA strands.[79] The binding between the peptide and the ATTTC sequence in a DNA strand is supported by available data, confirming the appearance of this particular sequence in the promoter region of the telomerase gene.[80] Such interaction of the Ala-Glu-Asp-Gly peptide and the ATTTC sequence can likely explain the geroprotective properties of Epitalon.[81]

Based on the analysis of physical and chemical characteristics of the DNA-peptide complex, a three-dimensional model of complimentary interaction between Ala-Glu-Asp-Gly and the ATTTC sequence was created (see Figure 20.1).[24,25,57] A synthetic nucleic acid preparation was used for studying the interaction between oligopeptides and double stranded DNA. It is a synthetic analogue of the binding site of transcription factors (TATA box, which is usually found in the binding sites of RNA-polymerase II) in the promoter regions of many eukaryotic genes.[82] On the surface of the major groove of the DNA synthetic preparation double helix a group of

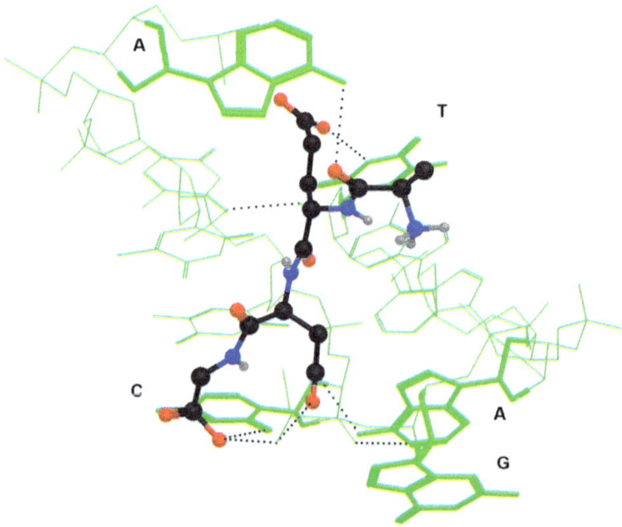

Figure 20.1 The interaction of the AEDG peptide with nitrogen bases of DNA (ATTTC sequence). The dotted line indicates hydrogen bonds between the atoms of peptides and DNA; bold green lines indicate nitrogen bases of DNA forming hydrogen bonds with the peptide. The DNA molecule is indicated in green; letters indicate nitrogen bases (A: adenine, T: thymine, G: guanine, C: cytosine). In the peptide molecule, blue is used for nitrogen atoms, red for oxygen atoms, grey for carbon atoms, and light grey for polar hydrogen atoms.

nucleobases occurs that can interact with the Ala-Glu-Asp-Gly groups. It was found that binding of six nucleotide pairs with TATATA of the DNthe A leading strand can be performed *via* an additional hydrogenous and one hydrophobic bond.[24,78] Thus, a regulatory peptide is believed to be able to bind with a complementary site on the gene promoter region, causing local separation of strands and thereby initiating the process of RNA polymerase gene transcription.

20.8 Conclusion

Physiologically active peptides, including short peptides, represent biologically active compounds that can modulate various cellular and molecular processes. Peptide compounds are essential for invention of new pharmaceutical products.[83] Peptide preparations are highly active, non-toxic and have no side effects, which comprise their main advantages as pharmaceuticals. Short peptides possess pronounced anticarcinogenic and geroprotective properties. In experimental studies in animals the peptides revealed the ability to decrease the risk of spontaneous and induced neoplasia and to enhance lifespan by 20–40%.[14] In general, these properties are determined by the peptides' capability to influence the immune system of the organism, thus preventing aging.[21,54] The peptides possess pronounced antioxidant potential: Vilon reduces the ROS level in *D. melanogaster* mitochondria; Epitalon inhibits the chemoluminescence level and enhances general antioxidant activity in mice blood serum.[14] Epitalon also has an inhibitory effect on the level of age-related chromosome aberrations in mice.[22] Short peptides activate heterochromatin in the cytoblasts of elderly patients and promote activation of genes repressed as a consequence of age-associated heterochromatinization of the euchromatic region of chromosomes.[19] Recognition of the short peptides' ability to influence the expression of various genes was essential for understanding of their role in the aging processes.[23]

Small peptides (di-, tri- and tetra-peptides) revealed the capability of complementary interaction with the DNA-specific binding sites on the promoter segment of genes, inducing separation of double helix strands and RNA polymerase activation. Discovery of the phenomenon of peptide activation of gene transcription allows determination of the mechanism to maintain physiological functions, which is based on the complementary interaction of DNA and regulatory peptides.[24,79]

Application of peptide bioregulators in humans for preventive purposes resulted in a significant restoration of the main physiological functions and a substantial mortality decrease in different age groups for a period of 6–12 years.[26]

Further investigation of the mechanisms of peptide geroprotective action can likely provide new avenues for peptidergic regulation of aging, prevention of premature aging, age-associated pathology and an increase in the period of active human longevity.

References

1. R. L. Walford, *Fed. Proc.*, 1974, **33**, 2020.
2. G. Goldstein, M. Scheid, U. Hammerling, D. Schlesinger, H. D. Niallan and E. A. Boyse, *Proc. Natl. Acad. Sci. U. S. A.*, 1975, **72**, 11.
3. E. Hannappel, S. Davoust and B. L. Horecker, *Proc. Natl. Acad. Sci. U. S. A.*, 1982, **82**, 1708.
4. A. Hershko, A. Ciechanover and I. A. Rose, *Proc. Natl. Acad. Sci. U. S. A.*, 1979, **76**, 3107.
5. A. Hershko, A. Ciechanover, H. Heller, A. L. Haas and I. A. Rose, *Proc. Natl. Acad. Sci. U. S. A.*, 1980, **77**, 1783.
6. V. T. Ivanov, A. A. Karelin, M. M. Philippova, I. V. Nazimov and V. Z. Pletnev, *Biopolymers*, 1997, **43**, 171.
7. S. Karlin and S. F. Altschul, *Proc. Natl. Acad. Sci. U. S. A.*, 1990, **87**, 2264.
8. V. Kh. Khavinson, *Neuroendocrinol. Lett.*, 2002, **23**, 144.
9. V. N. Anisimov and V. Kh. Khavinson, *Biogerontology*, 2010, **11**, 1399.
10. J. P. Kirkwood, *Mutat. Res.*, 1989, **219**, 1.
11. D. E. Harrison and T. N. Roderick, *Exp. Gerontol.*, 1997, **32**, 65.
12. D. Harman, *Ann. N. Y. Acad. Sci.*, 1994, **717**, 1.
13. M. K. Shigenaga, T. M. Hogen and B. N. Ames, *Proc. Natl. Acad. Sci. U. S. A.*, 1994, **91**, 10771.
14. V. N. Anisimov, *Molecular and Physiological Mechanisms of Aging*, Nauka, St. Petersburg, 2008.
15. V. A. Bohr and R. M. Anson, *Mutat. Res.*, 1995, **338**, 25.
16. K. A. Yeiding, *Perspect. Biol. Med.*, 1974, **17**, 201.
17. E. Giolloto, A. Mottura, L. De Carli and F. Nuzzo, *Exp. Cell Res.*, 1978, **113**, 415.
18. M. A. de Arce, *Hum. Genet.*, 1981, **57**, 83.
19. T. Lezhava, *Human Chromosomes and Aging: From 80 to 114 Years*, Nova Biomedical, 2006, p. 1147.
20. V. N. Anisimov, I. G. Popovich and M. A. Zabezhinski, *Methods Mol. Biol.*, 2013, **1048**, 145.
21. V. Kh. Khavinson and V. N. Anisimov, *Peptide Bioregulators and Aging*, Nauka, St. Petersburg, 2003.
22. S. V. Rosenfeld, E. F. Togo, V. S. Mikheev, I. G. Popovich, V. Kh. Khavinson and V. N. Anisimov, *Bull. Exp. Biol. Med.*, 2002, **13**, 274.
23. S. V. Anisimov, K. R. Bokeler, V. Kh. Khavinson and V. N. Anisimov, *Bull. Exp. Biol. Med.*, 2002, **133**, 293.
24. V. Kh. Khavinson, S. I. Tarnovskaya, N. S. Linkova, V. E. Pronaeva, L. K. Shataeva and P. P. Vakutseni, *Bull. Exp. Biol. Med.*, 2013, **154**, 403.
25. V. Kh. Khavinson, *Adv. Gerontol.*, 2014, **4**, 337.
26. O. V. Korkushko, V. Kh. Khavinson and V. B. Shatilo, *Pineal Gland: Ways for Correction in Aging*, Nauka, St. Petersburg, 2006.
27. V. G. Morozov and V. Kh. Khavinson, US Pat., 5070076, 1991.
28. V. G. Morozov and V. Kh. Khavinson, *Int. J. Immunopharmacol.*, 1997, **19**, 501.

29. V. A. Alexandrov, V. G. Bespalov, V. G. Morosov, V. Kh. Khavinson and V. N. Anisimov, *Carcinogenesis*, 1996, **17**, 1931.
30. V. Kh. Khavinson, B. I. Kuznik and G. A. Ryzhak, *Adv. Gerontol.*, 2013, **3**, 225.
31. V. N. Anisimov, V. Kh. Khavinson and V. G. Morozov, *Mech. Aging Dev.*, 1982, **19**, 245.
32. N. P. Napalkov, G. N. Iakovlev, V. N. Anisimov, V. G. Morozov and V. Kh. Khavinson, *Vopr. Onkol.*, 1988, **34**, 515.
33. V. Kh. Khavinson, V. G. Morozov, V. I. Semenova, O. V. Chaika and G. A. Ryzhak, RF Patent, 2163129, 2001.
34. V. N. Anisimov and V. Kh. Khavinson, in *Aging Interventions and Therapies*, ed. S. I. S. Rattan, World Scientific, 2005, vol. 7, pp. 127–146.
35. V. Kh. Khavinson, V. G. Morozov and V. N. Anisimov, *Experimental Studies of the Pineal Preparation Epithalamin*, Belin Heidelberg, Springer-Verlag, 2001.
36. V. N. Anisimov, M. N. Ostroumova and V. M. Dil'man, *Bull. Exp. Biol. Med.*, 1980, **89**, 723.
37. V. N. Anisimov, G. I. Miretskii, V. G. Morozov and V. Kh. Khavinson, *Bull. Exp. Biol. Med.*, 1982, **94**, 80.
38. V. M. Dilman, V. N. Anisimov, M. N. Ostroumova, V. Kh. Khavinson and V. G. Morozov, *Exp. Pathol.*, 1979, **17**, 539.
39. V. Kh. Khavinson and S. V. Mylnikov, *Dokl. Biol. Sci.*, 2000, **373**, 370.
40. V. N. Anisimov, V. Kh. Khavinson and V. G. Morozov, *Biogerontology*, 2000, **1**, 55.
41. V. Kh. Khavinson, V. G. Morozov, V. V. Malinin and S. V. Sery, US Pat., 6642201, 2003.
42. V. Kh. Khavinson and V. N. Anisimov, *Dokl. Biol. Sci.*, 2000, **372**, 261.
43. V. Kh. Khavinson, V. N. Anisimov, N. Y. Zavarzina, M. A. Zabezhinskii, O. A. Zimina, I. G. Popovich, A. V. Shtylik, V. V. Malininand and V. G. Morozov, *Bull. Exp. Biol. Med.*, 2000, **130**, 687.
44. V. Kh. Khavinson, US Pat., 6727227, 2004.
45. D. R. Cox and D. Oakes, *Analysis of Survival Data*, Chapman and Hall, London, 1996, p. 201.
46. N. D. Goncharova, V. Kh. Khavinson and B. A. Lapin, *Pineal Gland and Age-Related Pathology (Mechanisms and Correction)*, Nauka, St. Petersburg, 2007.
47. R. J. Reiter, C. M. Craft and J. E. Johnson, *Endocrinology*, 1981, **109**, 1205.
48. J. Hansen, *Cancer Causes Control*, 2006, **17**, 531.
49. R. G. Stivens, *Cancer Causes Control*, 2006, **17**, 501.
50. V. N. Anisimov, N. Y. Zavarzina, M. A. Zabezhinski, I. G. Popovich, O. A. Zimina, A. V. Shtylick, A. V. Arutjunyan, T. I. Oparina, V. M. Prokopenko, A. I. Mikhalski and A. I. Yashin, *J. Gerontol., Ser. A.*, 2001, **56**, B311.
51. I. A. Vinogradova, A. V. Bukalev, M. A. Zabezhinski, A. V. Semenchenko, V. Kh. Khavinson and V. N. Anisimov, *Bull. Exp. Biol. Med.*, 2007, **144**, 825.
52. I. A. Vinogradova, A. V. Bukalev, M. A. Zabezhinski, A. V. Semenchenko, V. Kh. Khavinson and V. N. Anisimov, *Bull. Exp. Biol. Med.*, 2008, **145**, 472.

53. G. M. Yakovlev, V. S. Novikov, V. S. Smirnov, V. Kh. Khavinson and V. G. Morozov, *Mechanisms of Bioregulation*, Nauka, St. Petersburg, 1992.
54. V. G. Morozov, V. Kh. Khavinson and V. V. Malinin, *Peptide Thymometics*, Nauka, St. Petersburg, 2000.
55. V. Kh. Khavinson, *Neuroendocrinol. Lett.*, 2002, **23**, 144.
56. V. N. Anisimov, A. V. Arutjunyn and V. Kh. Khavinson, *Neuroendocrinol. Lett.*, 2001, **22**, 9.
57. V. Kh. Khavinson, A. Yu. Solov'ev, D. V. Zelinskii, L. K. Shataeva and B. F. Vaniushin, *Adv. Gerontol.*, 2012, **2**, 277.
58. M. O. Yuneva, N. V. Guseva and A. A. Boldyrev, *Adv. Gerontol.*, 2000, **4**, 147.
59. V. Kh. Khavinson, T. A. Lezhava, J. R. Monaselidze, T. A. Jokhadze, N. A. Dvalishvili, N. K. Bablishvili and S. V. Trofimova, *Neuroendocrinol. Lett.*, 2003, **24**, 329.
60. T. A. Lezhava, V. Kh. Khavinson, J. R. Monaselidze, T. A. Jokhadze, N. A. Dvalishvili, N. K. Bablishvili and S. Barbakadze, *Biogerontology*, 2004, **5**, 73.
61. R. S. Hawley and T. Arbel, *Cell*, 1993, **72**, 301.
62. V. Kh. Khavinson, T. A. Lezhava and V. V. Malinin, *Bull. Exp. Biol. Med.*, 2004, **137**, 78.
63. M. O. Olson, M. Dundr and A. Szebeni, *Trends Cell Biol.*, 2000, **10**, 189.
64. V. N. Anisimov, V. Kh. Khavinson, M. Provinciali, I. N. Alimova, D. A. Baturin, I. G. Popovich, M. A. Zabezhinski, E. N. Imyanitov, R. Mancini and C. Franceschi, *Int. J. Cancer*, 2002, **101**, 7.
65. V. Kh. Khavinson, I. E. Bondarev and A. A. Butyugov, *Bull. Exp. Biol. Med.*, 2003, **135**, 590.
66. V. Kh. Khavinson, I. E. Bondarev, A. A. Butyugov and T. D. Smirnova, *Bull. Exp. Biol. Med.*, 2004, **137**, 503.
67. V. Kh. Khavinson and V. V. Malinin, *Gerontological Aspects of Genome Peptide Regulation*, Karger AG, Basel (Switzerland), 2005.
68. B. I. Kuznik, N. S. Linkova, S. I. Tarnovskaya and V. Kh. Khavinson, *Adv. Gerontol.*, 2013, **26**, 38.
69. V. Kh. Khavinson, V. G. Morozov, V. V. Malinin, T. B. Kazakova and E. A. Korneva, *Bull. Exp. Biol. Med.*, 2000, **130**, 898.
70. O. V. Korkushko, V. Kh. Khavinson, G. M. Butenko and V. B. Shatilo, *Peptide Preparations of Thymus and Pineal Gland in Prevention of Accelerated Aging*, Nauka, St. Petersburg, 2002.
71. K. H. Cooper, *JAMA*, 1968, **203**, 135.
72. O. V. Korkushko, V. Kh. Khavinson, V. B. Shatilo and I. A. Antonyk-Sheglova, *Bull. Exp. Biol. Med.*, 2011, **151**, 366.
73. V. Kh. Khavinson, B. I. Kuznik and G. A. Ryzhak, *Adv. Gerontol.*, 2014, **4**, 346.
74. L. I. Fedoreyeva, I. I. Kireev, V. Kh. Khavinson and B. F. Vanyushin, *Biochemistry*, 2011, **76**, 1210.
75. A. Schumacher, in *The New molecular and Medical Genetics*, ed. T. O. Tollefsbol, Elsevier Inc., Amsterdam, 2011, pp. 405–422.

76. J. T. Bell, P. C. Tsai, T. P. Yang, R. Pidsley, J. Nisbet, D. Glass, M. Mangino, G. Zhai, F. Zhang and A. Valdes, *et al.*, *PLoS Genet.*, 2012, **8**, e1002629.
77. G. Hannum, J. Guinney, L. Zhao, L. Zhang, G. Hughes, S. Sadda, B. Klotzle, M. Bibikova, J. B. Fan and Y. Gao, *et al.*, *Mol. Cell*, 2013, **49**, 359.
78. V. Kh. Khavinson, A. Yu. Solovyov and L. K. Shataeva, *Bull. Exp. Biol. Med.*, 2008, **146**, 624.
79. V. Kh. Khavinson, L. K. Shataeva and A. A. Chernova, *Neuroendocrinol. Lett.*, 2005, **26**, 237.
80. V. Wick, D. Zubon and G. Hagen, *Gene*, 1999, **232**, 97.
81. V. Kh. Khavinson, I. E. Bondarev, A. A. Butyugov and T. D. Smirnova, *Bull. Exp. Biol. Med.*, 2004, **137**, 503.
82. A. J. Warren, *Curr. Opin. Struct. Biol.*, 2002, **12**, 107.
83. M. Lebl and A. Houghten, *Peptides: The Wave of the Future*, American Peptide Society, San Diego, 2001, p. 1147.

CHAPTER 21

HDAC Inhibitors: A New Avenue in Anti-Aging Medicine

E. G. PASYUKOVA[a]* AND A. M. VAISERMAN[b]

[a]Laboratory of Genome Variation, Institute of Molecular Genetics of RAS, Kurchatov sq. 2, Russia; [b]Laboratory of Epigenetics, Institute of Gerontology, Vyshgorodskaya st. 67, Kiev 04114, Ukraine
*E-mail: egpas@img.ras.ru

21.1 Introduction

Gradual loss of physiological functions accompanied by decreasing fertility and increased risk of mortality with advancing age is recognized as age-related senescence, a process immanent to most living beings. Whether senescence may be prevented and/or postponed by certain approaches is a matter of utmost importance in today's world. Recent advances in biogerontology and an increasing number of pharmacological and dietary interventions suggested to have the anti-aging and life-extending effects[1,2] give hope that senescence may be effectively combated in the near future.

In recent years, epigenetic regulatory mechanisms have become increasingly appreciated as central to a variety of age-associated processes, such as cellular and organismal senescence, genomic instability, and tumorigenesis.[3–5] Epigenetic modifications provide a mechanism of heritable but reversible changes in gene function that occur without the change in the primary DNA sequence due to alterations in chromatin structure. Normally, throughout the life span, the epigenetic processes are both

RSC Drug Discovery Series No. 57
Anti-aging Drugs: From Basic Research to Clinical Practice
Edited by Alexander M. Vaiserman

Published by the Royal Society of Chemistry, www.rsc.org

finely tuned and influenced by multiple environmental cues that can be "remembered" due to the changes in the epigenome;[4] both internal and external attenuation of epigenetic processes may affect the normal rate of aging. The global changes in chromatin structure and certain local epigenetic modifications in the promoter regions of several specific genes including tumor suppressor genes are among the key age-associated epigenetic processes.[6] At the same time, the epigenetic patterns can be significantly destroyed along with the development of various diseases. Epigenetic dysregulation has been shown to be implicated in a wide variety of age-related chronic diseases, such as decline of immune function, atherosclerosis, type 2 diabetes, cancer, and neurodegenerative and psychiatric diseases.[7–10]

The main epigenetic mechanisms include DNA methylation, modifications of histones that package the DNA, and microRNA regulatory pathways. It is noteworthy that, unlike genetic changes (mutations) that cannot be restored, epigenetic aberrations are reversible and can be relatively easily corrected through nutritional and pharmacological interventions or due to certain physical factors and environmental exposures.[5,11] The potential reversibility of epigenetic aberrations makes them attractive targets for therapeutic drug development.[9] In the last several years, a novel class of drugs targeting epigenetic pathways ("epigenetic drugs") has been proposed. Among others, the members of superfamilies of histone deacetylases (HDACs) are currently considered as highly promising targets for epigenetic drugs with health-beneficial and anti-aging effects; in this context, HDAC inhibitors (HDACIs) are regarded as potential therapeutics.[12–15]

At the present time, invertebrates have been recognized as useful models for development of human diseases and are widely used in screens of agents with potential anti-aging properties.[16–21] For example, the nematode *Caenorhabditis elegans* and the fruit fly *Drosophila melanogaster* have high potential for such studies, the main advantages being their relatively short life span, ease of maintenance, sequences of the full genomes, availability of a large variety of environmental and genetic manipulations and stocks containing altered genes. In addition, the currently accepted view is that pathways are substantially conserved in a wide variety of species, from invertebrates to humans.[19] Highly specific and precise age-related invertebrate models may facilitate determining which HDACIs may extend longevity and which genes are implicated in these effects, as well as provide important information about the genetic basis of aging. The critical issue still remains whether the mechanisms of action of HDACIs are similar between invertebrates and mammals.

The main focus of this chapter is a review of the literature describing the health-beneficial and life-extending effects of inhibitors of HDAC activity, with an emphasis on data obtained using *D. melanogaster* as a model species.

21.2 Role of Histone Modification in Epigenetic Regulation

There are two major epigenetic mechanisms influencing gene expression throughout the eukaryotic life cycle, including aging: methylation of DNA and modification of histones.[22] Genomic DNA in eukaryotic cells is assembled in nucleosomes due to interaction with two histone H2A and histone H2B dimmers and a tetramer of histones H3 and H4; nucleosomes interact with the linker histone H1. The highly conserved core histones contain lysine-rich N-terminal tails that are able to undergo various covalent post-translational modifications, *i.e.* acetylation, phosphorylation, ubiquitylation, biotinylation, sumoylation, and others.[23] These histone modifications alter the histone-DNA interaction and create a "histone code" that coordinates the recruitment of transcription factors and polymerases.[12] Among all known histone modifications, acetylation has the highest potential to induce chromatin unfolding as it neutralizes the electrostatic interaction between the histone and the negatively charged DNA, making it more accessible to the transcriptional apparatus.[24] Overall, histone acetylation and deacetylation play a crucial role in modifying chromatin structure and thus regulating gene expression, cell proliferation, migration and apoptosis, as well as immune functions and angiogenesis.[25]

Many recent studies revealed a role of chromatin modification in aging. The heterochromatin loss model of aging proposed by Villeponteau[26] suggests that heterochromatin domains are set up early in embryogenesis but then are gradually lost with aging, which results in aberrant gene expression associated with old age. An association between chromatin silencing and life span has been revealed in various experimental models including yeast, *Caenorhabditis elegans*, mice and *Drosophila melanogaster*.[5] A dramatic reorganization of chromosomal regions with age in fruit flies was found in a whole genome study.[27] An overall decline of the active chromatin marks, such as H3K4me3 and H3K36me3, as well as a significant decrease in the enrichment of the repressive heterochromatin H3K9me3 and heterochromatin protein 1 (HP1) marks at pericentric heterochromatin loci, have been found with age. Such extensive alterations in repressive chromatin state were associated with age-related changes in gene expression.[28] Alterations in the structure and functions of genes encoding HDACs such as SIRT1/SIR2 and RPD3 are also known to be involved in the extension of lifespan in different organisms.[29,30] In *D. melanogaster*, genetic alterations in *sir2* and *Rpd3* and nutrition have distinct but interacting effects on longevity,[31,32] which emphasizes both the role of HDACs as overall modifiers of the metabolic status of an organism and the significance of this status in life span control.

Histone acetylation is controlled by a dynamic counterbalance between activities of histone acetyl transferases (HATs) and HDACs.[33] HATs catalyze the transfer of the acetyl moiety from acetyl coenzyme A to the ε-amino groups of histone lysine residues, thereby neutralizing the positive charge of the histone tails and reducing their affinity for DNA. This results in a more open chromatin state and greater access of DNA to transcription factors. HDACs, on the contrary, catalyze the removal of acetyl groups from lysine residues of histone tails, resulting in a more condensed, transcriptionally repressive chromatin conformation.[34] In normal cells, there is a fine balance between acetylation and deacetylation of histones.[35] Histone acetylation is associated with an open chromatin and activation of gene expression, while histone deacetylation is associated with closed chromatin and repression of transcription. In recent years, increasing evidence has been accumulated that HDACs play key roles in diverse biological processes such as cell proliferation, apoptosis, inflammation and others.[36]

Though modifying histones and chromatin structure is the predominant function of HDACs, they are able to modify non-histonal proteins.[37] Some of these proteins are transcription factors and others are regulatory proteins, which also determines the role of HDACs in regulation of gene expression profiles.

The HDACs are grouped into classes depending on sequence homology to the yeast original enzymes and domain organization. Class I, II and IV HDACs depend on a Zn^{2+} ion in their catalytic center. Class III HDACs (sirtuins, SIRTs) differ from other HDACs in that they use NAD^{+} as the cofactor.[34,38] In mammals, including humans, Class I includes HDACs1–3 and HDAC8, Class II includes HDACs4–7 and HDACs9–10, Class III includes SIRTs1–7, and, finally, Class IV is represented by a single member, HDAC11.[39] In other species, the composition of the main HDAC classes may be different. For example, in *D. melanogaster*, Class I includes RPD3 and dHDAC3, Class II includes dHDAC4 and dHDAC6, Class III includes dSIR2, dSIRT2, dSIRT4, dSIRT6, and dSIRT7,[40] and Class IV includes HDAC11.[41] HDAC inhibitors that target the Class I, II and IV HDACs and SIRTs are different; the latter will not be covered in this review.

21.3 Life Span-Modulating Effects of HDAC Inhibitors in Animal Models

Among the chemicals affecting HDAC activity, HDAC inhibitors (HDACIs) seem to be the most promising in the field of geroscience. Since the level of transcription of many genes, primarily metabolic and biosynthetic ones, is known to decrease with age,[42] the restoration of the transcriptional activity *via* HDACIs could likely delay the age-related functional decline. Moreover, HDAC inhibition can cause up-regulation of genes involved in response to stress and inflammation – pathways generally involved in the regulation of longevity.[43] Life span-modulating effects of HDACIs have been studied mostly in invertebrate experimental models such as *D. melanogaster*.

In recent years, experimental research has emerged on the life-extending potential of synthetic HDACIs, although several natural compounds, such as trichostatin A extracted from *Streptomyces hydrocsopicus*, sulforaphane contained in broccoli, curcumin extracted from turmeric and garlic-derived

Table 21.1 Most widely used HDACIs.

Class	Compound name	HDAC specificity class
Short-chain fatty acids	Phenylbutyrate (PBA)	I, II
	Sodium butyrate (SB)	I, II
	Valproic acid (VPA)	I, II
Hydroxamic acids	Trichostatin A (TSA)	I, II, IV
	Vorinostat (suberoylanilide hydroxamic acid, SAHA)	I, II, IV
	Givinostat (ITF2357)	I, II
	Abexinostat (PCI-24781)	
	Belinostat (PXD101)	I, II, IV
	Panobinostat (LBH589)	I, II, IV
	Resminostat (4SC-201)	I, II, IV
	Quisinostat (JNJ-26481585)	I, II, IV
Cyclic peptides	Depsipeptide (romidepsin)	I
	Apicidin	I, II
Benzamides	Entinostat (MS-275)	I, II
	Mocetinostat (MGCD0103)	I

diallyl disulfide, seem to be very promising as well. HDACIs include four chemical classes: cyclic peptides, hydroxamic acids, short chain fatty acids and synthetic benzamides, and they substantially vary in biological activity, structure and specificity.[44] The most commonly used HDACIs are listed in Table 21.1. In *D. melanogaster*, each HDAC was shown to regulate transcription of a unique set of genes and to have a distinct pattern of temporal expression.[45] Furthermore, a differential sensitivity of HDACs to HDACIs has been shown.

The research findings supporting the anti-aging and life-extending properties of HDACIs are reviewed in the subchapters below.

21.3.1 Phenylbutyrate

Sodium 4-phenylbutyrate (PBA) was shown to inhibit class I and II HDACs, which lead to elevated gene expression, reduced cellular proliferation, induction of apoptosis, and enhanced cell differentiation in neoplastic cell populations.[46]

The dose-dependent life-extending potential of the sodium salt of PBA in *D. melanogaster* was demonstrated by Kang and co-authors.[47] Feeding of flies with PBA resulted in a substantial extension of both mean and maximal life span by up to 30–50% regardless of the fly's genetic background, without diminution of locomotor activity and resistance to stress. This result was not due to caloric restriction, known to extend life span in different model organisms, or due to the decrease in reproductive activity. Treatment for a limited period, either early or late in adult life, has also been found to have potential to extend the flies' longevity, possibly by stimulating repair mechanisms and/or inhibiting the accumulation of damages.[47] The effects of PBA were also accompanied by marked changes in the levels of acetylation of histones H3 and H4 and either down- or up-regulation of several hundreds of genes, as was evident from the DNA microarray-based global transcriptional analysis. The general trend was up-regulation of genes involved in detoxification and chaperone activity, including several genes that have previously been found to be involved in life span determination in *D. melanogaster*, and down-regulation of genes involved in different metabolic pathways. These findings support the hypothesis that life span extension may be caused by overall generalized changes in epigenetic regulation.[48]

21.3.2 Sodium Butyrate

In several studies (Table 21.2), life-extending capacity was also shown for sodium butyrate (SB), a short chain fatty acid having HDAC inhibition activity and known to markedly influence the processes of cell growth, differentiation and apoptosis in both normal and transformed cells.[49,50] In *D. melanogaster*, an increase of both mean and maximum life span by 25.8% and 11.5%, respectively, was observed due to one-off treatment with SB.[51] Later, an increase in mean and/or maximum life span and a decrease in mortality rate after SB treatment were observed by other authors.[52–55] SB at concentrations varying from 10 to 40 mM demonstrated the potential to increase life span, whereas SB treatment at higher doses (more than 100 mM) decreased longevity.[51,52] In some cases[51,55] effects depended on whether the line used was short- or long-lived. The life-extending effects obtained were unlikely due to the decreased reproductive investment, because no reduction in reproductive activity (fecundity) was revealed in SB-treated female flies.[52] In some cases, life span improvement was accompanied by an increase in locomotor activity[55] (see also Figure 21.1), which is often considered as a marker of health and aging.[56]

Treatment with SB caused elevated acetylation levels at histone H3,[51,57–59] whereas the level of acetylation of histone H4 remained unchanged.[59] Histone H3 with elevated acetylation levels was found at the promoter of the *hsp22*,[57] *hsp70*,[58] and *hsp26*[59] genes. Also, SB affected the structure of chromatin at the site of cytogenetic location of the *hsp70* gene on the polythene chromosome.[60] Accordingly, elevated levels of expression of *hsp22*, *hsp26*, and *hsp70* genes were found in SB-treated flies.[51,57–60] According to Zhao and co-authors,[51,57] these findings suggest that the alterations in histone acetylation and, thereafter, the expression of chaperone genes, may contribute to the life-extending effects of SB and other HDACIs in *D. melanogaster*.

Other mechanisms, however, may also be contributing. In recent research by St Laurent and co-authors,[54] treatment with SB-supplemented food rescued the early mortality of flies with the pesticide rotenone-induced Parkinson's disease. In this model, SB was selected as a therapeutic candidate because it is known to be able to correct the disrupted HDAC activity in Parkinson's disease and other neurodegenerative disorders. The SB-mediated rescue of rotenone-induced Parkinson's disease was associated with elevated dopamine levels in the fly brain. At the same time, treatment

Table 21.2 Phenotypic and functional changes induced by SB treatment.[a]

Strain/model	Stage	Phenotypic changes	Functional changes
Oregon-R[52]	Larvae and adult	Increased MLS in both sexes	ND
	Adult	Increased male MLS	ND
Oregon-R[53]	Larvae	Increased male MLS and MaxLS; increased female MaxLS	Up-regulation of inducible expression of *sir2* gene
Oregon-R (unpublished data)	Adult	Increased male MLS; no effect on locomotion	ND
w[1118] (unpublished data)	Adult	Increased male MLS; increase in locomotion in 40 and 50 day old males	ND
Short-lived iso4 line[51]	Larvae	Increased MLS and MaxLS in a sex-pooled population	Hyperacetylation of core histone H3; elevated levels of expression of *hsp22* and *hsp70* genes
	Larvae and adult	No effect on life span	Hyperacetylation of core histone H3
Long-lived iso2 line[51]	Larvae	No effect on life span	Hyperacetylation of core histone H3; elevated levels of expression of *hsp22* and *hsp70* genes
	Larvae and adult	No effect on life span	Hyperacetylation of core histone H3
Normal-lived Ra strain[55]	Transition/senescent span	Decreased mortality rate and increased MLS	ND
Long-lived La strain[55]	Entire adult life span or healthspan	Decreased MLS	ND
	Adult	Decreased MLS	ND
Sin3A[lof] (the model of Parkinson disease)[54]	First 5 days of adult life	Rescued locomotor impairment and early mortality	Unchanged deficiency in the tyrosine hydroxylase mRNA level; however, elevated dopamine levels in the brain
Canton-S[58]	Larvae	ND	H3 hyperacetylation in the promoter and coding regions of *hsp70* gene; up-regulation of basal and inducible *hsp70* expression
NS[60]	Larvae	ND	Modification of chromatin structure at the site of cytogenetic location of *hsp70* gene; elevated level of transcription of *hsp70*
NS[57]	Larvae	ND	H3 hyperacetylation in the promoter and coding regions of *hsp22* gene; up-regulation of basal and inducible expression of *hsp22*
Canton-S[59]	Larvae	ND	H3 hyperacetylation in the promoter region of *hsp26* gene; decreased level of basal transcription and increased level of inducible transcription of *hsp26*

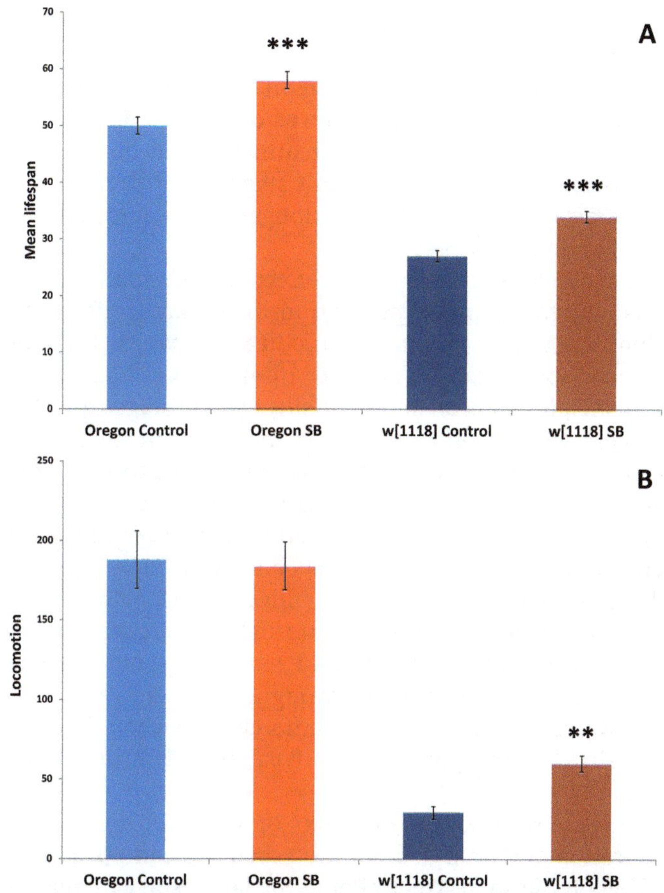

Figure 21.1 Effects of SB on the lifespan and locomotion of *D. melanogaster* males of two genotypes. (A) Mean lifespan ± SE. (B) Mean locomotion ± SE of 50 day-old flies. Lifespan and locomotion were measured according to ref. 104. Results of two independent life span measurements are combined (unpublished data). $*p < 0.05$, $**p < 0.01$, $***p < 0.001$.

with SB did not improve the deficiency in serotonin content, tyrosine hydroxylase (the rate-limiting enzyme for dopamine biosynthesis) mRNA levels and in SOD activity in rotenone-treated insects.

Up-regulation of inducible expression of *sir2* gene was observed after SB treatment.[52] SIR2 is known to be involved in the extension of life span in *D. melanogaster*.[30,31] Activation of Class III HDAC, SIR2, following inhibition of Class I–II HDACs by SB might represent an interesting example of complex interactions between HDACs involved in life span control. According to our preliminary unpublished results, an RNA-seq analysis of transcriptomes of SB-treated and control w^{1118} males demonstrated that the following functional gene sets were associated with SB treatment: (i) male fertility; (ii) regulation of immune system; (iii) regulation of response to stress, including JNK signaling. Accordingly, up-regulation of *foxo*, *hep* and other genes involved in life span control was revealed (unpublished data).

The effects of SB on life span depended on the stage and/or age when treatment was applied. Increased life span was observed after treatment at the larval stage in all studies,[51–53] whereas treatment at the both larval and adult stage or exclusively throughout the adult stage either decreased life span,[55] or increased it[52,53] or had no effect[51] on longevity. Thereto, SB effects on life span were shown to be sex-specific.[52,53]

To explain the inconsistencies in SB effects on life span described above, a hypothesis suggesting phase separation in the adult life of fruit fly and other gradually aging organisms into a health span, a transition phase, and a senescent span[61] can be applied. In analysis conducted in different model organisms, it has been shown that these life stages are characterized by different gene expression patterns. The health span is characterized by a tightly regulated gene expression pattern, which leads to a maximized tissue function and to a minimized inflammatory and other damage response; the transition phase is characterized by a gradual decline of the cellular regulatory capacity, and the senescent span is characterized by a gradual deregulation of the gene expression pattern.[62] Indeed, McDonald and co-authors[55] demonstrated a decrease in mortality rate and an increase in life span when flies were fed with SB during transition or senescent spans, but a decrease in life span when SB was administered throughout the entire adult life span or health span only. Similar results were demonstrated when flies were fed with curcumin during health span, senescent span and throughout the entire adult life span.[63]

To summarize, a wide variety of effects of SB on life span were observed. Stage and duration of SB supplementation seem to be important with respect to the SB effect on life span. Treatment at the larval stage increased longevity in most experiments. This fact indicates that SB might indeed affect epigenetic mechanisms of life span extension based on modification of histones. Treatment with SB at the adult stage had a positive effect on longevity only in some cases, in particular, though not exclusively, when SB was supplied later in life and could prolong active transcription of genes essential for maintaining health span. Short/normal-lived strains were more sensitive to SB treatment than long-lived strains. It must be taken into account that different authors used various genotypes, which could also contribute to the observed differences in SB effects. Contradictions between different experiments could, at least partially, also be explained by the sex-specificity of SB effects. In some experiments, life span was measured in a mixed population of males and females, and this could substantially bias the final result. However, despite the complexity and partial inconsistency of results, SB demonstrated a high potential as a life-extending agent.

The life-extending capacity of HDACIs, such as D-beta-hydroxybutyrate, to modulate aging and promote life span was also recently reported in *Caenorhabditis elegans*.[64] Supplementation with this agent extended the worm's mean life span by approximately 20%. In addition, it increased worm thermotolerance, prevented glucose toxicity, delayed Alzheimer's amyloid-beta toxicity and decreased Parkinson's alpha-synuclein aggregation. Interestingly, D-beta-hydroxybutyrate did not extend life span in a genetic model of dietary restriction, indicating that it is likely functioning through a similar mechanism.

21.3.3 Trichostatin A

Trichostatin A (TSA) is another widely used HDACI that demonstrates a broad spectrum of epigenetic activities, including inhibition of the cell cycle since the beginning of the growth stage and promotion of the expression of apoptosis-associated genes. TSA is recognized as a promising anticancer drug candidate. Possible mechanisms of action of this compound are induction of terminal differentiation, cell cycle arrest and apoptosis in different cancer cell lines, and thereby inhibition of tumorigenesis.[65]

The epigenetic and phenotypic effects of TSA treatment are very similar to those shown for SB treatment (Table 21.3). In *D. melanogaster*, an increase of both mean and maximum life span was observed due to both one-off and continuous treatment with 10 mkM TSA.[51,66] TSA treatment was effective both at the larval[51] and adult[66] stages and influenced the longevity of both short- and long-lived *D. melanogaster* lines, but to different extents.[51] Life span improvement affected both males and females, and in some cases was accompanied by an increase in locomotor activity (Figure 21.2).

These life-extending effects induced by the TSA treatment were accompanied by the hyperacetylation of core histone H3 in the promoter and coding regions of some chaperone genes, such as *hsp22*, *hsp26* and *hsp70*, along with up-regulation, in most cases, of both basal and inducible expression of these genes.[51,57–60,66] Modified chromatin morphology at the locus of *hsp22* was also revealed.[66] The authors suggested that the expression of chaperones can reduce the level of accumulation of damage, stimulate the repair mechanisms, and improve the cell stress resistance to create cellular and physiological environments that are favorable for longevity.

We performed an RNA-seq analysis of transcriptomes of TSA-treated and control w^{1118} males. According to our preliminary results, the following functional gene sets were associated with the differential expression in control and TSA-treated flies: (i) DNA replication; (ii) cell fate determination, differentiation and development of various organ systems, and (iii) mitochondria function and ATP synthesis. Surprisingly, up-regulation of many genes involved in development of the nervous system, heart and cuticula was revealed in TSA treated males in association with increased life span (unpublished data).

To summarize, the effects of TSA on life span seem more consistent than the effects of SB. TSA was shown to affect life span of both short/normal- and

Table 21.3 Phenotypic and functional changes induced by TSA treatment.[a]

Strain	Stage	Phenotypic changes	Functional changes
Canton-S[66]	Adult	Increased MLS and MaxLS in both sexes	Modified chromatin morphology at the locus of *hsp22* gene; increased *hsp22* transcription
Oregon-R (unpublished data)	*Adult*	Increased male MLS; no effect on locomotion	ND
w^{1118} (unpublished data)	*Adult*	Increased male MLS; increase in locomotion in 30, 40, and 50 day old males	ND
Short-lived iso4 line[51]	Larvae	Increase of MLS by one-off treatment; increase of MLS and MaxLS by continuous treatment	Hyperacetylation of core histone H3; elevated levels of expression of hsp22 and hsp70 genes
Long-lived iso2 line[51]		Increase of MLS by continuous treatment	
Canton-S[58]	Larvae	ND	H3 hyperacetylation in the promoter and coding regions of hsp70 gene; up-regulation of basal and inducible hsp70 expression
NS[60]	Larvae	ND	Modification of chromatin structure at the site of cytogenetic location of hsp70 gene; elevated level of transcription of hsp70 gene
NS[57]	Larvae	ND	H3 hyperacetylation in the promoter and coding regions of hsp22 gene; up-regulation of basal and inducible expression of hsp22 gene
Canton-S[59]	Larvae	ND	H3 hyperacetylation in the promoter region of *hsp26* gene; decreased level of basal transcription and increased level of inducible transcription of *hsp26* gene

[a]MLS: mean life span, MaxLS: maximum life span; NS: not specified; ND: not determined.

long-lived strains, and the stage of TSA supplementation seems to be less important compared to for SB treatment. The life span-modulating effects of TSA were also found to be sex-dependent. Similarly to SB, TSA demonstrated high potential as a life-extending agent. However, in our experiments, two HDACIs affected transcription of different sets of genes, with TSA treatment affecting transcription of much more genes and in a greater extent compared to SB treatment.

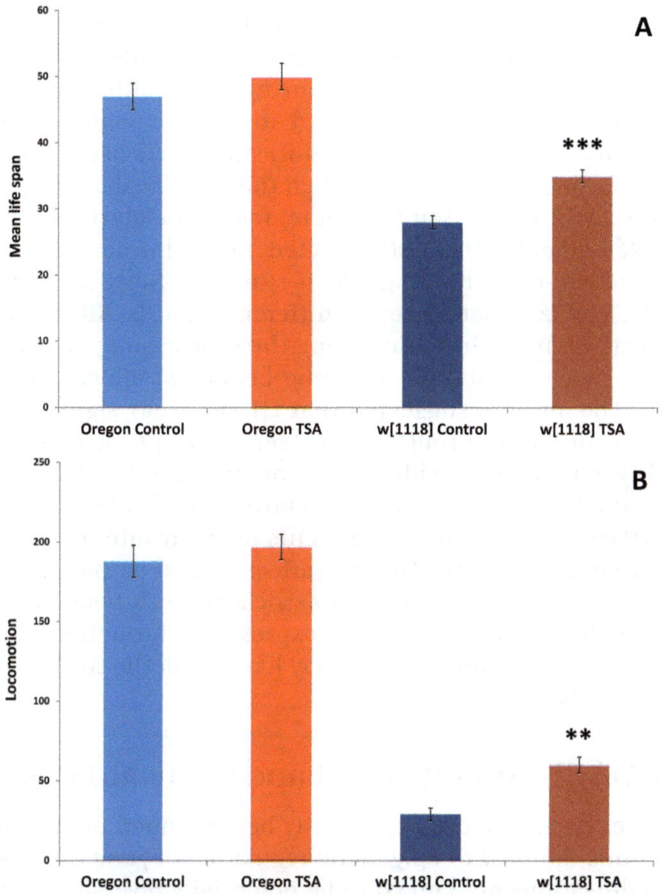

Figure 21.2 Effects of TSA on the lifespan and locomotion of *D. melanogaster* males of two genotypes. (A) Mean lifespan ± SE. (B) Mean locomotion ± SE of 50 day-old flies. Lifespan and locomotion were measured according to ref. 106. Results of two independent life span measurements are combined (unpublished data). $*p < 0.05$, $**p < 0.01$, $***p < 0.001$.

21.3.4 Suberoylanilide Hydroxamic Acid (SAHA)

One more HDACI that was shown to be able to extend life in fruit flies is suberoylanilide hydroxamic acid (SAHA). In *in vitro* studies, SAHA was found to have similar effects to SB, although at much lower effective doses.[25] This compound is known to induce growth arrest in transformed cells,[67] and it was shown to be effective in preventing Huntington's disease in various animal models.[68]

In a recent study by McDonald and co-authors,[55] the effects of administration with SAHA throughout *D. melanogaster* health span, transition phase, and senescent span were studied. Treatment with SAHA during the transition or senescent spans resulted in decreased mortality rate and extended longevity compared to the control, while supplementation during the entire adult life span or during the health span only led to decreased longevity in the normal-lived strain. The analysis of mortality curves indicated that there were no significant effects of the SAHA administration until the age of ~50 days. When the long-lived strain was administered with SAHA by the same scheme, mostly deleterious effects were detected. Remarkably, the SAHA-treated normal-lived *D. melanogaster* strain showed late-life extending effects similar to those seen in the same study for SB. The fact that these two different HDACIs, SB and SAHA, had similar effects on mortality rate during the senescent span indicates the similarity of mechanisms that underlie beneficial effects for this class of HDACIs. The authors suggested that HDACIs may significantly influence the mortality rate throughout the senescent phase by reducing the vulnerability of treated individuals, in a manner similar to that of dietary restriction. Indeed, as was mentioned above, genetic alterations in genes encoding HDACs and nutrition regiments partially interact in the course of longevity control.[31,32] HDACIs may affect several pathways involved in regulating gene expression patterns associated with healthy aging. The induction of these patterns of gene expression throughout senescence when they are not normally present may likely underlie the life-extending effects of HDACIs.

21.4 HDACIs in Preclinical and Clinical Trials

A lot of hope in geroscience is currently being pinned on pharmacological compounds targeted to epigenetic regulators of gene expression. Epigenetic modifications are known to be potentially reversible; this feature makes them attractive targets for pharmacological intervention. Over the past few years, a series of medications have been developed targeted to epigenetic regulators, including modulators of HDACs, HATs, DNA methyltransferases, and noncoding miRNAs, with potential effects against various types of disorders.[69,70] Several modulators of HDAC activity (primarily, HDACIs), among other drugs targeting epigenetic machinery, have been recently examined in human clinical trials, and some have been proposed as promising anti-ageing drug candidates.[13] However, as most HDACIs lack specificity,[71] their wide applicability is still questionable. Serious efforts are currently aimed at finding class-selective and isoform-selective HDACIs.[72]

HDACIs are expected to have clinical potential in preventing and/or treating many chronic pathological conditions, including cancer, cardiovascular disorders, metabolic and neurodegenerative disorders such as Parkinson's, Alzheimer's and Huntington's diseases, violated immune response, inflammation and arthritis. The research findings supporting the therapeutic properties of HDACIs in curing age-related pathologies are reviewed in the sub-sections below.

21.4.1 Cancer

The onset and progression of various cancers involve substantial dysregulation of HDAC activity. The antitumor effects of HDACIs are suggested to be attributed to both transcriptional repression of proto-oncogenes and transcriptional reactivation of silent tumor suppressor genes.[73] Their effects may also be mediated by the regulation of DNA repair, inducing cell cycle arrest and apoptosis, inhibiting angiogenesis and long-term stimulation of immune response.[74] HDACIs are considered to be very promising candidates in cancer treatments since these agents preferentially kill neoplastic cells and are relatively non-toxic to normal cells,[75] though the molecular mechanisms of this selectivity remain to be elucidated.

A wide range of HDACIs are emerging as promising anticancer pharmaceuticals.[75–77] HDACIs such as belinostat, panobinostat, SAHA and FK228,[72] as well as TSA, sodium butyrate, vorinostat, valproic acid and romidepsinor,[78] showed substantial activity in both haematological and solid tumors in different tissues. In the last few years, HDACIs have undergone a rapid phase of clinical development in different cancer types, either as monotherapy or combined with other anticancer modalities. To date, three HDACIs have been approved by the FDA for the treatment of cutaneous/peripheral T-cell lymphoma,[79] and four HDACIs, namely vorinostat, belinostat, romidepsin and panobinostat, have been approved by the FDA for the treatment of hematologic cancers.[80] Many other HDACIs are at different stages of clinical development for the treatment of hematological malignancies and solid tumors.[79]

21.4.2 Metabolic and Cardiovascular Pathology

The FDA's approval of HDACIs as anticancer agents has provided the motivation for using these medicines as treatment options for non-malignant diseases. The beneficial outcomes of HDAC inhibition were obtained in treatment of various types of inflammatory, neurodegenerative and cardiovascular disorders.[80] In particular, experimental evidence has indicated that inhibitors of Class I HDACs can attenuate the development of cardiac hypertrophy and preserve cardiac function in several small animal models.[81] In addition, HDACIs have been found to be beneficial in preventing myocardial infarction, hypertension, atherosclerosis, vascular calcification, supraventricular arrhythmia, cardiac remodeling, fibrosis, and neointima formation.[82] The putative mechanisms mediating beneficial effects of HDACIs on the heart function include suppression of oxidative stress and inflammation, enhancement of cardiac protein aggregate clearance and autophagic flux, as well as inhibition of MAP kinase signaling.[83] In addition, since HDACIs were reported to promote β-cell proliferation, differentiation and function, and positively affect late diabetic microvascular complications, HDAC inhibition was proposed as a novel treatment strategy for type 2 diabetes.[84]

21.4.3 Neurodegenerative Diseases

Accumulating evidence indicates that histone acetylation plays a crucial role in the etiology of neurodegenerative disorders. Several recent studies have highlighted the importance of HDACs and modifications of histone acetylation in Alzheimer's disease,[85] neuronal memory, learning, synaptic plasticity and neural regeneration.[86] This is not surprising since neurodegenerative diseases such as amyotrophic lateral sclerosis, polyglutamine-related diseases, as well as Parkinson's and Alzheimer's disease are known to be accompanied by transcriptional dysfunctions, leading to neuronal death.[87]

HDACIs show great promise to combat ageing-associated neurodegenerative diseases[70,88–90] and to ameliorate the symptoms of cognitive decline, post-traumatic stress disorder and depression.[91] Some studies demonstrate that HDACIs may be neuroprotective by regulating memory and synaptic dysfunctions in both *in vitro* and *in vivo* models of this pathology.[85] HDACIs were also reported to cause beneficial effects in both *in vitro* and *in vivo* models of Parkinson's disease. For example, a HDAC1/2 isoform-specific inhibitor, K560, was recently found to protect against pharmacologically induced neuronal death and to mitigate experimental Parkinson's disease in both *in vitro* and *in vivo* models of this disease.[92] Other HDACIs, such as SAHA and SB, were shown to improve memory function in the mouse model of Alzheimer's disease.[93,94] The potential mechanisms underlying these effects include maintenance of histone acetylation homeostasis and transcriptional activation of neuronal survival genes.[95] In the last few years, clinical trials have been initiated to examine the effectiveness of HDACIs in patients with Parkinson's disease. The loss of functional activity of HATs is likely a common mechanism related to the impairment of the chromatin acetylation status throughout the lifetime of neurons. The therapeutic potential of HAT activators in the treatment of neurodegenerative disorders has been established in preclinical studies. Substantial neuroprotective properties were revealed for one of the HATs termed cAMP response element binding protein (CREB)-binding protein (CBP), and also for several other HATs that were shown to be essential for processes of neuronal plasticity and memory formation.[87]

21.4.4 Inflammatory Disorders

Non-specific HDACIs also demonstrated anti-inflammatory effects in both *in vitro* and *in vivo* models.[96,97] Recently, evidence was obtained for the role of the NF-κB signal transduction pathway in mediating the effects of HDACIs on inflammatory responses.[98] The important point is that such effects were reported at concentrations that were 10–100-fold lower than the concentrations required for the anti-cancer effects of these compounds. Clinical application of these substances for treating inflammatory diseases is, however, hampered due to their low specificity and a wide variety of HDACs that they affect throughout the body.[99]

21.5 Conclusion

Environmental, life style and genetic interventions have clearly proven to be effective in prolonging life span in experimental animals.[100] Modifications in the epigenome are currently supposed to largely underlie the life-extending interventions.[101,102] Epigenetic changes that are associated with senescence may be slowed down or even reverted by several life style factors, such as diet or exercise. A variety of compounds, both chemical and natural, including HDACIs, also demonstrated beneficial effects on aging and longevity presumably due to the fine-tuning of epigenetic regulation.[103] Importantly, very accurate and precise fine-tuning is crucial for this purpose since any unbalancing in HDAC activity, similarly to unbalanced consumption of vitamins, antioxidants, or hormones, may result in disruption of mechanisms controlling homeostasis. In a living organism, the regulation of transcriptional and metabolic networks is a highly coordinated and orchestrated process affecting vital complex traits, such as development, reproduction, survival, and aging. Epigenetic regulation appears to be one of the central players modulating these traits, aging among others.[48,104] In this context, nonspecific HDACIs, which may potentially influence the expression of thousands of genes, including those that are involved in aging, can prove to be quite effective for further anti-aging treatments. For the same reason, the use of HDACIs requires caution. A reasonable compromise may be achieved by the use of tissue-, stage-, and HDAC-specific inhibitors, and serious efforts are currently aimed at finding these HDACIs.[105]

Acknowledgements

The authors would like to thank Oksana Zabuga for assistance in preparing the manuscript. This work was supported by a joint grant of the Russian Foundation for Basic Research and the State Fund for Fundamental Research of Ukraine (EGP and AV, project no. 11-04-90478-Ukr_f_a) and State budget of Russian Federation (EGP).

References

1. L. J. Dominguez, M. Barbagallo and J. E. Morley, *Aging Male*, 2009, **12**, 13.
2. V. K. Kapoor, J. R. Dureja and R. Chadha, *Drug Discovery Today*, 2009, **14**, 899.
3. U. Muñoz-Najar and J. M. Sedivy, *Antioxid. Redox Signaling*, 2011, **14**, 241.
4. J. D. Boyd-Kirkup, C. D. Green, G. Wu, D. Wang and J. D. Han, *Epigenomics*, 2013, **5**, 205.
5. J. G. Wood and S. L. Helfand, *Front. Genet.*, 2013, **4**, 274.
6. J. P. Issa, *Crit. Rev. Oncol. Hematol.*, 1999, **32**, 31.

7. M. Berdasco and M. Esteller, *Aging Cell*, 2012, **11**, 181.
8. A. Brunet and S. L. Berger, *J. Gerontol., Ser. A*, 2014, **69**(suppl. 1), S17.
9. R. Cacabelos and C. Torrellas, *Int. J. Mol. Sci.*, 2015, **16**, 30483.
10. A. Vaiserman, *Clin. Epigenet.*, 2015, **7**, 96.
11. T. O. Tollefsbol, *Cancer Treat. Res.*, 2014, **159**, 257.
12. M. New, H. Olzscha and N. B. La Thangue, *Mol. Oncol.*, 2012, **6**, 637.
13. A. M. Vaiserman and E. G. Pasyukova, *Front. Genet.*, 2012, **3**, 224.
14. R. Benedetti, M. Conte and L. Altucci, *Antioxid. Redox Signaling*, 2014, **23**, 99.
15. H. Olzscha, S. Sheikh and N. B. La Thangue, *Crit. Rev. Oncog.*, 2015, **20**, 1.
16. E. Rockenstein, L. Crews and E. Masliah, *Adv. Drug Delivery Rev.*, 2007, **59**, 1093.
17. U. B. Pandey and C. D. Nichols, *Pharmacol. Rev.*, 2011, **63**, 411.
18. M. Konsolaki, *Expert Opin. Drug Discovery*, 2013, **8**, 1503.
19. M. Lucanic, G. J. Lithgow and S. Alavez, *Ageing Res. Rev.*, 2013, **12**, 445.
20. D. S. Dwyer, E. Aamodt, D. Cohen and E. A. Buttner, *Front. Pharmacol.*, 2014, **5**, 177.
21. R. A. Kobet, X. Pan, B. Zhang, S. C. Pak, A. S. Asch and M. H. Lee, *Biomol. Ther.*, 2014, **2**, 371.
22. C. Huidobro, A. F. Fernandez and M. F. Fraga, *Mol. Aspects Med.*, 2013, **34**, 765.
23. T. Kouzarides, *Cell*, 2007, **128**, 693.
24. G. E. Zentner and S. Henikoff, *Nat. Struct. Mol. Biol.*, 2013, **3**, 259.
25. Q. Zhou, C. L. Dalgard, C. Wynder and M. L. Doughty, *BMC Neurosci.*, 2011, **12**, 50.
26. B. Villeponteau, *Exp. Gerontol.*, 1997, **32**, 383.
27. J. G. Wood, S. Hillenmeyer, C. Lawrence, C. Chang, S. Hoiser, W. Lightfoot, E. Mukherjee, N. Jiang, C. Schorl, A. S. Brodsky, N. Neretti and S. L. Helfand, *Aging Cell*, 2010, **9**, 971.
28. K. Larson, S.-J. Yan, A. Tsurumi, J. Liu, L. Zhou, K. Gaur, D. Guo, T. H. Eickbush and W. X. Li, *PLoS Genet.*, 2012, **8**, e1002473.
29. B. Rogina, S. L. Helfand and S. Frankel, *Science*, 2002, **298**, 1745.
30. S. Frankel, T. Ziafazeli and B. Rogina, *Exp. Gerontol.*, 2011, **46**, 391.
31. B. Rogina and S. L. Helfand, *Proc. Natl. Acad. Sci. U. S. A.*, 2004, **101**, 15998.
32. S. Frankel, J. Woods, T. Ziafazeli and B. Rogina, *Aging (Albany NY)*, 2015, **7**, 1112.
33. M. H. Kuo and C. D. Allis, *Bioessays*, 1998, **20**, 615.
34. O. Witt, H. E. Deubzer, T. Milde and J. Oehme, *Cancer Lett.*, 2009, **277**, 8.
35. J. R. Davie and V. A. Spencer, *J. Cell Biochem. Suppl.*, 1999, **32–33**, 141.
36. H. P. Chen, Y. T. Zhao and T. C. Zhao, *Crit. Rev. Oncog.*, 2015, **20**, 35.
37. M. A. Glozak, N. Sengupta, X. Zhang and E. Seto, *Gene*, 2005, **363**, 15.
38. S. Imai, C. M. Armstrong, M. Kaeberlein and L. Guarente, *Nature*, 2000, **403**, 795.
39. D. D. Leipe and D. Landsman, *Nucleic Acids Res.*, 1997, **25**, 3693.

40. K. T. Chang and K. T. Min, *Ageing Res. Rev.*, 2002, **1**, 313.
41. L. Gao, M. A. Cueto, F. Asselbergs and P. Atadia, *J. Biol. Chem.*, 2002, **277**, 25748.
42. L. Seroude, T. Brummel, P. Kapahi and S. Benzer, *Aging Cell*, 2002, **1**, 47.
43. N. Kourtis and N. Tavernarakis, *EMBO J.*, 2011, **30**, 2520.
44. K. C. Lakshmaiah, L. A. Jacob, S. Aparna, D. Lokanatha and S. C. Saldanha, *J. Cancer Res. Ther.*, 2014, **10**, 469.
45. Y. Cho, A. Griswold, C. Campbell and K. T. Min, *Genomics*, 2005, **86**, 606.
46. T. Iannitti and B. Palmieri, *Drugs R&D*, 2011, **11**, 227.
47. H.-L. Kang, S. Benzer and K. T. Min, *Proc. Natl. Acad. Sci. U. S. A.*, 2002, **99**, 838.
48. A. Vaiserman, *Ageing Res. Rev.*, 2011, **10**, 413.
49. E. Buommino, D. Pasquali, A. A. Sinisi, A. Bellastalla, F. Morelli and S. Metafora, *J. Mol. Endocrinol.*, 2000, **24**, 83.
50. S. Khan and G. B. Jena, *Chem.-Biol. Interact.*, 2014, **213**, 1.
51. Y. Zhao, H. Sun, J. Lu, X. Li, X. Chen, D. Tao, W. Huang and B. Huang, *J. Exp. Biol.*, 2005, **208**(4), 697.
52. A. V. Vaiserman, A. K. Koliada, N. M. Koshel, A. V. Simonenko and E. G. Pasyukova, *Adv. Gerontol.*, 2012, **25**, 126[In Russian].
53. A. M. Vaiserman, N. M. Koshel, O. G. Zabuga, A. K. Kolyada, N. V. Roshina and E. G. Pasyukova, *Adv. Gerontol.*, 2013, **26**, 111[In Russian].
54. R. St Laurent, L. M. O'Brien and S. T. Ahmad, *Neuroscience*, 2013, **246**, 382.
55. P. McDonald, B. M. Maizi and R. Arking, *Exp. Gerontol.*, 2013, **48**, 240.
56. A. L. Ridgel and R. E. Ritzmann, *Ageing Res. Rev.*, 2005, **4**, 23.
57. Y. Zhao, J. Lu, H. Sun, X. Chen, W. Huang, D. Tao and B. Huang, *Biochem. Biophys. Res. Commun.*, 2005, **326**, 811.
58. Y. Zhao, X. Chen, H. Sun, Z. G. Yuan, G. L. Ren, X. X. Li, J. Lu and B. Huang, *Cell Res.*, 2006, **16**, 566.
59. Y. Zhao, J. Lu, H. Sun and B. Huang, *Mol. Cell. Biochem.*, 2007, **306**, 1.
60. T. Chen, H. Sun, J. Lu, Y. Zhao, D. Tao, X. Li and B. Huang, *Arch. Biochem. Biophys.*, 2002, **408**, 171.
61. R. Arking, J. Novoseltseva, D. S. Hwangbo, V. Novoseltsev and M. Lane, *J. Gerontol., Ser. A*, 2002, **57**, B390.
62. R. Arking, in *Life Span Extension: Single Cell Organisms to Man*, ed. C. Sell, A. Lorenzini and H.M. Brown-Borg, Humana Press, Springer, Dordrecht, 2009, p. 59.
63. R. Arking, *Invertebr. Reprod. Dev.*, 2015, **59**(suppl. 1), 28.
64. C. Edwards, J. Canfield, N. Copes, V. Rehan, D. Lipps and P. C. Bradshaw, *Aging (Albany NY)*, 2014, **6**, 621.
65. T. Vanhaecke, P. Papeleu, G. Elaut and V. Roqiers, *Curr. Med. Chem.*, 2004, **11**, 1629.
66. D. Tao, J. Lu, H. Sun, Y. Zhao, Z. G. Yuan, X. Li and B. Huang, *Acta Biochim. Biophys. Sin.*, 2004, **36**, 618.
67. D. Yin, J. M. Ong, J. Hu, J. C. Desmond, N. Kawamata, B. M. Rknda, K. L. Black and H. P. Koeffler, *Clin. Cancer Res.*, 2007, **3**, 1045.

68. J. S. Steffan, L. Bodai, J. Pallos, M. Poelman, A. McCampbell, B. L. Apostol, A. Kazantsev, E. Schmidt, Y. Z. Zhu, M. Greenwald, R. Kurokawa, D. E. Housman, G. R. Jackson, J. L. Marsh and L. M. Thompson, *Nature*, 2001, **413**, 739.
69. A. O. Arguelles, S. Meruvu, J. D. Bowman and M. Choudhury, *Drug Discovery Today*, 2015, **21**, 499.
70. R. Cacabelos, *Drug Dev. Res.*, 2014, **75**, 348.
71. S. K. Pirooznia and F. Elefant, *Front. Cell. Neurosci.*, 2013, **7**, 30.
72. N. Ma, Y. Luo, Y. Wang, C. Liao, W. C. Ye and S. Jiang, *Curr. Top. Med. Chem.*, 2016, **16**, 415.
73. Y. Boumber and J. P. Issa, *Oncology (Williston Park)*, 2011, **25**(220), 228.
74. G. R. Leggatt and B. Gabrielli, *Immunol. Cell Biol.*, 2012, **90**, 33.
75. R. W. Johnstone, *Nat. Rev. Drug Discovery*, 2002, **1**, 287.
76. M. Slingerland, H. J. Guchelaar and H. Gelderblom, *Anticancer Drugs*, 2014, **25**, 140.
77. A. C. West and R. W. Johnstone, *J. Clin. Invest.*, 2014, **124**, 30.
78. P. Chun, *Arch. Pharmacal Res.*, 2015, **38**, 933.
79. M. Mottamal, S. Zheng, T. L. Huang and G. Wang, *Molecules*, 2015, **20**, 3898.
80. S. Yoon and G. H. Eom, *Chonnam Med. J.*, 2016, **52**, 1.
81. A. Abend and I. Kehat, *Pharmacol. Ther.*, 2015, **147**, 55.
82. G. H. Eom and H. Kook, *Pharmacol. Ther.*, 2014, **143**, 168.
83. B. S. Ferguson and T. A. McKinsey, *J. Mol. Cell. Cardiol.*, 2016, **83**, 14.
84. D. P. Christensen, M. Dahllöf, M. Lundh, D. N. Rasmussen, M. D. Nielsen, N. Billestrup, L. G. Grunnet and T. Mandrup-Poulsen, *Mol. Med.*, 2011, **17**, 378.
85. X. Lu, L. Wang, C. Yu, D. Yu and G. Yu, *Front. Cell. Neurosci.*, 2015, **9**, 226.
86. S. A. Ganai, M. Ramadoss and V. Mahadevan, *Curr. Neuropharmacol.*, 2016, **14**, 55.
87. B. R. Selvi, J. C. Cassel, T. K. Kundu and A. L. Boutillier, *Biochim. Biophys. Acta*, 2010, **1799**, 840.
88. T. Abel and R. S. Zukin, *Curr. Opin. Pharmacol.*, 2008, **8**, 57.
89. E. Hahnen, J. Hauke, C. Trankle, I. Y. Eyupoglu, B. Wirth and I. Blumcke, *Expert Opin. Invest. Drugs*, 2008, **17**, 169.
90. A. Mai, D. Rotili, S. Valente and A. G. Kazantsev, *Curr. Pharm. Des.*, 2009, **15**, 3940.
91. J. Penney and L. H. Tsai, *Sci. Signaling*, 2014, **7**, re12.
92. C. J. Choong, T. Sasaki, H. Hayakawa, T. Yasuda, K. Baba, Y. Hirata, S. Uesato and H. Mochizuki, *Neurobiol. Aging*, 2016, **37**, 103.
93. N. Govindarajan, R. C. Agis-Balboa, J. Walter, F. Sananbenesi and A. Fischer, *J. Alzheimer's Dis.*, 2011, **26**, 187.
94. E. Benito, H. Urbanke, B. Ramachandran, J. Barth, R. Halder, A. Awasthi, G. Jain, V. Capece, S. Burkhardt, M. Navarro-Sala, S. Nagarajan, A. L. Schütz, S. A. Johnsen, S. Bonn, R. Lührmann, C. Dean and A. Fischer, *J. Clin. Invest.*, 2015, **125**, 3572.
95. S. Sharma and R. Taliyan, *Parkinson's Dis.*, 2015, **2015**, 303294.

96. P. V. Licciardi, K. Ververis, M. L. Tang, A. El-Osta and T. C. Karagiannis, *Curr. Mol. Med.*, 2013, **13**, 640.
97. S. Schotterl, H. Brennenstuhl and U. Naumann, *Crit. Rev. Oncog.*, 2015, **20**, 139.
98. F. J. Dekker, T. van den Bosch and N. I. Martin, *Drug Discovery Today*, 2014, **19**, 654.
99. M. D. Cantley and D. R. Haynes, *Inflammopharmacology*, 2013, **21**, 301.
100. J. Jylhava, *Curr. Pharm. Des.*, 2014, **20**, 6058.
101. A. M. Vaiserman, *Rejuvenation Res.*, 2008, **11**, 39.
102. T. A. Rando and H. Y. Chang, *Cell*, 2012, **148**, 46.
103. F. Uchiumi, T. Oyama and K. Ozaki, *Pharm. Anal. Acta*, 2012, **3**, 7.
104. M. G. Bacalini, S. Friso, F. Olivieri, C. Parazzini, C. Giuliani, M. Capri, A. Santoro, C. Franceschi and P. Garagnani, *Mech. Ageing Dev.*, 2014, **136–137**, 101.
105. P. Gupta, R. C. Reid, A. Iyer, M. J. Sweet and D. P. Fairlie, *Curr. Top. Med. Chem.*, 2012, **12**, 1479.
106. N. V. Roshina, A. V. Simonenko, A. V. Krementsova, M. V. Trostnikov and E. G. Pasyukova, *Aging (Albany NY)*, 2014, **6**, 1076.

Section VI

Social Context

CHAPTER 22

Human Life Extension: Opportunities, Challenges, and Implications for Public Health Policy

ILIA STAMBLER[a]

[a]Department of Science, Technology and Society, Bar Ilan University, Ramat Gan 52900, Israel
*E-mail: ilia.stambler@gmail.com

22.1 Introduction: The Diverse Aspects of Life Extension Promotion as a Part of Health Promotion

The prospect of human life extension, understood either as a substantial increase in life expectancy or an increase in the human species-specific lifespan may have a profound and wide ranging impact on science and society. The issues that are involved may be highly complex and diverse, yet many of them may be classified into a limited number of categories. These would basically concern: (1) the scientific and technological feasibility of human life extension; (2) its individual and social desirability; and (3) normative actionable suggestions for research and public policy, building on and synthesizing

RSC Drug Discovery Series No. 57
Anti-aging Drugs: From Basic Research to Clinical Practice
Edited by Alexander M. Vaiserman

Published by the Royal Society of Chemistry, www.rsc.org

the first two groups, for the facilitation of potential beneficial scenarios and mitigation of negative scenarios related to human life extension.

The first group of questions concerning the *feasibility* of the accomplishment of life extension inquires: is it even theoretically and technologically possible? Is not the human lifespan a strongly genetically and evolutionary determined feature and thus any discussions about its possible significant modifications and improvements would be meaningless or even unscientific? How can a significant human life extension be even practically verifiable, given the large time periods needed to test and ascertain truly life-extending interventions? Or conversely, is human lifespan modifiable to a significant and verifiable degree? If yes, to what degree? What dominant research paradigms would be most productive for achieving this goal?[1] What technological, empirical or experimental approaches are most promising toward promoting human life extension, for example, reductionist *vs.* holistic,[2] invasive *vs.* minimally-invasive, therapeutic *vs.* preventive, *etc.*?[3,4] How can these diverse and numerous trends be identified, extrapolated, classified and rated? What key factors and analytic tools should be employed? What research priorities should be chosen? How will human life extension affect the incidence and severity of aging-related and life course-related diseases and disabilities? These questions are vastly complex. The current work cannot presume even to begin to comprehensively posit, let alone answer them. Many of those scientific questions have been raised elsewhere. The current chapter will only endeavor to present a brief overview, in the "Feasibility" section.

The second group of questions and challenges concerns the *desirability* of the accomplishment of a significant life extension for the individual and for the society, provided it will some day become possible through scientific intervention. How will the possibility of a significant human life extension affect the current perceptions of life-sustaining treatments?[5] Will such treatments still be considered as morally imperative, if they could be applied for prolonged periods of time? Will significant life extension lead to individual and social stagnation and perpetuation of undesirable psychological and communal patterns?[6,7] Will it lead to insufficient and inequitable distribution of resources, which can be harmful for social sustainability and stability as a whole, even though possibly beneficial for some individuals or select groups of individuals? Based on the existing historical and demographic trends, and plausible extrapolations and forecasts, how empirically grounded are the fears of shortage of resources and inequality due to life extension, or are they purely theoretical *a priori* constructs? Are the existing trends likely to continue or be amended?[8,9] What scenarios related to life extension are more likely—the dystopian (or nearly dystopian) or utopian (or nearly utopian)? These questions will be the focus of the "Desirability" section.

This work will provide a brief overview of the first two groups of questions. Those questions have been relatively widely debated in the past. Yet the thrust of this work will be mainly on the "normative" group of questions and challenges, as probably the most practically significant and urgent for a broad public, academic and political debate. What should we do, given the possibility and desirability of a significant human life extension? If human life extension is scientifically possible, how can this possibility be realized? If it is desirable, how can this desire be fulfilled? Should we patiently wait until effective life-extending therapies become conveniently available and affordable, or should we actively endeavor to make them available and affordable as soon as possible? How? Or conversely, if it is neither possible nor desirable, should we abandon or even ban this pursuit?

Generally, this work will support the position that significant human life extension is in principle possible and desirable, for the individual and for the society. The argument would be quite straightforward. Indeed, it may be argued that it is generally desirable and demonstrably possible to live a long and healthy life. The real questions arise in the normative and specific domain: What actions exactly should be taken to become "healthy, wealthy, wise (and long-lived)"? Who should undertake those actions? And who will enjoy the results of those actions? Could there be undesirable side effects to those actions? In case of a likely negative scenario, what should be done to prevent it? In case of a likely positive scenario, how could it be achieved sooner and for a greater number of people? In summary, assuming that significant human life extension is scientifically possible and socially desirable, and that its implications are either demonstrably positive or, in case of a negative forecast, they may be corrected—what practical implications should these determinations have for individual actions and public policy in particular social environments?

The above normative questions translate into specific and urgent questions and challenges for public health and research policy. Should, for example, more funds be dedicated to life extension research? In what way should such funds be designated and appropriated? In what ways should healthcare, life insurance and pension systems be made ready and abiding for a significant increase in life expectancy or extended lifespan? What should be the role of the pharmaceutical, biotechnological and medical technology industries? What will be the say of the scientific and humanitarian academic institutions, public associations, political organizations? Furthermore, given the significance of the deteriorative aging process for the development of chronic aging-related diseases, should medical research and treatment be reoriented to emphasize this significance?[10,11] Should a greater emphasis be given to basic, empirical, applied, engineering, environmental, or other approaches to human life extension? How can we ensure that life extending technologies will also be healthspan extending technologies? What should be the regulatory frameworks for potential life-extending, healthspan-extending and anti-aging treatments that would on the one hand prevent the emergence of unjustified claims, and on the other provide incentives for rapid and widely available beneficial therapeutic developments? Furthermore, in case technologies for healthy life extension are developed, how can society ensure their wide and equitable distribution? How could the practically inevitable gaps in access to life-extending technologies be bridged, within a particular society or between societies? What social institutions will make these decisions and how will they enact them?

Given the rapid population aging and the increasing incidence and burden of aging-related diseases, on the pessimistic side, and the rapid development of medical technologies, on the optimistic side, these become critical social challenges and vital questions of social responsibility. This work will generally argue for the urgent adoption of policies designed to promote biomedical research of aging and aging-related diseases to improve the health and longevity of the elderly population, including increased funding, incentives and institutional support for such research.[12] Yet various specific ramifications, concerns and challenges of such policies will be discussed in greater detail in the "Normative action" section.

22.2 Scientific and Technological Implications and Challenges: Is Human Life Extension Scientifically and Technologically Feasible?

The main question in this discussion can be summarized as follows: Is a significant human lifespan extension feasible? Should not we consider it as just a "pipe dream," an instance of "wishful thinking," a "science fiction" scenario or some other form of entertaining but impractical fantasy? An answer to this question also largely determines the answers to the questions of desirability and normative action. (After all, why desire or pursue something that is impossible?) Yet, theoretical considerations and empirical evidence indicate that the prospect is at least in principle feasible. Insofar as life is a material and pliable process, it can in principle be manipulated, either in the direction of shortening or lengthening. Similarly, the aging process, the main limiting factor for the human lifespan, is also material and pliable, and thus can be aggravated or ameliorated. Ameliorating the aging process appears to be a necessary condition for extending human lifespan. As the German hygienist Christoph Wilhelm Hufeland (1762–1836) contended in his book *Macrobiotics or the Art of Prolonging Human Life* (1796), the art of life prolongation mainly consists of avoiding "things that shorten life," and "endeavoring to avoid or remove the causes of disease."[13] The degenerative aging process is evidently the main "thing that shortens life" and the main "cause of disease." Thus, logically, we should "endeavor to avoid or remove" it first and foremost. There is indeed a growing consensus that degenerative aging is the main risk factor or even the main underlying root cause for most age-related diseases, such as cancer, type 2 diabetes, heart disease, neurodegenerative diseases, chronic obstructive pulmonary disease, and also a major aggravating factor for communicable infectious diseases. Hence, in order to truly effectively combat these diseases, to achieve healthy longevity, there is an essential need to treat the "root" of the problem, rather than to attempt to treat the various "branches" (particular diseases) or "leaves" (symptoms) separately.[10–12] (Still, the exact quantitative weight of aging as a non-specific risk factor or root cause of diseases needs to be elaborated on large sample sizes.[14,15])

Yet, this appreciation of the critical role of aging in pathogenesis, and the proactive therapeutic approach to aging that derives from it, seem not yet to have become a common knowledge and common practice. Thus, even such authoritative studies as the Global Burden of Disease (GBD), even though generally acknowledging that the incidence of chronic diseases increases with age, do not consider aging as a risk factor, let alone as a diagnosable and treatable medical condition (though these studies include such risk factors as "injuries by pedal cycle vehicles").[16] Generally, the methodology for estimating risk factors appears to be flawed. Thus in the GBD assessments, the risk of death from various factors can exceed hundreds of percent (when in fact it should be no more than 100%). The inability to quantitatively estimate the weight of aging as a contributing factor to diseases may be a part of the methodological challenge. Yet, critically, increasing the recognition of degenerative aging as a source of disease may be a major societal and perceptual challenge that may be a prerequisite for any significant increase in research and development directed toward human life extension.

The fact that degenerative aging is not considered by many as problematic will be addressed in greater detail in the following section on the "desirability" of life extension. Yet, the unwillingness to perceive it as a medical problem may largely derive from the fact that aging is often perceived as unyielding and "natural." To rephrase the old saying, "when there is no way [to ameliorate aging], there is no wish [to treat it]." Are the aging processes really modifiable? An answer is made difficult by the fact that there is still no commonly accepted clinical definition of aging, or a commonly shared theory of aging,[17] or commonly accepted criteria to evaluate changes in this process or the effectiveness of interventions into it.[18,19] The World Health Organization (WHO) in its *World Report on Ageing and Health* (October 1, 2015)[20] provides the following definition of aging (the original references are included in the quote):

"The changes that constitute and influence ageing are complex.[21] At a biological level, ageing is associated with the gradual accumulation of a wide variety of molecular and cellular damage.[22,23] Over time, this damage leads to a gradual decrease in physiological reserves, an increased risk of many diseases, and a general decline in the capacity of the individual. Ultimately, it will result in death. But these changes are neither linear nor consistent, and they are only loosely associated with age in years."

Furthermore, according to the WHO's report, "healthy ageing" is determined by "intrinsic capacity", which is (somewhat vaguely) defined as "the composite of all the physical and mental capacities that an individual can draw on" and that needs to be "improved." Such general definitions may fit a large number of working programs on aging. Yet, there seems to be little agreement in selecting the main "culprits of senescence" or evidence-based methods of its diagnosis and intervention into it.

The WHO is beginning to recognize the methodological and terminological challenge. Thus the WHO's *Global Strategy and Action Plan on Ageing and Health (GSAP)—2016–2020* (November 2015) includes "Strategic objective 5: Improving measurement, monitoring and research on Healthy Ageing," with a clause: "5.1: Agree on ways to measure, analyse, describe and monitor Healthy Ageing." The WHO desires that a "consensus should be reached on common terminology and on which metrics, biological or other markers, data collection measures and reporting approaches are most appropriate" (GSAP, Section 95).[24] The WHO even appears to be beginning to recognize the modifiable nature of aging, and the need to increase research in the field. Thus according to GSAP (Section 105), "finally, better clinical research is urgently needed on the etiology of, and treatments for, the key health conditions of older age... This could also be extended to include possible interventions to modify the underlying physiological and psychological changes associated with ageing." Still, this emerging recognition has not yet reached the level of practicable clinical and research guidelines, apparently in a large measure due to the deficit of agreement on the main definitions, criteria and targets for intervention.

Yet several classification systems have been proposed that prioritize several sets of basic aging processes as the main determinants for various sets of aging-related diseases. Accordingly, by intervening into these specific basic aging processes by specific types of biomedical technologies, it is hoped to reduce the general ill health due to aging and enhance the healthy lifespan. Crucially, there are already tangible proofs of principle for the possibility of intervention and modification of these processes, and the corresponding modifications of the lifespan. Several such examples of the arrays of presumed major determinants and countermeasures of aging can be cited. For example, in the SENS program (Strategies for Engineered Negligible Senescence), the "seven deadly things" with their corresponding countermeasures ("the seven vital virtues"?) include:[25] (1) *Damage from cell loss and tissue atrophy*. That is hoped to be countered by adding stem cells and tissue engineering (RepleniSENS). Of course, the effort to replenish aging cells and tissues is not a prerogative of the SENS approach. This is in fact the subject of the entire field of "regenerative medicine."[26,27] If perfected, this could become an effective preventative aid for virtually all age-related diseases, and there are already proofs of principles for a variety of therapies, from diabetes (replenishment of beta-cells)[28] through neurodegenerative diseases (enhancing neurogenesis)[29] and retinal diseases (retinal cell regeneration)[30] to heart disease (cardiomyocytes replenishment)[31] to infectious diseases (regeneration of the thymus and replenishment of naïve T cells).[32,33] (2) *Nuclear (epi-) mutations leading to cancer*. These should be neutralized by the removal of telomere-lengthening machinery (OncoSENS). This is still a very hypothetical branch of the SENS program. On the contrary, a large number of studies have worked on enhancing "telomere lengthening machinery" to improve tissue regeneration, with some promising results.[34,35] (3) *Mutant mitochondria*. These should be backed up by allotopic expression of 13 proteins in the nucleus (MitoSENS). There are advances in this area as well, together with various other means to address aging-related diseases of the mitochondria.[36,37] (4) *Death-resistant cells*. These should be removed by targeted ablation (ApoptoSENS). This is also not an exclusive SENS prerogative. The promising work with "senolytics"—or pharmacological substances capable of removing senescent cells, thereby reducing cell senescence-related pathologies—as well as various genetic engineering and immunological means to eliminate senescent cells, have been advancing.[38,39] Analogous approaches have been used against "death resistant" cancer cells.[40] (5) *Tissue stiffening*. That is intended to be prevented by compounds breaking Advanced Glycation End-products—the "AGE-breakers" (GlycoSENS)—and by tissue engineering. Such means could ameliorate heart disease and other forms of sclerosis, and there have been considerable advances in the development of the AGE-breakers.[41] Though with this approach, as probably with all the other interventions, caution is advised.[42] (6) *Extracellular aggregates*. These are to be cleaned up by immunotherapeutic clearance (AmyloSENS). Amyloid clearance technologies have been developing as well, though their clinical benefits are yet to be ascertained.[43,44] (7) *Intracellular aggregates.* These should be dissolved by novel lysosomal hydrolases (LysoSENS). Special emphasis in current research has been made on stimulation of autophagy, the mechanism for lysosomal degradation of dysfunctional cellular components.[45]

In other classification systems, the culprits and countermeasures are somewhat different, though there are considerable overlaps between the sets of basic aging processes and their countermeasures in the different systems. For example, at the 2013 US National Institutes of Health (NIH) Geroscience Summit, the following seven priority research areas have been identified:[46,47] (1) *Adaptation to Stress*. In this area, probably the best known "universal anti-stress" system involves activation of Sirtuin proteins, acting to stabilize the DNA and improve energy metabolism.[48,49] (2) *Epigenetics*. Epigenetic rejuvenation, for example using demethylating agents, small interfering RNAs or micronutrients, either for the entire organism or for individual tissues, has also been a burgeoning field.[50,51] (3) *Inflammation*. Anti-inflammatory medications have been widely tested to diminish aging-related degenerative pathologies, such as neurodegenerative pathologies, and to extend lifespan in animal models.[52] But pro-inflammatory effects have also been shown to be important for tissue regeneration.[53] (4) *Macromolecular Damage*. This may include AGE-breaking means mentioned above.[41] This may also refer to various DNA repair enhancing means, with potential beneficial effects ranging from neurodegenerative diseases to cancer.[54] (5) *Metabolism*. This is a very broad area. With specific reference to energy metabolism, diverse means to improve mitochondrial function and cellular respiration are being developed, from pharmacological[55,56] through genetic engineering[37] to physical means.[57] Some of the aging-related molecular signaling pathways have been under special scrutiny, especially those modifiable by drugs, proving the possibility of pharmacological intervention into the aging process. Some of the "star" metabolic pathways have involved the mTOR enzyme (the Mechanistic Target of Rapamycin, modified by the drug Rapamycin)[58] and IGF-1 (Insulin Growth Factor 1, presumably modifiable by the drug metformin).[59] (6) *Proteostasis* or maintaining protein homeostasis mainly involves pharmacological and enzymatic means to maintain protein stability as well as a means to remove protein aggregates "clogging" cell machinery.[60] (7) *Stem cells/regeneration* is also present in this classification system, as likely a necessary component in any system endeavoring to address degenerative aging processes and a host of accompanying diseases.[61,62]

There are several other examples of similar classification approaches, prioritizing research of major sets of aging processes. Thus, another authoritative classification system enumerates nine tentative hallmarks that represent common denominators of aging in different organisms, with a special emphasis on mammalian aging.[63] These hallmarks include: (1) genomic instability, (2) telomere attrition, (3) epigenetic alterations, (4) loss of proteostasis, (5) deregulated nutrient sensing, (6) mitochondrial dysfunction, (7) cellular senescence, (8) stem cell exhaustion, (9) altered intercellular communication. Some further classification sets could be added, such as substance balances of various kinds, including nutrient, microelement, redox and pH balances, involving the phenomena of over-mineralization or demineralization, the effects of various dietary restrictions, rest and activity regimens, and more.[64,65] Furthermore, the anti-aging and life-extending processes and interventions do not necessarily need to be chemical or biological, but can also be physical, in particular as relates to various resuscitation technologies as applied to the elderly, such as hypothermia and suspended animation,[66] oxygenation,[67] electromagnetic stimulation.[68] But the general concept prevails, namely, that degenerative aging is determined by a limited and definable number of basic processes, mainly at the molecular and cellular level, and these processes may be treated, with an accumulating body of evidence proving this principle.[69]

Such classification systems tend to focus on the cellular and molecular levels, with relatively little attention paid to the systemic regulatory level of aging. Addressing the regulatory mechanisms may perhaps be the next challenge of life extension research and development.[15,70] That is to say, having created the necessary technological tools to tackle the basic molecular and cellular mechanisms of aging, it may then become necessary to learn to coordinate, dose and calibrate the use of those tools for the entire organism. Despite the challenges, the examples above showcase some of the existing proofs of practical technological feasibility of intervention into aging processes and lifespan modification. The primary kinds of evidence and sources of hope are: the successful cases of life extension experimentally achieved in animal models and the development of new intervention techniques, based on the ever better elucidation of the mechanisms of aging.

But there are more sources of hope, as well as potential scientific challenges. On the basic theoretical level, there is no law in nature that sets a strict insurmountable limit to the lifespan of any organism. This is demonstrated by the existence of non-aging and slowly aging life-forms and the constant evolutionary adaptations of the lifespan even for humans, according to particular changing environmental and genetic conditions.[71,72] Still, the practical applicability of those demonstrations is limited. Another source of hope is the persistent increase in life expectancy around the world, with the current increases in the "developing world" being much faster and larger than in the "developed world."[73] An ever increasing proportion of the life-expectancy rise is attributable to advances of biomedical technology, rather than mere hygiene.[74] Still, closing the gaps in life expectancy and in access to medical technologies, within particular societies and between societies, remains a grand challenge.

The very rapid development of biomedical technologies in itself constitutes probably the strongest source of hope mingled with concerns (mainly the concerns over safety, efficacy and availability). Consider the amount of progress made, for example, since the positing of the cellular theory of immunity by the founder of gerontology Elie Metchnikoff in 1882 until the beginning of the synthesis of the first prototypes of artificial immune cells recently.[75] Aging and longevity research has always been an integral part of this progress and, moreover, several important biomedical technologies and therapies, such as probiotic diets, hormone replacement therapy and cell therapy, were born out of aging and longevity research.[76] In fact, the recent progress in biomedical technology has been so vast and rapid that some authors spoke of "exponential acceleration" of technological development, due to technologies' convergence and cross-fertilization, improved communication and computational capabilities.[77] Yet, even with less optimistic and uncertain forecasts, assuming the speed of technological development to continue at least as fast as it was for the past century and a half, and at least for a comparable time in the future—we may expect dramatic improvements in biomedical technological capabilities and their distribution. Of course, reaching truly effective, safe and widely available anti-aging and life-extending capabilities, may still be a long way off. Their actual achievement, as well as their safety, efficacy and affordability, especially at the initial stages of application, may remain some of the main potential challenges. Still, the principal feasibility of a significant human life expectancy and lifespan extension by scientific and technological means appears to be evident. But do we want this extension, if it were possible? What would we use it for and who would use it? The scientific and technological feasibility assessment opens the door for the ethical desirability assessment.

22.3 Implications and Challenges for the Individual and the Society: Is Life Extension a Desirable Goal?

The question of ethical desirability essentially boils down to the question: Is healthy life extension a good thing? The question may seem rhetorical, and an answer to it obvious. Yet, quite surprisingly (at least for the proponents of healthy longevity), for decades and centuries, strong opposition has been expressed to the very idea of increasing longevity. The opposition has been frequent among philosophers, and even among physicians and researchers of aging. There has been a strong tendency among well-established physicians and scholars to consider aging as inexorable and therefore "normal," and to see the lifespan as fixed and immutable. Accordingly, any attempts to "meddle" with the aging process or to extend the lifespan would be considered foolish, futile and even somehow unethical.

For example, the British philosopher Thomas Robert Malthus (1766–1834), in *An Essay on the Principle of Population* (1798), expounded on the "fallacy" of the "conjecture concerning the organic perfectibility of man, and the indefinite prolongation of human life" (1798) and presumed "the very great additional weight that an increase in the duration of life would give to the argument of population [burden]."[78] In nineteenth century Germany, the "pessimistic philosophies" of Arthur Schopenhauer (*The World as Will and Idea,* 1844), Eduard Hartmann (*The Philosophy of the Unconscious*, 1870) and Philipp Mainländer (*The Philosophy of Redemption*, 1872) propounded on the saturation with life and called to abandon the pursuit of life prolongation. Quite coherently with the pessimistic philosophy, Mainländer committed suicide at the age of 34 (1841–1876).[1] In 1905, the renowned Canadian physician William Osler (1849–1919) spoke of the "uselessness of men above sixty years of age."[79] In a more recent period, the resistant attitude is exemplified, among many others, by the works of Morris Fishbein (1889–1976), the editor of the *Journal of the American Medical Association* (from 1924 to 1950) and a sworn enemy of all "quackery." In his books *The Medical Follies* (1925) and *The New Medical Follies* (1927), he encapsulated the therapeutic fashions in America at the beginning of the twentieth century, which were presented as quack or foolish. "Rejuvenation" or attempts to intervene into the aging processes were included among such follies. Fishbein believed aging to be utterly immutable: "the tissues of the senile can no more be rejuvenated than can the elasticity of a worn-out pair of suspenders." According to the unyielding nature of aging, he claimed, people "may now confidently look forward under all ordinary circumstances to reaching the age of fifty to fifty-five years"[80] and "there has been, however, but little average prolongation of life beyond the age of seventy, and there is not the slightest scientific reason to believe that there ever will be."[81] The British philosopher and mathematician Bertrand Russell (1872–1970) in "The Menace of Old Age" (1931) was greatly worried by the prospect that "every increase in medical skill is bound to make the world more and more conservative." Hence, he proposed "to prevent all researches calculated to prolong the life of the very old."[82] And moreover, in "How to grow old" (1944), he maintained that "in an old man... the fear of death is somewhat abject and ignoble."[83]

Later on, Norbert Wiener (1894–1964), the author of the theory of cybernetic regulation in living organisms and machines, was deeply concerned by the potential consequences of life extension: "Consoling as the suggestion may seem at first sight, it is in reality very terrifying, and above all for the doctors. For if one thing is clear, it is that humanity as such could not long survive the indefinite prolongation of all lives which come into being" (1964).[84] Further, the Australian immunologist, the Nobel Laureate in medicine of 1960 and the author of the "intrinsic mutagenesis" theory of aging (1974),

Frank Macfarlane Burnet (1899–1985) "doubt[ed] very much whether anything worthwhile would be gained by extending the human life span beyond its present bracket of 70 to 100 years—and that if we wanted this extension of life, I am deeply sceptical about our chance of ever achieving it" and furthermore, "death in the old should be accepted as something always inevitable and sometimes as positively desirable" (1974).[85] Leonard Hayflick (b. 1928), the author of the "cell division limit" theory of aging, claimed that "no intervention will slow, stop, or reverse the aging process in humans" (2004).[86] And even if it were somehow possible, "the problems created by having the power to arrest or even slow the aging process could be enormous and damaging to both the individual and society in general" (1994).[87] In 2001, the US Presidential advisor on bioethics Leon Kass stated that "the finitude of human life is a blessing for every human individual, whether he knows it or not."[88] Quite recently, in 2014, the oncologist and White House advisor for health policy Ezekiel Emanuel expressed the "hope to die at 75" and argued that "society and families—and you—will be better off if nature takes its course swiftly and promptly."[89] The list can continue. (For a more extensive overview, see Ilia Stambler, *A History of Life Extensionism in the Twentieth Century*, 2014.[1]) Of course, it is necessary to note that the pursuit of life extension has also been a persistent and highly respectable medical tradition, upheld, among others, by a founder of modern hygiene—Christoph Wilhelm Hufeland (1762–1836), the founder of therapeutic endocrinology—Charles-Édouard Brown-Séquard (1817–1894), the founder of geriatrics—Ignatz Leo Nascher (1863–1944), the founder of gerontology—Elie Metchnikoff (1845–1916) and many more.[1,90] Still, the stream opposed to the possibility and desirability of life extension has been strong and influential.

Often the opposing arguments distort the terms, arguing against "immortality" or "indefinite life extension" as a way to imply the undesirability of *any* significant lifespan or even healthspan extension, or the futility of the development of medical technologies for lifespan and healthspan extension. Usually, the arguments against extending longevity are quite standard. Here the word "often" may refer to virtually all the works cited above and many others. These arguments are also refutable in standard ways, as has been discussed in the relevant ethical literature.[5,6,91–93] Some of these "golden standards" are presented below, in the form of "Frequently Asked Questions" (FAQ). Indeed, almost any person, anywhere in the world, reflecting for a short time on the possibilities of human life extension, comes up with most of these concerns, and if reflecting or debating a little longer arrives at most of the refutations. The questions and answers below may provide a short summary of such debates.

Many people worry that extending longevity would prolong human suffering, tantamount to implying that death is a solution to suffering. For the proponents of healthy life extension, death can never be perceived as a solution to any inconvenience. Suffering is not inevitable. Human beings have the ability to actively influence their fate and alleviate their hardships. And essentially, the desire to extend life does not imply a desire to prolong anguish and disease, but a desire to prolong health and well being (increasing the healthspan).

It is often implied that extending longevity would lead to extending boredom. Proponents of life extension argue to the opposite. For them, extended life also implies extended ability to learn and change. The sense of boredom does not necessarily depend on the period, and often comes and goes periodically. And generally, the feeling of boredom does not seem a sufficient reason to abandon the pursuit of life. And if it is (for some people)—their choices are in their hands, and should not diminish the choices and chances of others.

Some people ponder whether extending longevity would make human life meaningless, tantamount to saying that death gives meaning to human life. Life-extensionists counter that life may carry a meaning of its own, independent of death. It is difficult or even impossible to place a temporal limit on the meaning, love and enjoyment of life. Human beings are entitled to choose a prolonged existence, and that choice and pursuit alone may give their life meaning.

A persistent warning is that increasing longevity would stop progress and make individuals and societies stagnant. Yet, the opposite seems to be more likely, as the potential for learning will be increased by longer life-spans. Such a prolonged "cultural adaptation" may be sufficient and necessary for the survival of the society. Moreover, rationally controlled development and care for the survival of the weak may be more advantageous for progress than blind and cruel Darwinian selection.

Often it is assumed and stated that aging and death from aging are natural and inevitable phenomena. Contingently, their acceptance as natural and inevitable gives comfort in facing them. For the believers in life extension by scientific means, these assumptions make little sense. Concerning the inexorable "natural" limit to the human life, however comforting a reconciliation with death may be, it should not replace an active quest for life preservation. And more importantly, almost never a particular cause of death is completely "inevitable" but is always due to some identifiable material agent, and thus potentially a subject to prevention or amelioration. There is no limit "set in stone" to either the lifespan or the healthspan.

The acceptance of the "naturalness" and inevitability of degenerative aging and of the shortness of the lifespan due to it, has been so engrained that it has become a part of the individual and collective identity. What would then happen to our individual and collective identities in case of extending lives, either substantially or radically? Would we still be "us"? Furthermore, would the incessant transformations of the body and mind during prolonged existence even permit us to speak of a preservation of identity? A possible answer is that, during a prolonged life history, there may be a continuity of human existence. Or else, some "core" personal and communal patterns may be preserved, while various extensions and additions to them may develop in time.[1,94]

Perhaps the most frequent type of worry relates to the future availability of resources due to life extension. The common assumption is that "there will never be enough for everybody." This assumption has taken the form of two major related concerns: "longevity will only be available for the rich" and "overpopulation will happen due to extending longevity." Referring to the availability of resources, a very strong and persistent apprehension has been about the potentially unequal and selective access to life-extending technologies. Of all the possible concerns and challenges of human life extension, this is probably one of the most likely and disturbing, seeing the present inequalities in the access to health care.[95] Would then the extension of life only be made accessible for the rich and powerful? Would such preferential access for select groups be justifiable, continuous or inevitable? Would not such a fundamental disparity in the ability to survive threaten the very fabric of social coherence that would be fraught with constant resentment and strife? It has been asserted that the inability to provide a good to all people should not prevent providing it to some people.[6] Yet, such assertions may offer little consolation to people doomed to an early death by their social status. The inequality of access to medical means and technologies, and hence the unequal opportunity for lifespan and healthspan extension, appears to be a real danger. This danger is already here, manifesting in the present unequal access to health care, and is not necessarily reserved to future technologies. This danger needs to be recognized and a wide and equitable sharing of medical technologies, both the present and emerging ones, needs to become a primary social objective.

When addressing this concern, the upper class life-extensionists often reassure that the life-extending treatments will eventually be made cheaper as the technologies develop, and they will "trickle down" to the poor from the rich. Moreover, the rich may allow such treatments to the poor as they are interested in maintaining an "active and healthy workforce." Hence, in this type of social agreement, for the poor, a chance to obtain the treatments may only be contingent on their utility as "workforce," and if they have no such utility (for example, if the labor needs are already fulfilled, also from robotics), there are absolutely no incentives and no obligations to provide them with the life-extending treatments. Hence, at least for the initial stages of therapy development, the following options may be available for people of lesser means: (1) Wait patiently until the therapies "become cheaper" and/or "trickle down" from the rich; (2) fight for the right for access (perhaps also violently); or (3) advocate for universal public research, development and distribution programs for life-extending therapies, which will also give the public strong entitlement to such therapies. The third option appears preferable. Yet, in any case, the inequality of access does not seem to be a reason to hinder the emergence of new medical technologies, but only to intensify their development. The sooner they emerge, the faster they will likely become available for the people, hopefully for all.

An additional fear related to the availability of resources is that rising longevity would lead to a shortage of resources for the global population as a whole due to its unsustainable increase (also commonly known as "the problem of overpopulation due to life extension"). Yet, it must be argued that the term "overpopulation" does not simply relate to the number of people on a certain territory. Rather, it indicates the degree of availability of resources, especially food, for people at that territory. And, based on the available evidence and trends of development, scarcity of resources should not be anticipated as a result of increasing longevity. It was already calculated in the 1960s by the Agricultural Economics Research Institute, Oxford, that the agricultural productivity, even at that time, would be more than sufficient to feed 45 billion people. Since that time, agricultural capabilities in developed countries have increased dramatically, way ahead of increases in life expectancy or population.[96,97] The technological capabilities are here to feed the world. Then, why are there still famines? It often happens because of mismanagement or because the right technologies are not applied.[98] But technologies generally, or life-extending technologies in particular, should not be considered a cause of overpopulation or shortage of resources. On the contrary, in wealthy, technologically advanced countries, with high life expectancy, there are hardly any signs of "overpopulation" or shortage of resources. "Overpopulation" is often the problem of poorer, "developing" countries that overcompensate for high mortality (low life expectancy) with high birth rates, and that have limited access to medical and technological means to provide for the population increase. Hence, also in those countries, the way to combat overpopulation may be by *increasing* life expectancy, and the concomitant quality of life, medical and technological capabilities, not by decreasing them. Indeed, longevity (life expectancy) is an indispensable part of the Human Development Index (HDI), and it correlates with and synergistically reinforces the HDI's other parts, such as education and income per capita.[9] One may argue that even at diminishing resources, the prolongation of human life may be valuable and desirable. Yet, the most likely concomitant of extended longevity is rather abundance and not scarcity, as the same types of technologies that improve agricultural, technological and medical capabilities, are also instrumental for increasing the lifespan and healthspan.

And yet another very commonly perceived challenge is whether increasing life quantity would mean decreasing life quality for the population? In other words, wouldn't we have "too many old sick people"? Here too it may be argued that the perception of a human life as a "liability" to the person or to the society is questionable, and the preservation of life may be desirable even at some loss of life quality. Yet, it must be emphasized that the improvement in life quantify is often (though not always) inseparable from the improvement in life quality. A robust organism (similar to a robust machine) can operate efficiently and for longer periods of time. The same measures that enhance health, also enhance longevity. A good example is centenarians, who enjoy both exceptional longevity as well as quality of life, and preserved mental and physical ability, almost to the end of their lives.[99,100] Still, there is an evidently increasing incidence of aging-related diseases, following increasing life expectancy (partly because of not treating the underlying root causes of aging-related diseases). Yet, this increasing incidence is not a reason to stop biomedical research and development, especially for the amelioration of aging-related degeneration—the main cause of disease and disability in the aged—but to intensify this research and development. The advancement of this research and development is perhaps the only practical means to alleviate aging-related suffering and improve healthy and productive longevity for the elderly population. Some examples and grounds for the feasibility of success of this mission are discussed in the previous section. In summary, it is the extension of the human healthspan (healthy and productive lifespan) and not just of the lifespan that is pursued in the research and development of new medical means and technologies.

Responsible and active research and development will also help address another frequent and rather legitimate worry that the emerging anti-aging and life-extending therapies may not be as effective as anticipated or that they may be even unsafe, at least at their initial stages of development and application. The efficacy and safety of any new medical treatment are essential scientific and public concerns and they need to be addressed through rigorous study, through the development of and adherence to strict scientific criteria for efficacy and safety. Compliant with such criteria, new anti-aging and life-extending therapies may be highly desirable and beneficial commodities.

22.4 Normative Action: What Should We Do?

Given the feasibility and desirability, we enter the realm of normative suggestions and actions. What is it exactly that we need to do to achieve something that we desire and may have a chance to achieve, if not for ourselves then for our loved ones? What should we do to facilitate the emergence and availability of life-extending therapies? These are critical questions of public policy, in particular healthcare and research policy, and they need to be raised in the public arena. But these are also questions of personal choice and action. From independent individual actions, the movement for healthy longevity may grow, eventually allowing it a more visible place in the public policy debate. Some of the individual actions may include increasing personal knowledge on longevity science by various access means. Such increased interest and knowledge, when combined, may raise the demand for therapies that may in turn improve the offer. There are now extensive possibilities to join others with a similar interest, ranging from discussions with friends, to more formal live and online study groups, to joining networks and public associations of supporters of longevity science. There are now also expanding possibilities to participate, volunteer and assist in research, donate to or join academic and public organizations involved in longevity research and advocacy, including possibilities of "crowd-sourcing" and "crowd-funding." Through such networks and organizations, the individual may assist in healthy longevity promotion, as well as help advance legislation and public policies and programs supportive of longevity research. Perhaps most importantly, anyone could endeavor to practice a healthy, life-prolonging life-style, to improve one's chances to benefit from effective, safe and accessible life-extending technologies whenever they may arrive. These may be some of the actions that could be undertaken by individuals, essentially independent from any state or political structures.

Yet, obviously, specific regulatory, organizational and policy frameworks will be indispensable for any effort to achieve healthy life extension for the population. It may yet be too early to provide any specific regulatory and policy recommendations toward this achievement. To provide more thorough recommendations, the issue still needs to be raised more strongly in public, academic and political discourse. Yet, some preliminary recommendations may be offered. These may include increased funding, incentives and institutional support for research and development specifically directed toward alleviation of the aging process and for healthy life extension. Some preliminary recommendations are given in the position paper of the International Society on Aging and Disease (ISOAD), entitled "The Critical Need to Promote Research of Aging and Aging-related Diseases to Improve Health and Longevity of the Elderly Population."[12] Below some of the suggestions of that position paper are briefly discussed, with specific reference to funding, incentives and institutional support.

22.4.1 Funding

First of all, it will be necessary to ensure a significant increase of governmental and non-governmental funding for fundamental and goal-directed (translational) research in preventing the degenerative aging processes, and the associated chronic non-communicable diseases and disabilities, with the explicit purpose of extending healthy and productive life. The importance of increasing funding for biomedical research to increase its yield should be obvious. Yet, it is often tacitly implied, often by the lay public and policy makers, that fundamental and translational biological research of aging is somehow wasteful or inherently dangerous, or that the scientists already have "more than enough" and should not ask for more, or that the research money should be better spent on causes other than "aging-related" ill health (as if there are such "aging-unrelated" causes). An example of this attitude is no less than the UN "Sustainable Development Goals (SDG)—until 2030" adopted in September 2015.[101] The Sustainable Development Goal—SDG 3 "Ensure healthy lives and promote well-being for all at all ages" mandates: "By 2030, reduce by one third *premature* mortality from non-communicable diseases through prevention and treatment" (3.4., emphasis added) thus implying that "mature" mortality is somehow acceptable. This clause omits or does not explicitly mention the aged and the debilitating processes of aging underlying the non-communicable diseases (the formulation "for all ages" itself makes the problem of aging inconspicuous, not prioritized). Though, arguably, it is only by prevention and treatment of the underlying deteriorative aging processes that the goal of a significant reduction of mortality from non-communicable diseases could ever be achieved. This clause also implies the need to establish criteria for the distinction between "mature" and "premature" mortality, which are currently absent. Moreover, the SDG3 Clause 3.b mandates that the global community should "Support the research and development of vaccines and medicines for the communicable and non-communicable diseases that primarily affect developing countries, provide access to affordable essential medicines and vaccines." Apparently this undervalues the support for research of aging-related diseases that presumably primarily affect the "developed" (also known as "high income") countries, thus implying both that the aging plagues of the developed countries are not a research priority and that those plagues are irrelevant for the "developing" ("low income") countries, which is far from being the case. As a result of such a dismissive attitude, biomedical research of aging is seldom even considered as a budget item, either at the international, national or institutional level. Indicatively, as of 2016, the entire proposed budget for the World Health Organization's "Ageing and Health" program was $13.5M, out of the about $4.4 billion total WHO budget (0.3%).[102] This attitude should change if the scientific research of aging is to advance and produce positive results. Increasing research funding should become an explicit and emphatic point of advocacy.

Specifically, enhanced funding may require the dedication of a defined percentage of budget within relevant ministries, such as ministries of health and/or science, particularly in the divisions involved in research and treatment of non-communicable chronic diseases. There could also be a special legislation that would mandate a specific percentage of the profits of commercial pharmaceutical, biotechnology and medical technology companies to such research and development. The main purpose of such allocations within relevant ministries, governmental and non-governmental bodies would be to establish relevant research grant programs on a competitive as well as goal-directed basis. It may also be advisable to mandate incremental or factorial increases of such funding (say doubling every 2 or 5 years, or negotiating increases at defined periods of time). Such mandated long-term increases would, first of all, require the knowledge or establishment of the baseline for such funding (such a baseline is often absent), and secondly it would posit the commitment to continued investments in this research and development.

In practical terms, such increases in funding would necessitate painstaking work of research advocates with the relevant decision makers and stakeholders, also engaging the support of the broader community. The advocates would need to determine the agencies from which funding could be allocated to aging research, find out the possible procedural means to achieve these allocations, and establish contacts to negotiate and eventually achieve them. Presently, most aging research institutions are hardly in the position to hire professional lobbyists or materially support advocacy and public education organizations. The scientists are often simply not aware or dismissive of the benefits of targeted advocacy, and if they are aware of those benefits, they seldom have the time or resources to dedicate to advocacy or public education. But somebody has to do this work.

22.4.2 Incentives

Part of the promotion of life extension research could be accomplished not merely by increasing the amounts of financial investments put into the research, but by optimally effective management of the financial investments, combining financial and non-financial rewards for the advancement of the field. This optimization would necessitate the developing and adopting of legal and regulatory frameworks that give incentives for the relevant goal-directed biomedical research and development. Such incentives should accelerate the development, registration, administration and accessibility of drugs, medical technologies and other therapies that will effectively and evidentially ameliorate the aging processes and associated diseases and extend healthy life.

One of the primary specific requirements for developing the incentives would be to establish the criteria for the efficacy and safety of geroprotective (anti-aging) therapies. Such commonly agreed criteria are presently lacking. Yet, they appear to be absolutely necessary in order to set up the goals and define the merits that are to be rewarded or incentivized. There has recently been an intensifying discussion among longevity researchers and advocates about the need to recognize *the degenerative aging process as a treatable medical condition,* which would include the systemic aging-related factors that contribute to diseases and frailty.[103,104] Such recognition may accelerate research, development and distribution of therapies in several aspects: (1) the general public would be encouraged to actively demand and intelligently apply aging-ameliorating therapies; (2) the pharmaceutical and medical technology industries would be incentivized to develop and bring effective aging-ameliorating therapies and technologies to the market; (3) health insurance, life insurance and healthcare systems would obtain a new area for reimbursement, which may induce them and their subjects to promote healthy longevity; and (4) regulators and policy makers would be moved to increase investments of public funds into healthy life extension research and development. Yet, it appears that the primary necessary condition for the degenerative aging process to be recognized as a diagnosable and treatable medical condition and therefore an indication for research, development and treatment, is to develop evidence-based diagnostic criteria and definitions for degenerative aging and for the efficacy and safety of potential means against it. Without such scientifically grounded and clinically applicable criteria, the discussions about "ameliorating" or even "curing" degenerative aging will be mere slogans.

Such criteria are explicitly requested by major regulatory bodies. Thus, the WHO International Classification of Diseases (ICD-10) currently does include a category on "senility," synonymous with "old age" and "senescence" (carrying the code R54), but there are not yet any general symptoms, clinical definitions or test cases of this condition.[105] Furthermore, WHO's *Global Strategy and Action Plan on Ageing and Health (GSAP)—2016–2020* (November 2015) includes "Strategic objective 5: Improving measurement, monitoring and research on Healthy Ageing," with a clause "5.1: Agree on ways to measure, analyse, describe and monitor Healthy Ageing" (Section 95), which recognizes the need for such agreed measures.[24] Furthermore, major regulatory authorities, such as the US Food and Drug Administration (FDA) and the EU European Medicines Agency (EMA) have struggled for the inclusion of elderly subjects in all clinical trials and are beginning to search for a clinically applicable definition of the aging process. Thus, the EMA has been continuously searching for a consensus definition of age-related "frailty" and for criteria for effective and safe interventions against frailty, as well as for the accurate general assessment of the medication needs of older persons.[106,107] The direction at the US FDA appears to be similar. Here too the need for the inclusion of older subjects in all clinical trials and the necessity for devising specific criteria for their diagnostic and therapeutic assessment are recognized.[108] Yet, apparently, these needs have not yet been addressed satisfactorily. There is still no mandatory inclusion of elderly subjects in clinical trials, and no agreed criteria for their diagnostic and therapeutic evaluation, either in the EU or the US.

Nonetheless, a major advance recently occurred with the FDA. In November 2015, the FDA approved the "TAME" study—"Targeting Aging with Metformin," testing the ability of metformin (a well known anti-diabetic medication) to reduce or postpone multiple age-related diseases and dysfunctions.[109] Apparently this is the first time that a regulatory agency has approved a trial to intervene into the basic aging process (predominantly glycation) with the aim of reducing aging-related multimorbidity. Yet, it must be emphasized that the study does not test the effects on "*aging*" as such (for which there is presently no agreed formal or clinical definition or criteria), but on various age-related diseases and dysfunctions (which can be diagnosed in the clinic and which together are named "multimorbidity" or "comorbidity"). Yet, essentially, there is no agreed formal or clinical definition and criteria for multimorbidity either. There is still the need for an agreed and rigorous methodology to evaluate either aging itself or age-related multimorbidity or frailty as treatable medical conditions, within the EMA, FDA and other regulatory agencies. The achievement of such criteria, or at least a massive consultation process for their development, appear to be major policy requirements for the progress of life extension science.[15,110]

Integrally related to the issue of devising diagnostic criteria for degenerative aging and for the efficacy and safety of anti-aging and life-extending interventions is the facilitation of *in silico* and animal testing. While *in silico* testing is not commonly practiced in biogerontology, animal testing often faces severe public perception and regulatory hurdles, and not just in relation to anti-aging testing. There is also a need to provide guidelines for the ethical safety-enhanced human testing of anti-aging and life-extending therapies, which are currently rather absent. A special regulatory status may be sought for anti-aging and lifespan-extending therapies, which is currently also absent. In this regard, some existing regulatory frameworks may be "adopted." For example, as mentioned above, "age-related multimorbidity," "frailty" and "functional decline" may be less controversial, and better regulated and accepted targets for intervention than "aging." "Adjuvant therapy" (*i.e.* "supportive/additional" therapy) may be another "acceptable" term to describe the treatment of the aging organism as a whole, addressing the root causes of aging-related diseases rather than their particular symptoms. Indeed, in November 2015, the FDA approved an adjuvant therapy (the compound MF59, made with squalene oil, developed by Novartis) for a flu vaccine to boost the immune response in older persons. This development goes beyond "a drug against a disease" model, but seeks an appropriate regulatory framework to support the underlying health of older persons, using "adjuvant therapy."[111] The concepts of "life-saving therapies" and "life-extending therapies," which are already well established in major regulatory environments, mainly in relation to life-threatening conditions,[112] may also serve to address and advance anti-aging and lifespan extending therapies.

There is a special need to provide a shortened approval pathway for therapies with a high level of evidence for efficacy and safety in preclinical and early clinical trials. The approval may also need to be facilitated for cases of advanced degenerative and seemingly futile conditions. But here again, there is the problem of the deficit of consensus criteria for defining "enhanced efficacy and safety," as well as criteria for "advanced degeneration" and "*seemingly* futile" conditions. It may be both more scientifically justifiable and ethically acceptable to test and apply anti-aging interventions in their intended target population—the older frail persons, rather than the younger and healthier people who may exhibit entirely different responses and may not really need such treatments, even under the heading of "prevention." Yet how can the degree of "senile degeneration" be precisely gauged? For the "seemingly futile conditions," apparently, the criteria, methodology and terminology from critical and intensive care medicine may need to be examined.[113] The existing legal frameworks governing the conditions whose treatment is considered "futile" may be reconsidered in order to allow for the use of novel, less well tested therapies in severe cases to give the patients, and potentially others suffering from the same conditions, an improved chance to "live with dignity" rather than to "die with dignity."

It has also been a commonly voiced opinion that in order to accelerate biomedical progress generally, and the progress of anti-aging and life-extending therapies in particular, regulation on the development and use of such therapies should be generally softened, to allow for the proliferation of new ideas and methods.[114] The concept of "conditional approval" of therapies has been advanced, which would presumably make it easier for new therapies to enter the market and would reserve a greater share of research for the "post-market analysis" (*i.e.* after the medicines have already been sold and used).[115] A considerable number of patients, mainly the wealthy ones, seek to try new therapies in countries with particularly permissive regulatory requirements as a form of "medical tourism."[116] Personal ("do-it-yourself") testing is becoming increasingly popular.[117] There may be some logic in the argument for easy regulation. The developing and making available of new therapies has become notoriously costly and lengthy, in a considerable measure due to regulatory hurdles, among other reasons.[118,119] In many cases, there is a need to try for a chance. On the other hand, we may not wish people (including ourselves) to assume the role of mishandled guinea pigs. Some patients may become privileged gullible test subjects for their own money (if they have money). And others may become expendable unprotected test subjects (when they have no money). Both situations appear ethnically undesirable and may involve a considerable and unjustified risk to the patients' health and well being, though possibly with a good "profit margin" and "development potential" for the producers and suppliers of the new medications. Some balanced position needs to be found. Part of the answer may again lie in the development of strict scientific criteria for the diagnosis of the aging process and for the effectiveness and safety of interventions against it. Following the development of such evidence-based criteria, it may be easier to stall the dissemination of quack nostrums as well as to facilitate the availability of truly promising therapies. In other words, such criteria may help improve regulation, not discard it. This issue too should become a subject of broad academic and political discussion.

Another issue that apparently needs to be given much thought in advance is a normative procedure to make potential anti-aging and life-extending therapies universally accessible, rather than preferentially or exclusively available to the rich or to some other privileged social categories (unrelated to their medical indications). Some programs of public support for therapy research and development coupled with public entitlement to those therapies, when they are available, may be considered. There might be explored a kind of "longevity research bonds" (reminiscent of, but not identical to, government-issued "longevity bonds"[120]). In the past, various government-issued bonds and obligations programs helped to win wars, though the bond holders often never saw their worth back. Such issues need to be weighed. There may be other suggestions, for example, some special allocations within the public health systems. These and other options and ideas to make therapies accessible when they emerge need to be given a thorough prospective deliberation by the public, decision makers and academia.

22.4.3 Institutional Support

In any case, and in any discussion of health care research, funding and regulation, anti-aging and lifespan-extending therapies need to be included as an integral part. Special recognition, status and benefits need to be granted to commercial and public entities engaged in such research and development, on a par with any other branch of innovative biomedical science, or perhaps even higher due to the great importance and promise of the field. This essentially means strengthening the institutional basis of anti-aging and life extension science. There should be a greater thrust for the establishment and expansion of national and international coordination and consultation structures, programs and institutions to advance the research, development and education on the biology of aging and associated diseases. There needs to be a stronger institutional framework for the development of clinical guidelines to modulate the aging processes and associated aging-related diseases and to extend the healthy and productive lifespan for the population.

As a part of the stronger institutional support, aging research also needs a better place in academia. There appears to be an urgent need to establish or reinforce the specialty of Biogerontology and courses in Biogerontology as a common part of university curricula. Regretfully, this discipline is seldom a part of the curricula. The current curricula in life and health sciences around the world, very often, simply omit aging and longevity from processes of biological development. Furthermore, many biology textbooks do not include aging and dying, not to mention longevity, among the processes of life. The science of aging and longevity, and adjacent areas of study, need to become an entrenched part of education at every level, not just because of the scientific value of this subject, but also because of its great practical significance for society. In fact, the WHO "Global Strategy and Action Plan on Aging and Health (GSAP)" directly requests member states to "ensure competencies on ageing and health are included in the curricula of all health professionals" (of course, it should be stressed that knowledge of the biology of aging is one of such indispensable gerontological competencies).[24] Yet, this requirement is very far from implementation. The researchers of aging and longevity need to have a say in the development and dissemination of regimens for the extension of healthy longevity, based on the best available evidence, as a part of authoritative health recommendations. Such guidelines for healthy longevity for the public are commonly lacking. Very simply, researchers of aging and longevity need places and positions that would allow them to do their work. Such work places, which would be involved primarily and not tangentially with biomedical aging research, are quite few even in the developed world, and are almost absent in the "developing" or "low income" world.[121] There is a vital need to establish more and more cooperative centers of excellence for fundamental, translational and applied studies, alongside centers for strategic analysis, forecast, education and policy development on aging and longevity research, at academic institutes and various governmental and supra-governmental agencies.

The common rationale for these tentative policy recommendations is to reduce the burden of the aging process on the economy and to alleviate the suffering of the aged and the grief of their loved ones. It may be hoped that, if granted sufficient support, these measures can increase the healthy life expectancy for the elderly, extend their period of productivity and their interaction with society, and enhance their sense of enjoyment, purpose, equality and valuation of life. Thus, in light of the great need for and promise of human life extension, it may be considered a societal duty, especially of the professionals in biology, medicine, health care, economy and socio-political organizations, to strongly recommend greater investments, incentives and institutional support for research and development dealing with the understanding of mechanisms of human biological aging and translating these insights into effective, safe, affordable and universally available health and life-extending technologies and treatments. Given the feasibility and desirability of healthy human life extension, the normative "thing to do" would be simply "to do," to become proactive for the advancement of the field, to study and support the field, to realize the challenges facing the field, as well as its vital promises, and to contribute to overcoming the challenges and fulfilling the promises. It may be hoped that the present work will further contribute to the realization of this duty.

References

1. I. Stambler, *A History of Life-Extensionism in the Twentieth Century*, Longevity History, Rishon Lezion, 2014, http://www.longevityhistory.com/ (last accessed March 2016).
2. A. M. Vaiserman, *Front. Genet.*, 2014, **5**, 446.
3. I. Stambler, *Aging Dis.*, 2015, **6**(2), 76.
4. *The Future of Aging: Pathways to Human Life Extension*, ed. G. M. Fahy, M. D. West, L. S. Coles and S. B. Harris, Springer, New York, 2010.
5. R. Veatch, *Death, Dying, and the Biological Revolution. Our Last Quest for Responsibility*, Yale University Press, New Haven, 1977, pp. 293–305.
6. J. Harris, *Ann. N. Y. Acad. Sci.*, 2004, **1019**, 527.
7. C. Overall, *Aging, Death, and Human Longevity: A Philosophical Inquiry*, University of California Press, Berkeley CA, 2003.
8. E. Stokstad, *Science*, 2005, **309**, 102.
9. I. Stambler, *Curr. Aging Sci.*, 2014, **6**, 25.
10. M. J. Rae, R. N. Butler, J. Campisi, A. D. N. J. de Grey, C. E. Finch and M. Gough, *et al.*, *Sci. Transl. Med.*, 2010, **2**, 40cm21.
11. D. P. Goldman, D. Cutler, J. W. Rowe, P. C. Michaud, J. Sullivan, D. Peneva and S. J. Olshansky, *Health Aff.*, 2013, **10**, 1698.
12. K. Jin, J. W. Simpkins, X. Ji, M. Leis and I. Stambler, *Aging Dis.*, 2015, **6**(1), 1.
13. *Hufeland's Art of Prolonging Life*, ed. E. Wilson, Lindsay & Blakiston, Philadelphia, 1867, pp. 147, 165, originally C. W. Hufeland, *Makrobiotik; oder, Die Kunst das menschliche Leben zu verlängern*, Jena, 1796.
14. D. Blokh and I. Stambler, *Aging Dis.*, 2015, **6**, 196.
15. D. Blokh and I. Stambler, *Prog. Neurobiol.*, 2016, DOI: http://dx.doi.org/10.1016/j.pneurobio.2016.03.005.
16. R. Lozano, M. Naghavi, K. Foreman, S. Lim, K. Shibuya and V. Aboyans, *et al.*, *Lancet*, 2013, **380**, 2095.
17. Z. A. Medvedev, *Biol. Rev.*, 1990, **65**, 375.
18. T. Flatt, *Front. Genet.*, 2012, **3**, 148.
19. A. D. N. J. de Grey, *Rejuvenation Res.*, 2013, **16**, 89.
20. World Health Organization, *World Report on Aging and Health*, 2015, http://www.who.int/ageing/en/ (last accessed March 2016).
21. T. B. Kirkwood, *Nature*, 2008, **451**, 644.
22. C. J. Steves, T. D. Spector and S. H. Jackson, *Age Ageing.*, 2012, **41**, 581.

23. S. Vasto, G. Scapagnini, M. Bulati, G. Candore, L. Castiglia and G. Colonna-Romano, *et al.*, *Front. Biosci.*, 2010, **2**, 392.
24. World Health Organization, *Global Strategy and Action Plan on Ageing and Health (GSAP) - 2016-2020*, 2015, http://www.who.int/ageing/global-strategy/en/ (last accessed March 2016).
25. A. D. N. J. De Grey and M. Rae, *Ending Aging. The Rejuvenation Breakthroughs That Could Reverse Human Aging in Our Lifetime*, St. Martin's Press, New York, 2007.
26. A. Atala, *Curr. Aging Sci.*, 2008, **1**(2), 73.
27. G. Orlando, S. Soker, R. J. Stratta and A. Atala, *Cold Spring Harbor Perspect. Med.*, 2013, **3**(8), a015693.
28. Y. Cheng, H. Kang, J. Shen, H. Hao, J. Liu, Y. Guo, Y. Mu and W. Han, *Sci. Rep.*, 2015, **5**, 11703.
29. V. Galvan and K. Jin, *Clin. Interventions Aging*, 2007, **2**(4), 605.
30. C. M. Ramsden, M. B. Powner, A. J. Carr, M. J. Smart, L. da Cruz and P. J. Coffey, *Development*, 2013, **140**(12), 2576.
31. G. D'Uva, A. Aharonov, M. Lauriola, D. Kain, Y. Yahalom-Ronen and S. Carvalho, *et al.*, *Nat. Cell Biol.*, 2015, **17**(5), 627.
32. N. Bredenkamp, C. S. Nowell and C. C. Blackburn, *Development*, 2014, **141**(8), 1627.
33. Y.-H. Youm, T. L. Horvath, D. J. Mangelsdorf, S. A. Kliewer and V. D. Dixit, *Proc. Natl. Acad. Sci. U. S. A.*, 2016, **113**(4), 1026.
34. M. Jaskelioff, F. L. Muller, J. H. Paik, E. Thomas, S. Jiang and A. C. Adams, *et al.*, *Nature*, 2011, **469**(7328), 102.
35. C. Bär and M. A. Blasco, *F1000Research*, 2016, **5**, 89.
36. C. Bonnet, V. Kaltimbacher, S. Ellouze, S. Augustin, P. Bénit and V. Forster, *et al.*, *Rejuvenation Res.*, 2007, **10**(2), 127.
37. E. A. Schon and S. DiMauro, *Curr. Med. Chem.*, 2003, **10**, 2523.
38. D. J. Baker, T. Wijshake, T. Tchkonia, N. K. LeBrasseur, B. G. Childs, B. van de Sluis, J. L. Kirkland and J. M. van Deursen, *Nature*, 2011, **479**, 232.
39. D. J. Baker, B. G. Childs, M. Durik, M. E. Wijers, C. J. Sieben and J. Zhong, *et al.*, *Nature*, 2016, **530**, 184.
40. M. Leisegang, T. Kammertoens, W. Uckert and T. Blankenstein, *J. Clin. Invest.*, 2016, **126**(3), 854.
41. R. Nagai, D. Murray, T. O. Metz and J. W. Baynes, *Diabetes*, 2012, **61**(3), 549.
42. J. Saarikangas and Y. Barral, *eLife*, 2015, **4**, e06197.
43. V. P. Kramp and P. Herrling, *Neurodegener. Dis.*, 2011, **8**, 44.
44. E. M. Reiman, J. B. Langbaum, A. S. Fleisher, R. J. Caselli, K. Chen and N. Ayutyanont, *et al.*, *J. Alzheimer's Dis.*, 2011, **26**(suppl. 3), 321.
45. B. Levine, M. Packer and P. Codogno, *J. Clin. Invest.*, 2015, **125**(1), 14.
46. Healthspan Campaign, *NIH Geroscience Interest Group (GSIG) Releases Recommendations from the October 2013 Advances in Geroscience Summit*, 2013, http://www.healthspancampaign.org/2014/02/27/nih-geroscience-interest-group-gsig-releases-recommendations-october-2013-advances-geroscience-summit/ (last accessed March 2016).

47. B. K. Kennedy, S. L. Berger, A. Brunet, J. Campisi, A. M. Cuervo and E. S. Epel, *Cell*, 2014, **59**, 709.
48. Y. Kanfi, S. Naiman, G. Amir, V. Peshti, G. Zinman, L. Nahum, Z. Bar-Joseph and H. Y. Cohen, *Nature*, 2012, **483**, 218.
49. M. Kodali, V. K. Parihar, B. Hattiangady, V. Mishra, B. Shuai and A. K. Shetty, *Sci. Rep.*, 2015, **5**, 8075.
50. A. Brunet and S. L. Berger, *J. Gerontol., Ser. A*, 2014, **69**(suppl. 1), S17.
51. M. Manukyan and P. B. Singh, *Genes Cells*, 2012, **17**(5), 337.
52. C. He, S. K. Tsuchiyama, Q. T. Nguyen, E. N. Plyusnina, S. R. Terrill and S. Sahibzada, *et al.*, *PLoS Genet.*, 2014, **10**(12), e1004860.
53. M. Karin and H. Clevers, *Nature*, 2016, **529**, 307.
54. V. Gorbunova, A. Seluanov, Z. Mao and C. Hine, *Nucleic Acids Res.*, 2007, **35**(22), 7466.
55. V. P. Skulachev, V. N. Anisimov, Y. N. Antonenko, L. E. Bakeeva, B. V. Chernyak and V. P. Erichev, *et al.*, *Biochim. Biophys. Acta*, 2009, **1787**(5), 437.
56. A. L. Orr, L. Vargas, C. N. Turk, J. E. Baaten, J. T. Matzen and V. J. Dardov, *et al.*, *Nat. Chem. Biol.*, 2015, **11**(11), 834.
57. J. N. Kheir, L. A. Scharp, M. A. Borden, E. J. Swanson, A. Loxley and J. H. Reese, *et al.*, *Sci. Transl. Med.*, 2012, **4**(140), 140ra88.
58. A. Richardson, V. Galvan, A. L. Linc and S. Oddo, *Exp. Gerontol.*, 2015, **68**, 51.
59. V. N. Anisimov and A. Bartke, *Crit. Rev. Oncol. Hematol.*, 2013, **87**, 201.
60. J. Labbadia and R. I. Morimoto, *F1000Prime Rep.*, 2014, **6**, 7.
61. J. L. Olson, A. Atala and J. J. Yoo, *Chonnam Med. J.*, 2011, **47**(1), 1.
62. Y. Zhang, A. Desai, S. Y. Yang, K. B. Bae, M. I. Antczak and S. P. Fink, *et al.*, *Science*, 2015, **348**, aaa2340.
63. C. López-Otín, M. A. Blasco, L. Partridge, M. Serrano and G. Kroemer, *Cell*, 2013, **153**, 1194.
64. J. Santos, F. Leitão-Correia, M. J. Sousa and C. Leão, *Oxid. Med. Cell. Longevity*, 2016, **2016**, 4010357.
65. K. Muradian, *Biogerontology*, 2013, **14**(6), 687.
66. R. Bellamy, P. Safar, S. A. Tisherman, R. Basford, S. P. Bruttig and A. Capone, *et al.*, *Crit. Care Med.*, 1996, **24**(suppl. 2), S24.
67. G. G. Rogatsky and A. Mayevsky, *Undersea Hyperbaric Med.*, 2007, **34**(2), 75.
68. Y. P. Gerasimenko, D. C. Lu, M. Modaber, S. Zdunowski, P. Gad and D. G. Sayenko, *et al.*, *J. Neurotrauma*, 2015, **32**(24), 1968.
69. L. Fontana, B. K. Kennedy, V. D. Longo, D. Seals and S. Melov, *Nature*, 2014, **511**, 405.
70. A. A. Cohen, *Biogerontology*, 2016, **17**(1), 205.
71. M. R. Rose, *Evolutionary Biology of Aging*, Oxford University Press, Oxford, 1991.
72. R. Cutler, *Proc. Natl. Acad. Sci. U. S. A.*, 1975, **72**(11), 4664.
73. World Health Organization, *World Health Statistics 2014: Large Gains In Life Expectancy*, 2014, http://www.who.int/mediacentre/news/releases/2014/world-health-statistics-2014/en/ (last accessed March 2016).

74. S. Kunitz, in *Companion Encyclopedia of the History of Medicine*, ed. W. F. Bynum and R. S. Porter, Routledge, London, 2001, pp. 1693–1711.
75. I. Stambler, *Adv. Gerontol.*, 2015, **5**(4), 201.
76. I. Stambler, *Rejuvenation Res.*, 2014, **17**, 297.
77. R. Kurzweil, *The Singularity Is Near: When Humans Transcend Biology*, Penguin Books, New York, 2005, pp. 371–372.
78. T. R. Malthus, *An Essay on the Principle of Population, as it Affects the Future Improvement of Society with Remarks on the Speculations of Mr. Godwin, M. Condorcet, and Other Writers*, J. Johnson, London, 1798.
79. W. Osler, *Sci. Am.*, March 25, 1905, reproduced in S. Hall, *Senescence, the Last Half of Life*, D. Appleton & Company, New York, 1922, pp. 3–5.
80. M. Fishbein, *The Medical Follies. An Analysis of the Foibles of Some Healing Cults, Including Osteopathy, Homeopathy, Chiropractic, and the Electronic Reactions of Abrams, with Essays on the Antivivisectionists, Health Legislation, Physical Culture, Birth Control, and Rejuvenation*, Boni and Liveright, New York, 1925, p. 169.
81. M. Fishbein, *The New Medical Follies. An Encyclopedia of Cultism and Quackery in These United States, with Essays on the Cult of Beauty, The Craze for Reduction, Rejuvenation, Eclecticism, Bread and Dietary Fads, Physical Therapy, and a Forecast as to the Physician of the Future*, Horace Liveright, New York, 1927, p. 234.
82. B. Russell, *Mortals and Others, American Essays 1931-1935, Volumes I and II*, Routledge Classics, London, 2009, pp. 18–20.
83. B. Russell, *Portraits from Memory: And Other Essays*, Simon and Schuster, New York, 1956, pp. 50–53.
84. N. Wiener, *God and Golem, Inc. A Comment on Certain Points where Cybernetics Impinges on Religion*, The MIT Press, Cambridge MA, 1964, pp. 66–67.
85. F. M. Burnet, *The Biology of Aging*, Auckland University Press, Auckland NZ, 1974, pp. 63, 66.
86. L. Hayflick, *J. Gerontol., Ser. A*, 2004, **59**(6), B573.
87. L. Hayflick, *How and Why we Age*, Ballantine Books, New York, 1994, pp. 336–338.
88. L. R. Kass, *First Things*, 2001, **113**, 17.
89. E. J. Emanuel, *The Atlantic*, October 2014.
90. I. Stambler, *Front. Genet.*, 2015, **6**, 00202.
91. R. A. Miller, *Milbank Q.*, 2002, **80**(1), 155.
92. S. Sethe and J. P. de Magalhães, in *Ethics, Health Policy and (Anti-) Aging: Mixed Blessings*, ed. M. Schermer and W. Pinxten, Springer, Dordrecht, 2012, pp. 173–188.
93. F. Fuchs-Simonstein, *Self-evolution: The Ethics of Redesigning Eden*, Yozmot, Tel Aviv, 2004.
94. I. Stambler, *J. Evol. Technol.*, 2010, **21**, 13.
95. A. Deaton, *J. Econ. Lit.*, 2003, **41**(1), 113.
96. C. Clark, in *Man and His Future: A CIBA Foundation Volume*, ed. G. Wolstenholme, Little, Brown and Co., Boston, 1963, pp. 23–35.

97. Food and Agriculture Organization of the United Nations, *FAOSTAT*, 2013, http://faostat.fao.org/site/567/DesktopDefault.aspx?PageID=567#ancor (last accessed March 2016).
98. P. Walker, *Famine Early Warning Systems: Victims and Destitution*, Earthscan Publications Ltd., London, 1989.
99. S. Milman and N. Barzilai, *Cold Spring Harbor Perspect. Med.*, 2015, **6**(1), a025098.
100. M. A. Faria, *Surg. Neurol. Int.*, 2015, **6**, 49.
101. United Nations, *Transforming Our World: The 2030 Agenda for Sustainable Development*, 2015, https://sustainabledevelopment.un.org/ (last accessed March 2016).
102. World Health Organization, *Sixty-Eighth World Health Assembly: Proposed Programme Budget 2016-2017*, 2015, http://apps.who.int/gb/ebwha/pdf_files/WHA68/A68_7-en.pdf (last accessed March 2016).
103. A. Zhavoronkov and B. Bhullar, *Front. Genet.*, 2015, **6**, 326.
104. S. Bulterijs, R. S. Hull, V. C. E. Björk and A. G. Roy, *Front. Genet.*, 2015, **6**, 205.
105. World Health Organization, *International Statistical Classification of Diseases and Related Health Problems 10th Revision (ICD-10)-WHO Version 2016*, 2016, http://apps.who.int/classifications/icd10/browse/2016/en#R54 (last accessed March 2016).
106. European Medicine Agency, *Medicines for Older People*, 2015, http://www.ema.europa.eu/ema/index.jsp?curl=pages/special_topics/general/general_content_000249.jsp (last accessed March 2016).
107. M. Cesari, R. Fielding, O. Bénichou, R. Bernabei, S. Bhasin and J. M. Guralnik, *et al.*, *J. Frailty Aging*, 2015, **4**(3), 114.
108. US Food and Drug Administration, *Guidance for Industry, E7 Studies in Support of Special Populations: Geriatrics, Questions and Answers*, 2012, http://www.fda.gov/ScienceResearch/SpecialTopics/RunningClinicalTrials/GuidancesInformationSheetsandNotices/ucm219488.htm (last accessed March 2016).
109. S. S. Hall, *Science*, 2015, **349**(6254), 1275.
110. A. Moskalev, E. Chernyagina, V. Tsvetkov, A. Fedintsev, M. Shaposhnikov, V. Krut'ko, A. Zhavoronkov and B. K. Kennedy, *Aging Cell*, 2016, **15**(3), 407.
111. R. Preidt, *HealthDay News*, November 25, 2015, http://consumer.healthday.com/infectious-disease-information-21/flu-news-314/fda-approves-first-flu-shot-with-added-ingredient-to-boost-immune-response-705627.html (last accessed March 2016).
112. N. Olchanski, Y. Zhong, J. T. Cohen, C. Saret, M. Bala and P. J. Neumann, *Expert Rev. Pharmacoecon. Outcomes Res.*, 2015, **15**(6), 931.
113. J. L. Bernat, *Neurocrit. Care*, 2005, **2**(2), 198.
114. Y. Luo, G. Iyengar and V. Venkatasubramanian, *PLoS One*, 2016, **11**(3), e0150343.
115. Y. Fujita and A. Kawamoto, *Clin. Pharmacol. Ther.*, 2016, **99**(1), 26.
116. R. A. Charo, *N. Engl. J. Med.*, 2016, **374**(10), 901.

117. J. A. Greene, *N. Engl. J. Med.*, 2016, **374**(4), 305.
118. United States Government Accountability Office, *New Drug Development: Science, Business, Regulatory and Intellectual Property Issues Cited as Hampering Drug Development Efforts*, 2006, http://www.gao.gov/products/GAO-07-49 (last accessed March 2016).
119. S. Morgan, P. Grootendorst, J. Lexchin, C. Cunningham and D. Greyson, *Health Policy*, 2011, **100**(1), 4.
120. J. R. Brown and P. R. Orszag, *J. Risk Insur.*, 2006, **73**(4), 611.
121. A. Ahmad and S. Komai, *J. Am. Geriatr. Soc.*, 2015, **63**(6), 1283.

Subject Index